SCALE ISSUES IN
HYDROLOGICAL MODELLING

ADVANCES IN HYDROLOGICAL PROCESSES

Series Editors M. G. Anderson
D. E. Walling

Terrain Analysis and Distributed Modelling in Hydrology

Edited by K. J. Beven and I. D. Moore

Scale Issues in Hydrological Modelling

Edited by J. D. Kalma and M. Sivapalan

SCALE ISSUES IN HYDROLOGICAL MODELLING

Edited by

J. D. KALMA

Department of Civil Engineering and Surveying,
University of Newcastle, Callaghan, NSW, Australia

and

M. SIVAPALAN

Centre for Water Research, University of Western Australia, Perth, WA, Australia

JOHN WILEY & SONS

Chichester · New York · Brisbane · Toronto · Singapore

The papers in this volume were originally published in
Hydrological Processes — An International Journal, volume 9,
issues 3–4 (237–482) and issues 5–6 (483–728), 1995.

Other Wiley Editorial Offices

John Wiley & Sons, Inc., 605 Third Avenue,
New York, NY 10158-0012, USA

Jacaranda Wiley Ltd, 33 Park Road, Milton,
Queensland 4064, Australia

John Wiley & Sons (Canada) Ltd, 22 Worcester Road,
Rexdale, Ontario M9W 1L1, Canada

John Wiley & Sons (SEA) Pte Ltd, 37 Jalan Pemimpin #05-04,
Block B, Union Industrial Building, Singapore 2057

British Library Cataloguing in Publication Data

A catalogue record for this book is available from the British Library

ISBN 0-471-95847-6

Typeset in 10/12pt Times.
Printed and bound in Great Britain by BPC Wheatons Ltd, Exeter.

CONTENTS

DEDICATION

IAN DONALD MOORE 1951–1993

On 28 September 1993 Ian Moore, our friend and colleague, passed away after a brief illness, only a few weeks before the Robertson workshop, of which he was a convenor. At the time of his death, Ian held the prestigious Jack Beale Chair at the Australian National University, Canberra. Prior to this he had held academic and research positions at the University of Kentucky, CSIRO Division of Water and Land Resources, and the University of Minnesota.

Ian Moore's research in the last 10 years focused on erosion mechanics, hydrological modelling, dryland salinisation, and water quality. Throughout his scientific career, Ian was concerned with the hydrological implications of spatial heterogeneity in soils, landscapes and vegetation, and with issues of scale. He devised topographic indices that are used to describe the spatial patterns of soil wetness and erodibility, as well as the linkage between soil moisture and vegetation patterns.

Ian was an outstanding teacher, as well as inspiring supervisor and mentor to numerous postgraduate students. Above all he was a good friend and colleague to many, as evidenced by the many tributes which spoke of his energy, talents, and his generosity and personal warmth. Ian will be missed greatly by all those he influenced and touched as a friend, colleague and teacher. He enriched our lives. This volume in *Advances in Hydrological Processes* is dedicated to the memory of Professor Ian Donald Moore.

CONTRIBUTORS

Roni Avissar, Department of Meteorology and Physical Oceanography, Rutgers University, Cook Campus, PO Box 231, New Brunswick, NJ 08903, U.S.A.

Larry Band, Department of Geography, University of Toronto, Toronto, Ontario M5S 1A1, CANADA.

Bryson Bates, CSIRO, Division of Water Resources, Perth Laboratory, Floreat Park, Private Bag, Wembley, W.A. 6014, AUSTRALIA.

Keith J. Beven, Centre for Research on Environmental Systems and Statistics, Institute of Environmental and Biological Sciences, University of Lancaster, Lancaster, LA1 4YQ, U.K.

Günter Blöschl, Institut für Hydraulik, Gewässerkunde und Wasserwirtschaft, Technische Universität Wien, Karlsplatz 13/223, A-1040 Vienna, AUSTRIA.

Bobby H. Braswell, Institute for the Study of Earth, Oceans and Space, Engineering Research Building, University of New Hampshire, Durham, NH 03824-3525, U.S.A.

David R. Dawdy, Hydrologic Consultant, #3055 23rd Avenue, San Francisco, CA 94132, U.S.A.

Francois De Troch, Laboratory of Hydrology and Water Management, University of Gent, Coupure Links 653, B-9000 Gent, BELGIUM.

Bill Dietrich, Department of Geology and Geophysics, University of California at Berkeley, Berkeley, CA 94720, U.S.A.

Maurice J. Duncan, NIWA Climate, National Institute for Water and Atmospheric Research Ltd, Box 8602, Christchurch, NEW ZEALAND.

Jay Famiglietti, Department of Geological Sciences, University of Texas at Austin, Austin, TX 78712, U.S.A.

John J. Finnigan, CSIRO, Centre for Environmental Mechanics, GPO Box 821, Canberra, ACT 2601, AUSTRALIA.

Wolfgang-Albert Flügel, Geographische Institut, Abteilung Geohydrologie, Meckenheimer Allee 166, D-53115 Bonn, GERMANY.

Filippo Giorgi, National Center for Atmospheric Research, University Corporation of Atmospheric Research, PO Box 3000, Boulder, Colorado, CO 80307-3000, U.S.A.

Rodger Grayson, Centre for Environmental Applied Hydrology, Department of Civil and Environmental Engineering, University of Melbourne, Parkville, VIC 3052, AUSTRALIA.

Vijay K. Gupta, Centre for the Study of Earth from Space/CIRES, and Department of Geological Sciences, University of Colorado at Boulder, Boulder, CO 80309, U.S.A.

Yeboah Gyasi-Agyei, Department of Civil Engineering and Surveying, University of Newcastle, Rankin Drive, Newcastle, NSW 2308, AUSTRALIA.

Jörg M. Hacker, Flinders Institute for Atmospheric and Marine Sciences, Flinders University of South Australia, GPO Box 2100, Adelaide, SA 5001, AUSTRALIA.

Thomas Hatton, CSIRO, Division of Water Resources, Canberra Laboratory, GPO Box 1666, Canberra, ACT 2601, AUSTRALIA.

Mei-Ling Hsu, Department of Geography, University of California at Berkeley, Berkeley, CA 94720, U.S.A.

Jetse D. Kalma, CSIRO, Division of Water Resources, Canberra Laboratory, GPO Box 1666, Canberra, ACT 2601, AUSTRALIA. Now at: Department of Civil Engineering and Surveying, University of Newcastle, Rankin Drive, Newcastle, NSW 2308, AUSTRALIA.

Jumpei Kubota, Department of Environmental Science and Natural Resources, Tokyo University of Agriculture and Technology, Fuchu, Tokyo 183, JAPAN.

George Kuczera, Department of Civil Engineering and Surveying, University of Newcastle, Rankin Drive, Newcastle, NSW 2308, AUSTRALIA.

Thomas J. Lyons, Atmospheric Science, School of Biological and Environmental Sciences, Murdoch University, Murdoch, WA 6150, AUSTRALIA.

Marco Mancini, DIIAR, Politecnico Di Milano, Milano, ITALY.

Kooiti Masuda, Department of Geography, Tokyo Metropolitan University, Tokyo, JAPAN.

Hitoshi Matsuyama, Center for Climate System Research, University of Tokyo, Tokyo, JAPAN.

David Montgomery, Department of Geological Sciences and Quarternary Research Center, University of Washington, Seattle, WA 98195, U.S.A.

Ian D. Moore, Centre for Resource and Environmental Studies, Australian National University, Canberra, ACT 2601, AUSTRALIA (now deceased).

Katumi Musiake, Institute of Industrial Science, University of Tokyo, 7-22-1 Roppongi, Minato-ku, Tokyo 106, JAPAN.

Taikan Oki, Institute of Industrial Science, University of Tokyo, 7-22-1 Roppongi, Minato-ku, Tokyo 106, JAPAN.

Andrew Pitman, School of Earth Sciences, Macquarie University, North Ryde, NSW 2109, AUSTRALIA.

Michael R. Raupach, CSIRO, Centre for Environmental Mechanics, GPO Box 821, Canberra, ACT 2601, AUSTRALIA.

Robert Reiss, Department of Geology and Geophysics, University of California at Berkeley, Berkeley, CA 94720, U.S.A.

Justin Robinson, Centre for Water Research, Department of Environmental Engineering, University of Western Australia, Nedlands, WA 6009, AUSTRALIA.

Richard Silberstein, Centre for Water Research, Department of Environmental Engineering, University of Western Australia, Nedlands, WA 6009, AUSTRALIA.

Murugesu Sivapalan, Centre for Water Research, Department of Environmental Engineering, University of Western Australia, Nedlands, WA 6009, AUSTRALIA.

Richard C. G. Smith, Leeuwin Centre for Earth Sensing Technologies, Floreat, WA 6019, AUSTRALIA.

John Snell, Centre for Water Research, Department of Environmental Engineering, University of Western Australia, Nedlands, WA 6009, AUSTRALIA.

Peter A. Troch, Laboratory of Hydrology and Water Management, University of Gent, Coupure Links 653, B-9000 Gent, BELGIUM.

Garry Willgoose, Department of Civil Engineering and Surveying, University of Newcastle, Rankin Drive, Newcastle, NSW 2308, AUSTRALIA.

Eric F. Wood, Water Resources Program, Department of Civil Engineering and Operations Research, Princeton University, Princeton, NJ 08544, U.S.A.

Ross Woods, Centre for Water Research, Department of Environmental Engineering, University of Western Australia, Nedlands, WA 6009, AUSTRALIA. On leave from: National Institute for Water and Atmospheric Research Ltd, Box 8602, Christchurch, NEW ZEALAND.

Hsin-I Wu, Biosystems Research Group, Department of Industrial Engineering, Texas A&M University, College Station, TX 77843-3131, U.S.A.

Huang Xinmei, Atmospheric Science, School of Biological and Environmental Sciences, Murdoch University, Murdoch, WA 6150, AUSTRALIA.

ALPHABETICAL LIST OF REVIEWERS

Avissar, Roni (Rutgers)
Barry, Andrew (Western Australia)
Beven, Keith (Lancaster)
Blöschl, Günter (Tech. Univ. Vienna)
Choudhury, Bhaskar (NASA)
De Troch, Francois (Gent)
Dolman, Han (Wageningen)
Goodrich, David (USDA, Tucson)
Gupta, Vijay (Colorado, Boulder)
Gyasi-Agyei, Yeboah (Newcastle)
Held, Alex (CSIRO)
Hutchinson, Michael (ANU)
Kalma, Jetse (CSIRO)
Kirkby, Mike (Leeds)
Kuczera, George (Newcastle)
Loague, Keith (UC, Berkeley)
Maidment, David (Texas at Austin)
Mesa, Oscar (Medellin)
Nemani, Ramakrishna (Montana)
Pitman, Andrew (Macquarie)
Quinn, Paul (Lancaster)
Robinson, Justin (Western Australia)
Short, David (CSIRO)
Sivapalan, Murugesu (Western Australia)
Smith, Roger (USDA)
Stagnitti, Frank (Deakin U.)
Tarboton, David (Utah State)
Turner, Jeff (CSIRO)
Vörösmarty, Charles (New Hampshire)
Willgoose, Garry (Newcastle)
Wood, Eric (Princeton)
Wyrwoll, Karl-Heinz (Western Australia)

Band, Larry (Toronto)
Bates, Bryson (CSIRO)
Binley, Andrew (Lancaster)
Brutsaert, Wilfried (Cornell)
Cleugh, Helen (CSIRO)
Dietrich, Bill (UC, Berkeley)
Famiglietti, James (Texas at Austin)
Grayson, Rodger (Melbourne)
Gutknecht, Dieter (Tech. Univ. Vienna)
Hatton, Thomas (CSIRO)
Howard, Alan (Virginia)
Ibbitt, Richard (NIWA Ltd., NZ)
Kelliher, Francis (Landcare Res., NZ)
Koster, Randal (NASA)
Lettenmaier, Dennis (Washington)
Lyons, Thomas (Murdoch)
McNaughton, Keith (Hort Res., NZ)
Montgomery, David (Washington)
Parlange, Marc (UC, Davis)
Puente, Carlos (UC, Davis)
Rabin, Robert (NOAA/NSSL)
Running, Steven (Montana)
Silberstein, Richard (Western Australia)
Smith, James (Princeton)
Snell, John (Western Australia)
Suarez, Max (NASA)
Troch, Peter (Gent)
Viney, Neil (Western Australia)
Wetzel, Peter (NASA)
Wilson, John (CSIRO)
Woods, Ross (Western Australia)
Ying Fan (MIT)

PREFACE

This volume brings together the proceedings of a special three-day workshop on Scale Issues in Hydrological/Environmental Modelling held at Ranelagh House in Robertson, NSW, Australia, between November 30 and December 2, 1993. This was the third in a series of workshops organised with similar objectives in mind, the previous ones having taken place in Caracas, Venezuela in January 1982, and at Princeton University, U.S.A., in November 1984.

The objectives of the Robertson workshop were to discuss recent progress in research on scale issues in hydrological and environmental modelling, and to develop appropriate research strategies for linking model parameterisations across a range of spatial and temporal scales. By all accounts, the workshop was a great success in raising awareness about scale problems in hydrology, and in offering a number of alternative approaches towards their eventual resolution. The collection of papers appearing in this volume reflects much of the current thinking in scale-related research.

ACKNOWLEDGEMENTS

We are grateful for the financial and organisational support provided by the following Australian and international organisations: the Land and Water Resources Research and Development Corporation, the Australian Department of Industry, Technology and Regional Development, ALCOA of Australia Ltd., the Water Authority of Western Australia, the Water Research Foundation of Australia, the Centre for Water Research (University of Western Australia), CSIRO Division of Water Resources, the Centre for Resource and Environmental Studies (Australian National University), and the International Association of Hydrological Sciences. The smooth running of the workshop owed a lot to the hard work of a number of volunteers — our sincere thanks to them, and especially to Vicki Baas-Becking.

We thank all those who made presentations at the workshop, and contributed papers to this volume. We take this opportunity to pay a special tribute to our panel of reviewers, the full list of which is given on p. xii, for assisting us with the review process. Special thanks are also due to Professor Malcolm Anderson, Editor of *Hydrological Processes* and of *Advances in Hydrological Processes*, for agreeing to publish these papers, and for offering assistance and encouragement at all times. We are thankful to Professor Eric Wood of Princeton University for his help with the organisation of the workshop, and for assistance in the editorial process.

As convenors of the workshop, and editors of these proceedings, we could not but feel the excitement and hard work that went into the various presentations at the workshop, and the contributions included in this volume. We sincerely hope that the publication of these proceedings will give the same inspiration to the readers, that completing these proceedings gave us as convenors, contributors, and editors.

December 1994
Jetse D. Kalma
Canberra, AUSTRALIA

Murugesu Sivapalan
Perth, AUSTRALIA

1

SCALE PROBLEMS IN HYDROLOGY: CONTRIBUTIONS OF THE ROBERTSON WORKSHOP

MURUGESU SIVAPALAN

Centre for Water Research, Department of Environmental Engineering, The University of Western Australia, Nedlands, WA 6009, Australia

AND

JETSE D. KALMA*

Division of Water Resources, Commonwealth Scientific and Industrial Research Organisation, Canberra Laboratory, GPO Box 1666, Canberra, ACT 2601, Australia

INTRODUCTION

Hydrologists have made impressive gains in work leading to the understanding and quantification of individual hydrological processes and in a variety of environments. However, theories of many processes such as infiltration, evaporation, overland flow, sediment transport and subsurface water movement have been, and continue to be, developed at small space–time scales. Our record in implementing these theories towards the development of predictive models at much larger space and time scales has not been equally impressive. The extrapolation of the theories of non-linear hydrological processes to large-scale, three-dimensional natural systems such as large drainage basins, floodplains and wetlands continues to pose serious problems.

A number of recent reviews have identified the 'scale problem' as a major unresolved problem in hydrology (e.g. NRC, 1991). It figures prominently in research towards the construction of appropriate models for addressing the management of land and water resources, and of models of the behaviour of large ecological systems. New analytical formulations for atmosphere–soil–vegetation interactions, and methods for their experimental verification, are currently being sought to assist in the development of hydrological models of extensive land areas. Such models are urgently needed in regional water resource planning, in the amelioration of water quality degradation in large river basins, in the validation of general circulation models (GCMs) and in the prediction and/or interpretation of the hydrological impacts of global climate change.

The previous workshops held at Caracas and at Princeton were strongly influenced by, and revolved around, the geomorphological unit hydrograph of Rodriguez-Iturbe and Valdes (1979), and other related concepts. Scale research has since become considerably broader, no doubt influenced by a number of significant technological and scientific developments, and new research imperatives. These are: the widespread availability and use of digital elevation models (DEMs) and geographical information systems (GISs), the increasing use of remotely sensed data, the need to address the effects of large-scale anthropogenic changes at the land surface and possible interactions with the atmosphere, and the realization of the importance of land surface hydrological processes in GCMs of the atmosphere.

The papers in this special issue address many, but not all, of these emerging issues. Specifically, several papers address the vexed question of how to handle spatial heterogeneity, the existence or otherwise of natural or preferred time and space scales, and approaches to finding linkages between scales of state variables, parameters and conceptualizations. Although a clear picture is emerging about the nature and extent of the 'scale problem', there is, as yet, no consensus on the solution to the problem. The need for continued and sustained research on scale issues is therefore self-evident.

---

* Present address: Department of Civil Engineering and Surveying, University of Newcastle, Callaghan, NSW 2308, Australia.

SUMMARY OF THE CONTRIBUTIONS

The first paper by Blöschl and Sivapalan aims to provide a much needed review of scale issues in hydrology, by organizing the vast amount of relevant publications into a number of major themes. It attempts to give precise definitions for a number of widely used terms and concepts in scale research. Looseness of terminology, present in many papers even in this special issue, makes communication difficult, especially across disciplinary boundaries. Blöschl and Sivapalan then address two alternative approaches to scaling: a model-oriented approach focusing on the scaling of state variables, model parameters, inputs and conceptualizations, and a more holistic approach which involves using dimensional analysis and similarity concepts, and fractal and multi-fractal analyses. In each instance considerable advances have been achieved, and many of these are reviewed in the paper.

The remaining papers in this special issue are summarized below under six headings, covering hydrological process descriptions ranging from a few metres to a few hundred kilometres.

Spatial heterogeneity and scale: REA and statistical scaling analyses

A critical question faced by many surface hydrologists is whether or not there are certain preferred time and space scales at which simple conceptualizations of hydrological responses may be feasible. The search for such preferred scales has intensified following the introduction of the representative elementary area (REA) concept by Wood *et al.* (1988). The REA promises a spatial scale at which process representations can remain simple and distributed catchment behaviour can be represented without the apparently undefinable complexity of local heterogeneity. The REA concept is further examined and clarified in the two papers by Woods *et al.* and Blöschl *et al.*

Motivated by the REA concept, Woods *et al.* investigate the relationship between spatial heterogeneity and spatial scale based on distributed measurements of runoff in a river basin, rather than the model outputs which were used by Wood *et al.* (1988). Woods *et al.* show that for small subcatchments (less than about 1 km^2 in area), the variance of specific discharge between subcatchments of similar sizes decreases with area more quickly than might be expected if the catchments are mere random samples. At larger scales, however, the variance decreases in a way that is consistent with sampling from a stationary random field. These results from measured streamflow data are reinforced by an analysis of topographic data. Such behaviour is interpreted by Woods *et al.* as evidence supporting the presence of 'organization'.

Blöschl *et al.* re-examine the evidence for the REA concept, clarifying its meaning and its utility for distributed rainfall–runoff modelling. They repeat the REA analysis of Wood *et al.* (1988), but with a different model of runoff generation, and include runoff routing as well. They use a nested catchment approach which they conclude to be more appropriate for the investigation of the existence of 'separation of scales'. Their results indicate that the REA size is strongly governed by the correlation length of precipitation, and by storm duration. Blöschl *et al.* claim that the existence and size of an REA will be catchment and application specific, and in any case the existence of an REA is not a prerequisite for simple representations of catchment response.

The next two papers, by Wood, and by Gupta and Dawdy, present two applications of theories of statistical self-similarity, i.e. concepts of simple scaling and multiscaling, to characterise the heterogeneity of hydrological variables across a broad range of scales.

Wood investigates the scaling behaviour of soil moisture, a fundamental hydrological state variable which has an important impact on the partitioning of rainfall and energy at the land surface. He conducts both REA type analyses of the type presented above and data analyses using techniques of statistical self-similarity. Wood also carries out an investigation of scaling using measures of information content (i.e. entropy). These analyses are carried out on a distributed model output, on remotely sensed soil moisture and also on field measured soil hydraulic parameters. The results obtained indicate that the threshold area (scale) at which statistical representations can replace the actual patterns of variability at smaller scales may be of the order of 5–10 km^2. This is larger than the REA size of 1 km^2 suggested by Wood *et al.* (1988); Wood suggests that this may be due to increased watershed size and consequent large-scale variations in the hydrological variables. The analysis of statistical self-similarity for soil moisture showed that soil moisture does not obey simple scaling, but rather multiscaling.

Gupta and Dawdy present the results of statistical scaling analyses carried out in the context of regional flood frequency relations. They focus on the scaling exponents in the power law relationship between flood quantiles and drainage areas, finding evidence for both simple scaling (i.e. the scaling exponent does not vary with return period) and multiscaling (i.e. the scaling exponent varies systematically with return period), and they then proceed to investigate the underlying physical causes for the differences in scaling behaviour. The results suggest that snowmelt-generated floods exhibit simple scaling, whereas rainfall-generated floods exhibit multiscaling. Gupta and Dawdy also investigate the physical causes for the multiscaling behaviour in rainfall-generated floods, and find that the scaling behaviour in small basins, smaller than a critical drainage area, is determined by basin response, and that of larger basins is determined by precipitation input.

Terrain attributes, terrain-based modelling and GIS

The greatest single advance in hydrological modelling in the past decade has probably been the availability and use of digitized topographic data, and to a lesser extent other terrain attributes such as vegetation, soils and land use. Although the availability of DEMs has revolutionized hydrology, their use is still undergoing rapid development, and present a number of 'scale' problems. Similarly, GISs, are now widely used, often uncritically, ignoring a number of scale problems associated with their use. The next set of papers by Gyasi-Agyei *et al.*, Dietrich *et al.*, Band and Moore, Flügel, and Famiglietti *et al.* address some of these issues.

The paper by Gyasi-Agyei *et al.* examines the effects of map scale, and the vertical resolution of DEMs on a number of hydrologically relevant geomorphological parameters. Gyasi-Agyei *et al.* estimate a number of geomorphological parameters, e.g. Horton order ratios, fractal dimension, etc., and the topographic wetness index, using high resolution DEMs of both natural and synthetic catchments. For this study, as the surrogate for map scale, they use the constant threshold area normally used to extract channel networks from DEMs. They find that many of the parameters, e.g. the Horton order ratios, are dependent on the map scale and, therefore, cannot be used to reliably distinguish between catchments. On the other hand, a change in vertical resolution of the DEMs does not have a significant impact on many of the important parameters.

Although DEMs are fairly easy to construct, other important pieces of information such as soil depth and soil properties are not so easy to obtain. For this reason, there have been efforts to link soil characteristics to topographic position, based on the premise that both are products of co-evolution in terms of catchment morphology. Dietrich *et al.* present a physically based deterministic model of the spatial variation of soil depth. The soil depth model solves for the mass balance betwen soil production from the underlying bedrock and the divergence of diffusive soil transport. Model predictions of soil depth correspond well with field observations and are strongly correlated with topographic curvature. This model is then used with a model of shallow slope stability that includes the effects of root cohesion and vertically varying saturated hydraulic conductivity, and is used to predict the locations of shallow landslides; these are suggested as potential sites of stream channel initiation.

Band and Moore focus on the use of GIS techniques to extend hydrological models to larger areas, with particular emphasis given to the impact of changing data resolution on the estimation of surface attributes across the landscape. Change of scale or the resolution of spatial data sets involves the loss of information at the higher spatial frequencies, leading to significant biases, due to the strong non-linearity of many hydrological processes. Band and Moore review and assess the various methods for estimating and combining critical land surface attributes, and the structure of the watershed, using GIS tools. They indicate that the overlay process in GIS, as currently practised, is a poor sampling strategy as it involves the combination of incompatible data sets.

Flügel presents a GIS-based method for regionalization of catchment behaviour which is able to conserve the inherent heterogeneity of catchment attributes. It is based on 'hydrological response units' (HRUs) which are distributed unit areas within a catchment characterized by common land use and physiographic properties such as precipitation, topography, soils and geology. The HRUs by definition can be assumed to produce 'similar' hydrological responses, a property which allows them to be used as a vehicle for aggregating different parts of the catchment for modelling. However, this aggregation method sidesteps the question of scale by ignoring the natural heterogeneity of parameters and processes *within* the HRUs (see Band and Moore).

Famiglietti *et al.* are concerned with the space–time variability of water and energy cycles at continental scales, and the process controls on such variability. They develop a framework, based on principal component analysis of climate model output, for understanding the relative roles of soil moisture, precipitation, evapotranspiration, runoff and snow depth in controlling the spatial variability of the water balance over the continental USA. Their analyses lead to measures of *hydroclimatological similarity* which can be used to divide the whole continent into subcontinental-scale regions having 'similar' hydroclimatologies. A classification scheme based on such notions of similarity differs from classical GIS overlay techniques because it is based on the dominant modes of variability rather than specific indices such as vegetation or seasonal wetness.

Linking state variables and parameters across scales, without regard to conceptualizations

The next set of papers by Kalma *et al.*, Willgoose and Kuczera, and Snell and Sivapalan, address the question of finding linkages across scales in terms of state variables, and parameter values, *without* regard to conceptualizations. Whereas Kalma *et al.* investigate linking a state variable between scales, Willgoose and Kuczera, and Snell and Sivapalan, investigate linking parameter values.

Kalma *et al.* develop a methodology for the 'upscaling in a single step' (see Blöschl and Sivapalan for definition) of point-scale soil moisture measurements to a catchment-scale soil moisture state variable, expressed in terms of a scaled soil moisture store. They accomplish this by investigating the organized behaviour of a large number of long-term soil moisture measurements in an experimental catchment. Influenced by these observations, they apply a quasi-distributed/lumped model of hydrological response (the VIC model, see Sivapalan and Woods), which incorporates a conceptual form of such within-catchment heterogeneity. They show that this model is capable of reproducing not only the runoff at the catchment outlet, but also the variation of the catchment-scale soil moisture storage, *as well as* or slightly better than, a lumped model (i.e. the SFB model), which does not explicitly include the heterogeneity. The advantage of the quasi-distributed model, however, is that it also allows the disaggregation of predicted soil moisture storage to a distribution of point-scale soil moisture values.

Willgoose and Kuczera address the question of linking parameters across scales with respect to Hortonian overland flow. They show that kinematic wave overland flow parameters estimated from small plot experiments cause considerable biases when extrapolated to produce catchment-scale predictions. Similarly, model parameters obtained by calibration using catchment-scale rainfall–runoff data cannot be used to make predictions at small plot scales. Willgoose and Kuczera argue that this is due to the inadequacy of the kinematic wave approximation at large scales, which is caused by branching and braiding of overland flow, hydraulic jumps behind obstructions and backwater effects of debris dams, and subcritical/supercritical flow transitions. On the other hand, they report that infiltration parameters can indeed be adequately calibrated from small plot studies.

The paper by Snell and Sivapalan addresses the question of relating the hydraulic geometry parameters of individual channels to that of a channel network. Their work has been motivated by the geomorphological instantaneous unit hydrograph, or GIUH, of Rodriguez-Iturbe and Valdes (1979). The GIUH is a linear construct, and although linearity may be an adequate approximation for large catchments, empirical evidence suggests that small catchments are highly non-linear, a scale problem which cannot be fully addressed by the GIUH approach. To overcome the limitations of the GIUH, Snell and Sivapalan introduce the concept of a *meta-channel*, which is formed by collapsing the stream network into a single channel. The hydraulic properties of this meta-channel are estimated by conserving the spatial distributions (in terms of flow distance) of mass and mechanical energy dissipation between the stream network and the meta-channel. Snell and Sivapalan present the underlying principles, and an application of the meta-channel concept towards the construction of its effective hydraulic geometry.

Linking conceptualizations across scales: distributed modelling and similarity analyses

The next four papers by Beven, Hatton and Wu, Kubota and Sivapalan, and Robinson and Sivapalan, address the problem of linking conceptualizations between scales from a modelling perspective, given that the so-called 'aggregation approach' (see Beven), whereby a model that is applicable at small scales is applied at large scales using 'effective' parameter values, is an inadequate approach to the scale problem. In addition,

Robinson and Sivapalan, and Troch *et al.*, present applications of dimensional and similarity analyses as alternative approaches to the scaling of conceptualizations.

Beven proposes an alternative 'disaggregation approach' wherein the heterogeneity of hydrological processes are represented in terms of joint distribution functions which retain the essential non-linearities of the system while using the linear operation of integration to predict the areal fluxes at the scale of interest. He stresses that such a subgrid parameterization approach to modelling is, at any scale, dependent on the data available with which to calibrate the model, and recommends an analysis of predictive uncertainty to estimate the value of different types of data. Beven illustrates these points using a simple 'patch' model.

Hatton and Wu deal with the extrapolation of measurements of water use by individual trees to that for a *stand of trees*, which is a critical step in linking plant physiology and hydrology. They present the development of a non-linear relationship connecting tree water use to leaf area index, average solar irradiance and soil water potential, which can be used as a convenient scaling relationship to estimate stand water use. The derivation of this relationship is based on the concepts of hydrological equilibrium and ecological field theory. With a knowledge of the spatial distributions of soil water potential, solar irradiance and leaf areas, the actual scaling of tree water use measurements to the catchment scale appears straightforward.

As a solution to the problem of scaling conceptualizations, Kubota and Sivapalan and Robinson and Sivapalan, present two examples of what they call a 'disaggregation–aggregation approach' in which distribution functions are first used to represent sub-grid scale hydrological heterogeneity. The effects of such heterogeneity are then aggregated using the assumed distribution functions. This methodology is similar to the approach advocated by Beven, and makes the implicit assumption that distribution functions, rather than the actual spatial patterns, are sufficient to characterize subgrid heterogeneity, which presumably would hold beyond a certain threshold area, for example the REA of Wood *et al.* (1988).

By following the disaggregation–aggregation approach, Kubota and Sivapalan develop catchment-scale descriptions governing subsurface flow in steep forested catchments and have obtained results comparable with relationships obtained from field measurements. Robinson and Sivapalan follow a similar strategy and using a distributed model of runoff generation by both infiltration excess and saturation excess, they are able to obtain a lumped, physically based and catchment-scale model that describes both the extent of saturated areas, and the average infiltration capacity over the unsaturated areas during storm events. This lumped model, expressed in the form of a one-step equation, is able to predict the storm response for particular storms *as well as* the distributed model, but at a fraction of the computational cost.

Notions of scale invariance—for example, notions of geometric, kinematic and dynamic similarity—are widespread in science and engineering and have been responsible for some of the great advances in hydraulics. Wood, and Gupta and Dawdy, earlier presented the application of theories of *statistical* self- similarity, using concepts of simple scaling and multiscaling, to characterize the heterogeneity of hydrological variables across a broad range of scales. Robinson and Sivapalan, on the other hand, present an application of *deterministic* similarity analysis (see Blöschl and Sivapalan) using their physically based, lumped hydrological model. By expressing the model in terms of three non-dimensional *similarity parameters* they show that the *hydrological similarity* of catchments, in respect of their runoff generation potential, can be expressed very well by these similarity parameters.

The paper by Troch *et al.* also addresses the issue of hydrological similarity, but in catchments dominated by subsurface stormflow. Troch *et al.* apply a physically based model of the stream network morphology in these catchments which uses information about geology, climate and topography. Based on this model they develop relationships between drainage characteristics, e.g. drainage density, and the average water-table depth of the catchment. They suggest that the relationships relating a number of drainage characteristics, i.e. drainage density, stream spacing and channel geometry, to average water-table depth, can be considered to be catchment-scale *similarity relationships,* which can be used for comparative catchment studies. However, unlike Robinson and Sivapalan, the similarity has been expressed in terms of *measurable* and *catchment-scale* variables.

Land surface–atmosphere interactions: modelling and measurements at large catchment to regional scales

In the papers summarised so far, the spatial scale of interest has been small enough so that feedbacks between the land surface and atmosphere can be ignored in the models discussed. Climatic inputs such as rainfall and potential evaporation can then be prescribed externally. Increasingly, however, it is becoming clear that such feedbacks need to be considered explicitly for some of the large-scale hydrological models and during the development of land surface parameterization schemes for global and regional climate models.

The approaches to use in this instance depend on the length scale of the heterogeneities present in the landscape. Here we can consider two cases: (i) heterogeneities having length scales at which the surface energy fluxes can be assumed to be mainly vertical, and which can be parameterized by integration over assumed distribution functions characterizing such heterogeneity; and (ii) heterogeneities at length scales at which the interactions between 'patches' in the horizontal direction have to be explicitly included. The papers reviewed in this section, namely, Raupach and Finnigan, Silberstein and Sivapalan, Xinmei et al. (1), and Xinmei et al. (2) deal with scale issues related to such interactions between neighbouring patches over heterogeneous landscapes.

Raupach and Finnigan provide a comprehensive review of scale issues in land surface–boundary layer interactions, and present two guidelines for scaling: (i) the need to preserve model complexity between scales; and (ii) the fact that fluid mixing naturally smooths out manifestations of landscape heterogeneity. They next present two flux-matching criteria for upscaling models of land–air fluxes: (i) that surface fluxes average linearly; and (ii) that model form be preserved between scales. Using these criteria, they show that the Penman–Monteith equation provides a model which leads to physically consistent flux-matching rules for the upscaling of surface resistances. The next part of their paper presents new tests of a hypothesis that regionally averaged energy balances over land surfaces are insensitive to the scale of heterogeneities, being very similar in both the microscale and macroscale ranges. They review other evidence which suggests that the mesoscale range behaves similarly; however, questions remain about the consequences of clouds and precipitation.

Silberstein and Sivapalan present results based on simulations with a land surface water and energy balance model which is coupled to a convective boundary layer model. The coupled model is used to study the effects of land surface heterogeneity at the patch scale, and the averaging effects over multiple patches, which is handled without explicitly modelling the advection between neighbouring patches. The difference between lumping the entire mosaic of patches into a single patch, with assumed effective properties, as opposed to representing the land surface as a combination of patches acting in parallel, is also discussed using the results of the model simulations. Their model simulations demonstrate that significant biases can be generated in modelling catchment output if the heterogeneity effects are not fully accounted for. For example, the fluxes from a 'homogenized' surface with catchment average land surface properties (e.g. soil moisture) can be significantly higher than those from a heterogeneous surface with implicitly modelled inter-patch interactions.

Xinmei et al. (1) report on NOAA-AVHRR satellite measurements of land surface parameters, such as albedo, canopy resistance, leaf area index and fractional vegetation cover, over contrasting surfaces (native forest and agricultural crops) on both sides of the 'vermin-proof fence' in the south-west of Western Australia. The land surface parameters are evaluated against independent surface measurements and used as input into a numerical model to simulate the energy exchange between the surface and the overlying atmosphere. The simulation results are validated against detailed aircraft observations over both natural and agricultural vegetation.

Xinmei et al. (2) describe the meteorological impact of replacing native perennial vegetation with annual agricultural crops. Based on analysis of satellite data and the use of the one-dimensional boundary layer model described above, Xinmei et al. (2) suggest that convective mixing over the area cleared of forest cover is no longer able to reach the lifting condensation level for a significant period of the year. This implies a decrease in convective cloud formation and a reduction in the convective enhancement of rainfall. This paper presents further experimental evidence for the type of inter-patch interactions discussed by Raupach and Finnigan, and Silberstein and Sivapalan.

Water balance estimation at large catchment to continental scales is an extremely difficult task using traditional point measurements of rainfall, runoff, groundwater levels, etc., as of these, only runoff can be measured at the scale of interest with any degree of accuracy. With increasing availability of atmospheric data, the *atmospheric water balance* method is becoming an alternative method for water balance estimation for continental scale river basins. Oki et al. present encouraging results from their atmospheric water balance investigations for the Chao Phraya River basin in Thailand ($178\,000$ km^2) and for 70 other large river basins. Such analyses, in addition to being useful for the validation of continental-scale hydrological models, may help provide a vantage point to observe global water circulations, and the global teleconnections between human activities in one part of the world and the consequences in other parts of the world. They may also be useful towards the derivation of large-scale hydrological response functions, e.g. functions relating soil moisture storage to surface and subsurface runoff, and to evapotranspiration, which can then be incorporated in land surface parameterizations in global climate models.

Land surface parameterizations for climate models

Although hydrologists need to work together with climatologists, meteorologists and others to collect appropriate data sets, and to characterize and deal with spatial heterogeneities at all length scales, the incorporation of the effects of land surface heterogeneities at length scales greater than about 100 km remains a great challenge to *meteorologists*. This is the conclusion of the review paper by Avissar. Avissar identifies five land surface characteristics that need to be specified accurately in atmospheric models. These are stomatal conductance, soil surface wetness, surface roughness, leaf area index and albedo. Avissar argues that, for the first four of these, their spatial variability must also be considered in land surface schemes. When the length scale of the spatial variability of any of these characteristics is of the order of 100 km, they can result in mesoscale circulations which need to be considered explicitly. Spatial variability at smaller scales may, however, be parameterized.

The next two papers by Sivapalan and Woods, and by Pitman deal with the issue of handling the subgrid scale heterogeneity in land surface parameterization schemes. The paper by Sivapalan and Woods is a preliminary attempt to develop an efficient parameterization of the effects of spatial heterogeneity of rainfall and soil moisture. Sivapalan and Woods use a generalization of the simple bucket model which is efficient computationally and in terms of data requirements, and use it to investigate the effects of heterogeneity of soil moisture and rainfall, on runoff generation, evapotranspiration and long-term water balance. They show that the non-inclusion of spatial heterogeneity can lead to considerable biases in the land surface hydrological response, and proceed to present a parsimonious methodology to predict the space–time evolution of subgrid scale soil moisture heterogeneity.

The paper by Pitman is concerned with handling vegetation heterogeneity, and specifically with determining whether vegetation heterogeneity is important at the spatial scales of GCMs, whereas it is generally recognized that it is important at smaller scales (see Avissar). Pitman investigates this by using a land surface scheme in a stand-alone mode and driven by prescribed climate forcing, and by examining the sensitivity of model output to the treatment of vegetation heterogeneity. He concludes that the land surface response depends non-linearly on the method of aggregation of the vegetation parameters, suggesting that more care is needed in data aggregation, and calls for improved land surface parameterizations which incorporate such heterogeneity.

FUTURE CHALLENGES

There have been tremendous advances in the research into scale issues in hydrological modelling in the 10 years since the last workshop at Princeton. Many of these advances have been reflected in the papers presented in this special issue.

Use of digital terrain information, e.g. topography, has become very widespread. The beginnings of a physical correlation between topographic position and other terrain attributes such as soil depth can already be seen, although a lot more work remains to be carried out. Work on terrain attributes and stream network analysis should strive to incorporate the valuable information that could be provided by the underlying geology. The use of GIS is becoming widespread; however, there are fundamental scale issues associated with GIS which are not well understood. A number of other new concepts have been presented in these papers (e.g. deterministic and statistical self-similarity, entropy, meta-channel, atmospheric water balance, 'smart' buckets, etc.) and these will play important roles in future research.

The search for natural or preferred spatial and temporal scales has intensified, judging from the number of papers devoted to the REA concept. The concept of 'organization' has been presented in a preliminary manner, and may be related to the concept of the REA. The inappropriateness of 'effective' parameters when dealing with heterogeneous processes is now widely accepted, as a number of papers have suggested; however, there is no consensus about possible approaches to the problem of linking parameters, state variables and conceptualizations across scales, although a number of approaches are being tried. This is the area where the greatest advances are needed for modelling purposes.

It is an interesting, but disappointing, commentary on hydrology that the pace of development of concepts and theories have outstripped the pace of growth of spatial databases and measurements to validate these theories. Despite our best efforts, only a few papers presented at the workshop addressed the question of measurements. It is extremely important that this imbalance be redressed if we are to make substantial progress in our modelling efforts.

Research on greenhouse-induced climate change is providing a degree of urgency to fundamental research on the modelling of hydrological processes at large scales, not only to improve the land surface schemes in the current generation of atmospheric models, but also to simulate the impacts of any climatic changes on catchment and regional scale water balances. However, it should be noted that the scale problem in hydrology is, and has always been, a fundamental problem for hydrologists, regardless of the new urgency provided by the difficulties with global climate models. For example, hydrologists are increasingly called upon to develop water balance models at large catchment scales for water resource management purposes.

It is clear that there is still a wide gulf separating the length scales at which hydrologists and meteorologists traditionally work. There is some narrowing of this gap, especially between hydrologists and boundary layer meteorologists when dealing with the effects of land surface heterogeneity. However, in many other instances, for expediency, this gap has been ignored, and theories and models only applicable at the scales of a few metres have been used uncritically at the scale of the GCM grid square. A more gradual approach, which integrates small-scale processes, where possible, to develop lumped models, which are then verified with measurements at the larger scale, would appear to be preferable. Clearly, this cannot be achieved overnight and requires sustained research over a long period of time.

From the hydrologists' point of view, it would be wiser to gain considerable experience in building models at large catchment to continental scales, before attempting to build models at the GCM scale. Indeed, runoff at the catchment outlet remains the most reliable and areally integrated measurement of hydrological response presently available, and thus the large catchment perspective is extremely valuable. However, we cannot point to many attempts nor successes in hydrological modelling at large catchment scales. Work in this important area has been renewed with vigour, as part of the international Global Water and Energy Experiment (GEWEX).

Finally, traditional hydrologists must be concerned about the huge gap between hydrological theory and practice when it comes to scaling issues. In spite of the tremendous advances in hydrological theories related to scaling, these have not contributed adequately to the way hydrologists solve traditional problems in applied hydrology. Theoretical research in hydrology, some critics might claim, is carried out in a vacuum. They might also claim, perhaps unfairly, that we find it more rewarding and challenging to tackle the problems of the GCMs than to attempt to bridge the gap between hydrological theory and (engineering) practice. Hydrologists have a duty to respond to these important challenges.

REFERENCES

Gupta, V. K., Rodriguez-Iturbe I., and Wood, E. F., (Ed) 1986. *Scale Problems in Hydrology*, Reidel, Dordrecht. 246pp.
NRC, National Research Council 1991. *Opportunities in the Hydrologic Sciences*. National Academy Press, Washington, DC. 348pp.
Rodriguez-Iturbe, I., and Gupta, V. K. (Guest Eds) 1983. *Scale Problems in Hydrology, J. Hydrol. (Spec. Issue)*, **65**, 1–257.
Rodriguez-Iturbe, I., and Valdes, J. B. 1979. 'The geomorphologic structure of hydrologic response'. *Wat. Resour. Res.*, **15**, 1409–1420.
Wood, E. F., Sivapalan, M., Beven, K. J., and Band, L. E. 1988. 'Effects of spatial variability and scale with implications to hydrologic modeling', *J. Hydrol.*, **102**, 29–47.

2

SCALE ISSUES IN HYDROLOGICAL MODELLING: A REVIEW

G. BLÖSCHL*

Centre for Resource and Environmental Studies, The Australian National University, Canberra City, ACT 2601, Australia

AND

M. SIVAPALAN

Centre for Water Research, Department of Environmental Engineering, University of Western Australia, Nedlands, 6009, Australia

ABSTRACT

A framework is provided for scaling and scale issues in hydrology. The first section gives some basic definitions. This is important as researchers do not seem to have agreed on the meaning of concepts such as scale or upscaling. 'Process scale', 'observation scale' and 'modelling (working) scale' require different definitions. The second section discusses heterogeneity and variability in catchments and touches on the implications of randomness and organization for scaling. The third section addresses the linkages across scales from a modelling point of view. It is argued that upscaling typically consists of two steps: distributing and aggregating. Conversely, downscaling involves disaggregation and singling out. Different approaches are discussed for linking state variables, parameters, inputs and conceptualizations across scales. This section also deals with distributed parameter models, which are one way of linking conceptualizations across scales. The fourth section addresses the linkages across scales from a more holistic perspective dealing with dimensional analysis and similarity concepts. The main difference to the modelling point of view is that dimensional analysis and similarity concepts deal with complex processes in a much simpler fashion. Examples of dimensional analysis, similarity analysis and functional normalization in catchment hydrology are given. This section also briefly discusses fractals, which are a popular tool for quantifying variability across scales. The fifth section focuses on one particular aspect of this holistic view, discussing stream network analysis. The paper concludes with identifying key issues and gives some directions for future research.

INTRODUCTION

This review is concerned with scale issues in hydrological modelling, with an emphasis on catchment hydrology.

Hydrological models may be either predictive (to obtain a specific answer to a specific problem) or investigative (to further our understanding of hydrological processes) (O'Connell, 1991; Grayson *et al.*, 1992). Typically, investigative models need more data, are more sophisticated in structure and estimates are less robust, but allow more insight into the system behaviour. The development of both types of model has traditionally followed a set pattern (Mackay and Riley, 1991; O'Connell, 1991) involving the following steps: (a) collecting and analysing data; (b) developing a conceptual model (in the researcher's mind) which describes the important hydrological characteristics of a catchment; (c) translating the conceptual model

---

* Also at: Institut für Hydraulik, Gewässerkunde und Wasserwirtschaft, Technische Universität Wien, Vienna, Austria.

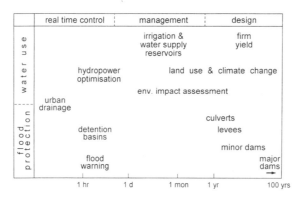

Figure 1. Problem solutions required at a range of time-scales

into a mathematical model; (d) calibrating the mathematical model to fit a part of the historical data by adjusting various coefficients; (e) and validating the model against the remaining historical data set.

If the validation is not satisfying, one or more of the previous steps needs to be repeated (Gutknecht, 1991a). If, however, the results are sufficiently close to the observations, the model is considered to be ready for use in a predictive mode. This is a safe strategy when the conditions for the predictions are similar to those of the calibration/validation data set (Bergström, 1991). Unfortunately, the conditions are often very different, which creates a range of problems. These are the thrust of this paper.

Conditions are often different in their space or time *scale*. The term *scale* refers to a characteristic time (or length) of a process, observation or model. Specifically, processes are often observed and modelled at short-time scales, but estimates are needed for very long time-scales (e.g. the life time of a dam). Figure 1 gives some examples of real-time control, management and design for which estimates of hydrological models are required. The associated time-scales range from minutes to hundreds of years. Similarly, models and theories developed in small space-scale laboratory experiments are expected to work at the large scale of catchments. Conversely, sometimes large-scale models and data are used for small-scale predictions. This invariably involves some sort of extrapolation, or equivalently, transfer of information across scales. This transfer of information is called *scaling* and the problems associated with it are *scale issues*.

In the past few years scale issues in hydrology have increased in importance. This is partly due to increased environmental awareness. However, there is still a myriad of unresolved questions and problems. Indeed, '. . . the issue of the linkage and integration of formulations at different scales has not been addressed adequately. Doing so remains one of the outstanding challenges in the field of surficial processes' (NRC, 1991: 143).

Scale issues are not unique to hydrology. They are important in a range of disciplines such as: meteorology and climatology (Haltiner and Williams, 1980; Avissar, 1995; Raupach and Finnigan, 1995); geomorphology (de Boer, 1992); oceanography (Stommel, 1963); coastal hydraulics (deVriend, 1991); soil science (Hillel and Elrick, 1990); biology (Haury *et al.*, 1977); and the social sciences (Dovers, 1995). Only a few papers have attempted a review of scale issues in hydrology. The most relevant papers are Dooge (1982; 1986), Klemeš (1983), Wood *et al.* (1990), Beven (1991) and Mackay and Riley (1991). Rodríguez-Iturbe and Gupta (1983) and Gupta *et al.* (1986a) provide a collection of papers related to scale issues and Dozier (1992) deals with aspects related to data.

This paper attempts to provide a framework for scaling and scale issues in hydrology. The first section gives some basic definitions. This is important as researchers do not seem to have agreed on what notions such as scale or upscaling exactly mean. The second section discusses heterogeneity and variability in catchments, which is indeed what makes scale issues so challenging. The third section addresses the linkages across scales from a modelling point of view. Specifically, different approaches are discussed for linking state variables, parameters, inputs and conceptualizations across scales. This section also deals with distributed parameter models, which are one way of linking conceptualizations across scales. The fourth section addresses the linkages across scales from a more holistic perspective, dealing with dimensional analysis and similarity concepts. The main difference to the modelling point of view is that dimensional analysis and similarity concepts deal with complex processes in a much simpler fashion. This section also briefly discusses fractals, which are a popular tool for quantifying variability across scales. The fifth section focuses on one particular aspect of this holistic view discussing stream network analysis. The paper concludes with identifying key issues and gives some directions for future research.

THE NOTION OF SCALES AND DEFINITIONS

Hydrological processes at a range of scales

Hydrological processes occur at a wide range of scales, from unsaturated flow in a 1 m soil profile to floods in river systems of a million square kilometres; from flashfloods of several minutes duration to flow in aquifers over hundreds of years. Hydrological processes span about eight orders of magnitude in space and time (Klemeš, 1983).

Figure 2 attempts a classification of hydrological processes according to typical length and time scales. Shaded regions show characteristic time–length combinations of hydrological activity (variability). This type of diagram was first introduced by Stommel (1963) for characterizing ocean dynamics and was later adopted by Fortak (1982) to atmospheric processes. Since then it has been widely used in the atmospheric sciences (e.g. Smagorinsky, 1974; Fortak, 1982). The shaded regions in Figure 2 can be thought of as regions of spectral power (in space and time) above a certain threshold. Stommel (1963: 572) noted, 'It is convenient to depict these different components of the spectral distribution of sea levels on a diagram (Stommel's Figure 1) in which the abscissa is the logarithm of period, P in seconds, and the ordinate is the logarithm of horizontal scale, L in centimeters. If we knew enough we could plot the spectral

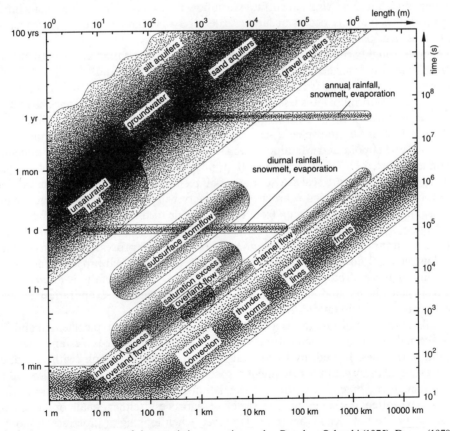

Figure 2. Hydrological processes at a range of characteristic space–time scales. Based on Orlanski (1975), Dunne (1978), Fortak (1982) and Anderson and Burt (1990) with additional information from the authors

distribution quantitatively by contours on this period-wavelength plane.' In a hydrological context, the shaded regions for a particular process in Figure 2 could be determined, at least conceptually, by plotting a characteristic time-scale (e.g. scale of maximum spectral power of a discharge record or, alternatively, response time of a catchment) versus a characteristic length scale (e.g. square root of catchment area).

Figure 2 is based on both data and heuristic considerations. Precipitation is one of the forcings driving the hydrological cycle. Precipitation phenomena range from cells (associated with cumulus convection) at scales of 1 km and several minutes, to synoptic areas (frontal systems) at scales of 1000 km and more than a day (Austin and Houze, 1972; Orlanski, 1975). Many hydrological processes operate — in response to precipitation — at similar length scales, but the time-scales are delayed. The time delay increases as the water passes through the subsurface and clearly depends on the dominant runoff mechanisms (Pearce et al., 1986). Consider, for example, a small catchment of, say, 1 km^2 size. Infiltration excess (i.e. Horton overland flow) response (such as often occurs in an arid climate during high rainfall intensities) is very fast (<30 minutes). Saturation excess (i.e. saturation overland flow) response (e.g. in humid climates and thin soils) is typically slower because the building up of a saturated layer delays the runoff response. Subsurface stormflow is often significantly slower, with response times of a day or longer for the same catchment size. Finally, groundwater-controlled flows are associated with time-scales from months to hundreds of years (Dunne, 1978; 1983; Zuidema, 1985; Anderson and Burt, 1990).

As in the case of atmospheric processes, different hydrological processes occur at different length scales (Figure 2). Runoff generation associated with rainfall intensities exceeding infiltration capacities (which produces infiltration excess/Horton overland flow) is a 'point phenomenon' and can, as such, be defined at a very small length scale. Saturation excess runoff (i.e. saturation overland flow) is an integrating process and needs a certain minimum catchment area to be operative. This is because, typically, the main mechanism for raising the groundwater table (which in turn produces saturation overland flow) is *lateral* percolation above an impeding horizon. Also, subsurface stormflow needs a certain minimum catchment area to be operative. Channel flow typically occurs at larger scales above a channel initiation area up to the length scales of the largest river basins.

Figure 2 suggests a roughly constant ratio of characteristic length and time-scales for a given process over a range of scales. This ratio is termed the *characteristic velocity* (Haltiner and Williams, 1980; Blöschl et al., 1995). For atmospheric processes this characteristic velocity is of the order of 10 m/s, for channel flow it is 1 m/s and for subsurface stormflow it is less than, say, 0·1 m/s. In Figure 2, the slopes have been selected as flatter than 1:1, which means a slight increase in the characteristic velocity with scale. This is consistent with Orlanski's (1975) definition of atmospheric scales, a slight tendency of channel flow velocities to increase with catchment scale (Leopold and Maddock, 1953) and the response times for catchments of different sizes (Anderson and Burt, 1990). It is speculated that, physically, this may be related to slightly decreased resistances (in a general sense) with scale. Also, the runoff response of a catchment is the combined effect of a number of processes. The combined characteristic velocity will reflect the relative contributions of individual processes. Specifically, when moving from the local scale to the regional scale, the increasing importance of channel flow will, typically, translate into increasing characteristic velocities with scale. In Figure 2 this means that the trace of the space–time relationship of combined processes (e.g. unsaturated flow, subsurface stormflow, channel flow) will be flatter than that of the individual processes shown.

Another forcing of the hydrological cycle is solar radiation. Consequently, a number of hydrological processes show a clear diurnal and annual cycle. Such processes include evaporation, snowmelt, and — depending on climate — precipitation (Gutknecht, 1993).

One of the striking features of Figure 2 is a certain relationship between typical length and time-scales for a given process. Small length scales tend to be associated with small time-scales and the same applies to large length and time-scales. Indeed, by looking at the plot of a hydrograph (with the time-scale given) we can roughly estimate the size of the catchment associated with it. For example, a slim and peaky hydrograph suggests a small catchment. Because of the relationship between length and time-scales of hydrological processes, we often just refer to 'the scale' of a process (either small or large) and implicitly assume that this relates to both length and time. However, when looking at processes in more detail, this is not always so. For example, unsaturated flow at the site scale (1 m) can be associated with time-scales

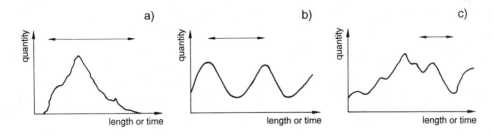

Figure 3. Three alternative definitions of process scale in space l (and time t). (a) Spatial extent (duration); (b) space (time) period; and (c) integral scale or correlation length (time)

from hours to weeks. To examine the scale of a process more closely, we need to first define, exactly, the notion of 'scale'. Specifically, we will first discuss the 'process scale' and then the 'observation scale'. The 'process scale' is the scale that natural phenomena exhibit and is beyond our control. On the other hand, we are free to choose the 'observation scale', within the constraints of measurement techniques and logistics.

Process scale

Characteristic *time-scales* of a hydrological process can be defined as (Figure 3): (a) the lifetime ($=$ duration) (for intermittent processes such as a flood); (b) the period (cycle) (for a periodic process such as snowmelt); and (c) the correlation length ($=$ integral scale) (for a stochastic process exhibiting some sort of correlation) (Haltiner and Williams, 1980; Dooge, 1982; 1986; Klemeš, 1983; Dagan, 1986; Stull, 1988). Similarly, characteristic *space scales* can be defined either as spatial extent, period or integral scale, depending on the nature of the process. Specifically, for a random function $w(x)$ that is stationary such that its covariance $C_w(x_1, x_2)$ depends on $r = x_1 - x_2$ rather than x_1 and x_2, the integral scale is defined as

$$I_w = \int \frac{C_w(r)\mathrm{d}r}{\sigma_w^2} \tag{1}$$

where x is either space or time, r is space (or time) lag and σ_w^2 is the variance. In other words, the integral scale refers to the average distance (or time) over which a property is correlated (Dagan, 1986). 'Correlation length' has two main usages: the first is identical with that of the integral scale, i.e. the average distance of correlation. The second refers to the maximum distance of correlation (de Marsily, 1986). Clark (1985) discusses the relationship between a number of different scale definitions.

The various definitions for scale (lifetime, period, correlation length) are often used interchangeably and it is not always clear, in a particular case (of, say, a stochastic *and* periodic phenomenon), which one is used. One justification for doing this is the relatively small difference between the above definition compared with the total range in scale of eight orders of magnitude. Indeed, the term 'scale' refers to a rough indication of the order of magnitude rather than to an accurate figure. In flood frequency analysis recurrence probabilities are sometimes thought of as a kind of scale. For example, a 100 year flood refers to a recurrence probability of 0·01 in any one year. This definition of scale is related to the interval *between* events (see crossing theory and the definition of runs, e.g. Yevjevich, 1972: 174).

Some hydrological processes exhibit (one or more) preferred scales, i.e. certain length (or time) scales are more likely to occur than others. These preferred scales are also called the *natural scales*. In a power spectrum representation preferred scales appear as peaks of spectral variance (Padmanabhan and Rao, 1988). Scales that are less likely to occur than others are often related to as a *spectral gap*. This name derives from the power spectrum representation in which the spectral gap appears as a minimum in spectral variance (Stull, 1988; 33). The existence of a spectral gap is tantamount to the existence of a *separation of scales* (Gelhar, 1986). Separation of scales refers to a process consisting of a small-scale (or fast) component superimposed on a much larger (or slow) component with a minimum in spectral variance in between. In meteorology, for example, microscale turbulence and synoptic scale processes are often separated by a spectral gap which manifests itself in a minimum in the power spectrum of windspeeds at the frequency of 1/hour.

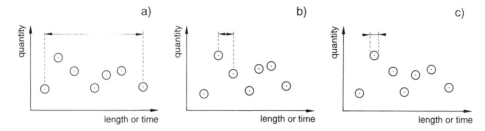

Figure 4. Three alternative definitions of measurement scale in space l (and time t). (a) Spatial (temporal) extent; (b) spacing (= resolution); and (c) integration volume (time constant)

In the time domain, many hydrological processes (such as snowmelt) exhibit preferred time-scales of one day and one year with a spectral gap in between. Clearly, this relates to the periodicity of solar radiation (Figure 2). In the space domain, there is no clear evidence for the existence of preferred scales. Precipitation, generally, does not seem to exhibit preferred scales and/or spectral gaps (e.g. Gupta and Waymire, 1993). Wood *et al.* (1988) suggested that catchment runoff may show a spectral gap at 1 km^2 catchment area. More recent work by Blöschl *et al.* (1995), however, indicates that both the existence and size of a spectral gap is highly dependent on specific catchment properties and climatic conditions. Also, runoff data (Woods *et al.*, 1995) do not support the existence of a spectral gap in runoff.

Observation scale

The definition of the 'observation scale' is related to the necessity of a finite number of samples. Consequently, 'observation scale' in space and time can be defined (Figure 4) as: (a) the spatial (temporal) extent (= coverage) of a data set; (b) the spacing (= resolution) between samples; or (c) the integration volume (time) of a sample.

The integration volume may range from 1 dm^3 for a soil sample to many square kilometres (i.e. the catchment size) for discharge measurements. The integration time (i.e. the time constant) is often a property of a particular instrument. Typically, space resolutions are much poorer than time resolutions in hydrology. This is seen as one of the major obstacles to fast progress in scale-related issues. Obviously, the particular observation scale chosen dictates the type of the instrumentation, from detailed sampling of a soil profile to global coverage by satellite imagery (Dozier, 1992).

Process versus observation scale

Ideally, processes should be observed at the scale they occur. However, this is not always feasible. Often the interest lies in large-scale processes while only (small-scale) point samples are available. Also, hydrological processes are often simultaneously operative at a range of scales.

Cushman (1984; 1987), in a series of papers, discussed the relationship between process and observation scale and pointed out that sampling involves filtering. Figure 5 gives a more intuitive picture of the effect of

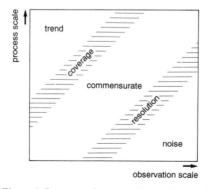

Figure 5. Process scale versus observation scale

sampling: processes larger than the coverage appear as trends in the data, whereas processes smaller than the resolution appear as noise. Specifically, the highest frequency that can be detected by a given data set of spacing d is given by the Nyquist frequency f_n (Jenkins and Watts, 1968; Russo and Jury, 1987)

$$f_n = \frac{1}{2d} \qquad (2)$$

Modelling (working) scale

Yet another scale is the modelling (or working) scale. The modelling scales generally agreed upon within the scientific community are partly related to processes (Figure 2) and partly to the applications of hydrological models (Figure 1).

In space, typical modelling scales are (Dooge, 1982; 1986):

the local scale (1 m);
the hillslope (reach) scale (100 m);
the catchment scale (10 km);
and the regional scale (1000 km).

In time, typical modelling scales are:

the event scale (1 day);
the seasonal scale (1 yr);
and the long-term scale (100 yrs).

Unfortunately, more often than not, the modelling scale is much larger or much smaller than the observation scale. To bridge that gap, 'scaling' is needed.

Definition of upscaling, downscaling and regionalization

To scale, literally means 'to zoom' or to reduce/increase in size. In a hydrological context, *upscaling* refers to transferring information from a given scale to a larger scale, whereas *downscaling* refers to transferring information to a smaller scale (Gupta *et al.*, 1986a). For example, measuring hydraulic conductivity in a borehole and assuming it applies to the surrounding area involves upscaling. Also, estimating a 100 year flood from a 10 year record involves upscaling. Conversely, using runoff coefficients derived from a large catchment for culvert design on a small catchment involves downscaling (Mein, 1993). *Regionalization*, on the other hand, involves the transfer of information from one catchment (location) to another (Kleeberg, 1992). This may be satisfactory if the catchments are similar (in some sense), but error-prone if they are not (Pilgrim, 1983). One of the factors that make scaling so difficult is the heterogeneity of catchments and the variability of hydrological processes.

NATURE OF HETEROGENEITY AND VARIABILITY IN SPACE AND TIME

Heterogeneity at a range of scales

Natural catchments exhibit a stunning degree of heterogeneity and variability in both space and time. For example, soil properties and surface conditions vary in space, and vegetation cover, moisture status and flows also vary in time. The term 'heterogeneity' is typically used for media properties (such as hydraulic conductivity) that vary in space. The term 'variability' is typically used for fluxes (e.g. runoff) or state variables (e.g. soil moisture) that vary in space and/or time. Heterogeneity and variability manifest themselves at a range of scales.

Figure 6a illustrates subsurface spatial heterogeneity in a catchment. At the local scale, soils often exhibit macropores such as cracks, root holes or wormholes. These can transport the bulk of the flow with a minimum contribution of the soil matrix (Beven, 1981; Germann, 1990). For the macropore flow to become operative, certain thresholds in precipitation intensity and antecedent moisture may need to be met (Kneale and White, 1984; Germann, 1986). At the hillslope scale, preferential flow may occur through

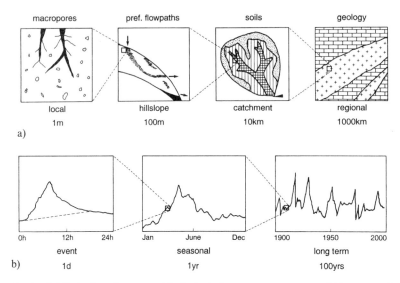

Figure 6. Heterogeneity (variability) of catchments and hydrological processes at a range of (a) space scales and (b) time-scales

high conductivity layers and pipes (Jones, 1987; Chappell and Ternan, 1992) and water may exfiltrate to the surface as return flow (Dunne, 1978). Figure 7 shows an example of flow along such a high conductivity layer in a catchment in the Austrian Alps. The location of the springs in Figure 7 indicates that the layer is parallel to the surface. Heterogeneity at the catchment scale (Figure 6a) may relate to different soil types and properties. Typically, valley floors show different soils as hillslopes and ridges (Jenny, 1980). At the regional scale, geology is often dominant through soil formation (parent material) and controls on the stream network density (von Bandat, 1962; Strahler, 1964).

Similarly, variability in time is present at a range of scales (Figure 6b). At the event scale the shape of the runoff wave is controlled by the characteristics of the storm and the catchment. At the seasonal scale, runoff is dominated by physioclimatic characteristics such as the annual cycle in precipitation, snowmelt and evaporation. Finally, in the long term, runoff may show the effects of long-term variability of precipitation (Mandelbrot and Wallis, 1968), climate variability (Schönwiese, 1979), geomorphological processes (Anderson, 1988) and anthropogenic effects (Gutknecht, 1993).

Types of heterogeneity and variability

The classification of heterogeneity (variability) is useful as it has a major bearing on predictability and scaling relationships (Morel-Seytoux, 1988). In fact, such a classification may serve as a predictive framework (White, 1988). Here, we will follow the suggestion of Gutknecht (1993), which is consistent with the definitions of 'scale' given earlier (Figure 3).

Hydrological processes may exhibit one or more of the following aspects: (a) *discontinuity* with discrete zones (e.g. intermittency of rainfall events; geological zones) — within the zones, the properties are relatively uniform and predictable, whereas there is disparity between the zones; (b) *periodicity* (e.g. the diurnal or annual cycle of runoff), which is predictable; and (c) *randomness*, which is not predictable in detail, but predictable in terms of statistical properties such as the probability density function.

At this stage it might be useful to define a number of terms related to heterogeneity and variability. *Disorder* involves erratic variation in space or time similar to randomness, but it has no probability aspect. Conversely, *order* relates to regularity or certain constraints (Allen and Starr, 1982). *Structure*, on the other hand, has two meanings. The first, in a general sense, is identical with order (e.g. 'storm structure'; Austin and Houze, 1972; geological 'structural facies', Anderson, 1989). The second, more specific, meaning is used in stochastic approaches and refers to the moments of a distribution such as mean, variance and correlation length (Hoeksema and Kitanidis, 1985). 'Structural analysis', for example, attempts to estimate these moments (de Marsily, 1986). *Organization* is often used in a similar way to 'order', but tends to relate to a more complex form of regularity. Also, it is often closely related to the function and the formation (genesis) of the system (Denbigh, 1975). These aspects are illustrated in more detail in the ensuing discussion.

Organization

Catchments and hydrological processes show organization in many ways, which include the following examples. (a) Drainage networks embody a 'deep sense of symmetry' (Rodríguez-Iturbe, 1986). This probably reflects some kind of general principle such as minimum energy expenditure (Rodríguez-Iturbe *et al.*, 1992b) and adaptive landforms (Willgoose *et al.*, 1991). Also, Rodríguez-Iturbe (1986) suggested that Horton's laws (Horton, 1945; Strahler, 1957) are a reflection of this organization. (b) Geological facies (Walker, 1984) are discrete units of different natures which are associated with a specific formational process (Anderson, 1989; 1991; Miall, 1985). These units are often 'organized' in certain sequences. For example, in a glacial outwash sequence, deposits close to the former ice fronts mostly consist of gravel, whereas those further away may consist mainly of silt and clay. (c) Austin and Houze (1972) showed that precipitation patterns are organized in clearly definable units (cells, small mesoscale areas, large mesoscale areas, synoptic areas), which may be recognized as they develop, move and dissipate. (d) Soils tend to develop in response to state factors (i.e. controls) such as topography (Jenny, 1941; 1980). Different units in a catchment (e.g. 'nose', 'slope' and 'hollow'; Hack and Goodlett, 1960; England and Holtan, 1969; Krasovskaia, 1982) may have a different function and are typically formed by different processes. Soil catenas (i.e. soil sequences along a slope) are a common form of organization of soils in catchments (Milne, 1935; Moore *et al.*, 1993a; Gessler *et al.*, 1993).

Organization versus randomness

Randomness is essentially the opposite of organization (Dooge, 1986). Given that 'randomness is the property that makes statistical calculations come out right' (Weinberg, 1975: 17) and the high degree of organization of catchments illustrated in the preceding sections, care must be exercised when using stochastic methods. It is interesting to follow a debate between field and stochastic hydrogeology researchers on exactly this issue: Dagan (1986) put forward a stochastic theory of groundwater flow and transport and Neuman (1990) proposed a universal scaling law for hydraulic conductivity, both studies drawing heavily on the randomness assumption. Two comments prepared by Williams (1988) and Anderson (1991), respectively, criticized this assumption and emphasized the presence of organized discrete units as formed by geological processes. Specifically, Williams (1988) pointed out that the apparent disorder is largely a consequence of studying rocks through point measurements such as boreholes. This statement is certainly also valid in catchment hydrology. It is clear that for 'enlightened scaling' (Morel-Seytoux, 1988), the identification of organization is a key factor. The importance of organization at the hillslope and catchment scale has been exemplified by Blöschl *et al.* (1993), who simulated runoff hydrographs based on organized and random spatial patterns of hydrological quantities. Although the covariance structure of the organized and the random examples were identical, the runoff responses were vastly different. Similar effects have been shown by Kupfersberger and Blöschl (1995) for groundwater flow. It is also interesting to note that hydrological processes often show aspects of both organization and randomness (Gutknecht, 1993; Blöschl *et al.*, submitted).

LINKAGES ACROSS SCALES FROM A MODELLING PERSPECTIVE

Framework

Earlier in this paper, the term 'scaling' was defined as transferring information across scales. What, precisely, does information mean in this context? Let $g\{s; \theta; i\}$ be some small-scale conceptualization of hydrological response as a function of state variables s, parameters θ and inputs i, and let $G\{S; \Theta; I\}$ be the corresponding large-scale description. Now, the information scaled consists of state variables, para-

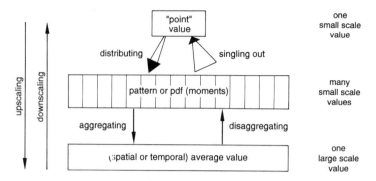

Figure 8. Linkages across scales as a two-step procedure

meters and inputs as well as the conceptualization itself

$$s \leftrightarrow S; \quad \theta \leftrightarrow \Theta; \quad i \leftrightarrow I; \quad g\{s; \theta; i\} \leftrightarrow G\{S; \Theta; I\} \tag{3}$$

However, in practice, often only one piece of information out of this list is scaled and the others are, implicitly, assumed to hold true at either scale. It is clear that this is not always appropriate. Also, the definition of a large-scale parameter Θ is not always meaningful, even if θ and Θ share the same name. The same can be the case for state variables. For example, it may be difficult (if not impossible) to define a large-scale hydraulic gradient in an unsaturated structured soil.

Upscaling, typically, consists of two steps. For illustration, consider the problem of estimating catchment rainfall from one rain gauge (or a small number of rain gauges), i.e. upscaling rainfall from a dm^2 scale to a km^2 scale. The first step involves *distributing* the small-scale precipitation over the catchment (e.g. as a function of topography). The second step consists of *aggregating* the spatial distribution of rainfall into one single value (Figure 8). Conversely, downscaling involves *disaggregating* and *singling out*. In some instances, the two steps are collapsed into one single step, as would be the case of using, say, station weights in the example of upscaling precipitation. However, in other instances only one of the steps is of interest. This has led to a very unspecific usage of the term 'upscaling' in published work referring to either distributing, aggregating, or both. In this paper we will use the terms as defined in Figure 8.

Scaling can be performed either in a deterministic or a stochastic framework. In a deterministic framework, distributing small-scale values results in a spatial (or temporal) pattern which is aggregated into one 'average' value. In a stochastic framework, distributing results in a distribution function (and covariance function), often characterized by its moments, which is subsequently aggregated. The stochastic approach has a particular appeal as the detailed pattern is rarely known and distribution functions can be derived

Table I. Examples for linkages across scales. Abbreviations: swe = snow water equivalent

	State variables	Parameters	Inputs	Conceptualizations
Distributing	Wetness index swe = f (terrain)	Kriging conductivities Soil type mapping	Spline interpolation of rainfall	Distributed model
Singling out	Trivial	Trivial	Trivial	Trivial
Aggregation	Mostly trivial	Effective conductivity = geometric mean	Trivial	Perturbation approach
Disaggregation	Wetness index and satellite data		Mass curves Daily → hourly rainfall	
Upscaling in one step	Snow index stations		Thiessen method depth–area curves	
Downscaling in one step		Runoff coefficient for culvert design		

more readily. On the other hand, deterministic methods have more potential to capture the organized nature of catchments, as discussed earlier in this paper.

Methods of scaling greatly depend on whether state variables, parameters, inputs or conceptualizations are scaled. Although parameters often need to be scaled in the context of a particular model or theory, inputs and state variables can in many instances be treated independently of the model. Table I gives some examples that are of importance in hydrology. Distributing information is usually addressed by some sort of interpolation scheme such as kriging. Singling out is always trivial as it simply involves selecting a subset of a detailed spatial (or temporal) pattern that is known. Aggregating information is trivial or mostly trivial for state variables (e.g. soil moisture) and inputs (e.g. rainfall) as the aggregated value results from laws such as the conservation of mass. It is not trivial, however, to aggregate model parameters (e.g. hydraulic conductivity) as the aggregated value depends on the interplay between the model and the model parameters. Disaggregating information is often based on indices such as the wetness index. Methods for upscaling/downscaling in one step are often based on empirical regression relationships. Finally, scaling conceptualizations refers to deriving a model structure from a model or theory at another scale. Recall that no scaling is involved for a model that is formulated directly at the scale at which inputs are available and outputs are required. Some of the examples in Table I are discussed in more detail in the following sections.

Distributing information

Distributing information over space or time invariantly involves some sort of interpolation. Hydrological measurements are, typically, much coarser spaced in space than in time, so most interpolation schemes refer to the space domain.

The classical interpolation problem in hydrology is the spatial estimation of rainfall from rain gauge measurements. A wide variety of methods has been designed, including: the isohyetal method; optimum interpolation/kriging (Matheron, 1973; Journel and Huijbregts, 1978; Deutsch and Journel, 1992); spline interpolation (Creutin and Obled, 1982; Hutchinson, 1991); moving polynomials; inverse distance; and others. The various methods have been compared in Creutin and Obled (1982), Lebel et al. (1987) and Tabios and Salas (1985), and a review is given in Dingman (1994: 120).

Other variables required for hydrological modelling include hydraulic conductivity (de Marsily, 1986), climate data (Hutchinson, 1991; Kirnbauer et al., 1994) and elevation data (Moore et al., 1991), for which similar interpolation methods are used.

In many instances the supports (i.e. measurements) on which the interpolation is based are too widely spaced and the natural variability in the quantity of interest is too large for reliable estimation. One way to address this problem is to correlate the quantity of interest to an auxiliary variable (i.e. a *covariate* or *surrogate*), whose spatial distribution can more readily be measured. The spatial distribution of the quantity is then inferred from the spatial distribution of the covariate.

In catchment hydrology topography is widely used as a covariate because it is about the only information known in a spatially distributed fashion. Precipitation has a tendency to increase with elevation on an event (Fitzharris, 1975) and seasonal (Lang, 1985) scale, but this is not necessarily so for hourly and shorter scales (Obled, 1990). A number of interpolation schemes have been suggested that explicitly use elevation, ranging from purely statistical (Jensen, 1989) to dynamic (Leavesley and Hay, 1993) approaches. Similarly, snow characteristics, though highly variable at a range of space scales, are often related to terrain (Golding, 1974; Woo et al., 1983a; 1983b). For example, Blöschl et al. (1991) suggested an interpolation procedure based on elevation, slope and curvature. The coefficients relating snow water equivalent to elevation were based on a best fit to snow course data, whereas the other coefficients were derived from qualitative considerations.

Topographic information has also been used to estimate the spatial distribution of soil moisture. The topographic wetness index was first developed by Beven and Kirkby (1979) and O'Loughlin (1981) to predict zones of saturation. The wetness index of Beven and Kirkby (1979) is based on four assumptions:

1. The lateral subsurface flow-rate is assumed to be proportional to the local slope $\tan \beta$ of the terrain. This implies kinematic flow, small slopes ($\tan \beta \approx \sin \beta$) and that the water-table is parallel to the topography.

2. Hydraulic conductivity is assumed to decrease exponentially with depth and storage deficit is assumed to be distributed linearly with depth (Beven, 1986).
3. Recharge is assumed to be spatially uniform.
4. Steady-state conditions are assumed to apply, so the lateral subsurface flow-rate is proportional to the recharge and the area a drained per unit contour length at point i.

Introducing the wetness index w_i at point i

$$w_i = \ln\left(\frac{\bar{T} \cdot a}{T_i \cdot \tan \beta}\right) \tag{4}$$

where T_i is the local transmissivity and $\bar{T}$ is its average in the basin, gives a simple expression for the storage deficit S_i at point i

$$S_i = \bar{S} + m \cdot (\bar{w} - w_i) \tag{5}$$

$\bar{S}$ and $\bar{w}$ are the averages of the storage deficit and the wetness index over the catchment, respectively, and m is an integral length which can be interpreted as the equivalent soil depth. For a detailed derivation, see Beven and Kirkby (1979), Beven (1986) and Wood et al. (1988). A similar wetness index has been proposed by O'Loughlin (1981;1986) which, however, makes no assumption about the shape of hydraulic conductivity with depth.

The wetness index w_i increases with increasing specific catchment area and decreasing slope gradient. Hence the value of the index is high in valleys (high specific catchment area and low slope), where water concentrates, and is low on steep hillslopes (high slope), where water is free to drain. For distributing the saturation deficit S_i over a catchment, Equation (5) can be fitted to a number of point samples. However, the predicitive power of the wetness index has not yet been fully assessed. Most comparisons with field data were performed in very small catchments. Moore et al. (1988) found that the wetness index [Equation (4)] explained 26–33% of the spatial variation in soil water content in a 7·5 ha catchment in NSW, Australia, whereas a study in a 38 ha catchment in Kansas (Ladson and Moore, 1992) suggested an explained variation of less than 10%. Burt and Butcher (1985) found that the wetness index explained 17–31% of the variation in the depth to saturation on a 1·4 ha hillslope in south Devon, UK. There also seems to be a dependence of the wetness index on the grid size used for its calculation, which may become important when using the wetness index in larger catchments (Moore et al., 1993b; Vertessy and Band, 1993; Band and Moore, 1995).

A range of wetness indices has been suggested which attempts to overcome some of the limitations of Equation (4). For example, Barling et al. (1994) developed a quasi dynamic wetness index that accounts for variable drainage times since a prior rainfall event. This relaxes the steady-state assumption. Other developments include the effect of evaporation as controlled by solar radiation (e.g. Moore et al., 1993c). A comprehensive review is given in Moore et al. (1991). Based on the rationale that in many landscapes pedogenesis of the soil catena (Milne, 1935) occurs in response to the way water moves through the landscape (Jenny, 1941; 1980; Hoosbeek and Bryant, 1992), terrain indices have also been used to predict soil attributes (McKenzie and Austin, 1993). For example, Moore et al. (1993a) predicted quantities such as A horizon thickness and sand content on a 5·4 ha toposequence in Colorado where the explained variances were about 30%. Gessler et al. (1993) reported a similar study in south-east Australia.

There is a range of other covariates in use for distributing hydrological information across catchments. Soil type or particle size distributions are the traditional indices to distribute soil hydraulic properties (Rawls et al., 1983; Benecke, 1992; Williams et al., 1992). General purpose soil maps are widely available in considerable detail, such as the STATSGO database developed by the US Department of Agriculture (Reybold and TeSelle, 1989). Unfortunately, the variation of soil parameters (such as hydraulic conductivity) within a particular soil class is often much larger than variations between different soils (Rawls et al., 1983; McKenzie and MacLeod, 1989). Part of the problem is that the spatial scale of variation of hydraulic properties tends to be much smaller (<10 m) than the resolution of most soil maps (Bridges, 1982).

Alternatively, geophysical data have been used for estimating quantities of interest to hydrological modelling (e.g. Rubin et al., 1992; Copty et al., 1993). Mazac et al. (1985) discuss the factors affecting the relationship between the electrical and hydraulic properties of aquifers. For example, an increased clay content tends to decrease both the electrical resistivity and hydraulic conductivity. Kupfersberger and Blöschl (1995) showed that auxiliary (geophysical) data are particularly useful for identifying high hydraulic conductivity zones such as buried stream channels. There is also a range of indices supporting the interpolation of information related to vegetation. For example, Hatton and Wu (1995) used the leaf area index (LAI) for interpolating measurements of tree water use. Further covariates include satellite data, particularly for the spatial estimation of precipitation, snow properties, evapotranspiration and soil moisture (Engman and Gurney, 1991). However, a full review of this is beyond the scope of this paper.

An exciting new area is that of using indicators (Journel, 1986) to spatially distribute information. Indicators are binary variables which are often more consistent with the presence of zones of uniform properties (e.g. clay lenses or geological zones at a larger scale) than the ubiquitous log-normal distributions (Hoeksema and Kitanidis, 1985). They are also more consistent with the type and amount of information usually available. Indicator-based methods have been used to estimate conductivity in aquifers (Brannan and Haselow, 1993; Kupfersberger, 1994) and they have also been used to interpolate rainfall fields (Barancourt et al., 1992). In the rainfall example, the binary patterns represent areas of rainfall/no rainfall.

Aggregating model parameters

The study of the aggregation of model parameters involves two questions. (a) Can the microscale equations be used to describe processes at the macroscale? (b) If so, what is the aggregation rule to obtain the macroscale model parameters with the detailed pattern or distribution of the microscale parameters given.

The macroscale parameters for use in the microscale equations are termed *effective parameters*. Specifically, an effective parameter refers to a single parameter value assigned to all points within a model domain (or part of the domain) such that the model based on the uniform parameter field will yield the same output as the model based on the heterogeneous parameter field (Mackay and Riley, 1991). Methods to address the questions (a) and (b) make use of this definition by matching the outputs of the uniform and heterogeneous systems. If an adequate match can be obtained, an effective parameter exists. The aggregation rule is then derived by relating the effective parameter to the underlying heterogeneous distribution. Methods used include analytical approaches (e.g. Gutjahr et al., 1978), Monte Carlo simulations (e.g. Binley et al., 1989) and measurements (e.g. Wu et al., 1982). Effective parameters are of clear practical importance in distributed modelling, but aggregation rules are of more conceptual rather than of practical importance, either in a deterministic or a stochastic framework. In a deterministic framework, the aggregation rule yields an effective parameter over a certain domain with the pattern of the detailed parameter given. However, in practical applications there is rarely enough information on the detailed pattern available to use the aggregation rule in a useful way. In a stochastic framework, the aggregation rule yields the spatial moments of the effective parameter with the spatial moments of the microscale parameter given. This has been used to determine element spacings or target accuracy for calibration (Gelhar, 1986), but in practice other considerations such as data availability are often more important. Another limitation is the assumption of disordered media properties on which such approaches are often based (e.g. Rubin and Gómez-Hernández, 1990). If some sort of organization is present (e.g. preferential flow; Silliman and Wright, 1988), the same aggregation rules cannot be expected to hold.

What follows is a review of effective parameters and aggregation rules for a number of processes that are of importance in catchment hydrology. These include saturated flow, unsaturated flow, infiltration and overland flow.

The concept of effective hydraulic conductivity has been widely studied for saturated flow. Consider, in a deterministic approach, uniform (parallel flow lines) two-dimensional steady saturated flow through a block of porous medium made up of smaller blocks of different conductivities. It is easy to show that for an arrangement of blocks in series the effective conductivity equals the harmonic mean of the block values. Similarly, for an arrangement of blocks in parallel, the effective conductivity equals the arithmetic mean.

In a stochastic approach, the following results were found for steady saturated flow in an infinite domain without sinks (i.e. uniform flow).

(a) Whatever the spatial correlation and distribution function of conductivity and whatever the number of dimensions of the space, the average conductivity always ranges between the harmonic mean and the arithmetic mean of the local conductivities (Matheron, 1967, cited in de Marsily, 1986). Specifically, for the one-dimensional case, the effective conductivity K_{eff} equals the harmonic mean K_H as it is equivalent to an arrangement of blocks in series

$$\text{One-dimensional}: \quad K_{eff} = K_H \tag{6}$$

(b) If the probability density function of the conductivity is log-normal (and for any isotropic spatial correlation), Matheron (1967) and Gelhar (1986) showed that for the two-dimensional case the effective conductivity K_{eff} equals the geometric mean K_G.

$$\text{Two-dimensional}: \quad K_{eff} = K_G \tag{7}$$

Matheron also made the conjecture that for the three-dimensional case the effective conductivity K_{eff} is

$$\text{Three-dimensional}: \quad K_{eff} = K_G \cdot \exp(\sigma_{\ln K}^2/6) \tag{8}$$

where $\sigma_{\ln K}^2$ is the variance of $\ln K$. This result is strongly supported by the findings of Dykaar and Kitanidis (1992) using a numerical spectral approach. Dykaar and Kitanidis (1992) also suggested that Equation (8) is more accurate than the small perturbation method (Gutjahr et al., 1978) and the imbedding matrix method (Dagan, 1979) when the variances $\sigma_{\ln K}^2$ are large.

(c) If the probability distribution function is not log-normal (e.g. bimodal), both the numerical spectral approach and the imbedding matrix method seem to provide good results, whereas the small perturbation method is not applicable (Dykaar and Kitanidis, 1992).

Unfortunately, steady-state conditions and uniform flow are not always good assumptions (Rubin and Gómez-Hernández, 1990). For transient conditions, generally, no effective conductivities independent of time may be defined (Freeze, 1975; El-Kadi and Brutsaert, 1985). Similarly, for bounded domains and flow systems involving well discharges the effective conductivity is dependent on pumping rates and boundary conditions (Gómez-Hernández and Gorelick, 1989; Ababou and Wood, 1990; Neuman and Orr, 1993). It is interesting to note that the effective conductivity tends to increase with increasing dimension of the system for a given distribution of conductivity. This is consistent with intuitive reasoning: the higher the dimension the more degrees of freedom are available from which flow can 'choose' the path of lowest resistance. Hence, for a higher dimension, flow is more likely to encounter a low resistance path which tends to increase the effective conductivity.

For unsaturated flow in porous media, generally, there exists no effective conductivity that is a property of the medium only (Russo, 1992). Yeh et al. (1985) suggested that the geometric mean is a suitable effective parameter for certain conditions, but Mantoglou and Gelhar (1987) showed that the effective conductivity is heavily dependent on a number of variables such as the capillary tension head. In their examples, an increase in the capillary tension head from 0 to 175 cm translated into a decrease in the effective conductivity of up to 10 orders of magnitude. Mantoglou and Gelhar (1987) also demonstrated significant hysteresis in the effective values. Such hysteresis was produced by the soil spatial variability rather than the hysteresis of the local parameter values.

Similarly to unsaturated flow, for infiltration there is no single effective conductivity in the general case. However, for very simple conditions, effective parameters may exist. Specifically, for ponded infiltration (i.e. all the water the soil can infiltrate is available at the surface) Rogers (1992) found the geometric mean of the conductivity to be an effective value using either the Green and Ampt (1911) or the Philip's (1957) method. This is based on the assumption of a log-normal spatial distribution of hydraulic conductivity. For non-zero time to ponding, Sivapalan and Wood (1986), using Philip's (1957) equation, showed that effective parameters do not exist, i.e. the point infiltration equation does no longer hold true for spatially variable conductivity.

Overland flow is also a non-linear process so, strictly speaking, no effective parameter exists for the general case. However, Wu et al. (1978; 1982) showed that reasonable approximations do exist for certain cases. Wu et al. investigated the effect of the spatial variability of roughness on the runoff hydrographs for an experimental watershed facility surfaced with patches or strips of butyl rubber and gravel. Wu *et al.* concluded that an effective roughness is more likely to exist for a low contrast of roughnesses, for a direction of strips normal to the flow direction and for a large number of strips. They also suggested that the approximate effective roughness can be estimated by equating the steady-state surface water storage on the hypothetical uniform watershed to that on the non-uniform watershed. In a similar analysis, Engman (1986) calculated effective values of Manning's *n* for plots of various surface covers based on observations of the outflow hydrograph. Abrahams *et al.* (1989) measured distribution functions of flow depths and related them to the mean flow depth.

All these studies analysed a single process only. If a number of processes are important, it may be possible to define approximate effective parameters, but their relationship with the underlying distribution is not always clear. Binley *et al.* (1989) analysed, by simulation, three-dimensional saturated–unsaturated flow and surface runoff on a hillslope. For high-conductivity soils, they found effective parameters to reasonably reproduce the hillslope hydrograph, although there was no consistent relationship between the effective values and the moments of the spatial distributions. For low-conductivity soils, characterized by surface flow domination of the runoff hydrograph, single effective parameters were not found to be capable of reproducing both subsurface and surface flow responses. At a much larger scale, Milly and Eagleson (1987) investigated the effect of soil variability on the annual water balance. Using Eagleson's (1978) model they concluded that an equivalent homogeneous soil can be defined for sufficiently small variance of the soil parameters.

The use of effective parameters in microscale equations for representing processes at the macroscale has a number of limitations. These are particularly severe when the dominant processes change with scale (Beven, 1991). Often the dominant processes change from matrix flow to preferential flow when moving to a larger scale. Although it may be possible to find effective values so that the matrix flow equation produces the same output as the preferential flow system, it does so for the wrong reasons and may therefore not be very reliable. A similar example is the changing relative importance of hillslope processes and channel processes with increasing catchment size. Ideally, the equations should be directly derived at the macroscale rather than using effective parameters. However, until adequate relationships are available, effective parameters will still be used.

Disaggregating state variables and inputs

The purpose of disaggregation is, given the average value over a certain domain, to derive the detailed pattern within that domain. Owing to a lack of information for a particular situation, the disaggregation scheme is often based on stochastic approaches. Here, two examples are reviewed. The first refers to disaggregating soil moisture in the space domain and the second refers to disaggregating precipitation in the time domain.

Disaggregating soil moisture may be required for estimating the spatial pattern of the water balance as needed for many forms of land management. The input to the disaggregation procedure is usually a large-scale 'average' soil moisture. This can be the pixel soil moisture based on satellite data, an estimate from a large-scale atmospheric model or an estimate derived from a catchment water balance. One problem with this 'average' soil moisture value is that the type of averaging is often not clearly defined. Nor does it necessarily represent a good estimate for the 'true' average value in a mass balance sense. Although the associated bias is often neglected, there is a substantial research effort underway to address the problem. One example is the First International Satellite Land Surface Climatology Project (ISLSCP) Field Experiment (FIFE) (Sellers *et al.*, 1992). Guerra *et al.* (1993) derived a simple disaggregation scheme for soil moisture and evaporation. The scheme is based on the wetness index and additionally accounts for the effects of spatially variable radiation and vegetation. It is important to note that the wetness index [Equation (4)] along with Equation (5) can be used for both disaggregating and distributing soil moisture. Equation (5) can, in fact, be interpreted as a disaggregation operator (Sivapalan, 1993), yielding the spatial pattern of saturation deficit with the areal average of saturation deficit given. In a similar application, Sivapalan and Viney (1994a,b) disaggregated soil moisture stores from the 39 km^2 Conjurunup catchment in Western Australia to 1–5 km^2 subcatchments, based on topographic and land use characteristics. Sivapalan and

Viney (1994a,b) tested the procedure by comparing simulated and observed runoff for individual subcatchments.

Disaggregating precipitation in time is mainly needed for design purposes. Specifically, disaggregation schemes derive the temporal pattern of precipitation within a storm, with the average intensity and storm duration given. The importance of the time distribution of rainfall for runoff simulations has, for example, been shown by Woolhiser and Goodrich (1988). Mass curves are probably the most widely used concept for disaggregating storms. Mass curves are plots of cumulative storm depths (normalized by total storm depth) versus cumulative time since the beginning of a storm (normalized by the storm duration). The concept is based on the recognition that, for a particular location and a particular season, storms often exhibit similarities in their internal structure despite their different durations and total storm depths. The time distribution depends on the storm type and climate (Huff, 1967). For example, for non-tropical storms in the USA maximum intensities have been shown to occur at about one-third of the storm duration into the storm, whereas for tropical storms they occurred at about one-half of the storm duration (Wenzel, 1982; USDA-SCS, 1986). Mass curves have recently been put into the context of the scaling behaviour of precipitation (Koutsoyiannis and Foufoula-Georgiou, 1993). A similar method has been developed by Pilgrim and Cordery (1975), which is a standard method in Australian design (Pilgrim, 1987). A number of more sophisticated disaggregation schemes (i.e. stochastic rainfall models) have been reported (e.g. Woolhiser and Osborne, 1985) based on the eary work of Pattison (1965) and Grace and Eagleson (1966). A review is given in Foufoula-Georgiou and Georgakakos (1991).

Another application of disaggregating temporal rainfall is to derive the statistical properties of hourly rainfall from those of daily rainfall. The need for such a disaggregation stems from the fact that the number of pluviometers in a given region is always much smaller than that of daily read rain gauges. Relationships between hourly and daily rainfall are often based on multiple regression equations (Canterford *et al.*, 1987).

Up/downscaling in one step

There are a number of methods in use for up/downscaling that do not explicitly estimate a spatial (or temporal) distribution. This means that these methods make some implicit assumption about the distribution and skip the intermediate step in Figure 1. An example is the Thiessen (1911) method for estimating catchment rainfall. It assumes that at any point in the watershed the rainfall is the same as that at the nearest gauge. The catchment rainfall is determined by a linear combination of the station rainfalls and the station weights are derived from a Thiessen polygon network. Another example is depth–area–duration curves. These are relationships that represent the average rainfall over a given area and for a given time interval and are derived by analysing the isohyetal patterns for a particular storm (WMO, 1969; Linsley *et al.*, 1988). Depth–area–duration curves are scaling relationships as they allow the transfer of information (average precipitation) across space scales (catchment sizes) or time-scales (time intervals).

If recurrence intervals or return periods (Gumbel, 1941) are included in the definition of scale, the wide area of hydrological statistics relates to scale relationships. Up/downscaling then refers to the transfer of information (e.g. flood peaks) across time-scales (recurrence intervals).

Linking conceptualizations

Linking conceptualizations across scales can follow either an upward or a downward route (Klemeš, 1983). The upward approach attempts to combine, by mathematical synthesis, the empirical facts and theoretical knowledge available at a lower level of scale into theories capable of predicting processes at a higher level. Eagleson (1972) pioneered this approach in the context of flood frequency analysis. This route has a great appeal because it is theoretically straightforward and appears conceptually clear. Klemeš (1983), however, warned that this clarity can be deceptive and that the approach is severely limited by our incomplete knowledge and the constraints of mathematical tractability (Dooge, 1982; 1986). A well-known example is that of upscaling Darcy's law (a matrix flow assumption), which can be misleading when macropore flow becomes important (White, 1988; Beven, 1991). Klemeš (1983) therefore suggested adopting the downward approach, which strives to find a concept directly at the level of interest (or higher) and then looks for the steps that could have led to it from a lower level. It is clear that the 'depth of inference' (i.e.

the range of scales over which a concept can be inferred) is much more limited in the downward case. There-fore, ideally, the upward and downward search should be combined to form the basis of testable hypotheses (Klemeš, 1983). We will give examples for the two approaches from different areas in hydrology.

The example for the upward approach relates to deriving a macroscale equation for saturated flow in a stochastic framework based on the small perturbation approach (Bakr et al., 1978; Gelhar, 1986). The approach assumes that the local hydraulic conductivity K and local piezometric head H are realizations of a stochastic process and composed of two components

$$K = \bar{K} + k \qquad H = \bar{H} + h \tag{9}$$

where $\bar{K}$ and $\bar{H}$ are the large-scale components and k and h are the local fluctuations. In other words, $\bar{K}$ and $\bar{H}$ are assumed to be smooth functions in space and correspond to mean quantities whereas k and h are realizations of a zero-mean stochastic process. It is clear that the decomposition suggested depends on the scale of the problem examined. For example, what represents the mean at a small-scale laboratory model might be viewed as a fluctuation on a large-scale problem. Combining Equation (9) with the flow equation gives, after a series of assumptions (Bakr et al., 1978), an equation for the mean behaviour. This equation is similar to the local flow equation, but involves additional terms (e.g. representing the covariances of heads and conductivities) which need to be parameterized. This way it gives a new equation, especially if the new additional terms are larger in magnitude than the original terms (Cushman, 1983).

An example of the downward approach has been given by Klemeš (1983). Klemeš related monthly pre-cipitation to monthly runoff in a 39 000 km² basin in Canada. Klemeš found a poor relationship and went back to hypothesize about the reasons. He consequently included the effect of evaporation, gravity storage and tension storage in steps. At each step he tested the hypothesis by examining the data and could finally separate the effects of gravity and tension storage. It is important that these results have been arrived at by analysis rather than by postulating them *a priori*. Along similar lines, Gutknecht (1991b) derived a flood routing model for an Austrian basin. Based on a multiple linear storage model, large events were consis-tently overestimated. Gutknecht (1991b) hypothesized about the reasons and included the effects of inun-dations for floods higher than a certain stage. He then tested this hypothesis by examining the data and the forecasts were consistently improved. A number of model components were included in steps (in case they turned out to be important), which allowed model complexity to be minimized.

Distributed parameter hydrological models.

Nature of distributed parameter models. Distributed parameter models attempt to quantify the hydrological variability that occurs at a range of scales by subdividing the catchment into a number of units (i.e. subareas). These units may either be so-called hydrological response units (e.g. Leavesley and Stannard, 1990), subcatchments (Sivapalan and Viney, 1994a,b), hillslopes (e.g. Goodrich, 1990), contour-based elements (e.g. Grayson et al., 1992) or, for convenience, square grid elements (e.g. Abbott et al., 1986). In such an approach, processes with a characteristic length scale smaller than the grid/element size are assumed to be represented implicitly (= parameterized), whereas processes with length scales larger than

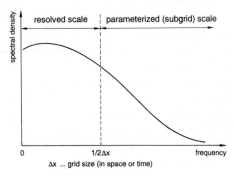

Figure 9. Schematic power spectrum of a hydrological process. Processes with a characteristic scale larger than the grid size are resolved whereas processes with a characteristic scale smaller than the grid size are parameterized as subgrid variability

the grid size are represented explicitly by element to element variations (Smagorinsky, 1974). This is shown in Figure 9, based on a schematic representation of the spectrum of some hydrological process. For example, such a spectrum might be arrived at by, conceptually, sampling infiltration rates along a transect in a catchment and transforming these into the frequency domain. The spectrum in Figure 9 indicates that large-scale processes (low frequency) have more spectral power (= variance) than small-scale processes, which is typical of many hydrological phenomena. It is therefore reasonable to explicitly represent the large-scale processes.

The representation of processes within a unit (i.e. element) involves (a) local (or site) scale descriptions and (b) some assumptions on the variability within the unit. Distributed parameter hydrological models often represent local phenomena (e.g. site-scale infiltration) in considerable detail, while the variability within a unit (i.e. subgrid variability) is often neglected. To drive the models for each unit, input variables (e.g. precipitation) need to be estimated for each element. This involves some sort of interpolation between observations. Conversely, the response from individual elements is coupled by routing routines. Distributed parameter catchment models have been reviewed by Goodrich and Woolhiser (1991) and Moore et al. (1991) and recent contributions are given in Beven (1992) and Rosso (1994).

Many arguments have been put forward in favour of distributed models (e.g. Beven et al., 1980). Although distributed models never lived up to their expectations in terms of their performance in predicting runoff (Loague, 1990; Obled, 1990), this has been early recognized by some workers (Freeze and Harlan, 1969). Bergström (1991) presented an excellent discussion on modelling philosophy in hydrology in general and model complexity in particular. Clearly, the optimum model complexity depends on the nature of a specific problem. There is a wide range of problems where bulk models cannot do the job and distributed parameter models are called for. These include those where spatially distributed estimates of runoff are required and/or a high degree of process understanding is needed (Bergström, 1991).

Unfortunately, distributed models invariably suffer from a number of limitations (Beven, 1989; Grayson et al., 1992; 1993; Kirnbauer et al., 1994). These include (a) the extreme heterogeneity of catchments, which makes it difficult to accurately define element to element variations and subgrid variability and (b) the large number of model parameters, which makes model calibration and evaluation very difficult.

Both aspects are scale-related issues. The quantification of element to element variations has been examined earlier in this paper (distributing information). The representation of subgrid variability and model evaluation are now being discussed.

Subgrid variability. There are three approaches for quantifying the variability of hydrological processes within a computational element (grid cell).

The first approach assumes that the parameters and processes are uniform within each element and that the local (small-scale) descriptions apply to the whole element. The local parameters are then replaced by effective parameters (see section on aggregating model parameters). However, effective parameters do not always exist, particularly when the processes are non-linear.

The second approach uses distribution functions rather than single values. This has advantages for non-linear systems, but complicates identifiability significantly. Goodrich (1990) used distribution functions of saturated hydraulic conductivity for representing variability on a hillslope as a part of a catchment model. A hillslope was decomposed into a number of non-interacting strips in parallel and the values for each strip were assigned according to a log-normal distribution. In a similar fashion, Famiglietti (1992) modelled the spatial variability of evaporation by a distribution function based on the wetness index [Equation (4)]. The distribution of the index was discretized into a number of intervals and the local model was applied to each interval. These and similar techniques are clearly related to methods of distributing information.

The third approach parameterizes subgrid variability without explicitly resorting to the local equations. An excellent example is the parameterization of overland flow in rills suggested by Moore and Burch (1986). In this parameterization individual rills are not modelled explicitly, but rather the lumped effect of a larger number of rills is represented by a power law of the form

$$R = \xi A^m \tag{10}$$

where R is the hydraulic radius, A is the cross-sectional area and ξ and m are parameters. It is important to

Figure 7. Preferential flow along a high conductivity layer. Löhnersbach catchment in the Austrian Alps. Photo courtesy of R. Kirnbauer and P. Haas

note that the information on the detailed geometry of the rills is replaced by only two parameters. This is reminiscent of Darcy's law, where the detailed information on pore geometry is replaced by one lumped parameter (i.e. hydraulic conductivity). The exponent m is unity for sheet flow and 0.5 for trapezoidal or parabolic geometries. For most natural surfaces m lies between these limits (Foster *et al.*, 1984; Parsons *et al.*, 1990; Willgoose and Riley, 1993). The representation of overland flow also illustrates well the partitioning of topographic variability into resolved and parameterized variability (Figure 9) as used in some models (Grayson *et al.*, in press). The resolved variability (i.e. element to element variability) is represented by the shapes, slopes and aspects of individual elements and their relative configuration. The parameterized (i.e. subgrid) variability is represented by the lumped Equation (10). As long as these two representations are consistent, the element size should have little effect on the results. However, this is not always so. Willgoose and Riley (1993), for example, showed that the parameters of Equation (10) can vary significantly with catchment size (see also Willgoose and Kuczera, 1995).

If there is a minimum in the power spectrum of the process to be modelled (Figure 9), i.e. a spectral gap, the quantification of subgrid variability becomes easier. Specifically, the same parameterization can be used for a range of element sizes as long as the element size falls into the spectral gap. This is discussed in more detail in Blöschl *et al.* (1995).

Model evaluation. The difficulty in evaluating and calibrating distributed models on the basis of catchment runoff (e.g. Blöschl *et al.*, 1994) lead Bathurst and Cooley (in press) to remark 'A complicating feature of multiple parameter models is the possibility that apparently equally satisfactory simulations can be achieved with different combinations of physically realistic parameter values, the change in one parameter being compensated for by a change in another.' In other words, runoff, being an integrated value, cannot easily identify the high-frequency component of spatial parameter distributions. One way to address this problem is to use subcatchment runoff for cross-checking (e.g. Sivapalan and Viney, 1994a,b). However, in practice the number of subcatchments for which data are available is often very limited. Also, groundwater levels and tensiometer readings within the catchment have been used (Koide and Wheater, 1992), but these are invariably point values and not always representative. As an alternative it has been suggested to use spatial *patterns* of state variables to assess the accuracy of models within a catchment (Blöschl *et al.*, 1994). Patterns have a very high *space* resolution as opposed to a very high *time* resolution of time series such as hydrographs. One notable example is Moore and Grayson (1991), who evaluated a distributed parameter model on the basis of saturation patterns (i.e. zones of surface saturation). Their analyses were based on a $2 \, m^2$ laboratory catchment. Moore and Grayson (1991) demonstrated that accurate runoff simulations at the catchment outlet do not necessarily imply accurate simulations of distributed catchment response. In a similar fashion, Blöschl *et al.* (1991) used snow cover depletion patterns to evaluate a distributed hydrological model in a $10 \, km^2$ Alpine catchment. They showed that the patterns allowed the assessment of the accuracy of individual model components (such as radiative exchange) and discrimination between alternative model assumptions (Blöschl *et al.*, 1994). It is clear that distributed model evaluation is the key to progress in distributed modelling. However, significantly more data are needed in the future. This particularly refers to reliable measurements of soil moisture patterns, which are ideal for the evaluation of distributed models.

DIMENSIONAL ANALYSIS AND SIMILARITY

Dimensional techniques

Dimensional techniques are powerful for dealing with complex physical problems, as they can potentially describe these systems by very simple relationships. Dimensional techniques have been responsible for some of the great advances in hydraulics (Fischer *et al.*, 1979) and yet have only been used sparingly in catchment hydrology (Dooge, 1986).

Dimensional techniques can be useful in a number of ways. They can be used to define similarity relationships (between two catchments or between a catchment and a scale model) in the same fashion as the Froude and Reynolds numbers have been used in hydraulics. Further, they can be used to establish relationships that are valid over a wide range of scales in the same fashion as the Moody diagram (e.g. Chow *et al.*, 1988) is being used for pipe network design. Also, they are powerful for data reduction. This makes them appealing for regional flood frequency analysis and regionalized loss estimation (NRC, 1988; Wood *et al.*, 1990; Beven, 1991) as they allow for classifying data into zones of similar behaviour. Finally, they have potential for characterizing the dominant hydrological mechanisms operating on a specific catchment.

The foundation of all dimensional techniques rests in the concept of similarity. Similarity exists between two systems whenever the characteristics of one system can be related to the corresponding characteristics of another system by a simple conversion factor, called the scale factor (Langhaar, 1951). Three types of similarity are possible in physical systems: geometric, kinematic or dynamic (Tillotson and Nielsen, 1984). Geometric similarity relates to the size relationship between two systems. When two catchments of different size have similar shapes they are considered to be geometrically similar. Kinematic similarity applies when the ratios of velocities in the two systems are constant multiples of each other. When the instantaneous unit hydrographs of two catchments are related by a constant scale factor, they can be said to be kinematically similar. One example of kinematic similarity relationships in catchment hydrology is the geomorphologic unit hydrograph (Rodríguez-Iturbe and Valdés, 1979). However, it is well known that small catchments behave differently from large catchments due to the effects of spatial heterogeneity and changing processes, among other things, and dynamic similarity may need to be invoked to describe these.

In hydraulics, dynamic similarity is said to exist when the forces in one system are a constant multiple of the equivalent forces in the other. For example, suppose two flow systems depend on inertial and viscous forces only. They are dynamically similar if the ratio of viscous to inertial forces (the Reynolds number) is identical between the two systems. In general, dynamic similarity reflects a balance of two dynamic quantities. This can be a balance of forces, a balance of energy, of mass, of momentum, etc. In catchment hydrology the term dynamic similarity may be ambiguous in meaning, as it is not perfectly clear whether we would consider ratios that we know clearly capture the 'dynamics' of catchment responses as truly representing dynamic similarity. Examples are the ratio of surface to subsurface runoff or the ratio of hillslope response times to stream network response times. Strictly speaking, only kinematic quantities are involved in both examples. Clearly, more careful thought needs to be given to a definition of similarity appropriate to catchment hydrology.

All methods used to determine scale factors can be grouped into three main classes: dimensional analysis, similarity analysis and functional normalization (Tillotson and Nielsen, 1984). Dimensional anlaysis and similarity analysis allow the definition of geometric, kinematic or dynamic similarity, whereas functional normalization does not. In this section we give brief definitions of these techniques and examples of their applications in hydrological modelling and flood estimation.

Dimensional analysis. Dimensional analysis is a simplification process by which the number of dimensional quantities used to describe a system is reduced to a fewer number of non-dimensional quantities called π terms. The number of non-dimensional π terms needed to completely define a system is given by the Buckingham pi theorem (Buckingham, 1914). Note that dimensional analysis can be carried out even when the laws or equations governing the physical system are unknown. In fact, it is a particularly useful technique in such instances.

The systematic procedure by which dimensional analysis should be carried out for any selected physical problem is presented in many standard texts (Fischer *et al.*, 1979; Stull, 1988). This involves two main steps: (a) select the variables relevant to the problem and organize the variables into non-dimensional groups; and (b) perform an experiment, or gather the relevant data from previous experiments, to determine the values of the non-dimensional groups and determine relationships between the non-dimensional groups based on the empirical data. The first step can potentially give dimensionless numbers (such as the Froude number), which could be used to design scale experiments. The second step, in case consistent relationships can be found, is even more useful as it allows the establishment of 'universal' relationships such as those represented in the Moody diagram.

Similarity analysis. As in dimensional analysis, similarity analysis seeks to (a) organize variables into non-dimensional groups and to (b) determine relationships between the groups (Hellums and Churchill,

1961; Tillotson and Nielsen, 1984). The second step, i.e. determining relationships between the groups, follows along the lines of dimensional analysis and can be carried out by field (or laboratory) experiments and/or numerical simulations. However, the first step, i.e. the organization of variables into non-dimensional groups, is different. Specifically, unlike dimensional analysis, similarity analysis requires that physical laws or equations governing the system of interest are known. The idea in similarity analysis is to rewrite the partial differential equations in non-dimensional form, which gives the dimensionless groups. Similarity analysis is more powerful than dimensional analysis (Miller, 1990). For example, it can handle more than one parameter with a given unit (e.g. more then one quantity of unit length) and it can also handle dimensionless quantities, which is not the case with dimensional analysis. A classical example of similarity analysis in soil physics is the work of Miller and Miller (1956).

Functional normalization. Unlike the two physically based dimensional techniques discussed earlier, functional normalization (Tillotson and Nielsen, 1984) is an empirical method. It begins with a set of empirical relationships between two or more variables such as soil moisture characteristic relations. The objective then is to coalesce all such empirical relationships in the set into one, or a few, reference curve(s) that describe the set as a whole. Functional normalization is therefore a simplification process which is empirical and does not necessarily have any physical justification.

Hydrological similarity of catchment responses

Dimensional analysis of catchment form and dynamics. Strahler (1964) suggested that catchments with common underlying geology, climate and lithology contain a high degree of geometrical similarity. 'Horton's laws' of network morphology (Horton, 1945; Strahler, 1957) (which are discussed later in this paper), describe a form of geometrical similarity. Note that these are not physical laws, but rather empirical laws and the process by which they are derived is an example of functional normalization.

Strahler (1964) made the first attempt at applying dimensional analysis to examine the similarity of catchment form based on its dynamics. Through application of the Buckingham pi theorem, Strahler suggested that the normalized drainage density of catchments (denoted by HD) can be related to three physically based non-dimensional groupings, or similarity parameters

$$HD = \Pi(Q_r K, \ Q_r \rho H / \mu, \ Q_r^2 / Hg) \tag{11}$$

where Q_r is runoff intensity (rainfall intensity minus infiltration capacity), K is an erosion proportionality factor defined as the ratio of erosion intensity to erosion force (a measure of erodibility of the ground surface), D is the drainage density, H is the vertical relief representing the potential energy of the system, ρ is the density of water, μ is viscosity and g is the acceleration due to gravity. $Q_r K$ is called the Horton number, which expresses the relative intensity of erosion processes within the catchment. $Q_r \rho H / \mu$ is a form of the Reynolds number in which Q_r takes the place of the velocity and H is the characteristic length. Finally, Q_r^2 / Hg is a form of the Froude number. The reduction of the seven dimensional variables to four non-dimensional groups, shown in Equation (11), 'focuses attention upon dynamic relationships, simplifies the design of controlled empirical observations, and establishes conditions essential to the validity of comparisons of models with prototypes' (Strahler, 1964: 4–71). However, the actual functional form of Π in Equation (11) has not yet been determined. Also, it is not clear whether the non-dimensional terms are measurable in practice.

Similarity analysis of runoff generation. As runoff generation is a complex phenomenon, involving tremendous spatial heterogeneity, it has been difficult to generate universally applicable concepts and theories to describe runoff generation. Sivapalan *et al.* (1987) applied similarity analysis to gain insights into the processes of runoff generation. The starting point was a physically based model of runoff generation, based on TOPMODEL concepts (Beven and Kirkby, 1979), which included both infiltration excess and saturation runoff generation.

The first step in the similarity analysis involved recasting the model equations into non-dimensional forms. All variables in the model equations were first transformed to non-dimensional form by dividing each by a fixed arbitrary reference quantity of the same dimension. These were then substituted into the model equations. The resulting equations were then rearranged to their non-dimensional forms in such a way that the coefficient of the leading term of every equation, often the left-hand side of the equation, was equal to unity. The similarity analysis along these lines by Sivapalan *et al.* (1987) resulted in the identification of five non-dimensional similarity parameters and three auxiliary variables. It was believed that these parameters embody the interrelationships of topography, soil and rainfall that lead to similar catchment responses. However, the identification of these non-dimensional groupings is only the first step in similarity analysis. The second, and perhaps even more important, step is the derivation of relationships among the similarity parameters, i.e. similarity relationships.

Larsen *et al.* (1994) carried out an extension of the similarity analysis of Sivapalan *et al.* (1987), which resulted in the derivation of a number of similarity relationships. Their study was carried out on a number of small agricultural catchments in the eastern wheat belt of Western Australia. The study concentrated on the relative dominance of two alternative mechanisms of runoff generation in the region: infiltration excess and saturation excess. To quantify this, Larsen *et al.* (1994) defined a runoff measure, denoted by R, as the ratio of the volume of simulated saturation excess runoff to the volume of simulated combined runoff, integrated over a hypothetical year. If $R = 1$, then all runoff is of the saturation excess type, whereas if $R = 0$, then all runoff is of the infiltration excess type. According to Larsen *et al.*'s similarity theory, the dependence of R on topography, soils and rainfall, and their spatial variability, can be expressed in terms of five non-dimensional similarity parameters and two auxiliary variables, as follows:

$$R = \Psi\{p^*, t_d^*; f^*, K_0^*, B^*, C_{V, K_0}, C_{V, \ln(a/\tan\beta)}\} \tag{12}$$

where p^* and t_d^* are auxiliary climatic parameters relating to the mean rainfall intensity and mean storm duration, and $f^*, K_0^*, B^*, C_{V, K_0}, C_{V, \ln(a/\tan\beta)}$ are the non-dimensional similarity parameters (f^* is a measure that is inversely related to the storage capacity of the soil profile; K_0^* is a measure of the infiltration capacity of the soil; B^* is related to the pore size distribution; C_{V, K_0} and $C_{V, \ln(a/\tan\beta)}$ are the coefficients of variation (in space) of K_0^* and the wetness index, respectively). The purpose of similarity analysis is to derive Equation (12) based on a combination of field or laboratory data and simulations.

By means of sensitivity analysis with their model, and analysis of field data from seven small agricultural catchments in the eastern wheat belt, Larsen *et al.* (1994) showed that only the first four parameters in Equation (12) were important for predicting R (i.e. predicting which runoff generation process dominated in these catchments). Larsen *et al.* (1994) then carried out extensive simulations, with artificial catchments derived by varying the four critical similarity parameters. Each simulation resulted in a value of R which could then be used to infer the complex similarity relationship among p^*, t_d^*, f^* and K_0^*. For

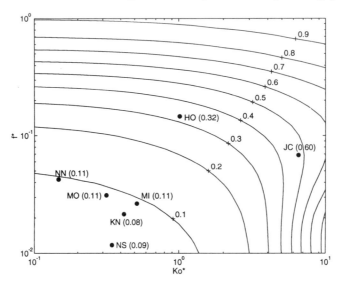

Figure 10. Contours of R values (ratio of simulated saturation excess runoff volume to simulated total runoff volume) as a function of K_0^* (a measure of infiltration capacity) and f^* (a measure that is inversely related to the storage capacity of the soil). The points represent actual catchments. From Larsen *et al.* (1994)

verification, Larsen *et al.* (1994) also estimated the values of R for each of the seven experimental (actual) catchments.

Larsen *et al.* (1994) presented their results in the form of contour plots of the variation of R in $K_0^*-f^*$ space and $p^*-t_d^*$ space. Figure 10 is an example of the similarity relationships so obtained, and presents the variation of R in $K_0^*-f^*$ space. The R values for the seven actual catchments are presented within brackets alongside their positions on the diagram. Both the parameters f^* and K_0^* have physical meanings which are directly related to the two runoff generation mechanisms. In Figure 10, the bottom left-hand corner, for example, relates to deep soils (small f^*) of low infiltration capacity (small K_0^*), hence runoff is mainly of the infiltration excess type. This is clearly consistent with physical reasoning. The contours in Figure 10 are similarity relationships between the parameters f^* and K_0^* for catchments in the eastern wheat belt region. The R values for the actual catchments are consistent with the values predicted by the similarity theory (contour lines). This demonstrates that catchment responses can indeed be scaled using the similarity parameters f^* and K_0^*.

Robinson and Sivapalan (1995) have further extended the work of Larsen *et al.* (1994). They investigated the regionalization, based on the same concepts of similarity, of a lumped catchment-scale runoff prediction equation which included both infiltration excess and saturation excess mechanisms of runoff generation. One of the advantages of this approach is that the regionalization is based on runoff generating mechanisms and field soil properties that, at least in principle, can be measured in the field. In reality, such field estimation is not a trivial exercise.

Regionalization of flood frequency.

Empirical flood frequency analysis: functional normalization

One of the practical motivations for the investigation of hydrological similarity is the widely perceived need to improve the physical basis for the grouping of catchments for regional flood frequency analysis. Ideally, regionalization of flood frequency should be based on the actual flooding mechanisms operating on catchments within the region. However, in current practice, such regionalizations are carried out mainly on empirical grounds with only a few exceptions (e.g. Kölla, 1987). The empirical procedures used for this exercise belong to the category of functional normalization. This is because they are aimed at collapsing a large number of empirical relationships (flood magnitude against probability of exceedance for different catchment types and sizes) into one or a few regional relationships(s).

Empirical regionalization of flood frequency involves two steps: forming catchment groupings, and coalescing all of the flood frequency curves within each group into one characteristic regional curve. The latter is usually done by scaling the flood peaks by the mean annual flood. This procedure is known as the index flood method (Flood Studies Report, 1975; NRC, 1988).

Very sophisticated techniques are presently available for grouping catchments based on objective statistical criteria (Wallis, 1988; Wiltshire, 1985). An excellent example of the application of these techniques for regional flood frequency analysis is provided by work carried out in New Zealand, as summarized in McKerchar and Pearson (1989) and Pearson (1993).

Model-based derived flood frequency: similarity analysis

The use of similarity analysis in combination with the derived flood frequency approach was pioneered by Eagleson (1972). Considerable work has been carried out since by Wood and Hebson (1986) and Hebson and Wood (1986), who applied similarity analysis to gain insights into the effects of catchment size (in relation to storm size) on the flood frequency curve. However, much of this work was concerned with runoff routing rather than runoff generation.

Sivapalan *et al.* (1990) extended the derived flood frequency work of Wood and Hebson (1986) and Hebson and Wood (1986) by explicitly considering aspects of runoff generation. Their derived flood frequency model combined the runoff generation model of Sivapalan *et al.* (1987) and a generalized geomorphologic unit hydrograph (based on Horton order ratios) to handle partial area runoff generation. The combined model was rewritten into a non-dimensional form by adopting the methods of similarity analysis used by Sivapalan *et al.* (1987). Using this model Sivapalan *et al.* (1990) carried out extensive simulations to investigate the sensitivity of the flood frequency characteristics to the similarity parameters.

An important conclusion of the work of Sivapalan *et al.* (1990) was that there is indeed a connection

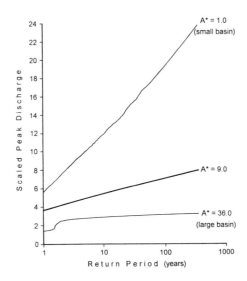

Figure 11. Sensitivity of the flood frequency distribution to the scaled catchment area A^* from Sivapalan *et al.* (1990)

between the shape of the flood frequency curve and the relative dominance of the mechanisms of runoff generation. In view of the groupings of the eastern wheat belt catchments obtained by Larsen *et al.* (1994) in terms of the dominant mechanisms of runoff generation, i.e. the ratio R, any possible connection between the empirical and physically based groupings of the catchments merits very careful examination. Hebson and Wood (1986) and Sivapalan *et al.* (1990) also showed that the ratio of the spatial correlation length of the storms to the catchment size was critical to the shape of the flood frequency curve. This is illustrated by Figure 11, taken from Sivapalan *et al.* (1990), which indicates that the shape of the flood frequency curve changes with catchment scale, becoming steeper with decreasing catchment size (relative to storm size). Current work by Gupta and Dawdy (1995; 1994) suggests that these conclusions are consistent with a theoretical framework of regional flood frequency based on multi-scaling arguments. Further investigation of these connections is worthwhile as they may have wide-ranging practical significance.

Self-similarity and fractals

Fractals, like other similarity concepts, can potentially describe complex phenomena by a minimum of parameters, which makes them an appealing concept. Fractals are based on the recognition that variability exists at a range of scales. In fact, fractals allow the relationship of variability between different scales to be quantified. The main use of fractal concepts in hydrology is in determining this relationship. This can be the quantification of rainfall variability as a function of time- (or space) scale or the quantification of soil parameters as a function of space scale. Once the relationship is derived for certain conditions it is possible to extrapolate the rainfall (or soil parameter) variability to a larger or smaller time- (or space) scale. Similar properties of fractals are used in analyses of the structure of stream networks. One specific application is the definition of a critical scale at which the natural phenomena (e.g. topography) deviate from the 'ideal' fractal relationship. This seems to have potential to discriminate between hillslope and channel processes (Tarboton *et al.*, 1991; Montgomery and Foufoula-Georgiou, 1993). It has also been attempted to use fractals to *classify* landscapes (Klinkenberg and Goodchild, 1992), soils and other variables.

Fractals are particularly convincing by their visual appearance (such as simulated clouds), which often resembles natural phenomena (e.g. Lovejoy and Schertzer, 1985). However, this may be deceptive as fractals, generally, are non-physical approaches and care must be exercised in their interpretation. It will be a challenge to investigate to which degree fractals can be compatible with physical reasoning (Gupta and Dawdy, 1995; 1994).

What are fractals? Simple fractals are sets with certain properties independent of scale (Voss, 1988; Cutler, 1993). Fractals can be either deterministic or random. In the case of deterministic (geometric) fractals, the property independent of scale is the geometric shape of the set. In the case of random (statistical) fractals, these properties independent of scale are certain statistical properties of that set (Feder, 1988). The selection of the statistical properties that are required to be independent of scale depends on whether an ordered set or an unordered set is considered. Ordered sets are sets where a quantity is associated with a location (in space or time), whereas unordered sets are simply collection of objects. Specifically, unordered sets are said to be simple random fractals if their probability density function (pdf) is independent of scale (e.g. area–perimeter relationships; Lovejoy, 1982; Rodríguez-Iturbe *et al.*, 1992a), whereas ordered sets (i.e. mathematical processes) are said to be simple random fractals if properties such as the pdf of their *increments* are independent of scale (see Mandelbrot and Van Ness, 1968; Klinkenberg and Goodchild, 1992). We will detail the requirements for ordered sets to be simple random fractals as these are most commonly used in hydrology. Usually, one of the following three relationships is examined.

(a) Scaling of the increments. In the first description, the *pdf of the increments* of a process is required to be independent of scale (Mandelbrot and Van Ness, 1968; Lovejoy and Schertzer, 1989). Such a pdf is called, equivalently, scaling, self-similar or hyperbolic. The condition of independence of scale is satisfied by a power law

$$\{Z(x+h) - Z(x)\} \stackrel{d}{=} \lambda^{-H}\{[Z(x+\lambda h) - Z(x)]\} \tag{13}$$

where h is the spacing (lag) between two points x and $x + h$; λ is the scaling parameter; H is a scaling exponent; $Z(x + h) - Z(x)$ is the increment (fluctuation) of the property for a given spacing h and point x and the equality $\{\cdot\} \stackrel{d}{=} \{\cdot\}$ indicates equality of pdfs. Equation (13) compares two pdfs. One is derived from a set of increments with a small spacing. For example, this may be a time series of rainfall Z with a spacing (time step) h of, say, five minutes. The other pdf is derived from a set of increments with a larger spacing (λh), which might be a time series of rainfall with hourly time steps. In practical applications, the latter is often arrived at by resampling the former. The ratio of the spacings is the scaling parameter λ ($\lambda = 12$ in the above example).

(b) Semivariogram. In the second description, the *variance of the increments* of a process (i.e. the semivariogram) is required to be independent of scale (Mandelbrot and Van Ness, 1968; Mandelbrot, 1977)

$$\gamma(h) = \lambda^{-2H}\gamma(\lambda h) \tag{14a}$$

and

$$\gamma(h) = 0\cdot5E\{[Z(x+h) - Z(x)]^2\} \tag{14b}$$

where $\gamma(h)$ is the semivariogram function and $E\{\ \}$ refers to the expectation (Oliver and Webster, 1986). Clearly, Equation (14) is a relaxation of Equation (13) as only variance is required to scale rather than the complete pdf.

(c) Power spectrum. In the third description the *power spectral density S* of a process is required to be independent of scale (Mandelbrot and Van Ness, 1968; Voss, 1985)

$$S(\lambda f) = \lambda^{-\beta}S(f) \tag{15}$$

where f is the frequency and β is a scaling exponent. Equations (14) and (15) are different representations of the same information. 'Roughly speaking', a power law in both Equations (14) and (15) is consistent (Carlson, 1986: 162ff; Voss, 1988) and Equation (14) is equivalent to Equation (15) with a linear relationship between β and H (e.g. Voss, 1988). Strictly speaking, however, the consistency is not very clear and the relationship between H and β appears to be non-linear (Gallant *et al.*, 1994). Part of the problem is that the power spectrum Equation (15) is undefined as fractals are non-stationary.

The power laws Equations (13), (14) and (15) appear as straight lines in double logarithmic plots. The goodness of fit of data to that straight line is generally used as the one criterion to judge if a given data set may be approximated by a fractal or not (Feder, 1988; Tarboton *et al.*, 1988). However, it is notoriously easy to fit double logarithmic relationships to data (Lewis, 1995). Also, the fractal approximation is often limited by a lower and an upper cutoff (Feder, 1988).

There are a variety of other techniques that are used to determine if a set is a simple random fractal or not (Gallant *et al.*, 1994) and the value of the scaling exponent. These techniques include the divider method (Mandelbrot, 1967) and the rescaled range analysis (Mandelbrot and Wallis, 1969).

Using various techniques, a substantial amount of geophysical data (including rainfall and streamflow) has been examined (e.g. Hurst, 1951; Mandelbrot and Wallis, 1968; 1969; Lettenmaier and Burges, 1977; Lovejoy, 1982; Skoda, 1987; Klinkenberg and Goodchild, 1992). Most analyses showed strong evidence for values of the scaling exponent H [in Equation (13) or (14)] to be greater than 0·5 (typically between 0·7 and 0·8). For a process in which any event is dependent only on the preceding event (a Markov process), the exponent H is 0·5, whereas $H > 0·5$ indicates long-term persistence (the Hurst phenomenon). The existence of long-term persistence would question traditional statistical methods in hydrology, particularly those related to design. However, it is not yet clear how robust the methods used in these analyses really are. Specifically, work by Mesa and Poveda (1993) questioned their reliability. Based on a so-called 'GEOS' diagram which they deemed more reliable, Mesa and Poveda found no evidence that H is significantly different from 0·5. More work seems to be needed on that issue.

The preceding section dealt with simple random fractals. Simple fractals have a number of drawbacks, particularly for rainfall modelling (Lovejoy and Schertzer, 1989). Firstly, they involve only two parameters (slope and intercept of the straight line in the double logarithmic plot) and are thus very special. Secondly, they cannot cope with zero rainfall and require thresholding of the fractal. One way of simultaneously overcoming both drawbacks is to use 'multifractals' (multiple scaling).

Multifractals are more general than simple fractals. They consist of intertwined fractal subsets with different scaling exponents (Feder, 1988). Also, they exhibit a very singular limiting behaviour with values everywhere almost surely zero, and all of the mass is concentrated into singularities of various orders. Applications in the context of atmospheric turbulence, rainfall and streamflow have been suggested by Oboukhov (1962), Gupta and Waymire (1990; 1993), Lovejoy and Schertzer (1991), Sreenivasan (1991) and Gupta and Dawdy (1994, 1995). Multifractals involve *multiplicative* modulation of the small scales by the large, as opposed to *adding* random elements at different scales (as is the case for simple fractals). Also, multifractals are defined for *measures*, whereas simple fractals are defined for *sets* (Falconer, 1990). One convenient model of multifractals is the binomial multiplicative cascade (e.g. Gupta and Waymire, 1993). Multifractals can be characterized by their singularity spectrum $f(\alpha)$ (Halsey *et al.*, 1986; Ijjasz-Vasquez *et al.*, 1992), which describes the scaling behaviour of the singularities of a process rather than that of the process itself. Ijjasz-Vasquez *et al.* (1992) argued that the singularity spectrum is 'useful as a first step toward the development of multiplicative cascade models of energy dissipation similar to models implemented in studies of turbulence (Menevau and Sreenivasan, 1987; Chhabra *et al.*, 1989; Menevau *et al.*, 1990).' However, Chhabra *et al.* (1989) indicated that extracting underlying multiplicative processes from the singularity spectrum may be very difficult. Wavelets (e.g. Daubechies, 1992) may have some potential (Tarboton, pers. comm.). It will be a challenge for future research to adequately address this identification problem. It will also be a challenge, once the underlying process is known for a given catchment, to translate it into information of practical relevance (Smith, 1992; Gupta and Dawdy, 1994, 1995).

STREAM NETWORK ANALYSIS

Importance of stream network analysis

Stream network analysis is one area in which dimensional techniques and similarity are particularly important and widely used. This is because dimensional techniques and similarity concepts are particularly powerful for dealing, in a simple way, with a range of complex processes such as those involved in forming stream channels. The original motivation of characterizing stream networks for hydrological applications was the hope that synthetic unit hydrographs can be derived for ungauged catchments based on properties of the stream network. Although this has never been successful for practical applications, it has been useful in understanding the scaling behaviour of catchment response. In other words, stream network analysis provides investigative rather than predictive tools. A further potential and future goal is that the spatial variability of soils might be inferred from stream networks. This is based on the rationale that the processes involved in the formation of stream networks and soils are coupled by a number of feedback mechanisms (Beven *et al.*, 1988; Gupta *et al.*, 1986b).

To describe a network completely, two pieces of information are needed. These are (a) the topology (i.e. map view and elevation) and (b) the cross-sectional geometry and hydraulic properties. Information on both aspects is needed for models such as the geomorphologic unit hydrograph or the metachannel.

Quantitative description of the topology of stream networks

Early work on the quantification of stream networks relied on an ordering system devised by Horton (1945) and later improved by Strahler (1957), which highlighted the natural convergence inherent to stream networks. Following the analysis of a large number of stream networks, Horton (1945) and Schumm (1956) presented a set of empirical relationships, known as *Horton's laws*, which were based on this ordering system. They found that, in a region of uniform geology and climate, the ratios of number of streams, length of streams, area of streams and slopes of streams between successive orders are approximately constant, i.e. independent of order. These constants are generally known as the Horton order ratios, and are given by

$$\text{Bifurcation ratio} \qquad R_B = N_{w-1}/N_w \qquad (16a)$$

$$\text{Length ratio} \qquad R_L = L_w/L_{w-1} \qquad (16b)$$

$$\text{Area ratio} \qquad R_A = A_w/A_{w-1} \qquad (16c)$$

$$\text{Slope ratio} \qquad R_S = S_{w-1}/S_w \qquad (16d)$$

where N_w is the number of streams of order w, L_w is the mean length of streams of order w, A_w is the mean area contributing to the streams of order w and S_w is the mean slope of streams of order w.

The Horton order ratios [Equations (16)] are dimensionless parameters that are often thought to collectively represent scaling relationships for catchments of different sizes. However, it has been shown (Kirchner, 1993) that almost all possible networks have Horton order ratios similar to those derived for natural catchments. This is considered by Kirchner to imply that Horton's 'laws' are artefacts of stream ordering techniques and as such express little or nothing as regards either the uniqueness of individual networks or indeed the scaling between catchments of different sizes with their attendant networks. Moreover, the scaling of network properties using Horton order ratios becomes even more difficult in regions of changing or non-uniform lithology (the erodibility and infiltration capacity of soils), geology (structural forms) and climate (Horton, 1945; Chorley *et al.*, 1984).

Several other quantitative measures of stream networks have also been proposed, which do not assume the Strahler ordering. Rather, they express the cumulative properties of the network geometry in terms of the flow distance s, or altitude h, from the outlet, in effect collapsing the network into a single 'effective' channel. Some of the more commonly used functions based on the flow distance s or altitude h are: (1) the width function, $N(s)$, which expresses the number of links occurring at a flow distance s from the outlet (Lee and Delleur, 1976; Kirkby, 1976; Mesa and Mifflin, 1986); (2) the distance–area function, $a(s)$, which describes the local area draining per unit length of channel at a flow distance s from the outlet (Snell and Sivapalan, 1994a,b); (3) the link concentration function, $N(h)$, which expresses the number of links occurring at an altitude h above the outlet (Gupta *et al.*, 1986b); and (4) the hypsometric curve (Strahler, 1957), denoted by $A(h)$, describing the cumulative area of the catchment above a given altitude h.

In the past, the methods for estimating the Horton order ratios, and other geomorphological functions, relied heavily on topographic maps and used manual procedures, which are time consuming. With the advent of digital elevation models (DEMs), there has been a proliferation of automatic procedures for processing DEMs, extracting channel networks and estimating the various geomorphological functions and parameters (Band, 1986; Jenson and Domingue, 1988; Band and Moore, 1995). All of these methods define the streams on a DEM as consisting of all points with accumulated drainage area above some threshold value, called a 'support area'.

The choice of support area is often made in a subjective manner as insufficient guidance is available based on physical considerations. There is considerable debate as to whether the use of a support area is the best method to define channel initiation, considering the spatial non-uniformity of geology, soils and rainfall over large geographical regions. As yet, the problem is far from resolved and work is continuing, using both theoretical and field studies, to find the most objective way to define the initiation of stream channels (Tarboton *et al.*, 1991; Dietrich and Dunne, 1993). For example, Montgomery and Foufoula-Georgiou (1993) have shown that a constant critical support area, the method most commonly used at present, is more appropriate for depicting the hillslope/valley transition than for identifying channel heads and suggest that a slope-dependent (i.e. non-constant) critical support area is both theoretically and empirically more appropriate for defining the extent of channel networks. Resolution of this problem is important as it is becoming clear that geomorphological attributes such as order ratios and the width and area functions are heavily dependent on the support area used to extract stream networks from DEMs (Lee and Delleur, 1976; Helmlinger *et al.*, 1993; Snell and Sivapalan, 1994b; Gyasi-Agyei *et al.*, 1995).

Some recent work has attempted to characterize stream network morphology in terms of fractals and multifractals as a way to better quantify the scaling in geomorphological attributes (Tarboton *et al.*, 1988; 1989; Rosso *et al.*, 1991). In particular, most of this work was geared towards the estimation of fractal exponents and to the study of how these exponents can be related to geomorphological laws such as the Horton's laws. Much work is now in progress to build models of hydrological response based on a fractal or multi-fractal characterization of stream networks (Marani *et al.*, 1991; Rinaldo *et al.*, 1991). Although these methods are mathematically sophisticated, it is unlikely that significant progress can be achieved without accounting for the effects of geology, lithology and climate. For example, geology is clearly a dominant control of network structure (e.g. von Bandat, 1962; Drury, 1987). This is probably best known for drainage density (the ratio of total stream length to basin area). Drainage densities can vary more than three orders of magnitude between, say, limestone strata in Germany ($< 0.28 \, \text{km/km}^2$; Kern, 1994) and badlands developed on clay in New Jersey (about $700 \, \text{km/km}^2$; Strahler, 1964). It is clear that such differences are also very important for more complex stream network characteristics. Also, mainly because of the differences *between* regions, drainage density is a very useful parameter for practical hydrological applications such as regional flood frequency (or low flow) analysis. Here, as in the case of sophisticated stream network theories, the rationale is the complex interplay and interdependence between runoff processes and stream network evolution (Troch *et al.*, 1995). For future research it may be more rewarding to look at the differences *between* different network types as related to geology and climate. Such analyses are more likely to give us a clue on feedback mechanisms and may be more useful for practical applications.

Hydraulic geometry

Hydraulic geometry refers to the quantitative expression of the relationships between the hydraulic characteristics of a channel cross-section and the flow through the channel. Initial work on hydraulic geometry was performed by Leopold and Maddock (1953), who derived a number of empirical relationships based on field measurements. They expressed these relationships in the following way:

$$v \propto Q^m; \quad d \propto Q^f; \quad w \propto Q^b; \quad S \propto Q^z; \quad n \propto Q^y \tag{17}$$

where v is the velocity of flow, d is the mean depth of the flow, w is the surface flow width, S is the energy slope (approximately equal to stream bed slope), n is the Manning friction coefficient and the exponents m, f, b, z and y are constants. These relationships apply to the variations of the hydraulic characteristics with flow, either at a single site, or in the downstream direction (under the assumption that all cross-sections are experiencing flow having an identical probability of exceedance). The exponents m, f, b, z and y would be different for these two instances. Because of the direct connection of discharge to catchment area, these downstream hydraulic geometry relationships can be viewed as scaling relationships. However, caution has to be exercised in such treatment of hydraulic geometry as changes in hydraulic geometry due to floodplain formation and river meandering, often found at large scales, have not been incorporated into these relationships. For example, in large lowland rivers, channels can actually narrow downstream even as the total runoff increases.

Typical values found by Leopold and Maddock (1953) for the exponents for at a site hydraulic geometry were $m = 0.34$, $f = 0.40$ and $b = 0.26$ ($m + f + b = 1$ since $Q = v \cdot d \cdot w$). For downstream hydraulic geometry, the values were $m = 0.1$, $f = 0.4$ and $b = 0.5$. Leopold and Langbein (1962) carried out a theoretical analysis of hydraulic geometry based on considerations of entropy and obtained roughly similar values. The small value of m obtained for downstream hydraulic geometry implies that, whereas flow velocity varies considerably with the flow at a single site (i.e. varies in time), it remains approximately constant in the downstream direction (i.e. is almost constant in space). This result has been independently confirmed by a number of workers (Pilgrim, 1977). However, the power law forms of the hydraulic geometry relationships derived by Leopold and Maddock (1953) have been shown to be simplistic in nature, both as regards the at a site and downstream hydraulic geometries they are trying to express. At a site geometry exponents display a functional relationship with discharge expressed by Richards (1982) in log-quadratic form. Furthermore, Ferguson (1973) and Knighton (1974; 1975) show that at a site hydraulic geometry is multivariate rather than bivariate with the sedimentology of the banks playing an important part in the determination of channel width. Downstream hydraulic geometry is also complicated by factors such as spatial variability in the magnitude and frequency of events, especially with regard to upstream and downstream reaches, together with systematic variation at the local scale and the multivariate nature of the controls on channel geometry. Reviews on factors determining the hydraulic geometry relationships can be found in Richards (1982), Knighton (1984) and Hey (1988).

Rodríguez-Iturbe et al. (1992b) showed that the combination of three principles of energy expenditure can be used to derive the most important structural characteristics observed in stream networks, including downstream hydraulic geometry variation, the downstream variation of stream slope as a function of catchment area and Horton order ratios. The three principles used by Rodríguez-Iturbe et al. (1992b) are: (1) the principle of minimum expenditure in any link of the network; (2) the principle of equal energy expenditure per unit area of channel anywhere in the network; and (3) the principle of minimum total energy expenditure in the network as a whole. However, some of the assumptions, such as a constant rating curve across the catchment, appear to be speculative and need further examination. Chang (1979; 1982), Yang (1976) and Yang and Song (1979) have foreshadowed the work of Rodríguez-Iturbe et al. (1992b) by their use of variational principles regarding minimum stream power per unit length and minimum rate of energy dissipation in deriving hydraulic geometries of rivers which have reached stable, equilibrium values, i.e. are in regime.

Geomorphologic instantaneous unit hydrograph

The concept of a geomorphologic instantaneous unit hydrograph (GIUH) was introduced by Rodríguez-Iturbe and Valdés (1979), based on Lee and Delleur (1972; 1976), and was later generalized by Gupta et al. (1980) and Rinaldo et al. (1991). The original development was geared towards deriving synthetic unit hydrographs for ungauged catchments, but its main use has since been to understand the scaling behaviour of catchment response rather than practical applications. The GIUH is built on the following premises:

1. The GIUH of a catchment can be taken to be the probability distribution of arrival times at the outlet due to a unit impulse of excess rainfall which is uniform across the catchment.
2. The distribution of arrival times is strongly governed by the distribution of water flow pathways from the sources of runoff generation to the catchment outlet.
3. The pathways can be characterized by two fundamental properties. These are (i) a probability that a droplet falling into the catchment will follow this pathway and (ii) a characteristic length representing the distance the droplet will travel from the source of the pathway to the outlet.
4. A pathway is generic — it does not define a path between a specific point on the catchment to the outlet, but rather it defines a set of such points whose individual paths are 'similar', a consequence of the postulates attributed to Shreve (1966; 1967).
5. Each pathway is associated with a pdf of the travel times along that pathway.

Current formulations of the GIUH differ in the manner in which the network morphology is used to characterize the possible pathways and to estimate the probabilities and lengths associated with these pathways. They also differ in the parameterizations assumed to describe the travel time distributions in each pathway. Two alternative approaches have been proposed to characterize the distribution of pathways in terms of network morphology (Snell and Sivapalan, 1994a): (i) methods based on Strahler ordering (Rodríguez-Iturbe and Valdés, 1979; Gupta et al., 1980; Rinaldo et al., 1991); and (ii) methods based on the width and area functions (Lee and Delleur, 1976; Kirkby, 1976; Mesa and Mifflin, 1986; Beven and Wood, 1993; Snell and Sivapalan, 1994a).

In the methods based on Strahler ordering, the concept of a pathway becomes a set of transitions between the initial order of a water droplet (the order of the stream into which the droplet is initially injected) and higher and higher ordered streams until the outlet is eventually reached. Rodríguez-Iturbe and Valdés (1979) and Gupta et al. (1980) estimated the pathway probabilities and path lengths in terms of the Horton order ratios. In this formulation the GIUH suppresses the unique structure of an individual network in favour of the expected characteristics of the ensemble of networks that are similar in terms of the Horton order ratios (Surkan, 1969; Beven et al., 1988). Rinaldo et al. (1991) showed that, with complete knowledge of the morphology and topology of a Strahler ordered network, these parameters can also be estimated, without recourse to Horton order ratios.

In the methods based on the width or area function, the concept of a pathway as a set of states is not meaningful and consequently we are no longer concerned about the probabilities of transitions from lower to the higher states. Instead, the normalized area function (following division by catchment area), represents the probability of a droplet being sourced at that flow distance from the outlet. It can therefore be considered to be the probability of a droplet finding its way to the outlet through a pathway whose length is that distance.

A number of alternative parameterizations have been suggested to describe travel time distributions for individual flow pathways. Each of these corresponds to an approximate conceptualization of the underlying hydraulics of flow in river channels. These are: (i) dirac delta function, corresponding to pure translation routing (Kirkby, 1976; Gupta et al., 1986b); (ii) exponential distribution, corresponding to a linear store, or series of linear stores (Rodríguez-Iturbe and Valdés, 1979; Gupta et al., 1986b); (iii) the semi-Gaussian Green's function (Mesa and Mifflin, 1986; Troutman and Karlinger, 1985), which corresponds to the linearized diffusion wave equation; (iv) the gamma distribution (Gupta et al., 1986b; van der Tak and Bras, 1990); and (v) the solution to the linearized dynamic wave equation (Harley, 1967; Troutman and Karlinger, 1985).

The GIUH based on Strahler ordering (Gupta et al., 1980; Rinaldo et al., 1991) and that based on the normalized area function (Mesa and Mifflin, 1986; Snell and Sivapalan, 1994a) can be written as

$$\text{Strahler ordering}: \quad f(t) = \sum_{\gamma \in \Gamma} p(\gamma) \cdot h(L_\gamma, t) \tag{18a}$$

$$\text{Area function}: \quad f(t) = \int_0^\infty p(s) \cdot h(s, t) \, ds \tag{18b}$$

In Equation (18a), $f(t)$ is the GIUH, γ denotes a particular flow pathway based on Strahler ordering, Γ is the set of all such pathways, $p(\gamma)$ and L_γ are, respectively, the associated pathway probability and pathway length. In Equation (18b), s denotes the flow distance from the outlet and $p(s)$, where $p(s) = a(s)/A$, is the normalized area function. In both equations $h(x, t)$ is an appropriate travel time distribution for a selected flow pathway.

It is clear from the above [Equation (18)] that the two GIUH approaches are indeed comparable, provided the same function $h(x, t)$ is used in both instances. In one instance the result is an integral (area function approach), whereas in the other it is in the form of a finite sum (Strahler ordering). Snell and Sivapalan (1994a) estimated the GIUHs for two natural catchments using the two different approaches and using the semi-Gaussian Green's function of Mesa and Mifflin (1986) to parameterise $h(x, t)$. They found that the approach based on Strahler ordering produced a GIUH which was substantially different from that produced by the area function approach. This is because Strahler ordering narrows down the number of possible pathways (to five or 10, for example) which narrows down the possibilities, while the area function allows an infinite number of pathways.

As the GIUH is either parameterized in terms of the Horton order ratios (which are dimensionless quantities) or the normalized area function (which is also dimensionless), the GIUH can be viewed as a scaling relationship and can be used to link the responses of catchments of different sizes. For example, Rodríguez-Iturbe and Valdés (1979) showed that the non-dimensional peak, q_p^*, and time to peak, t_p^*, of the GIUH can be approximated by

$$q_p^* = \frac{q_p L_\Omega}{v} = 1 \cdot 31 R_L^{0 \cdot 43} \tag{19a}$$

$$t_p^* = \frac{t_p v}{L_\Omega} = 0 \cdot 44 R_L^{-0 \cdot 38} (R_B / R_A)^{0 \cdot 55} \tag{19a}$$

where L_Ω is the length of the highest order stream in the network (the characteristic length scale) and v is a velocity parameter. Thus two catchments would be similar, in terms of their GIUH and regardless of their size, if R_L and R_B/R_A are the same. Equations (19a) and (19b) will then serve as the scaling relationships.

The GIUH approach described so far has ignored the processes of runoff generation on the hillslopes, implicitly assuming that runoff is generated uniformly over the entire catchment area and that hillslope travel times are negligible compared with the travel times within the channel network. Sivapalan et al. (1990) have generalized the GIUH of Rodríguez-Iturbe and Valdés (1979) to incorporate partial area runoff generation. Lee and Delleur (1976), Kirkby (1976) and Mesa and Mifflin (1986) have presented a general framework which can be used to extend the area function approach so as to incorporate hillslope response functions.

Another limitation is that the GIUH is a linear construct, underlain by the assumptions of constant velocity and independence of flow pathways. Although this may be an adequate approximation for large catchments, empirical evidence suggests that small catchments arer highly non-linear, with the non-linearity decreasing with increasing catchment size (Minshall, 1960). This may be a consequence of the fact that channel processes are not as important in small catchments. The GIUH, being a linear model, may not therefore be applicable to small catchments. The quantification of the non-linearity with decreasing catchment size, and the investigation of the sources of non-linearity (hillslopes versus channel network) are important problems that cannot be addressed by the GIUH approach.

Meta-channel concept

As a way of overcoming such limitations of the GIUH approach, Snell and Sivapalan (1995) made a first step with the introduction of their *meta-channel* concept. They form their meta-channel by collapsing the stream network on the basis of flow distance and estimate the hydraulic properties of the new meta-channel by conserving the spatial distributions (in terms of flow distance) of mass, momentum and energy between the natural network and the meta-channel.

Their development of the meta-channel precludes the *a priori* need to assume linearity of channel behaviour. Rather, physically based governing equations, such as the St. Venant equations, can be used to model the flow. Computationally, the modelling of the hydraulics is much simpler as there is only a single channel to model, whose hydraulic characteristics are already known. It is also fairly straightforward to integrate more realistic hillslope response functions with the meta-channel approach. The disadvantage is that the meta-channel approach cannot yield an analytical solution. However, it does give a more general, non-linear solution which can be used as a benchmark against which the accuracy of more approximate solutions can be tested. More specifically, the metal-channel approach permits the development of scaling relationships that hold for a wider range of catchment sizes, as compared with the GIUH.

FUTURE RESEARCH

This paper has defined a framework for scale issues in hydrology. As a starting point, some basic definitions and fundamental issues of heterogeneity were discussed. Then two alternative approaches to scaling were suggested. The first is a model-oriented approach and focuses on the scaling of state variables, model parameters, inputs and conceptualizations. The second (i.e. dimensional analysis and similarity concepts) deals with complex processes in a much simpler fashion. In each of these areas considerable advances have been achieved in the last years. There are, however, a number of key issues that need more attention in the future:

1. Organization in catchments has been identified as one of the keys to 'enlightened scaling'. New and innovative techniques for quantifying organization are needed.
2. The space resolution of hydrological measurements is typically much poorer than the time resolution. What is needed is high-resolution spatial data of quantities such as soil moisture. Also, the data should be representative over the soil profile (or at least over a depth of several decimetres), rather than just over a thin layer at the soil surface. These patterns are needed for defining scale relationships, identifying organization and evaluating distributed parameter models.
3. Current work on stream network analysis has often neglected geologic information. Geology is clearly a dominant control of network structure and should therefore play a central part in the analysis. It may be more rewarding to search for relationships between geology and network characteristics than to refine theories that are sophisticated but disregard the most obvious controls.
4. Theoretical achievements in hydrological scaling have been considerable over the last few years and decades, but, often, they have not contributed much to solve the problems of engineering hydrology. There is a definite need for more sophisticated techniques in practice and we trust that adapting current theoretical concepts to the engineering practice is a feasible endeavour. Indeed, for future scale research in hydrology it will be one of the most rewarding challenges to bridge the gap between theory and practice.

ACKNOWLEDGEMENTS

The authors thank the Fonds zur Förderung der wissenschaftlichen Forschung, Vienna, Project No. J0699-PHY for financial support. We are grateful to two anonymous reviewers for a number of valuable comments and to J. Snell for assistance towards significant improvements to the section on stream network analysis. The first author also thanks D. Gutknecht for the many useful discussions on organization in catchments.

REFERENCES

Ababou, R. and Wood, E. F. 1990. 'Comment on "Effective groundwater model parameter values: Influence of spatial variability of hydraulic conductivity, leakance, and recharge" by J. J. Gómez-Hernández and S. M. Gorelick', *Wat. Resour. Res.*, **26**, 1843–1846.
Abbott, M. B., Bathurst, J. C., Cunge, J. A., O'Connell, P. E., and Rasmussen, J. 1986. 'An introduction to the European Hydrological System — Système Hydrologique Européen, "SHE", 2: structure of a physically-based, distributed modelling system', *J. Hydrol.*, **87**, 61–77.
Abrahams, A. D., Parsons, A. J., and Luk, S. H. 1989. 'Distribution of depth of overland flow on desert hillslopes and its implications for modelling soil erosion', *J. Hydrol.*, **106**, 177–184.
Allen, T. F. H., and Starr, T. B. 1982. *Hierarchy*. The University of Chicago Press, Chicago. 310 pp.
Anderson, M. G. (Ed.) 1988. *Modelling Geomorphological Systems*. Wiley, Chichester. 458 pp.
Anderson, M. G. and Burt, T. P. 1990. 'Subsurface runoff' in Anderson, M. G. and Burt, T. P. (Eds), *Process Studies in Hillslope Hydrology*. Wiley, Chichester. pp. 365–400.
Anderson, M. P. 1989. 'Hydrogeologic facies models to delineate large-scale spatial trends in glacial and glaciofluvial sediments', *Geol. Soc. Am. Bull.*, **101**, 501–511.
Anderson, M. P. 1991. 'Comment on "Universal scaling of hydraulic conductivities and dispersivities in geologic media" by S. P. Neuman', *Wat. Resour. Res.*, **27**, 1381–1382.
Austin, P. M. and Houze, R. A. 1972. 'Analysis of the structure of precipitation patterns in New England', *J. Appl. Meteorol.*, **11**, 926–935.
Avissar, R. 1995. 'Scaling of land–atmosphere interactions: an atmospheric modelling perspective', *Hydrol. Process.*, **9**, 679–695.
Bakr, A. A., Gelhar, L. W., Gutjahr, A. L., and MacMillan, J. R. 1978. 'Stochastic analysis of spatial variability in subsurface flows. 1. Comparison of one- and three-dimensional flows', *Wat. Resour. Res.*, **14**, 263–271.
Band, L. E. 1986. 'Topographic partition of watersheds with digital elevation models', *Wat. Resour. Res.*, **22**, 15–24.
Band, L. E. and Moore, I. D. 1995. 'Scale: landscape attributes and geographical information systems', *Hydrol. Process.*, **9**, 401–422.
Barancourt, C., Creutin, J. D., and Rivoirard, J. 1992. 'A method for delineating and estimating rainfall fields', *Wat. Resour. Res.*, **28**, 1133–1144.
Barling, R. D., Moore, I. D., and Grayson, R. B. 1994. 'A quasi-dynamic wetness index for characterising the spatial distribution of zones of surface saturation and soil water content', *Wat. Resour. Res.*, **30**, 1029–1044.
Bathurst, J. C. and Cooley, K. R. 'Use of the SHE hydrological modelling system to investigate basin response to snowmelt at Reynolds Creek, Idaho', *J. Hydrol.*, in press.

Benecke, P. 1992. 'Vorhersagbarkeit der Wasserbindungs- und der Wasserleitfähigkeitsfunktion an bodenkundlichen Substratmerkmalen' in Kleeberg, H.-B. (Ed.), *Regionalisierung in der Hydrologie, DFG-Mitt. XI*. VCH Verl. ges., Weinheim. pp. 221–239.

Bergström, S. 1991. 'Principles and confidence in hydrological modelling', *Nordic Hydrol.*, **22**, 123–136.

Beven, K. 1981. 'Kinematic subsurface stormflow', *Wat. Resour. Res.*, **17**, 1419–1424.

Beven, K. 1986. 'Runoff production and flood frequency in catchments of order n: an alternative approach' in Gupta, V. K., Rodríguez-Iturbe, I., and Wood, E. F. (Eds), *Scale Problems in Hydrology*. D. Reidel, Dordrecht. pp. 107–131.

Beven, K. 1989. 'Changing ideas in hydrology — the case of physically based models', *J. Hydrol.*, **105**, 157–172.

Beven, K. J. 1991. 'Scale considerations' in Bowles, D. S. and O'Connell, P. E. (Eds), *Recent Advances in the Modeling of Hydrologic Systems*. Kluwer, Dordrecht. pp. 357–371.

Beven, K. (Ed.) 1992. 'Future of distributed modelling', *Hydrol. Proc.*, **6**, 253–268.

Beven, K. J. and Kirkby, M. J. 1979. 'A physically-based variable contributing area model of basin hydrology', *Hydrol. Sci. Bull.*, **24**, 43–69.

Beven, K. and Wood, E. F. 1993. 'Flow routing and the hydrological response of channel networks' in Beven, K. and Kirkby, M. J. (Eds), *Channel Network Hydrology*. Wiley, Chichester. pp. 99–128.

Beven, K., Warren, R., and Zaoui, J. 1980. 'SHE: towards a methodology for physically-based distributed forecasting in hydrology', *Proc. Oxford Symp., IAHS Publ.*, **129**, 133–137.

Beven, K., Wood, E. F., and Sivapalan, M. 1988. 'On hydrological heterogeneity — catchment morphology and catchment response', *J. Hydrol.*, **100**, 353–375.

Binley, A., Beven, K., and Elgy, J. 1989. 'A physically based model of heterogeneous hillslopes. 2. Effective hydraulic conductivities', *Wat. Resour. Res.*, **25**, 1227–1233.

Blöschl, G., Kirnbauer, R., and Gutknecht, D. 1991. 'Distributed snowmelt simulations in an Alpine catchment. 1. Model evaluation on the basis of snow cover patterns', *Wat. Resour. Res.*, **27**, 3171–3179.

Blöschl, G., Gutknecht, D., Grayson, R. B., Sivapalan, M., and Moore, I. D. 1993. 'Organisation and randomness in catchments and the verification of hydrologic models', *EOS, Trans. Am. Geophys. Union*, **74**, 317.

Blöschl, G., Gutknecht, D., and Kirnbauer, R. 1994. 'On the evaluation of distributed hydrologic models' in Rosso, R., Peano, A., Becchi, I., Bemporad, G. A. (Eds.), *Advances in Distributed Hydrology. Proceedings of a Workshop held in Bergamo, Italy, June 1992*. Water Resources Publications, in press.

Blöschl, G., Grayson, R. B., and Sivapalan, M. 1995. 'On the representative elementary area (REA) concept and its utility for distributed rainfall-runoff modelling', *Hydrol. Process.*, **9**, 313–330.

Blöschl, G., Gutknecht, D., and Grayson, R. B. 'On spatial organisation and randomness in hydrology', *Wat. Resour. Res.*, submitted.

Brannan, J. R. and Haselow, J. S. 1993. 'Compound random field models of multiple scale hydraulic conductivity', *Wat. Resour. Res.*, **29**, 365–372.

Bridges, E. M. 1982. 'Techniques of modern soil survey' in Bridges, E. M. and Davidson, D. A. (Eds), *Principles and Applications of Soil Geography*. Longman, London, New York. pp. 29–57.

Buckingham, E. 1914. 'On physically similar systems: Illustrations of the use of dimensional equations', *Phys. Rev.*, **4**, 345–376.

Burt, T. P. and Butcher, D. P. 1985. 'Topographic controls of soil moisture distributions', *J. Soil Sci.*, **36**, 469–486.

Canterford, R. P., Pescod, N. R., Pearce, H. J., and Turner, L. H. 1987. 'Design intensity–frequency–duration rainfall' in Pilgrim, D. H. (Ed.), *Australian Rainfall and Runoff*. The Institution of Engineers, Barton, ACT. pp. 15–40.

Carlson, A. B. 1986. *Communication Systems*. 3rd edn. McGraw Hill, New York. 686pp.

Chang, H. H. 1979. 'Minimum stream power and river channel patterns', *J. Hydrol.*, **41**, 303–327.

Chang, H. H. 1982. 'Mathematical model for erodible channels', *J. Hydraul. Div. ASCE*, **108**, (HY5), 678–689.

Chappell, N. and Ternan, L. 1992. 'Flow path dimensionality and hydrological modelling', *Hydrol. Process.*, **6**, 327–345.

Chhabra, A. B., Jensen, R. V., and Sreenivasan, K. R. 1989. 'Extraction of underlying multiplicative processes from multifractals via the thermodynamic formalism', *Phys. Rev. A*, **40**, 4593–4611.

Chorley, R. J., Schumm, S. A., and Sugden, D. E. 1984. *Geomorphology*. Methuen, London. 605 pp.

Chow, V. T., Maidment, D. R., and Mays, L. W. 1988. *Applied Hydrology*. McGraw-Hill, New York. 572 pp.

Clark, W. 1985. 'Scales of climate impacts', *Climatic Change*, **7**, 5–27.

Copty, N., Rubin, Y., and Mavko, G. 1993. 'Geophysical–hydrological identification of field permeabilities through Bayesian updating', *Wat. Resour. Res.*, **29**, 2813–2825.

Creutin, J. D. and Obled, C. 1982. 'Objective analyses and mapping techniques for rainfall fields: an objective comparison', *Wat. Resour. Res.*, **18**, 413–431.

Cushman, J. H. 1983. 'Comment on "Three-dimensional stochastic analysis of macrodispersion in aquifers" by L. W. Gelhar and C. L. Axness', *Wat. Resour. Res.*, **19**, 1641–1642.

Cushman, J. H. 1984. 'Unifying the concepts of scale, instrumentation, and stochastics in the development of multiphase transport theory', *Wat. Resour. Res.*, **20**, 1668–1676.

Cushman, J. H. 1987. 'More on stochastic models', *Wat. Resour. Res.*, **23**, 750–752.

Cutler, C. D. 1993. 'A review of the theory and estimation of fractal dimension', *Tech. Rep. Ser. STAT-93-06*, Department of Statistics and Actuarial Science, University of Waterloo, Ontario, 107 pp.

Dagan, G. 1979. 'Models of groundwater flow in statistically homogeneous porous formations', *Wat. Resour. Res.*, **15**, 47–63.

Dagan, G. 1986. 'Statistical theory of groundwater flow and transport: pore to laboratory, laboratory to formation and formation to regional scale', *Wat. Resour. Res.*, **22**, 120S–134S.

Daubechies, I. 1992. *Ten Lectures on Wavelets*. SIAM, Philadelphia. 357 pp.

de Boer, D. H. 1992. 'Hierarchies and spatial scale in process geomorphology: a review', *Geomorphology*, **4**, 303–318.

de Marsily, G. 1986. *Quantitative Hydrogeology*. Academic Press, San Diego. 440 pp.

Denbigh, K. G. 1975. 'A non-conserved function for organized systems' in Kubát, L. and Zeman, J. (Eds), *Entropy and Information in Science and Philosophy*. Elsevier, Amsterdam. pp. 83–92.

Deutsch, A. and Journel, A. G. 1992. *GSLIB, Geostatistical Software Library and User's Guide*. Oxford University Press, New York, Oxford. 340 pp.

deVriend, H. J. 1991. 'Mathematical modelling and large-scale coastal behaviour', *J. Hydraul. Res.*, **29**, 727–740.

Dietrich, W. E. and Dunne, T. 1993. 'The channel head' in Beven, K. and Kirkby, M. J. (Eds), *Channel Network Hydrology*. Wiley, Chichester. pp. 175–219.

Dingman, S. L. 1994. *Physical Hydrology*. Macmillan, New York. 575 pp.

Dooge, J. C. I. 1982. 'Parameterization of hydrologic processes' in Eagleson, P. S. (Ed.), *Land Surface Processes in Atmospheric General Circulation Models*. Cambridge University Press, London. pp. 243–288.

Dooge, J. C. I. 1986. 'Looking for hydrologic laws', *Wat. Resour. Res.*, **22**, 46S–58S.

Dovers, S. R. 1995. 'A framework for scaling and framing policy problems in sustainability', *Ecol. Econ.*, **12**, 93–106.

Dozier, J. 1992. 'Opportunities to improve hydrologic data', *Rev. Geophys.*, **30**, 315–331.

Drury, S. A. 1987. *Image Interpretation in Geology*. Allen and Unwin, London. 243 pp.

Dunne, T. 1978. 'Field studies of hillslope flow processes' in Kirkby, M. J. (Ed.), *Hillslope Hydrology*. Wiley, Chichester. pp. 227–293.

Dunne, T. 1983. 'Relation of field studies and modeling in the prediction of storm runoff', *J. Hydrol.*, **65**, 25–48.

Dykaar, B. B. and Kitanidis, P. K. 1992. 'Determination of the effective hydraulic conductivity for heterogeneous porous media using a numerical spectral approach. 2. Results', *Wat. Resour. Res.*, **28**, 1167–1178.

Eagleson, P. S. 1972. 'Dynamics of flood frequency', *Wat. Resour. Res.*, **8**, 878–898.

Eagleson, P. S. 1978. 'Climate, soil, and vegetation (in 7 parts)', *Wat. Resour. Res.*, **14**, 705–776.

El-Kadi, A. and Brutsaert, W. 1985. 'Applicability of effective parameters for unsteady flow in nonuniform aquifers', *Wat. Resour. Res.*, **21**, 183–198.

England, C. B. and Holtan, H. N. 1969. 'Geomorphic grouping of soils in watershed engineering', *J. Hydrol.*, **7**, 217–225.

Engman, E. T. 1986. 'Roughness coefficients for routing surface runoff', *J. Irrig. Drainage Div. Proc. ASCE*, **112**, 39–53.

Engman, E. T. and Gurney, R. J. 1991. 'Recent advances and future implications of remote sensing for hydrologic modeling' in Bowles, D. S. and O'Connell, P. E. (Eds), *Recent Advances in the Modeling of Hydrologic Systems*. Kluwer, Dordrecht. pp. 471–495.

Falconer, K. J. 1990. *Fractal Geometry: Mathematical Foundations and Applications*. Wiley, Chichester. 288 pp.

Famiglietti, J. S. 1992. 'Aggregation and scaling of spatially-variable hydrological processes: local, catchment-scale and macroscale models of water and energy balance, *PhD Dissertation*, Princeton University, 207 pp.

Feder, J. 1988. *Fractals*. Plenum Press, New York and London. 283 pp.

Ferguson, R. I. 1973. 'Channel pattern and sediment type', *Area*, **5**, 38–41.

Fischer, H. B., List, E. J., Koh, R. C. Y., Imberger, J., and Brooks, N. H. 1979. *Mixing in Inland and Coastal Waters*. Academic Press, New York, 483 pp.

Fitzharris, B. B. 1975. 'Snow accumulation and deposition on a westcoast midlatitude mountain', *PhD Thesis*, University of British Columbia, Vancouver 367 pp.

Flood Studies Report 1975. *Vol. I — Hydrological Studies*. Natural Environment Research Council, London. 570 pp.

Fortak, H. 1982. *Meteorologie*. Dietrich Reimer, Berlin. 293 pp.

Foster, G. R., Huggins, L. F., and Meyer, L. D. 1984. 'A laboratory study of rill hydraulics: I. Velocity relationships', *Trans. Am. Soc. Agric. Engin.*, **27**, 790–796.

Foufoula-Georgiou, E. and Georgakakos, K. P. 1991. 'Hydrologic advances in space–time precipitation modeling and forecasting' in Bowles, D. S. and O'Connell, P. E. (Eds), *Recent Advances in the Modeling of Hydrologic Systems*. Kluwer, Dordrecht. pp. 47–65.

Freeze, R. A. 1975. 'A stochastic-conceptual analysis of one-dimensional groundwater flow in nonuniform homogeneous media', *Wat. Resour. Res.*, **11**, 725–741.

Freeze, R. A. and Harlan, R. L. 1969. 'Blueprint for a physically-based, digitally simulated hydrologic response model', *J. Hydrol.*, **9**, 237–258.

Gallant, J. C., Moore, I. D., Hutchinson, M. F., and Gessler, P. E. 1994. 'Estimating fractal dimension of profiles: a comparison of methods', *Math. Geol.*, **26**, 455–481.

Gelhar, L. W. 1986. 'Stochastic subsurface hydrology from theory to applications', *Wat. Resour. Res.*, **22**, 135S–145S.

Germann, P. F. 1986. 'Rapid drainage response to precipitation', *Hydrol. Process.*, **1**, 3–14.

Germann, P. F. 1990. 'Macropores and hydrologic hillslope processes' in Anderson, M. G. and Burt, T. P. (Eds), *Process Studies in Hillslope Hydrology*. Wiley, Chichester. pp. 327–367.

Gessler, P. E., McKenzie, N. J., Hutchinson, M. F., and Moore, I. D. 1993. 'Soil–landscape modelling in southeastern Australia: scale relationships' in Kalma, J., Sivapalan, M., and Wood, E. (Eds), *Scale Issues in Hydrological/Environmental Modelling, Proc. Workshop Robertson, CRES*. Australian National University, Canberra. p. 9.

Golding, D. L. 1974. 'The correlation of snowpack with topography and snowmelt runoff on Marmot Creek basin, Alberta', *Atmosphere*, **12**, 31–38.

Gómez-Hernández, J. J. and Gorelick, S. M. 1989. 'Effective groundwater model parameter values: influence of spatial variability of hydraulic conductivity, leakance, and recharge', *Wat. Resour. Res.*, **25**, 405–419.

Goodrich, D. C. 1990. 'Geometric simplification of a distributed rainfall–runoff model over a large range of basin scales', *PhD Thesis*, Univ. Arizona, Tucson, 361 pp.

Goodrich, D. C. and Woolhiser, D. A. 1991. 'Catchment hydrology', *Rev. Geophys.* (suppl.), April, 202–209.

Grace, R. A., and Eagleson, P. S. 1966. 'The synthesis of short-time-increment rainfall sequences', *Rep. No. 91*, MIT, Hydrodynamics Laboratory 105 pp.

Grayson, R. B., Moore, I. D., and McMahon, T. A. 1992. 'Physically-based hydrologic modelling: 2. Is the concept realistic?', *Wat. Resour. Res.*, **26**, 2659–2666.

Grayson, R. B., Blöschl, G., Barling, R. D., and Moore, I. D. 1993. 'Process, scale and constraints to hydrological modelling in GIS' in Kovar, K. and Nachtnebel, H. P. (Eds), *Applications of Geographic Information Systems in Hydrology and Water Resources Management (Proc. Vienna Symp., April 1993)*, *IAHS Publ.*, **211**, 83–92.

Grayson, R. B., Blöschl, G., and Moore, I. D. 'Distributed parameter hydrologic modelling using vector elevation data: THALES and TAPES-C' in Singh, V. P. (Ed.), *Computer Models of Watershed Hydrology*. CRC Press, in press.

Green, W. H. and Ampt, G. A. 1911. 'Studies on soil physics, Part I. The flow of air and water through soils', *J. Agric. Sci.*, **4**, 1–24.

Guerra, L., Moore, I. D., Kalma, J. D., and Hofstee, C. 1993. 'Predicting spatially distributed evaporation using terrain, soil and land cover information' in Bolle, H.-J., Feddes, R. A., and Kalma, J. D. (Eds), *Exchange Processes at the Land Surface for a Range of Space and Time Scales, IAHS Publ.*, **212**, 611–618.

Gumbel, E. J. 1941. 'The return period of flood flows', *Ann. Math. Stat.*, **12**, 163–190.

Gupta, V. K. and Dawdy, D. R. 'Regional analysis of flood peaks: multiscaling theory and its physical basis', in Rosso, R., Peano, A., Becchi, I., Bemporad, G. A. (Eds.), *Advances in Distributed Hydrology. Proceedings of a Workshop held in Bergamo, Italy, June 1992*. Water Resources Publications, pp. 149–168.

Gupta, V. K. and Dawdy, D. R. 1995. 'Some physical implications of regional variations in the scaling exponents of flood quantiles', *Hydrol. Process.*, **9**, 347–361.

Gupta, V. K. and Waymire, E. C. 1990. 'Multiscaling properties of spatial rainfall and river flow distributions', *J. Geophys. Res.*, **95**(D3), 1999–2009.

Gupta, V. K. and Waymire, E. C. 1993. 'A statistical analysis of mesoscale rainfall as a random cascade', *J. Appl. Meteorol.*, **32**, 251–267.

Gupta, V. K., Waymire, E., and Wang, J. R. 1980. 'A representation of an instantaneous unit hydrograph from geomorphology', *Wat. Resour. Res.*, **16**, 855–862.

Gupta, V. K., Rodríguez-Iturbe, I., and Wood, E. F. (Eds) 1986a. *Scale Problems in Hydrology*. D. Reidel, Dordrecht. 246 pp.

Gupta, V. K., Waymire, E., and Rodríguez-Iturbe, I. 1986b. 'On scales, gravity and network structure' in Gupta, V. K., Rodríguez-Iturbe, I., and Wood, E. F. (Eds), *Scale Problems in Hydrology*. D. Reidel, Dordrecht, pp. 159–184.

Gutjahr, A. L., Gelhar, L. W., Bakr, A. A., and MacMillan, J. R. 1978. 'Stochastic analysis of spatial variability in subsurface flows, 2, evaluation and application', *Wat. Resour. Res.*, **14**, 953–959.

Gutknecht, D. 1991a. 'On the development of "applicable" models for flood forecasting' in Van de Ven, F. H. M., Gutknecht, D., Loucks, D. P., and Salewicz, K. A. (Eds), *Hydrology for the Water Management of Large River Basins (Proc. Vienna Symp., August 1991), IAHS Publ.*, **201**, 337–345.

Gutknecht, D. 1991b. 'Computer aided modelling for operational forecasting systems', *Ann. Geophys. Suppl.* **9**, C480.

Gutknecht, D. 1993. 'Grundphänomene hydrologischer Prozesse', *Zürcher Geogr. Schriften*, **53**, Geographisches Institut der Eidgenossischen Technischen Hochschule, Zurich, 25–38.

Gyasi-Agyei, Y., Willgoose, G., and De Troch, F. P. 1995. 'Effects of vertical resolution and map scale of digital elevation models on geomorphologic parameters used in hydrology', *Hydrol. Process.*, **9**, 363–382.

Hack, J. T. and Goodlett, J. G. 1960. 'Geomorphology and forest ecology of a mountain region in the Central Appalachians', *US Geol. Surv. Prof. Pap.*, **347**, 66pp.

Halsey, T. C., Jensen, M. H., Kadanoff, L. P., Procaccia, I., and Shraiman, B. I. 1986. 'Fractal measures and their singularities: the characterization of strange sets', *Phys. Rev. A*, **33**, 1141–1151.

Haltiner, G. J. and Williams, R. T. 1980. *Numerical Prediction and Dynamic Meteorology*. Wiley, New York. 477 pp.

Harley, B. M. 1967. 'Linear routing in uniform open channels', *M. Eng. Sci. Thesis*, Department of Civil Engineering, National University of Ireland.

Hatton, T. J. and Wu, H. I. 1995. 'Scaling theory to extrapolate individual tree water use to stand water use', *Hydrol. Process.*, **9**, 000–000.

Haury, L. R., McGowan, J. A., and Wiebe, P. H. 1977. 'Patterns and processes in the time–space scales of plankton distributions' in Steele, J. H. (Ed.), *Spatial Pattern in Plankton Communities*. Plenum Press, New York and London. pp. 277–327.

Hebson, C. S. and Wood, E. F. 1986. 'A study of the scale effects in flood frequency response' in Gupta, V. K., Rodríguez-Iturbe, I., and Wood, E. F. (Eds), *Scale Problems in Hydrology*. D. Reidel, Dordrecht. pp. 133–158.

Hellums, J. D. and Churchill, S. W. 1961. 'Dimensional analysis and natural circulation', *Chem. Eng. Progr. Symp. Ser.*, **57**, 75–80.

Helmlinger, K. R., Kumar, P., and Foufoula-Georgiou, E. 1993. 'On the use of digital elevation model data for Hortonian and fractal analyses of channel networks', *Wat. Resour. Res.*, **29**, 2599–2613.

Hey, R. D. 1988. 'Mathematical models of channel morphology' in Anderson, M. G. (Ed.), *Modelling Geomorphological Systems*. Wiley, Chichester. pp. 99–125.

Hillel, D., and Elrick, D. E. 1990. *Scaling in Soil Physics: Principles and Applications*. Soil Science Society of America, Vol. 25. 122 pp.

Hoeksema, R. J., and Kitanidis, P. K. 1985. 'Analysis of the spatial structure of properties of selected aquifers', *Wat. Resour. Res.*, **21**, 563–572.

Hoosbeek, M. R. and Bryant, R. B. 1992. 'Towards the quantitative modeling of pedogenesis — a review', *Geoderma*, **55**, 183–210.

Horton, R. E. 1945. 'Erosional development of streams and their drainage basins: hydrophysical approach to quantitative morphology', *Geol. Soc. Am. Bull.*, **38**, 275–370.

Huff, F. A. 1967. 'Time distribution of rainfall', *Wat. Resour. Res.*, **3**, 1007–1018.

Hurst, H. E. 1951. 'Long-term storage capacity of reservoirs', *Trans. Am. Soc. Civ. Engin.*, **116**, 770–808.

Hutchinson, M. F. 1991. 'The application of thin plate smoothing splines to continent-wide data assimilation' in Jasper, J. D. (Ed.), *Data Assimilation Systems, BMRC Res. Rep. 27*, Bureau of Meteorology, Melbourne: 104–113.

Ijjasz-Vasquez, E. J., Rodríguez-Iturbe, I., and Bras, R. L. 1992. 'On the multifractal characterization of river basins', *Geomorphology*, **5**, 297–310.

Jenkins, G. M. and Watts, D. G. 1968. *Spectral Analysis and its Applications*. Holden-Day, San Francisco. 525 pp.

Jenny, H. 1941. *Factors of Soil Formation*. McGraw Hill, New York. 281 pp.

Jenny, H. 1980. *The Soil Resource*. Springer, New York. 377 pp.

Jensen, H. 1989. 'Räumliche Interpolation der Stundenwerte von Niederschlag, Temperatur und Schneehöhe', *Zürcher. Geogr. Schriften*, **35**, Geographisches Institut der Eidgenössischen Technischen Hochschule, Zurich, 70 pp.

Jenson, S. K. and Domingue, J. O. 1988. 'Extracting topographic structure from digital elevation data for geographic information system analysis', *Photogr. Engin. Remote Sensing*, **54**, 1593–1600.

Jones, J. A. A. 1987. 'The effect of soil piping on contributing areas and erosion patterns', *Earth Surf. Process. Landforms*, **12**, 229–248.

Journel, A. G. 1986. 'Constrained interpolation and qualitative information — the soft kriging approach', *Math. Geol.*, **18**, 269–287.

Journel, A. G. and Huijbregts, C. J. 1978. *Mining Geostatistics*. Academic Press, London. 600 pp.

Kern, K. 1994. *Grundlagen naturnaher Gewässergestaltung*. Springer, Berlin. 256 pp.

Kirchner, J. W. 1993. 'Statistical inevitability of Horton's laws and the apparent randomness of stream channel networks', *Geology*, **21**, 591–594.

Kirkby, M. J. 1976. 'Tests of the random network model and its applications to basin hydrology', *Earth Surf. Process.*, **1**, 197–212.

Kirnbauer, R., Blöschl, G., and Gutknecht, D. 1994. 'Entering the era of distributed snow models', *Nordic Hydrol.*, **25**, 1–24.

Kleeberg, H.-B. (Ed.) 1992. *Regionalisierung in der Hydrologie. DFG-Mitt. XI*. VCH Verl. ges., Weinheim. 444 pp.

Klemeš, V. 1983. 'Conceptualisation and scale in hydrology', *J. Hydrol.*, **65**, 1–23.

Klinkenberg, B. and Goodchild, M. F. 1992. 'The fractal properties of topography: a comparison of methods', *Earth Surf. Process. Landforms*, **17**, 217–234.

Kneale, W. R., and White, R. E. 1984. 'The movement of water through cores of a dry (cracked) clay-loam grassland topsoil', *J. Hydrol.*, **67**, 361–365.

Knighton, A. D. 1974. 'Variation in width–discharge relation and some implications for hydraulic geometry', *Bull. Geol. Soc. Am.*, **85**, 1059–1076.

Knighton, A. D. 1975. 'Variations in at-a-station hydraulic geometry', *Am. J. Sci.*, **275**, 186–218.

Knighton, D. 1984. *Fluvial Forms and Processes*. Edward Arnold, London. 218 pp.

Koide, S. and Wheater, H. S. 1992. 'Subsurface flow simulation of a small plot at Loch Chon, Scotland', *Hydrol. Process.*, **6**, 299–326.

Kölla, E. 1987. 'Estimating flood peaks from small rural catchments in Switzerland', *J. Hydrol.*, **95**, 203–225.

Koutsoyiannis, D. and Foufoula-Georgiou, E. 1993. 'A scaling model of a storm hyetograph', *Wat. Resour. Res.*, **29**, 2345–2361.

Krasovskaia, I. 1982. 'Hypothesis on runoff formation in small watersheds in Sweden', *FoU-notiser No. 19*, SMHI, Norrköping.

Kupfersberger, H. 1994. 'Integrating different types of information for estimating aquifer transmissivities with the sequential indicator cosimulation method' in Dracos, T. H. and Stauffer, F. (Eds), *Transport and Reactive Processes in Aquifers (Proc. 5 IAHR Symp. Zurich)*, Balkema, Rotterdam pp. 165–170.

Kupfersberger, H. and Blöschl, G. 1995. 'Estimating aquifer transmissivities — on the value of auxiliary data', *J. Hydrol.*, **165**, 85–99.

Ladson, A. R., and Moore, I. D. 1992. 'Soil water prediction on the Konza Prairie by microwave remote sensing and topographic attributes', *J. Hydrol.*, **138**, 385–407.

Lang, H. 1985. 'Höhenabhängigkeit der Niederschläge' in *Der Niederschlag in der Schweiz, Beiträge zur Geologie der Schweiz, Nr. 31*. Verl. Kümmerlyt. Frey, Bern. pp. 149–157.

Langhaar, H. L. 1951. *Dimensional Analysis and Theory of Models*. Wiley, New York. 166 pp.

Larsen, J. E., Sivapalan, M., Coles, N. A., and Linnet, P. E. 1994. 'Similarity analysis of runoff generation processes in real-world catchments', *Wat. Resour. Res.*, **30**, 1641–1652.

Leavesley, G. H. and Hay, L. E. 1993. 'A nested-model approach for investigating snow accumulation and melt processes in mountainous regions', *EOS, Trans. Am. Geophys. Union*, **74**, 237.

Leavesley, G. H. and Stannard, L. G. 1990. 'Application of remotely sensed data in a distributed-parameter watershed model' in Kite, G. W. and Wankiewicz, A. (Eds), *Proc. Workshop on Applications of Remote Sensing in Hydrology, Saskatoon, February 1990*. pp. 47–64.

Lebel, T., Bastin, G., Obled, C., and Creutin, J. D. 1987. 'On the accuracy of areal rainfall estimation: a case study', *Wat. Resour. Res.*, **23**, 2123–2134.

Lee, M. T. and Delleur, J. W. 1972. 'A program for estimating runoff from Indiana watersheds, 3, analysis of geomorphologic data and a dynamic contributing area model for runoff estimation', *Tech. Rep. 24*, Purdue Univ. Water Resour. Res. Center, Lafayette 144 pp.

Lee, M. T. and Delleur, J. W. 1976. 'A variable source area model of the rainfall–runoff process based on the watershed stream network', *Wat. Resour. Res.*, **12**, 1029–1036.

Leopold, L. B. and Langbein, W. B. 1962. 'The concept of entropy in landscape evolution', *US Geol. Surv. Prof. Pap.*, **500-A**, A20 pp.

Leopold, L. B. and Maddock, T. 1953. 'The hydraulic geometry of stream channels and some physiographic implications', *US Geol. Surv. Prof. Pap.*, **252**, 57 pp.

Lettenmaier, D. P. and Burges, S. J. 1977 'Operational assessment of hydrologic models of longterm persistence', *Wat. Resour. Res.*, **13**, 113–124.

Lewis, A. 1995. 'Scale in spatial environmental databases', *PhD Thesis*, Centre for Resource and Environmental Studies, Australian National University, Canberra.

Linsley, R. K., Kohler, M. A., and Paulhus, J. L. H. 1988. *Hydrology for Engineers*. McGraw Hill, New York. 492 pp.

Loague, K. 1990. 'R-5 Revisited. 2. Reevaluation of a quasi-physically based rainfall-runoff model with supplemental information', *Wat. Resour. Res.*, **26**, 973–987.

Lovejoy, S. 1982. 'The area–perimeter relations for rain and cloud areas', *Science*, **216**, 185–187.

Lovejoy, S. and Schertzer, D. 1985. 'Generalized scale invariance in the atmosphere and fractal models of rain', *Wat. Resour. Res.*, **21**, 1233–1250.

Lovejoy, S. and Schertzer, D. 1989. 'Comment on "Are rain rate processes self-similar?" by B. Kedem and L. S. Chiu', *Wat. Resour. Res.*, **25**, 577–579.

Lovejoy, S. and Schertzer, D. 1991. 'Multifractal analysis techniques and rain and cloud fields from 10^{-3} to 10^6 m' in Schertzer, D. and Lovejoy, S. (Eds), *Scaling, Fractals and Non-linear Variability in Geophysics*. Kluwer, Dordrecht. pp. 111–144.

Mackay, R. and Riley, M. S. 1991. 'The problem of scale in the modelling of groundwater flow and transport processes' in *Chemodynamics of Groundwaters, Proc. Workshop November 1991, Mont Sainte-Odile, France*. EAWAG, EERO, PIR "Environment" of CNRS, IMF Université Louis Pasteur Strasbourg. pp. 17–51.

Mandelbrot, B. B. 1967. 'How long is the coast of Britain? Statistical self-similarity and fractional dimension', *Science*, **156**, 636–638.

Mandelbrot, B. B. 1977. *Fractals: Form, Chance, and Dimension*. Freeman, San Francisco. 365 pp.

Mandelbrot, B. B. and Van Ness, J. W. 1968. 'Fractional Brownian motions, fractional noises and applications', *SIAM Rev.*, **10**, 422–437.

Mandelbrot, B. B. and Wallis, J. R. 1968. 'Noah, Joseph, and operational hydrology', *Wat. Resour. Res.*, **4**, 909–918.

Mandelbrot, B. B. and Wallis, J. R. 1969. 'Robustness of the rescaled range R/S in the measurement of noncyclic long run statistical dependence', *Wat. Resour. Res.*, **5**, 967–988.

Mantoglou, A. and Gelhar, L. W. 1987. 'Capillary tension head variance, mean soil moisture content, and effective specific soil moisture capacity of transient unsaturated flow in stratified soils', *Wat. Resour. Res.*, **23**, 47–56.

Marani, A., Rigon, R., and Rinaldo, A. 1991. 'A note on fractal channel networks', *Wat. Resour. Res.*, **27**, 3041–3049.

Matheron, G. 1967. 'Eléments pour une théorie des milieux poreux', cited in de Marsily (1986).

Matheron, G. 1973. 'The intrinsic random functions and their applications', *Adv. Appl. Prob.*, **5**, 438–468.

Mazac, O., Kelly, W. E., and Landa, I. 1985. 'A hydrogeophysical model for relations between electrical and hydraulic properties of aquifers', *J. Hydrol.*, **85**, 1–19.

McKenzie, N. J. and Austin, M. P. 1993. 'A quantitative Australian approach to medium and small scale surveys based on soil stratigraphy and environmental correlation', *Geoderma*, **57**, 329–355.

McKenzie, N. J. and MacLeod, D. A. 1989. 'Relationships between soil morphology and soil properties relevant to irrigated and dryland agriculture', *Aust. J. Soil. Res.*, **27**, 235–258.

McKerchar, A. I. and Pearson, C. P. 1989. 'Flood frequency in New Zealand', *Publ. No. 20*, Hydrology Centre, Christchurch, 87 pp.

Mein, R. 1993. 'Flood hydrology', *Catchword, May 1993*, Cooperative Centre for Catchment Hydrology, Monash University, Clayton, Victoria.

Menevau, C. and Sreenivasan, K. R. 1987. 'Simple multifractal cascade model for fully developed turbulence', *Phys. Rev. Lett.*, **59**, 1424–1427.

Menevau C., Sreenivasan, K. R., Kailasnath, P., and Fan, M. S. 1990. 'Joint multifractal measures: theory and applications to turbulence', *Phys. Rev. A*, **41**, 894–913.

Mesa, O. J. and Mifflin, E. R. 1986. 'On the relative role of hillslope and network geometry in hydrologic response' in Gupta, V. K., Rodríguez-Iturbe, I., and Wood, E. F. (Eds), *Scale Problems in Hydrology*. D. Reidel, Dordrecht, pp. 1–18.

Mesa, O. J. and Poveda, G. 1993. 'The Hurst effect: the scale of fluctuations approach', *Wat. Resour. Res.*, **29**, 3995–4002.

Miall, A. D. 1985. 'Architectural-element analysis: a new method of facies analysis applied to fluvial deposits', *Earth-Sci. Rev.*, **22**, 261–308.

Miller, E. E. 1990. 'Scaling in soil physics — introduction' in Hillel, D. and Elrick, D. E. (Eds), *Scaling in Soil Physics: Principles and Applications*. Vol. 25. Soil Science Society of America. pp. xvii–xxi.

Miller, E. E. and Miller, R. D. 1956. 'Physical theory of capillary flow phenomena', *J. Appl. Phys.*, **27**, 324–332.

Milly, P. C. and Eagleson, P. S. 1987. 'Effects of spatial variability on annual average water balance', *Wat. Resour. Res.*, **23**, 2135–2143.

Milne, G. 1935. 'Some suggested units of classification and mapping particularly for East African soils', *Soil. Res.*, Berlin, **4**, 183–198.

Minshall, N. E. 1960. 'Predicting storm runoff on small experimental watersheds', *J. Hydraul. Div. Am. Soc. Civ. Engin.*, **86**, (HY8) 17–38.

Montgomery, D. R. and Foufoula-Georgiou, E. 1993. 'Channel network source representation using digital elevation models', *Wat. Resour. Res.*, **29**, 3925–3934.

Moore, I. D. and Burch, G. J. 1986. 'Sediment transport capacity of sheet and rill flow: application of unit stream power theory', *Wat. Resour. Res.*, **22**, 1350–1360.

Moore, I. D. and Grayson, R. B. 1991. 'Terrain based prediction of runoff with vector elevation data', *Wat. Resour. Res.*, **27**, 1177–1191.

Moore, I. D., Burch, G. J., and Mackenzie, D. H. 1988. 'Topographic effects on the distribution of surface soil water and the location of ephemeral gullies', *Trans. Am. Soc. Agric. Engin.*, **31**, 1098–1107.

Moore, I. D., Grayson, R. B., and Ladson, A. R. 1991. 'Digital terrain modelling: a review of hydrological, geomorphological, and biological applications', *Hydrol. Process.*, **5**, 3–30.

Moore, I. D., Gessler, P. E., Nielsen, G. A., and Peterson, G. A. 1993a. 'Soil attribute prediction using terrain analysis', *Soil Sci. Soc. Am. J.*, **57**, 443–452.

Moore, I. D., Lewis, A., and Gallant, J. C. 1993b. 'Terrain attributes: estimation methods and scale effects' in Jakeman, A. J., Beck, M. B., and McAleer, M. (Eds), *Modelling Change in Environmental Systems*. Wiley, Chichester 189–214.

Moore, I. D., Turner, A. K., Wilson, J. P., Jenson, S. K., and Band, L. E., 1993c. 'GIS and land surface-subsurface process modeling' in Goodchild, M. F., Parks, B. O., and Steyaert, L. T. (Eds), *Geographic Information Systems and Environmental Modeling*. Oxford University Press, Oxford. pp. 196–230.

Morel-Seytoux, H. J. 1988. 'Soil–aquifer–stream interactions — a reductionist attempt toward physical-stochastic integration', *J. Hydrol.*, **102**, 355–379.

Neuman, S. P. 1990. 'Universal scaling of hydraulic conductivities and dispersivities in geologic media', *Wat. Resour. Res.*, **26**, 1749–1758.

Neuman, S. P. and Orr, S. 1993. 'Prediction of steady state flow in nonuniform geologic media by conditional moments: exact nonlocal formalism, effective conductivities, and weak approximations', *Wat. Resour. Res.*, **29**, 341–364.

National Research Council (NRC) 1988. *Estimating Probabilities of Extreme Floods: Methods and Recommended Research*. National Academy Press, Washington. 141 pp.

National Research Council (NRC) 1991. *Opportunities in the Hydrologic Sciences*. National Academy Press, Washington. 348 pp.

Obled, Ch. 1990. 'Hydrological modeling in regions of rugged relief' in Lang, H. and Musy, A. (Eds), *Hydrology in Mountainous Regions. I — Hydrological Measurements; the Water Cycle (Proc. Lausanne Symp., August 1990), IAHS Publ.*, **193**, 599–613.

Oboukhov, A. 1962. 'Some specific features of atmospheric turbulence', *J. Fluid Mech.*, **13**, 77–81.

O'Connell, P. E. 1991. 'A historical perspective' in Bowles, D. S. and O'Connell, P. E. (Eds), *Recent Advances in the Modeling of Hydrologic Systems*. Kluwer, Dordrecht. pp. 3–30.

Oliver, M. A. and Webster, R. 1986. 'Semi-variograms for modelling the spatial pattern of landform and soil properties', *Earth Surf. Process. Landforms*, **11**, 491–504.

O'Loughlin, E. M. 1981. 'Saturation regions in catchments and their relation to soil and topographic properties', *J. Hydrol.*, **53**, 229–246.

O'Loughlin, E. M. 1986. 'Prediction of surface saturation zones in natural catchments by topographic analysis', *Wat. Resour. Res.*, **22**, 794–804.

Orlanski, I. 1975. 'A rational subdivision of scales for atmospheric processes', *Bull. Am. Meteorol. Soc.*, **56**, 527–530.

Padmanabhan, G. and Rao, A. R. 1988. 'Maximum entropy spectral analysis of hydrologic data', *Wat. Resour. Res.*, **25**, 1519–1533.

Parsons, A. J., Abrahams, A. D., and Luk, S.-H. 1990. 'Hydraulics of interrill overland flow on a semi-arid hillslope, southern Arizona', *J. Hydrol.*, **117**, 255–273.

Pattison, A. 1965. 'Synthesis of hourly rainfall data', *Wat. Resour. Res.*, **1**, 489–498.

Pearce, A. J., Stewart, M. K., and Sklash, M. G. 1986. 'Storm runoff generation in humid headwater catchments. 1. Where does the water come from?', *Wat. Resour. Res.*, **22**, 1263–1272.

Pearson, C. P. 1993. 'Regional flood frequency analysis for small New Zealand basins. 2. Flood frequency groups', *J. Hydrol. (New Zealand)*, **30**, 77–92.

Philip, J. R. 1957. 'The theory of infiltration, 1–7', *Soil Sci.*, **83–85**.

Pilgrim, D. H. 1977. 'Isochrones of travel time and distribution of flood storage from a tracer study on a small watershed', *Wat. Resour. Res.*, **13**, 587–595.

Pilgrim, D. H. 1983 'Some problems in transferring hydrological relationships between small and large drainage basins and between regions', *J. Hydrol.*, **65**, 49–72.

Pilgrim, D. H. (Ed.) 1987. *Australian Rainfall and Runoff*. The Institution of Engineers, Barton, ACT. 374 pp.

Pilgrim, D. H. and Cordery, I. 1975. 'Rainfall temporal patterns for design floods', *J. Hydraul. Div. ASCE*, **101**, 81–95.

Raupach, M. R. and Finnigan, J. J. 1995. 'Scale issues in boundary layer — meteorology: surface energy balances in heterogenous terrain', *Hydrol. Process.*, **9**, 589–612.

Rawls, W. J., Brakensiek, D. L., and Miller, N. 1983. 'Green–Ampt infiltration parameters from soils data', *J. Hydraul. Engin.*, **109**, 62–70.

Reybold, W. U. and TeSelle, G. W. 1989. 'Soil geographic data bases', *J. Soil Wat. Conserv.*, Jan–Feb, 28–29.

Richards, K. 1982. *Rivers — Form and Process in Alluvial Channels*. Methuen, London. 361 pp.

Rinaldo, A., Marani, A., and Rigon, R. 1991. 'Geomorphologic dispersion', *Wat. Resour. Res.*, **27**, 513–525.

Robinson, J. S. and Sivapalan, M. 1995. 'Catchment-scale runoff generation model by aggregation and similarity analysis', *Hydrol. Process.*, **9**, 555–574 .

Rodríguez-Iturbe, I. 1986. 'Scale problems in hydrologic processes' in Shen, H. W., Obeysekera, J. T. B., Yevjevich, V., and Decoursey, D. G. (Eds), *Multivariate Analysis of Hydrologic Processes*. Colorado State University. pp. 51–65.

Rodríguez-Iturbe, I. and Gupta, V. K. (Eds) 1983. 'Scale problems in hydrology', *J. Hydrol.*, **65** (spec. issue).

Rodríguez-Iturbe, I. and Valdés, J. B. 1979. 'The geomorphologic structure of hydrologic response', *Wat. Resour. Res.*, **15**, 1409–1420.

Rodríguez-Iturbe, I., Ijjász-Vásquez, E. J., Bras, R. L., and Tarboton, D. G. 1992a. 'Power law distributions of discharge mass and energy in river basins', *Wat. Resour. Res.*, **28**, 1089–1093.

Rodríguez-Iturbe, I., Rinaldo, A., Rigon, R., Bras, R. L., Marani, A., and Ijjász-Vásquez, E. 1992b. 'Energy dissipation, runoff production and the 3-dimensional structure of river basins', *Wat. Resour. Res.*, **28**, 1095–1103.

Rogers, A. D. 1992. 'The development of a simple infiltration capacity equation for spatially variable soils', *B.E. Thesis*, Univ. of West. Aust., Nedlands. 64 pp.

Rosso, R., Peano, A., Becchi, I., Bemporad, G. A. (Eds.) 1994. 'Advances in distributed hydrology' in *Proceedings of a Workshop held in Bergamo, Italy, June 1992*. Water Resources Publications, 416 pp.

Rosso, R., Bacchi, B., and La Barbera, P. 1991. 'Fractal relation of mainstream length to catchment area in river networks', *Wat. Resour. Res.*, **27**, 381–387.

Rubin, Y. and Gómez-Hernández, J. J. 1990. 'A stochastic approach to the problem of upscaling of conductivity in disordered media: theory and unconditional numerical simulations', *Wat. Resour. Res.*, **26**, 691–701.

Rubin, Y., Mavko, G., and Harris, J. 1992. 'Mapping permeability in heterogeneous aquifers using hydrologic and seismic data', *Wat. Resour. Res.*, **28**, 1809–1816.

Russo, D. and Jury, W. A. 1987. 'A theoretical study of the estimation of the correlation scale in spatially variable fields 1. stationary fields', *Wat. Resour. Res.*, **23**, 1257–1268.

Russo, D. 1992. 'Upscaling of hydraulic conductivity in partially saturated heterogeneous porous formation', *Wat. Resour. Res.*, **28**, 397–409.

Schönwiese, C. D. 1979. *Klimaschwankungen*. Springer, Berlin.

Schumm, S. A. 1956. 'Evolution of drainage systems and slopes in Badlands at Perth Amboy, New Jersey, *Geol. Soc. Am. Bull.*, **67**, 597–646.

Sellers, P. J., Hall, F. G., Asrar, G., Strebel, D. E., and Murphy, R. E. 1992. 'An overview of the First International Satellite Land Surface Climatology Project (ISLSCP) Field Experiment (FIFE)', *J. Geophys. Res.*, **97**(D17), 18 345–18 371.

Shreve, R. L. 1966. 'Statistical law of stream numbers', *J. Geol.*, **74**, 17–37.

Shreve, R. L. 1967. 'Infinite topologically random networks', *J. Geol.*, **75**, 178–186.

Silliman, S. E. and Wright, A. L. 1988. 'Stochastic analysis of paths of high hydraulic conductivity in porous media', *Wat. Resour. Res.*, **24**, 1901–1910.

Sivapalan, M. 1993. 'Linking hydrologic parameterisations across a range of scales: hillslope to catchment to region' in Bolle, H.-J., Feddes, R. A., and Kalma, J. D. (Eds), *Exchange Processes at the Land Surface for a Range of Space and Time scales, IAHS Publ.*, **212**, 115–123.

Sivapalan, M. and Viney, N. R. 1994a. 'Large scale catchment modelling to predict the effects of land use and climate', *Water, J. Aust. Wat. Wastewat. Assoc.*, **21**, 33–37.

Sivapalan, M. and Viney, N. R. 1994b. 'Application of a nested catchment model for predicting the effects of changes in forest cover'. In: Ohta, T. (Ed.) *Proc. Int. Symp. Forest Hydrology*, Univ. Tokyo, Tokyo. IUFRO, pp. 315–322.

Sivapalan, M. and Wood, E. F. 1986. 'Spatial heterogeneity and scale in the infiltration response of catchments' in Gupta, V. K., Rodríguez-Iturbe, I., and Wood, E. F. (Eds), *Scale Problems in Hydrology*. D. Reidel, Dordrecht. pp. 81–106.

Sivapalan, M., Beven, K., and Wood, E. F. 1987. 'On hydrologic similarity, 2. A scaled model of storm runoff production', *Wat. Resour. Res.*, **23**, 2266–2278.

Sivapalan, M., Wood, E. F., and Beven, K. J. 1990. 'On hydrologic similarity. 3. A dimensionless flood frequency model using a generalized geomorphologic unit hydrograph and partial area runoff generation', *Wat. Resour. Res.*, **26**, 43–58.

Skoda, G. 1987. 'Fractal dimension of rainbands over hilly terrain', *Meteorol. Atmos. Phys.*, **36**, 74–82.

Smagorinsky, J. 1974. 'Global atmospheric modeling and the numerical simulation of climate' in Hess, W. N. (Ed.), *Weather and Climate Modification*. Wiley, New York. pp. 633–686.

Smith, J. A. 1992. 'Representation of basin scale in flood peak distribution', *Wat. Resour. Res.*, **28**, 2993–2999.

Snell, J. D. and Sivapalan, M. 1994a. 'On geomorphological dispersion in natural catchments and the geomorphological unit hydrograph', *Wat. Resour. Res.*, **30**, 2311–2323.

Snell, J. D., and Sivapalan, M. 1994b. 'Threshold effects in geomorphological parameters extracted from DEM's', *Trans. Jpn Geomorph. Union*, **15A**, 67–93.

Snell, J. D. and Sivapalan, M. 1995. 'Application of the meta-channel concept: Construction of the meta-channel hydraulic geometry for a natural catchment', *Hydrol. Process.*, **9**, 485–506.

Sreenivasan, K. R. 1991. 'Fractals and multifractals in fluid turbulence', *Annu. Rev. Fluid Mech.*, **23**, 539–600.

Stommel, H. 1963. 'Varieties of oceanographic experience', *Science*, **139**, 572-576.

Strahler, A. N. 1957. 'Quantitative analysis of watershed geomorphology', *Trans. Am. Geophys. Union*, **38**, 913–920.

Strahler, A. N. 1964. 'Quantitative geomorphology of drainage basins and channel networks' in Chow, V. T. (Ed.), *Handbook of Applied Hydrology*. McGraw-Hill, New York. pp. 4.39–4.76.

Stull, R. B. 1988. *An Introduction to Boundary Layer Meteorology*. Kluwer Academic, Dordrecht. 666 pp.

Surkan, A. J. 1969. 'Synthetic hydrographs: effects of network geometry', *Wat. Resour. Res.*, **5**, 112–128.

Tabios, G. Q. and Salas, J. D. 1985. 'A comparative analysis of techniques for spatial interpolation of precipitation', *Wat. Resour. Bull.*, **21**, 365–380.

Tarboton, D. G., Bras, R. L., and Rodríguez-Iturbe, I. 1988. 'The fractal nature of river networks', *Wat. Resour. Res.*, **24**, 1317–1322.

Tarboton, D. G., Bras, R. L., and Rodríguez-Iturbe, I. 1989. 'Scaling and elevation in river networks', *Wat. Resour. Res.*, **25**, 2037–2051.

Tarboton, D. G., Bras, R. L., and Rodríguez-Iturbe, I. 1991. 'On the extraction of channel networks from digital elevation data' in Beven, K. J. and Moore, I. D. (Eds), *Terrain Analysis and Distributed Modelling in Hydrology*. Wiley, Chichester. pp. 85–106.

Thiessen, A. H. 1911. 'Precipitation averages for large areas', *Monthly Weather Rev.*, **39**, 1082–1084.

Tillotson, P. M. and Nielsen, D. R. 1984. 'Scale factors in soil science', *Soil Sci. Soc. Am. J.*, **48**, 953–959.

Troch, P. A., De Troch, F. P., Mancini, M., and Wood, E. F. 1995. 'Stream network morphology and storm response in humid catchments', *Hydrol. Process.*, **9**, (at press).

Troutman, B. M. and Karlinger, M. R. 1985. 'Unit hydrograph approximations assuming linear flow through topologically random networks', *Wat. Resour. Res.*, **21**, 743–754.

USDA-SCS (US Department of Agriculture Soil Conservation Service) 1986. 'Urban hydrology for small watersheds', *Tech. release 55*, USDA-SCS, Washington.

van der Tak, L. D. and Bras, R. L. 1990. 'Incorporating hillslope effects into the geomorphologic instantaneous unit hydrograph', *Wat. Resour. Res.*, **26**, 2393–2400.

Vertessy, R. and Band, L. 1993. 'Cross comparison of terrain analysis methods for hydrologic simulations' in Kalma, J., Sivapalan, M., and Wood, E. (Eds), *Scale Issues in Hydrological/Environmental Modelling, Proc. Workshop Robertson, CRES*. Australian National University, Canberra. p. 16.

von Bandat, H. F. 1962. *Aerogeology*. Gulf Publishing, Houston. 350 pp.

Voss, R. F. 1985. 'Random fractals: characterisation and measurement' in Pynn, R. and Skjeltorp, A. (Eds), *Scaling Phenomena in Disordered Systems*. Plenum Press, New York. pp. 1–11.

Voss, R. F. 1988. 'Fractals in nature: from characterization to simulation' in Pietgen, H. and Saupe, D. (Eds), *The Science of Fractal Images*. Springer, New York. pp. 21–70.

Walker, R. G. (Ed.) 1984. *Facies Models*. Geological Association of Canada, Toronto. 317 pp.

Wallis, J. R. 1988. 'Catastrophes, computing and containment: living with our restless habitat', *Spec. Sci. Technol.*, **11**, 295–315.

Weinberg, G. M. 1975. *An Introduction to General Systems Thinking*. Wiley, New York. 279 pp.

Wenzel, H. G. 1982. 'Rainfall for urban stormwater design' in Kibler, D. F. (Ed.), *Urban Storm Water Hydrology. Wat. Resour. Monogr. 7*. American Geophysical Union, Washington, pp. 35–64.

White, I. 1988. 'Measurement of soil physical properties in the field' in Steffen, W. L. and Denmead, O. T. (Eds), *Flow and Transport in the Natural Environment: Advances and Applications*. Springer, Berlin. pp. 59–85.

Willgoose, G. and Kuczera, G. 1995. 'Estimation of sub-grid scale kinematic wave parameters for hillslopes', *Hydrol. Process.*, **9**, 469–482.

Willgoose, G. and Riley, S. 1993. 'Scale dependence of runoff and the hydrology of a proposed mine rehabilitation' in Institution of Engineers (Ed.), *Towards the 21st Century, Hydrology and Water Resources Symposium, Newcastle, 1993*. The Institution of Engineers, Barton, ACT. pp. 159–164.

Willgoose, G. R., Bras, R. L., and Rodríguez-Iturbe, I. 1991. 'A physically based coupled network growth and hillslope evolution model: 1. Theory', *Wat. Resour. Res.*, **27**, 1671–1684.

Williams, R. E. 1988. 'Comment on "Statistical theory of groundwater flow and transport: pore to laboratory, laboratory to formation, and formation to regional scale" by Gedeon Dagan', *Wat. Resour. Res.*, **24**, 1197–1200.

Williams, J., Ross, P., and Bristow, K. 1992. 'Prediction of the Campbell water retention function from texture, structure, and organic matter' in Van Genuchten, M. Th., Leij, F. J., and Lund, L. J. (Eds.), *Proc. Int. Workshop on Indirect Methods for Estimating the Hydraulic Properties of Unsaturated Soils*. Univ. California, Riverside 427–441.

Wiltshire, S. E. 1985. 'Grouping basins for regional frequency analysis', *Hydrol. Sci. J.*, **30**, 151–159.

World Meteorological Organisation (WMO) 1969. 'Manual for depth–area–duration analysis of storm precipitation', *WMO No. 237, Tech. Pap.*, **129**, 1–31.

Woo, M.-K., Heron, R., Marsh, P., and Steer, P. 1983a. 'Comparison of weather station snowfall with winter snow accumulation in High Arctic basins', *Atmosphere–Ocean*, **21**, 312–325.

Woo, M.-K., Marsh, P., and Steer, P. 1983b. 'Basin water balance in a continuous permafrost environment' in *Permafrost, Fourth International Conference*. National Academy Press, Washington. pp. 1407–1411.

Wood, E. F. and Hebson, C. S. 1986. 'On hydrologic similarity. 1. Derivation of the dimensionless flood frequency curve', *Wat. Resour. Res.*, **22**, 1549–1554.

Wood, E. F., Sivapalan, M., Beven, K., and Band, L. 1988. 'Effects of spatial variability and scale with implications to hydrologic modeling', *J. Hydrol.*, **102**, 29–47.

Wood, E. F., Sivapalan, M., and Beven, K. 1990. 'Similarity and scale in catchment storm response', *Rev. Geophys.*, **28**, 1–18.

Woods, R. A., Sivapalan, M., and Duncan, M. J. 1995. 'Investigating the representative elementary area concept: an approach based on field data', *Hydrol. Process.*, **9**, 291-312.

Woolhiser, D. A. and Goodrich, D. 1988. 'Effect of storm rainfall intensity patterns on surface runoff', *J. Hydrol.*, **102**, 335–354.

Woolhiser, D. A. and Osborne, H. B. 1985. 'A stochastic model of dimensionless thunderstorm rainfall', *Wat. Resour. Res.*, **21**, 511–522.

Wu, Y.-H., Yevjevich, V., and Woolhiser, D. A. 1978. 'Effects of surface roughness and its spatial distribution on runoff hydrographs', *Colo. State Univ., Fort Collins, Colo., Hydrol. Pap.*, **96**, 47 pp.

Wu, Y.-H., Woolhiser, D. A., and Yevjevich, V. 1982. 'Effects of spatial variability of hydraulic resistance of runoff hydrographs', *J. Hydrol.*, **59**, 231–248.

Yang, C. T. 1976. 'Minimum unit stream power and fluvial hydraulics', *J. Hydraul. Div. ASCE*, **102**(HY7), 919–934.

Yang, C. T. and Song, C. C. S. 1979. 'Theory of minimum rate of energy dissipation', *J. Hydraul. Div. ASCE*, **105**(HY7), 769–784.

Yeh, T. C. J., Gelhar, L. W., and Gutjahr, A. L. 1985. 'Stochastic analysis of unsaturated flow in heterogeneous soils. 1. Statistically isotropic media', *Wat. Resour. Res.*, **21**, 447–456.

Yevjevich, V. 1972. *Stochastic Processes in Hydrology*. Water Resources Publications, Fort Collins. 276 pp.

Zuidema, P. K. 1985. *Hydraulik der Abflussbildung während Starkniederschlägen*. Mitt. VAW 79, ETH Zürich, 150 pp.

3

INVESTIGATING THE REPRESENTATIVE ELEMENTARY AREA CONCEPT: AN APPROACH BASED ON FIELD DATA

ROSS WOODS* AND MURUGESU SIVAPALAN

Centre for Water Research, Department of Environmental Engineering, University of Western Australia, Nedlands, WA 6009, Australia

MAURICE DUNCAN

NIWA Freshwater, National Institute for Water and Atmospheric Research Ltd, Box 8602, Christchurch, New Zealand

ABSTRACT

Changing the scale of observation or averaging has a significant, but poorly understood, impact on the apparent variability of hydrological quantities. The representative elementary area (REA) concept is used as a motivation for measuring inter-storm streamflow and calculating wetness index distributions for the subcatchments of two small study areas in New Zealand. Small subcatchments are combined to provide larger scale samples, and then the variance of specific discharge between similar sized subcatchments is calculated. For small subcatchments (area less than ~ 1 km^2) this variance is found to decrease with area more quickly than might be expected if the catchments were random samples. Such behaviour is tentatively interpreted as evidence supporting the concept of 'organization'. At larger scales, variance between catchments decreases in a way that is consistent with sampling from a stationary random field. The results from the streamflow data are reinforced by an analysis of topographic data for the two study areas, although some questions remain open.

Both the flow and topographic data support the idea that it is possible to find an averaging scale where the variability between catchments is sufficiently small for a 'distribution function' approach to be used in distributed rainfall–runoff modelling. Consistent estimates of the scale at which the study areas become stationary (0·5 km^2 for Little Akaloa, 2 km^2 for Lewis) are obtained using both flow and topographic data. The data support a pragmatic REA concept which allows meaningful averages to be formed: this may be a useful base for further conceptual developments, but it is not appropriate for a classical continuum approach. Further conceptual development combined with field measurement and computer simulation are still required for the REA to have operational impacts. In particular, it is not clear which models are appropriate for use at the REA scale.

INTRODUCTION

The complex interaction between spatial scale and spatial variability is widely perceived as a substantial obstacle to progress in hydrology. Numerous workers, including Dooge (1986; 1988), Klemes (1983; 1986), Beven (1989; 1991) and the National Research Council (US) (1991), have made this point. The work of Dunne (1983), Klemes (1983), DeCoursey (1991) and O'Loughlin (1990) all suggest that a closer link between measurement, modelling and theoretical development is the best way forward. Here we address the problem by using detailed streamflow and topography data to assess spatial variability at multiple spatial scales, for a single time-scale. Observations of the interaction between variability and scale are a vital ingredient for any related theoretical or conceptual developments in hydrology, if we are to progress beyond the stage of conjecture to well-founded concepts.

---

* Also at NIWA Freshwater, New Zealand.

Coping with variability: the representative elementary area concept

The approach used here has been motivated by the representative elementary area (REA) concept proposed by Wood *et al.* (1988; 1990) in their search for an appropriate spatial scale (the REA) at which a simple description of the rainfall–runoff process could be obtained. Once the size of the REA is known, the catchment being studied could presumably then be disaggregated into REA-sized subcatchments (the inputs and parameter values would need to be specified at the REA scale) and a simple parameterization developed. Their approach does not show which parameterization to use, but instead suggests a spatial scale at which a simple one might be found.

Wood *et al.* (1988) suggest that the actual patterns of variability within small catchments are important in determining their hydrological response, even if all the catchments are considered as being drawn from the same (hypothetical) statistical distribution. Taking this view, the differences between small catchments are caused by inadequate sampling. One solution is to use large enough catchments to properly sample this small-scale variability, so that the catchments will have sufficiently similar response functions, provided there is no larger scale source of variability (which must be resolved explicitly).

Looking for a representative elementary area

The scale at which the variability of hydrological response (between catchments) falls to an acceptably low level (and possibly rises again) is taken by Wood *et al.* (1988) as the method of detection for the REA. The REA is thus a 'good' scale at which to take samples of catchment runoff generation, assuming that the primary objective of sampling is to infer the mean catchment runoff generation. They also give a definition of REA as the 'smallest discernible point which is representative of the continuum'. Both of these descriptions of the REA are subjective to some extent.

Wood *et al.* (1988; 1990) investigated the REA concept by computer simulation, using spatially variable topography and synthetic rainfall and hydraulic conductivity fields with TOPMODEL, a runoff generation model. They found that the degree of scatter between groups of subcatchments decreased with increasing subcatchment size, and that the mean response among subcatchments appeared to stabilize at about 1 km^2. Blöschl *et al.* (1995) have suggested that the method of plotting the results as moving means may have influenced the size of the REA; using more points in each 'window' will smooth out differences between window means and produce a smaller REA. Wood *et al.* (1988) did not detect any increase in the variability of hydrological response between subcatchments at larger areas, possibly because their largest samples were not large enough to detect non-stationarity in the topography data (the rainfall and hydraulic conductivity fields were approximately stationary at the 17 km^2 scale). If all the subcatchments of a sufficiently large area are producing almost identical runoff, in spite of the differences in actual detailed patterns of topography, soils and rainfall between the subcatchments, we can reasonably conclude that only a summary description of the topography, soils and rainfall is necessary, i.e. that the 'distribution function' approach (Beven, 1991) is valid at that scale.

The primary conclusion of Wood *et al.* (1988), that an REA does exist in the context of runoff generation in catchments, has aroused considerable interest among hydrologists, given the claimed benefits that accrue if an REA exists. There was no evidence presented to suggest that their results are any more than a consequence of the 'law of large numbers' (Garbrecht, 1991); provided the population is stationary, we would expect larger samples to be closer to one another. What Wood *et al.* (1988) show, in effect, is that the runoff generation population appears to be stationary, given their assumptions about the spatial variability of moisture deficit, soils and rainfall. It is also possible that the variability was decreasing at a rate other than that suggested by statistical sampling theory, indicating some more complex spatial structure, but they did not address this question.

Measurements of spatial variability in hydrology

Regardless of whether or not the 1 km^2 REA was actually present in the work of Wood *et al.* (1988; 1990), the REA concept provides a motivation for other work because it highlights the interaction between scale and variability. Debates about the validity of their work can be kept in perspective by remembering that they analysed the output of a computer program, rather than a set of measurements. This study attempts to answer the question: what do measurements show about the interaction of spatial scale and spatial variability?

Measurements used in studies of spatial variability in hydrology are typically 'point scale' measurements such as saturated hydraulic conductivity and rainfall (e.g. Thauvin and Lebel, 1991). The term 'point scale' is relative: the point scale is infinitesimal when compared with the scale of the problem under study. Having measurements only at the point scale does not prevent the construction of larger scale estimates using spatial interpolation techniques such as surface fitting or kriging. However, what it does prevent is the understanding of the effect of scale. Interpolation techniques implicitly assume that scale is unimportant within the domain being studied. Decisions about the appropriateness of interpolation are purely a matter of judgement and experience.

Measurements of streamflow are inherently spatial averages, so one useful approach for studying spatial variability and scale is to measure the flows at many points along the mainstem of a stream and associate the increase in flow with the catchment area draining to the stream between gauging points. Previous workers (Anderson and Burt, 1978; Mosley, 1979; Huff et al., 1982; Burt and Arkle, 1986; Cooper, 1990; Mulholland et al., 1990) have used a similar approach, but often with the aim of associating the variations in flow per unit area with some other catchment feature, such as topography or water quality. For this study we are interested purely in the effects of scale on the variability of streamflow and topography.

Genereux et al. (1993) briefly considered the testing of the REA concept using streamflow data and found that the runoff rate increased with catchment area throughout the (rather narrow) range (30–40 ha) covered by their data. They found bedrock geology to be a dominant control on runoff generation and concluded that 40 ha was insufficient to obtain a representative sample of the geology. Wolock (1993) studied the effect of catchment size on the sample moments of distributions of a topographic wetness index within the 112 km^2 Sleepers River catchment. He found that variability between catchments decreased as catchment size increased from 0·01 to 1 km^2, and that there was little variability among catchments in the range 1–100 km^2.

Overview of this work

Here we report measurements of streamflow per unit area of catchment and calculations of a topographic wetness index, covering scales from 0·01 to 50 km^2. The approach has several important features in the context of previous measurements of spatial variability: we measure flow, a quantity hydrologists often need

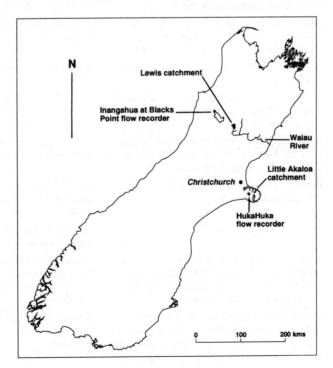

Figure 1. Location map, showing study sites and nearby flow recording sites within South Island, New Zealand

to predict and model; measurements are spatially integrated quantities, not point scale data; the measurements completely cover a contiguous region; and measurements can be combined to provide larger scale samples.

To test further the results obtained using the limited range of spatial scales covered by our flow data, we study the effect of scale on the variability of topography. We also investigate to what extent topography can be considered as a correlated random field and look at the possibility that some further spatial structure is important in determining patterns of hydrological response. Finally, we discuss some implications of detecting a REA on future developments in hydrological modelling.

Table I. Comparative features of the two study areas. Data from field observation, Bowen (1964), Suggate (1973), Ministry of Works and Development (1979), Department of Survey and Land Information (1983; 1989), New Zealand Meteorological Service (1986)

Feature	Little Akaloa	Lewis River
Channel description	Single thread, steep	Single thread, steep
Channel morphology	Pool and riffle, with runs in lower reaches	Pool, riffle and run
Bed material	Boulders and large cobbles in upper reaches, small cobbles and gravel in lower reaches, sand at the last (tidal interface) cross-section	Large cobbles, large boulders protruding at some sites, bedrock pools up to 2 m deep
Longitudinal slope	0·046 on average, ranging from 0·1 in upper reach to 0·027 in lowest 1500 m.	0·02, relatively even throughout
Channel network	Main stem is joined by lower order streams in a herring-bone pattern, Horton order of main stem is 2	Main stem is joined by lower order streams in a herring-bone pattern, Horton order of main stem is 3
Gauging cross-sections	2–8 m wide (mean 4·1 m) 0·16–0·68 m deep (mean 0·35 m)	3·8–17·7 m wide (mean 9·3 m) 0·29–1·12 m deep, (mean 0·57 m)
Catchment area	14·1 km^2 at downstream site	52·4 km^2 at downstream site
Catchment description	Hilly, with valley slopes from 18 to 35°, bottom quarter of catchment has alluvial terraces up to 300 m wide at seafront	Steep and mountainous, with 60% of valley slopes 25–35° and some steeper
Elevation range	Sea level to 738 m above sea level (mean 364 m)	640 to 1870 m above sea level (mean 1109 m)
Lithology	Valley dissects andesitic and basaltic lava flows and tephra deposits, lower slopes overlain by loess	Strongly indurated greywacke and argillite
Soils	Southern yellow–grey and yellow brown soils, on lower slopes of true left side of valley, soils are on deep loess	Steepland yellow brown soils, developing into podzols and gley soils on flatter sites
Vegetation	Pasture, with a little scrub, woodlot and remnant native bush	Beech forest on lower slopes, with tussock grassland above and bare rock on highest tops
Rainfall	1163 mm annual rainfall	3000 mm annual rainfall, with snow on tops in winter
Hydrology	Highest flows in winter and spring, under saturated conditions, freshes about once a month, lower flows in summer, though floods can occur at any time	Highest flows in spring and early summer, with rain adding to melting winter snow, freshes almost weekly, lower flows in summer
Likely runoff generation mechanisms	Mixed, with subsurface flow possible in yellow brown soils, but infiltration excess during heavy rain on yellow grey soils and saturation excess in valley bottoms	Subsurface stormflow dominant, with limited saturation excess in narrow valley bottoms

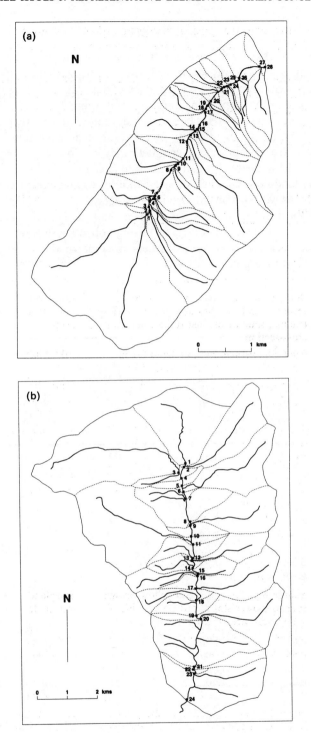

Figure 2. Stream channels, site positions and subcatchment boundaries. (a) Little Akaloa Stream and (b) Lewis River

STUDY SITES

There are two study sites: Little Akaloa Stream is on the north-east side of Banks Peninsula, South Island, New Zealand (Figure 1); Lewis River drains southwards from the main divide of the South Island's Southern Alps, 130 km north-west of Christchurch, and is a tributary of the Waiau River (Figure 1). The streams were chosen predominantly on catchment shape (long and thin) and channel network. They had to be headwater streams so the flows at the upstream sites would be small, and the potential contributions to main stem flow by tributaries would be large enough to significantly increase the main stem flow. Tributary size had to be relatively even so when flows from them were aggregated, they would not be dominated by disparate sized tributaries. The rivers had to be wadeable for gauging and to have reasonable gauging sites above and below major tributaries. Uniform land use was a factor, as we did not want differing land use affecting the flows. Access to stream gauging sites was also an important consideration, so the large number of sites could be gauged by a small number of people in one day. Both rivers had roads running close to them and allowed ready access over grazed paddocks (Little Akaloa) or through open beech forest or along tributaries to the main stem. A steep walled rock gorge prevented access to some of the lower reaches of the Lewis. A description of the two catchments is given in Table I.

Table II. Contributing areas for flow gauging sites

Little Akaloa Stream			Lewis River		
Site number	Area of immediate upstream catchment (km^2)	Total area upstream of site (km^2)	Site number	Area of immediate upstream catchment (km^2)	Total area upstream of site (km^2)
1	4·87	4·87	1	4·40	4·40
2	0·55	5·42	2	2·98	7·38
3	0·04	5·46	3	7·89	15·26
4	1·06	6·52	4	0·49	15·75
5	0·08	6·59	5	0·37	16·12
6	0·26	6·85	6	3·02	19·14
7	0·34	7·19	7	0·06	19·21
8	0·50	7·69	8	2·46	21·67
9	0·23	7·92	9	2·75	24·42
10	0·06	7·92	10	0·58	25·00
11	0·58	8·56	11	6·27	31·26
12	0·28	8·84	12	1·04	32·30
13	1·21	10·05	13	1·33	33·63
14	0·06	10·11	14	0·66	34·29
15	0·45	10·56	15	0·63	34·92
16	0·54	11·09	16	0·99	35·91
17	0·70	11·80	17	2·90	38·81
18	0·22	12·02	18	0·91	39·72
19	0·01	12·04	19	0·99	40·72
20	0·17	12·20	20	3·07	43·78
21	0·34	12·54	21	3·98	47·76
22	0·12	12·66	22	0·70	48·46
23	0·12	12·79	23	0·08	48·54
24	0·13	12·92	24	3·87	52·40
25	0·19	13·11			
26	0·24	13·34			
27	0·34	13·68			
28	0·40	14·08			

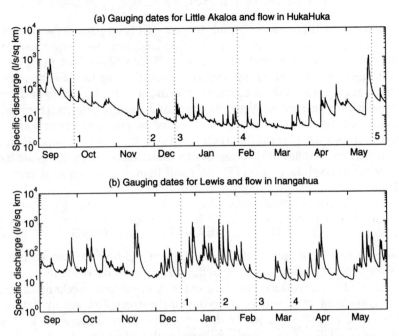

Figure 3. Flow records for nearby flow recording sites, with gauging days indicated by vertical lines. (a) Little Akaloa Stream and (b) Lewis River

METHOD OF STREAMFLOW GAUGING

Gauging sites were chosen on both streams to be upstream and downstream of observable tributaries. There are 28 sites on Little Akaloa Stream and 24 sites on the Lewis River, as shown in Figure 2. The precise positioning of each cross-section was chosen to provide the best cross-section for flow gauging. However, the choice of a suitable cross-section can depend on the flow, so the gaugers used their discretion on some occasions to move to a nearby better cross-section, but without significantly altering the contributing area between sites. (Note that site 3 for the Lewis River is on a large tributary, but for analysis a dummy site was created on the mainstream; the flow for this site was calculated to match the inflow from the tributary.)

Flow was gauged using Pygmy, small Ott and Gurley current meters. At least 15 verticals (i.e. locations across a cross-section) were measured and usually 20, although some cross-sections on Little Akaloa Stream were so narrow (2 m) that even 15 verticals were very close together. Point velocities were measured at one point in each vertical (0·6 of the depth, measured from the surface). A team of four experienced gaugers was used on each gauging day, with two to three extra staff to book gaugings. Each gauging took approximately 45 minutes, including the time to walk to the next site and set up. The total labour requirement for data collection was approximately 600 hours.

Table III. Summary of gaugings

Little Akaloa Stream			Lewis River		
Gauging day	Date of gauging	Flow at most downstream site (l/s)	Gauging day	Date of gauging	Flow at most downstream site (l/s)
1	29 September 1992	428·0	1	21 December 1992	1436
2	26 November 1992	131·8	2	20 January 1993	4024
3	17 December 1992	98·3	3	17 February 1993	1772
4	4 February 1993	77·0	4	16 March 1993	1033
5	21 May 1993	298·0			

Table II shows the catchment area contributing to each of the streams between each pair of adjacent gauging sites. Areas for both streams were obtained by using landmarks to position gauging sites on 1:50 000 scale maps (Department of Survey and Land Information, 1983; 1989) and then using topographic contours to manually draw catchment boundaries. These boundaries were then digitized to obtain subcatchment areas.

For the flows at adjacent sites measured at different times during the same day to be directly comparable, it is necessary to assume that the flow is near steady throughout the gauging day. All measurements were made during flow recessions and the rate of recession was checked by measuring river stage against three local datums for each stream (sites 2, 11, 17 for Little Akaloa and sites 4, 13, 24 for Lewis). Figure 3 shows the flow record at neighbouring continuous flow measurement sites; the dates of the gaugings are indicated by vertical lines. The location of the HukaHuka Stream (12 km^2) and Inangahua River (234 km^2) flow recorders are shown in Figure 1.

RESULTS OF STREAMFLOW GAUGING

There were five gaugings carried out for Little Akaloa Stream and four for Lewis River during the period September 1992 to May 1993. The dates and gauged flows at the most downstream site on each day for each stream are shown in Table III. Gauging days for each stream are four to six weeks apart on average. The days were chosen several weeks in advance, with back-up days if the streamflow was likely to change quickly (i.e. during or just after rain). The observed within-day variations in river stage for each gauging day were negligible (typically 3–8 mm fall in river stage, compared with typical flow depths of 300–500 mm) for the purpose of treating the gauging as simultaneous along the stream.

Downstream variation in flow and lateral inflow

Figure 4 shows the gauged flow in the mainstream, plotted against upstream contributing area, at each site on each gauging day for both streams. The estimated 95% confidence limits are shown by vertical bars (reflecting sampling errors due to both the limited sampling time of approximately 40 seconds for each velocity measurement and the limited number of sampling points within each vertical and across the section). A number of decreases in flow with increasing area are apparent in Figure 4, as well as some unlikely looking increases in flow, particularly when the catchment area between sites is considered. It is possible for flow to decrease as we move downstream—for example, because of infiltration into the stream bed or because of measurement error. As some reaches show consistent decreases for all gaugings, whereas others only have decreases for some gaugings, both of these reasons are likely to have affected the data presented here, mainly the data from Little Akaloa.

If we subtract the flow at one site from the flow at the next downstream site, we have an indirect measurement of the extra flow from the catchment area draining to the stream between the two gauging sites. Dividing the increase in flow (ΔQ_{ij}^k) between gauging sites i and j on the kth gauging day by the contributing area (ΔA_{ij}) gives a specific discharge ($q_{ij}^k = \Delta Q_{ij}^k / \Delta A_{ij}$) for that 'catchment' on that day. To carry out an REA analysis of these data, specific discharge (rather than flow) is used as a measure of hydrological response. [The variability of flow is dominated by variations in catchment area (see Figure 4), so flow is not an appropriate measure of the hydrological response.] Figure 5 plots the specific discharge for each subcatchment between adjacent sites. The specific discharge values (q_{ij}^k) are normalized so that gaugings from different days plot together: specific discharge for each subcatchment is divided by the specific discharge for the whole catchment on that day (q_{0N}^k, where the '$0N$' subscript indicates the complete catchment drained by the most downstream gauging site) to obtain a normalized specific discharge ($\bar{q}_{ij}^k = q_{ij}^k / q_{0N}^k$). Decreases in flow between gauging sites are plotted as negative specific discharges on Figure 5. The very high yielding subcatchments are balanced by others which do not contribute sufficiently to overcome channel infiltration and measurement errors.

Figure 4. Gauged flow at all sites, with vertical bars showing 95% error bounds, for each gauging day. (a) Little Akaloa Stream and (b) Lewis River

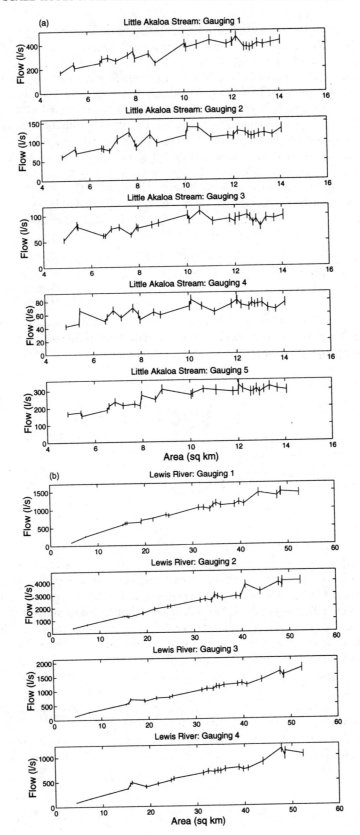

ANALYSIS OF STREAMFLOW GAUGING

Inflows at larger scales

To study the effect of catchment area on the variability of specific discharge, we first need flows measured at larger scales than those shown in Table II (central column). We can sum the inflows from *adjacent* subcatchments and add their areas. The ratio of combined flow to combined area is the specific discharge from this larger catchment. We can obviously continue this process, combining three or four or more adjacent catchments to obtain larger areas. The resulting collection of various sized regions are similar in character to those in Wood *et al.* (1988), where they used various 'seeding' thresholds to subdivide Coweeta catchment to varying levels of detail.

Suppose, for simplicity, we had eight subcatchments labelled A–H. Then one approach might be to progressively accumulate in pairs, giving the list A, B, C, D, E, F, G, H, A + B, C + D, E + F, G + H, A + B + C + D, E + F + G + H, A + B + C + D + E + F + G + H. To study the effect of subcatchment scale on

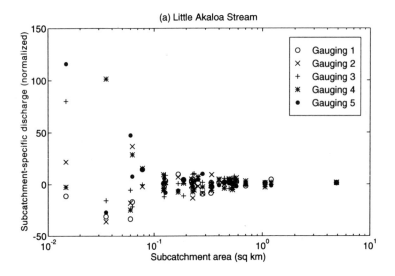

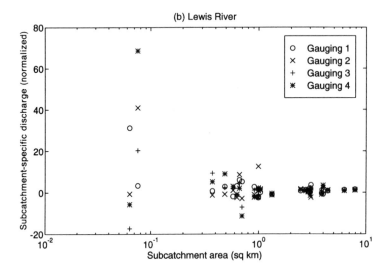

Figure 5. Specific discharge (normalized by catchment specific discharge) for the subcatchments between each pair of adjacent sites, for each gauging day, plotted against subcatchment area. (a) Little Akaloa Stream and (b) Lewis River

variability, we will sort a list such as this by area, and group the list into 'windows' for calculating the variance between catchments of similar sizes. Clearly, there will be a problem if, for example, catchment A is noticeably larger than catchments B–H, because then A will be a significant contributor to all the catchments in the last window (largest subcatchments) and the variance for that window will have little statistical meaning because the samples are not independent.

An improved method for combining the subcatchments is to start with all 28 subcatchments (24 for Lewis), and find the pair of adjacent subcatchments whose combined area is smallest. Combine them to form a new subcatchment. Now repeat the process, working on a reduced list where the new subcatchment replaces its two components. Continue combining adjacent subcatchments in this way until the list is exhausted: every subcatchment created during the process can be used for analysis. This method of combining adjacent subcatchments is less prone to combining adjacent regions with very disparate areas. It produces a list whose largest subcatchments are relatively independent of one another.

Figure 6 shows the specific discharge plotted against the catchment area for each gauging day at each site, including all the combined subcatchments (55 for Little Akaloa and 48 for Lewis). Figure 6 is similar in concept to Figure 4a of Wood *et al.* (1988) and has a similar general appearance [note that we show individual data points, and have not used moving means as in Wood *et al.* (1988)]. The range of specific discharge values decreases sharply as we increase the spatial scale.

Error analysis

It is relevant to note that this decrease in range with increasing area seems unlikely to be related to any effect of flow magnitude (and hence catchment area) on flow measurement error. The estimated relative error in a flow gauging is predominantly determined by the expected random error in the velocity measurements and the extent to which the cross-section was fully sampled. As we used the same gauging strategy (sampling time and number of verticals) for all sites, regardless of flow, we conclude that the relative error in flow is independent of flow magnitude. Thus the relative error in calculating the difference between two flows is not affected by the size of the flows, and so we would not expect any effect of flow magnitude (and hence contributing area) on the uncertainty of the increase in flow. A second possible source of error is in the determination of contributing area from topographic maps. There is a minimum resolution associated with the scale of the maps (1:50 000), so that relative errors in estimating the contributing area could be larger for the smallest areas (less than 0.05 km^2). For this reason we place less weight on those data points with very small areas; the expected decrease in variance is still apparent without them.

It is tempting at this stage to estimate the REA scale by eye, as in Wood *et al.* (1988): perhaps $1.5-2 \text{ km}^2$ for Little Akaloa Stream and $2-3 \text{ km}^2$ (or is it 5 km^2?) for Lewis River. Clearly, this a rather subjective approach: we will delay estimation of the REA until a more objective approach is developed. Certainly there is no evidence in Figure 6 to suggest any separation of scales, although from Figure 4a, Little Akaloa Stream does receive a larger proportion of its inflow in the upper 10 km^2, and much less in the last 4 km^2.

Variance of sample means for a changing sample size

An important question to ask is: what would a sceptic expect to find in these data, i.e. what is the null hypothesis? If we think of the list of subcatchments as random samples of various sizes, from a well-defined parent population, then one obvious choice is to assume that runoff generation is a random process. It is a standard result from sampling theory (Mood *et al.*, 1974) that the variance of a sample mean decreases as $(1/n)$, where n is the number of random samples, i.e.

$$\text{var}(\bar{X}_n) = \sigma^2/n \tag{1}$$

where $\bar{X}_n$ is the sample mean of a sample of size n and σ^2 is the variance of the population from which the sample is drawn. An analogous result holds for random fields (Vanmarcke, 1984): in the two-dimensional case the variance of $\bar{X}_A$, (the sample mean whose averaging area is A) decreases as $(1/A)$, i.e.

$$\text{var}(\bar{X}_A) = \sigma^2/A \tag{2}$$

The development of the REA concept given by Beven *et al.* (1988) places the REA scale as intermediate

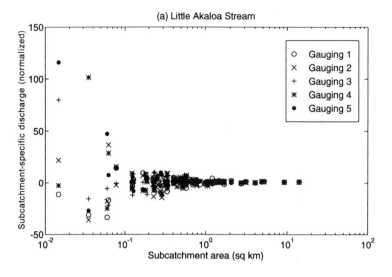

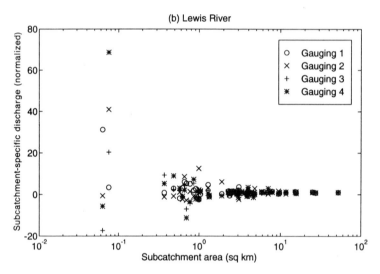

Figure 6. Specific discharge (normalized by catchment specific discharge) for subcatchments including combined subcatchments, for each gauging day, plotted against subcatchment area. (a) Little Akaloa Stream and (b) Lewis River

between the correlation scale of soil properties (typical correlation distances of 10–100 m) and the scale at which non-stationarity of soil, rainfall or morphology becomes significant. Equation (2) can be valid in the presence of correlation, provided that the averaging area A is significantly larger than the correlation scale. We would expect our measurements to be consistent with Equation (2) for some of the area range between these two correlation scales.

If there are two widely separated correlation length scales in the field from which we sample, then as the area increases (for scales well beyond the smaller correlation length), the sample variance should decrease more slowly than Equation (2), as the influence of large-scale variability starts to appear. It is possible that the variance will reach a minimum value and then rise, but once the sampling scale becomes larger than that of the large-scale variability, then the variance will fall again, ultimately to a level even lower than the local minimum. This increase in variance is theoretically detectable using the approach of Wood *et al.* (1988; 1990), but we are not aware of any documented case for hydrological variables. The use of nested subcatchments suggested by Blöschl *et al.* (1995) may be a more sensitive method for detecting a separation of scales.

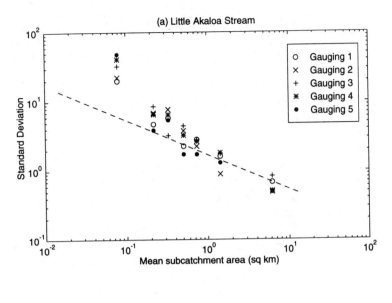

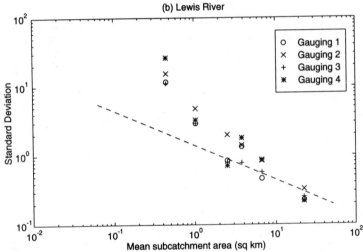

Figure 7. Standard deviation of normalized specific discharge, for each gauging day, plotted against subcatchment area with logarithmic axes, and showing the theoretical line for sampling from a stationary random field. (a) Little Akaloa Stream and (b) Lewis River

Lumley and Panofsky (1964; 41) give a mathematical example of this variance behaviour for a simplified problem, with an approximate expression for the minimum variance and the scale at which it occurs. In their case they assumed the random field was the sum of two independent stationary random processes, one of which had a much smaller correlation length than the other. The variance due to the small-scale process decreased as $(1/A)$ as we would expect from Equation (2), whereas the variance due to the larger scale process grew as A^4, because they approximated the slowly varying process by parabolic arcs. An application of their approach to this problem is not appropriate without independent information on the correlation scales of our problem.

Application

We are now in a position to see what these data say regarding the REA concept. Sort the list of all subcatchments by area, for each stream. Now take the ordered subcatchments in samples of, say, eight: they will all have roughly the same area. Each sample of eight is a sample of specific discharge at a particular scale (the average area of the eight) and we can calculate its sample variance or standard deviation. The sample standard deviation is plotted against the average area for that group in Figure 7 using logarithmic axes. Superimposed on the data is a line derived from Equation (2), with σ chosen to have the curve pass through the cluster of points with the largest area because it is at the largest area that the population appears to be stationary. (Other choices of σ will translate the curve vertically.)

Standard deviation does indeed generally decrease with increasing area, as was expected from Figure 6. Note that there is some flexibility in choosing the appropriate area to associate with each data point because of the range of catchment areas used within each window. Here we have used the arithmetic mean area from each sample.

For both streams, the variance decreases much more quickly at small scales than is predicted by equation (2). However, as we noted earlier, the high variance at the very smallest scales may be caused by errors in defining catchment area (for catchment areas less than ~ 0.05 km^2). Thus for Little Akaloa, with its small subcatchments, the observation is tentative and is re-evaluated later in this paper using topographic data. Subcatchments with larger areas have a variance which decreases at the rate predicted by Equation (2), indicating approximately stationary behaviour at scales up to ~ 10 km^2.

As the area increases, the variance does not fully stabilize or increase for either of the study sites. Without a minimum in variance, it is not possible to establish an objective value for the REA for these catchments at this stage. We might choose to detect an REA by choosing an acceptably low level of variance in relation to the mean (for these data, if the standard deviation is required to be less than the mean, we require areas greater than ~ 2 km^2). This is a significant improvement over not knowing anything about spatial variability, but it remains a subjective decision. Alternatively, we could detect an REA at the area where the variance first starts to behave consistently with Equation (2). This would give an REA of ~ 0.5 km^2 for Little Akaloa and ~ 2 km^2 for Lewis River. It has the advantage that it does not depend so strongly on defining an 'acceptable' level of variance in relation to the mean. Instead, it is the smallest scale at which changes in variability between catchments are due purely to accumulating smaller independent catchments. Of course with these different choices we must remember that we are effectively redefining the REA and opening the way for a confusion with the 'classical' definition of Wood *et al.* (1988).

ORGANIZATION

We now address the question: 'what could have caused the variance to drop so much more at small areas than is predicted by Equation (2)?' A faster than expected drop in variance suggests that larger regions are not simply a random collection of smaller regions. To put it another way, perhaps there is 'organization' (Denbigh, 1975; Blöschl *et al.*, 1993) at scales larger than $0.5-2$ km^2 which is not present at smaller scales. Without providing precise definitions, Denbigh (1975) distinguishes organization ('an assembly of parts and sub-parts which are interconnected') from order ('the extent to which any actual specimen of an entity corresponds to the ideal or pattern to which it is compared'). The concepts of organization and 'function' are closely related (a system is *organized* to carry out some *function*): we might describe catchments as organized collections of hillslopes, channels, etc. whose function is the drainage of a region. We do not exploit the concept of organization in any depth here, but simply use it as a label, a way of asking 'as convergent regions have the function of draining that region through a relatively narrow outlet, are they organized, i.e. is their spatial layout not purely random? At what scales is this organization present?'

To test whether organization could possibly be associated with the sharp fall in variance seen in Figure 7, we now turn to topography as a source of detailed spatial data which may also have this organization. Following Blöschl *et al.* (1993), if organization is unimportant, then there should be little difference between the actual topography and a correlated random field in a plot of variance against area. We take the 80th percentile (P_{80}) of the sample distribution of the wetness index, $\log(a/\tan\beta)$, as a hydrological characteristic of any subcatchment (almost identical results were obtained using the sample mean in place of P_{80}). We do not consider any sampling regions other than (non-nested) subcatchments, because it seems unlikely that arbitrary regions will have organization.

TOPOGRAPHIC DATA PROCESSING

Point elevation data along contours were obtained for the Little Akaloa and Lewis catchments and a rectangular grid of elevation values at 25 m intervals was obtained using the bicubic spline technique of Inoue (1986). Sinks in the elevation data were filled and flow direction and accumulated area values were calculated for each grid point using the methods of Jenson and Domingue (1988). Values of $\log(a/\tan\beta)$ were calculated at each grid point (Figure 8a and 8c), using a multiple direction algorithm similar to that described by Quinn *et al.* (1991).

Sample distributions of $\log(a/\tan\beta)$ were formed for subcatchments similar to those shown in Figure 2. This time, the subcatchments were generated using the topographic data by starting at the most downstream location used for flow gauging and 'climbing' up the main stream, always choosing as upstream the grid point with maximum contributing area upstream of the current point. Along this main stream, a new subcatchment was 'seeded' whenever the accumulated area had fallen by at least 32 pixels ($0.02\,\text{km}^2$) since the last subcatchment was seeded. This produced 103 subcatchments for Little Akaloa and 241 for Lewis River, considerably more than were available from the flow gauging data (28 and 24, respectively).

Correlated random fields with the same statistics (mean, variance, correlation length) as the actual $\log(a/\tan\beta)$ fields were generated (Figure 8b and 8d) using the turning bands method (Mantoglou and Wilson, 1981). Subcatchment sample distributions of the random field were then obtained using the topographically defined subcatchments from above.

ANALYSIS OF TOPOGRAPHY RESULTS

Using the same procedure as for the flow data, for each of Little Akaloa and Lewis, the subcatchments were combined to form larger subcatchments (making a total of 203 subcatchments for Little Akaloa, 479 for Lewis) and then sorted by area and grouped into windows of eight. For each subcatchment, a value of P_{80} was extracted from the sample distribution and for each window the standard deviation of P_{80} was calculated. Figure 9 shows how the sample values and standard deviations between subcatchments vary with subcatchment area for $\log(a/\tan\beta)$ and the random field, for both Little Akaloa and Lewis. The straight lines on Figure 9 are derived from Equation (2), with the underlying variance chosen so that the line interpolates the random field data at larger scales. At the very smallest scales the random field data are below the line because the field is correlated at small scales.

Here the null hypothesis is that there is no difference between the $\log(a/\tan\beta)$ and random fields. At scales larger than 0.4–$0.8\,\text{km}^2$ (Little Akaloa) and $2\,\text{km}^2$ (Lewis), the variance of the $\log(a/\tan\beta)$ and random fields are similar, and they decrease at the rate predicted by Equation (2) (with no increase in variance detected at larger scales). However, at smaller scales there appears to be a distinction between the data from the $\log(a/\tan\beta)$ and random fields. In both study areas, the variance between subcatchments is lower at small scales for $\log(a/\tan\beta)$ than it is for the random field. The difference in variance is less pronounced for areas smaller than $0.1\,\text{km}^2$.

DISCUSSION

The interpretation of the sharp fall in variance for the flow data as a consequence of some form of 'organization' is one possible explanation, although the same behaviour is not shown in the topographic data. More fine scale flow data, with greater accuracy, would be needed to confirm the flow data result. Further work is also required to find out why the variance falls so slowly for $\log(a/\tan\beta)$ in intermediate-sized areas.

If the assumptions of random samples and stationarity are valid, a decrease in variance can be expected in many situations: for instance, the same catchment mapped at two different scales may show decreasing variance in both instances. We might be tempted to say this suggests different REA values and that REA is a scale-dependent quantity. If we do select the REA as the scale at which the standard deviation falls to, say, one-tenth of its value at the smallest resolution in a given study, then it will indeed be scale-dependent and of

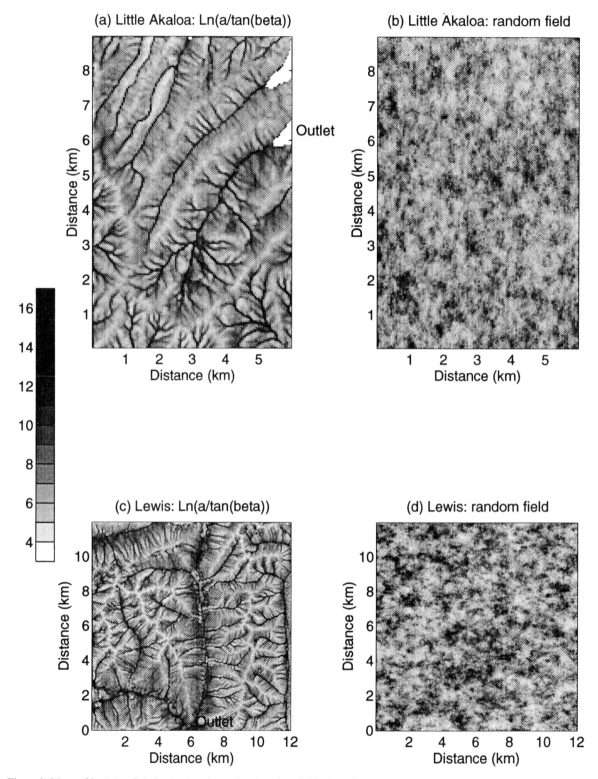

Figure 8. Maps of log(a/tanβ) index (a, c) and correlated random fields (b, d) for areas surrounding Little Akaloa Stream and Lewis River

little general use. The point to note is that we must define the REA consistently in all instances, as, for example, the scale at which standard deviation between catchments is 10% of the mean (note that the expected value of the mean is independent of scale). With such a definition, any scale dependence of the REA is unlikely, although this is not to say that only one REA is possible, as it is conceivable that there could be three well-separated scales of variability.

The use of instantaneous, inter-storm flow measurements to investigate the REA concept was a deliberate choice to maximize the possibility of detecting any REA 'effect'. Our results do not necessarily apply to the peak runoff response of the same catchments, nor to the seasonal or annual runoff response. If alternative 'responses' had been used in the present paper (e.g. peak runoff instead of inter-storm specific discharge), a different pattern of changing variability with scale could have emerged, and different inferences might then have been drawn about the REA concept. In particular, even if our measurements of inter-storm specific discharge had been made in the Coweeta or Kings Creek catchments used by Wood *et al.* (1988, 1990), that data could not necessarily be used for evaluation of the results found by their rainfall–runoff simulations for two reasons. Firstly, those rainfall–runoff simulations estimated instantaneous runoff (i.e. neglecting the effects of routing on spatial variability), rather than streamflows and, secondly, there is no guarantee that the same REA will apply for both storm and inter-storm responses.

The possible importance of time-scale to REA has been suggested by the simulation study of Blöschl *et al.* (1995). Note that the time-scale influence is in addition to the possible differences between the variability of instantaneous peak and interstorm responses. The first step in resolving the influence of time-scale is the identification of a limited number of meaningful time-scales (e.g. duration of rainfall event and inter-storm period, characteristic response times of a catchment to rainfall and evaporative forcing). Then perhaps it will be meaningful to extend the set of 'hydrological responses' to include streamflow summed over the rainfall-event time-scale or inter-storm timescale, in addition to instantaneous peak or inter-storm responses. This need not lead to an endless list of different responses for which separate REA values are needed.

For the present, we prefer to treat the REA as a quantity which may depend on the type of hydrological response in question, although it is worth seeking a unifying principle to place all such REA values into a single context. For example, in a region dominated by saturation excess runoff, it may be that the REA for peak runoff is always determined by a scale at which the channel network 'shape' changes only slowly, while the REA for storm total runoff is determined instead by spatial variability of a topography-based pre-storm wetness index and the REA for annual runoff by the spatial variability of annual rainfall.

The search for 'organization' in topography has led to an interesting result. Using two separate data sources (flow and topography data), the population becomes effectively stationary at similar scales, i.e. 0.5 km^2 (Little Akaloa) and 2 km^2 (Lewis) for the each of the study areas. For larger scales, the variance decreases according to Equation (2) in both study areas. This suggests that those scales are the smallest at which 'simple' behaviour will be found in each study area: there are complexities in the spatial patterns at smaller scales. As with all localized field studies, we must bear in mind that these results are not universal: it is not yet possible to extrapolate these scales to other regions.

We still lack any proper definition or description of this quality termed 'organization'. It is associated with the ideas of unity or a common theme among many components; Denbigh (1975) expands on this point. We might take the common theme to be adaptation to upstream conditions, or the need to provide a continuous flow path without internal sinks. In this way, 'organization' is identified with the accumulation of area into subcatchments, so that all points in the plane have a drainage path which leads to the edge of the region.

From this simplistic point of view, any catchment displays organization and we do not obtain any useful information on the effects of scale. One avenue to try is a simplified description of topographic features: is it possible to show that sufficiently large subcatchments in each study area are all made up of the same distribution of fundamental building blocks (valleys, interfluves, etc. with particular sizes, slopes, curvatures, etc.) linked by a channel network? With this approach, can some of the major differences between catchments be captured in the differences between these distributions?

For larger areas, there is a continuing shortage of evidence regarding the effects of large-scale variability on REA, due perhaps to the small catchments studied. The fact that no increase in variability has been detected using the Wood *et al.* (1988; 1990) approach serves to remind us that their REA is not associated with a

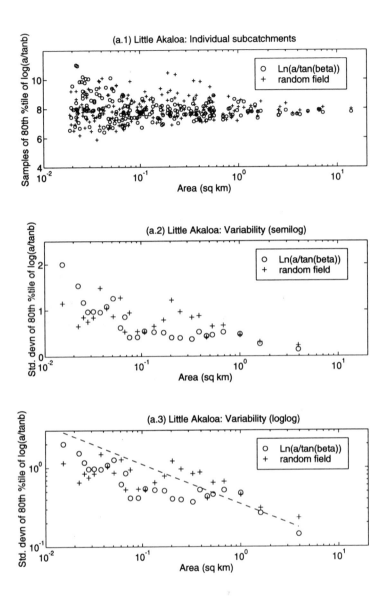

Figure 9. Mean and standard deviation (shown with both natural and logarithmic scales) of 80th centile of sample distributions from log(a/tanβ) index and correlated random fields for both study areas, plotted against subcatchment area (logarithmic scale) and showing the theoretical line for sampling from a stationary random field. (a) Little Akaloa Stream and (b) Lewis River

separation of scales, but rather with the variance between catchments (of similar size, chosen without regard for location) falling to an acceptably small value. Even if this approach had detected a separation of scales, the fact that the catchments are not adjacent to one another means that we cannot assume the associated continuum is differentiable.

A practical next step would be a large area ($\sim$1000 km^2), high detail study based on digital topographic data, although this would be at the risk of ignoring other significant variables such as rainfall, geology and vegetation. The availability of radar reflectivity data on rainfall, with very detailed temporal data, as well as useful spatial resolution (though not usually down to the finest scales used here) provides an opportunity to use realistic large-scale variability, rather than relying on random field generators. The preferred, but more

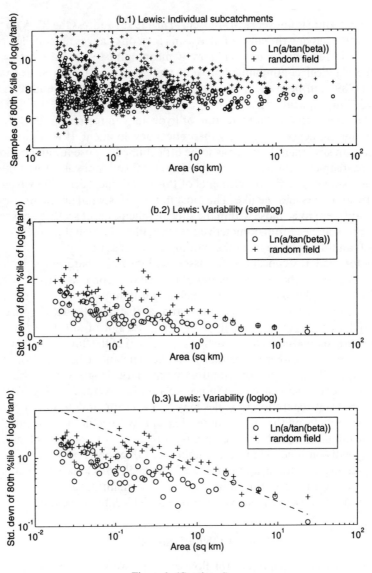

Figure 9. (Continued)

expensive, option would be to expand the present measurement programme to cover a larger area in the same detail, possibly including temporal variability, as well as expanding the range of variables to include rainfall and some indicator of moisture status at scales larger than 'point' scale.

Where does the REA lead?

By following the approach of Wood *et al.* (1988) we have shown that their concept of REA does have some meaning for field data. A significant question remains, however: what should we do once we find an REA? Although Wood *et al.* (1988) discuss the classical definition of representative elementary volume (REV) and state that the REA is exactly analogous to the REV, their work does not appear to lead to the classical path usually associated with the search for a continuum scale. Wood *et al.* (1988) did not take larger and larger scale averages about a common point, with the aim of formulating differential equations at the REA scale, relating spatial and temporal derivatives of quantities defined at that scale. Rather, they investigated the question of how to choose a scale at which spatial averages are meaningful. Although it is possible that their work could lead to a 'reformulation of the physics' (Short *et al.*, 1993), the development path to be followed from their averaging procedure is not yet clear. Blöschl *et al.* (1995) address the REA question more along the classical continuum path, studying the effect of increasing area on changes in modelled runoff rate, for sets of nested subcatchments. It remains to be seen how a meaningful set of differential equations might be formulated using the classical continuum approach, and what continuum scale state variables will be appropriate. Beven (1989) can be read as an argument that existing continuum concepts are inadequate for hydrological modelling applications.

The geometry of catchments is a defining feature of hydrology in regions drained by channels, and seems likely to have a profound influence on any new theoretical development. If we think of the REA as being a subcatchment of a certain size, then there seems to be little immediate use for any notion of spatial gradient between adjacent subcatchments for a state variable (e.g. total soil moisture stored in each subcatchment). With our present understanding, adjacent elements of this size do not 'drive' flow from one to another, via inter-element pressure gradients and the like. The most important spatial gradients for the determination of runoff generation would seem to be those within the subcatchment, i.e. at scales smaller than the REA. If this is so, then there is a pressing need to parameterize these small-scale spatial gradients (rather than trying to resolve the fine detail) and thus escape from the tyranny of small scale.

It is not necessary for the REA concept or the associated idea of separation of scales to be applicable in every hydrological setting. It may be simply a matter of using it for problems where it will apply. For example, annual rainfall may have length scales which are much larger than those of topography and soils, although hourly rainfall may not show such a clear separation. In that instance, the REA concept would be useful in developing physically based models of annual water yield, but not of hourly storm runoff. Similarly, the distributions of the length and nature (surface versus subsurface) of the flow paths in a catchment are a result of long-term climatic forcing (weathering, erosion). As the time and length scales in this forcing are fairly large, we might also expect the flow path distribution to have large length scales. This is because after a very long time all points in the catchment have been subject to a similar average forcing. There is still a small-scale variability, of course, but this separation of scales may allow a simpler description of the runoff generation process by allowing us to focus on the distributions. The ability to define specific puzzles and then separate those puzzles into meaningful, tractable pieces seems to be a key to unlocking their secrets. Without this separation, hydrology will continue to have more unknowns than observables.

Finally, there is a pressing need for meaningful quantities which reflect the essence of what is going on at the REA scale and also for constitutive relationships between these quantities at these larger scales. This need will only be satisfied if workers consciously strive to combine field data collection, modelling and theoretical development.

CONCLUSIONS

By using standard measurement techniques for flow gauging, we have collected a data set of inter-storm specific discharge which covers a useful range of spatial scales. Our analysis of the data shows that variance between catchments decreases as catchment area increases, but the rate of decrease is not constant across all scales. The qualitative behaviour of variability in relation to scale from the Wood *et al.* (1988, 1990) simulation studies is clearly supported by the data. Similarly to Wood *et al.* (1988, 1990), we did not detect any increase in variance at large areas. At small areas, the variance of specific discharge decreased faster than would be expected for a random sample, and at large areas the variance decreased at a rate consistent with sampling a stationary random field.

The fast decrease in variance is interpreted as possible evidence of 'organization', and this is investigated by an equivalent analysis of a topographic wetness index, where it is shown that a spatial structure more complex than correlation length is needed to mimic the important features of small-scale patterns shown in nature. Consistent estimates of the scale at which the study areas become stationary (0.5 km^2 for Little Akaloa, 2 km^2 for Lewis) are obtained using both flow and topographic data. Given the limited underlying theory, we caution against extrapolating these scales to other regions.

The steady rate of decrease in variance at large areas suggests that large-scale variability is not present within the catchments, most likely because the catchments are not large enough. If further studies of either field data or model output are carried out, they should use larger areas, with realistic large-scale variability.

Detecting the REA is not an end in itself: it is useful to hydrology if it leads to better ways of describing some essentials of catchment behaviour. The REA concept will only have a significant impact if workers go on to define meaningful hydrological variables and relationships at that scale. For this to succeed, a combined field, modelling and theoretical approach is essential.

ACKNOWLEDGEMENTS

We thank all the NIWA staff who were involved in the flow gauging and data processing, the New Zealand Department of Survey and Land Information for licensing the digital topographic data, and Guenter Blöschl, for several fruitful discussions, and for introducing us to 'organization'. This research was partially funded by the New Zealand Foundation for Research, Science and Technology, under contract CO1319. This document is Centre for Water Research reference ED 830 RW.

REFERENCES

Anderson, M. G., and Burt, T. P. 1978. 'The role of topography in controlling throughflow generation', *Earth Surf. Process. Landforms*, **3**, 331–344.

Beven, K. J. 1989. 'Changing ideas in hydrology—the case of physically based models', *J. Hydrol*, **105**, 157–172.

Beven, K. J. 1991. 'Scale considerations' in Bowles, D. S. and O'Connell, P. E. (Eds), *Recent Advances in the Modelling of Hydrologic Systems*. Kluwer Academic, Dordrecht, pp. 357–371.

Beven, K. J., Wood, E. F., and Sivapalan, M. 1988. 'On hydrological heterogeneity—catchment morphology and catchment response', *J. Hydrol*, **100**, 353–375.

Blöschl, G., Gutknecht, D., Grayson, R. B., Sivapalan, M., and Moore, I. D. 1993. 'Organisation and randomness in catchments and the verification of distributed hydrologic models', *EOS Suppl.*, Oct 26, 317.

Blöschl, G., Grayson, R. B., and Sivapalan, M. 1995. 'On the representative elementary area (REA) concept and its utility for distributed rainfall-runoff modelling', *Hydrol. Process.*, **9**, 313–330.

Bowen, F. E. 1964. *Sheet 15 Buller (1st edn) Geological Map of New Zealand 1:250000*, Department of Scientific and Industrial Research, Wellington.

Burt, T. P. and Arkle, B. P. 1986. 'Variable source areas of stream discharge and their relationship to point and non-point sources of nitrate pollution', *IAHS Publ.*, **157**, 155–164.

Cooper, A. B. 1990. 'Nitrate depletion in the riparian zone and stream channel of a small headwater catchment', *Hydrobiologia*, **202**, 13–26.

DeCoursey, D. G. 1991. 'Mathematical models: research tools for experimental watersheds' in Bowles, D. S., and O'Connell, P. E. (Eds), *Recent Advances in the Modelling of Hydrologic Systems*, Kluwer Academic, Dordrecht, pp. 591–612.

Denbigh, K. G. 1975. 'A non-conserved function for organized systems' in Kubat, L. and Zeman, J. (Eds), *Entropy and Information in Science and Philosophy*. Elsevier, Amsterdam.

Department of Survey and Land Information 1983. *NZMS 260 N36, AKAROA*, Edition 1, Department of Survey and Land Information, Wellington.

Department of Survey and Land Information 1989. *Infomap 260 M31, LEWIS*, and *Infomap 260 M32, BOYLE*, Edition 1, Department of Survey and Land Information, Wellington.

Dooge, J. C. I. 1986. 'Looking for hydrologic laws', *Wat. Resour. Res.*, **22**, 46S–58S.

Dooge, J. C. I. 1988. 'Hydrology past and present', *J. Hydr. Res.* **26**, 5–26.

Dunne, T. 1983. 'Relation of field studies and modelling in the prediction of storm runoff', *J. Hydrol*, **65**, 25–48.

Garbrecht, J. 1991. 'Effects of spatial accumulation of runoff on watershed response', *J. Environ. Qual.*, **20**, 31–35.

Genereux, D. P., Hemond, H. F., and Mulholland, P. J. 1993. 'Spatial and temporal variability in streamflow generation on the West Fork of Walker Branch Watershed', *J. Hydrol*, **142**, 137–166.

Huff, D. D., O'Neill, R. V., Emaneul, W. R., Elwood, J. W., and Newbold, J. D. 1982. 'Flow variability and hillslope hydrology', *Earth Surf. Process. Landforms*, **7**, 91–94.

Inoue, H. 1986. 'A least squares smooth fitting for irregularly spaced data: finite- element approach using the cubic B-spline basis', *Geophysics*, **51**, 2051–2066.

Jenson S. K., and Domingue, J. O. 1988. 'Extracting topographic structure from digital elevation data for geographic information systems', *Photogr. Engin. Remote Sensing*, **54**, 1593–1600.

Klemes, V. 1983. 'Conceptualization and scale in hydrology', *J. Hydrol*, **65**, 1–23.

Klemes, V. 1986. 'Dilettantism in hydrology: transition or destiny?', *Wat. Resour. Res.*, **22**, 177S–188S.

Lumley, J. L., Panofsky, H. A. 1964. *The Structure of Atmospheric Turbulence*. Wiley, New York.

Mantoglou, A., and Wilson, J. L. 1981. 'Simulation of random fields with the turning bands method', *Technical Report 264*, Ralph M. Parsons Laboratory, MIT, Cambridge.

Ministry of Works and Development 1979. *Our Land Resources—a Bulletin to Accompany New Zealand Land Resource Inventory Worksheets*, Government Printer, Wellington.

Mood, A. M., Graybill, F. A., and Boes, D. C. 1974. *Introduction to the Theory of Statistics*. McGraw-Hill, Tokyo.

Mosley, M. P. 1979. 'Streamflow generation in a forested watershed, New Zealand', *Wat. Resour. Res.*, **15**, 795–806.

Mulholland, P. J., Wilson, G. V., and Jardine, P. M. 1990. 'Hydrogeochemical response of a forested watershed to storms: effects of preferential flow along shallow and deep pathways', *Wat. Resour. Res.*, **26**, 3021–3026.

National Research Council (US) 1991. *Opportunities in the Hydrologic Sciences*. National Academy Press, Washington, DC.

New Zealand Meteorological Service 1986. 'Climate map series 1:2000000 part 6: annual rainfall', *New Zealand Meteorological Service Misc. Publ. 175*, Ministry of Transport, Wellington.

O'Loughlin, E. M. 1990. 'Perspectives on hillslope research' in Anderson, M. G., and Burt, T. P. (Eds), *Process Studies in Hillslope Hydrology*. Wiley, Chichester.

Quinn, P., Beven, K., Chevallier, P., and Planchon, O. 1991. 'The prediction of hillslope flow paths for distributed hydrological modelling using digital terrain models', *Hydrol. Proc.*, **5**, 59–79.

Short, D. L., Crapper, P. F., and Kalma, J. D. 1993. 'Moving between scales in surface hydrology: the need to reformulate physics', *IAHS Publ.*, **212**, 503–511.

Suggate, R. P. 1973. *Sheet 21, Christchurch (2nd edn) Geological Map of New Zealand 1:250000*, Department of Scientific and Industrial Research, Wellington.

Thauvin, V. and Lebel, T. 1991. 'EPSAT–Niger study of rainfall over the Sahel at small time steps using a dense network of raingauges', *Hydrol. Process.*, **5**, 251–260.

Vanmarcke, E. 1984. *Random Fields, Analysis and Synthesis*. MIT Press, Cambridge.

Wolock, D. M. 1993. 'Effects of topography and subbasin size on simulated flow pathways in Sleepers River watershed, Vermont', *EOS Suppl.*, Oct 26, 291.

Wood, E. F., Sivapalan, M., Beven, K. J., and Band, L. 1988. 'Effects of spatial variability and scale with implications to hydrologic modelling', *J. Hydrol.*, **102**, 29–47.

Wood, E. F., Sivapalan, M., and Beven, K. J. 1990. 'Similarity and scale in catchment storm response', *Rev. Geophys.*, **28**, 1–18.

4

ON THE REPRESENTATIVE ELEMENTARY AREA (REA) CONCEPT AND ITS UTILITY FOR DISTRIBUTED RAINFALL–RUNOFF MODELLING

G. BLÖSCHL*

Centre for Resource and Environmental Studies, The Australian National University, Canberra City, ACT 2601, Australia

R. B. GRAYSON

Centre for Environmental Applied Hydrology, University of Melbourne, Parkville, 3052, Australia

AND

M. SIVAPALAN

Centre for Water Research, University of Western Australia, Nedlands, 5009, Australia

ABSTRACT

Since the paper of Wood *et al.* (1988), the idea of a representative elementary area (REA) has captured the imagination of catchment modellers. It promises a spatial scale over which the process representations can remain simple and at which distributed catchment behaviour can be represented without the apparently undefinable complexity of local hetero-geneity. This paper further investigates the REA concept and reassesses its utility for distributed parameter rainfall–runoff modelling. The analysis follows Wood *et al.* (1988) in using the same topography and the same method of generating parameter values. However, a dynamic model of catchment response is used, allowing the effects of flow routing to be investigated. Also, a 'nested catchments approach' is adopted which better enables the detection of a minimum in variability between large- and small-scale processes. This is a prerequisite of the existence of an REA.

Results indicate that, for an impervious catchment and spatially invariant precipitation, the size of the REA depends on storm duration. A 'characteristic velocity' is defined as the ratio of a characteristic length scale (the size of the REA) to a characteristic time-scale (storm duration). This 'characteristic velocity' appears to remain relatively constant for different storm durations. Spatially variable precipitation is shown to dominate when compared with the effects of infiltration and flow routing. In this instance, the size of the REA is strongly controlled by the correlation length of precipitation. For large correlation lengths of precipitation, a separation of scales in runoff is evident due to small-scale soil and topographic variability and large-scale precipitation patterns. In general, both the existence and the size of an REA will be specific to a particular catchment and a particular application. However, it is suggested that a separation of scales (and therefore the existence of an REA), while being an advantage, is not a prerequisite for obtaining simple representations of local heterogeneity.

INTRODUCTION

Distributed parameter hydrological models are being increasingly used in investigations of spatial scale and catchment heterogeneity as well as general rainfall–runoff applications (e.g. Goodrich, 1990; Grayson *et al.*, 1993, in press; Rosso *et al.*, 1994; Sivapalan and Viney, 1994). A critical problem in the application of these models is the choice of element size, which must be able to represent the heterogeneity of the catchment

---

*Now at: Institut für Hydraulik, Gewässerkunde und Wasserwirtschaft, Technische Universität Wien, Vienna, Austria.

response, be compatible with the algorithms used in the model, and must enable definition of the model parameters. The choice of element size relates to the representation of spatial heterogeneity of catchment response in general, and of the model parameters in particular. The representative elementary area (REA) work of Wood *et al.* (1988) was an initial attempt to determine at what scale, if any, the description of catchment properties becomes simple. This relates directly to an 'ideal' or 'preferred' element size for distributed parameter catchment modelling and will depend on the effect on catchment response of the heterogeneity of soils, topography and precipitation at various spatial scales.

Since the initial work of Wood *et al.* in 1988, the idea of an REA has captured the imagination of catchment modellers. It promises a spatial scale over which the process representations can remain simple and at which distributed catchment behaviour can be represented without the apparently undefinable complexity of local heterogeneity.

Wood *et al.*'s original work was based on a modified version of TOPMODEL (Beven and Kirkby, 1979) and the topography of the Coweeta catchment in North Carolina. Wood *et al.* (1988) used TOPMODEL to compute the total runoff volume for a one hour period of rainfall, but did not include dynamic effects such as the routing of 'runoff waves'. Spatially correlated stationary random fields of soil hydraulic conductivity and precipitation were used to investigate the spatial characteristics of runoff. For each realization of soil and rainfall characteristics, Wood *et al.* (1988) determined the runoff volume from 148 subcatchments. These runoff volumes were ranked on the basis of subcatchment size, irrespective of their relative position in the catchment. The average of a 15 element filter, moving in steps of five, was plotted versus area (expressed as number of pixels). These plots were then used to determine the REA, defined as the area where the curve joining points of increasing subcatchment area flattened out. This was shown by Wood *et al.* (1988) to be approximately 1 km^2. The choice of different rainfall correlation lengths varying from 125 to 1250 m or spatially invariant precipitation did not significantly change this result. They concluded that the REA was strongly influenced by topography and that the length scales of soils and rainfall characteristics played only a secondary part.

In a more recent discussion of the REA concept, Beven (1991) highlighted a number of aspects of the original work that require further investigation. These included the influence of larger correlation lengths for rainfall, the effects of flow routing on the volume and timing of runoff, and the influence of geology and large-scale synoptic effects.

The objective of this paper is to investigate further the utility of the REA concept. Specifically, it goes beyond the original work of Wood *et al.* (1988) in the following ways:

1. Sets of nested subcatchments are used here, whereas Wood *et al.* (1988) used averaging over non-nested subcatchments.
2. A dynamic model of catchment response is used, allowing the effects of flow routing to be investigated.
3. Rainfall correlation length is explored over a wider range and rainfall duration is varied.
4. The utility of the REA concept for distributed parameter rainfall–runoff modelling is reassessed.

The reasons for using nested catchments are discussed in detail later but, briefly, it is because it enables the detection of a separation of scales between large- and small-scale variability.

The paper is separated into nine sections. The first is a discussion of the theoretical considerations that underlie the concept of the REA. The second and third present the modelling assumptions used in the subsequent simulations and the methods used to test for the existence of an REA. The following sections present the results of the simulations starting from the simplest case of spatially uniform parameters, then introducing spatially variable parameters in steps. The last two sections are a discussion and conclusion, respectively, including a review of the utility of the REA concept.

CONTINUUM ASSUMPTIONS, REV AND REA

Heterogeneity in natural catchments is of stunning complexity and so far has virtually defied detailed description and/or measurement. Many branches of science which deal with similar heterogeneity adopt a 'continuum approach' and ignore the local (microscale) heterogeneity for modelling purposes, replacing the real system by a fictitious continuous medium at the macroscale. For example, in groundwater hydrology, the detailed patterns of pore structure at the microscopic level are replaced by a continuous field of porosity at the macroscopic level. Porosity is a coefficient which describes the lumped effect of the microscopic pattern (Hubbert, 1956; Wheatcraft and Cushman, 1991). Bear and Bachmat (1990) suggest the following advantages of the continuum approach: (a) no knowledge of the microscale patterns is required; (b) the continuous medium is differentiable; and (c) the continuum (potentially) represents measurable quantities.

The continuum can be arrived at either by ensemble averaging or by space averaging. For the ensemble average, the medium is assumed to be a realization of a random process. A property such as porosity can then be defined, at a given point in space, as the average over all possible realizations of the same process (deMarsily, 1986; Cushman, 1987). The space average can be a volume or area average and defines a property, at a given point in space, as the average property of a certain volume (or area) of material surrounding it. Two area averaging methods have been suggested in the context of continuum assumptions in catchment hydrology: Gupta *et al.* (1986) proposed elevation bands in high relief terrain, whereas Wood *et al.* (1988; 1990) suggested subcatchments. Here we will follow Wood *et al.* (1988).

In principle, an arbitrary elementary area (AEA) of *any* size can be selected as an averaging area for passing from the microscopic to the macroscopic level (Bear and Bachmat, 1990: 16; Stull, 1988: 420). It is likely, however, that different AEAs will yield different averaged values for each quantity of interest, which makes model parameterization difficult. The question then is whether we can determine an AEA where the averaged variables become insensitive to variations in the size of the AEA. Hubbert (1956; 227) addressed this very problem in the context of defining macroscopic porosity at a particular point in a porous medium:

> About this point we take a finite volume element ΔV, which is large as compared with the grain or pore size of the rock. Within this volume element the average porosity is defined to be $\bar{f} = \Delta V_f / \Delta V$ where ΔV_f is the pore volume within ΔV. We then allow ΔV to contract about the point P and note the value of $\bar{f}$ as ΔV diminishes. If we plot $\bar{f}$ as a function of ΔV (Hubbert's Fig. 5), it will approach smoothly a limiting value as ΔV diminishes until ΔV approaches the grain or pore size of the solid. At this stage $\bar{f}$ will begin to vary erratically and will ultimately attain the value of either 1 or 0, depending upon whether P falls within the void or the solid space.

It is important to note that Hubbert (and also Bear, 1972: 16, who based his ideas on those of Hubbert), considered one fixed reference point with a volume of changing size around it. Bear (1972) denoted, as the representative elementary volume (REV), the order of magnitude where 'f (*porosity*) approaches smoothly a limiting value' (i.e. varies only smoothly with changing volume). In direct analogy, Wood *et al.* (1988), coined the term Representative Elementary Area (REA) in catchment hydrology.

The existence of a scale where macroscopic properties vary smoothly is tantamount to the existence of a 'separation of scales' (Gelhar, 1986). 'Separation of scales' refers to a process consisting of a small-scale (or fast) component superimposed on a much larger (or slow) component. The scales are called 'separated' if there is a minimum in the power spectrum. In meteorology, for example, microscale turbulence and synoptic scale processes are often separated by a 'spectral gap' which shows up as a minimum in the power spectrum of wind speed. Typically, this spectral gap is at a frequency of 1/hour; both higher and lower frequencies are much more common (Stull, 1988). Although Wood *et al.* (1988) did not use this terminology, an REA would exist if there was a 'spectral gap' (or equivalently a separation of scales) in catchment variability.

The application of the continuum approach to distributed hydrological modelling requires the deterministic representation of large-scale variability (as different values in different model elements) and the parameterization of the small-scale processes within an element (see, e.g. Kirnbauer *et al.*, 1994). Variability within an element is often referred to as 'sub-grid' variability. Wood *et al.* (1988) suggested the use of distribution functions for parameterizing the sub-grid variability of infiltration. Provided a separation of scales exists, the same parameterization of sub-grid processes can be used for a range of element sizes. This becomes possible when the element size falls into the spectral gap or, in other words, when the element size is of the order of magnitude of the size of the REA. A small change in element size does not then significantly change the relative contributions of sub-grid and large-scale variability.

It should be emphasized that the existence of a spectral gap (i.e. the existence of an REA) is not a prerequisite for the parameterization of processes within an element. Small-scale processes can be parameterized for any element size. The advantage of the existence of a spectral gap is that the same parameterization can be used for a range of element sizes and so would be more generally applicable. If such a gap existed and was 'universal', then there would be a basis for developing generalized macroscale modelling approaches.

METHODS

The simulation exercise in this paper uses the distributed parameter model known as THALES (Grayson *et al.*, 1992), which is based on the terrain analysis software TAPES-C (Moore and Grayson, 1991). The general framework of the model is described in the following, as are differences from the model as presented in Grayson *et al.* (1992; in press).

Following the work of Wood *et al.* (1988), a 30 m Digital Elevation Model (DEM) of the 17 km² Coweeta catchment in North Carolina is used. Elevations range from 680 m to 1600 m. From the DEM, contours with variable spacing from 5 to 40 m are derived and used to determine the flow paths. Adjacent flow paths and contours define elements. The resulting element network contains 6206 elements with an average element size of 50 by 50 m (Figure 1). Model parameters are assumed to be constant within each element. Runoff is produced via infiltration excess or saturation excess mechanisms and flows downslope. The infiltration equation of Philip (1957) is used, based on the parameter description of Wood *et al.* (1988). Surface runoff is described as overland sheet flow and is routed using the kinematic wave assumptions with a Manning's *n* of 0·4. The infiltration of surface runoff in downslope elements (runon) is allowed in these simulations and the channel network is defined by a threshold upslope contributing area of 10 ha. Elements with areas greater than this threshold are assumed to contain channels with a width of 5 m and a Manning's *n* of 0·1. Although in the original version of THALES, a soil depth above an impermeable boundary is defined, for the simulations

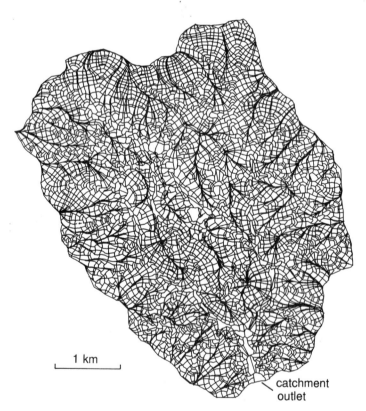

1 km

catchment
outlet

Figure 1. Element network for the Coweeta topography, North Carolina, USA. The elements are defined by adjacent flow paths and contours

Table I. Parameter sets used in the simulations

	Manning's n overland flow	Manning's n channel flow	(Average) hydraulic conductivity (mm/h)	Correlation length of conductivity (m)	Distribution of initial sat. deficit	Initial saturated area (%)	Storm duration	(Average) intensity of precip. (mm/h)	Correlation length of precipitation (m)
Figure 2abc	0·4	0·1	5	100	Topogr.	30	1 h	5	Invar.
Figure 4	0·4	0·1	5	100	Topogr.	30	1 h	5	Invar.
Figure 5	0·4	no chann.	Impervious	Impervious	Impervious		1 h	5	Invar.
Figure 6a	0·4	no chann.	Impervious	Impervious	Impervious		12 min	5	Invar.
Figure 6b	0·4	no chann.	Impervious	Impervious	Impervious		2 h	5	Invar.
Figure 7	0·04	no chann.	Impervious	Impervious	Impervious		12 min	5	Invar.
Figure 8a	0·4	no chann.	5	100	Topogr.	30	12 min	5	Invar.
Figure 8b	0·4	no chann.	5	100	Topogr.	70	12 min	5	Invar.
Figure 9a	0·4	0·1	5	100	Topogr.	30	12 min	5	Invar.
Figure 9b	0·4	0·1	5	100	Topogr.	30	1 h	5	Invar.
Figure 10a	0·4	0·1	5	100	Topogr.	30	1 h	5	125
Figure 10b	0·4	0·1	5	100	Topogr.	30	1 h	5	2500
Figure 10c	0·4	0·1	5	100	Topogr.	30	1 h	5	625
Figure 11a	0·4	0·1	5	100	Topogr.	30	1 h	5	125
Figure 11b	0·4	0·1	5	100	Topogr.	30	1 h	5	2500
Figure 12 I	0·4	0·1	Impervious	Impervious	Impervious		1 h	5	2500
Figure 12 V	0·4	0·1	5	100	Topogr.	30	1 h	5	2500
Figure 12 T	0·4	0·1	5	Invar.	Topogr.	30	1 h	5	2500
Figure 12 R	0·4	0·1	5	Invar.	Rand.field	30	1 h	5	2500

Abbreviations: sat. deficit = saturation deficit; precip. = precipitation; no chann. = no channels; invar. = spatially invariant; topogr. = based on topographic wetness index; invar. = spatially invariant; and rand. field = random field.

in this paper, the 'saturation deficit' concept of Beven and Kirkby (1979) is used to ensure consistency with the work of Wood *et al.* (1988). A simulation time step of two minutes is used.

The use of THALES as a basis for the simulations in this paper introduces some differences from the work presented by Wood *et al.* (1988). The model is dynamic and both channel and overland flow are actually routed though the catchment so runoff peaks and storm volumes can be defined. This introduces the ability to represent the influence of channelized versus overland flow.

Spatially, model and input parameters are assumed to be either constant, or stationary correlated random fields. Specifically, in the case of precipitation, the correlation length refers to a double exponential correlation function while in the case of the hydraulic conductivity it refers to a Bessel type correlation function. The exception is the initial saturation deficit which is either represented by a random field of exponential correlation structure or by the topographic wetness index of Beven and Kirkby (1979). The methods and parameters used strictly follow the work of Wood *et al.* (1988, Appendix A and B) unless where otherwise indicated. Precipitation is assumed to be a rain block of 12 min, 1 hour or 2 hour duration. Table 1 gives the parameter sets used in the simulations.

EFFECT OF AVERAGING

The analysis of Wood *et al.* (1988) involved the ranking of runoff volumes according to subcatchment size and subsequent filtering. Our concern is that this method of averaging might influence the definition and size of the REA. This section presents an analysis similar to that of Wood *et al.* (1988) and explores the effect of different averaging methods.

Simulations were performed for uniform precipitation, spatially variable hydraulic conductivity and topographically defined initial saturation deficit based on the $\ln(a/\tan\beta)$ index of Beven and Kirkby (1979) where a is the upslope contributing area per unit contour length and β is the local slope. Subcatchments were chosen using the following procedure: ten ridge points were chosen at random from the element network and the flow paths from each of these points to the catchment outlet were determined. The upslope contributing area for each element on the flow path defined a subcatchment. Figure 3 shows some of the subcatchments defined by this procedure for one flow path. This resulted in 161 subcatchments, which is close to the 148 used by Wood *et al.* (1988).

The flow volumes for each subcatchment were ranked according to subcatchment size in an identical manner to that of Wood *et al.* (1988). The same filtering procedure based on a window of 15 subcatchments, moving in steps of five, was used resulting in 30 classes for the 161 subcatchments. The resulting average flow volumes are plotted versus subcatchment area (Figure 2a). Figure 2a is very similar to Figure 4a in Wood *et al.* (1988), with the response flattening out above approximately 1 km^2.

Figure 2b shows the same simulation results, but with 1394 subcatchments selected for averaging using a larger number of flow paths. The same number of filtering classes as in Figure 2a were used (30 in each instance). For the 1394 subcatchments, this corresponds to a window of 135 subcatchments moving in steps of 45. Clearly, the window to window variations are smoothed out when a sufficient number of subcatchments per window are selected. The shape of the curve, with lower volumes in small catchments, reflects the influence of the topographic index on saturation deficit with large deficits for the small subcatchments near the ridges (i.e. lower $\ln(a \tan \beta)$).

Given the smoothing effect of the number of subcatchments in a window on average volumes, it may be more consistent with the requirement for a 'minimum of variance', to examine the variability within each window. Figure 2c shows the variance in the windows used to construct Figure 2a, which indeed is much lower above a threshold area of 1 km^2 than for smaller subcatchments. However, the meaning of such a quantity in the context of distributed rainfall–runoff modelling is unclear, as it compares catchments of similar size, irrespective of their relative position or distance apart. In fact, any approach that ignores the relative position of subcatchments will be unable to detect an increase in variance at large scales. This is illustrated by the following example. Two small adjacent catchments differ only due to small-scale variability. Two small catchments, separated by a large distance, differ due to both the small-scale variability as well as any large-scale variability present in the landscape. If, on the other hand, we compare two large catchments, small-scale

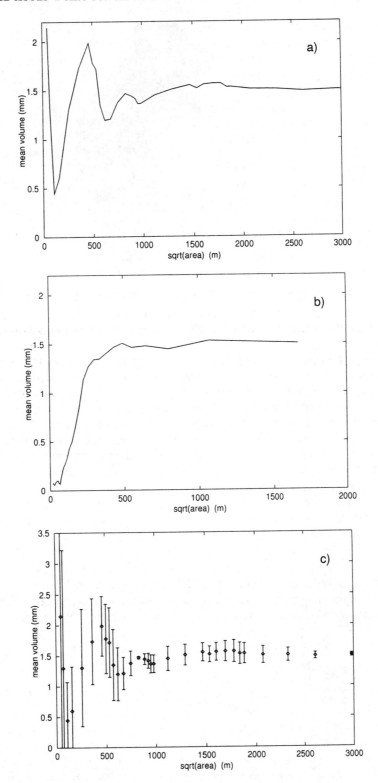

Figure 2. Mean runoff volume versus catchment size for spatially variable soils and constant precipitation. (a) Filtering based on a window of 15 subcatchments moving in steps of five, resulting in 30 classes as in Figure 4a of Wood *et al.* (1988). (b) Filtering based on a window of 135 subcatchments moving in steps of 45, resulting in 30 classes. (c) Standard deviations within windows for (a)

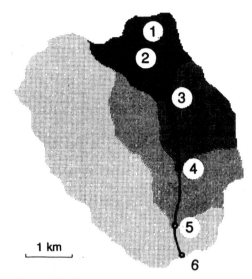

Figure 3. Example of a set of nested catchments. Numbers indicate some of the subcatchment outlets and refer to those in Figure 4.

variability is averaged out within each catchment and we have variability due only to large-scale effects. It is therefore apparent that large catchments will *never* appear more variable than widely spaced small catchments. To show true changes in variability and detect a spectral gap, adjacent areas must be compared.

The averaging method should therefore consider one fixed reference point and a domain of variable size around it. In catchment hydrology, this is a set of nested catchments, as illustrated in Figure 3. In this example, the fixed reference point is a ridge point at the top of Figure 3 and the 'domain of variable size' is defined by subcatchments of increasing size. This approach is analogous to that of Hubbert (1956) and Bear (1972), used in their definition of the REV. The difference from their analysis is that the nested catchments are grouped asymmetrically around the reference point, rather than symmetrically as in the work of Hubbert (1956) and Bear (1972). This is because water flows downhill. The outlets of the subcatchments lie on a flow path from the ridge point to the catchment outlet.

The 'averaged properties' are peaks and volumes of the hydrographs at the respective subcatchment outlets. These are plotted versus the square root of area for the same conditions as Figure 2a in Figure 4. The

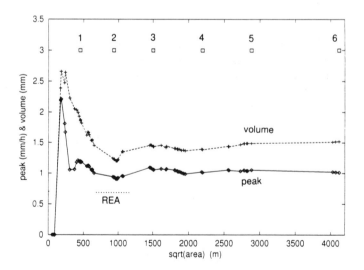

Figure 4. Runoff volume and peak versus subcatchment size for one set of nested catchments as shown in Figure 3. Spatially variable soils and constant precipitation

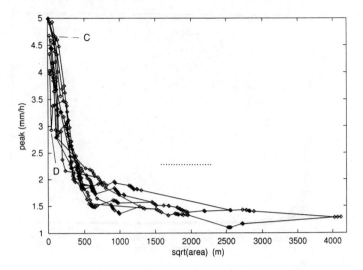

Figure 5. Peak flow versus subcatchment area for 10 sets of nested catchments. Impervious conditions, no channels, Manning's $n = 0.4$, one hour rainblock. Precipitation is spatially invariant. The dotted line suggests the approximate size of the REA. C refers to a convergent element and D refers to a divergent element

use of length units rather than area for the x-axis allows closer examination of small scales and is more in keeping with the notion of 'length scales'. Following the approach of Hubbert (1956) and Bear (1972), the REA is described as the scale where $|dq/da|$ is small, with q being the hydrograph peak or volume and a the subcatchment area. For Figure 4 this occurs around 600 to 1200 m as indicated by the dotted line.

It is important to note that Figure 4 shows the results of one set of nested catchments without any filtering. This set of nested catchments can be interpreted as a realization of topography. In the following sections, 10 such sets (realizations) are presented in each graph. Runoff peaks will be used rather than runoff volume because they better capture the dynamics of runoff processes and may be more relevant to practical applications.

EFFECT OF TOPOGRAPHY ON ROUTING AND THE 'CHARACTERISTIC VELOCITY'

This section examines the effect of topography for impervious conditions and spatially invariant precipitation of variable duration.

Figure 5 shows the results for a one hour rainblock of 5 mm/h intensity. As in Figure 4, peak flow at one subcatchment outlet plots as a marker in Figure 5 and nested subcatchments are joined by lines. Peaks are dampened significantly with increasing flow length (subcatchment area). There is substantial scatter between the individual sets of nested catchments, which is largely a consequence of convergent and divergent flow across the hillslopes. Examples of convergent and divergent elements (i.e. subcatchment outlets) are marked in Figure 5 by the letters C and D, respectively. The convergent element is 50 m long and its width decreases from 119 to 56 m downhill. Consequently, the peak flow is significantly higher (4·7 mm/h) than the average of other elements with the same upslope contributing area (3·6 mm/h). The width of the divergent element increases from 7 to 28 m downhill and gives a peak of 2·9 mm/h as opposed to the average of 4·4 mm/h. These peak flows are also affected by the convergence/divergence of the upslope neighbouring elements, which have similar shapes to the elements examined. The curves flatten out at around 2000 m, indicating the length scale of the REA.

Figure 6 illustrates the effect of storm duration using the same rainfall intensity of 5 mm/h. A shorter storm duration of 12 min decreases the length scale of the REA significantly (to 300 m; Figure 6a), whereas an increase in storm duration increases the length scale of the REA to 4000 m or larger (Figure 6b). This indicates some relationship between length scale and storm duration. If we choose the square root of the REA as a characteristic length scale and the storm duration as a characteristic time-scale, a 'characteristic velocity' can

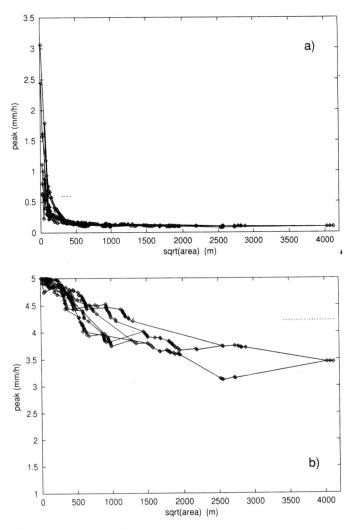

Figure 6. Peak flow versus subcatchment area for 10 sets of nested catchments. Impervious conditions, no channels, Manning's $n = 0·4$. (a) 12 minute rainblock; and (b) two hour rainblock. Precipitation is spatially invariant

be defined. In the case of the simulation in Figure 6, this is about 2 km/h This characteristic velocity is expected to be either related to wave celerity or particle speed, so a decreased roughness should increase the characteristic velocity V and consequently the characteristic length scale L (i.e. $\sqrt{\text{REA}}$) for a given characteristic time-scale T according to:

$$L = V * T$$

Figure 7 shows this effect. Using a 12 minute storm (as in Figure 6a), but smoother surface conditions (Manning's $n = 0·04$ rather than $n = 0·4$) gives a characteristic length scale of about 700 to 1000 m, which corresponds to a characteristic velocity of 3·5 to 5 km/h. This is more than double the value of the rougher case in Figure 6a ($n = 0·4$).

EFFECT OF INFILTRATION

In this section, runoff can be produced by saturation excess and infiltration excess mechanisms. Hydraulic conductivity is spatially variable (with an average of 5 mm/h and a correlation length of 100 m as in Wood

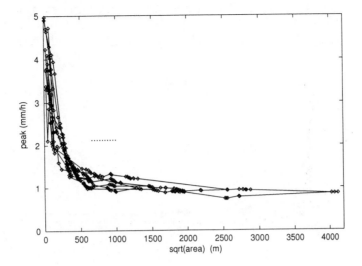

Figure 7. Peak flow versus subcatchment area for 10 sets of nested catchments. Impervious conditions, no channels, 12 minutes rain-block, Manning's $n = 0.04$. Precipitation is spatially invariant (same as Figure 6a, but lower roughness)

et al., 1988), but precipitation is spatially invariant. A 12 minute rainblock of 5 mm/h is used and the initial saturation deficit is based on the topographic wetness index.

Figure 8a presents the results for initial conditions as used in Wood *et al.* (1988), where 30% of the area is saturated at the beginning of the storm ($m = 30$ mm and $\overline{S} = 15.3$ mm in Wood *et al.*, Appendix A1). Allowing spatially variable infiltration introduces significant scatter between the sets of nested catchments (as compared with Figure 6a) and increases the length scale of the REA to about 800–1200 m. The saturated part of the catchment behaves as an 'impervious surface' for runoff generation, so a larger saturated area is expected to reduce the length scale of the REA, closer to that of the impervious case in Figure 6a. This is borne out by the simulations shown in Figure 8b, which have a 70% saturated area at the beginning of the storm ($m = 30$ mm and $\overline{S} = -34.7$ in Wood *et al.*, Appendix A1). Figure 8b indicates a length scale of the REA of about 500–600 m.

EFFECT OF CHANNELS

This section examines the effect of assuming channelized flow in elements which have an upslope contributing area larger than 10 ha, superimposed on the effects of topography and infiltration.

Figure 9a presents the results for a 12 minute rainblock of 5 mm/h intensity. As compared with Figure 8a (sheet flow only), the length scale of the REA is increased. This might be related to the higher conveyance of the channels (due to the geometry and the smoother surface condition), which should increase the characteristic velocity. However, using a one hour rainblock does not increase the length scale of the REA to the same extent (Figure 9b).

EFFECT OF SPATIALLY VARIABLE PRECIPITATION

For ease of comparison with the original work of Wood *et al.* (1988), a one hour rainblock is used in the following analyses. Spatially variable precipitation fields with correlation lengths of 125, 625 m and 2500 m are used. Spatially variable conductivity, infiltration and channel flow are represented as in Figure 9b.

Figure 10a shows results for 10 realizations of precipitation with correlation lengths of 125 m for one set of nested catchments. As compared with Figure 9b (spatially invariant precipitation), peaks are much higher for small catchments. The curves flatten out at about 1000 m. Using a correlation length of 2500 m changes the results dramatically (Figure 10b). There is no single scale where the curves flatten out. Indeed, Figure 10b

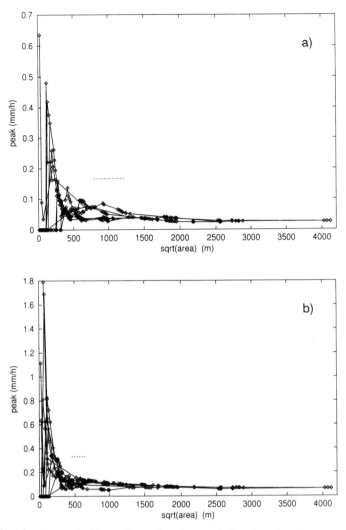

Figure 8. Peak flow versus subcatchment area for 10 sets of nested catchments. Infiltration allowed, no channels, 12 minute rainblock, Manning's $n = 0.4$. (a) Initial saturated area is 30%. (b) Initial saturated area is 70%. Precipitation is spatially invariant and soils are spatially variable

suggests that there is both variability at the small scale (< 500 m) and at the large scale (> 2500 m) with little variability in between. It is important to note that each of the curves in Figure 10b should be inspected individually and the minimum in variability appears where $|dq/da|$ is small (dq/da is the change in peak flow with change in subcatchment area for a set of nested subcatchments). Using a correlation length which is between the two extremes (Figure 10c) gives length scales between those of Figure 10a and 10b as would be expected. Similar results are found for other sets of nested catchments (Figure 11).

To illustrate the combined effect on the REA of different assumptions of infiltration and precipitation, two realizations of precipitation are investigated in more detail for one set of nested catchments. Figure 12 shows results of using a correlation length of 2500 m for precipitation and either:

- impervious surface conditions (marked I in Figure 12),
- spatially variable conductivity and topographically defined initial saturation deficit (marked V in Figure 12),
- spatially invariant conductivity and topographically defined initial saturation deficit (marked T in Figure 12),
- spatially invariant conductivity and random initial saturation deficit (marked R in Figure 12).

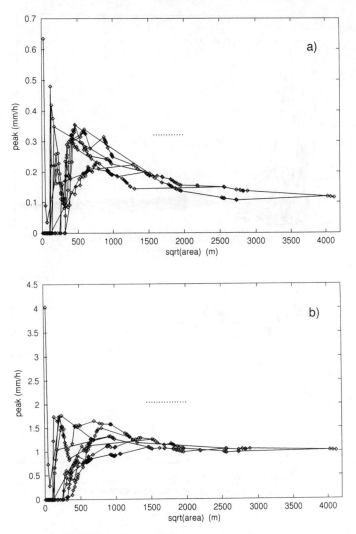

Figure 9. Peak flow versus subcatchment area for 10 sets of nested catchments. Infiltration allowed, 30% initial saturated area, Manning's $n = 0.4$, channelized flow for upslope contributing areas > 10 ha. (a) 12 minute rainblock; and (b) one hour rainblock. Precipitation is spatially invariant and soils are spatially variable

The 'random initial saturation deficit' was generated as a stationary random field with the same distribution function and covariance function as the topographically defined initial saturation deficit. Figure 12 indicates that the different assumptions on infiltration markedly affect the magnitude of the runoff peaks. However, the change of flow peaks with scale (and consequently the size of the REA) is dominated by the precipitation pattern. Again, Figure 12 shows small-scale variability below 500 m and large-scale variability above 2000 m.

DISCUSSION

Peak flows in nested catchments are used to examine the effects of various assumptions on the REA.

For the impervious case, storm duration is shown to be dominant. A characteristic velocity is defined which relates the length scale of the REA to storm duration. It is about 2 km/h and suggests a larger REA for longer storm durations. In the context of distributed parameter rainfall–runoff modelling, this finding is not surprising. Essentially, it means that coarser temporal resolutions in precipitation are consistent with larger element sizes. In fact, many rainfall–runoff modelling applications use, say, time-scales of minutes for

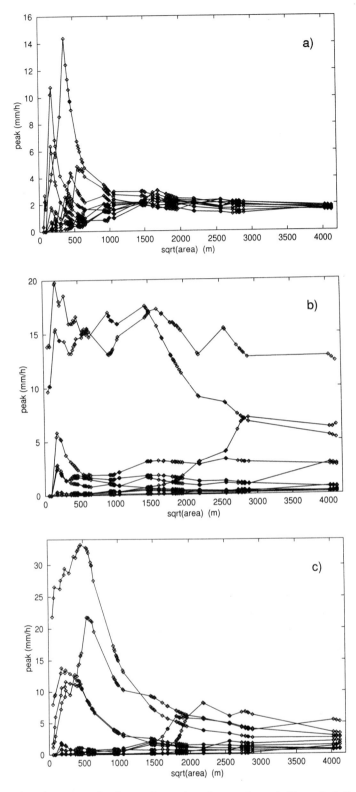

Figure 10. Peak flow versus subcatchment area for the set of nested catchments shown in Figure 3. Infiltration characteristics and channels as in Figure 9. Ten realizations of spatially variable precipitation of one hour duration with correlation lengths of (a) 125 m; (b) 2500 m; and (c) 625 m

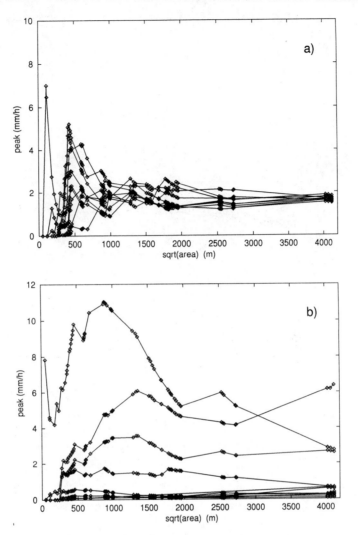

Figure 11. As Figure 10 for another set of nested catchments. Correlation length of precipitation is (a) 125 m; and (b) 2500 m. The highest intensity realization is not shown in (b)

catchments of several hectares, whereas time steps of hours or days are typically used for catchments of tens or hundreds of square kilometres.

Also, the notion of a characteristic velocity is known from meteorology. For many atmospheric processes, the characteristic velocity, relating time and length scales of the processes, is about 10 m/s. Physically, this may correspond to particle velocities (such as the typical vertical velocities in convective cells) or wave celerities (such as the celerities of large scale frontal systems) (Haltiner and Williams, 1980; Fortak, 1982; Stull, 1985). The much smaller characteristic velocities found here (2 km/h = 0·6 m/s) are clearly consistent with the perception of overland flow in catchments.

It is interesting to note that the effect of the characteristic velocity seems not to be related to one particular catchment size. Simulation runs (not shown here) using the Coweeta topography scaled up by a factor of 10 (i.e. area = 1700 km^2), without changing the slopes, gave very similar characteristic velocities to the real topography for the same value of surface roughness.

Another way of thinking about this is to consider the 'area of influence' upstream of any point in the catchment, i.e. the actual contributing area of any point at peak discharge. The results in Figure 6 indicate that, for a given roughness (or wave speed) this area depends on the rainfall duration. That is to be expected as a longer storm will increase the contributing area at peak discharge. If the roughness is reduced, a similar

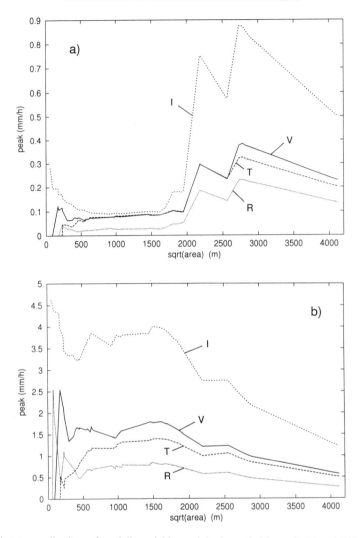

Figure 12. As Figure 10, but two realizations of spatially variable precipitation only [shown in (a) and (b)]; Correlation length of precipitation is 2500 m in both instances. The different graphs relate to different assumptions on infiltration (see text)

effect to that of increasing storm duration is expected as the wave speed increases and therefore the contributing area at peak discharge increases. Figure 7 illustrates this point.

When allowing for infiltration and channel flow, the effect of storm duration shows up less clearly and seems to be largely dominated by the former. Specifically, for the parameter sets used here, the results suggest that the individual effects of infiltration and channel flow are of the same order of magnitude as those of routing a one hour rainblock on an impervious catchment. Therefore, superposition of these processes gives a combined length scale of the REA, which is similar to that of the individual processes. This is about 1000–2000 m.

Introducing spatially variable precipitation clearly dominates over routing and infiltration. It also dominates over storm duration, as results not shown here indicate. The size of the REA seems to be directly related to the correlation length of precipitation. For very large correlation lengths, however, the results indicate both large-scale and small-scale variability, with little variability in between. Figure 12 shows that different assumptions about infiltration do not affect the large-scale variation, but they significantly affect the small-scale variability. Clearly, the small-scale variability is related to the effect of variable soil parameters and topography, whereas the large-scale variability is due to precipitation patterns. A minimum in variance in between indicates that a 'separation of scales' in runoff is present.

The following example is used to show the implications of a separation of scales for distributed catchment modelling. Assume that a distributed rainfall–runoff model for the catchment and the conditions used here is to be set up. It is then prudent to choose the element size of the model somewhere between 500 and 2000 m. The large-scale variability due to precipitation would be represented explicitly as element to element variations. The small-scale variability due to soils and topography would be parameterized as subgrid variability. For example, the parameterization may be performed by a distribution function approach (e.g. Beven, 1991). The advantage of the existence of a separation of scales now is that the same distribution function for parameterizing the small-scale variability can be used for any element size between 500 and 2000 m. If no separation of scales existed in the above example, the catchment might still be modelled with an element size of, say, 2000 m and the small-scale variability could still be parameterized by a distribution function. However, when the element size was then changed to, say, 500 m, the distribution function representing small-scale variability would also have to change.

Following Wood *et al.* (1988), stationary random fields have been used in this study for simulating the spatial variation of rainfall and soil properties. The stationarity assumption (i.e. no large-scale variability) is the reason for most graphs shown in this paper levelling out at a certain scale. The same applies to the case of spatially constant soil properties and precipitation as these also assume no large-scale variability. Most of this paper was therefore concerned with the *size* of the REA and its controls rather than with the *existence* of an REA. However, hydrological processes are rarely stationary. It is now being argued that precipitation may be a multiscale phenomenon (or be fractal) and therefore have no preferred scale (e.g. Lovejoy and Schertzer, 1985; Kumar and Foufoula-Georgiou, 1993). There is also an increasing awareness that geological controls are operative at a multitude of scales (Gelhar, 1986; Wheatcraft and Cushman, 1991). Given the dominant controls of the correlation lengths of soils and precipitation on both the existence and the size of the REA shown in this paper, it may be that, generally, an REA will not exist. Although a spectral gap in runoff (and hence an REA) may exist for a particular catchment it seems that the size of the REA will be highly catchment-specific. Conceptually then, one universal size of an REA or one universal 'optimum element size' seems unlikely, although research will continue to determine if the REA concept is valuable at a regional level.

The existence of a spectral gap in runoff (and hence of an REA) is *not* a requirement for parameterizing sub-grid processes. For distributed rainfall–runoff modelling in the future, the question of how to parameterize may become more important than the question of an optimum scale at which to parameterize.

CONCLUSIONS

1. It is suggested that nested catchments be used for testing the existence and size of an REA. The length scale of the REA is defined as the scale where, with increasing area, changes of peak flows (or flow volumes) become small for one set of nested catchments. The existence of an REA is equivalent to the presence of separation of scales in runoff. The approach is consistent with the original definition of the REV of Hubbert (1956) and Bear (1972). It is shown to be capable of detecting a minimum in variability between large- and small-scale processes.

2. For impervious catchments of a given roughness, and spatially invariant precipitation, the size of the REA is shown to be controlled by a characteristic velocity. The characteristic velocity is defined as the ratio of a characteristic length scale (the size of the REA) and a characteristic time-scale (storm duration). The characteristic velocity is about 2 km/h for the conditions examined here.

3. Infiltration and channel flow exhibit less obvious controls on the REA. When allowing for these processes, length scales of the REA are of the order of 1000 m for the Coweeta catchment topography and conditions simulated in this study.

4. Spatially variable precipitation is shown to dominate over other processes considered. The size of the REA is strongly controlled by the correlation length of precipitation. The results indicate a separation of scales in runoff for large correlation lengths of precipitation and the conditions considered here. In this instance, large-scale variability is due to precipitation, whereas small-scale variability is related to soil characteristics and topography.

5. The simulations show that the size of the REA depends on many factors, including storm duration and variability, flow routing and infiltration characteristics. It is therefore apparent that the size of the REA

will be specific to a particular catchment and a particular application. There is no evidence for one universal size of an REA or one universal 'optimum element size' in the context of distributed rainfall–runoff modelling.

6. Following Wood *et al.* (1988), this study has used stationary random fields for simulating the spatial variation of rainfall and soil properties. Multiscale input fields might be more realistic. Given the dominant effect of the correlation lengths on the REA, it seems unlikely that for multiscale controls, a spectral gap (and hence an REA) should exist. This requires further study.

7. The existence of a spectral gap in runoff (and hence of an REA) is not a requirement for parameterizing sub-grid processes. It simply makes any parameterization more generally applicable. There remains a major challenge in deriving parameterizations for sub-grid variability and for distributed rainfall–runoff modelling in the future, the question of how to parameterize may be more important than the question of an optimum scale at which to parameterize.

ACKNOWLEDGEMENTS

The authors thank the Fonds zur Förderung der Wissenschaftlichen Forschung, Vienna, project no. J0699-PHY for financial support. The authors also thank Keith Beven and Ross Woods for their useful comments on the manuscript.

REFERENCES

Bear, J. 1972. *Dynamics of Fluids in Porous Media*. American Elsevier, New York. 764pp.
Bear, J., and Bachmat, Y. 1990. *Introduction to Modeling of Transport Phenomena in Porous Media*. Kluwer, Dordrecht. 533pp.
Beven, K. 1991. 'Scale considerations' in Bowles, D.S. and O'Connell, P.E. (Eds), *Recent Advances in the Modeling of Hydrologic Systems*. Kluwer, Dordrecht. pp.357–371.
Beven, K., and Kirkby, M. 1979. 'A physically based variable contributing area model of basin hydrology', *Hydrol. Sci. Bull.*, **24**, 43–69.
Cushman, J. H. 1987. 'More on stochastic models', *Wat. Resour. Res.*, **23**, 750–752.
deMarsily, G. 1986. *Quantitative Hydrogeology*. Academic Press, London. 440pp.
Fortak, H. 1982. *Meteorologie*. Dietrich Reimer, Berlin. 293pp.
Gelhar, L. W. 1986. 'Stochastic subsurface hydrology from theory to applications', *Wat. Resour. Res.*, **22**, 135S–145S.
Goodrich, D. C. 1990. 'Geometric simplification of a distributed rainfall-runoff model over a range of basin scales', *PhD Thesis*, The University of Arizona. 361pp.
Grayson, R. B., Moore, I. D., and McMahon, T. A. 1992. 'Physically based hydrologic modeling, 1. A terrain-based model for investigative purposes', *Wat. Resour. Res.*, **28**, 2639–2658.
Grayson, R. B., Blöschl, G., Barling, R. D., and Moore, I. D. 1993. 'Process, scale and constraints to hydrologic modelling in GIS' in *HydroGIS93, Proc. Vienna Symp., April 1993*. Kovar, K. and Nachtnebel, H. P. (Eds) *IAHS Publ.*, **211**, 83–92.
Grayson, R. B., Blöschl, G., and Moore, I. D. Distributed parameter hydrologic modelling using vector elevation data: THALES and TAPES-C. in Singh, V. P. (Ed), *Computer Models of Watershed Hydrology*. CRC Press, Boca Raton, in press.
Gupta, V. K., Waymire, E., and Rodríguez-Iturbe, I. 1986. 'On scales, gravity and network structure in basin runoff' in Gupta, V.K., Rodríguez-Iturbe, I., and Wood, E.F. (Eds), *Scale Problems in Hydrology*, Reidel, Dordrecht. pp.159–184.
Haltiner, G. J., and Williams, R. T. 1980. *Numerical Prediction and Dynamic Meteorology*. Wiley, Chichester 477pp.
Hubbert, M. K. 1956. 'Darcy's law and the field equations of the flow of underground fluids', *Trans. Am. Inst. Min. Met. Engln.*, **207**, 222–239.
Kirnbauer, R., Blöschl, G., and Gutknecht, D. 1994. 'Entering the era of distributed snow models', *Nordic Hydrol.*, **25**, 1–24.
Kumar, P. and Foufoula-Georgiou, E. 1993. 'A multicomponent decomposition of spatial rainfall fields, 2. Self-similarity in fluctuations', *Wat. Resour. Res.*, **29**, 2533–2544.
Lovejoy, S., and Schertzer, D. 1985. Generalized scale invariance and fractal models of rain. *Wat. Resour. Res.*, **21**, 1233–1250.
Moore, I. D. and Grayson, R. B. 1991. 'Terrain-based catchment partitioning and runoff prediction using vector elevation data', *Wat. Resour. Res.*, **27**, 1177–1191.
Philip, J. R. 1957. 'The theory of infiltration, 1–7', *Soil Sci.*, **83, 84**, and **85**.
Rosso, R., Peano, A., Becchi, I., and Bemporad, G. A. 1994. 'Advances in distributed hydrology' in *Proceedings of a Workshop held in Bergamo, Italy, 25–26 June 1992*. Water Resources Publications, Littleton, CO, 416 pp.
Sivapalan, M. and Viney, N. R. 1994. 'Large scale catchment modelling to predict the effects of land use and climate', *Water, J. Austr. Water and Wastewater Assoc.*, **21**, 33–37.
Stull, R. B. 1985. 'Predictability and scales of motion', *Bull. Am. Meteorol. Soc.*, **66**, 432–436.
Stull, R. B. 1988. *An Introduction to Boundary Layer Meteorology*. Kluwer, Dordrecht. 666pp.
Wheatcraft, S. W. and Cushman, J. H. 1991. 'Hierarchical approaches to transport in heterogeneous porous media', *Rev. Geophys., Suppl.*, 263–269.
Wood, E. F., Sivapalan, M., Beven, K., and Band, L. 1988. 'Effects of spatial variability and scale with implications to hydrologic modeling', *J. Hydrol.*, **102**, 29–47.
Wood, E. F., Sivapalan, M., and Beven, K. 1990. 'Similarity and scale in catchment storm response', *Rev. Geophys.*, **28**, 1–18.

5

SCALING BEHAVIOUR OF HYDROLOGICAL FLUXES AND VARIABLES: EMPIRICAL STUDIES USING A HYDROLOGICAL MODEL AND REMOTE SENSING DATA

ERIC F. WOOD

Water Resources Program, Princeton University, Princeton, NJ 08544, USA

ABSTRACT

The effects of small-scale heterogeneity in land surface characteristics on the large-scale fluxes of water and energy in the land-atmosphere system has become a central focus of many of the climatology research experiments. The acquisition of high resolution land surface data through remote sensing and intensive land–climatology field experiments (such as HAPEX and FIFE) has provided data to investigate the interactions between microscale land–atmosphere interactions and macroscale models. One essential research question is how to account for the small-scale heterogeneities and whether 'effective' parameters can be used in the macroscale models. To address this question of scaling, it is important to carry out modelling studies by analysing the spatial behaviour of process-based, distributed land–atmospheric models and available data from land surface climate experiments such as those designed under ISLSCP (e.g. FIFE and BOREAS) and HAPEX (e.g. HAPEX-MOBILY, HAPEX-SAHEL) and GEWEX (e.g. GCIP) and from smaller scale remote sensing experiments. Using data from FIFE'87 and WASHITA'92, a soil moisture remote sensing experiment, analyses are presented on how the land surface hydrology during rain events and between rain event varies; specifically, runoff during rain events, evaporation between rain events and soil moisture. The analysis with FIFE'87 data suggests that the scale at which a macroscale model becomes valid, the representative elementary scale (REA), is of the order of 1·5–3 correlation lengths, which for the land processes investigated appear to be about 750–1250 m. For the Washita catchment data, analysis of soil-based infiltration data supports an REA of this spatial scale, but model derived and remotely sensed soil moisture data appear to suggest a larger scale. Statistical self-similarity is investigated to further understand how soil moisture scales over the Washita catchment and to provide a basis for macroscale models.

INTRODUCTION

The complex heterogeneity of the land surface through soils, vegetation and topography, all of which have different length scales, and their interaction with meteorological inputs that vary with space and time, results in fluxes of energy and water whose scaling properties are unknown. Research into land–atmospheric interactions suggests a strong coupling between land surface hydrological processes and climate (Charney *et al.*, 1977; Walker and Rowntree, 1977; Shukla and Mintz, 1982; Sud *et al.*, 1990). The issue of 'scale interaction' for land surface–atmospheric processes has emerged as one of the critical unresolved problems for the parameterization of climate models (WCRP, 1991).

Understanding the interaction between scales increases in importance when the apparent effects of surface heterogeneities on the transfer and water and energy fluxes are observed through remote sensing and intensive field campaigns such as HAPEX and FIFE (Sellers *et al.*, 1988). The ability to parameterize macroscale models based on field experiments or remotely sensed data has emerged as an important research question for programmes such as the Global Energy and Water Experiment (GEWEX) or the Earth Observing System (EOS). It is also important for the parameterization of the macroscale land surface hydrology necessary in climate models, which have grid scale lengths ranging from 40 to over

250 km, and it is crucial to our understanding of the representation of sub-grid variability in these macroscale models.

In fact, the acquisition of high resolution land surface data through remote sensing and intensive land–climatology field experiments (such as HAPEX and FIFE) has provided data to investigate the interactions between microscale land–atmospheric interactions and macroscale models. One essential research question is how to account for the small-scale heterogeneities and whether 'effective' parameters can be used in the macroscale models. The current scientific thinking on this issue is mixed. For example, Sellers *et al.* (1992) claim that analysis of the FIFE data supports the view that land–atmospheric models are almost scale-invariant, a conclusion also reached by Noilhan (pers. comm.) using HAPEX–MOBILHY data. Counter arguments have been made by Avissar and Pielke (1989), who found that heterogeneity in land characteristics resulted in sea breeze-like circulations, and significant differences in surface temperatures and energy fluxes across the patches. The results of analyses presented in Wood (1994) suggest that soil moisture is the critical variable that controls the non-linear behaviour of land–atmospheric interactions, and the effect is most pronounced when soil moisture heterogeneity is such that part of the domain is under soil–vegetation control and part is under atmospheric control.

From a modelling perspective, it is important to establish the relationship between spatial variability in the inputs and model parameters, the scale being modelled and the proper representation of the hydrological processes at that scale. Figure 1 shows a schematic model for modelling over a range of scales. Let us consider this figure in the light of the terrestrial water balance, which for a control volume may be written as

$$\left\langle \frac{\partial S}{\partial t} \right\rangle = \langle P \rangle - \langle E \rangle - \langle Q \rangle \tag{1}$$

where S represents the moisture in the soil column, E evaporation from the land surface into the atmosphere, P the precipitation from the atmosphere to the land surface and Q the net runoff from the control volume. The spatial average for the control volume is noted by $\langle \quad \rangle$.

Equation (1) is valid over all scales and only through the parameterization of individual terms does the water balance equation become a 'distributed' or 'lumped' model. By 'distributed' model, we mean a model which accounts for spatial variability in inputs, processes or parameters; examples ranging from

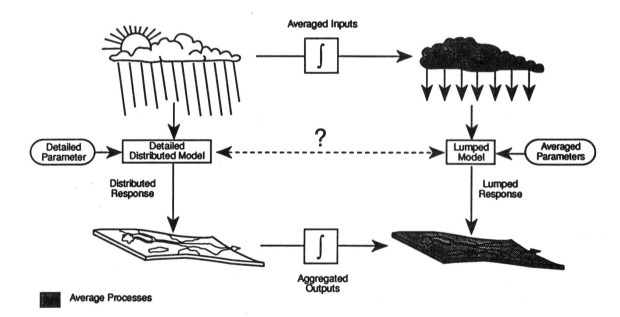

Figure 1. Schematic of aggregation and scaling in hydrological modelling

deterministic numerical models (Binley *et al.*, 1989; Paniconi and Wood, 1992) through models in which the patterns of variability are represented statistically — examples being models such as TOPMODEL (Beven and Kirkby, 1979) and its variants (see Wood *et al.*, 1990) in which topography and soil play an important part in the distribution of water within the catchment.

By a 'lumped' model we mean a model that represents the catchment (or control volume) as being spatially homogeneous with regard to inputs and parameters. There is a large number of hydrological water balance models of varying complexity that do not consider spatial variability. These range from the well known unit hydrograph and its variants, the water balance models of Eagleson (1978) to complex atmospheric–biospheric models being proposed for general circulation models [examples being the biosphere atmosphere transfer scheme (BATS) of Dickinson *et al.* (1984) and the simple biosphere model (SiB) of Sellers *et al.* (1986)].

The terrestrial water balance, including infiltration, evaporation and runoff, has been revealed to be a highly non-linear and spatially variable process. However, little progress has been made in relating the observed small-scale complexity that is apparent from field and remote sensing experiments to models and parameterizations at large scales. Why is this so? Basically, it appears that current research suggests two competing concepts: (i) the idea that sub-grid processes have a significant effect on processes at large scales and that the non-linearity in small-scale processes prevents simple scaling (Avissar and Pielke, 1989; Famiglietti and Wood, 1995b); and (ii) the competing claim suggesting that analysis of field data from FIFE (Sellers *et al.*, 1992) or HAPEX-MOBILY (Noilhan, pers. comm.) imply that land–atmospheric models are almost scale-invariant, and that 'effective' parameter approaches can be used in macroscale models.

Little progress has been made in relating the observed small-scale complexity in soil moisture and evaporation to models and parameterizations at large scales. It is this relationship that is the subject of this paper, which includes an initial attempt to clarify the concepts of aggregation, scaling and effective parameters as used in land surface models. We start by reviewing results for the aggregation of hydrological responses, first investigated through modelling experiments by Wood *et al.* (1988). The hydrological modelling results will be compared with results derived from field and remote sensing data. We will then consider the issue of 'effective' parameters and finally present some new results on the scaling behaviour of soil moisture and the implications for relating water balance terms (such as soil moisture) across spatial scales.

AGGREGATION OF HYDROLOGICAL RESPONSES

Surface hydrological fluxes are now known to result from a complexity of mechanisms. For example, during a particular storm different mechanisms may generate runoff from different parts of a catchment and during interstorm periods spatial differences in soil moisture may result in differences in evaporation and transpiration rates. As reviewed in Wood *et al.* (1990), the runoff generation mechanisms include runoff due to rainfall on areas of low permeability soils (referred to as the infiltration excess mechanism), or from rainfall on areas of soil saturated by a rising water-table even in high permeability soil (referred to as saturation excess runoff generation). These saturated contributing areas expand and contract during and between storm events.

As first shown by Beven and Kirkby (1979), variations in topography play a significant part in the spatial variation of soil moisture within a catchment, setting up spatially variable initial conditions for both runoff from rainstorms and evaporation during interstorm periods. Beven and Kirkby (1979) were one of the first to develop a saturated storm response model (TOPMODEL). This model has been further expanded to include the above mechanisms (see Beven, 1986a, 1986b; Sivapalan *et al.*, 1987). A complete description of the models, incorporating spatial variability in topography and soils, is provided in Wood *et al.* (1990) and will not be repeated here.

During interstorm periods, topography plays an important part in the downslope redistribution of soil moisture and, with soil properties, sets up the initial conditions for evaporation. The maximum evaporation rate is that rate demanded by atmospheric conditions, referred to as the potential rate, and

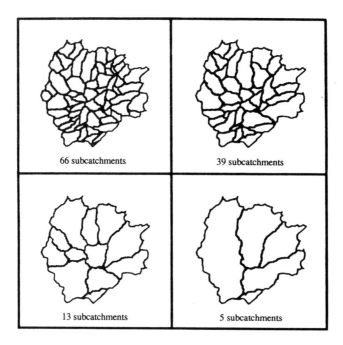

Figure 2. Natural subcatchment divisions for Kings Creek, Kansas

this rate is met if the soil column can deliver the moisture to the surface. Rates lower than the potential rate will be at a 'soil controlled' rate to be determined by soil properties and soil moisture levels. The model with both storm and interstorm processes is fully described in Famiglietti (1992) and Famiglietti and Wood (1994).

The water balance model described in Famiglietti *et al.* (1992) was applied to the Kings Creek catchment of the FIFE area in Kansas. Figure 2 shows the division of the $11 \cdot 7 \, km^2$ catchment into subcatchments — the number ranging from five to 66 depending on the scale. All subcatchments represent hydrologically consistent units in that runoff flows out of the subcatchment through one flow point, and that the surface runoff flux across the other boundaries is zero.

For a rainfall storm on 4 August 1987, the average runoff for the subcatchments was calculated for two times and plotted in Figure 3 against subcatchment area measured in pixels. Each pixel is $900 \, m^2$. Notice that the runoff, Q, is rescaled by the average precipitation, P. The same type of plot was constructed for selected times during an interstorm period that extended from 18 July to 31 July 1987 and is presented as Figure 4. The behaviour of the catchment shows that at small scales there is extensive variability in both storm response and evaporation. This variability appears to be controlled by variability in soils and topography whose length scales are of the order of $10^2-10^3 \, m$ — the typical scale of hillslope. With increased scale, the increased sampling of hillslopes leads to a decrease in the difference between subcatchment responses. At some scale, the variance between hydrological responses for catchments of the same scale should reach a minimum (Wood *et al.*, 1990). Wood *et al.* (1988) suggest that this threshold scale represents a representative elementary area (REA), which is proposed to be a fundamental building block for hydrological modelling. As defined in Wood *et al.* (1988; 1990): 'The REA is the critical scale at which implicit continuum assumptions can be used without explicit knowledge of the actual patterns of topographic, soil, or rainfall fields. It is sufficient to represent these fields by their statistical characterization.'

The implication of the REA concept is that patterns representing inputs and parameters can be replaced by their distributional (statistical) representation. It must also be recognized that this transition in scale, where pattern no longer becomes important but only its statistical representation, may occur at different spatial scales for different processes, may vary temporally for the same parameter (except during wet or

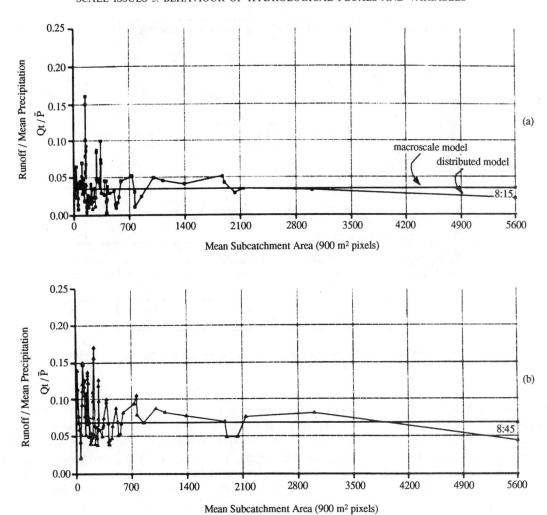

Figure 3. Comparison of storm runoff generated from the distributed model and from the macroscale water balance model for two time intervals on 4 August 1987. (a) 08.15 am, and (b) 08.45 am

dry periods) and may vary with the scale of the 'macroscale' model. Much of the research to date may indicate what the REA scale for catchment modelling is, with scales ranging from $0 \cdot 1$ to $25 \, km^2$.

By inspecting Figures 3 and 4, based on modelling at $11 \cdot 7 \, km^2$ catchment, it appears that the size of the REA is of the order of $1 \, km^2$ (about 1000–1200 pixels, each of which are $900 \, m^2$). Thus the results suggest that at scales larger than this REA scale it would be possible to model the responses using a simplified macroscale model based on the statistical representation of the heterogeneities in topography, soils and hydrological forcings (rainfall and potential evaporation).

Using the statistical distribution of the topographic–soil index, we can determined the fraction of the catchment simulated to be saturated due to the local soil storage being full. These areas will generate saturation excess runoff at the rate $\bar{p}$, the mean rainfall rate. For that portion of the catchment where infiltration occurs, the local expected runoff rate at time t, m_q, can be calculated as the difference between the mean rainfall rate, $\bar{p}$, and the local expected infiltration rate, m_g. This implies that m_q and m_g are *conditioned* on a topographic–soil index whose *statistical distribution* is central to the REA macroscale model. The difference between averaged rainfall and infiltration can be expressed as

$$m_q[t \mid \ln aT_e/T_i \tan \beta] = \bar{p} - m_g[t \mid \ln aT_e/T_i \tan \beta] \qquad (2)$$

Modeled Interstorm Evaporation Following Rain Ending 01:30 July 18 1987

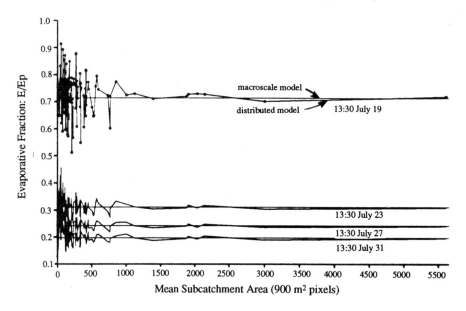

Figure 4. Comparison of inter-storm evapotranspiration from the distributed model and from the macroscale water balance model for four times during the 18–31 July 1987 inter-storm period

where $\ln(aT_e/T_i \tan \beta)$ is the topographic–soil index for a location i in the catchment and is a function of a, the contributing area upslope to i; $\tan \beta$, the local slope angle; T_i, the soil transmissivity at i; and $\ln(T_e)$, the catchment average of $\ln(T_i)$. The full development of the topographic–soil index is provided in Beven and Kirkby (1979), Beven (1986a; 1986b), Sivapalan *et al.* (1987) and Wood *et al.* (1990). Both the local expected runoff rate and the local expected infiltration rate are (probabilistically) conditioned on the topographic–soil index, $\ln(aT_e/T_i \tan \beta)$. The runoff production from the catchment is found by integrating, usually numerically, the conditional rate over the statistical distribution of topographic–soil index. Figure 3 also gives results for the macroscale model along with the distributed model. As the macroscale model is scale-invariant, it appears as a straight line in Figure 3.

In a similar way, a macroscale evaporation model is developed for inter-storm periods. As stated earlier, topography plays an important part in the interstorm redistribution of soil moisture. Variations in soil properties and topography leads to variations in soil moisture and the initial conditions for the evaporation calculations. For those portions of the catchment for which the soil column can deliver water at a rate sufficient to meet the potential evapotransipiration or atmospheric demand rate, E_p, the actual rate E equals E_p; otherwise, the rate will be at a lower soil controlled rate E_s. Within the TOPMODEL framework, locations with the same value of the topographic–soil index will respond similarly, implying a macroscale model conditioned on that index. The macroscale model can be written as

$$m_E[t \,|\, \ln(aT_e/T_i \tan \beta)] = \min\{m_{E_s}[t \,|\, \ln(aT_e/T_i \tan \beta)], \bar{E}_p(t)\} \qquad (3)$$

where m_E refers to the mean evaporation rate at locations in the catchment with the same index, m_{E_s} refers to the mean soil controlled rate and $\bar{E}_p$ to the spatially average potential or atmospheric demand rate. Figure 4, which compares the evaporation rates from the distributed model across the range of scales for Kings Creek, also includes the derived rates from the macroscale evaporation model. As in Figure 3, the macroscale model is scale invariant and appears as a straight line.

VALIDATION OF THE REPRESENTATIVE ELEMENTARY AREA CONCEPT

The results from the aggregation simulations suggest that the concept of a threshold scale, which has been referred to as the REA, may be valid. The reported work to date (Wood *et al.*, 1988; 1990; Famiglietti and Wood, 1995) suggest approximately 1–2 km^2 as being the critical spatial scale where the REA concept holds. This spatial scale was arrived at empirically, through simulation using models based on the soil–topographic index, and it is important to consider other independent measurements and analyses that may support this spatial scale.

What data could be used for validation? As evaporation and runoff are closely linked to soil properties, an analysis of soil infiltration variability should provide further insight into the REA scale. Farajalla and Vieux (1995) analysed the relationship between the variability in Green–Ampt infiltration parameters and the scale of the underlying soil properties for the Little Washita basin, a 525 km^2 experimental USDA catchment in the southern great plains region of the USA. The analysis used soil texture data from the Soil Conservation Service (SCS) (Bogard *et al.*, 1978; Moffatt, 1973; Mobeley and Brinley, 1967) which estimated 64 different soil series with 162 soil phases (Farajalla and Vieux, 1995). The analysis applied the techniques of Ahuja *et al.* (1989) and Rawls *et al.* (1983) to estimate the Green–Ampt parameters from the soil classification data. Farajalla and Vieux (1995) then estimated the information content (entropy) to identify the spatial scale (resolution) which captures the essential variability of the infiltration parameters. To do this, they determined the change in entropy, defined as

$$I = -\sum_{i=1}^{N} P_i \log(P_i) \tag{4}$$

where N is the number of intervals in the histogram for the random variable and P_i is the probability of the variable falling into interval i. Farajalla and Vieux (1995) estimated the entropy for the infiltration parameters based on the SCS soil data for the Little Washita. Figure 5, from Farajalla and Vieux (1995), gives the entropy results for derived soil parameters (hydraulic conductivity, suction head, porosity and infiltration depth). These data suggest a spatial scale about 1 km^2 for the REA.

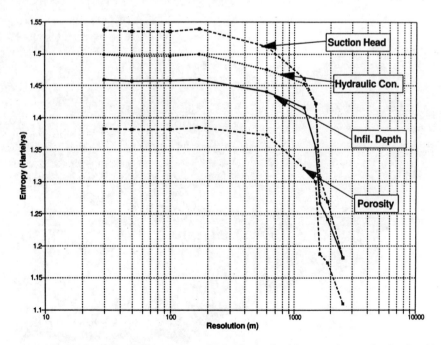

Figure 5. Entropy versus resolution for derived soil parameters: hydraulic conductivity, porosity, suction head and infiltration depth (from Farajalla and Vieux, 1995)

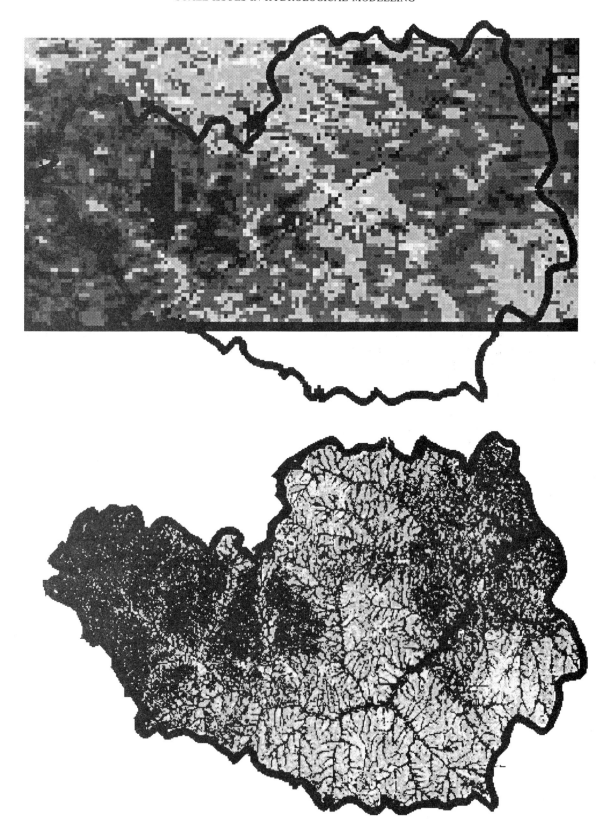

Another hydrological variable of interest is soil moisture. Determining the scale at which patterns of soil moisture can be replaced by its statistical representation is important for macroscale land surface parameterizations used in climate models. Such models need to represent subgrid heterogeneity in soil moisture to improve grid-average evapotranspiration, which is a function of soil moisture. Inclusion of sub-grid soil moisture variability is most important when there is a mix of soil and atmospheric controls on evapotranspiration within the grid (as further described in Wood, 1994).

A source of data for investigating the soil moisture REA scale is remotely sensed soil moisture data. Patterns of soil moisture are more complex as soil moisture integrates variability in rainfall and vegetation (as it transpires and depletes soil moisture) with the hydrological response of the catchment — in particular, variability in soils and topography. Thus we would expect that the REA for soil moisture would be larger than for soil properties.

During June 1992, there was a multi-sensor airborne remote sensing campaign in the Little Washita catchment (WASHITA'92) organized by the Hydrological Sciences Branch of NASA and USDA. As part of that campaign, passive microwave (L-band) data were collected from the electronically scanned thin array radiometer (ESTAR) instrument flown on NASA's C-130 aircraft. The resolution of the ESTAR data is 200 m. Brightness temperatures based on L-band have been used successfully to estimate surface (0–5 cm) soil moisture (Schmugge, 1983; Engman, 1990; Jackson, 1993). The remove sensing field campaign was carried out from 10–18 June 1992, with the catchment conditions being very wet at the beginning followed by an eight day dry down with no rain. The experimental period was also simulated using the water and energy model (TOPLATS) of Famiglietti and Wood (1994). The soil moisture fields for 10 June 1992 (Figure 6) and for 14 June 1992 (Figure 7) are shown with the top portion of these figures corresponding to the ESTAR estimated fields and the bottom portion corresponding to the model-derived fields. The deep sandy soils in the central portion of the basin are clearly evident in both the ESTAR and model-derived fields, reflecting in lower soil moisture values.

Using both the ESTAR-based soil moisture estimated and the TOPLATS model estimates a similar analysis was carried out to that shown in Figures 3 and 4: the catchment was divided into subcatchments and the average subcatchment surface soil moisture estimated from the ESTAR data or simulated using the hydrological model. Figure 8 demonstrates the average surface soil moisture with subcatchment scale for the ESTAR-based estimates. For Figure 8, the averaging was performed for the catchment area shown in Figures 6 and 7. Similar results, not shown, were found for the hydrological model-derived soil moisture fields. Looking at Figure 8, it appears that variance in catchment-average surface soil moisture falls with an increasing spatial scale, with a significant reduction at spatial scales of the order of 5 to $10 \, km^2$. This suggests that the REA scale for soil moisture may be larger than for soil properties. It also may be dependent on the scale of the domain being investigated, or that at scales larger than those investigated previously, there exists 'residual' variability that reflects larger scale variability in soils, topography and vegetation. To investigate this further, the model-derived soil moisture fields, shown in the bottom portion of Figures 6 and 7, were averaged over two large subcatchment areas, one corresponding to sandy soils and lower soil moistures and one corresponding to loam soils and wetter soil moistures. The results, presented in Figure 9, suggest an REA of about $5 \, km^2$, but, more interestingly, the drier large subcatchment (top portion of the figure) had a much lower residual variability at larger scales than the wetter subcatchment area (lower portion of the figure). This spatial and temporal variability in the implied REA scale is not inconsistent with our REA concept, which only states that patterns can be replaced by a distributional representation at the scale of macroscale models.

In addition, the entropy for these soil moisture fields was estimated using Equation (4). The soil moisture fields were the ESTAR-derived soil moisture (at a 200 m resolution) and the hydrological model-derived soil moisture, based on the TOPLATS model (Famiglietti and Wood, 1994) with 30 m topography. Figure 10 illustrates the results for 10 June 1992 (a day with high soil moisture) and 18 June 1992 (a drier day after an

Figure 6. Estimated soil moisture fields for the Little Washita for 10 June 1992: top figure based on ESTAR (Jackson, pers. comm.); bottom figure based on hydrological modelling. Dark areas represent higher volumetric soil moisture values. The range of soil moisture can be estimated from Figures 8 and 9

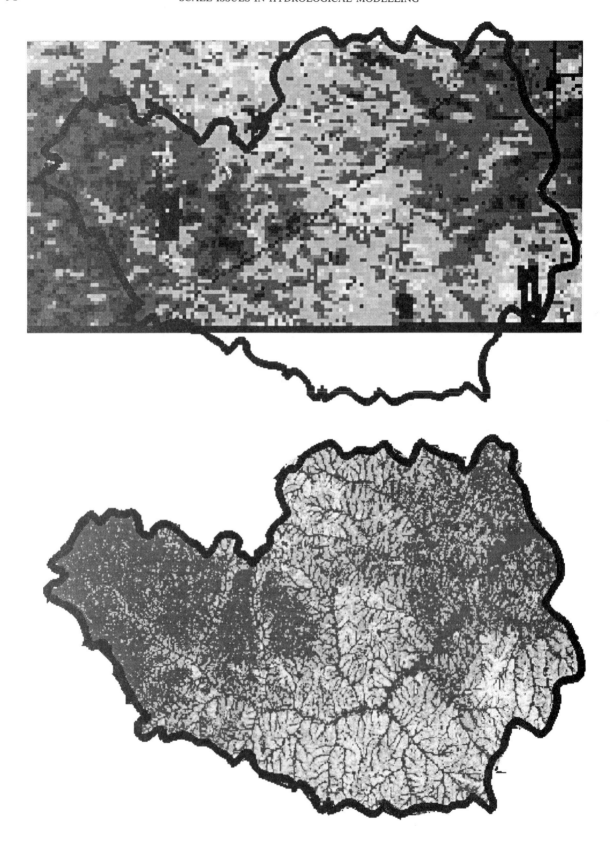

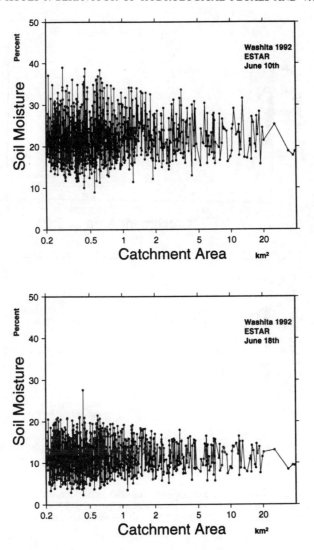

Figure 8. Variation in ESTAR-derived average soil moisture with subcatchment area. The area within the Little Washita as shown in the top portion of Figures 6 and 7 were used in the analysis

eight day dry down). The results from the model-derived soil moisture show that the entropy starts to fall off at about 300 m and that the decay in entropy is similar to that from the ESTAR data (the increase in entropy in the model-derived data with the increase in scale from 30 to about 90 m is unexplained at this time). A tentative conclusion from these results is that the ESTAR-derived surface soil moisture, with a resolution of 200 m, may have insufficient resolution to determine important small-scale variability in soil moisture. To further explore this, the theory of statistical self-similarity was applied to the hydrological model-derived soil moisture.

THEORY OF SIMPLE AND MULTI-SCALING

To further investigate the scaling behaviour in soil moisture, the technique of statistical self-similarity

Figure 7. Estimated soil moisture fields for the Little Washita for 14 June 1992: top figure based on ESTAR (Jackson, pers. comm.); bottom figure based on hydrological modelling. Dark areas represent higher volumetric soil moisture values. The range of soil moisture can be estimated from Figures 8 and 9

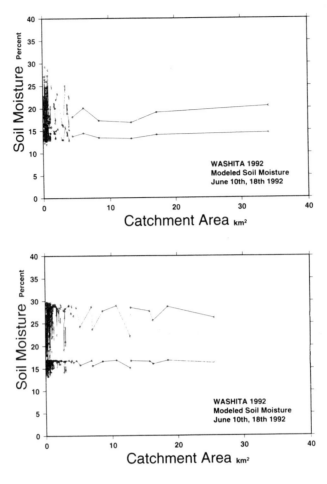

Figure 9. Variation in hydrological model-derived average soil moisture with subcatchment area. Top figure for the middle, drier portion of the basin and the lower figure for a wetter (western) portion. See Figure 6 or 7 for spatial variation in soil moisture. Upper lines in both figures refers to 10 June 1992 and the lower lines to 14 June 1992

can be applied; see Gupta and Waymire (1990) for an overview in rainfall and river flows and Gupta and Dawdy (1994) for its application to flood peaks. Following Gupta and Waymire (1990) and Gupta and Dawdy (1994), consider the simple scaling of a variable $X(A)$, where A refers to a spatial scale parameter and where the parametric dependence of X on A implies statistical homogeneity within a region. Let $\lambda > 0$ be a dimensionless scalar equal to the ratio (A_i/A_j), for areas i and j. Then simple scaling is defined as

$$X(A_i) \overset{d}{=} \left(\frac{A_i}{A_j}\right)^{\theta} X(A_j) \tag{5}$$

where $\overset{d}{=}$ means that the probability distribution of $X(A_i)$ is the same as that of $(A_i/A_j)^{\theta} X(A_j)$. Equation (5) yields a scale-invariant structure such that the probability distribution of $X(A_i)$ can be determined for the area A_i from the distribution of $X(A_j)$ having area A_j. As explained by Gupta and Waymire (1990), the parameter θ is referred to as the scaling exponent, a functional parameter which can take either positive or negative values.

From Equation (5), simple scaling implies that the moments between $X(A_i)$ and $X(A_j)$ are related by

$$E[X^n(A_t)] = \left(\frac{A_i}{A_j}\right)^{n\theta} E[X^n(A_j)] \tag{6}$$

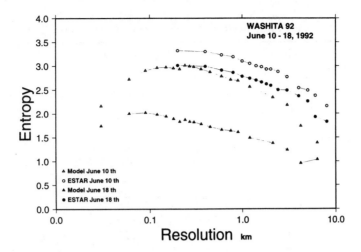

Figure 10. Entriopy versus resolution for hydrological model-derived and ESTAR-derived soil moisture for 10 June and 18 June 1992 from the Little Washita catchment

for $n = 1, 2, \ldots$ and $E[.]$ denotes expectation. Taking logarithms of both sides of Equation (6) results in a log–log linear equation for the moments that can be expressed as

$$\log\{E[X(\lambda)^n]\} = \theta_n \log \lambda + a_n \qquad (7)$$

where a_n, the intercept and θ_n the slope [which in Equation (6) is $n\theta$] can be estimated via regression. As pointed out by Gupta and Dawdy (1994), simple scaling implies that the coefficients of variation C_v are equal for $X(A_i)$ and $X(A_j)$.

To test whether soil moisture obeys simple scaling, model-derived soil moisture for the Little Washita catchment during the WASHITA'92 experiment period was analysed with respect to a scale parameter $\lambda = (A_i/A_j)$, defining the highest resolution of the model (30 m) as the 'unit area' so $A_j = 1$ and $\lambda = A_i$. This empirical analysis did not require any specific mathematical assumptions on the data except that $E[X_\lambda^n] < \infty$ for the order n of the moments analysed. Specifically, the average soil moisture over a pixel of side $\lambda^{0.5}$ was determined and the moments calculated at each scale as λ increased. Within the Little Washita catchment, a square rectangle of 512×512 pixels was extracted. Then rectangles of sizes 128×128, 64×64, 32×32, etc. corresponding to λ values of 16, 64, 256, 1024, etc., respectively, were analysed. Thus by varying the scale parameter λ we can calculate the statistical moments of average, model-derived soil moisture over spatial regions varying from 30×30 m to $15\cdot36 \times 15\cdot36$ km — about three orders of magnitude in area.

The slope and intercept of Equation (7) were estimated by regression for each moment order n. Table I shows the results for the 10 June 1992 simulation (a wet day) and for the 18 June 1992 (after a one week dry down). In Table I, the scaling exponent θ is reported. The last column of Table I gives the correlation coefficient of the log–log linear relationship of Equation (7). The results are plotted in Figure 11. It is clear that the linearity is extremely good up to the sixth-order moment, as required for simple scaling. However, the computed values of the slope, θ_n, shows that $\theta_n < 0$ for all the moment orders analysed (as would be expected), but the relationship between the moment order, n, and the slope is non-linear, which violates simple scaling.

It has been shown (Gupta and Waymire, 1990; 1993; Gupta and Dawdy, 1994) that a variety of hydrological processes (e.g. rainfall and flood peaks) do not follow simple scaling. Gupta and Waymire (1990) presented the framework of multi-scaling to accommodate departure from simple scaling. Basically multi-scaling results in the slope of Equation (7) being expressed as a function of the moment order, i.e. $\theta(n)$. Further research is required to understand how to parameterize $\theta(n)$, which will probably depend on soil characteristics and average soil moisture.

Table I. Intercepts and scaling exponent for model-derived soil moisture from WASHITA'92

Moment	Intercept	Scaling exponent	Correlation coefficient
10 June 1992 data			
1	0·528	−0·001	0·905
2	1·049	−0·021	0·996
3	1·567	−0·035	0·997
4	2·087	−0·044	0·998
5	2·612	−0·048	0·998
6	3·143	−0·050	0·905
18 June 1992 data			
1	0·493	−0·002	0·928
2	0·987	−0·006	0·993
3	1·480	−0·009	0·996
4	1·972	−0·012	0·997
5	2·464	−0·014	0·997
6	2·957	−0·016	0·928

The use of the hydrological model-derived soil moisture yields new insights into the scaling properties of soil moisture. From the comparisons carried out to date, it appears that the model captures the essential spatial variability observed in the remotely sensed data. In addition, the model can investigate the spatial variability at scales smaller than those observed by the ESTAR instrument. This suggests that model-derived soil moisture fields can be used to develop a consistent theory for the scaling of soil moisture, which can be verified at large scales (and low resolution) by remote sensing instruments such as the ESTAR. With such a scaling theory, soil moisture fields determined by low resolution passive sensors on satellites (resolution at scales of 1–10 km) can be statistically extended to smaller scales using a correctly parameterized Equation (7).

These model results yield intriguing new insights into the scaling of soil moisture. Before we can further understand the scaling behaviour of soil moisture, it would be useful to investigate the scaling behaviour for the remotely sensed soil moisture data. Unfortunately, the domain size for the collected ESTAR data precludes the sort of analysis carried out for the modelled area. One aspect of the above analysis that is an important by-product is a further understanding of the data requirements for soil moisture scaling research. It is expected through new experiments during 1994 with the Shuttle Imaging Radar (SIR-C) remotely sensed data from larger domains can be used to further explore the scaling issues alluded to in the modelling study.

Figure 11. Departure from simple scaling in the growth of the scaling exponent with respect to moment order using hydrological model-derived soil moisture for 10 June and 18 June 1992 from the Little Washita catchment

SUMMARY

This paper reviews a number of studies that investigated the transition in hydrological responses during storm and inter-storm periods as scale was increased in the presence of spatial variability. The results from FIFE'87 indicate that the macroscale models that preserve the statistical characterization of the small-scale variability in the hydrological controls (topography and soils) can accurately represent both storm and inter storm water fluxes (Famiglietti and Wood, 1994). An important research objective in scaling hydrological variables is the determination of the threshold scale where statistical representations of smaller (sub-grid) areas can replace actual patterns of variability. Using model-derived surface flux data from the $11 \cdot 7 \, \text{km}^2$ Kings Creek catchment with the FIFE'87 experiment, this threshold appears to be 1–$2 \, \text{km}^2$; a scale supported by an analysis of infiltration parameters, derived from soil classification data, on the $525 \, \text{km}^2$ Little Washita watershed in Oklahoma by Farajalla and Vieux (in press).

An analysis of remotely sensed soil moisture data and model-derived soil moisture for the Little Washita catchment has suggested that spatial scale for the threshold may be of the order of 5–$10 \, \text{km}^2$; this increased threshold scale from the earlier studies reflects the increased watershed size and the regional variations [with length scales of $O(10) \, \text{km}$]. An analysis of statistical self-similarity using model-derived soil moisture showed that soil moisture does not obey simple scaling theory, but multi-scaling. The analysis showed a strong relationship between the estimated statistical moments and spatial scale over which the soil moisture was averaged. This suggests great promise for scaling soil moisture — namely, combining soil moisture scaling theory with large-scale estimates from macroscale hydrological models and/or low resolution soil moisture satellite sensors to determine the statistical behaviour of sub-grid soil moisture. Such statistical representations are critical to the determination of surface fluxes, which may be controlled through atmospheric demand for moisture or limited by the ability of the soil–vegetation system to deliver water to the atmosphere (Wood, 1994). The results presented here are based on the most extensive data sets currently available for such studies; nonetheless, as discussed in the text, such studies need to be expanded over a wider range of catchment and climatic scales to further verify the concepts presented here.

ACKNOWLEDGEMENTS

I thank Dom Thongs and Mark Zion for their help in preparing the data and model runs for the Little Washita catchment results, and Ralph Dubayah, of the University of Maryland, who ran the scaling analysis. The research presented here was supported by NASA grants NAG5-1628, NAS5-31719 and NOAA grant NA36GP0419. This support is gratefully acknowledged.

REFERENCES

Ahuja, L.R., Cassel, D., Bruce, R., and Barnes, B. 1989. 'Evaluation of spatial distribution of hydraulic conductivity using effective porosity data', *Soil Science*, **148**, 404–411.

Avissar, R. and Pielke, R.A. 1989. 'A parameterization of heterogeneous land surface for atmospheric numerical models and its impact on regional meteorology', *Monthly Weather Rev.*, **117**, 2113–2136.

Beven, K.J. 1986a. 'Runoff production and flood frequency in catchments of order n: an alternative approach', in Gupta, V.K., Rodriguez-Iturbe, and Wood, E.F. (Eds), *Scale Problems in Hydrology*. Reidel, Hingham. pp. 107–131.

Beven, K.J. 1986b. 'Hillslope runoff processes and flood frequency characteristics', in Abrahams, A.D. (Ed.), *Hillslope Processes*. Allen and Unwin, London. pp. 187–202.

Beven, K.J. and Kirkby, M.J. 1979. 'A physically-based variable contributing area model of basin hydrology', *Hydrol. Sci. Bull.*, **24**, 43–69.

Binley, A.M., Elgy, J. and Beven, K.J. 1989. 'A physically-based model of heterogeneous hillslopes', *Wat. Resour. Res.*, **25**, 1219–1226.

Bogard, V., Fielder, A. and Meinders, H. 1978. *Soil Survey of Grady County, Oklahoma*. USDA Soil Conservation Service.

Charney, J., Quirk, W., Chow, S. and Kornfield, J. 1977. 'A comparative study of the effects of albedo change on drought in semi-arid regions', *J. Atmos. Sci.*, **34**, 1366–1385.

Dickinson, R.E., Henderson-Sellers, A., Kennedy, P.J. and Wilson, M.F. 1986. 'Biosphere–atmosphere transfer scheme (BATS) for the NCAR community climate model', *NCAR Tech. Note*, 69 pp.

Eagleson, P.S. 1978. 'Climate soil and vegetation', *Wat. Resour. Res.*, **14**, 705–776.

Engman, E.T. 1990. 'Process in microwave remote sensing of soil moisture', *Can. J. Remote Sensing*, **16**, 6–14.

Farajalla, N. and Vieux, B. 1995. 'Capturing the essential variability in distributed hydrological modeling: infiltration parameters', *Hydrol. Process.*, **9**, 55–68.

Famiglietti, J. S. 1992. 'Aggregation and scaling of spatially-variable hydrological processes: local, catchment scale and macroscale models of water and energy balance', *PhD Dissertation*, Department of Civil Engineering and Operations Research, Princeton University, Princeton.

Famiglietti, J. S. and Wood, E. F. 1990. 'Evapotranspiration and runoff from large land areas: land surface hydrology for atmospheric general circulation models', in Wood, E. F. (Ed.), *Land Surface–Atmospheric Interactions for Climate Models: Observations, Models, and Analyses*. Kluwer Academic, Dordrecht. pp. 179–204.

Famiglietti, J. S. and Wood, E. F. 1994. 'Multi-scale modeling of spatially-variable water and energy balance processes', *Wat. Resour. Res.*, **30**, 3061–3078.

Famiglietti, J. S. and Wood, E. F. 1995. 'Effects of spatial variability and scale on areal-average evapotranspiration', *Wat. Resour. Res.*, **31**, 699–712.

Famiglietti, J. S., Wood, E. F., Sivapalan, M., and Thongs, D. J. 1992. 'A catchment scale water balance model for FIFE', *J. Geophys. Res.*, **97**(D17), 18 997–19 008.

Gupta, V. K. and Dawdy, D. R. 1994. 'Regional analysis of flood peaks: multi scaling theory and its physical basis', in *Advances in Distributed Hydrology*. Water Resources Publications, Fort Collins.

Gupta, V. K. and Waymire, E. 1990. 'Multiscaling properties of spatial rainfall and river flow distributions', *J. Geophys. Res.* **95**(D3), 1999–2009.

Gupta, V. K. and Waymire, E. 1993. 'A statistical analysis of meso-scale rainfall and river flow distributions', *J. Appl. Meteorol.*, **32**, 251–267.

Jackson, T. J. 1993. 'Measuring surface soil moisture using passive microwave remote sensing', *Hydrol. Process.* **7**, 139–152.

Mobley, H. and Brinley, R. 1978. *Soil Survey of Comanche County, Oklahoma*. USDA — Soil Conservation Service.

Moffatt, H. 1978. *Soil Survey of Caddo County, Oklahoma*. USDA — Soil Conservation Service.

Paniconi, C. and Wood, E. F. 1993. 'A detailed model for simulation of catchment scale subsurface hydrological processes', *Wat. Resour. Res.*, **29**, 1601–1620.

Rawls, W., Brakensiek, D., and Miller, N. 1983. 'Green-Ampt infiltration parameters from soil data', *ASCE J. Hydr. Eng.*, **109**, 62–70.

Schmugge, T. J. 1983. 'Remote sensing of soil moisture: recent advances', *IEEE Trans. Geosci. Remote Sens.*, **GE-21**, 336–344.

Sellers, P. J., Mintz, Y., Sud, Y. C., and Dalcher, A. 1986. A simple biosphere model (SiB) for use within general circulation models', *J. Atmos. Sci.*, **43**, 505–531.

Sellers, P. J., Hall, F. G., Asrar, G., Strebel, D. E., and Murphy, R. E. 1988. 'The first ISLSCP field experiment (FIFE)', *Bull. Am. Meteorol. Soc.*, **69**, 22–27.

Sellers, P. J., Heiser, M. D., and Hall, F. G. 1992. 'Relations between surface conductance and spectral vegetation indices at intermediate (100 m^2 to 15 km^2) length scales', *J. Geophys. Res.*, **97**(D17), 19 033–19 059.

Shukla, J. and Mintz, Y. 1982. 'The influence of land-surface evapotranspiration on earth's climate', *Science*, **215**, 1498–1501.

Shukla, J., Nobre, C., and Sellers, P. J. 1990. 'Amazon deforestation and climate change', *Science*, **247**, 1322–1325.

Sivapalan, M., Beven, K. J., and Wood, E. F. 1987. 'On hydrological similarity: 2. A scaled model of storm runoff production', *Wat. Resour. Res.*, **23**, 2266–2278.

Sud, Y. C., Sellers, P., Chow, M. D., Walker, G. K., and Smith, W. E. 1990. 'Influence of biosphere on the global circulation and hydrologic cycle — a GCM simulation experiment', *Agric. Forestry Meteorol.*, **52**, 133–188.

Walker, J. M. and Rowntree, P. R. 1977. 'The effect of soil moisture on circulation and rainfall in a tropical model', *Q. J. R. Meteorol. Soc.*, **103**, 29–46.

Wood, E. F., 1991. 'Global scale hydrology: advances in land surface modeling', *Rev. Geophys. Suppl.*, **29**, 193–201.

Wood, E. F. 1994. 'Scaling, soil moisture and evapotransipiration in runoff models', *Adv. Wat. Resour.*, **17**, 25–34.

Wood, E. F., Sivapalan, M., Beven, K. J., and Band, L. 1988. 'Effects of spatial variability and scale with implications to hydroligic modeling', *J. Hydrol.*, **102**, 29–47.

Wood, E. F., Sivapalan, K. J., Beven, K. J., 1990. 'Similarity and scale in catchment storm response', *Rev. Geophys.*, **28**, 1–18.

World Climate Research Programme 1992. 'The scientific plan for the GEWEX continental scale international project', *WCRP-67 WMO/TD-No. 461*, WCRP, Geneva. 65 pp.

6

PHYSICAL INTERPRETATIONS OF REGIONAL VARIATIONS IN THE SCALING EXPONENTS OF FLOOD QUANTILES

VIJAY K. GUPTA

Center for the Study of Earth From Space/CIRES, and Department of Geological Sciences, University of Colorado, Boulder, CO 80309, USA

AND

DAVID R. DAWDY

Hydrologic Consultant, 3055 23rd Avenue, San Francisco, CA 94132, USA

ABSTRACT

The concepts of simple scaling and multiscaling provide a new theoretical framework for the study of spatial or regional flood frequency relations and their underlying physical generating mechanisms. In particular, the scaling exponents in the power law relationship between flood quantiles and drainage areas contain a 'basic signature of invariance' regarding the spatial variability of floods, and therefore suggest different hypotheses regarding their physical generating mechanisms. If regional floods obey simple scaling, then the slopes do not vary with return periods. On the other hand, if regional floods obey multiscaling, then the slopes vary with return periods in a systematic manner. This premise is expanded here by investigating the empirical variations in the scaling exponents in three states of the USA: New York, New Mexico and Utah. Distinct variations are observed in the exponents among several regions within each state. These variations provide clear empirical evidence for the presence of both simple scaling and multiscaling in regional floods. They suggest that snowmelt-generated floods exhibit simple scaling, whereas rainfall-generated floods exhibit multiscaling. Results from a simple rainfall–runoff experiment, along with the current research on the spatial scaling structure of mesoscale rainfall, are used to give additional support to these physical hypotheses underlying two different scaling structures observed in floods. In addition, the rainfall–runoff experiment suggests that the behaviour of the flood exponents in small basins is determined by basin response rather than precipitation input. This finding supports the existence of a critical drainage area, as has been reported for the Appalachia flood data in the USA, such that the spatial variability in floods in basins larger than the critical size is determined by the precipitation input, and in basins smaller than the critical size is determined by the basin response.

INTRODUCTION

Scale problems arise naturally in seeking connections among physical processes at disparate scales of aggregation and disaggregation in space and time. The classic problem of prediction from ungauged basins, and a growing emphasis on spatially distributed hydrological modelling in river basins within the context of numerical climate models, have brought into focus the issues of aggregation and disaggregation. Therefore, the recognition and importance of scale issues in hydrology has grown enormously within the last decade. In particular, we want to know what properties change and what properties remain invariant under scale change. For this, proper notions of scale transformation and scale invariance are needed. Scale invariance notions in terms of geometric and dynamic similarity are widespread in science and engineering. However, the presence of several spatial scales due to nesting of different sized sub-basins within a basin, or different sized basins within a geographical region, requires that notions of 'self-similarity' are formulated and tested. Indeed, geometric self-similarity has permeated the broad scientific literature under the popular name of 'fractals'

(Schroeder, 1991). It is also finding a natural place in describing river network topology and geometry within the context of the Horton–Strahler ordering scheme (Peckham, 1995). In addition, the presence of fluctuations and variability across a broad range of space and time-scales is bringing different notions of 'statistical self-similarity' into focus. For example, new and important forms of self-similarity, e.g. statistical simple scaling, multiscaling and multifractal cascade geometry, have been identified in a wide variety of physical contexts; see, for example, Feder (1988), Falconer (1990) and Schroeder (1991).

Ideas of statistical simple scaling and multiscaling provide a natural framework to begin to understand the physical structure of floods (Gupta and Waymire, 1990; Smith, 1992; Gupta et al., 1994) and other hydrological processes within a region (Schertzer and Lovejoy, 1987; Gupta and Waymire, 1989; 1993). Many examples from physics show that the scaling exponents contain the key physical information about a process. We expand on this premise here by formulating two basic hypotheses about floods. Firstly, there is a critical drainage area, as has been observed in the Appalachia flood data (Smith, 1992), such that scaling in floods in basins larger than the critical size is determined by the precipitation input, whereas in basins smaller than the critical size scaling is determined by the basin response. Secondly, in basins larger than the critical, the snowmelt-generated floods exhibit simple scaling, whereas the rainfall-generated floods exhibit multi-scaling. Most of the paper is devoted to testing these two hypotheses through (i) empirical analyses of regional flood data and (ii) a simple rainfall–runoff simulation experiment. Current research on the multiscaling structure of rainfall fields at the mesoscale is briefly summarized to give further support to these hypotheses.

The early study by Cadavid (1988) showed that floods in the south-eastern and the Appalachia regions of the USA did not obey simple scaling. Gupta and Waymire (1990) proposed that the framework of multiscaling be explored in describing floods in those regions where simple scaling does not hold. In this vein, a maximum likelihood test of simple scaling versus multiscaling by Smith (1992) rejected the presence of simple scaling in the Appalachian flood data. By contrast, the index flood assumption, which has been widely used in research on regional flood frequency analysis for over 30 years, is connected to simple scaling (Smith, 1992; Gupta et al., 1994).

In the existing published work, deviations from the index flood assumption are generally attributed to a lack of 'homogeneity' in the flood data [National Research Council (NRC), 1988]. Gupta et al. (1994) have argued that this notion of homogeneity is ad hoc and is not useful in regional flood frequency studies. They have defined floods to be regionally homogeneous if the drainage area serves as the indexing set for the random field of peak flows. Assuming homogeneity, the theory of statistical simple scaling and under certain conditions the multiscaling theory predict that flood quantiles are log–log linear with respect to drainage areas. The behaviour of the slopes, called scaling exponents, with respect to return periods contains a 'basic signature of invariance' in the spatial variability of floods. If regional floods obey simple scaling, then the slopes do not vary with return periods. On the other hand, if regional floods obey multiscaling, then the slopes vary with return periods in a systematic manner (Gupta et al., 1994).

The behaviour of the scaling exponents described here can be inferred from the empirical analyses of flood data which have been carried out routinely for over 30 years by the United States Geological Survey (USGS). For example, the USGS uses the 'quantile regression approach' (Benson, 1962) as a standard tool for the analysis of regional flood data, and a wealth of regional information about floods is contained in the USGS regional flood frequency reports (Waltemeyer, 1986; Lumia, 1991). Unfortunately, in the absence of a general theoretical framework to interpret data, these empirical studies have had little impact on research on regional flood frequency. In this paper we use the results from the USGS reports to examine the behaviour of the exponents of drainage areas with respect to return periods in a power law relationship between quantiles and drainage areas, and interpret this behaviour to support the physical hypothesis regarding snowmelt- versus rainfall-generated floods.

The second section gives a self-contained non-technical review of the quantile–drainage area relationships predicted by the simple and multiscaling theories under the assumption of spatial statistical homogeneity of annual peak flows (Gupta et al., 1994). The next section summarizes and interprets empirical results from three different states of the USA to support the physical hypothesis regarding differences in the scaling exponents of snowmelt- versus rainfall-generated floods. These analyses are taken from the USGS regional flood frequency reports. A simple rainfall–runoff experiment is then used to support our hypothesis that scaling exponents in small basins are determined by the basin response rather than the precipitation input. We then discuss recent progress in space–time rainfall modelling to suggest that the physical basis of multiscaling in rainfall-generated floods in large basins is related to the presence of multiscaling in spatial rainfall. We conclude with some remarks for future directions in this new line of research in regional flood frequency analysis and distributed hydrology.

A BRIEF REVIEW OF QUANTILE REPRESENTATIONS IN SIMPLE AND MULTISCALING THEORIES

The current research on regional flood frequency analysis is based on the index flood assumption, which implies that the coefficient of variation (CV) of floods is a constant and does not vary with drainage area. This constancy of CV is generally regarded as the definition of 'homogeneity' (NRC, 1988: 38). Therefore, observed deviations in the flood data from the index flood assumption are generally attributed to a lack of homogeneity. In our view, this notion of homogeneity is *ad hoc* and is not very useful. The definition of regional statistical homogeneity of floods by Gupta *et al.* (1994) is based on the formulation that a drainage network serves as an indexing set in describing the spatial random field of peak flows. The invariance of the probability distributions of peak flows under translation on this indexing set defines statistical homogeneity in a network. This definition is similar in spirit to how stationarity or homogeneity is defined in published works on stochastic processes. Within the context of statistical scaling theories used here, simple scaling implies a constant CV, whereas multiscaling does not. However, applications of both these theories to flood frequency have been based on the assumption of statistical homogeneity. We show a simple test of regional homogeneity in flood data which is based on the empirical quantile regression analyses contained in the USGS flood frequency reports.

There are two important scaling theories that are relevant to regional flood frequency analysis. The first of these, known as the simple scaling theory, is closely related to the widely used index flood assumption. Typically, the generating mechanism underlying a simple scaling stochastic processes is 'additive'. Some well-known examples of simple scaling include the Brownian motion and the fractional Brownian motion processes (Feder, 1988). It can be simply shown (see Gupta *et al.*, 1994) that if the regional floods obey simple scaling then (i) the quantile–drainage area relationship is log–log linear and (ii) the slope in this relationship is a constant and does not vary with the return period. To state this mathematically, let $Q(A)$ be a stochastic process denoting the annual peak flows from a basin of drainage area A. Here, the drainage area serves as a scale parameter and the dependence of $Q(A)$ on A alone represents the assumption of statistical homogeneity in regional floods. If the floods obey simple scaling, then the pth flood quantile defined as

$$P[Q(A) > q_p(A)] = p \tag{1}$$

varies with drainage area as

$$q_p(A) = c(p)A^\theta \tag{2}$$

where the coefficient $c(p)$ depends on the probability of exceedance p, or the return period $T_r = 1/p$, but the scaling exponent θ does not. We will show later that Equation (2), which exhibits both of the properties of simple scaling listed above, is observed to hold in some of the regional flood data.

By contrast, many other empirical analyses of floods given in in the next section exhibit the log–log linearity described earlier, but violate the constancy of the scaling exponent θ. This property of the regional floods can be described by another type of a scaling invariance property called multiscaling. It can be mathematically motivated by specifying how it differs from simple scaling (Gupta and Waymire, 1990). Specifically, simple scaling representation in terms of statistical moments implies that (i) the statistical moments $E[Q^j(A)]$, $j = 1,2,\ldots$ are log–log linear in A and (ii) the slopes $s(j) = j\theta$ of these equations are linear in j. Early analyses of the rainfall and the peak flow data showed that although they obeyed the log–log linearity of the moments, they did not obey the linearity of the slope $s(j)$ as a function of j. Now, if $s(j)$ is non-linear, then it can either be concave or convex in j. Gupta and Waymire (1990) argued that the concavity of slope implies a decrease in spatial variability with an increase in the scale of the underlying process, whereas the convexity of slope implies an increase in spatial variability with scale. This results in two distinct mathematical representations of a multiscaling process.

Under the assumption that floods are spatially homogeneous and multiscaling, Gupta *et al.* (in press) have derived theoretical expressions relating flood quantiles with drainage areas for the class of log-Levy models, of which the log-normal is a special case. Some preliminary tests by them show that the log-normal model describes the Appalachia flood data better than several other models from the log-Levy class. Even though log-normality or any other parametric assumption is not the focus of this paper, we give expressions of flood quantiles for the log-normal multiscaling model to briefly illustrate the theory, i.e.

$$\log q_p(A) = (a_0 - \mu_0 \log A) - \left[b^2 - \sigma_0^2 \log \left(\frac{A}{A_0} \right) \right]^{1/2} w_{1-p}, \qquad A_c < A < A_0 \tag{3}$$

and

$$\log q_p(A) = (\mu_1 \log A - a_1) - \left[\sigma_1^2 \log \left(\frac{A}{A^0} \right) + b^2 \right]^{1/2} w_{1-p}, \qquad A^0 < A < A_c \tag{4}$$

where $(a_i, \mu_i, \sigma_i^2 > 0; i = 0, 1)$, $b > 0$, are statistical parameters, and w_{1-p} is the $(1 - p)$th quantile of a normal random variable W with a mean of zero and a variance of unity. The interval (A_c, A_0) denotes the range over which Equation (3) holds, and (A^0, A_c) the range over which Equation (4) holds. The critical drainage area below which Equation (3) does not hold, but Equation (4) holds, is denoted by A_c. Note that the non-linear terms within brackets in Equations (3) and (4) always remain positive. Unlike simple scaling, we must decide which one of these two representations is appropriate for a given range of drainage areas. In the papers by Smith (1992) and Gupta *et al.* (1994), it is observed that for the Appalachia flood data the multiscaling corresponding to Equation (3) only holds for basins larger than about 20 square miles (52 km^2). By contrast, in smaller basins, the applicability of Equation (4) on both empirical and physical grounds has so far remained unclear. We present some results on this issue in the fourth section, which deals with the role of the basin response on scaling exponents in small basins via rainfall–runoff simulations.

It can be seen that neither of these two expressions exhibits log–log linearity between quantiles and drainage areas. However, when $b^2 + \sigma_0^2 \log(A_0) \gg \sigma_0^2 \log(A)$, the second term on the right-hand side of Equation (3) can be expanded in a Taylor series to yield log–log linearity between quantiles and drainage areas. This can be expressed as

$$q_p(A) = c(p) A^{\theta(p)} \tag{5}$$

The exponent $\theta(p)$ and the intercept $c(p)$ depend on the probability of exceedance p, and are given by

$$\log c(p) = a_0 - b_0^{1/2} w_{1-p} \tag{6}$$

$$\theta(p) = -\mu_o + \frac{\sigma_0^2}{2b_0^{1/2}} w_{1-p} \tag{7}$$

where $b_0 = b^2 + \sigma_0^2 \log(A_0)$. By definition of a quantile, as p becomes smaller $(1 - p)$ increases and w_{1-p} decreases. Therefore, it follows from Equation (7) that $\theta(p)$ decreases with p. This means that bigger floods have a smaller exponent than smaller floods. We will use this important property to infer the presence of multiscaling in the flood data analysed in the next section. We conclude this section with several remarks regarding the multiscaling theory:

1. The quantile expression given in Equation (3) has been used by Gupta and Dawdy (1994) to suggest connections between the multiscaling theory and the process-based similarity theory of Sivapalan *et al.* (1990) and a physical interpretation of two of the four statistical parameters appearing in Equation (3). Further theoretical details about multiscaling can be found in Gupta and Waymire (1990), Smith (1992) and Gupta *et al.* (1994).

2. Smith (1992) and Gupta *et al.* (1994) have given some results on the estimation of parameters in the multiscaling model using the Appalachian flood data. Therefore, in this paper we do not deal with the

issues of parameter estimation and goodness of fit of different models. In addition, Smith (1992) has given a maximum likelihood statistical inference test of simple versus multiscaling for the log-normal model.

3. It is important to note that both Equations (2) and (5) obey log–log linearity between quantiles and drainage areas. However, unlike Equation (2), the exponent in Equation (5) decreases as floods become large. This feature cannot be exhibited by either the simple scaling theory or the index flood assumption. It was first reported by Dawdy (1961) and was instrumental in the adoption of the quantile regression method by the USGS. In this sense the multiscaling theory provides the first theoretical underpinnings of the empirically observed variation in the exponents. These variations for several regions of the USA are given in the next section.

USE OF USGS QUANTILE REGRESSION RELATIONS TO GAIN PHYSICAL UNDERSTANDING OF VARIATION IN THE SCALING EXPONENTS

Our main objective here is to investigate to what extent the regional flood data discriminate between the simple and multiscaling theories reviewed in the previous section, and support the hypothesis that snowmelt-generated floods exhibit simple scaling, whereas the rainfall-generated floods exhibit multiscaling. The USGS performs regional flood frequency analyses on a state by state basis throughout the USA. As the USGS data and their analyses contain a wealth of regional information on flood response to precipitation input, they should serve as a major component in developing a physically based understanding of regional flood frequency behaviour. As a start towards that attempt, three USGS state reports for New York (Lumia, 1991), New Mexico (Waltemeyer, 1986) and Utah (Thomas and Lindskov, 1983) are used in this section. New York is divided into eight regions with an average area of 6200 square miles (16 000 km^2), New Mexico into eight regions with an average area of 15 200 square miles (39 000 km^2) and Utah into six regions with an average area of 14 150 square miles (37 000 km^2). We will focus on two types of analyses. The first is concerned with the issue of regional homogeneity and the second with a test of the variation of scaling exponents with respect to return periods, and their physical interpretations. Further tests of these hypotheses should serve as an important guide in the development of a general theory of regional flood frequency analysis.

The quantile regression method being used by the USGS consists of fitting a log-Pearson III probability distribution to flood data at each station in a region and then using those distributions to compute the quantile discharges corresponding to 2, 5, 10, 25, 50, 100 and 500 year return periods. The logarithms of these computed quantile discharges are then regressed against the logarithms of several physical variables, e.g. drainage area, mean elevation, mean rainfall, basin shape and many others, in a stepwise multiple regression analysis. The equations derived for regional flood frequency analysis often include other variables in addition to area. The 'standard error of estimate (SE)' is the square root of the residual variance which a regression equation does not explain [see Equation (10)]. However, the USGS reports do not give the SE, but instead derive a 'standard error of prediction (SP)' in terms of percentages. These two are related as

$$\frac{1}{2}[(10^{SE} - 1) + (1 - 10^{-SE})] = \frac{SP}{100} \tag{8}$$

Clearly, the SE can be computed from the SP and vice versa using Equation (8). We will use the SE for testing the issue of regional homogeneity and the variation of scaling exponents with respect to return periods. Even though the values of the SP given in the report are derived from the SE, the latter is preferable as a direct measure of the unexplained variability by the regression equation in the log-space. It is advisable to refer to the state reports referenced for a variety of other details not given here.

Test of regional homogeneity

As explained in the previous section, Gupta et al. (1994) have discussed at length the issue of regional homogeneity and have defined floods to be regionally homogeneous if they are indexed by drainage area alone and nothing else. They suggested a simple, but preliminary, criterion as a test of homogeneity on the basis of regional quantile analyses given in the flood frequency reports. We will illustrate this test for the state of Utah.

The USGS divides every state into several geographical regions and regression equations for each quantile in a region generally contain drainage area and several other variables. In some states, e.g. New York (Lumia, 1991), the report also contains regression equations using drainage area alone. In any event, if they are not given for a region, it is easy to use the USGS quantile data to derive the regression equations containing area alone. The basic idea behind the test is to compare the two sets of standard errors of estimate from the two sets of regression equations. The first set, denoted as SE1, is from equations given in the USGS reports containing several descriptor variables. The second set, denoted by SE2, is from regression equations containing drainage area alone. If the difference between SE1 and SE2 is 'nominal', then a region is homogeneous. The idea clearly is that if SE1 and SE2 are close to each other, then one does not explain more variance by adding other physical descriptors over and above the drainage area. By contrast, if SE1 and SE2 differ substantially, say (SE2 − SE1)/SE2 is greater than 0·15, then the region appears to be non-homogeneous. The value of 0·15 is *ad hoc*, and is chosen here only to illustrate a methodology. Certainly, rigorous statistical tests of homogeneity which are physically based will have to be developed in the future.

For most of the quantiles in the six regions of Utah, the USGS used the drainage area and the mean elevation in their regression equations (Thomas and Lindskov, 1983). We selected five of these seven regions, which are shown in Table I. As the drainage areas based regression equations are not available, they are derived for this study. Table I gives the two sets of the standard errors of estimate, SE1 and SE2, explained earlier. SE1 was computed from the standard error of prediction given in the report. The starred numbers in Table I indicate those quantiles where the USGS used area alone and, therefore, SE1 = SE2. The USGS did not compute regressions for the 500 year floods, but provided the 500 year peaks for each station. Thus the comparison between SE1 and SE2 for the 500 year flood quantile is not shown in Table I.

For the second, fourth and fifth regions in Table I, the reduction in SE1 over SE2 is nominal. Therefore, all of these three regions may be taken to be approximately homogeneous. The reduction in SE1 over SE2 for the first region is by about 20%, and this region is probably not homogeneous. However, for the Uinta basin region the standard error SE2 is reduced by 30% or more for most of the quantiles, and this region certainly does not seem to be homogeneous. These inferences make physical sense because the mean basin elevation, probably as a surrogate for mean annual snowfall, explains some of the variability in the data. Therefore, the standard errors are reduced.

This test shows that in some regions the drainage area is indeed sufficient as an indexing parameter for the flow random field, whereas in others it is not. Clearly, in those regions other variables contain additional physical information which should be used in the development of a theory of regional flood frequency. Exactly how this information can be used, other than to derive empirical prediction equations, is not clear at this time, and this constitutes an important topic of further research; see the comments in Gupta *et al.* (1994) with respect to extending this theory to non-homogeneous regions.

Empirical analyses and interpretations of scaling exponents

The second set of tests is concerned with the variation of scaling exponents with respect to return periods in the quantile–drainage area relations, and with physical interpretations of this behaviour. As drainage area is

Table I. Comparison of two sets of standard errors of prediction, SE1 and SE2, in a test of homogeneity for five regions of Utah

Quantiles	Northern mountains high elevation		Northern mountains low elevation		Uinta Basin S		High plateaus		Low plateaus	
	SE1	SE2	SE1	SE2	SE1	SE2	SE1	SE2	SE1	SE2
Q2	0·185	0·22	0·31	0·31	0·325	0·34	0·27**	0·27	0·34	0·37
Q5	0·16	0·20	0·29	0·29	0·27	0·33	0·22**	0·22	0·29	0·33
Q10	0·155	0·20	0·29	0·29	0·26	0·36	0·22	0·22	0·275	0·32
Q25	0·155	0·20	0·28	0·30	0·27	0·40	0·23	0·24	0·26	0·31
Q50	0·155	0·20	0·29	0·30	0·285	0·43	0·255	0·26	0·26	0·31
Q100	0·16	0·20	0·29	0·31	0·30	0·46	0·28	0·29	0·27	0·31

Table II. Comparison of least-squares and geometric mean estimates of scaling exponents for three different patterns in the standard errors of estimate

| Quantiles | New Mexico Region-4 | | | Utah | | | | | |
| | | | | Northern mountains high elevation | | | Uinta Basis | | |
	SE2	LSEXP	GMEXP	SE2	LSEXP	GMEXP	SE2	LSEXP	GMEXP
Q2	0·45	0·67	0·71	0·22	0·90	1·02	0·34	0·37	0·57
Q5	0·34	0·59	0·65	0·20	0·89	0·99	0·33	0·31	0·54
Q10	0·28	0·55	0·59	0·20	0·88	0·98	0·36	0·29	0·55
Q25	0·23	0·50	0·53	0·20	0·87	0·97	0·40	0·27	0·59
Q50	0·21	0·465	0·49	0·20	0·87	0·97	0·43	0·25	0·62
Q100	0·19	0·44	0·46	0·20	0·86	0·97	0·46	0·24	0·65
Q500	0·19	0·41	0·43	0·21	0·86	0·97	0·49	0·24	0·68

the only variable used in this investigation, it implies that all of the regions considered here are assumed to be homogeneous.

All of our slope estimates are based on least-squares regressions for direct comparison with the USGS equations, which are obtained from the generalized least-squares method. An estimate of the scaling exponent for each quantile from a regression equation introduces a 'regression effect' into the estimate. This can potentially bias the results and thereby our interpretations of simple versus multiscaling. To understand the nature of this problem, let $Y = a + bX$ denote a regression equation. Then the least-squares estimate of b is given by

$$b = \frac{\sigma(Y)}{\sigma(X)} \rho(X, Y) \tag{9}$$

and the standard error of estimate is given by

$$SE = \sigma(Y)[1 - \rho^2(X, Y)]^{1/2} \tag{10}$$

where denotes σ the standard deviation and ρ denotes the correlation coefficient (see Parzen, 1960:387). The term $[1 - \rho^2(X, Y)]^{1/2}$ is called the regression effect. If $\rho = 1$, then both the regression effect and SE are zero, and $b = \sigma(Y)/\sigma(X)$ gives the slope estimate without the regression effect. However, in general $\rho \neq 1$. For a fixed $\sigma(Y)$, larger SE means a smaller ρ and therefore a smaller slope b due to the regression effect. This discussion implies that if the standard errors within a region are about the same for different flood frequencies, then each regression equation explains about the same amount of variance and the regression effect does not influence the relative values of the exponents in the quantile relations. Then presence of uniformly large standard errors for all the quantiles will reduce the magnitude of the exponents, but will not change their relation to each other. In view of Equations (2), (5) and (7), it is mainly the relation of slopes to each other that we care about the most, as that will enable us to infer whether simple or multiscaling is being suggested in a region.

To further investigate the role of regression effect in the slope computations, we considered the line of organic correlation investigated by Hirsch and Gilroy (1984). As the slope of Y regressed on X is b, and the slope of X regressed on Y is $1/b$, it follows from Equation (9) that a division of the estimate of b by that of $1/b$ cancels out ρ. Therefore, taking the geometric mean of this ratio gives a slope estimate which has no regression effect. This property also makes this approach well-suited to the extension of hydrological records. We compared the exponents of the quantile relations estimated by these two methods for three regions. These regions shown in Table II are selected so that in the first region the standard error of estimate (SE2) decreases with flood frequency; it remains essentially constant in the second and in the third region it increases.

In the first two regions, the difference between the least-squares exponents (LSEXP) and the geometric

Table III. Exponents and standard errors of estimate in flood quantile–drainage area relations for four New York regions

Quantiles	Region 3: Hudson Valley		Region 7: Central Lake Ontario		Region 2: Adirondacks		Region 4: Catskills	
	EXP	SE2	EXP	SE2	EXP	SE2	EXP	SE2
Q2	0·700	0·15	0·828	0·25	0·816	0·17	0·914	0·16
Q5	0·687	0·15	0·829	0·25	0·803	0·17	0·887	0·16
Q10	0·686	0·16	0·831	0·25	0·796	0·18	0·872	0·17
Q25	0·687	0·17	0·834	0·26	0·789	0·18	0·856	0·18
Q50	0·689	0·18	0·837	0·26	0·786	0·18	0·846	0·19
Q100	0·691	0·19	0·841	0·27	0·783	0·18	0·839	0·20
Q500	0·698	0·21	0·849	0·28	0·778	0·18	0·827	0·22

mean exponents (GMEXP) is not greater than 10% in magnitude and the pattern of their decrease with respect to the flood frequency in each region is very similar. This suggests multiscaling in region 1, whereas simple scaling is suggested in region 2. By contrast, the Uinta basin, with the largest standard errors and with the standard errors increasing with quantiles, gives contradictory results from the two approaches. Therefore, any interpretation of those exponents in terms of simple or multiscaling is open to question.

In view of this analysis, our criterion for data interpretation will be to compare the standard errors of estimate (SE2) for each quantile in a region. If the standard errors remain constant or decrease or are uniformly small, then we will interpret whether the corresponding slope behaviour suggests simple or multiscaling and discuss what physical generating mechanism underlies this behaviour. Of course, more rigorous statistical inference tests of simple versus multiscaling, preferably non-parametric, need to be developed in the future; see Smith (1992) for a maximum likelihood test of a simple versus multiscaling log-normal model.

The first state for consideration is New York. Lumia (1991) gives the full regression equations as well as the equations based on drainage area alone. Table III shows the empirically estimated scaling exponents (EXP) and SE2 for four regions of New York. The SE2 values for all these regions were computed from the errors of prediction using Equation (8). Note that in each region SE2 is small. Thus, as per our earlier discussion, the regression effect does not affect the variation of exponents relative to each other.

Regions within New York with uniform scaling exponents are region 3, the Lower Hudson River Valley, and region 7, south of the centre of Lake Ontario. Region 7 has exponents which actually increase slightly with the return period. Region 2, the Adirondack Province, has exponents which decrease slightly with return

Table IV. Exponents and the standard error of estimate in flood quantile–drainage area relations for four New Mexico regions

Quantiles	Region 1 north-east plains		Region 5 northern mountains		Region 2 north-west plateau		Region 4 south-east plains	
	EXP	SE2	EXP	SE2	EXP	SE2	EXP	SE2
Q2	0·56	0·35	0·912	0·34	0·52	0·38	0·671	0·45
Q5	0·55	0·31	0·920	0·32	0·47	0·33	0·591	0·34
Q10	0·55	0·30	0·924	0·32	0·44	0·31	0·546	0·28
Q25	0·55	0·30	0·929	0·34	0·41	0·30	0·498	0·23
Q50	0·55	0·31	0·933	0·35	0·39	0·30	0·465	0·21
Q100	0·56	0·32	0·936	0·37	0·37	0·30	0·436	0·19
Q500	0·581	0·32	0·940	0·39	0·365	0·32	0·408	0·19

Table V. Exponents and standard error of estimate in flood quantile–drainage area relations for five Utah region

Quantiles	Northern mountains high elevation		Northern mountains low elevation		Uinta Basin S		High Plateaus		Low plateaus	
	EXP	SE2	EXP	SE2	EXP	SE2	EXP	SE2	EXP	SE2
Q2	0·898	0·22	0·741	0·31	0·371	0·34	0·800	0·27	0·506	0·37
Q5	0·887	0·20	0·738	0·29	0·314	0·33	0·740	0·22	0·438	0·33
Q10	0·881	0·20	0·736	0·29	0·289	0·36	0·710	0·22	0·404	0·32
Q25	0·873	0·20	0·732	0·30	0·267	0·40	0·679	0·24	0·369	0·31
Q50	0·868	0·20	0·730	0·30	0·254	0·43	0·660	0·26	0·347	0·31
Q100	0·863	0·20	0·727	0·31	0·244	0·46	0·643	0·29	0·328	0·31
Q500	0·858	0·21	0·725	0·32	0·236	0·49	0·628	0·32	0·311	0·32

period, and region 4, the Catskills, decreases more then other regions in New York (Lumia, 1991). The approximate constancy of the exponents for regions 3, 7 and 2 indicates that simple scaling holds, whereas multiscaling is suggested for region 4.

Maximum seasonal snowfall occurs in the Adirondacks and adjacent Tug Hill, in regions 1 and 2 and in the vicinity of region 7. 'In mid-March as much as 10 in. (25·4 cm) of water content can still remain in the snowpack of the Adirondack Mountains and in the highlands to the east of Lake Ontario.' (Lumia, 1991). The domination of peak discharges by seasonal snowmelt might account for the nearly constant exponents for regions 2 and 7. Why region 3 should have such constant exponents and region 4 have such varying exponents is not understood.

The second state considered is New Mexico. Table IV shows four USGS defined regions. Quantile relations for Q2 to Q500 for regions 1 and 2 are as given in Waltemeyer (1986). Relations for the other two regions were computed by us based on the USGS data. Notice once again that SE2 satisfies the criterion discussed earlier, and therefore the variability or the lack of it of the exponents within a region is not due to the regression effect.

Table IV shows different characteristics in terms of scaling exponents for these regions. Region 5 is a mountainous region and is the headwaters of the Canadian River. Not only are the exponents relatively constant, but the runoff is almost directly related to the drainage area, as all the exponents are greater then 0·9. Waltemeyer states 'in the northern mountain region, floods generally are produced from snowmelt runoff'. Region 1 is downstream from region 5, and its mainstream stations receive flows from region 5. The approximate constancy of the exponents indicates that simple scaling holds for both regions 1 and 5. This supports the hypothesis that in these two regions simple scaling is a property of snowmelt-generated flood peaks. Regions 2 and 4 have generally decreasing exponents with increasing return period of floods. These two regions as well as region 6 between them, which is not shown here, constitute a wide band across the middle of the state and exhibit similar decreases in exponents. Waltemeyer further states that 'Those (floods) in the plains, plateaus, valleys, and deserts generally are produced by rainfall'. The decreasing exponents suggest multiscaling, which supports the hypothesis that, in these two regions with considerable thunderstorm activity, multiscaling is a property of rainfall-generated flood peaks.

The third state considered is Utah (Thomas and Lindskov, 1983). The Utah data shown in Table V are derived from station data provided by the USGS. All five regions have exponents which tend to decrease with the return period. However, the two mountainous regions have the least variability in exponents, a difference of 0·04 and 0·016, because they are most influenced by snowmelt peaks. For the northern mountains high elevation region, Thomas and Lindskov (1983) state that 'most of the floods result from snowmelt. Thunderstorm floods generally are smaller in magnitude than the annual snowmelt peaks.' For the Low Elevation Region they state that 'This is a mixed population flood area where floods may result from snowmelt, thunderstorms, or a combination. The infrequent thunderstorm floods usually will have a higher magnitude than the snowmelt floods'. The plateau regions have the greatest variability in exponent, a difference of 0·172 and 0·195, probably because they are more influenced by thunderstorm activity. According to Thomas and Lindskov, the high plateau region 'is a mixed population flood area and the infrequent thunderstorm floods can have greater magnitudes than snowmelt floods.' For the low plateaus region 'Summer thunderstorms produce most of the large magnitude floods. Snowmelt floods are rare and usually small.' For the Uinta Basin region 'the topography is mostly high plateaus cut by deeply incised streams. All four stream categories can be found in the region. This is a mixed population flood area and thunderstorms will produce the larger floods'. The Uinta Basin is in between the other two pairs of regions. However, as discussed earlier, due to the behaviour of SE2 for this basin, we cannot interpret the pattern of the Uinta exponents. So, except for Uinta, the other two pairs of regions in Utah support our hypothesis that simple scaling is exhibited by snowmelt-generated floods, whereas rainfall-produced floods exhibit multiscaling.

In summary, several observations can be made about the behaviour of the scaling exponents from the above field data:

1. There are similarities and differences in the exponents among states, as well as among regions within each state. Firstly, with regard to the regions having essentially uniform exponents, note that region 5 of New Mexico and the northern mountain high elevation region of Utah are similar, with an exponent of about 0·9. Regions 2 and 7 of New York are similar with an exponent of about 0·8. Region 3 of New York and the northern mountain low elevation region of Utah are similar with an exponent of about 0·7. All of these three groups of regions are different from each other and from region 1 of New Mexico, which has a uniform exponent of 0·56. As floods in all of these regions are largely generated by snowmelt runoff, the similarity and differences among these regions should be related to the causative physical factors such as spatial snowpack patterns (Shook *et al.*, 1993), basin geomorphology, soils and vegetation.
2. With regard to variable exponents, region 2 of New Mexico and the low plateau region of Utah are similar as their exponents vary from about 0·52 to 0·31. The remainder of the regions in these three states with variable exponents are different from each other. These differences should be related to spatial rainfall variability, basin geomorphology, soils and vegetation types.

EFFECT OF BASIN RESPONSE ON THE SCALING EXPONENTS: RESULTS FROM A PRELIMINARY RAINFALL–RUNOFF EXPERIMENT

Empirical analyses of regional flood data in the preceding section support the hypothesis that the major cause of variability or uniformity in scaling exponents via Equations (2) and (3) is related to precipitation variability, i.e. rainfall versus snowmelt. However, the basin response itself might have an effect on the scaling exponents. For example, larger basins average flows from many smaller basins contained within them. Non-synchronization of the peaks from the smaller basins should affect the peak flows for the larger basins. In addition, overbank flows attenuate peak flows, especially if they are large. A simple rainfall–runoff experiment was undertaken to study in a preliminary way the effects of these factors of the basin response on scaling exponents, and to investigate if the multiscaling representation Equation (4) can exhibit these effects.

A 'uniformly nested' basin was constructed using HEC-1 (Hydrologic Engineering Center, 1990). A series of 0·25 square mile (0·65 km^2) basins, each with a main channel length of 2500 ft (760 m) were constructed and combined to generate a 63·75 square miles (165·1 km^2) basin. Each internal link bifurcates into two, has a length of 2500 ft (760 m), and drains a 0·25 square mile area from its two sides. As a basin with m sources has $(2m - 1)$ links, it follows that $63·75 = 0·25(2m - 1)$. Solving for m gives 128 external links and 127 internal links. The peak flows were recorded in the downstream direction at various sub-basins having areas of 0·25 (0·65), 0·75 (1·94), 1·75 (4·53), 3·75 (9·71), 7·75 (20·07), 15·75 (40·79), 31·75 (82·23), and 63·75 square miles (165·1 km^2). Notice that the area of a downstream basin in square miles is obtained by multiplying the upstream basin area by 2 and then adding 0·25 to it. A random rainfall amount was generated from a probability distribution and was assigned uniformly over a five minute interval as a pulse to each sub-basin with no infiltration to study the effect of channel routing alone. Kinematic wave routing was then used to route the flows overland and then through the main channels. The overland flow length was kept at 200 ft (61 m) over all the 0·25 (0·65 km^2) sub-basins. The above procedure was repeated for several hundred rainfall input sequences.

We performed two types of analyses. In the first set of analysis we fixed a recurrence frequency and then observed how the ratios of flood discharge quantiles are related to the ratios of corresponding drainage areas in the downstream direction. The ratios of drainage areas were selected to be $0\cdot75/0\cdot25 = 3$, $1\cdot75/0\cdot75 = 2\cdot33$, $3\cdot75/1\cdot75 = 2\cdot14$, $7\cdot75/3\cdot75 = 2\cdot07$ and so on. The idea was to test how well the log–log linearity between quantiles and drainage areas holds. If the log–log linearity holds, as displayed in Equation (5), then the exponent of the drainage area ratios in the downstream direction should remain constant, as the recurrence frequency is fixed. We observed that the exponent of the drainage area did not stay constant, but increased in the downstream direction. For example, for a 10 year flood it was found that

$$\frac{q_{0\cdot1}(0\cdot75)}{q_{0\cdot1}(0\cdot25)} = \left(\frac{0\cdot75}{0\cdot25}\right)^{0\cdot78} \tag{11}$$

and

$$\frac{q_{0\cdot1}(7\cdot75)}{q_{0\cdot1}(3\cdot75)} = \left(\frac{7\cdot75}{3\cdot75}\right)^{0\cdot9} \tag{12}$$

However, the exponent did not continue to grow and began to stabilize around 0.91 for all drainage areas greater than 7·75 square miles.

For smaller basins the overland flow has the maximum effect in delaying the transmission of the flood wave to the channels. Therefore a smaller fraction of the drainage area contributes to the peak flows and this is reflected in a smaller exponent of the area ratios shown in Equation (11). As the drainage size increases, the channel network geometry and therefore the channel flows become more dominant. This is reflected in a larger exponent.

To investigate if the type of non-linearity displayed in Equations (11) and (12) can be exhibited by the non-linear term in Equation (4), we take a set of hypothetical values of the parameters in Equation (4) for illustration. These are $a_1 = \mu_1 = 0$, $\sigma_1^2 = 1$ and $b^2 - \log A^0 = 1$. Then the ratio of quantiles can be expressed as

$$\frac{q_p(A_1)}{q_p(A_2)} = \exp(-w_{1-p}\{[\log(A_1) + 1]^{1/2} - [\log(A_2) + 1]^{1/2}\}) \tag{13}$$

Numerical computation of the quantile ratio from Equation (13) can be used to express it as an area ratio to an exponent, say $(A_1/A_2)^{\beta(p)}$. The exponent corresponding to $A_1 = 3\cdot75$ and $A_2 = 1\cdot75$ for the 10 year flood i.e. when $w_{0\cdot9} = -1\cdot28$ is $0\cdot24$, whereas the value corresponding to $A_1 = 7\cdot75$ and $A_2 = 3\cdot75$ is $0\cdot21$. This shows a relative decrease in downstream direction. As the subsequent area ratios begin to stabilize around $2\cdot0$, the 'rate of decrease' in the exponent corresponding to $A_1 = 31\cdot75$ and $A_2 = 63\cdot75$ is about $0\cdot17$. These computations show that the non-linearity contained in Equation (4) does not quite exhibit the results of the simulation experiment. Some further computations carried out by us, but not shown here, suggest that the assumption of log-normality is responsible for this discrepancy. Therefore, a log-Levy exponent different from $\frac{1}{2}$ in Equation (4) would be necessary to exhibit an increase in the area exponent in the downstream direction. However, we do not pursue this interesting and potentially important issue any further in this paper and leave it as an open problem for further research.

In the second set of analyses different quantiles corresponding to different return periods were computed, and the exponents of drainage area ratios for different sub-basins were computed in a manner similar to the 10 year return period described earlier. We found that the exponents showed an increase with the return period. For example, for the area ratio of $0\cdot75/0\cdot25$, the exponent increased from $0\cdot78$ for the 10 year flood to $0\cdot88$ for the 100 year flood. As before, the exponent did not continue to grow and gradually stabilized around $0\cdot91$ for all drainage areas greater than 7·75 square miles (20·07 km^2). This behaviour of the area exponent with respect to return period is the opposite to that given by Equation (3). To exhibit this feature by the multiscaling representation Equation (4), for a fixed pair of drainage areas A_1 and A_2, express the right-hand side Equation of (13) by $(A_1/A_2)^{\beta(p)}$ as before. As $A_1/A_2 > 1$, it follows from Equation (13) that a larger value of the exponent $-w_{1-p}$ implies a larger value of $\beta(p)$ and vice versa. As an example, note that $w_{0\cdot9} = -1\cdot28$ and $w_{0\cdot99} = -2\cdot33$. This means that the 100 year flood has a bigger exponent than the 10 year flood. This argument is meant to show numerically that the multiscaling representation Equation (4) for small basins can qualitatively exhibit the feature observed in the simulation experiment.

Muskingum–Cunge routing was used next, with an assumption that overbank flows occurred at about the 50% quantile varying in a downstream direction. The Muskingum–Cunge routing produced results similar to those for the kinematic wave routing. The constant main link length might tend to produce the results shown here because the smaller channels combine in a synchronous manner. Therefore, the next experiment was to produce channel lengths with a uniform distribution from 1250 to 3750 ft (380 to 1140 m) in length. The resulting Muskingum–Cunge routing produced results which were almost identical with those for the constant channel lengths in terms of scaling exponents and were only slightly less in absolute magnitude of peak values.

The results from these kinematic wave and Muskingum–Cunge routing experiments using spatially uniform inputs reinforce the conclusions tentatively drawn from the study of quantile regression results discussed earlier, that the behaviour of the scaling exponents of flood peaks for large basins is mainly determined by the spatial variability of precipitation fields rather than by basin response. As discussed in the next section, current research on the spatial variability of the mesoscale rainfall shows that it has a multiscaling structure. This structure, given by Equation (3), results in a decrease of scaling exponents as the return period increases, which is qualitatively similar to the behaviour observed in rainfall-generated floods in the preceding section. By contrast, the variability of snowfall and snowmelt is spatially damped, which appears to make the scaling exponents uniform for larger basins.

The increase in scaling exponents with respect to return periods shown by our experiment seems to be empirically exhibited in small basins in the Appalachia flood data (Smith, 1992; Gupta *et al.*, 1994). However, this issue needs further tests because the Appalachia flood data from basins smaller than 20 square miles (52 km^2) exhibit a lot more scatter than the larger basins. Our routing experiment suggests that (i) a physical explanation of this behaviour of flood exponents in small basins lies in the structure of basin response and (ii) the multiscaling representation in Equation (4) can be used to model this behaviour of floods via basin response. The theory of geomorphic unit hydrograph via the width function may play a basic part here. This important issue needs to be explored further.

More investigation of real basins and of differences among basins in a region may show that basin response has some effect for large basins, but we do not expect this to be a significant factor. For example, Pitlick (1994) has independently carried out flood frequency analyses in five mountainous regions of the western USA. Most of the basins in each of the five regions analysed by him are larger than 50 km^2. Pitlick finds that 'regional variation in flood frequency distribution reflects largely the variability in precipitation amount and intensity rather than differences in physiography'. In summary, our findings suggest a physical basis for the basic difference in the scaling structure of floods between large and small basins. The basin response-dominated behaviour of the scaling exponents in small basins, and the input-dominated behaviour in large basins imply that there is a critical basin size which serves as a threshold in between these two sets (Gupta *et al.*, 1994). A computation of the threshold area involving the network bifurcation structure, the channel geometry and perhaps the scaling parameters of the precipitation fields constitute an important problem in regional flood frequency and in distributed hydrology.

DISCUSSION OF RELATIONSHIP BETWEEN SCALING IN SPATIAL RAINFALL AND FLOODS

Results in the two preceding sections suggest that we need to examine the spatial scaling structure of rainfall and snowcovers to physically understand the appearance of multiscaling and simple scaling in floods. In this section we will briefly explain that research on the spatial variability of rainfall supports our hypothesis that rainfall-generated floods inherit their multiscaling structure from rainfall. Similar investigations need to be carried out with respect to spatial variability of snowcovers and their scaling properties; see Shook *et al.* (1993) in this regard.

Past attempts at the stochastic modelling of spatial rainfall have evolved along two separate lines. The first group of models have used the observed geometry of the spatial hierarchy in rainfall intensities as their empirical basis. They assume the presence of distinct scales in the spatial organization of rainfall intensities. The early pioneering study on the mathematical theory of cluster point processes to describe this type of rainfall variability began with the work of LeCam (1961). Research along these lines during the last decade and a half has provided new physical insights into the empirical structure of space–time rainfall (see Gupta and Waymire, 1979; Waymire *et al*, 1984; Smith and Karr, 1985; Kavvas *et al.*, 1987; Cox and Isham, 1988). It has also guided research into developing numerical simulations to address issues of rainfall measurement (Valdes *et al.*, 1990) and in conducting further statistical tests of these theories (Phelan and Goodall, 1990). A major difficulty with the point process approach has been its parameterization, because each spatial scale is treated separately and thus has its own set of parameters. Therefore, these theories do not exhibit any scaling invariance. In addition, because these models can typically contain up to 10 parameters, the problems of parameter estimation and their physical interpretations become cumbersome.

The second line of theoretical investigations has been carried out under the assumption that rainfall possesses no characteristic spatial scales. In particular, these models applied the mathematical theory of statistical simple scaling for modeling rainfall (Lovejoy, 1982; Lovejoy and Mandelbrot, 1985; Lovejoy and Schertzer, 1985; Waymire, 1985). However, the assumption of simple scaling has been shown to lead to contradictions between the theoretical results and many empirical features of rainfall, such as the widely observed spatial intermittency or regions of zero rainfall (Kedem and Chiu, 1987; Gupta and Waymire, 1991).

These two sets of theories suggest that there may be another form of scaling invariance which is different from simple scaling, and which preserves the many empirical properties contradicted by simple scaling such as spatial intermittency. A solution to this problem has been addressed by Schertzer and Lovejoy (1987), Lovejoy and Schertzer (1990) and independently by Gupta and Waymire (1990; 1993). These investigations support a common correction to the previous theories based on the random cascade geometry. The spatial rainfall intensities generated by random cascades can be viewed as 'multifractal measures'. For the present purposes it is unnecessary to go into an explanation of the meaning of multifractal measures; the interested readers can refer to Gupta and Waymire (1993) for further details, and Holly and Waymire (1992) for a careful mathematical exposition of the random cascade theory. The ideas underlying random cascades originated in the statistical theories of fully developed turbulence (Kolmogorov, 1941, 1962; Mandelbrot, 1974; Kahane and Peyriere, 1976)

In the present context of floods it is most important to note that multiscaling is intimately connected to random cascades (Gupta and Waymire, 1993: 259). In particular, spatial rainfall generated by random cascades has the feature that the variability of average rainfall over a given region decreases as the scale or size of the region increases. It is precisely this decrease in the spatial variability which results in a decrease in scaling exponents as the return period increases. Equivalently, it is also reflected in a decrease in the coefficient of variation as the size of a region increases. Therefore, in large basins, multiscaling in spatial rainfall seems to be the most plausible physical mechanism for the presence of multiscaling in regional floods generated by rainfall. This issue needs to be investigated in depth to establish a firm physical connection between the scaling exponents of spatial rainfall and regional floods.

FINAL REMARKS

The results of simple scaling and multiscaling analyses of regional flood frequencies predict power law relations between flood quantiles and drainage areas under the assumption of statistical homogeneity (Gupta *et al.*, 1994). We have shown here that the empirically computed exponents of the drainage area show systematic variations within different states of the USA and within different regions of a state. These differences support our hypotheses regarding differences in the scaling structure of snow melt- versus rainfall-generated floods. Moreover, based on a simple rainfall–runoff experiment, and current research on the spatial variability of mesoscale rainfall, we have argued that variations in scaling exponents in small and large basins provide important insights about the differences in physical processes at work. There are many interesting and important questions still to be answered for the development of a general theory of regional flood frequency. Such a theory, among other things, can suggest improvements in the existing methods of regional flood frequency analysis, or can lead to new methods. Some of the questions which need to be addressed are:

1. What measures of space–time precipitation fields, e.g. averages over time or space, are best related to the

scaling properties of regional flood frequency analyses? In particular, what are the relationships between the scaling exponents of rainfall, or snowmelt, and those of regional floods?

2. The USGS often uses mean annual precipitation as a variable in flood frequency analyses in the eastern USA. In the west they often use rainfall intensity, i.e. T-hour, N-year values. This is based on empirical studies to determine statistical significance. How can this difference in response be quantified and applied to a general theory of regional floods based on scaling invariance?

3. All USGS regional flood frequency analyses define a plethora of basin characteristics and use those characteristics in multiple regressions to improve flood prediction. Drainage area invariably is the most and often the only significant variable. As discussed in the third section, many times other variables are found to be statistically significant and used. These types of empirical analyses suggest that the assumption of homogeneity in the theory of scaling invariance needs to relaxed. For example, where the elevation exerts a significant influence on regional floods, there the peak flows should be parameterized by drainage area A, and elevation z, i.e. $Q(A,z)$. This and other types of generalizations incorporating non-homogeneity in regional floods remain to be explored.

4. The difference in climate and the underlying causative and geometrical factors in basin development and geomorphology need to be quantified and incorporated into a general scaling theory of regional floods in small and large basins.

ACKNOWLEDGEMENTS

This research was supported in part by grants from NSF and NASA. We gratefully acknowledge the help of several USGS district offices who sent us their data and flood frequency reports, and the comments by three anonymous refrees on an earlier draft which led to major improvements in this paper.

REFERENCES

Benson, M. A. 1962. 'Factors influencing the occurrence of floods in a humid region of diverse terrain', *US Geol. Surv. Wat. Supply Prof. Pap.* **1580B**, 62.

Cadavid, E. 1988. 'Hydraulic geometry of channel networks: tests of scaling invariance' *MS Thesis*, University of Mississippi.

Cox, D. R., and Isham, V. 1988. 'A simple spatial–temporal model of rainfall', *Proc. R. Soc. London, Ser. A*, **415**, 317–328.

Dawdy, D. R. 1961. 'Variation of flood ratios with size of drainage area', *US Geol. Surv. Res.*, **424-C**, paper C36.

Falconer, K. 1990. *Fractal Geometry Mathematical Foundations and Applications*. Wiley, New York.

Feder, J. 1988. *Fractals.* Plenum Press, New York.

Frisch, U., and Parisi, G. 1985. 'Fully developed turbulence and intermittency' Ghil, M., Benzi, R., and Parisi, G. (Eds), *Turbulence and Predictability and Geophysical Dynamics and Climate Dynamics*. North Holland, Amsterdam. pp. 84–88.

Gupta, V. K., and Dawdy, D. R. 1994. 'Regional analysis of flood peaks: multiscaling theory and its physical basis' in *Advances in Distributed Hydrology*. Water Resources Publication, Highlands Ranch, Co. pp. 149–166.

Gupta, V. K., and Waymire, E. C. 1979. 'A stochastic kinematic study of subsynoptic space–time rainfall', *Wat. Resour. Res.* **15**, 637–644.

Gupta, V. K., and Waymire, E. C. 1989. 'Statistical self-similarity in river networks parameterized by elevation', *Wat. Resour. Res.*, **25**, 463–476.

Gupta, V. K., and Waymire, E. C. 1990. 'Multiscaling properties of spatial rainfall and river flow distributions', *J. Geophys. Res.*, **95**(D3), 1999–2009.

Gupta, V. K., and Waymire, E. C. 1991. 'On log normality and scaling in spatial rainfall averages' in Lovejoy, S. and Schertzer, D. (Eds), *Scaling, Fractals and Nonlinear Variability in Geophysics*. Kluwer, Hingham. pp. 175–183.

Gupta, V. K. and Waymire, E. C. 1993. 'A statistical analysis of mesoscale rainfall as a random cascade', *J. Appl. Meterol.*, **32**, 251–267.

Gupta, V. K., Mesa, O., and Dawdy, D. R. 1994. 'Multiscaling theory of flood peaks: regional quantile analysis', *Wat. Resour. Res.*, **30**(12), 3405–3421.

Hirsch, R. M., and Gilroy, E. J. 1984. 'Methods of fitting a straight line to data: examples in water resources', *Wat. Res. Bull.*, **20**, 705–712.

Holly, R., and Waymire, E. C. 1992. 'Multifractal dimensions and scaling exponents for strongly bounded random cascades', *Ann. Appl. Prob.*, **2**(4).

Hydrologic Engineering Center, US. Army Corps of Engineers, 1990. *HEC-1 Flood Hydrograph Package, User's Manual.*

Kahane, J. P., and Peyriere, J. 1976. 'Sur certaines martingales de Benoit Mandelbrot', *Adv. Math.*, **22**, 131–145.

Kavvas, M. L., Saquib, M. N., and Puri, P. S. 1987. 'On a stochastic description of the time–space behavior of extratropical cyclonic precipitation fields', *Stoch. Hydrol. Hydraul.*, **1**, 37–52.

Kedem, B., and Chiu, L.S. 1987. 'Are rainrate processes self-similar?', *Wat. Resour. Res.*, **23**, 1816–1818.

Kolmogorov, A. N. 1941. 'Local structure of tubulence in an incompressible liquid for very large Reynolds numbers', *C. R. (Dok.) Acad. Sci. URSS*, **30**, 301–305.

Kolmogorov, A. N. 1962. 'A refinement of previous hypothesis concerning the local structure of turbulence in a viscous inhomogeneous fluid at high Reynolds number,' *J. Fluid Mech.*, **13,** 82–85.

LeCam, L. 1961. 'A stochastic descriptions of precipitation in *4th Berkley Symposium on Mathematics Statistics and Probability*. Vol. 3. Univ. of California, Berkley. pp. 165–186.

Lovejoy, S. 1982. 'Area–perimeter relation for rain and cloud areas', *Science*, **216,** 185–187.

Lovejoy, S., and Mandelbrot, B. B. 1985. 'Fractal properties of rain and a fractal model', *Tellus*, **37A,** 209–232.

Lovejoy, S., and Schertzer, D. 1985. 'Generalized scale invariance in the atomsphere and fractal models of rain', *Wat. Resour. Res.*, **21,** 1233–1250.

Lumia, R. 1991. 'Regionalization of flood discharges for rural, unregulated streams in New York excluding Long Island', *USGS Wat. Resour. Invest. Rep.*, **90–4197**.

Mandelbrot, B. B. 1974. 'Intermittent turbulence in self-similar cascades: divergence of high moments and dimension of the carrier', *J. Fluid Mech.*, **62,** 331–358.

National Research Council 1988. *Estimating Probabilities of Extreme Floods: Methods and Recommended Research*. National Academy Press, Washington DC.

Parzen, E. 1960. *Modern Probability Theory and its Application*. Wiley, New York.

Peckham, S. 1995. 'Self-similarity in the 3-D geometry and dynamics of large river basins' *PhD Dissertation*, Program in Geophysics, University of Colorado, Boulder.

Phelan, M. J., and Goodall, C. R. 1990. 'An assessment of a generalized Waymire–Gupta–Rodriguez–Iturbe model for GARP Atlantic Tropical Experimental Rainfall', *J. Geophys. Res.*, **95**(D6), 7603–7615.

Pitlick, J. 1994. 'Relation between peak flows, precipitation, and physiography for five mountainous regions in the western USA', *J. Hydrol.*, **158,** 219–240.

Schertzer, D., and Lovejoy, S. 1987. 'Physical modeling and analysis of rain and clouds by anisotropic scaling multiplicative processes', *J. Geophys. Res.*, **92,** 9693–9714.

Schroeder, M. 1991. *Fractals, Chaos and Power Laws: Minutes from an Infinite Paradise*. Freeman, New York.

Shook, K., Gray, D. M., and Pomeroy, J. M. 1993. 'Geometry of patchy snowcovers' in *Proceedings of 50th Eastern Snow Conference and 61st Western Snow Conference, Quebec City, Quebec*.

Sivapalan, M., Wood, E. F., and Beven, K. J. 1990. 'On hydrologic similarity, 3 a dimensionless flood frequency model using a generalized geomorphic unit hydrograph and partial area runoff generation', *Wat. Resour. Res.*, **26,** 43–58.

Smith, J. A. 1992. 'Representation of basin scale in flood peak distributions', *Wat. Resour. Res.*, **28,** 2993–2999.

Smith, J. A. and Karr, A. F. 1985. 'Parameter estimation for a model of space-time rainfall', *Wat. Resour. Res.*, **21,** 1251–1257.

Thomas, B. E., and Lindskov, K. L., 1983. 'Methods for estimating peak discharge and flood boundaries of streams in Utah', *US Geol. Surv. Wat. Resour. Invest. Rep.*, **83–4129**.

Valdes, J. B., Nakamoto, S., Shen, S. P., and North, G. R., 1990. 'Estimation of multidimensional precipitation parameters by areal estimates of oceanic rainfall', *J. Geophy. Res.*, **95**(D3), 2101–2111.

Waltemeyer, S. D. 1986. 'Techniques for estimating flood-flow frequency for unregulated streams in New Mexico', *USGS Wat. Resour. Invest. Rep.*, **86–4104**.

Waymire, E. 1985. 'Scaling limits and self-similarity in precipitation fields', *Wat. Resour. Res.*, **21,** 1271–1281.

Waymire, E. C., Gupta, V. K., and Rodriguez-Iturbe, I. 1984. 'A spectral theory of rainfall intensity at the meso-β scale', *Wat. Resour. Res.*, **20,** 1453–1465.

7

EFFECTS OF VERTICAL RESOLUTION AND MAP SCALE OF DIGITAL ELEVATION MODELS ON GEOMORPHOLOGICAL PARAMETERS USED IN HYDROLOGY

YEBOAH GYASI-AGYEI*

Laboratory of Hydrology, Free University Brussels, Brussels, Belgium

GARRY WILLGOOSE

Civil Engineering and Surveying Department, The University of Newcastle, Newcastle, Australia

AND

FRANCOIS P. DE TROCH

Laboratory of Hydrology and Water Management, University of Ghent, Ghent, Belgium

ABSTRACT

The advent of digital elevation models (DEMs) has made it possible to objectively extract, calculate and store geomorphological parameters for hydrological modelling at several scales. For a grid-based DEM, the threshold area used to extract the channel network is analogous to the scale of the map produced. In addition to the map scale, the effects of the vertical resolution of the DEM on some frequently used geomorphological parameters in hydrology are examined using high-resolution DEMs of two natural and two artificial catchments. The vertical resolution was varied between 1 cm and 1 m, the most common vertical resolution of DEMs. At a fixed map scale, the mean absolute percentage error in the geomorphological parameters caused by a decrease in vertical resolution is within the range 0–5% for the medium-sized catchments and 0–10% for the small catchments studied. Although it is true that a change in vertical resolution may cause a change in the individual pixel slope, area and topographic index (area/slope), particularly in low relief terrain, their cumulative distributions do not show any significant change with the vertical resolution. The shape of the normalized width function is not very sensitive to the vertical resolution and the map scale. For small catchments order change may occur at different map scales for the different vertical resolution DEMs of the same catchment, causing a significant change in order-related parameters such as Horton ratios. It is suggested that the vertical resolution of the DEM of a catchment be considered satisfactory for most hydrological applications if the ratio of the average drop per pixel and vertical resolution is greater than unity. This ratio criterion could be used to define the minimum pixel area for reliable channel network definition for any given vertical resolution. The minimum pixel area places a lower bound on the horizontal resolution with which a channel network can be extracted from a DEM. These results could potentially be used to assess the adequacy for hydrological purposes of existing and proposed digital elevation databases.

INTRODUCTION

Contour (or topographic) maps and aerial photographs were the primary source of geomorphological information before the 1980s, the former being the most common source. However, it has been demonstrated by Coates (1958), Morisawa (1959), Coffman *et al.* (1972) and Montgomery and Foufoula-Georgiou (1993) that blue line networks from topographic maps miss a substantial proportion of first-, second- and even

---

* Present address: Department of Civil Engineering and Surveying, The University of Newcastle, Newcastle NSW 2308, Australia

third-order streams. Two or three topographic maps of different map scales may be available for a particular catchment and the geomorphological parameters estimated from these maps may lead to an erroneous conclusion about the scale effects on geomorphological parameters. For a grid-based digital elevation model (DEM) the variation of the threshold area (TA) used to extract the channel network produces an effect on the representation of the channel network extent that is analogous to the effect commonly associated with changes in map scale. As a result, DEMs make it possible to extract geomorphological parameters at several map scales in an objective way. This enables the identification of map scale independent geomorphological parameters which could be used in regression models for the prediction of the hydrological response characteristics of a catchment.

The mainstream length–area scaling exponent (MLASEM) has been estimated for natural catchments (Hack, 1957; Gray, 1961; Hjelmfelt, 1988; Robert and Roy, 1990) and related to the fractal dimension of channel networks (Mandelbrot, 1982; Hjelmfelt, 1988; Robert and Roy, 1990). Using three different map scales of the same catchment, Robert and Roy (1990) concluded that the MLASEM is much lower (0·546) at the largest scale than at the smallest scale (0·65). Mesa and Gupta (1987) established that the MLASEM decreases with an increase in catchment size. Indirect estimates of the fractal dimension of channel networks based on the logarithms of Horton ratios have been reported (Rosso et al., 1991; La Barbera and Rosso, 1989; Feder, 1988; Tarboton et al., 1988, 1990; Beer and Borgas, 1993). Some of these relationships have been found to be of little relevance to hydrological systems (Beer and Borgas, 1993). It is not the intention of this paper to estimate fractal dimensions, but only to show the fluctuations of some of the reported relationships with scale.

More recent research seeks to relate the discharge hydrograph to fundamental catchment characteristics that can be easily obtained from topographic maps or DEMs. A major difference between the approaches is the way in which the channels constituting the channel network are grouped. One approach expresses the instantaneous unit hydrograph (IUH) as a function of Horton ratios (e.g. Rodriguez-Iturbe and Valdes, 1979; Gupta et al., 1980; Hebson and Wood, 1982; Kirshen and Bras, 1983; Agnese et al., 1988; Rinaldo et al., 1991). The other models the IUH as a function of the magnitude–diameter distribution reflected in the width function (e.g. Gupta and Waymire, 1983; Troutman and Karlinger, 1985; 1986; Mesa and Mifflin, 1986; Naden, 1992).

The distribution of the topographic index (area/slope) is widely used in physically based runoff production models (e.g. Beven and Kirkby, 1979; O'Loughlin, 1986) and will also be examined in this paper.

The need to standardize the extraction procedure of geomorphological parameters and their screening before use cannot be overemphasized. Channel network extraction procedures from DEMs have been proposed by Tarboton et al. (1991), Montgomery and Dietrich (1989; 1992) and Montgomery and Foufoula-Georgiou (1993). However, there is no generally accepted procedure.

Helmlinger et al. (1993) examined the effects of TA on the morphometric and scaling (Horton ratios and fractal dimension) properties of channel networks. They found that TA considerably affects the estimation of these properties of a channel network and that reporting these values without reference to the TA used is meaningless. A similar analysis is carried out in this paper using smaller TA increments and a wider range of geomorphological parameters to identify map scale independent geomorphological parameters. Geomorphological parameters that are highly correlated with magnitude, which is used as a surrogate variable for TA, are also identified. Most importantly, the effects of the vertical resolution of the DEM data on geomorphological parameters, which have not yet been addressed in published work, are examined.

DEFINITION OF PERTINENT GEOMORPHOLOGICAL PARAMETERS

Many of the following definitions can be found in Horton (1945), Shreve (1966; 1967; 1969), Smart (1972) and Abrahams (1984). The structure of an idealized river network containing no lakes, islands, nor junctions of more than two streams at the same point may be represented by a trivalent planted tree, or channel network. The root of the tree, or the outlet of the channel network, corresponds to the point furthest downstream, sources are the points furthest upstream and the remaining nodes, or junctions, correspond to the points where two streams flow together. A link is a segment of the channel network between two successive junctions

Table I. Basic properties of the DEM data sets

Catchment	Pixel dimension (m)	Vertical resolution (cm)	Elevation			Number of	
			Max. (m)	Min. (m)	Relief (m)	Rows	Columns
Dyfi	50	10	907	6	901	719	600
MCK	20	1	206	113	93	37	60
S40*	1	1	20	10	10	40	40
S200*	1	1	52	10	42	200	200

* Units are dimensionless as generated by SIBERIA.

or between the outlet and the first junction upstream. The terms exterior link and interior link signify links terminating at their upstream ends and junctions, respectively. The magnitude of a link is the number of sources upstream of it, implying that an exterior link has a magnitude of unity and an interior link has a magnitude equal to the sum of the magnitudes of the two links joining at its upstream end. The magnitude of a

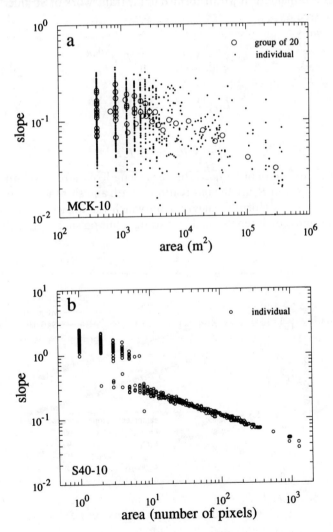

Figure 1. Plot of instantaneous slope versus contributing area

network is that of its outlet link. In these terms an idealized network of magnitude M will have $2M-1$ links, of which M are exterior links and $M-1$ are interior links.

A path is the shortest route between the outlet of a channel network and a source or a junction; thus the idealized channel network of magnitude M has $2M-1$ paths. The topological distance of a given source from the outlet of a channel network is the number of links, one of which is exterior, in the path joining these two points. The diameter of the network is the maximum (over all the sources) of the topological distances. The actual distance of a source from the outlet of the channel network is the sum of the link lengths in the path joining these two points, and the mainstream length of the network is the maximum of such distance. Drainage density is defined as the ratio of total channel length over the catchment area. Overland flow length is the length of flow of water over the ground surface before it becomes concentrated in definite stream channels and it is approximately equal to half the reciprocal of the drainage density.

The Strahler (1952) order, order for short, of a channel, k, is given mathematically, by

$$k = \text{maximum}\,[i,j, \text{integer}(1 + [i+j]/2)] \tag{1}$$

where i and j are the orders of the upstream streams and all exterior links have order unity. Given his ordering, Horton (1945) demonstrated three empirical laws: the law of stream numbers, the law of stream lengths and the law of stream slopes. Mathematically transformed to the framework of Strahler's (1952) ordering scheme, these laws may be expressed as

the law of stream numbers

$$N_i = N_1 R_b^{1-i} = N_1 R_b e^{-i \ln R_b} \tag{2}$$

the law of stream lengths

$$\bar{L}_i = \bar{L}_1 R_l^{i-1} = \bar{L}_1 R_l^{-1} e^{i \ln R_l} \tag{3}$$

the law of stream slopes

$$\bar{S}_i = \bar{S}_1 R_s^{1-i} = \bar{S}_1 R_s e^{-i \ln R_s} \tag{4}$$

where N_i is the number of streams of order i, $\bar{L}_i$ is the mean length of a stream of order i, $\bar{S}_i$ is the mean slope of streams of order i, and R_b, R_l and R_s are the bifurcation, length and slope ratios, respectively. These laws are often represented graphically as Horton's diagrams in which the logarithms of N_i, $\bar{L}_i$ and $\bar{S}_i$ are plotted against i. A geometric progression is represented by a straight line on such a plot. The various ratios are easily

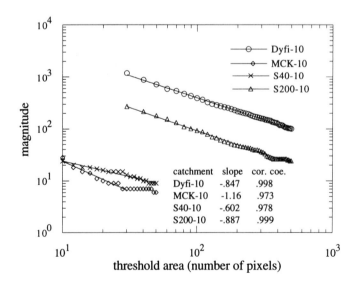

Figure 2. Power law scaling of magnitude with threshold area (regression coefficients are given in the legend)

Table II. Results of power law fit to magnitude and some geomorphological parameters

Parameter	D	D_d	ML	MLA	MLE	MLI
Dyfi-10						
Scaling exponent	0·687	0·456	−0·016	−0·543	−0·522	−0·560
Regression coefficient	0·996	0·997	−0·964	−0·999	−0·992	−0·999
S200-10						
Scaling exponent	0·583	0·436	−0·040	−0·569	−0·562	−0·577
Regression coefficient	0·996	0·996	−0·948	−0·994	−0·984	−0·994

D = Diameter; D_d = drainage density; ML = mainstream length; MLA = mean link length of all links; MLE = mean link length of exterior links; and MLI = mean link length of exterior links.

obtained from the slopes of the straight lines. Schumm (1956) proposed a Horton-type law for catchment area given by

$$\bar{A}_i = \bar{A}_1 R_a^{i-1} = \bar{A}_1 R_a^{-1} e^{i \ln R_a} \qquad (5)$$

where $\bar{A}_i$ is the average catchment area, including tributaries, of order i streams and R_a is the area ratio.

Empirical results indicate that the values for R_b normally range from 3 to 6, between 1·5 and 4 for R_l, between 1·5 and 3 for R_s and between 3 and 6 for R_a (Smart, 1968; 1972). Elsewhere, it has been observed that if the catchment's outlet link does not drain into a higher order link (incomplete channel network), the highest order channel may be far shorter than its true value and thus affect the length ratio (Gyasi-Agyei, 1993). In this instance the number of the lower order channels may fall short and thus affect the bifurcation ratio. The area ratio may also be affected, but the slope ratio normally does not change significantly as the slopes of links near the outlet are generally low. In some instances this may not be possible as the highest order channel may cross an edge of the DEM data set. Hence we neglected the values of the highest order channel during the calculation of the Horton ratios for orders five and higher. In general, the correlation improved when values of the highest order channel were neglected. The values of the bifurcation and length ratios showed a significant change.

The width function is defined as the number of links at a particular distance from the catchment's outlet. A normalized width function (NWF) is derived by dividing distances by the mainstream length and then scaling the ordinates to obtain a unit area under the NWF. Inevitably, the NWF depends on the sampling interval, particularly for catchments whose drainage density is not constant. To take into account each pixel of the

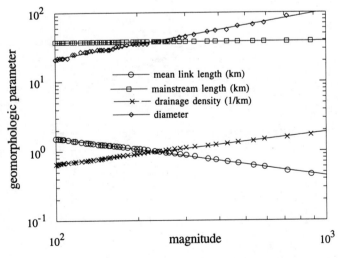

Figure 3. Power law scaling of diameter, drainage density, mean link length and mainstream length with magnitude for Dyfi-10

Table III. Mean and maximum (in parentheses) absolute percentage error caused in geomorpholigical parameters due to reduction in the vertical resolution. S200: 100-1 means reduction in vertical resolution from cm to 1 m of catchment S200. MCK: 100-1 means reduction in vertical resolution from cm to 1 m of catchment MCK

Geomorphological parameter	S200: 100-1	S200: 100-10	S40: 100-1	S40: 100-10	Dyfi: 10-1	MCK: 100-1	MCK: 100-10
Diameter	0·93 (9·09)	0·28 (9·09)	2·45 (20·0)	0·00 (0·00)	6·30 (18·2)	13·8 (22·2)	3·68 (12·5)
Magnitude	0·68 (5·71)	0·21 (3·23)	2·14 (20·0)	0·52 (11·1)	1·33 (3·74)	8·50 (14·3)	3·91 (12·5)
Drainage density	1·00 (1·69)	0·06 (0·24)	1·79 (4·26)	0·10 (0·82)	0·19 (1·00)	5·21 (8·86)	4·47 (8·22)
MLASEM	1·20 (2·88)	0·03 (1·01)	6·91 (78·4)	0·14 (6·92)	1·14 (8·40)	5·09 (10·5)	1·41 (2·51)
MLASEA	1·33 (8·88)	0·32 (4·08)	5·88 (63·7)	0·88 (19·6)	0·75 (2·40)	8·90 (17·7)	6·35 (13·1)
Horton bifurcation ratio	0·26 (2·28)	0·07 (0·96)	1·12 (9·49)	0·25 (5·33)	0·57 (2·55)	7·78 (14·3)	2·97 (12·5)
Horton length ratio	0·93 (5·85)	0·13 (1·29)	2·42 (13·2)	0·55 (17·6)	1·40 (7·07)	19·6 (52·2)	12·0 (56·9)
Horton area ratio	0·27 (2·50)	0·07 (1·59)	1·48 (13·4)	0·49 (17·4)	0·50 (1·81)	5·17 (9·09)	3·19 (9·45)
Horton slope ratio	0·98 (5·93)	0·30 (5·19)	0·00 (0·00)	0·00 (0·00)	1·06 (2·85)	26·7 (82·9)	11·8 (80·2)
Highest order length	3·74 (4·28)	0·98 (2·14)	0·00 (0·00)	0·00 (0·00)	1·02 (1·91)	339· (1975)	189· (1690)
Mean link length	1·54 (5·00)	0·24 (3·42)	3·54 (21·8)	0·50 (9·74)	1·45 (4·00)	6·36 (11·5)	6·54 (11·6)
Mean external link length	1·36 (8·60)	0·34 (6·09)	5·35 (45·4)	1·01 (28·7)	1·75 (5·57)	8·05 (26·5)	8·52 (18·1)
Mean internal link length	2·10 (5·52)	0·18 (3·49)	4·28 (31·6)	1·04 (38·9)	1·26 (3·80)	11·7 (29·9)	5·58 (12·9)
Mainstream length	1·66 (2·15)	0·09 (0·37)	0·43 (3·30)	0·04 (1·91)	0·38 (0·53)	3·87 (5·15)	2·08 (2·29)
Mean path length	0·42 (2·74)	1·09 (1·69)	3·05 (5·34)	0·12 (1·27)	1·07 (2·76)	4·32 (12·0)	3·20 (5·71)
Mean overland flow length	1·02 (1·89)	0·06 (0·83)	1·73 (3·95)	0·11 (1·14)	0·17 (0·54)	5·50 (9·70)	4·71 (8·92)
Normalized width function							
First moment	0·50 (1·13)	0·46 (0·52)	1·36 (3·38)	0·10 (0·48)	0·38 (0·74)	2·17 (3·88)	0·64 (1·95)
Second moment	0·46 (1·63)	0·66 (0·76)	3·28 (7·75)	0·20 (0·90)	0·82 (1·49)	3·31 (7·29)	1·68 (5·51)
Peak	3·21 (9·92)	0·61 (0·75)	9·06 (20·5)	0·20 (1·05)	4·20 (11·9)	14·3 (51·1)	16·5 (47·0)
Distance to peak	2·34 (17·4)	0·30 (3·95)	4·71 (19·7)	0·00 (0·00)	20·2 (136·)	30·6 (59·6)	9·32 (54·9)

Table IV. Some statistics of some geomorphological parameters

Geomorphological parameter	Minimum	Maximum	Mean	Standard deviation	Coefficient of variation
Dyfi-10					
MLASEM	0·539	0·702	0·569	0·028	0·049
MLASEA	0·753	0·861	0·802	0·033	0·041
Horton bifurcation ratio	3·53	4·41	3·97	0·203	0·051
Horton length ratio	1·83	2·17	2·01	0·101	0·050
Horton area ratio	3·60	4·86	4·15	0·332	0·080
Horton slope ratio	2·10	2·64	2·40	0·13	0·055
L_Ω (km)	10·5	20·9	11·4	2·904	0·254
Mainstream length (km)	37·4	38·6	37·9	0·392	0·010
Mean path length (km)	18·9	20·1	19·5	0·314	0·016
Normalized width function					
First moment	0·480	0·505	0·490	0·006	0·012
Second moment	0·288	0·308	0·295	0·005	0·016
Peak	2·25	2·70	2·44	0·101	0·040
Distance to peak	0·308	0·764	0·652	0·184	0·282
S200-10					
MLASEM	0·507	0·669	0·582	0·050	0·086
MLASEA	0·727	0·890	0·819	0·035	0·042
Norton bifurcation ratio	2·78	4·19	3·18	0·337	0·106
Horton length ratio	1·43	2·12	1·64	0·211	0·129
Horton area ratio	3·01	4·54	3·60	0·387	0·107
Horton slope ratio	1·27	1·74	1·39	0·122	0·088
L_Ω	19·1	121	74·1	51·3	0·692
Mainstream length*	271	305	283	9·83	0·035
Mean path length*	159	187	173	6·9	0·040
Normalized width function					
First moment	0·536	0·581	0·555	0·011	0·020
Second moment	0·340	0·385	0·340	0·012	0·035
Peak	1·76	2·38	2·05	0·182	0·089
Distance to peak	0·667	0·813	0·764	0·031	0·040

MLASEM = Mainstream length–area scaling exponent using only nodes on the mainstream; *MLASEA* = mainstream length = length–area scaling exponent using all nodes on the channel network; L_Ω = highest order channel length.
* Units are dimensionless as generated by SIBERIA.

channel network, a sampling interval equal to the minimum of the pixel dimension was used. The first and second moments about the origin (i.e. the catchment's outlet), the peak and distance to peak of the NWF are considered in the analysis.

DATA AND METHODOLOGY

For this study high-resolution DEM data sets must be available. The DEM data sets of two natural catchments, the Dyfi at Dyfi Bridge (471·3 km^2) and Middle Creek (0·44 km^2), and two artificial catchments, S40 and S200, are used.

The artificial catchments were generated by the SIBERIA model (Willgoose *et al.*, 1989; 1991a; 1991b). This model generates catchments with geomorphological characteristics similar to those of natural catchments, in particular the area–slope and area–slope–elevation relationships (Willgoose *et al.*, 1991c; Willgoose, 1994). With this model it is possible to completely control the spatial variation of erosivity on the catchment. However, the artificial catchments are of uniform erosivity with fluvial-dominated sediment transport.

The Dyfi catchment is situated in North Wales, UK, and drains south-westwards from the highest region

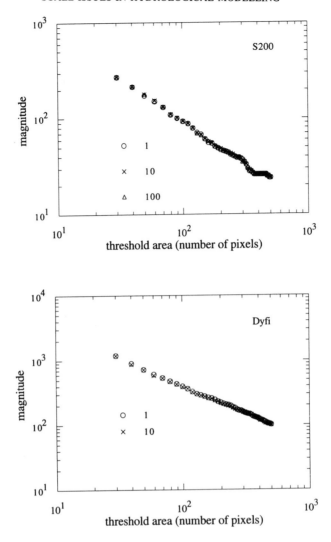

Figure 4. Sensitivity of power law scaling of magnitude with threshold area to vertical resolution. Results from DEMs of vertical resolution of m, dm and cm are labelled 1, 10 and 100, respectively

along the spine of the Cambrian Mountains. The DEM data of the Dyfi catchment were obtained from the Institute of Hydrology, Wallingford, UK, and is of good quality.

The raw elevation data of the Middle Creek catchment (MCK), Pokolbin region of the Hunter valley, Australias was obtained for a tourist development from a ground survey using stadia (Willgoose, 1994). This irregularly spaced data with an average spacing of 17 m was interpolated into a 20 m grid. The SIBERIA model was used to process the DEM data, producing depressionless DEM data, with the elevation data to the nearest centimetre.

For each DEM, lower vertical resolution DEMs were derived by successively truncating the last digit until the DEM to the nearest metre was obtained, as most of the commercially available DEM data are to the nearest metre. As an example, if a pixel in the highest resolution DEM has an elevation value of, say, 15·158 m, its value in centimetre resolution DEM would be 15·15 m and in metre resolution DEM would be just 15 m. So, in total, three DEM data sets of different vertical resolution were derived for the two simulated and the MCK catchments and only two for the Dyfi catchment. From here onwards extensions -1, -10, -100 associated with data or figures indicate vertical resolution truncated to the nearest metre, to the nearest decimetre and to the nearest centimetre, respectively. Table I provides the basic properties of the DEM data sets.

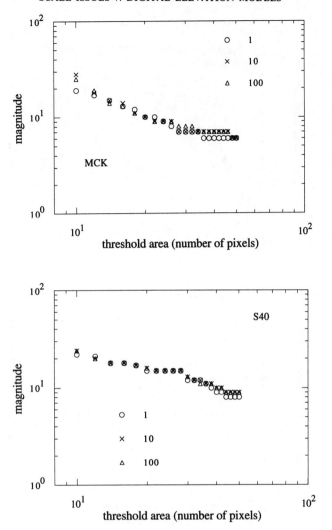

Figure 4. (Continued)

The Jenson and Domingue (1988) algorithms, available as public domain FORTRAN programs, were used to condition the DEM data sets, producing a depressionless DEM, a steepest descent single flow direction matrix based on eight cardinal directions and a flow accumulation matrix. Figure 1 is a plot of instantaneous slope (ratio of elevation difference between a pixel and the next in the steepest descent flow direction to the distance between them) versus contributing area of the catchments MCK-10 and S40-10. The homogeneity and heterogeneity in slope for a given area exhibited in S40 and MCK, respectively, are apparent.

Channel networks were extracted at TAs between 30 and 500 pixels in increments of 10 pixels for the Dyfi and S200, and between 10 and 50 pixels in increments of 2 pixels for the MCK and S40 catchments. The following geomorphological parameters were estimated at each TA for all the derived DEM data sets; magnitude (M); diameter (D); drainage density (D_d); mean link length of all links (MLA), of exterior links (MLE) and of interior links (MLI); mean path length (MPL); mainstream length–area scaling exponent using only nodes on the mainstream length ($MLASEM$) and using all nodes on the catchment ($MLASEA$); order (Ω); Horton bifurcation (R_b), length (R_l), area (R_a), and slope (R_s) ratios; first and second moments, peak and distance to peak of the normalized width function (NWF); highest order length (L_Ω); and the mainstream length (ML). FORTRAN programs have been developed to calculate these parameters for a given channel network.

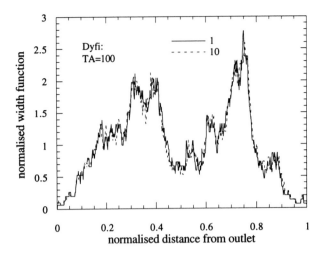

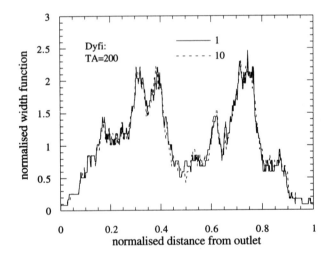

Figure 5. (a) Sensitivity of the normalized width function of vertical resolution (1 corresponds to m and 10 to dm resolutions)—Dyfi catchment.

RESULTS AND DISCUSSION

Figure 2 is a plot of TA versus magnitude for the four catchments. It is observed that magnitude exhibits a power law scaling with TA. Magnitude was therefore used as a surrogate variable for TA or map scale. At very high TAs an increase in TA may not cause a decrease in magnitude, resulting in a slight departure from the straight line at high TAs. Next, the parameters that are highly correlated with magnitude are determined. After screening of possible curve types, a power law was fitted to magnitude and mean link length, drainage density, diameter and mainstream length. These parameters show a positive or negative trend with magnitude. Typical results are shown in Table II for the Dyfi-10 and S200-10 catchments. It is observed in Table II that the mean link length (of all links, exterior links or interior links), drainage density and diameter are highly correlated with magnitude, with significant scaling exponents. Hence these parameters show power law scaling with TA as shown in Figure 3. Similar results were obtained by Helmlinger *et al.* (1993). Hence these parameters can be represented without undue loss of accuracy by magnitude. Mean overland flow length can also be represented by magnitude as it has one to one mapping with drainage density. As a result of very low scaling exponents, mainstream length can be considered as a map scale independent parameter,

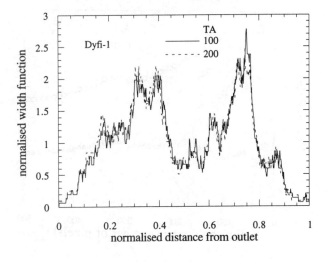

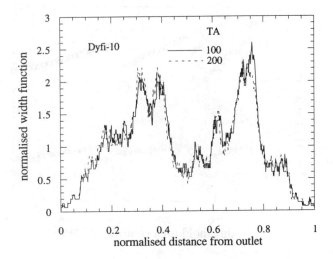

Figure 5. (b) Sensitivity of the normalized width function to threshold area (TA)—Dyfi catchment

supported by its low coefficient of variation (ratio of standard deviation to mean) given in Table IV. The other geomorphological parameters listed in Table III did not show any trend with magnitude.

To investigate the effects of the vertical resolution on the geomorphological parameters, the highest vertical resolution DEM was considered to be the 'truth'. The results obtained using this DEM were compared with those obtained from lower vertical resolution DEMs. Figure 4 depicts the log–log plot of magnitude versus TA for the four catchments using all the derived DEMs. It is seen that points of different vertical resolution DEMs of the same catchment cannot be distinguished. The correlation between magnitude and the parameters shown in Figure 3 did not change with change in the vertical resolution.

Channel networks of the same TA were extracted from the different DEMs of the same catchment. The absolute percentage error of a particular parameter caused by a decrease in vertical resolution was then calculated. Table III presents the mean and maximum absolute percentage error for the range of TAs analysed. For the larger catchments, Dyfi and S200, the mean errors fall below 5% and the maximum below 12%, with the exception of the distance to peak of the NWF of the MCK catchment. A maximum error of 18% in diameter of the Dyfi catchment was observed at TA of 440 pixels. A mean error in the range of 0–10% and a maximum error in the range 0–20% was observed for the smaller catchments, except diameter, bifurcation and length ratios, highest order channel length and peak and distance to peak of the NWF. The

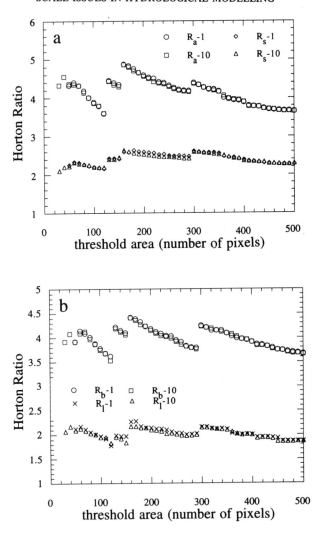

Figure 6. Vertical resolution and map scale effects on Horton ratios: Dyfi (a and b) and S200 (c and d) catchments. Results from DEMs of vertical resolution of m, dm and cm are labelled 1, 10 and 100, respectively

increase in the error for smaller catchments is primarily due to the complexity of the channel network, particularly at small TA.

Figure 5 shows the sensitivity of the NWF of the Dyfi catchment to vertical resolution and TA. In general, the shape of the NWF is not very sensitive to the vertical resolution and TA. This fact is reflected in the low coefficient of variation of the moments of the NWF, as seen in Table IV.

The variation of Horton ratios with scale is presented in Figure 6 for the Dyfi and S200 catchments. This figure illustrates that there are no significant effects of vertical resolution on the Horton ratios. For the small catchments studied, order change occurred at different TAs for the DEMs of the same catchment, causing a significant change in the Horton ratios at the same TA, as demonstrated in Figure 7 for the bifurcation ratio. The change in order is primarily due to the removal of key first-order channels which cause reductions in the order of the adjoining channels, which propagate downstream through the channel network. Table IV contains typical statistics of the Horton ratios for the Dyfi-10 and S200-10 catchments. It is seen that the range of the Horton ratios falls within the observed ranges of natural catchments.

For the catchments studied, the MLASEM varies between 0·5 and 0·7, with a mean of about 0·57, within the range of TAs analysed. This mean value is close to the value of 0·568 given by Gray (1961) for natural

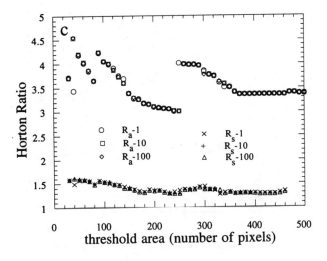

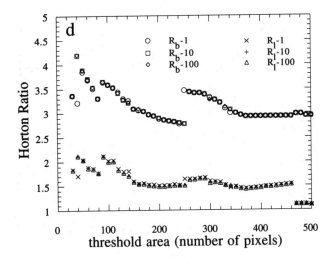

Figure 6. (Continued)

catchments. Hence if the fractal dimension is twice the MLASEM (Mandelbrot, 1982), then it would vary between 1 and 1·4 depending on the map scale used. Examining the variation of MLASEM for 23 catchments at three different map scales (1 : 2000; 1 : 50 000; and 125 000), Robert and Roy (1990) found that MLASEM is much lower (0·546) at the largest scale than at the smallest scale (0·65). Mesa and Gupta (1987) observed that the MLASEM remains rigid at 0·6 for catchments less than 20 000 km^2, changes suddenly to 0·5 for catchments between 20 000 km^2 and 25 000 km^2 and finally drops to 0·47 for larger catchments. The current work did not show any trend of the MLASEM with map scale. Estimation of the MLASEM using all the nodes on the catchment instead of only points on the mainstream channel produced a variation between 0·7 and 0·9 with a mean of 0·8.

The coefficient of variation of the Horton ratios, MLASEM and MLASEA, are fairly low (Table IV) and thus the mean could be used as a representative value of the catchment. However, the mean values may be different depending on the range and increments of TAs used in the analysis. As for the same catchment these parameters span at least half the possible range of values for natural catchments depending on map scale, it is difficult to use them to distinguish between catchments. At best they may be used as indices to test artificial catchments. They may be used effectively if the problem of identifying channel networks from DEMs is resolved.

Of the parameters examined, mainstream length and the mean path length appear to be the most

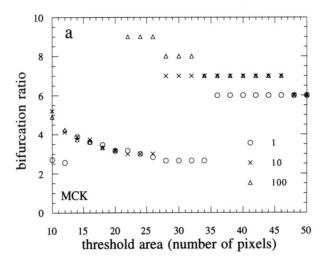

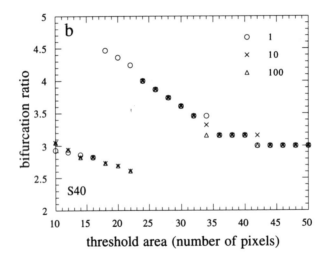

Figure 7. Change of bifurcation ratio (a and b) and order (c and d) with vertical resolution and threshold area. Results from DEMs of vertical resolution of m, dm and cm are labelled 1, 10 and 100, respectively

scale-independent as they have the lowest coefficient of variation (Table IV) and a wider range of possible values for natural catchments.

Some physically based runoff production models use the topographic index (area/slope), pioneered by Beven and Kirkby (1979) and O'Loughlin (1986), to determine locations on the catchment that are saturated during a storm event. Although it is true that a change in the vertical resolution may cause a change in the individual pixel slope and area (particularly in low relief terrain), the cumulative distribution of the area, slope and the topographic index do not show any significant change with the vertical resolution as depicted in Figure 8 for the small catchments. In Figure 8, the instantaneous slopes of all pixels on the catchments were used.

From the analysis carried out so far in this paper, the vertical resolution does not seem to be very important, at least down to the nearest metre. But how far can we truncate elevation values before the geomorphic statistics fall apart? To address this issue the DEM data were truncated to the nearest 10 m. Figure 9 depicts

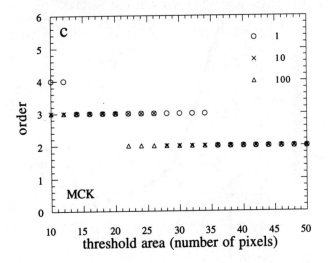

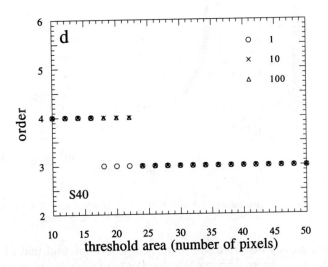

Figure 7. (Continued)

the channel network of the MCK catchment at four different vertical resolutions. It is seen that at a vertical resolution of 10 m the channel network of this catchment bears little resemblance to the higher resolution channel networks. The S200 catchment behaved similarly at a vertical resolution of 10 m. However, the channel network of the Dyfi catchment remained largely undistorted, even at a 10 m vertical resolution (Figure 10).

The drop per pixel in metres averaged over the whole catchment was estimated as 16.3, 3.1, 2.1 and 1.7, respectively, for the Dyfi, MCK, S200 and S40 catchments. These values were the same for the different vertical resolution DEMs. Note that a vertical resolution of 10 m is not possible for the S40 catchment as it has a relief of only 10 m. The drop is the elevation difference between a pixel and the next in the steepest descent flow direction. Only the Dyfi catchment has its average drop per pixel greater than 10 m and this explains why its channel network remained undistorted at the vertical resolution of 10 m. In fact, the channel network of the

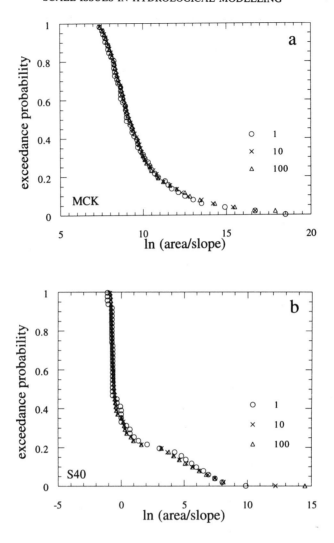

Figure 8. Vertical resolution effects on the distribution of Beven–Kirkby topographic index. Results from DEMs of vertical resolution m, dm and cm are labelled 1, 10 and 100, respectively

MCK catchment became grossly distorted at 5 m vertical resolution and that of Dyfi catchment at 50 m. Hence it appears that there is a threshold value less than unity of the ratio of the average drop per pixel and vertical resolution below which the channel network becomes unacceptably distorted. We suggest conservatively that the vertical resolution of the DEM of a catchment be considered satisfactory for most hydrological applications if the ratio of the average drop per pixel and vertical resolution is greater than unity.

The average drop per pixel of a catchment of a given vertical resolution will decrease with pixel area. Hence the ratio of the average drop per pixel and vertical resolution criterion could be used to define the minimum pixel area for reliable channel network definition for any given vertical resolution. We hypothesize that a steeper slope catchment would require a smaller minimum pixel area than a lower slope catchment. The minimum pixel area places a lower bound on the horizontal resolution with which a channel network can be extracted from a DEM.

CONCLUSIONS

The variation of the TA used to extract channel networks from DEMs produces an effect on the representation of the channel network extent that is analogous to the effects commonly associated with

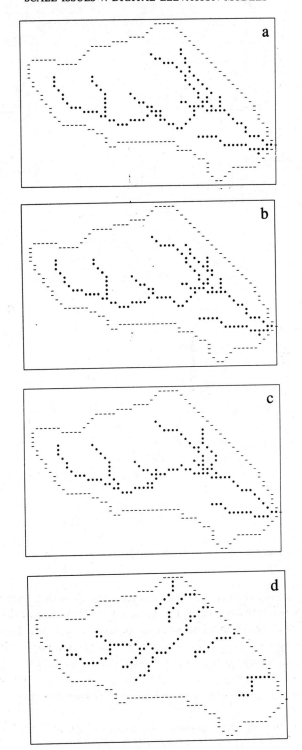

Figure 9. Channel networks of MCK catchment (TA = 20 pixels). Vertical resolution is (a) 0·01, (b) 0·1, (c) 1 and (d) 10 m

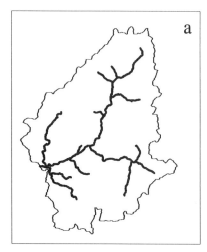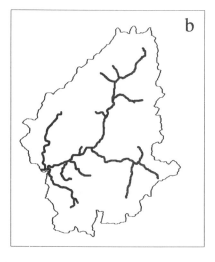

Figure 10. Channel networks of Dyfi catchment (TA = 5000 pixels). Vertical resolution is (a) 0·1 and (b) 10 m

changes in map scale. Hence TA can be used as a surrogate variable for map scale. The effects of map scale and the vertical resolution of the DEM on some frequently used geomorphological parameters in hydrology have been examined using high-resolution DEMs of two natural and two artificial catchments. Lower vertical resolution DEMs were derived by successively truncating the last digit of the elevation value until the DEM reaches a vertical resolution of a metre, the most common vertical resolution of commercially available DEMs.

As already demonstrated by Helmlinger *et al.* (1993), channel network magnitude, diameter, mainstream length, drainage density and mean link length exhibit power law scaling with map scale. It is shown that the channel network magnitude, which could be used as a surrogate variable for map scale, can represent mean link length, diameter, mainstream length and drainage density in regression models. The scaling exponent of the mainstream length is, however, very small. The other geomorphological parameters analysed did not show any trend with magnitude or map scale.

At a fixed map scale, the mean absolute percentage error in the geomorphological parameters caused by a decrease in vertical resolution is within the range 0–5% for the medium catchments and 0–10% for the small catchments studied. However, for small catchments order change may occur at different map scales for the different vertical resolution DEMs of the same catchment, causing a significant change in order-related parameters such as Horton ratios. It is observed that for the same catchment the Horton ratios and mainstream length–area scaling exponent may vary between more than half the range of observed values for natural catchments depending on the map scale. Until the problem of channel network extraction from the DEMs is resolved, these parameters cannot be used to distinguish between catchments. However, they may be used as indices to test artificial catchments. Of the geomorphological parameters examined, the mainstream length and the mean path length appear to be the most scale-independent.

A change in vertical resolution may cause a change in the individual pixel slope, area and topographic index (area/slope), particularly in low relief terrain, but their cumulative distributions do not show any significant change with the vertical resolution. The shape of the normalized width function is not very sensitive to vertical resolution and TA.

It is suggested conservatively that the vertical resolution of the DEM of a catchment be considered satisfactory for most hydrological applications if the ratio of the average drop per pixel and vertical resolution is greater than unity. This ratio criterion could be used to define the minimum pixel area for reliable channel network definition for any given vertical resolution. The minimum pixel area places a lower bound on the horizontal resolution with which a channel network can be extracted from a DEM. These results could potentially be used to assess the adequacy for hydrological purposes of existing and proposed digital elevation databases.

ACKNOWLEDGEMENTS

Comments by G. Kuczera were very helpful. We thank D. B. Boorman and D. Morris of the Institute of Hydrology, Wallingford, UK, for providing the Dyfi catchment DEM data set.

REFERENCES

Abrahams, A. D. 1984. 'Channel networks: a geomorphological perspective', *Wat. Resour. Res.*, **20**, 161–188.

Agnese, C., D'Asaro, F., and Giordana, G. 1988. 'Estimation of the time scale of the geomorphologic instantaneous unit hydrograph from effective streamflow velocity', *Wat. Resour. Res.*, **24**, 969–979.

Beer, T., and Borgas, M. 1993. 'Horton laws and the fractal nature of streams', *Wat. Resour. Res.*, **29**, 1475–1487.

Beven, K., and Kirkby M. J. 1979. 'A physically based, variable contributing area model of basin hydrology,' *Hydrol. Sci. Bull.*, **24**, 43–69.

Coates, D. R. 1958. 'Quantitative geomorphology of small drainage basins of southern Indiana', *Tech. Rep. 10*, Department of Geology, Columbia University, New York.

Coffman, D. M., Keller, E. A., and Melhorn, W. N. 1972. 'New topologic relationship as an indicator of drainage network evolution', *Wat. Resour. Res.*, **8**, 1497–1505.

Feder, J. 1988. *Fractals*, Plenum Press, New York.

Gray, D. M. 1961. 'Interrelationship of watershed characteristics', *J. Geophys. Res.* **66**, 1215–1223.

Gupta, V. K., and Waymire E. 1983. 'On the formulation of an analytical approach to hydrologic response and similarity at the basin scale', *J. Hydrol.*, **65**, 95–123.

Gupta, V. K., Waymire, E., and Wang, C. T. 1980. 'A representation of an instantaneous unit hydrograph from geomorphology', *Wat. Resour. Res.*, **16**, 855–862.

Gyasi-Agyei, Y. 1993. 'Geomorphologic investigations for catchment hydrologic response modelling using a digital elevation model', *PhD Dissertation*, Free University Brussels, 229pp.

Hack, J. T. 1957. 'Studies of longitudinal stream profiles in Virginia and Maryland', *US Geol. Surv. Prof. Pap.*, **292B**, 45–97.

Hebson, C., and Wood, E. F. 1982. 'A derived flood frequency distribution using Horton order ratios', *Wat. Resour. Res.*, **18**, 479–486.

Helmlinger, K. R., Kumar, P., and Foufoula-Georgiou, E. 1993. 'On the use of digital elevation model data for Hortonian and fractal analyses of channel networks', *Wat. Resour. Res.*, **29**, 2599–2613.

Hjelmfelt, A. T. Jr 1988. 'Fractals and the river-length catchment area ratio', *Wat. Resour. Bull.*, **24**, 455–459.

Horton, R. E. 1945. 'Erosional developments of streams and their drainage basins; hydrological approach to quantitative morphology', *Bull. Geol. Soc. Am.*, **56**, 275–370.

Jenson, S. K., and Domingue, J. O., 1988. 'Extracting topographic structure from digital elevation data for geographic information system analysis', *Photogr. Engin. Remote Sensing*, **54**, 1593–1600.

Kirshen, D. M., and Bras, R. L. 1983. 'The linear channel and its effects on the geomorphologic instantaneous unit hydrograph, *J. Hydrol.*, **65**, 175–208.

La Barbera, P., and Rosso, R. 1989. 'On the fractal dimension of stream networks', *Wat. Resour. Res.*, **24**, 735–741.

Mandelbrot, B. B. 1982. *The Fractal Geometry of Nature*. Freeman, New York.

Mesa, D. J., and Gupta V. K. 1987. 'On the main channel length-area relationship for channel networks', *Wat. Resour. Res.*, **23**, 2119–2122.

Mesa, D. J., and Mifflin, E. R. 1986. 'On the relative role of hillslope and network geometry in hydrologic response', in: Gupta, V. K., I. Rodriguez-Iturbe and E. F. Wood (Eds), *Scale Problems in Hydrology*. Reidel, Dordrecht. pp. 1–17.

Montgomery, D. R., and Dietrich, W. E. 1989. 'Source areas, drainage density, and channel initiation', *Wat. Resour. Res.*, **25**, 1907–1918.

Montgomery, D. R., and Dietrich, W. E. 1992. 'Channel initiation and the problem of landscape scale', *Science*, **255**, 826–830.

Montgomery, D. R., and Foufoula-Georgiou, E. 1993. 'Channel network source representation using digital elevation models', *Wat. Resour. Res.*, **29**, 3925–3934

Morisawa, M. E. 1959. 'Relation of quantitative geomorphology to stream flow in representative watersheds of the Appalachian plateau province', *Tech. Rep. 20*, Department of Geology, Columbia Univ., New York.

Naden, P. S. 1992. 'Spatial variability in flood estimation for large catchments; the exploitation of channel network structure', *Hydrol. Sci. J.*, **37**, 53–71.

O'Loughlin, E. M. 1986. 'Prediction of surface saturation zones in natural catchments by topographic analysis', *Wat. Resour. Res.*, **22**, 794–804.

Rinaldo, A., Marani, A., and Rigon R. 1991. 'Geomorphologic dispersion', *Wat. Resour. Res.*, **27**, 513–525.

Robert, A., and Roy, A. G. 1990. 'On the fractal interpretation of the mainstream length–drainage area relationship', *Wat. Resour. Res.*, **26**, 839–842.

Rodriguez-Iturbe, I. and Valdes J.B. 1979. The geomorphologic structure of hydrologic response', *Wat. Resour. Res.*, **15**, 1409–1420.

Rosso, R., Bacchi B., and La Barbera P. 1991. 'Fractal relation of mainstream length to catchment area in river networks', *Wat. Resour. Res.*, **27**, 381–387.

Schumm, S. A. 1956. 'The evolution of drainage systems and sloped in badlands at Perth Amboy, New Jersey,' *Bull. Geogr. Soc. Am.*, **67**, 597–646.

Shreve, R. L. 1966. 'Statistical laws of stream numbers', *J. Geol*, **74**, 17–37.

Shreve, R. L. 1967. 'Infinite topologically random channel networks', *J. Geol.*, **75**, 178–186.

Shreve, R. L. 1969. 'Stream length and areas in topological random channel networks', *J. Geol.* **77**, 397–414.

Smart, J. S. 1968. 'Statistical properties of stream lengths', *Wat. Resour. Res.*, **4**, 1001–1014.

Smart, J. S. 1972. 'Channel networks', *Adv. Hydrosci.*, **8**, 305–346.

Strahler, A. N. 1952. 'Hypsometry (area-altitude) analysis of erosional topography', *Bull. Geol. Soc. Am.* **63**, 117–1142.
Tarboton, D. G., Bras, R. L., and Rodriguez-Iturbe, I. 1988. 'The fractal nature of river networks', *Wat. Resour. Res.*, **24**, 1317–1322.
Tarboton, D. G., Bras, R. L., and Rodriguez-Iturbe, I. 1990. 'Comment on "On the fractal dimension of stream networks' by Paolo La Baebera and Renzo Rosso', *Wat. Resour. Res.*, **26**, 2243–2244.
Tarboton, D. G., Bras, R. L., and Rodriguez-Iturbe, I. 1991. 'On the extraction of channel networks from digital elevation models', *Hydrol. Process.*, **5**, 81–100.
Troutman, B. M., and Karlinger, M. R. 1985. 'Unit hydrograph approximations assuming linear flow through topological random channel network,' *Wat. Resour. Res.*, **21**, 743–754.
Troutman, B. M., and Karlinger, M. R. 1986. 'Averaging properties of channel networks using methods in stochastic branching theory; in: Gupta, V. K., I. Rodriguez-Iturbe and E. F. Wood (Eds), *Scale Problems in Hydrology*. Reidel, Dordrecht. pp. 185–216.
Willgoose, G. R. 1994. 'A physical explanation of an observed area–slope–elevation relationship for catchments with a declining relief', *Wat. Resour. Res.*, **30**, 151–159.
Willgoose, G. R., Bras, R. L., and Rodriguez-Iturbe, I. 1989. 'A physically based channel network and catchment evolution model', *TR322*, Department of Civil and Environmental Engineering, MIT, Boston.
Willgoose, G. R., Bras, R. L., and Rodriguez-Iturbe, I. 1991a. 'A physically based channel network and catchment evolution model, 1, theory', *Wat. Resour. Res.*, **27**, 1671–1684.
Willgoose, C. R., Bras, R. L., and Rodriguez-Iturbe, I. 1991b. 'A physically based channel network anc catchment evolution model, 2, nondimensionalisation and applications', *Wat. Resour. Res.*, **27**, 1685–1696.
Willgoose, G. R., Bras, R. L., and Rodriguez-Iturbe, I. 1991c. 'A physical explanation of an observed link area-slope relationship', *Wat. Resour. Res.* **27**, 1697–1702.

8

A PROCESS-BASED MODEL FOR COLLUVIAL SOIL DEPTH AND SHALLOW LANDSLIDING USING DIGITAL ELEVATION DATA

WILLIAM E. DIETRICH AND ROBERT REISS

Department of Geology and Geophysics, University of California, Berkeley, CA 94720, USA

MEI-LING HSU

Department of Geography, Univesity of California, Berkeley, CA 94720, USA

AND

DAVID. R. MONTGOMERY

Department of Geological Sciences, University of Washington, WA 98195, USA

ABSTRACT

A model is proposed for predicting the spatial variation in colluvial soil depth, the results of which are used in a separate model to examine the effects of root strength and vertically varying saturated conductivity on slope stability. The soil depth model solves for the mass balance between soil production from underlying bedrock and the divergence of diffusive soil transport. This model is applied using high-resolution digital elevation data of a well-studied site in northern California and the evolving soil depth is solved using a finite difference model under varying initial conditions. The field data support an exponential decline of soil production with increasing soil depth and a diffusivity of about $50\,cm^2/$ yr. The predicted pattern of thick and thin colluvium corresponds well with field observations. Soil thickness on ridges rapidly obtain an equilibrium depth, which suggests that detailed field observations relating soil depth to local topographic curvature could further test this model. Bedrock emerges where the curvature causes divergent transport to exceed the soil production rate, hence the spatial pattern of bedrock outcrops places constraints on the production law.

The infinite slope stability model uses the predicted soil depth to estimate the effects of root cohesion and vertically varying saturated conductivity. Low cohesion soils overlying low conductivity bedrock are shown to be least stable. The model may be most useful in analyses of slope instability associated with vegetation changes from either land use or climate change, although practical applications may be limited by the need to assign values to several spatially varying parameters. Although both the soil depth and slope stability models offer local mechanistic predictions that can be applied to large areas, representation of the finest scale valleys in the digital terrain model significantly influences local model predictions. This argues for preserving fine-scale topographic detail and using relatively fine grid sizes even in analyses of large catchments.

INTRODUCTION

In steep, soil-mantled landscapes, shallow landsliding of the soil can generate debris flows which scour low-order channels, deposit large quantities of sediment in higher order channels and, in urbanized settings, destroy property and kill people (e.g. Costa and Wieczorek, 1987; Selby, 1993). The practical significance of shallow landsliding has motivated many different kinds of approaches to mapping the potential hazard in a watershed (see review in Montgomery and Dietrich, 1994). One approach that seems particularly promising is to use digital elevation data and simple coupled hydrological and slope stability models to delineate those areas most prone to instability (Okimura and Kawatani, 1987; Dietrich *et al.*, 1992; 1993; Wu, 1993; Montgomery and Dietrich, 1994).

Soil thickness strongly affects relative slope stability, yet the spatial variation in soil thickness in land-slide-prone areas is rarely estimated (two exceptions are Okimura, 1989; DeRose *et al.*, 1991; 1993). Soils are typically thin to absent on sharply defined ridges and thickest in unchannelled valleys. Vegetation provides a root strength to the soil, and on steep lands with organic-rich, low-density soils, this root strength may dominate or provide a significant portion of the total strength of the material. Vegetation can root through thin soils into the underlying bedrock typically found on ridges and side slopes and provide considerable strength. In thick soils typical of unchannelled valleys, slope instability is favoured because failure planes can form below the rooting depth and the topography forces subsurface flow convergence and elevated pore pressures (e.g. Dietrich and Dunne, 1978; Reneau, 1988; Crozier *et al.*, 1990).

Land-use and climate change modify vegetation. The local soil depth must be known to understand the influence of changes in vegetation on slope stability, yet such information is rarely available and it is impractical to measure for even a modest sized watershed.

Furthermore, soil thickness affects the availability of soil moisture, the relative role of subsurface to overland flow (e.g. Dunne, 1978) and, therefore, the general hydrological response of a landscape. It introduces a spatially organized influence on runoff processes through its dependence on topography, yet a lack of detailed field information inhibits incorporating these effects in models. In addition, by breaking down bedrock into smaller erodible sized particles, soil generation strongly influences the rate of landscape evolution (e.g. Kirkby, 1985; Anderson and Humphrey, 1989). On a practical level, land management decisions are ultimately made on the local level and it would be desirable to account for the local influence of soil thickness on slope stability. Hence there would be considerable value (for these and other problems) in a model that predicts the general spatial distribution of soil depth across a landscape.

Two kinds of models have been proposed to predict the spatial pattern of soil characteristics. Process-based models are few and only one appears to have attempted to relate a theoretical prediction with field observations. Ahnert (1970) described the results of a computer simulation which solves for the local 'waste cover' thickness as a mass balance between waste production and slope-dependent transport removal. He estimated waste cover thickness variation along several hillslope profiles in North Carolina and found that he could explain with his model the variation in cover thickness with local slope and distance from the divide. The most thorough model is that proposed by Kirkby (1985). He developed a model for the evolution of regolith-mantle slopes that determines the spatial distribution of a 'soil deficit', or the amount of original parent material remaining on a hillslope, as influenced by rock type and climate. This model, although instructive about the influence of weathering on slope evolution, requires a large amount of hydrological, mechanical and geochemical information to be applied to a specific site. Kirkby also only considered the developed of a soil along a hillslope profile.

Numerous empirical models for estimating the spatial variation in soil attributes have been proposed, and Moore *et al.* (1993) offer an excellent brief review of these studies. Moore *et al.* (1993) and Gessler *et al.* (in press) also propose a new method using correlations between observed properties and analysis of digital terrain. Such an approach enables high spatial resolution and estimation of many soil attributes other than depth, but requires a large amount of field data. Also, because of its empirical nature it can only apply to areas where it has been developed. Such models, although very useful, have only limited value in providing a mechanistic explanation for the spatial distribution of soil thickness.

Here we present a simple model for predicting the spatial variation in colluvial soil thickness and then use this model in a coupled hydrological and slope stability model to examine the influence of root strength on the pattern of slope instability. Our soil thickness model is similar to that described by Ahnert (1970), but it is developed in such a way that the parameters can be estimated from field observations and it is applied to a real three-dimensional landscape rather than a hillslope profile. Where the topography is reasonably well capture in the digital elevation data, our model successfully predicts the extent of thick colluvial deposits in unchannelled valleys and identifies areas of significant bedrock outcrop. The model also suggests that soil depth quickly tends to a constant value on divergent slopes, leading to a testable hypothesis about the relationship between topographic curvature and soil depth. Modelled slope stability is strongly influenced by the root cohesion, with slope instability most likely in the steep unchannelled valleys with thick colluvial deposits, a result in agreement with field observations. In contrast with river systems which can drain very large areas, hillslopes are of finite extent, hence it is possible to apply the soil depth and slope stability models to local areas in large watersheds using fine scale grids.

THEORY FOR COLLUVIAL SOIL THICKNESS

Here we briefly describe a model for soil depth and show examples of its application to a site where we have mapped the pattern of thick colluvium and located shallow landslide scars. A more detailed development of the soil model can be found elsewhere (Dietrich *et al.* in prep.). The term 'soil' is used here to mean the surficial material mantling the underlying weathered or fresh bedrock and lacking relict rock structure. In thin soils it is roughly equivalent to the solum (the A and B horizons) and in areas of thick accumulations due to mass wasting it is equivalent to colluvium. As long as there is a gradient to the topography, however, all soils in this model are colluvial rather than residual.

This model is thought to be most applicable to unglaciated landscapes underlain by mechanically strong bedrock lacking a well-developed saprolite. We do not consider the role of chemical and physical breakdown of the underlying rock and its influence on soil generation in this model. Although such processes are certainly important, for simplicity we have chosen not to treat them explicitly in the model. Instead, we use a general soil production function proposed by others for which we have some field evidence. The soil production function we use may represent the role of biogenic processes in mechanically disrupting the underlying bedrock and converting it to a mobile soil layer. In the following we emphasize the role of such processes, but the general model does not specifically require only biogenic processes to occur.

On many hilly landscapes, the loose surface soil appears to be largely derived from the underlying bedrock by biogenic processes that dig into the bedrock and force pieces of it into the soil layer or onto the ground surface. In our experience in such landscapes, the transition from the soil to the underlying bedrock is usually abrupt. The most obvious example of this mixing of bedrock into the soil occurs by tree throw. We have observed in many environments, including in the Pacific Northwest, Puerto Rico and Australia, clear instances where the uplifted root wad of the fallen tree still contained bedrock with recognizable structures such as bedding. As the roots decay the bedrock will break into pieces, tumble to the ground and become part of the mobile soil layer. Other obvious agents of mixing of bedrock into the soil include the burrowing effects of animals and insects.

When these biogenic processes operate on an inclined surface, the presence of a downslope component of gravity presumably causes a net transport downslope at a rate roughly proportional to the gradient. Soil thickness is then the result of the dynamic balance between the downslope changes in the rate of transport and the production rate of soil (Figure 1). If solution processes play a minor part in mass transport, the conservation of mass equation for soil thickness, h, can be written as

$$\rho_s \frac{\partial h}{\partial t} = -\rho_r \frac{\partial e}{\partial t} - \nabla \cdot \rho_s \tilde{q}_s \tag{1}$$

where ρ_s and ρ_r are the bulk density of soil and rock, respectively, e is the elevation of the bedrock–soil interface and $\tilde{q}_s$ is the soil transport vector. The first term is the change in soil thickness with time, t. The second term is the rate of conversion of bedrock to soil due to lowering of the bedrock–soil interface and the last term is the divergence of soil transport.

To solve Equation (1) for the spatial variation in soil depth, we need a transport law for q_s and a soil production law. The simplest approach is to consider the case where hillslope processes can be represented by a purely slope-dependent transport law

$$\tilde{q}_s = -K\nabla z \tag{2}$$

in which K is a parameter equivalent to a diffusion coefficient with units of L^2/t and is assumed to be isotropic. Such a law has its origins in the works by Davis (1892) and Gilbert (1909), has been used in analytical and numerical models of landscape evolution (i.e. Culling, 1963; Kirkby, 1971; Koons, 1989; Anderson and Humphrey, 1989; Howard, 1994; see review in Fernandes and Dietrich, in prep.), and has some field evidence to support the simple linear dependence on gradient (McKean *et al.*, 1993). The

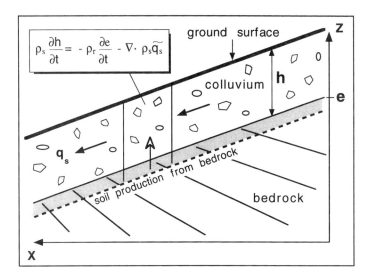

Figure 1. Balance of soil transport and production which controls local colluvial soil depth. In our study area, mass transport down-slope, q_s, of the entire active layer of the soil is caused primarily by biogenic processes acting on an inclined surface. The equation in the figure is Equation (1) in the text and all terms are defined there and shown graphically in the figure. The shaded area between the base of the soil at elevation e and the broken line is the amount of bedrock that would be converted to soil over some specified time interval. Note that $z = e + h$

diffusion coefficient in Equation (2) is not an arbitrary constant, but rather can be estimated by a variety of field methods, but when used in modelling its value is clearly scale-dependent, increasing in magnitude from hillslopes to mountain ranges (see review in Fernandes and Dietrich, in prep.). Equation (2) does not apply to runoff-driven transport responsible for valley evolution, nor is it apparently applicable to landsliding. It could be argued that the gradient term should be given by the sine of the slope (A. Howard, pers. comm.), but we will not consider that effect here.

The approach we will take is to apply Equation (2) to real landscapes represented by a digital elevation field in which diffusive transport processes predominate on the ridges and tend to fill the valleys between short periods of erosion by landsliding and gullying. In our experience such landscape processes typify (but are not limited to) unglaciated, hilly, mostly soil-mantled landscapes in humid to semi-arid climates where Horton overland flow is rare or absent and the underlying bedrock is mechanically strong. By ignoring the effects of river incision at the base of slopes and landsliding, we are in effect taking the digital landscape as given and solving for what the soil distribution should tend to be for the present topography under the assumption that landform change is sufficiently slow that soil depth tends to the local steady-state condition (of either constant depth or constant aggradation). The relatively narrow range of soil depth on hillslopes and the systematic thickening of soil in unchannelled valleys in our field sites support this assumption.

Despite its importance, we known of no field study that defines the production law for an area. Cox (1980) summarizes various theoretical expressions for this law, all of which are based on the assumption that the production rate is a function of the thickness of the soil. Since the original suggestion by Gilbert (1877) it has been assumed that the production rate is zero for soils greater than some depth and that for shallow soils the production rate increases, perhaps reaching a maximum when the bedrock is exposed or at some intermediate soil depth (Figure 2). It was the inference that the production rate reaches some maximum that led (thanks in large measure to Carson and Kirkby, 1972) to the now widely used terms, transport-limited (where there is a soil mantle) and weathering-limited (bedrock at the surface) landscapes. Ahnert (e.g. 1988) has offered the most specific relationships between weathering rate and 'cover thickness', reasoning that mechanical weathering decreases exponentially with cover thickness, but chemical weathering increases with increasing thickness, leading to a weathering rate that has a maximum at some finite cover thickness. An argument can be made that if mechanical disruption by biota plays a major

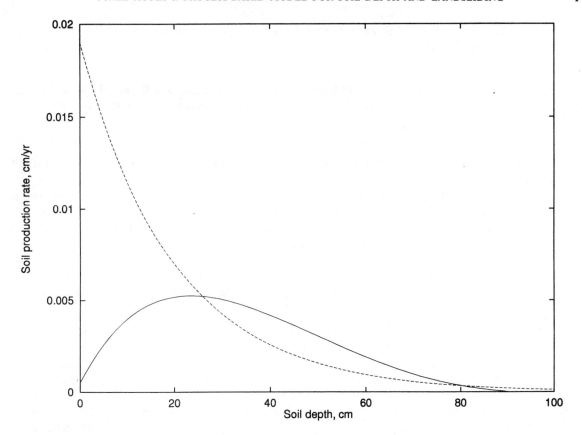

Figure 2. Two production functions estimated from field data. The exponential function was fitted to two data points: 0·0042 cm/yr at 30 cm and 0 at 150 cm. The other function, represented by a polynomial which gives a peak at about 25 cm, was set to a peak close to the 30 cm value for the exponential, but the intercept value is arbitrarily low and the production was assumed to go to zero at 90 cm. Note that the production rate is $-\partial e/\partial t$ and that at equilibrium on the ridges this rate is equal to $-K(\rho_s/\rho_r)\nabla^2 z$

part in converting in place bedrock to mobile soil, a similar 'humped' production curve may be appropriate. The frequency of contact and disruption of the colluvium–bedrock boundary should decrease as the soil thickens, hence production should tend to decrease with thickening soil. It is reasonable to suspect that sufficiently thin soil cannot support burrowing animals and that completely exposed bedrock has a lower rate of conversion to soil than that which is partly buried.

We have explored two general kinds of production laws, one which is a simple exponential decline with thickening soil, i.e. $-\partial e/\partial t = P_0 e^{-mh}$ (in which P_0 and m are empirical constants) and one in which the soil depth is a complex, bell-shaped function of h, with a maximum at some thin soil value (Figure 2). So, in general (assuming K and ρ_s are spatially constant), we can write for $f(h) = -(\partial e/\partial t)$

$$KV^2z = \frac{\partial h}{\partial t} - \frac{\rho_r}{\rho_s}f(h) \tag{3}$$

If the production rate reaches zero with increasing thickness, then valleys which receive convergent sediment transport will tend to fill and the local production rate will reach zero, leading to

$$KV^2z = \frac{\partial h}{\partial t} \tag{4}$$

The deposition rate in valleys is set by the diffusion coefficient and the topographic divergence; the total amount of soil deposited depends on how long aggradation has occurred since the last evacuation event (see Reneau, 1988; Reneau et al., 1989; 1990). This result is most applicable to unchannelled valleys.

On divergent slopes, i.e. $\nabla^2 z < 0$, if the soil depth reaches a time-independent value, then

$$K\nabla^2 z = -\frac{\rho_r}{\rho_s} f(h) \tag{5}$$

and if the production rate varies inversely with depth, sharply curved ridges will have the thinnest soils. Bedrock will appear at the surface, however, where the transport divergence exceeds the production rate

$$-K\nabla^2 z > \frac{\rho_r}{\rho_s} f(h) \tag{6}$$

Anderson and Humphrey (1989) point out this effect in a more complicated one-dimensional model which includes landsliding.

Once bedrock appears, the proposed transport law no longer applies because there is insufficient soil to satisfy the net soil flux. In our finite difference numerical model, for those elements where the transport rate from the element exceeds the sum of the soil thickness, the production amount and the upslope influx of soil, we reduce the transport rate to the total available from these three sources.

THEORY FOR THE INFLUENCE OF SOIL DEPTH ON SHALLOW SLOPE INSTABILITY

In earlier papers we have proposed simple theories for coupled shallow subsurface flow and landsliding of the soil mantle (i.e. Dietrich et al., 1986; 1992; 1993) which did not explicitly account for the influence of soil depth, unlike the innovative model proposed by Okimura and Kawatani (1987). Our previous model uses a steady-state runoff model to define the topographic control on pore pressure distribution and the resultant spatial pattern of slope instability. Soil cohesion is not included and the angle of internal friction and soil bulk density have been treated as spatially constant. We have shown that this very simple, one to two para-meter model fairly accurately delineates those parts of the landscape prone to shallow landsliding (Dietrich et al., 1993; Montgomery and Dietrich, 1994). The model, however, does not account for the effects of ve-getation change on slope instability and therefore cannot explicitly examine the effects on vegetation and slope stability. Also the model uses the assumption that the saturated conductivity is invariant with depth which is far from the case in natural landscapes. Here we take advantage of our soil model to estimate the effect of exponentially declining saturated conductivity and varying cohesive strength due to roots on the pattern of slope instability. An important question we will address is whether this much more complicated model offers substantial improvements in estimating the location of shallow landsliding.

We use the infinite slope model (e.g. Selby, 1993) which accounts for the strength contributed by roots as an apparent cohesion, C_r,

$$\rho_s g h \sin\theta \cos\theta = C_r + C_{sw} + (\rho_s g h \cos^2\theta - \rho_w g u \cos^2\theta)\tan\phi \tag{7}$$

in which g is gravity, θ is the hillslope angle, C_{sw} is the cohesion of the soil when it is wet, ρ_w is the density of water, ρ_s is the soil bulk density (including the mass contribution from soil moisture), u is the pore pressure and ϕ is the angle of internal friction. The vertical surcharge of vegetation is neglected and short-term changes in root strength due to land use are not considered here (see Sidle, 1992). This slope stability equa-tion is linked to the hillslope hydrology through the pore pressure term in which

$$u = h - y_w \tag{8}$$

and y_w is the distance below the surface to the water-table. As such Equation (8) only applies if $y_w \leqslant h$. When the water-table is below the bedrock–soil interface, then u is assumed to be zero. The wet soil bulk density varies from saturated values to moist values, but for simplicity we just use the saturated values. If Equation (8) is substituted into Equation (7) for the pore pressure term, then the equation can be solved for the ratio of the distance below the surface to the water-table, y_w and the soil depth, h

$$\frac{y_w}{h} = 1 - \frac{\rho_s}{\rho_w}\left[1 - \frac{1}{\tan\phi}\left(\tan\theta - \frac{C_r + C_{sw}}{h\rho_s g \cos^2\theta}\right)\right] \tag{9}$$

Failure will occur even when the water-table is below the bedrock interface when

$$\tan \theta \geqslant \frac{C_r + C_{sw}}{h \rho_s g \cos^2 \theta} + \tan \phi \tag{10}$$

and land this steep is unconditionally unstable. Slopes with sufficiently low gradient that they will not fail even when saturated, occurs if

$$\tan \theta \leqslant \frac{C_r + C_{sw}}{h \rho_s g \cos^2 \theta} + \tan \phi \left(1 - \frac{\rho_s}{\rho_w} \right) \tag{11}$$

Field evidence suggests that saturated conductivity declines exponentially through the soil and through the underlying bedrock, but at different rates (Montgomery, 1991), such that we can write

$$k = k_1 e^{-n_1 y \cos \theta} \qquad \text{for } y \leqslant h \tag{12}$$

and

$$k = k_2 e^{-n_2 y \cos \theta} \qquad \text{for } y > h \tag{13}$$

where k is the saturated conductivity (assumed to be isotropic) at vertical distance y below the surface and k_1 is the saturated conductivity at the ground surface and k_2 is this value when the bedrock values are projected to the ground surface. The exponents for these equations (n_1 and n_2) account for the decrease in saturated conductivity normal to the ground surface rather than vertically, and that is why the $\cos \theta$ is included. Following the approach used previously (Dietrich *et al.*, 1992; 1993, Montgomery and Dietrich, 1994), we solve for the steady-state runoff by subsurface flow parallel to the ground surface (hence given by the topographic gradient, $\sin \theta$), by integrating Equations (12) and (13) and multiplying by the assumed head gradient

$$qa = \left[\int_{y_w}^{y=h_0} k_1 e^{-n_1 y \cos \theta} \partial y \cos \theta + \int_{y=h_0}^{\infty} k_2 e^{-n_2 y \cos \theta} \partial y \cos \theta \right] b \sin \theta \tag{14}$$

in which q is the precipitation minus evapotranspiration which falls on the horizontal surface area, a, and flows across the unit contour length, b. The depth, h_0, is equal to the soil depth, h, unless the change in slope (from n_1 to n_2) occurs within the colluvium, which can happen in thick unchannelled valley-fills (Montgomery, 1991). In this latter case, h_0, is equal to h up to the depth where it changes and then it is fixed at that depth value.

Integrating Equation (14) and solving for y_w/h, gives

$$\frac{y_w}{h} = \frac{1}{-n_1 h \cos \theta} \ln \left(\frac{qan_1}{k_1 b \sin \theta} + e^{-n_1 h_0 \cos \theta} - \frac{k_2 n_1}{n_2 k_1} e^{-n_2 h_0 \cos \theta} \right) \tag{15}$$

A single equation coupling the slope stability and the hydrological model can then be obtained by setting Equations (15) and (9) equal, and solving for the ratio of effective rainfall rate to the saturated conductivity at the surface

$$\frac{q}{k_1} = \frac{b \sin \theta}{an_1} \left(e^{-n_1 \beta h \cos \theta} - e^{-n_1 h_0 \cos \theta} + \frac{k_2 n_1}{n_2 k_1} e^{-n_2 h_0 \cos \theta} \right) \tag{16}$$

where

$$\beta = 1 - \frac{\rho_s}{\rho_w} \left[1 - \frac{1}{\tan \phi} \left(\tan \theta - \frac{C_r + C_{sw}}{h \rho_s g \cos^2 \theta} \right) \right] \tag{17}$$

In this dimensionless form Equation (16) can easily be used in a digital terrain model in which topographic attributes, θ, a and b, can be measured and the differing values of the ratio of effective rainfall to infiltration rate needed for instability can be mapped. The local soil depth, h, is estimated from numerically solving Equation (3). Nine parameters must be specified to use Equation (16) once the topographic and soil depth field has been defined: h_0, n_1, n_2, k_2, k_1, $\tan \phi$, ρ_s, C_s and C_r. An argument can be made that these parameters should not be single valued, but should be assigned probability distributions (e.g. review in Mulder, 1991). Because field observations suggest that these parameters probably vary in space systematically rather than randomly (co-varying, for example, with soil depth), this would require assigning this spatial correlation structure, further complicating the model. For simplicity here we treat these parameters as spatially constant.

For a given landscape the strength properties contained in β determine the range of slopes susceptible to instability. The topographic term, $b\sin\theta/a$, in Equation (16) shows that for this steady-state model, lower gradient (but steep enough to fail), strongly convergent (low b/a) areas are least stable. The lower gradient is favoured because destabilizing pore pressures build up with less rainfall on lower gradient hillslopes. Intense precipitation much shorter in duration than that necessary for a steady-state response may favour instead steep side slopes as the least stable (Hsu and Dietrich, in prep.).

Equation (16) also shows that the amount of precipitation necessary for instability varies directly with the saturated conductivity at the ground surface, k_1, and inversely with the rate of decline of the conductivity in the soil, n_1. As in all subsurface flow problems, a meaningful saturated conductivity is difficult to define from field data, but has a large effect on the result. Because we have used k_1 to normalize our results, and because it appears as part of a ratio in the third term on the right-hand side, we do not need to know the exact value of k_1 unless we wish to judge the model in terms of whether it requires reasonable amounts of precipitation. Given that the model is steady state, which rarely, if ever, occurs in most natural storms, evaluating the precipitation rate required for instability may not be instructive: the primary result is the relative rating given by Equation (16) in its dimensionless form. Hence small q/k_1 means least stable and large q/k_1 is most stable. One reason, however, to attempt to assign specific values to the hydrological parameters, k_1, k_2, n_1, n_2, is to estimate whether the rainfall associated with a large q/k is so great as to indicate that the chances of instability are very low. This has obvious practical implications. We explore this problem in the application section.

Soil depth must be prescribed to estimate the role of cohesion in contributing strength to the soil; the deeper the soil the less significant the contribution from cohesion. The strong tendency for soils to be thin on narrow ridges and thick in convergent areas enhances the importance of steep unchannelled valleys as debris flow source areas (see review in Reneau and Dietrich, 1987b). Not only are unchannelled valleys typically mantled with a thicker colluvium which reduces the effectivness of root cohesion, but they are less stable because of the hydrological effects of topographic convergence (Dietrich *et al.*, 1986). Because of the thicker colluvium in unchannelled valleys these features, when they fail, produce larger, more destructive debris flows. Hence the coupled soil depth–slope stability model may be particularly useful in identifying debris flow hazards.

A test of the usefulness of Equation (16) is to apply it with and without a spatially varying soil depth to determine if, when applied to a real landscape, the range in depth and its spatial distribution has a primary influence on the location of shallow landsliding. If the topography dominates the location of shallow landsliding, then even a simpler model which we have previously used (Dietrich *et al.*, 1992; 1993; Montgomery and Dietrich, 1994) may be sufficient. This simpler model has the advantage of having only two parameters.

APPLICATION OF MODELS

In using Equations (3) and (16) in a digital terrain model, the first consideration should be whether the hydrological and erosion processes represented by these models actually occur in the landscape of interest. The soil depth model in the form given in Equation (3) assumes that over the time period sufficient to influence soil depth, the dominant hillslope transport process can be represented by a slope-dependent transport law. The slope stability model assumes that shallow subsurface flow parallel to the ground surface dictates the build-up of pore pressures and that instability involves just the weaker colluvial mantle. Also important is the quality of the digital elevation data. If the grid cells are large and therefore unable to portray either the local slope or the unchannelled valleys accurately, application of Equations (3) and (16) would seem unwarranted, and a simpler model such as that described in Montgomery and Dietrich (1994) would be more appropriate.

We use a grid-based rather than a contour-based model because it is substantially easier to apply these models and evolve the land surface over large basins. To reduce the grid artifacts, drainage area, slope and transport are determined in all eight possible directions. Programs available in ARC/INFO were used to grid the original data and to generate figures. Details of our finite difference numerical model are given elsewhere (Dietrich *et al.*, in prep.).

Field site

We apply these models to a small watershed in Tennessee Valley, Marin County, California, where we have approximately 10 m resolution digital elevation data, extensive field observations about runoff and erosion processes, and where we have already tested the simpler, depth-independent slope stability model (Montgomery and Dietrich, 1989; Montgomery, 1991; Dietrich *et al.*, 1992; 1993; Montgomery and Dietrich, 1994). The area is mostly underlain by tectonically deformed greywacke with some greenstone and chert. Colluvial soils mantle the landscape with thick deposits (several metres) in unchannelled valleys and thin soils on the ridges giving way to bedrock outcrops. Bedrock also crops out on steep side slopes and canyon bottoms. Grass and chaparral predominate in these hilly lands. There is clear evidence of biogenic processes playing a major part in mobilizing fractured bedrock into the shallow soil mantle and in causing the downslope transport of debris. Debris slide scars involving just the soil mantle are common. In the following we parameterize the model based on field observations and test the utility in explaining observed phenomena.

Soil depth

To apply Equation (3) to Tennessee Valley, we need an estimate of the diffusivity, K, the bedrock and soil density, and the production function. Reneau (1988) estimated diffusivity by solving Equation (2) for K by dividing a calculated flux of sediment required to in-fill unchannelled valleys (based on radiocarbon determined deposition rates) by the mean gradient of the adjacent source slopes. This gives a roughly Holocene averaged diffusivity. As reported in McKean *et al.* (1993), data supplied by Reneau (1988) for 34 unchannelled valleys in the coastal mountains of California, Oregon and Washington gave a mean diffusivity of $49 \pm 37 \text{ cm}^2/\text{yr}$, with no clear regional differences. The nearest and most similar sites to Tennessee Valley gave values of 54 and 44 cm^2/yr. Here we will use 50 cm^2/yr. The bedrock to soil density ratio is, according to data in Reneau (1988) about 1.7.

The production function is not known, but we have some guidance from field data of what might be reasonable. Bulk density profiles in the thick colluvial deposits of the unchannelled valleys of this area show a near-linear decrease with depth to about 1.5 m and then remain nearly constant (Reneau, 1988). These deposits accumulate from convergence transport caused by the diffusive-like effects of frequent biological disturbance, largely due to burrowing mammals in this area (e.g. Black and Montgomery, 1991). The depth profile may reflect the decrease in dilational effects of biogenic mixing (see Brimhall *et al.*, 1992), for discussion of this phenomenon), with penetration very rare below 1–1.5 m. Field observations of depth of rooting and animal burrows are congruent with this interpretation. Hence we suggest that the production of loose soil from the underlying weathered bedrock stops once a soil reaches this depth. At the two nearby sites where the diffusivity was determined, the average soil depth on the convex side slopes which serve as sources for the sediment that accumulates in the valleys is about 30 cm. Assuming the soil depth to remain constant during the period of accumulation in the valley axis we can estimate the bedrock to soil conversion rate from the net erosion recorded in the thickened deposits in the valley axis. For the two sites this value is 0.035 and 0.05 mm/yr. Here we assume that soil is produced at the rate of 0.0042 cm/yr from bedrock for a 30 cm thick layer of soil. By fitting an exponential function, $-\partial e/\partial t = P_0 e^{-mh}$, to the thick (no production at 150 cm) and thin soil production rates (0.0042 cm/yr at 30 cm) we obtain $P_0 = 0.019$ cm/yr and $m = 0.05$ (for a soil depth in cm). For depths greater than 100 cm, the productivity rate is assumed to be zero.

We could reason that maximum production should occur at some intermediate depth most favourable to biological activity and frequent contact with the underlying bedrock. In our study area, pocket gophers are common, and generate tunnels about 5 cm in diameter. As an alternative to the exponential function, we

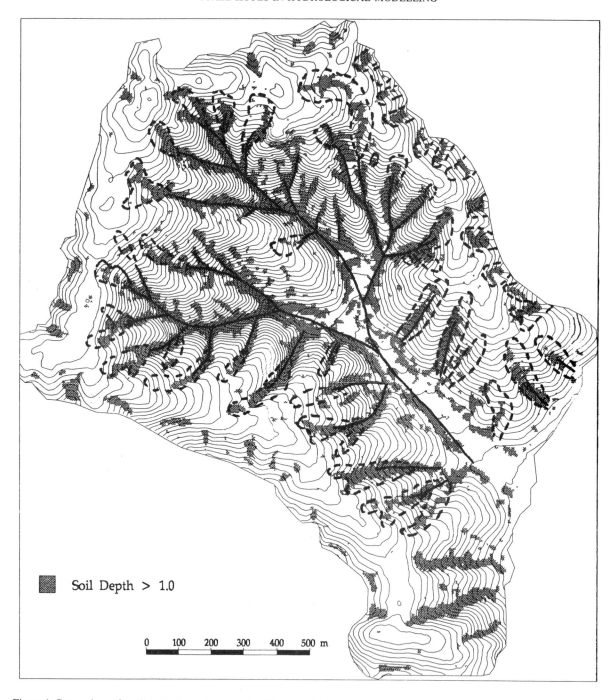

Figure 4. Comparison of predicted pattern of soil depth with areas of thick colluvium mapped reported by Montgomery and Dietrich (1989) on a different topographic base map. Shaded area are elements in Figure 3 with depth greater than 1 m. Channel network mapped in the field is shown as a solid line; some tributary channels do not connect to the main branch. Mapped boundaries of thick colluvium are shown with broken lines that mostly define nearly closed loops at the end or surrounding first-order channels

have selected a peak production at 25 cm (Figure 2). Given the lack of data to specify this function, we have made the peak production close to the observed value (at 30 cm depth) and assigned an arbitrarily low value at the zero depth intercept.

Numerical experiments show that the model is insensitive to the initial soil depth, so we selected 30 cm as a value close to that for much of the ridges. Radiocarbon dating of basal colluvium in this area indicates that deposition began at about 9000 to 15 000 years ago (Reneau *et al.*, 1990), hence a run time of 15 000 years was chosen to generate the soil depth. We found that the model gave consistent results if we used 100 year time steps or less. Hundreds of numerical experiments have been performed, but with limited space here we give only two examples.

Figure 3 shows the predicted pattern of soil depth after 15 000 years for a 5 m grid spacing for the exponential case. No tuning of any of the parameters has been performed. As observed in the field, narrow ridges have the thinnest soils and thick colluvium has accumulated in the valleys. This might be the pattern of soil depth if all incision due to water runoff ceased. Such processes present soil build-up in the larger valley bottoms and locally roughen the topography. We have not accounted for this effect. Figure 4 shows the mapped pattern of thick colluvium reported by Montgomery and Dietrich (1989) using a completely different and coarser base map before the digital elevation data were available. There is good correspondence between the mapped and predicted area of thick (greater than 1.0 m) colluvium. Also shown is the mapped channel network which, where it extends downslope of the colluvial valleys, is bedrock or alluvial-mantled. The edges of the area shown which lack the heavy lines were not mapped in 1989. In the centre of the map area, however, several small valleys are predicted to have thick colluvial deposits, but instead are mostly thinly mantled bedrock. This appears to be due to scour by shallow landsliding.

The alternative production law, with the peak production at 25 cm, gives substantially different results (Figure 5). The lower peak value causes many of the more sharply curved ridges to have bedrock at the surface — a result inconsistent with field observations. This effect could, of course, be eliminated if we put the peak up as high as the exponential intercept with zero depth, but this will not eliminate one other inconsistency. As explained by Carson and Kirkby (1972) for a similar production function, soil depths on the left-hand side of the peak of the production rate are not stable values. Without erosion, soil will simply progressively thicken to the point where production declines to zero. With soil erosion, thinning of the soil will decrease to induce less soil production, and this will lead to stripping of the soil to bedrock. Hence there should be no equilibrium soil depth values observed in the field less than the depth of peak production if the 'humped' production law is to apply. In our study site, soil depths between zero to 25 cm are common on ridges. Our modelling suggests that depth adjustments are rapid on thin soils, hence these thin soils are probably in local equilibrium (erosion approximately equals production). Therefore, if a peak production exists, it must occur at soil depths close to zero, rather than at a depth of 25 cm as chosen.

For this landscape the exponential function is the simplest and is determined empirically. We do not yet know when once the bedrock completely emerges at the surface whether the prodcution rate actually drops or stays at the high value given by the exponential projection to zero depth. For the problems examined here, this distinction is not essential.

In Figure 6, the rate of surface elevation change ($\partial z/\partial t = \partial h/\partial t + \partial e/\partial t$) is plotted against the divergence of sediment transport ($K\nabla^2 Z$) for each cell used to create Figure 3 (soil depth after 15 000 years). The data fall along two distinct relationships. In convergent areas (valleys), net deposition halts soil production ($\partial e/\partial t = 0$) once the depth exceeds 1 m. Equation (4) then applies, giving the linear relationship shown in Figure 6, with a slope of 1.0. In divergent areas (ridges), a balance developes between the production rate and the divergence of sediment transport, soil thickness is time dependent ($\partial h/\partial t = 0$) and Equation (5) then applies. The slope of this linear relationship, then, is ρ_r/ρ_s, the rock to soil bulk density ratio. This general tendency for the landscape to be divided into hillslopes with time-independent soil thickness and valleys with a net accumulation is quickly established (within several thousand years). This suggests that unless a significant recent landslide or a climatically driven change in the production rate or diffusivity has occurred, the soil thickness on divergent areas should be at equilibrium thickness values.

The model smoothes the initial topography, most rapidly removing the sharp curves (as long as the bedrock does not emerge). If the model is run for a period much longer than that dictated by the age of the colluvial fills (more than 30 000 years at our site), the topographic contours become very smooth and much of the local depth variation is eliminated. Although initial gridding of the digital elevation data introduces some artificial roughness which is removed by diffusive transport, we elected to use the value

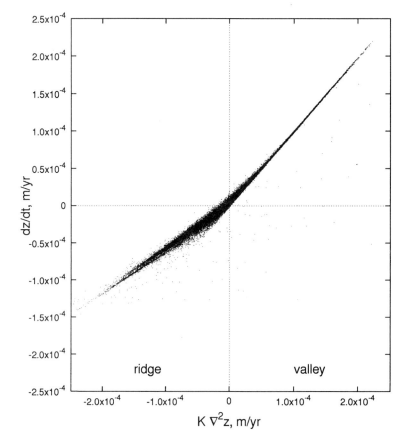

Figure 6. Plot of rate of vertical elevation change of the ground surface as a function of local topographic curvature, $\nabla^2 z$, times the diffusion coefficient after 15 000 years

of soil depth at 15 000 years with the initial topography to drive the slope stability model to retain as much of the present topographic influence on slope stability as possible.

Slope stability

To apply Equation (16), four strength parameters and five hydrological parameters must be estimated. Although the model would allow these parameters to be specified individually for each cell, unlike the soil depth model we have no theory or field evidence for the spatial structure of these parameters, so here we assign one value for the entire area. Based on sampling and testing reported elsewhere (Reneau and Dietrich, 1987a; 1987b; Dietrich *et al.*, 1993), we estimate ϕ to be 40°, the wet soil bulk density to be 2000 kg/m^3 and the soil cohesion to be zero. Root cohesion varies with vegetation type and we will examine the effect of varying cohesion on the predicted pattern of instability.

Values for the hydrological properties are estimated from falling head tests performed by D. Montgomery in Tennessee Valley area on piezometers in the thickened colluvium around a channel head (Montgomery and Dietrich, in press) and by C. Wilson in a nearby basin where tests were performed on shallow side slope colluvium, underlying bedrock and in deep colluvium and underlying bedrock in two hollows (Wilson and Dietrich, 1987; Wilson, 1988). Montgomery's data show the shallow colluvium in the hollow at his study site to have a saturated conductivity greater than 2×10^{-4} m/s and to decrease slowly with depth below the surface. Wilson's data suggest that the thinner side slope colluvium may be less conductive than in the hollow axis and that the bedrock immediately underneath the colluvium is about 10 times less conductive and declines with depth more rapidly. Wilson and Dietrich (1987) and

Wilson *et al.* (1989) demonstrate that although the bedrock is less conductive it contributes to storm flow in large rainfall events. Our best estimate, which combines these two data sets, is $k_1 = 2 \times 10^{-4}$ m/s, $k_2 = 4 \times 10^{-5}$ m/s, $n_1 = 0.5$ (1/m) and $n_2 = 1.4$ (1/m). In thick colluvium the bulk density decreases rapidly with depth and by about 1.5 m the saturated conductivity is not distinguishable from that of the bedrock. Hence we set h_0 equal to 1.5 m. There is considerable uncertainty about these parameters and we explore the effects of different parameter values.

Figure 7 shows the predicted and observed pattern of slope instability for the strength and hydrological parameters estimated above (values given in the figure legend). A logarithmic scale of q/k_1 is used to show the full range of possible values. All but 3 of the 43 scars fall within the instability zones with 80% at least partly touch an area within $\log(q/k_1)$ of less than -2.5, i.e. $q/k_1 = 0.00316$. For this instance in which we estimate $k_1 = 2 \times 10^{-4}$ m/s, the steady-state daily rainfall corresponding to $\log(q/k_1)$ of -3.1, -2.8, -2.5, -2.2 and -1.9 is 1.4, 2.7, 5.5, 10.9 and 21.8 cm/d, respectively. Note that if this estimate of k_1 is roughly correct, an extremely high steady-state rainfall is necessary to cause instability of the steep side slope and ridges.

Figure 8 shows a map of the relative amount of rainfall necessary to cause instability for four values of cohesion: zero, 1000, 5000 and 13 000 N/m^2. These values represent no vegetation, grass, brush and hardwood trees, respectively (e.g. Reneau and Dietrich, 1987a). This shows how root cohesion strongly affects the range of topographic settings where instability is likely to occur and the intensity of rainfall needed to cause instability. With decreasing root strength, areas of predicted instability spread up to the valleys and onto the ridges. The importance of root cohesion is not a new finding, but these maps appear to be the first portrayal of how realistic changes in root strength with vegetation type alters the location and relative instability of the land.

The four maps in Figure 8 may represent the effects of land-use and climate change. The highest root cohesion (13 000 N/m^2) would be associated with forest that might have covered this area during glacial period. The warmer, drier climate of the interglacials resulted in brush and grasslands (5000 N/m^2). Grazing and fire could reduce the root cohesion to chronically low values of 1000 N/m^2 and locally to values of zero.

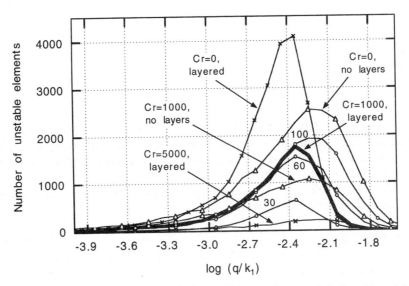

Figure 9. Number of elements that are predicted to have shallow landslides as a function of the logarithm of the net precipitation divided by the surface saturated conductivity for varying strength, soil depth and hydrological properties. The bold black line is the example shown in Figure 7. The two lines labelled with x's are the zero and 5000 N/m^2 cohesion example shown in Figure 8. The three lines labelled with circles show the instability distribution for constant soil depths of 30, 60 and 100 cm, but otherwise the same parameters as the case represented by the bold black line. The two lines labelled with triangles show the effect of having no contrast (no layers) between the soil and bedrock conductivities — that is, there is a single exponential decline in conductivity from the ground surface to infinity, for the case of $C_r = 0$ and for $C_r = 1000$ N/m^2.

Given the uncertainty about the parameters in Equation (16), we have performed a number of sensitivity analyses. Figure 9 summarizes some of our findings. This shows the number of elements that are unstable for a given value of $\log(q/k_1)$, with the heavy line being the example shown in Figure 7. Three findings are apparent in this figure. Firstly, if the soil is more conductive and varies less in its conductivity with depth than the underlying bedrock, then less precipitation is required to cause instability and the area of instability is much greater. Compare '$C_r = 0$, layered' with '$C_r = 0$, no layers', the latter being the case in which the saturated conductivity declines uniformly with depth across the soil–bedrock boundary and there is significant flow in the bedrock. The bold line and the line labelled '$C_r = 1000$, no layers' and marked with triangles is this same comparison for $C_r = 1000$ N/m². If there is no variation with depth ($k_1 = k_2$, $n_1 = n_2$) and n_1 is small ($n_1 < 2$), then deep flow in the bedrock occurs and instability is focused on the unchannelled valleys.

Secondly, soil depth has a large effect on the pattern of relative slope stability. Shallow soils (30 cm) are held in place even with minor root cohesion (1000 N/m²) and few elements are unstable. With increasing thickness, the areas of predicted instability spread up the ridges. Uncertainty in an appropriate mean soil depth has a large effect on the pattern of predicted instability. The soil depth model, on the other hand, systematically and correctly predicts thin soils on the ridges and thick in the valleys, which has the effect when combined with the influence of root cohesion of making ridges more stable because of the larger influence of root strength. There are relatively small differences between the model for a constant soil depth of 60 cm and the predicted spatially variable soil depth, although again we found that the instability is spread further up the ridges with the constant soil depth. In a practical application it would be difficult to claim that the slope instability based on the spatially variable soil depth and that based on a 60 cm soil depth are different. The challenge would be to estimate a representative average depth. If part of the hazard is defined by the potential volume that can be released, however, then the spatially uniform soil depth model is inferior because it will either underestimate by large amounts the volume in unchannelled valleys or it will greatly overestimate the side slope and ridge instability (by predicting too great a soil depth there).

Finally, Figures 8 and 9 clearly show the effect of varying root cohesion on relative stability. Much more than any other parameter, root strength can rapidly change due to local effects such as disease and due to large-scale effects such as land-use, fire or climate change.

DISCUSSION

We have attempted to formulate the soil depth and slope stability models on a physical basis with a minimum number of parameters and to make these parameters quantifiable from field observations or mechanical analysis. In the soil depth model, the diffusion coefficient can be estimated by a variety of field methods, as described in this paper. The soil production law has not previously been quantified, and here it is only very crudely estimated as an exponential function of soil depth. The soil production law subsumes all weathering processes inside the production function, ignoring chemical losses that occur during soil transport. No effects of velocity structure with depth are considered. If soil transport is by a linear slope-dependent processes, then our model suggests that because local equilibrium is quickly established, field measurements of soil depth and local curvature would define the shape of the soil production function [see Equation (5)]. A field study is underway to test this hypothesis.

The extent of bedrock exposure also places an important constraint on the production function and diffusivity. According to Equation (6), bedrock will crop out wherever

$$\nabla^2 z > \frac{1}{K}\frac{\rho_r}{\rho_s}\frac{\partial e}{\partial t}$$

that is, where the topographic curvature exceeds the production rate divided by the diffusivity. If the digital elevation data are of sufficiently high resolution to define the local curvature in areas of bedrock exposure, then this criteria can be used to set this ratio, and if the diffusivity can be estimated, then the maximum production rate can be determined.

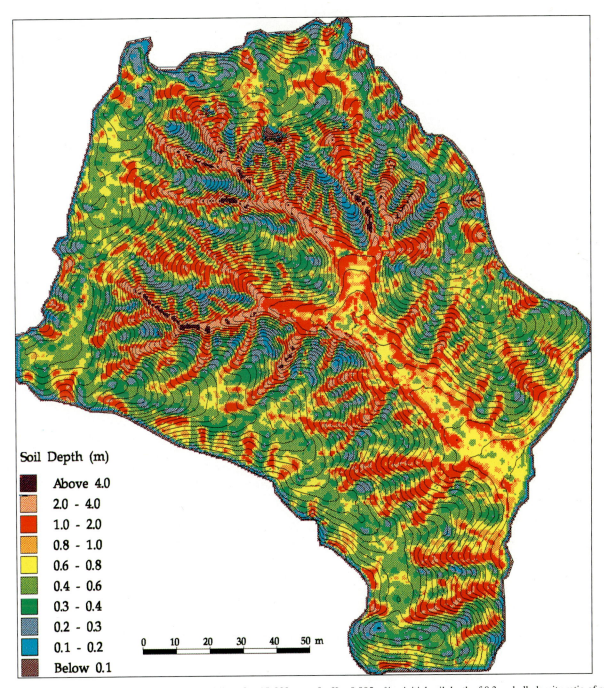

Figure 3. Predicted pattern of soil depth in Tennessee Valley after 15,000 years for K = 0.005 m²/yr, initial soil depth of 0.3 m, bulk density ratio of rock to soil of 1.7 and the exponential production function, i.e.$-\partial e/\partial t = 0.019$ cm/yr e $^{-0.05*\text{soil depth}}$. Calculations were performed in 100 year time steps for 5 m elements. Legend gives the soil depth categories in metres

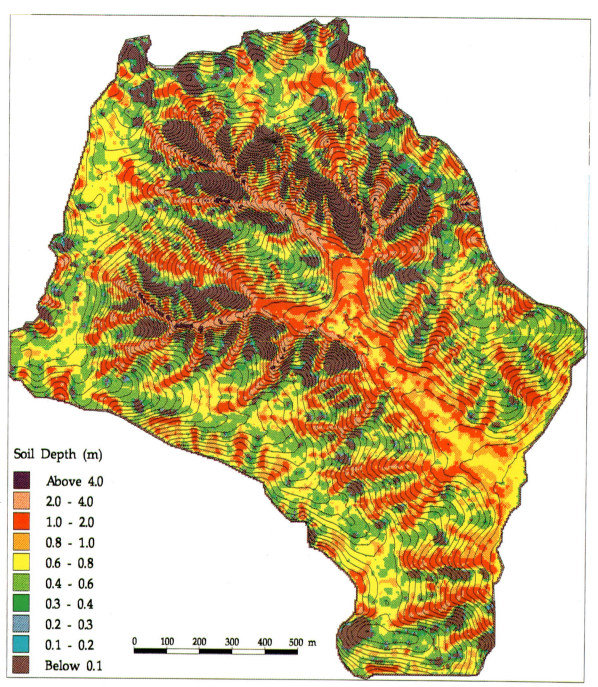

Soil Depth (m)

■	Above 4.0
■	2.0 - 4.0
■	1.0 - 2.0
■	0.8 - 1.0
■	0.6 - 0.8
■	0.4 - 0.6
■	0.3 - 0.4
■	0.2 - 0.3
■	0.1 - 0.2
■	Below 0.1

0   100   200   300   400   500 m

Figure 5. Predicted pattern of soil depth in Tennessee Valley after 15,000 years using the same parameters as in Figure 3, except the production function has a peak at 25 cm, as shown in Figure 2

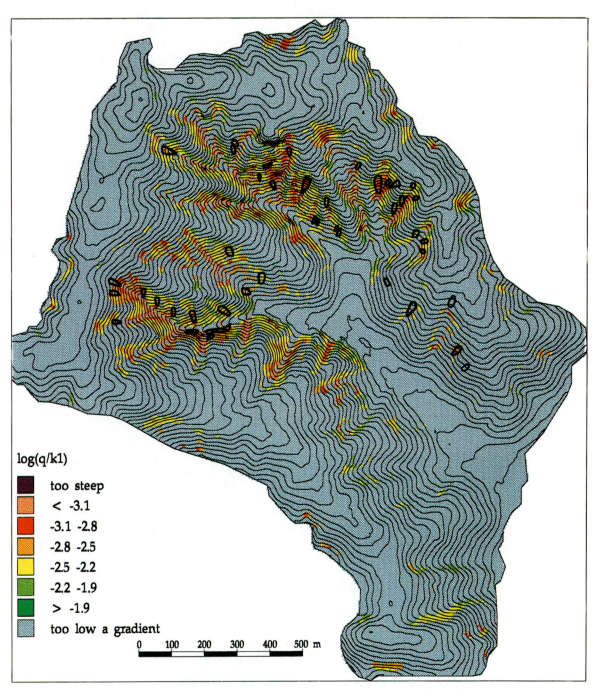

log(q/k1)

■ (dark maroon)	too steep
■ (light orange)	< -3.1
■ (red)	-3.1 -2.8
■ (orange)	-2.8 -2.5
■ (yellow)	-2.5 -2.2
■ (light green)	-2.2 -1.9
■ (green)	> -1.9
■ (grey)	too low a gradient

0   100   200   300   400   500 m

Figure 7. Predicted and observed pattern of slope instability in Tennessee Valley using Equation (16) in text. Soil depth is that shown in Figure 3. Angle of internal friction is 40°, root cohesion is 1000 N/m², $n_1 = 0.5$, $n_2 + 1.4$, $k_2/k_1 = 0.2$. Relative stability is shown by the logarithm of the ratio of effective precipitation to surface infiltration rate, with high negative values implying the small amounts of steady-rate rainfall necessary for instability. The black outlines are the approximate location of the debris slide scars. The scars are shown larger than their actual size to account for plotting uncertainty with regard to location

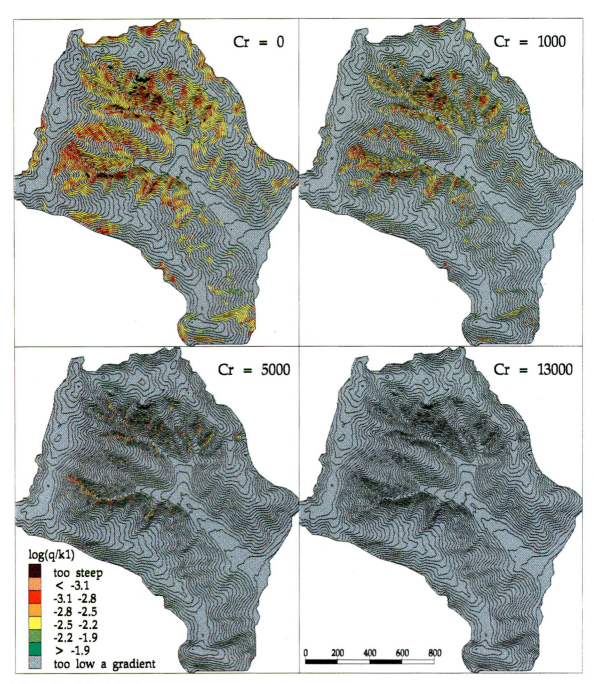

Figure 8. Comparison of the relative slope stability with changing root strength, C_r, shown in N/m². All other factors the same as in Figure 7 and this figure is repeated here without the landslide scars

Spatial variability in diffusion and production, however, is difficult to quantify. These rates probably vary with rock type, climate and the soil biota. Large differences in diffusion coefficient have been identified with rock type (e.g. McKean *et al.*, 1993), but climatic dependency is not well established (see review in Fernandes and Dietrich, in prep.).

If the results of the soil depth model are viewed from the perspective of a longer term landscape evolution, important questions are also raised. Figure 3 shows that the soil depth on ridges is predicted to vary widely across the landscape. Local equilibrium is established between production and erosion and production varies exponentially with depth. This implies that there is a large variation in rates of erosion and the land is far from morphological dynamic equilibrium — even though the soil depth is locally in equilibrium. Only when the ridges have a contact curvature, $\nabla^2 z$, will the erosion rate by diffusive transport be spatially uniform, hence the mapping of curvature from high-resolution digital data may prove a useful method to examine landscape equilibrium tendencies. Spatial variation in diffusivity, K, although certainly likely, would have to covary with curvature for the ridges at our study site to be eroding at a constant rate. This seems unlikely. Bedrock outcrops in a soil-mantled landscape may also indicate a non-steady-state landscape, because bedrock crops out either at locations of maximum erosion (exponential production law) or minimum erosion (maximum production at some finite depth). Changes in diffusivity or production rate with time could change the extent of bedrock outcrop and the spatial dependency of depth or curvature, but unless the model is incorrect, non-uniform curvature values for ridges imply a spatially variable erosion rate.

All the strength and hydrological parameters in the slope stability model can be obtained from field measurements or laboratory analysis. In practice, however, some of these parameters are difficult to define, particularly with regard to their spatial variation. Even for our small study area, we have extrapolated parameters from nearby areas and assumed no spatial variation in their magnitudes.

Our hydrological model for the case in which $k_1 = k_2$, $n_1 = n_2$ of Equation (14) is similar to that used in TOPMODEL (Beven and Kirkby, 1979), which can be written as

$$qa = km \tan\theta e^{-y/m} \tag{18}$$

in which, to convert to our model, $m = -1/n$, $y = y_w \cos\theta$ and $a = a/b$. Note, however, that we use the more physically correct $\sin\theta$ rather than $\tan\theta$, which is important on steep slopes. Also, Equation (14) allows us to examine, explicitly, the effect of conductivity contrasts between bedrock and soil. Our k values reported here are low for catchment scale applications of TOPMODEL (M. J. Kirkby, pers. comm.). Our estimates are based on field measurements and we suggest that the lower values may be more appropriate for hillslopes rather than entire catchments. Also, our n values are very high compared with those used in TOPMODEL (M. J. Kirkby, pers. comm.). Again, this may be a scale effect in that our estimates of n come from field measurements of conductivity on hillslopes and are used to predict only local hillslope response, not catchment scale runoff.

The study by Okimura (1989) offers an instructive comparison with our results. Guided by earlier modelling work using digital elevation data and a coupled slope stability and hydrological model which solves for the pore pressure necessary to cause instability as a function of time since the start of rainfall (Okimura and Kawatani, 1987), Okimura conducted a field study to document the spatial variation in potential failure layer thickness. This layer is identified on the basis of cone penetration results and therefore may not correspond to our definition of soil. Unlike our field observations on soil thickness and our modelling results, he found no systematic relationship with topography and attributed this to effects of past failures. Similar to our findings, however, he concluded that using the measured or estimated spatial distribution of potential failure depth gave very different results from assuming a model with a spatially constant estimated mean depth. He also showed that the number of elements classified as unstable increases significantly with increasing mean depth. Okimura's model does not consider the effects of exponential decline in conductivity with depth, nor the effects of flow in bedrock.

Despite the general success of models based on the infinite slope assumptions, hillslope failures in detail are not necessarily well represented by the infinite slope model, especially when root cohesion matters (e.g. Burroughs *et al.*, 1985). For example, the size of landslides varies with root strength and, therefore, vegetation (e.g. Reneau and Dietrich, 1987a). Also, field observations at sites of shallow failure often suggest that instability was initiated by erosion of the toe due to channel head advance. Local exfiltration gradients associated with bedrock fracture heterogeneity may be responsible for the timing and location of landslides (e.g. Wilson and Dietrich, 1987; Montgomery *et al.*, 1990). These local effects are not included in the infinite slope model and may be more important than the uncertainty in some of the model parameters.

Is the more complicated slope stability model proposed here a substantial improvement over our earlier, two-parameter model that does not include effects of soil depth variation, vertically varying conductivity or root strength? In terms of gaining insight regarding controls or slope stability, we suggest the answer is 'yes'. But with regard to the practical application to slope stability, the answer is less certain. Analysis by Montgomery and Dietrich (1994) of the same study site using our earlier simpler model yielded equally successful results in identifying failure sites. This is partly due to the low root cohesion in this area. When the root cohesion is large, the more complicated model will predict a much smaller area of potential instability than the earlier simpler model. The simpler model cannot be used to ask questions about effects of vegetation changes or differences in bedrock or soils (as influencing vertically varying conductivities). Field studies may eventually yield ranges of diffusivities and soil production laws and their environmental controls such that the soil model will be fairly easy to apply to a new area in a manner that correctly portrays the gross spatial variation in soil depth. Unfortunately, for essentially all practical purposes, the hydrological component of the slope stability model cannot be calibrated, as details of the subsurface conductivity field are rarely known and are not easily obtained.

The large influence of cohesion on the spatial extent of instability (Figures 8 and 9) presents a specific dilemma. Root cohesion is the most variable of all of the slope stability parameters. Fire, disease, land-use, the effects of extreme weather events may all cause local or watershed-wide rapid changes in root strength. How should a map of slope stability with a specific root cohesion be interpreted? For example, consider a long, weakly convergent 30° hillslope with dense vegetation on it which, due to root strength, is estimated to be stable even when saturated. Is it safe to build a house downslope? Is this a site that does not experience landsliding? Our geomorphic analysis might clearly say 'no', as the weakly convergent topography may be a partially infilled bedrock hollow where periodic landslides occur. Presumably the most cautious approach is to assume that at some time the root cohesion can be reduced to zero. There is a lot of land, however, between the area predicted to be unstable with no cohesion and that with a modest amount of cohesive strength which may have considerable value.

Further work is needed on how to incorporate the importance of root strength into land-use decisions. If the conservative approach of assuming no root strength is used (which can be done with the two-parameter model), then vast areas of land in steep hilly areas would be considered potentially unstable — and that is probably correct on geomorphic time-scales, but may be less easily defended on a management time-scale. An approach that has been explored is to treat root strength and other parameters probabilistically (e.g. Hammond *et al.*, 1992) or, in the case of timber harvest, a function of time since cutting (e.g. Sidle, 1992). Although the assignment of probability functions remains largely guesswork, an approach of this kind seems warranted in the context of land-use decisions.

Unlike rivers, which progressively drain larger areas with increasing total catchment area, hillslopes are of only local extent, even in large watersheds. Soil depth and shallow slope instability are strongly dictated by local topography. Therefore, it is desirable to retain the local physical basis of these models when applying them to large watersheds. The simplicity of these models allows this, although again it may be necessary to assume that parameters do not vary significantly over local scales. Hence, although it is desirable and probably necessary to develop ways of representing the hydrological response of large areas by means other than the summing up of all the detailed processes, for soil depth and slope stability analysis the limited spatial extent of hillslopes and the importance of local processes argues for retaining the fine scale analysis. If the goal is to model sediment flux, then perhaps the diffusivity, like saturated conductivity, can be treated as a scale-dependent parameter (e.g. Koons, 1989). A field calibrated transport law for soil flux from shallow landsliding is not available, hence analysis of scaling issues for this process seems premature.

CONCLUSIONS

The simple model for soil depth proposed here appears to explain the general tendency in hilly landscapes for the sharp convex ridges to have thin soil or bedrock outcrop and the swales or unchannelled valleys to be mantled with thick soils. The model also sheds some light on the likely form of a 'law' for the rate of conversion of bedrock to soil. Soil depth rapidly develops a locally stable value for a given curvature. There is no stable value for soil depths thinner than that at which maximum production occurs. Because thin soils are common in steep, sharp-crested hillslopes, this suggests that the maximum production rate of soil probably occurs at nearly zero soil depth. The shape of the production function for the case of soil transport by a linear slope-dependent diffusive process is given by the variation of soil depth with local topographic curvature. These model results suggest specific field observations that should lead to an enhanced understanding of soil production rates.

With the ability to predict soil depth, it is possible to use the infinite slope model to analyse the influence of root cohesion and subsurface conductivity structure on the spatial distribution of shallow landslide potential on real landscapes. Root cohesion affects the total area prone to slope instability and the relative amount of precipitation needed to cause landslides. The variation in saturated conductivity with depth partly controls the relative instability of hollows or unchannelled valleys versus steep side slopes. The less conductive the bedrock relative to that of the soil, the less stable the side slopes. Root cohesion increases the stability of side slopes relative to hollows. However, root cohesion, unlike other parameters affecting soil strength, can vary widely in time due to land use, climate or biological processes such as disease. This sensitivity may give it a dominant role in linking land-use and climate change to slope stability.

ACKNOWLEDGEMENTS

This paper benefited greatly from instructive reviews by M. J. Kirkby, A. D. Howard and A. Rinaldo, as well as two anonymous reviewers. M. J. Kirkby, A. D. Howard, L. Band, N. Fernandes and J. Kirchner also provided critical insights through conversations during various phases of the analysis reported here. Ian Moore offered valuable encouragement and is greatly missed. Kent Rich rescued and improved several of the illustrations. The project was supported by grants TFW FY92-010 and TFW FY94-004 through the CMER and SHAMW committees of the Washington State Timber/Fish/Wildlife agreement.

REFERENCES

Ahnert, F. 1970. 'A comparison of theoretical slope models with slopes in the field', *Z. Geomorphol. Suppl Bd*, **9**, 88–101.
Ahnert, F. 1988. 'Modelling landform change' in Anderson, M. G. (Ed.), *Modelling Geomorphological Systems*. Wiley, Chichester. pp. 375–400.
Anderson, R. S. and Humphrey, N. F. 1989. 'Interaction of weathering and transport processes in the evolution of arid landscapes' in Cross, T. A. (Ed.), *Quantitative Dynamic Stratigraphy*. Prentice Hall, Englewood Cliffs. pp. 349–361.
Beven, K. and Kirkby, M. J. 1979. 'A physically based variable contributing area model of basin hydrology', *Hydrol. Sci. Bull.*, **24**, 43–69.
Black, T. A. and Montgomery, D. R. 1991. 'Sediment transport by burrowing mammals, Marin County, California', *Earth Surf. Process Landforms*, **16**, 163–172.
Brimhall, G. H., Chadwick, O. A., Lewis, C. J., Compston, W., Dietrich, W. E., Power, M. E., Hendricks, D., and Bratt, J. 1992. 'Deformational mass transport and invasive processes in soil evolution', *Science*, **255**, 695–702.
Burroughs, E. R. Jr, Hammond, C. J., and Booth, G. D. 1985. 'Relative stability estimation for potential debris avalanche sites using field data' in Takei, A. (Ed.), *Proceedings of the International Symposium on Erosion, Debris Flow and Disaster Prevention*. Erosion Control Society, Tokyo. pp. 335–339.
Carson, M. A. and Kirkby, M. J. 1972. *Hillslope Form and Process*. Cambridge University Press, Cambridge. 475pp.
Costa, J. E. and Wieczorek, G. F. 1987. 'Debris flows/avalanches: process, recognition and mitigation', *Geol. Soc. Am. Rev. Engin. Geol.*, Vol. VII. Geological Society of America, Boulder. 239pp.
Cox, N. J. 1980. 'On the relationship between bedrock lowering and regolith thickness', *Earth Surf. Process.*, **5**, 271–274.
Crozier, M. J., Vaughn, E. E., and Tippett, J. M. 1990. 'Relative instability of colluvium-filled bedrock depressions', *Earth Sur. Process Landforms*, **15**, 329–339.
Culling, W. E. H. 1963. 'Soil creep and the development of hillside slopes', *J. Geol.*, **71**, 127–161.
Davis, W. M. 1892. 'The convex profile of badland divides', *Science*, **20**, 245.
DeRose, R. C., Trustrum, N. A., and Blaschke, P. M. 1991. 'Geomorphic change implied by regolith–slope relationships on steepland hillslopes, Taranaki, New Zealand', *Catena*, **18**, 489–514.

DeRose, R. C., Trustrum, N. A., and Blaschke, P. M. 1993. 'Post-deforestation soil loss from steepland hillslopes in Taranaki, New Zealand', *Earth Surf. Process Landforms*, **18**, 131–144.

Dietrich, W. E. and Dunne, T. 1978. 'Sediment budget for a small catchment in mountainous terrain', *Z. Geomorphol. Suppl. Bd.*, **29**, 191–206.

Dietrich, W. E., Wilson, C. J., and Reneau, S. L. 1986. 'Hollows, colluvium and landslides in soil-mantled landscapes' in Abrahams, A. (Ed.), *Hillslope Processes, Sixteenth Annual Geomorphology Symposium*. Allen and Unwin, London. pp. 361–388.

Dietrich, W. E., Wilson, C. J., Montgomery, D. R., McKean, J. and Bauer, R. 1992. 'Erosion thresholds and land surface morphology', *Geology*, **20**, 675–679.

Dietrich, W. E., Wilson, C. J., Montgomery, D. R. and McKean, J. 1993. 'Analysis of erosion thresholds, channel networks and landscape morphology using a digital terrain model', *J. Geology*, **101**, 161–180.

Gessler, P. E., Moore, I. D., McKenzie, N. J., and Ryan, P. J., Soil–landscape and spatial prediction of soil attributes', *Int. J. GIS*, in press.

Gilbert, G. K. 1877. *Report of the Geology of the Henry Mountains (Utah), US Geographical and Geological Survey of the Rocky Mountains Region*. US Government Printing Office, Washington, DC. 169pp.

Gilbert, G.K. 1909. 'The convexity of hilltops', *J. Geol.*, **17**, 344–350.

Hammond, C., Hall, D., Miller, S., and Swetik, P. 1992. 'Level 1 Stability Analysis (LISA) documentation for Version 2.0, *USDA, Forest Service, Intermountain Research Station, General Tech. rep. INT-285*, 190pp.

Howard, A. D. 1994. 'A detachment-limited model of drainage basin evolution', *Wat. Resour. Res.*, **30**, 2261–2285.

Kirkby, M. J. 1971. 'Hillslope process-response models based on the continuity equation', *Inst. Br. Geogr. Spec Publ.*, **3**, 15–30.

Kirkby, M. J. 1985. 'A model for the evolution of regolith-mantled slopes' in Woldenberg, M. J. (Ed.), *Models in Geomorphology*. Allen and Unwin, London. pp. 213–228.

Koons, P. O. 1989. 'The topographic evolution of collisional mountain belts: a numerical look at the Southern Alps, New Zealand', *Am. J. Sci.*, **289**, 1041–1069.

McKean, J. A., Dietrich, W. E., Finkel, R. C., Southon, J. R., and Caffee, M. W. 1993. 'Quantification of soil production and downslope creep rates from cosmogenic [10]Be accumulations on a hillslope profile', *Geology*, **21**, 343–346.

Montgomery, D. R. 1991. 'Channel initiation and landscape evolution', *Unpubl. PhD Dissertation*, University of California, Berkeley, 421pp.

Montgomery, D. R. and Dietrich, W. E. 1989. 'Channel initiation, drainage density and slope', *Wat. Resour. Res.*, **25**, 1907–1918.

Montgomery, D. R. and Dietrich, W. E. 1994. 'A physically based model for topographic control on shallow landsliding', *Wat. Resour. Res.*, **30**, 1153–1171.

Montgomery, D. R. and Dietrich, W. E. 'Hydrologic processes in a low-gradient source area', *Wat. Resour. Res.*, in press.

Montgomery, D. R., Dietrich, W. E., Torres, R., Anderson, S. P., Heffner, J. T., Sullivan, K. O., and Loague, K. M. 1990. 'Hydrologic experiments in a steep unchanneled valley: (1) experimental design and piezometric response', *EOS, Trans. Am. Geophys. Union*, **71**, 1342.

Moore, I. D., Gessler, P. E., Nielsen, G. A. and Peterson, G. A. 1993. 'Soil attribute prediction using terrain analysis', *Soil Sci. Soc. Am. J.*, **57**, 443–452.

Mulder, H. F. H. M. 1991. 'Assessment of landslide hazard', *Doctoral Thesis*, Faculty of Geographical Sciences, University of Utrecht, 150pp.

Okimura, T. 1989. 'Prediction of slope failure using the estimated depth of the potential failure layer', *J. Natural Disaster Sci.*, **11**, 67–89.

Okimura, T. and Kawatani, T. 1987. 'Mapping of the potential surface-failure sites on granite slopes' in Gardiner, V. (Ed.), *International Geomorphology 1986, Part I*. Wiley, Chichester. pp. 121–138.

Reneau, S. L. 1988. 'Depositional and erosional history of hollows: application to landslide location and frequency, long-term erosion rates, and the effects of climatic change', *Unpubl. PhD Dissertation*, University of California, Berkeley, 328 pp.

Reneau, S. L. and Dietrich, W. E. 1987a. 'Size and location of colluvial landslides in a steep forested landscape', *Proc. Int. Symp. Erosion and Sedimentation in the Pacific Rim, 3–7 August 1987, Corvallis, OR, USA, Int. Assoc. Hydrol. Sci. Bull. Publ. No. 165*, 39–48.

Reneau, S. L. and Dietrich, W. E. 1987b. 'The importance of hollows in debris flow studies' in Costa, J. E. and Wiezcorek, G. F. (Eds), *Debris Flows/Avalanches: Process, Recognition and Mitigation, Geol. Soc. Am. Rev. Engin. Geol.* Vol. VII. Geological Society of America, Boulder. pp. 165–180.

Reneau, S. L., Dietrich, W. E., Rubin, M., Donahue, D. J., and Jull, J. T. 1989. 'Analysis of hillslope erosion rates using data colluvial deposits', *J. Geol.*, **97**, 45–63.

Reneau, S. L., Dietrich, W. E., Donahue, D. J., and Jull, A. J. T. 1990. 'Late Quaternary history of colluvial deposition and erosion in hollows, Central California Coast Ranges', *Geol. Soc. Am. Bull.*, **102**, 969–982.

Selby, M. 1993. *Hillslope Materials and Processes*. Oxford University Press, Oxford. 289pp.

Sidle, R. C. 1992. 'A theoretical model of the effects of timber harvesting on slope stability', *Wat. Resour. Res.*, **28**, 1897–1910.

Wilson, C. J. 1988. 'Runoff and pore pressure development in hollows', *Unpubl. PhD Dissertation*, University of California, Berkeley, 284pp.

Wilson, C. J. and Dietrich, W. E. 1987. 'The contribution of bedrock groundwater flow to storm runoff and high pore pressure development in hollows', in *Proc. Int. Symp. on Erosion and Sedimentation in the Pacific Rim, 3–7 August 1987, Corvallis, OR, USA, Int. Assoc. Hydrol. Sci. Bull. Publ. No. 165*, 49–59.

Wilson, C. J., Dietrich, W. E. and Narasimhan, T. N. 1989. 'Predicting high pore pressures and saturation overland flow in unchannelled hillslope valleyes', in *Proc. Hydrol. and Wat. Resour. Symp.* Institute of Engineers Australia. pp. 392–396.

Wu, W. 1993. 'Distributed slope stability analysis in steep, forested basins', *Unpubl. PhD Dissertation*, Utah State University, 134pp.

9

SCALE: LANDSCAPE ATTRIBUTES AND GEOGRAPHICAL INFORMATION SYSTEMS

L. E. BAND

University of Toronto, Toronto, Canada

AND

I. D. MOORE

Australian National University

ABSTRACT

The roles and limitations of geographical information systems (GISs) in scaling hydrological models over heterogeneous land surfaces are outlined. Scaling is defined here as the extension of small-scale process models, which may be directly parameterized and validated, to larger spatial extents. A process computation can be successfully scaled if this extension can be carried out with minimal bias. Much of our understanding of land surface hydrological processes as currently applied within distributed models has been derived in conjunction with 'point' or 'plot' experiments, in which spatial variations and patterns of the controlling soil, canopy and meteorological factors are not defined. In these cases, prescription of model input parameters can be accomplished by direct observation. As the spatial extent is expanded beyond these point experiments to catchment or larger watershed regions, the direct extension of the point models requires an estimation of the distribution of the model parameters and process computations over the heterogeneous land surface. If the distribution of the set of spatial variables required for a given hydrological model (e.g. surface slope, soil hydraulic conductivity) can be described by a joint density function, $f(x)$, where $x = x_1, x_2, x_3, \ldots$ are the model variables, then a GIS may be evaluated as a tool for estimating this function. In terms of the scaling procedure, the GIS is used to replace direct measurement or sampling of $f(x)$ as the area of simulation is increased beyond the extent over which direct sampling of the distribution is feasible. The question to be asked is whether current GISs and current available spatial data sets are sufficient to adequately estimate these density functions.

INTRODUCTION

Spatially distributed models of watershed hydrological processes have been developed that incorporate the spatial patterns of terrain, soils and vegetation as estimated with the use of remote sensing and geographical information systems (GISs) (Band *et al.*, 1991; 1993; Famiglietti and Wood, 1991; 1994; Moore and Grayson, 1991; Moore *et al.*, 1993b; Wigmosta *et al.*, 1994). This approach makes use of various algorithms to extract and represent watershed structure from digital elevation data (e.g. Marks *et al.*, 1984; O'Callaghan and Mark, 1984; Band, 1986a; 1986b; 1989a,b; O'Loughlin, 1986; Jenson and Domingue, 1988; Moore *et al.*, 1988). Land surface attributes are mapped into the watershed structure as estimated directly from remote sensing imagery (e.g. canopy leaf area index), digital terrain data (slope, aspect, contributing drainage area) or from digitized soil maps, such as soil texture or hydraulic conductivity assigned by soil series. Two approaches to the spatial distribution of key surface variables have emerged. The first is to explicitly map surface attributes into a grid or flow element structure, such that water (and sediment) can be directly routed through the flow path network to compute the distribution of soil water and runoff production. A second approach has been to use functional relations between topographic and soil variables and the distribution of soil water following principles of hillslope hydrology to approximate the effects of soil water routing as embodied in TOPMODEL (Beven and Kirkby, 1979) and the steady-state version of TOPOG (O'Loughlin, 1981; 1986) and TAPES (Moore and Grayson, 1991). The latter approach obviates the need for explicit patterns of surface attributes for flow routing, reducing the parameterization requirements to the specification of the attribute density functions.

Two problems present themselves in the extension of these models to regional scales. The first is that much of the process understanding on which these models are built is based on physical interactions and mechanisms that often occur on length scales of metres or less. Current canopy evapotranspiration models assume a knowledge of soil profile properties at high vertical resolution, along with a fairly detailed knowledge of canopy and lower atmospheric conditions. Recent work has shown that simulated surface/atmosphere exchange and runoff production can be very sensitive to the parameter distribution functions specified, and replacement of parameter distributions by mean values often leads to significant bias (e.g. Avissar, 1992; Dolman, 1992; L'Homme, 1992; Band, 1993) under certain conditions. The key topographic and soil variables used in many distributed catchment models typically have length scales smaller than a hillslope or small catchment. At regional scales, however, the best available digital topographic data may not contain sufficient information to resolve flow patterns, and regional soil maps are highly generalized.

The second, related problem has to do with the sampling method that is used to estimate the joint distribution of surface attributes that is inherent in a standard GIS approach. Most GISs have been developed with a cartographic paradigm in which maps (mostly area–category maps) are input, information is combined by overlay analysis and maps are output. Unfortunately, with the exception of high-resolution digital topographic and remote sensing information, the input maps are often heavily generalized and produced at a variety of scales, such that the spatial information for the different surface attributes (soils, vegetation, terrain) are filtered to different levels. Typically, soil information is known with the least certainty and at the greatest level of generalization due to the difficulty of sampling, high-field variability and the general inability to remotely image the required soil properties. Simple GIS overlay of digital terrain data or remotely sensed vegetation data with digital soil maps often results in poor estimates of the co-occurrence of these variables. In terms of parameterization of soil–vegetation–atmosphere–transfer (SVAT) models, this problem is often manifested in the GIS combination of biophysically infeasible soil and vegetation associations. Working in water-limited areas, we often find that overlay analysis results in associations of high leaf area index (LAI) as determined by Thematic Mapper imagery with soils of very low available water capacity. In this instance, the generalized soil maps cannot show the locations of pockets of deeper soils or narrow riparian zones due to sampling and scale limitations. Although the overlay method does not need to be used, the more general availability of digital area–category maps for soils and vegetation, and the ease of cartographic overlay analysis in GIS, has resulted in its indiscriminate use, with little regard for how the entire process performs as a sampling strategy to estimate the joint occurrence of model parameters.

In comparison with GIS analysis of surface attribute information, field sampling is expensive, time consuming and can often only measure land surface variables at points or small plot levels (e.g. soil pits or forest plots.) However, direct field sampling can have the important property of measuring the important variables at the same or similar resolution, a property that may be lost when combining generalized map information produced at different scales or levels of support. If the models being run are sensitive to the covariation of model parameters below the resolution of the original mapping units, then it may be necessary to attempt to estimate what the variance and covariance of the significant model variables are within the initial mapping units and the simulation units, a process which cannot generally be performed with standard GIS techniques.

In this paper, methods for estimating and combining critical land surface attributes and the structure of the watershed with GIS tools are reviewed and assessed. The impact of data scale or resolution, and the standard GIS tools of processing and combining spatial data, are investigated in terms of the estimation of both the patterns and distributions of terrain, soil and vegetation attributes. We begin by reviewing some of the approaches that have been taken to deal with surface heterogeneity in hydrological models.

This discussion provides a basis for surface data requirements that may be provided by a GIS. The next section covers the GIS estimation of watershed structure and terrain variables from digital elevation models (DEMs) at different grid resolutions. This is followed by a review of the GIS and remote sensing derivation of vegetation attribute distributions and the estimation of soil properties over watershed areas. The performance of GIS as a paradigm for combining critical land surface attributes for hydro-logical models is assessed with respect to scale effects and the parameter sensitivity of a distributed hydroecological model. Within this context, methods for the estimation and incorporation of 'subgrid' parameter variance are discussed. Finally, given specific limitations of GISs to produce adequate data sets for many current models, the process of model development and its relations to data constraints is evaluated with respect to the performance and expectations of the entire modelling and data analysis exercise.

MODELLING APPROACHES TO LAND SURFACE HETEROGENEITY

Over the past two to three decades, the evolution of hydrological models has proceeded along two key trajectories: increased physical complexity and realism of the process description, and spatially distributing the representation of key hydrological processes. A combination of the two approaches has largely occurred just in the last decade with the advent of available spatial data sets. Before the advent of extensive data sets that could be used to describe the heterogeneity of the land surface, most hydrological models conceptually lumped landscape conditions into average or typical parameter estimates which were often calibrated to available rainfall–runoff data. In this instance it was recognized that the calibrated parameters behaved as 'effective' parameters that may be unique to the calibration conditions. The desire to develop more physically realistic, distributed models has been motivated partially by the desire to produce more general models that could confidently be used in situations for which no data is available and for forecasting change in hydrological behaviour due to a variety of land use or climate changes. An important part of this goal is to replace the dependence of models on calibrated 'effective' parameters with physically realistic process descriptions that use parameters that can be determined by the direct observation of land surface conditions.

The evolution of more physically realistic models of hydrological processes has been largely based on experiments carried out on small uniform field plots or under laboratory conditions. Application of the principles and theories learned under these conditions to larger, more complex watersheds necessitates a simplification of the model process description and some method of dealing with land surface hetero-geneity. Often, although not always, the greater the degree of heterogeneity acknowledged or represented by the modeller, the greater the degree of process simplification in recognition of the inability to estimate the necessary model variables or parameters. In this respect we may view a number of existing models as existing along a spectrum ranging from those that emphasize a physically complex description of processes while averaging or lumping landscape heterogeneity, to those models that emphasize a representation of the landscape heterogeneity while simplifying process representation. This is by no means a good categor-ization of all models, and the ability to determine more complex sets of spatially distributed surface variables and parameters is increasing with advances in remote sensing and spatial data processing.

As models become more distributed spatially experimental evidence indicates an unavoidable need for some effective or calibrated, rather than directly observed, parameters, no matter how physically realistic or simplified the model is (e.g. Loague and Freeze, 1985). These parameters may be estimated by experi-ence and (numerical) experimentation or by calibration with rainfall–runoff data. The need appears to arise from both the simplification of the process representation and also some degree of inability to actually measure or specify an adequate set of physically based parameters. Beven (1989) and others have discussed the need to retain some effective parameters because of limitations in both our understand-ing and representation of physical processes and in the spatial data available for parameter determination. Note that this need is not necessarily avoided by retaining more physically complex process representations in the models, as this places additional demands on data for adequate parameterization. The unfortunate consequence of relying on the calibrated 'effective' parameters in addition to the GIS derived variables is in the potential confusion of why a model may or may not appear to work relative to the variables used for calibration (e.g. rainfall–runoff.) As a check of the internal consistency, temporal or spatial patterns of model state or flux variables other than those used for calibration could be used as part of a validation exercise, such as the distribution of surface soil moisture, or an observed time series of evapotranspiration, although such data are much more difficult to collect than runoff data and are rarely available except in large field experiments. However, advances in remote sensing technology and the expansion of biophysical models to incorporate a greater number of variables (e.g. vegetation canopy states) will ease this process.

SVAT models designed for coupling with atmospheric circulation models are largely simplified from more detailed plot level models. In general, there is no available data to calibrate these models over large areas, with the possible exception of large field and remote sensing campaigns such as FIFE, BOREAS and HAPEX. The necessary aggregation of land surface conditions up to the level of the atmospheric grid cell involves a large loss of land surface information and concern has been expressed about potential bias in computed land surface exchange. This has spurred a significant amount of research into methods to incorporate subgrid resolution variability in the form of statistical descriptions of the surface. Entekhabi and Eagleson (1989) prescribed the forms of distribution functions of available soil water and then produced areal average evaporation and runoff production by integrating a hydrological model over these distributions, whereas Band *et al.* (1993) and Famiglietti and Wood (1991) derived soil water distribution functions on the basis of terrain variables. Avissar (1991; 1992) and others demonstrated the sensitivity of land surface energy and water exchange to different prescribed distributions of surface conductance with similar means and concluded that significant bias can be gained by not incorporating distributional information in model parameters. Similar conclusions have been observed in rainfall–runoff simulations with regard to the specification of saturated hydraulic conductivity or infiltration capacity (Freeze, 1980; Woolhiser *et al.*, 1990; Wood *et al.*, 1992).

The direct approach of incorporating surface heterogeneity can be cast as

$$E(p) = \int p(x)f(x)\,dx \tag{1}$$

where p is a process model, x is a vector of model parameters and f is the joint frequency distribution of x. Generally, it is too difficult to analytically integrate Equation (1) unless f and p are greatly simplified. Therefore, Equation (1) is often modified as a summation over a frequency table of the model parameters. Depending on the length of x, integration (or summation) of Equation (1) may be overly complex and alternative methods of incorporating surface heterogeneity should be found. Equation (1) can be altered to simulate p over discrete, homogeneous areas if the key driving variables of the model can be effectively partitioned in space. Often the components of the parameter vector x are strongly associated, such as soil, vegetation and topographic patterns. If an organizing principle in the landscape can be found and described, such as a recognizable catenary relation between soil and topographic position, or clear associations of forest stand composition and soils, f can be simplified, facilitating the integration process. Finally, other methods of approximating the effects of the surface heterogeneity may be found short of incorporating the full distribution function. One such method was demonstrated by Bresler and Dagan (1988) and Band *et al.* (1991), which uses a Taylor series expansion about the vector mean of model parameters

$$E(p) = p(\bar{x}) + \frac{1}{2}\sum_i \left(\sigma_{x_i}^2 \frac{\partial^2 p}{x_i^2} + \sigma_i \sigma_j \frac{\partial^2 p}{\partial x_i \partial x_j} + \dots \right) \tag{2}$$

In Equation (2) the distributional information required is reduced to the variance–covariance matrix of the key model variables. These methods assume p can be taken as a spatially independent process. If the behaviour of p is dependent on the surrounding terrain, this spatial dependence must either be directly solved for (e.g. routing of water and sediment) or incorporated into the function by some neighbourhood function or parameterization.

If we accept that distributional information on key model variables is important to incorporate, at least under certain conditions, then we must find efficient ways to estimate the forms of these distributions at scales and over areas appropriate to specific models. Each model will require different levels of information depending on its sensitivity to different variables. As an example, a model with a degree day snowmelt algorithm will not require the same topographic data as a model that resolves radiation exchange. Models that incorporate explicit routing of runoff and/or sediment require spatial patterns of topographic and soil attributes, whereas spatial patterns are not required for other runoff models that do not perform routing. Therefore, the discussion that follows regarding our current ability to measure and describe the distribution of surface attributes must be interpreted in terms of the needs of specific models. Although a number of studies have demonstrated that hydrological models are sensitive to the prescribed distributions of surface attributes, operational modelling requires that the distributions be estimated as appropriate forms for different types of landscapes and regions.

The ability to locate or store the spatial location of individual attribute values is generally either not feasible (e.g. saturated hydraulic conductivity) or is not required for effective simulation over large areas. In these instances, the limit to the locational accuracy of many important land surface variables may be a definable area (e.g. subwatershed, catchment or hillslope), which can be assigned a description of the variable distribution. Therefore, one important function a GIS can be used for is to spatially organize or partition the land surface into physically meaningful units (e.g. catchments, hillslopes, flow elements) for which the distributions of key model variables and parameters can be estimated and stored for simulation. The concept of the representative elementary area (REA) (Wood *et al.*, 1988; 1990) is attractive as it suggests a scale at which the land surface heterogeneity, and the effects of the heterogeneity on hydrological processes, stabilizes. If this scale exists such that the variances and covariances of key variables are invariant in land units above a certain threshold size, the specification of distributional information required in Equation (1) or (2) would be substantially facilitated. If such a concept is applicable to real landscapes, it presumably exists at different scales in different types of terrain. It may be possible to identify these scales using appropriate high-resolution data sets and GIS processing in conjunction with specific process models as demonstrated by Wood *et al.* (1988).

We conclude that GIS can be used for a number of functions in seeking to both structure and extend watershed models to larger scales. These include:

1. Determining the spatial patterns of surface attributes at scales or resolution appropriate to route water through the terrain or represent spatial dependence of certain processes (such as net erosion, run-on infiltration or advection).
2. Extracting and representing the watershed structure by the nested system of the stream network and associated hillslope and subcatchment drainage areas.
3. Sampling and constructing statistical descriptions of key surface variables including variance and covariance structures or joint distribution functions.
4. Determining the effects of scale on the distribution functions of key land surface variables and the correlation structures between these variables, and investigating the scaling behaviour of these distributions as they are sampled across different resolutions.
5. Optimal partitioning of the surface into sufficiently homogeneous land units minimizing within unit variance, or into functional watershed units at a scale corresponding to the REA (if it can be shown to exist).
6. Facilitating the storage and retrieval of observed biophysical data for comparison with a set of state and flux variables predicted by the model, such that modelled spatial and temporal patterns of variables not used for calibration can be assessed.

The ability of any GIS to perform these tasks is, of course, highly dependent on the available data sets, so our discussion in the following sections is with reference to currently available data and data that should shortly become available through advances in remote sensing and information processing technologies.

GIS DATA MODELS, SPATIAL GENERALIZATION AND AGGREGATION

Spatial information stored in a GIS has traditionally been represented with a cartographic model in one of two formats: vector or raster. Although there are numerous variations of data structures based on each of these spatial data formats, a general differentiation between them may be described in which vector formats use a node, line and area model to flexibly distinguish between non-overlapping areas (or polygons), whereas a raster model samples the occurrence of events over a regular, predetermined grid. The triangular irregular network (TIN) can be thought of as a vector rendition of a DEM as it represents a set of contiguous, non-overlapping areas (triangles) with uniform slope and aspect as determined by the density of elevation samples. The vector or polygon format has been referred to as the area–category model and is largely inherited from the format of paper maps produced by traditional cartographic techniques. Although not all area–category data sets in a GIS are produced by digitizing paper maps (they are often produced by vectorization of a classified raster image), the data model is that of a cartographic product that is fixed at a given scale and level of generalization as an exhaustive partition of space into discrete entities with no internal pattern or variance, although statistical information on internal variance may accompany attribute information.

It has been recognized that the use of the cartographic model for many types of environmental information results in data error and bias. Mark and Csillag (1989) discussed errors associated with the use of area–class maps to describe the spatial distribution of soils in a GIS and suggested that the use of sharp boundaries to delimit soil polygons produces unrealistically steep gradients and a sense of artificial accuracy in soil properties. The process of cartographic generalization can be viewed as a mode filter, in which the dominant surface property is assigned to areas delimited as a function of map scale and surface patterns. As map scale decreases (covers larger areas), the degree of spatial modal filtering increases. Unless small inclusions are considered to be important by the cartographer, their occurrence is not explicitly represented. Remote sensing derived information is often categorized and filtered with a minimum area operation as preparation for vectorization, mimicking the cartographic process and significantly reducing the data set information content.

If statistical information on polygon attribute occurrence is unknown or not maintained within the GIS, which is generally so, the combination of two or more generalized spatial data sets by polygon overlay may not maintain the spatial co-occurrence of critical surface attributes, especially when the data were drawn from sources of different scales. In this respect the cartographic model in GISs may be regarded as a poor sampling strategy for estimating the joint distribution function of surface attributes. Although this may appear as an obvious conclusion, it should be noted that the cartographic model is still the dominant model in current GIS, and standard practice is to digitize whatever existing paper map products are available regardless of the mix of scales this may produce. If the covariation of surface attributes is critical to a modelling strategy, it is essential that methods be developed that can improve the spatial information contained in these data sets.

Raster data formats are often preferred in many environmental simulation projects as they more readily lend themselves to a finite difference structure and are more compatible with remote sensing imagery and digital terrain data. As a local model is scaled up to larger areal extents, the volume of raster data expands rapidly at a fixed resolution and there is pressure to reduce resolution for the effects of both data volume and computational burden. As the resolution decreases, the surface is filtered by the process of aggregating larger areas into the cell unit area, which takes only one attribute value in the typical raster data model. The method of assigning the cell attribute value may be critical to the resulting attribute distributions (Moore et al., 1993c). If categorical attributes are aggregated, a mode filtering occurs, with results similar to the mode filtering in cartographic generalization as categories that typically occur in small areas are progressively lost. Aggregation with mean values tends to preserve the distribution mean, but reduce the variance. If subsequent spatial operations on the surface are carried out, the results of these operations may be substantially affected by the reduction in pixel to pixel variance. As an example, as the resolution of a DEM is decreased, the computed slopes also decrease as the elevation distribution approaches the mean elevation value. This will have secondary effects on the distribution of derivative geophysical products, such as direct beam radiation as the spherical distribution of the surface normal approaches the vector mean surface normal for the landscape, and consequently the patterns and distribution of potential evapo-transpiration. In addition, any catenary relations between soil, vegetation, terrain and microclimate that exist at the scales appropriate to the modelled processes may be lost.

Another problem with raster data that becomes more noticeable as the cell size is increased is the inflexibility of the grid compared with vector data sets. Whereas vector data sets provide primitive elements for zero-, one- and two-dimensional objects (points, lines and areas), raster data provide only a unit area primitive that must be used to approximate points (single cells), lines (chains of cells) and areas (connected component regions). The fixed dimension of the unit area makes computation of any geometric information very error prone, unless the object for which information is computed has dimensions many times the unit area dimension. The location and width of the stream channel and the topography of the area right around the channel can be critical determinants of runoff behaviour, but are typically of such small dimension (with the exception of very large rivers) that use of raster data processing to estimate these properties is nearly useless.

The effects cited here pertain to problems of aggregation and generalization of the surface in both vector and raster GIS. Therefore, scaling surface hydrological models by reduction in the resolution and aggregation of spatial information in either vector or raster models runs the risk of biasing results by progressively biasing the representation of the surface heterogeneity. The extent to which hydrological models using this aggregated information are biased depends on both the sensitivity of the model to parameter distributional information and the success of any attempts to incorporate descriptions of subgrid (or subpolygon) resolution statistical information.

SCALE AND RESOLUTION EFFECTS ON SURFACE ATTRIBUTES

Digital terrain analysis

Over the past decade numerous approaches have been developed for the automated extraction of watershed structure from grid DEM (for a review, see Tribe, 1992). The most widely used method for the extraction of stream networks that has emerged is to accumulate the contributing area upslope of each pixel through a tree or network of cell to cell drainage paths and then to prune the tree to a finite extent based on a threshold drainage area required to define a channel or to seek local morphological evidence in the terrain model that a channel or valley exists. The drainage area accumulation methods have also been adapted to produce the distribution of the relative wetness indices used in TOPMODEL, TOPOG and other distributed and quasi-distributed catchment models (Band and Wood, 1988; Moore *et al.*, 1988). Only recently have researchers shown interest in the impact of the various methods used for the computation of drainage area or the resolution and quality of the original terrain data on the derived terrain variables.

Stream network and watershed structures

The techniques used for the generation of the drainage path network by surface routing of drainage area and local identification of valley forms are ultimately dependent on a topographic signal generated in a local neighbourhood on the DEM, although a form of graph optimization is used in the watershed delineation routine in the GRASS GIS (USACERL, 1991). As the approach is used to extract watershed structure with increasingly lower resolution terrain data, higher frequency topographic information is lost as the larger sampling dimensions of the grids act as a filter. If watershed structural information is used to drive a hydrological model, the scaling behaviour and consistency of the derived stream network with grid dimension needs to be addressed. As an example, the use of the stream network for the geomorphological unit hydrograph computations (e.g. Rodriguez-Iturbe, 1993) requires information on network branching characteristics for state transition probabilities. If resampling of the terrain data to lower resolution results in scale dependencies in basic network statistics such as bifurcation, length and area ratios (R_b, R_l and R_a, respectively), the simulated hydrological behaviour of the basins may shift based on the source or sampling density of the DEM.

Helmlinger *et al.* (1993) have documented just such a dependency in three stream networks extracted from a DEM. In their study, it was shown that R_b, length ratios R_l and the estimated fractal dimension were not stable across changes in the area threshold required to define first-order channels, or as the resolution of the DEM is changed. They suggest that a slope–area threshold relation, following Dietrich *et al.* (1993) and Montgomery and Dietrich (1988), would be more appropriate for defining channel heads, although this relation must be known *a priori* for a region. The lack of stable scaling behaviour in R_b should not be surprising as numerous data sets show a concave upward plot of stream number–stream order plots (showing 'excess' lower order streams), even for data sets carefully field mapped for evidence of concentrated surface flow (e.g. Maxwell, 1960), and Shreve (1966) demonstrated almost three decades ago that this corresponds to the most probable trend of stream number–stream order relations assuming topologically random channel networks. These observations contradict the assumptions of LaBarbera and Rosso (1989) that R_b and R_l are constant within a basin and independent of scale. In this regard, a critical aspect for stream networks extracted from a DEM is the reliability of first-order channel definition at a given threshold area or resolution.

A more basic impact of DEM resolution would occur if the connectivity of significant portions of the watershed is altered as a result of DEM resolution. Figure 1 shows stream networks derived from DEM resolutions of 100 m and 1 km. In this example, within the Rocky Mountains of western Montana, the

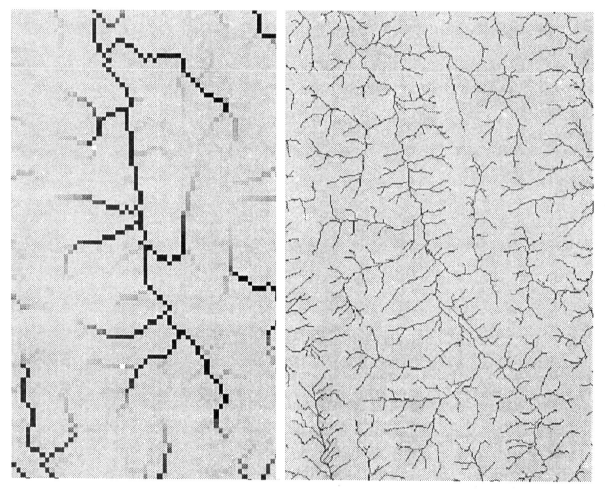

Figure 1. Drainage network for the South Fork of the Flathead River in Montana extracted from a 100 m and a 1 km DEM. The stream network is largely pruned moving to the lower resolution DEM without significant errors in connectivity

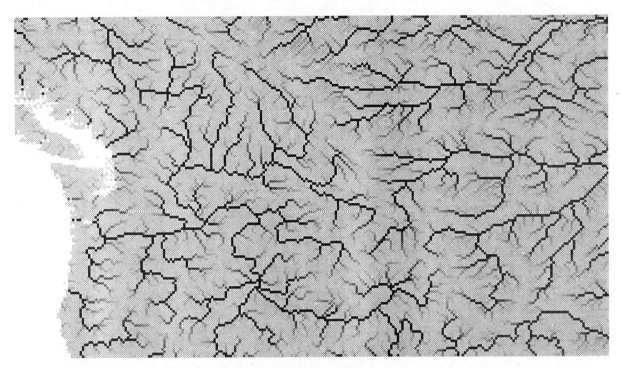

Figure 2. Drainage network for the Pacific Northwest and western Canada as extracted from ETOPO5 (10 arc minute) terrain data. Although most of the drainage systems are adequately represented, some topological errors occur due to low horizontal resolution. Note the disconnection of the Fraser River just upstream from its outlet

network structures appear only to be pruned without significant connectivity problems. Therefore, over at least the limited range of spatial resolutions used here and within a similar geomorphic province (alpine glaciated with large valleys), the topological structure may simply be filtered (pruned). However, more serious problems emerge with continental to global scale data sets, or in areas of lower relief. Figure 2 shows the drainage networks of North America as extracted from the ETOPO5 (10 arc minutes) data set. In this instance the horizontal and vertical resolution is clearly not sufficient to capture the proper flow direction in low gradient regions or where the watershed narrows (e.g. note some confusion between the Columbia and adjacent drainage networks). These errors would be minimized with higher resolution data, but it does illustrate limits with these methods applied to currently available continental to global data sets. Effects on flow routing through the grid cells of the DEM are discussed in the next section.

Representation of the watershed has expanded in the last few years to include the full surface of nested watersheds and hillslopes (or valley sides) draining into the channel network. Lammers and Band (1990) used a formal model of watershed geomorphology to identify explicitly linked stream links and hillslope areas from digital terrain data. An alternative model of the full surface was developed by O'Loughlin (1981; 1986) and Moore et al. (1988) of a fully connected graph of flow strips formed by the intersection of contour and slope lines. Each method allows the mapping of surface attributes into each landscape surface unit (hillslopes or flow elements) to facilitate distributed modelling of hydrological and ecological processes. The hillslope and stream channel method used by Lammers and Band (1990) provides an aggregation strategy by pruning the stream network and combining the individual hillslopes into larger, more complex units within which the internal variability of surface attributes and topography is statistically summarized.

One advantage of using functional hillslope units is that the surface variance within the units is significantly lower than comparably sized grid cells as the hillslopes follow the major topographic features retained at given scales (Band, 1989b). In mountainous regions in which differences in microclimate and

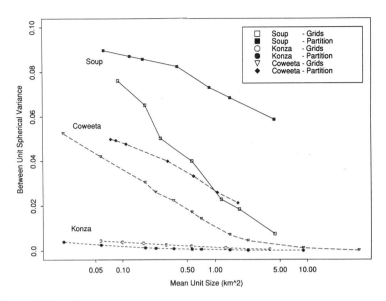

Figure 3. Spherical variance of the surface normal for landscapes represented by grid cells and hillslope facet partition for three different catchments. A greater amount of the land surface variance is maintained by a hillslope partition in the two steeper catchments (Soup and Coweeta) with no significant difference in the Konza Prairie site

associated soil and vegetation patterns are largely driven by solar illumination, the hillslope (or valley side) partition produces an effective method of organizing and representing surface heterogeneity (Band, 1991), especially compared with grid cells of similar size. Figure 3 shows the spherical variance of the surface normal computed for landscape partitions into hillslopes and grid cells of different sizes for three different terrain types. In these plots, Soup Creek in Montana is the steepest and most complex, followed by the Coweeta watershed in North Carolina and the Konza Prairie site (Swede Creek) in Kansas. As the boundaries of the hillslopes follow the main break points of the terrain (e.g. ridges, valley bottoms), there is significantly less smoothing and more of the terrain complexity is captured, with the exception of the Konza site, which has very gentle topography.

Terrain variables

Work has been carried out on the sensitivity of specific terrain variables to a shift in DEM resolution. Moore *et al.* (1991), Wolock and Price (1994), Quinn *et al.* (1991) and Band *et al.* (1993) have examined DEM resolution impact on the computed distribution of the topographic portion of the hydrological similarity index used in TOPMODEL (Beven and Kirkby, 1979), $\ln(a/\tan\beta)$, where a is the area drained per unit contour and β is the local surface gradient. As resolution decreases (cell size increases), β tends to drop as the terrain variance is filtered (in the limit, the world is flat, or at least spherical), whereas the distribution of a tends to shift towards larger values. The net result for $\ln(a/\tan\beta)$ is shown in Figure 4 for a watershed in western Montana for three different methods of computing the drainage area and the local slope (Band *et al.*, 1993). The methods used are: D8 (terminology from Moore *et al.* 1993a), in which all the drainage area is routed to the lowest adjacent cell; FR8, in which the drainage area is partitioned to all lower adjacent cells weighted by gradient; and the contour-based method of TOPOG (Moore *et al.*, 1988; O'Loughlin, 1990), which accumulates area through a network of flow elements defined by the intersection of the contours and a set of projected slope lines (orthogonals to the contours). In all three instances there is a shift of the distribution towards higher values as the resolution is decreased. The significance of these shifts needs to be assessed relative to the sensitivity of the models or methods using this information.

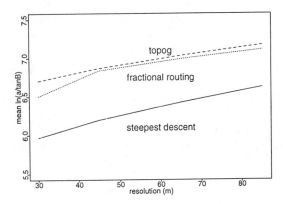

Figure 4. Trends in the mean value of the topographic index [$\ln(a/\tan\beta)$] extracted from Soup Creek in western Montana using three different computation methods over a range of DEM resolutions

Moore *et al.* (1993c) fitted a series of three-parameter log-normal distributions to the wetness index and examined the impact of DEM resolution on the distributional moments for three different watersheds. Although the threshold parameter increased monotonically with grid resolution, the scale and shape parameters increased up to grid resolution of 100 m, then varied irregularly. The lack of regular trends in the latter two parameters above 100 m was interpreted as indicating an approximate length scale to the wetness index, which makes physical sense considering the 100–200 m hillslope lengths in the study catchments. The resolution sensitivity within this range may be some cause for concern as it suggests that large-scale modelling exercises using data with as fine a resolution as 100 m do not capture much of the surface variance in wetness.

Much of the sensitivity to cell resolution for the grid-based methods is dependent on the inflexible definition of flow width, which defaults to the grid cell dimension or a function of grid dimension and the partition of flow to adjacent cells. This reflects the basic problem with a raster GIS of attempting to portray lines or continuously deforming areas with chains or groups of pixels. Alternatively, the TOPOG method also suffers a topological problem by treating streamlines as slope lines; flow cannot cross slope lines so all area is accumulated down the sides of the streamlines across fairly short downslope segments of contour. The large stream side elements are therefore assigned exaggerated drainage areas. It would be feasible to prescribe finite widths for the stream channels and route area and flow into the channel segments, but the scale of topographic information required may be an order of magnitude greater than is captured in the generalized contours that are typically available. It is interesting that both grid- and contour-based methods do not handle stream channels well due to topological inconsistencies and the lack of adequate resolution information to effectively resolve both hillslope and stream channel features, which may occur at very different scales. As much hydrological behaviour is dependent on the local environment around the channels, it is clear that adequate information may not be available by computing the more general topographic information available in either a grid- or vector-based DEM, but may need to be measured or prescribed independently, especially as grid or flow element size increases. Dietrich *et al.* (1993) have pointed out that an adequate prescription of the extent of stream channels (i.e. the location of channel heads) in a given landscape requires independent field assessment as the scale of the channel heads are below that resolvable on standardly available DEM.

Grayson *et al.* (1993) have pointed out that as the information content of a DEM drops below that necessary to capture the significant structure and patterns of the surface, the assumptions underlying distributed models that explicitly route water cannot be met. They argue that this indicates that many recent approaches to deterministic watershed simulation are little more than academic exercises as it is difficult to collect data on terrain and soil information to adequately characterize even very small, intensively studied research catchments. This interpretation is consistent with the recognition that available soil and vegetation information are generally inadequate to parameterize the vertical structure of the rooting zone and canopy to the extent required by many SVAT models over large areas. The gains in accuracy supposedly given by producing models that rely on more physically complex descriptions of soil, canopy or terrain structure and the processes that are modulated by their structures, may be nullified by the lack of sufficient data commensurate to the model sensitivity. Although the models may actually be more accurate representations of the basic hydrological processes (although this has been questioned) the practical availability of adequate databases to actuate these improvements cannot be ignored, especially by the model developer who may regard the collection of the required field data to be someone else's problem. This theme has received much discussion (e.g. Beven, 1989; Grayson et al., 1992a; 1992b; Goodrich and Woolhiser, 1994), but the envelope of what consitutes a model that is too data demanding for practical application is constantly shifting as the technology to collect high-resolution field data improves.

SOIL AND VEGETATION COVER

It is well known that the spatial distribution of soil water and the production of runoff are dependent on catchment topography, soil and vegetation canopy (or land cover) patterns. The spatial variability and association of terrain, soil and vegetation canopy properties have been well documented over a long period of time and covary in observable patterns (depending on the degree of disturbance) that may be identified at specific scales. The catena concept refers to the repeated patterns of soils and vegetation along topographic gradients. At scales of the individual hillslope, a catena may develop in response to the flux of water downslope, with consequent effects on weathering processes, sediment transport, and water and nutrient limitations. At the scale of a mountain block, the catena may be controlled by climatic variations in temperature, precipitation and radiation. There may exist a number of spatially nested covariance structures between the same set of land surface attributes. Therefore, the particular covariance structure or joint frequency distribution estimated for an area may be strongly dependent on the scale or resolution of sampling, and the spatial domain of interest.

Vegetation

Surface vegetation cover is generally derived within a GIS from three major sources: surface sampling (plot information); remote sensing estimation of canopy type and/or properties; and generalized vegetation maps. Remote sensing may be used to produce quantitative measures of vegetation cover, such as canopy leaf area index (LAI), although general algorithms not reliant on empirical calibration and ecosystem specific correction routines (Spanner et al., 1990; 1994; Nemani et al., 1993) are still lacking. In addition, forest canopy stem density and biomass (Waring and Peterson, 1994) can be empirically estimated in many instances. Remote sensing classification of stand type is often used to assign quantitative variables on a per category basis. In the latter case, the information content of the spatially variable surface reflectance spectra is generalized to produce a product analogous to generalized vegetation maps. Although it may be feasible to match the sampling support of the digital terrain analysis with a registered image product (e.g. 30 m USGS DEM with 30 m Thematic Mapper data), the generalization of the remote sensing information effectively degrades the support for the vegetation data relative to the terrain support. Various image processing or mapping conventions, including the concept of minimum mappable areas and cartographic generalization, tend to alter the original information both in terms of spatial and attribute characteristics. In many applications, the only information on vegetation available is either coarse resolution satellite data (best resolution >1 km) or small-scale vegetation maps. Any significant spatial covariance that exists between terrain variables and vegetation (e.g. vegetation clustering in riparian zones) would not be captured. In semi-arid zones, assignment of mean LAI over the full landscape may result in significant bias in runoff production, areal average evapotranspiration and other hydrological processes, as will be demonstrated in the following section.

As discussed earlier, physically complex canopy models which require detailed descriptions of canopy structure in multiple layers, including some idea of rooting depths and densities, are often difficult to extend beyond the level of small plots with much confidence. Generally, the form of the models and available data are such that evaluation of their performance is very difficult, and only aggregated model performance (e.g. catchment discharge) can be assessed. If some parameters are still calibrated, matching a discharge record does not necessarily indicate a correct representation of spatially distributed hydrological processes, just the local success of the calibration.

Soils

Soils information derived from a GIS are generally gathered in a similar manner to vegetation, with the exception that remote sensing often cannot provide critical volume information about the soil properties, especially if the soil is obscured by a vegetation canopy. Substantial progress has been made in estimating near-surface and profile soil water content with active and passive microwave sensors and in the estimation of hydraulic properties by model inversion (e.g. Entekhabi *et al.*, 1994). However, in general, soil spatial information is the least known of the land surface attributes relative to its well-known spatial variability that has been observed in many studies (Nielsen and Bouma, 1985). Consequently, whereas topographic and to an extent vegetation information may be sampled at reasonably high spatial resolution, soil information of the type required for hydrological modelling is either heavily generalized in terms of occurrence and spatial distribution, or is completely absent for many parts of the earth's surface. Although a soil scientist may recognize more detailed soil patterns in the field and may be able to describe the conditions in which soil inclusions occur within larger mapped soil bodies, the method of cartographic representation of the soils at a fixed scale and expense precludes explicit mapping. Even large-scale soil maps (e.g. <1 : 24 000) often cannot capture the inherent variability of critical soil properties at the spatial resolution demanded by the sensitivity of many models. Although some descriptions of the purity of the mapped soil bodies and the occurrence of soil inclusions may be given in a soil report or map legend, this information is typically not included in a GIS.

An additional problem is the high variability of soil hydraulic properties within soil types. It has been demonstrated that the form of hillslope hydrographs can be very sensitive to the distribution of soil hydraulic conductivities and characteristic functions (Freeze, 1980; Binley and Beven, 1989; Sharma *et al.*, 1987, Woolhiser *et al.*, 1990) within hillslope or flow element areas, often well below what can realistically be represented as variation in soil types on maps. Beven (1989) has stated that this level of knowledge of soil properties over the landscape may simply be 'unknowable' from the standpoint of hydrological modelling, although Beven (1981) had earlier argued that at least some of the unknown variation of soil properties might be inferred from known landform–soil relations. If the detailed patterns of these variables are in fact unavailable in any practical sense, then the models requiring this information may also be considered impractical for operational or predictive purposes (Beven, 1989; Grayson *et al.*, 1992b). For given model purposes, certain soil properties may be calibrated as 'effective' values, such as the concept of effective soil conductivity or transmissivity within a catchment (Quinn *et al.*, 1991) for use in TOPMODEL. However, Binley and Beven (1989), using a three-dimensional saturated/unsaturated flow model over hypothetical hillslopes, have shown that an effective single soil hydraulic conductivity may not always be definable or may be event-specific.

It may be possible in many instances to specify the distributional form of critical soil properties exclusive of spatial pattern. Woolhiser and Goodrich (1988) assumed a log-normal distribution of the saturated hydraulic conductivity over length scales of metres to incorporate the effects of soil variability into infiltration excess runoff computations. TOPMODEL (Beven and Kirkby, 1979) and TOPOG (O'Loughlin, 1981; 1986) rely on the co-occurrence of local terrain variables [$\ln(a/\tan\beta)$, discussed earlier], and soil transmissivity, T, given as L^2T^{-1}, the latter of which is rarely available at resolutions matching terrain data. Although T may vary significantly over the terrain, most applications of TOPMODEL have implicitly assumed that the composite similarity term, $\ln(aT_e/T_i\tan\beta)$, where T_i and T_e denote point and area integrated terms, respectively, is dominated by the topographic components (e.g. Wood *et al.*, 1990). However, this conclusion is predicated on the assumption of a lack of significant correlation between transmissivity and the terrain variables, and may also be a result of an underestimation of the actual field variation of T given the spatially filtered nature of map information and the great difficulty of directly collecting the needed information. Certainly, if soil and terrain variables are associated in

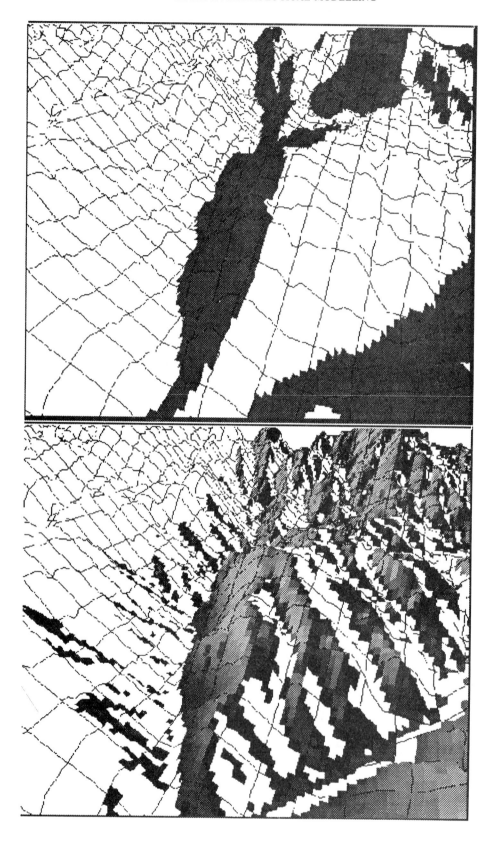

an observable catena, alternative techniques need to be found to both estimate and represent that covariation.

If the distribution of soil properties must be known in conjunction with other observable variables to make physical sense within a model, some other methods must be found to estimate the approximate covariation of the soil with more easily observable variables. To an extent, this is what most soil scientists do when they produce soil maps as the occurrence and extent of a given soil is typically inferred on the basis of observable landform and vegetation associations, although the cartographic model of representing these patterns in area–category maps limits the scale and information content of what is reported. Moore *et al.* (1993a) have pointed out that much of the variation in soils at the mesoscale (in hydrology, more than a small plot, less than a watershed) may be explained with key terrain and vegetation variables, corresponding to the soil–landform conceptual model that has been employed by soil scientists. Efforts to exploit these relationships by the automation of soil inference within a GIS context have been attempted by Skidmore *et al.* (1991), Moore *et al.* (1993b) and Zhu and Band (1994) using a mix of statistical methods, fuzzy set theory and expert systems. In all three of these instances, the results appear sufficiently encouraging, but will require additional work in the areas of GIS and in quantifying and representing knowledge of soil–landform–environment relations to become operational. The advantage to this approach rests on the ability to infer soil types and properties from more easily observable variables, including landform position, local terrain attributes and vegetation cover. If this information can be generated at high resolution by the methods of digital terrain analysis and remote sensing, then the barrier of scale built into the current area–category model of soil maps may be reduced, along with mitigating other problems related to inconsistent mapping, or even a lack of mapping in non-agricultural areas (e.g. mountainous terrain). In addition, as the inference of the soil attributes is based on terrain and vegetation attributes, the covariation of soil–terrain–vegetation systems can be maintained. These methods are not likely to produce soil patterns of sufficient spatial and attribute accuracy to match the high resolution flow element networks or grids used in current kinematic waves models, but may provide a better estimate of the spatial mean values of soil properties over a landscape (e.g. between ridge and valley, knob and swale). This landscape scale often corresponds to catenary relations that develop between soil, topographic and vegetation attributes, and the inferred joint distributions of these variables may provide the statistical information required within landforms or subcatchment areas that is generally not feasible to directly map. An important research question is the relative significance of soil variability at this scale (tens to hundreds of metres) to aggregate runoff production in catchments compared with the variability observed on length scales of metres, which probably cannot be produced by these inference methods, but may be estimated and incorporated on the basis of local sampling.

As an example of these approaches, soil patterns in the Lubrecht Experimental Forest in western Montana were inferred on the basis of local terrain and vegetation cover variables within an expert system with a fuzzy set inference engine (Zhu, 1994). Comparison of a detailed soil map prepared for the area with the inferred patterns over the terrain (Figure 5) shows that the fuzzy inference method was able to produce a more spatially detailed representation of the soil series patterns, with corresponding estimates of the inference certainty for the occurrence of any soil series at a given location. Field observations of soil series occurrence in this area showed that the fuzzy inference method was better able to capture the spatial soil pattern than the soil map as topographic features below the scale of the field mapping could be extracted from the DEM.

METEOROLOGICAL VARIABLES

The lack of adequate data on the spatial and temporal occurrence of precipitation is probably the single most vexing problem in hydrological modelling. In addition to precipitation data, insufficient information on temperature, radiation, wind, humidity and snowpack patterns can also be significant sources of error

Figure 5. Mapped and inferred distributions of a soil series occurring on south facing slopes in an area of western Montana. The inferred distribution values are fuzzy memberships for the soil series

for interstorm periods. The extent to which GIS techniques can estimate or interpolate the distribution of these driving variables from basic meteorological observations and association with landscape (terrain) attributes may significantly improve the general state of hydrological forecasting and explanation.

Over local complex terrain, the general principles of mountain meteorology can be used to extrapolate approximate micrometeorological conditions around the landscape from observations at a base station. Running *et al.* (1987) used simple corrections of solar illumination angles, temperature lapse rates and adjustments to atmospheric transmissivity to produce estimates of micrometeorological variables at different slopes, aspects and elevations. These methods were incorporated within a landscape feature model produced from digital terrain analysis (Band *et al.*, 1991) and extended by Moore *et al.* (1993d) to incorporate greater detail on terrain effects on insolation within a pixel-based system. Orographic effects on precipitation quantities are only crudely handled in these systems as they are based on isohyetal/elevation difference adjustments, which are only stable for long time steps (e.g. monthly, annual).

Another approach has been to use remote sensing data to estimate the distribution of surface meteorological variables (see Goward *et al.*, 1994). Although these methods are promising, they have not yet been demonstrated to be general, and are often limited in temporal and spatial resolution due to satellite resolution, fixed overpass time, repeat imaging time and weather conditions.

Methods to incorporate a dynamic orographic precipitation model with a digital terrain model have been produced (Hay *et al.*, 1993; Leavesley and Hay, 1993; Barros and Lettenmaier, 1994). These methods significantly improve the estimation of spatial precipitation fields over mountain areas over simple isohyetal methods by capturing event conditions by directly estimating the impact of the terrain on the dynamics of the atmosphere. In the absence of major terrain features, however, and with much smaller scale convectional events, the detailed variation of precipitation fields currently appears to defy deterministic modelling at scales important to local hydrological processes. It is expected that some of the current problems with spatial precipitation uncertainty will be eased in coming years (at least in portions of the world) with the expansion of rain-radar systems. However, length scales well below a kilometre have been observed for small-scale convectional activity which is, at least for the forseeable future, below our ability to measure on an operational basis either by precipitation guages or rain-radar systems. For large areas affected by precipitation processes with length scales well below our observational limits, a number of other methods have been developed to synthesize stochastic precipitation fields (Georgakakos and Kavvas, 1987). This is an area where GISs may not contribute substantially, but the GIS based modelling approaches will need to operate in conjunction with the set of stochastic precipitation models.

SENSITIVITY OF HYDROLOGICAL MODELS TO GIS RESOLUTION

To investigate the question of model sensitivity to various GIS techniques for scaling or aggregation of landscape attributes, it is necessary to work within the context of a given model's data requirements and sensitivity. Model sensitivity to input data error propagation can be evaluated to specify the form and acceptable limits of accuracy of input data sets describing land surface attributes. For certain hydrological and ecological processes that are strongly dependent on soil water status, including evapotranspiration and net photosynthesis, much of the process variations at sufficiently large spatial and temporal scales can be explained without any direct observations of soil water using landscape surrogates for water availability. Field *et al.* (1994) have shown that much of the variance in primary productivity due to soil water conditions can be estimated from normalized difference vegetation index (NDVI) measurements at seasonal time steps and continental to global scales. Nemani and Running (1989) have demonstrated the ability to infer relative canopy resistance at regional levels from the Advanced Very High Resolution Radiometer (AVHRR) measurements of NDVI and surface temperature. However, for other processes such as runoff production and for lower space and time steps, NDVI measurements tend to be more stable than soil water content, and the impact of soil water stress and variable source area runoff generation. For questions posed at these levels it is necessary to more directly address the spatial and temporal variability of these processes.

A few studies have explicitly investigated the impacts of varying the resolution of spatial data on simulated hydrological processes. Most of these studies have focused on varying the resolution of terrain data as it has become a commonly available data type at different resolutions and techniques for processing terrain data have become more widespread over the past few years. The chief attributes that have been addressed are surface slope and exposure for computation of radiation loads, and the hydrological similarity index, $\ln(a/\tan\beta)$, discussed earlier for computation of watershed runoff. In terms of radiation modelling, the primary terrain variable of interest is the surface normal vector, of which the distribution will determine the variance of insolation intensity at a given solar position. Although shading from adjacent terrain can also be important, its significance is generally well below that of the surface normal except in very steep terrain at lower solar elevations. Geophysical fluxes which are dependent on surface attitude, most significantly radiation, would follow similar trends in terms of variance scale dependency. Dubayah *et al.* (1991) constructed semivariograms of simulated insolation and found correlation lengths of approximately 200 m, indicating that sensors sampling at lower resolution such as AVHRR and (in the future) MODIS, would not capture any of the landscape variance in these and related surface patterns.

Quinn *et al.* (1991) investigated the sensitivity of TOPMODEL to the use of two different DEM resolutions and for estimation of $\ln(a/\tan\beta)$ for a small African catchment. The results indicated that catchment behaviour as predicted by TOPMODEL could be sensitive to the different approaches for terrain parameterization. Band *et al.* (in press) extended this work by investigating the sensitivity of an integrated hydroecological model, RHESSys (Regional HydroEcological Simulation System), which incorporates TOPMODEL coupled to a forest canopy model that computes evapotranspiration and net canopy photosynthesis. They found regular shifts in runoff behaviour in terms of hydrograph timing and low flow of a 15 km^2 watershed in the northern Rocky Mountains as the DEM resolution changed, although the snowmelt dominated hydrology in this area tended to mask or overshadow the differences in flow regime at the seasonal level. Although the effective TOPMODEL parameters (e.g. m, α; see Sivapalan *et al.*, 1987) were not recalibrated in this exercise to fit observed outflow, it is clear that such a recalibration of effective parameters would produce a degree of dependence on the terrain data resolution.

Figure 6 shows results of the same model comparing the use of AVHRR grid cells and the hillslope disaggregation of the watershed discussed earlier on spatially averaged evapotranspiration and net canopy photosynthesis (Band, 1993). Two variations of the hillslope disaggregation method were used, one treating within-hillslope variance with a TOPMODEL-based approach in which a simple statistical description of the catenary associations of topographic position, soil and vegetation (LAI) is given as sampled from high resolution data, and one treating each hillslope as a bucket with mean topographic, soil and vegetation parameters assigned. The AVHRR grids are also assigned mean surface attribute parameter values, although the (larger scale) catenary associations between soil and vegetation attributes with topographic slope and exposure are not maintained. Over the mountainous terrain of western Montana, AVHRR resolution grid cells significantly average a large proportion of the landscape terrain variance, effectively rendering the topography as rolling hills without the steep north- and south-facing slopes that produce very distinct radiation microenvironments and strongly smoothing variations in forest LAI. The results are such that there is no significant radiation limitation on evapotranspiration or photosynthesis during the growing season. Therefore, water is lost more rapidly by evapotranspiration early in the season compared with the hillslope disaggregation methods in which north-facing slopes maintain a lower, but more steady evapotranspiration rate through the season. Adding the simple catenary associations and TOPMODEL to the within-hillslope hydrology preserves a greater variance in available soil water and tends to buffer the impact of the drought as portions of the watershed remain wet in the summer. The general principle that is suggested by these simulations is that the incorporation of environmental heterogeneity tends to moderate the system response, with lower peak flux rates during non-limiting conditions (e.g. adequate moisture supply, temperature) and higher flux rates during the most limiting conditions, such as the summer drought. Further work with the same modelling approach has suggested that the partition of surface heterogeneity between and within land units over a range of scales (extent of the stream network and numbers of hillslopes) tends to produce stable simulation results of watershed runoff production, evapotranspiration and vegetation productivity as the overall variance is preserved. Therefore, the incorporation of some catenary associations of key model variables can be seen as producing important modulation of the system behaviour. The degree of significance of this effect of course

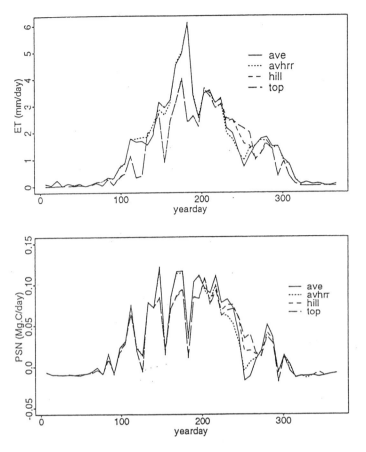

Figure 6. Simulated seasonal trends of evapotranspiration (ET) and net canopy photosynthesis (PSN) for a watershed in western Montana using four different methods of surface representation: (1) hillslope disaggregation with internal hillslope variance (TOP); (2) hillslope disaggregation with no internal hillslope variance (HILL); (3) a 1·1 km grid cell disaggregation of the watershed (AVHRR); and (4) fully lumped representation (avg). Figure from Band (1993), reproduced by permission, *Journal of Hydrology*.

depends on the actual heterogeneity of the landscape and the degree of spatially variable limitations that may occur, but there is a clear suggestion that GIS techniques should be used to develop an understanding and representation of spatial associations between model variables, rather than simply overlaying and aggregating individual spatial variable fields.

DISCUSSION

This paper has focused on the use of GIS techniques to extend hydrological models to larger areas. Emphasis was placed on the ability of GIS to estimate surface attribute patterns and distributions across the landscape and the impact of changing data resolution or scale on these estimations. The cartographic model, which has been the dominant data model in GIS, is based on automating map analysis and map combination to integrate disparate sources of spatial information. Although this was traditionally performed in a qualitative sense as a manual approach, with a large amount of human interpretation, in the GIS environment it has been translated to a straightforward combination of attribute values based on the georegistration of digital data sets, without the input and use of human intelligence as a filter. Owing to differences in generalization and original source resolution between the georegistered data sets, the information content of the overlays do not necessarily correspond. In this instance, overlay analysis can often be a poor sampling strategy unless special care is taken to assure the compatibility of input data sets or the use of other techniques to maintain the 'true' covariance structure of the critical surface attributes. In the latter instance, techniques have been developed to attempt to estimate difficult to sample attibutes such as soil properties from better known variables such as terrain and vegetation cover. A variety of different approaches has been taken, including statistical methods, expert systems and fuzzy set theory.

Change of scale or resolution of spatial data sets involves a loss of information at higher spatial frequencies. Polygon data structures are especially susceptible to this information loss due to the cartographic generalization process that focused on information at length scales based on the mapping scale, whereas raster data allow a greater sampling density. The strong non-linearity of many hydrological processes with respect to soil water content and regulating surface attributes can result in significant bias when the higher frequency variation is lost. Therefore, it is necessary to incorporate subgrid or within-polygon variance and covariance information, which can be taken as a representation of landscape catenary relations, but which most GIS projects currently do not provide. A distinction needs to be made between simple data combination by GIS overlay and the extraction and synthesis of the spatial statistical associations within an area.

GIS techniques are particularly effective in extracting and representing watershed structure from digital elevation data over at least a range of DEM resolution (dependent on terrain conditions). The aggregation of watershed structure and properties can be effectively carried out by partitioning the landscape into component catchments and hillslopes tributary to the drainage network and successively pruning the network to include only higher order or magnitude streams. This approach of representing the heterogeneity of the landscape with hydrologically functional land units (hillslope, subcatchments) and explicitly representing larger scale surface variance by the aggregation of information into within-land unit means can be usefully combined with simple statistical catenary descriptions of within-land unit heterogeneity. As the watershed description is increasingly aggregated into larger, fewer land units, more of the surface variance is shifted to the within-unit statistical information, preserving more of the total landscape variance compared with simple resolution reduction.

As Grayson *et al.* (1993) pointed out, GIS do not 'create' information. However, there appears to have developed an implicit reliance on GIS to provide information adequate to parameterize physically based distributed hydrological models, often at spatial resolution and accuracy levels that are unrealistic given the original source of spatial data. The overlay process in GIS, as currently practised, is often a very poor sampling strategy as it facilitates the combination of incompatible data sets. Although various techniques are being developed to improve the information content of poorly observed variables with associated better observed variables and to concentrate more on catenary patterns rather than simple GIS overlay, model development and operation still needs to be modest and more realistic in terms of matching model complexity and physical realism with the feasible accuracy and precision of available spatial data.

ACKNOWLEDGEMENTS

This paper benefited substantially from comments of two anonymous reviewers. Support from NOAA, NSERC, NASA and the CRC for Catchment Hydrology, CSIRO are gratefully acknowledged. Ian Moore died suddenly after a short illness after the paper was conceived and outlined. Ian was a leader in this field and pioneered the use of digital terrain analysis and GIS with hydrological modelling. I have attempted to represent his views as best I understood them, but any shortcomings or miscontructions in the paper are entirely my own.

REFERENCES

Avissar, R. 1991. 'A statistical–dynamical approach to parameterize subgrid-scale land surface heterogeneity in climate models' In Wood, E. F. (Ed.), *Land Surface–Atmosphere Interactions for Climate Modelling*. Kluwer Academic, Dordrecht. pp. 155–178.
Avissar, R. 1992. 'Conceptual aspects of a statistical–dynamical approach to represent landscape subgrid-scale heterogeneities in atmospheric models', *J. Geophys. Res.*, **97** (D3), 2729–2742.

Band, L. E. 1986a. 'Topographic partition of watersheds with digital elevation models', *Wat. Resour. Res.*, **22**, 15–24.
Band, L. E. 1986b. 'Analysis of drainage basin structure using digital elevation data' In *Proceedings of the Second International Symposium on Spatial Data Handling, July 1986*. International Geographical Union. pp. 437–450.
Band, L. E. 1989a. 'A terrain based watershed information system', *Hydrol. Process.*, **3**, 151–162.
Band, L. E. 1989b. 'Spatial aggregation of complex terrain', *Geogr. Anal.*, **21**, 279–293.
Band, L. E. 1991. 'Distributed parameterization of complex terrain' In Wood, E. F. (Ed.), *Land Surface–Atmosphere Interactions for Climate Modeling*. Kluwer Academic, Dordrecht. pp. 249–270.
Band, L. E. 1993. 'Effect of land surface representation on forest water and carbon budgets', *J. Hydrol.*, **150**, 749–772.
Band, L. E. and Wood, E. F. 1988. 'Strategies for large scale, distributed hydrologic modelling', *Appl. Math. Comput.*, **27**, 23–37.
Band, L. E., Peterson, D. L., Running, S. W., Coughlan, J. C., Lammers, R. B., Dungan, J. and Nemani, R. 1991. 'Forest ecosystem processes at the watershed scale: basis for distributed simulation', *Ecol. Modell.*, **56**, 151–176.
Band, L. E., Patterson, P., Nemani, R. and Running, S. W. 1993. 'Forest ecosystem processes at the watershed scale: incorporating hillslope hydrology', *Agric. Forest Meteorol.*, **63**, 93–126.
Band, L. E., Lammers, R. and Vertessey, R. 1994. 'The effect of different terrain representation schemes on watershed processes', *Z. Geomorphol.*, in press.
Barros, A. and Lettenmaier, D. P. 1994. 'Dynamic modelling of the spatial distribution of precipitation in remote mountainous areas', *Monthly Weather Rev.*, **121**, 1195–1214.
Beven, K. J. 1981. 'Comments on 'A stochastic-conceptual analysis of rainfall–runoff processes on a hillslope' by R. Allan Freeze', *Wat. Resour. Res.*, **17**, 431–432.
Beven, K. J. 1989. 'Changing ideas in hydrology — The case of physically-based models', *J. Hydrol.*, **105**, 157–172.
Beven, K. J. and Kirkby, M. J. 1979. 'A physically based, variable contributing model of basin hydrology', *Hydrol. Sci. Bull.*, **24**, 43–69.
Binley, A. and Beven, K. 1989. 'A physically based model of heterogeneous hillslopes. 2. Effective hydraulic conductivities', *Wat. Resour. Res.*, **25**, 1227–1233.
Bresler, E. and Dagan, G. 1988. 'Variability of yield of an irrigated crop and its causes, 1, Statement of the problem and methodology', *Wat. Resour. Res.*, **24**, 381–388.
Dietrich, W. E., Wilson, C. J., Montgomery, D. R. and McKean, J. 1993. 'Analysis of erosion thresholds, channel networks and landscape morphology using a digital terrain model', *J. Geol.*, **101**, 259–278.
Dolman, A. J. 1992. 'A note on areally averaged evaporation and the value of the effective surface conductance', *J. Hydrol.*, **138**, 583–589.
Dubayah, R., Dozier, J. and Davis, F. 1991. 'Topographic distribution of clear-sky radiation over the Konza Prairie, Kansas', *Wat. Resour. Res.*, **26**, 679–690.
Entekhabi, D. and Eagleson, P. S. 1989. 'Land surface hydrology parameterization for atmospheric general circulation models including subgrid scale spatial variability', *J. Climate*, **2**, 816–831.
Entekhabi, D., Nakamura, H. and Njoku, E. G. 1994. 'Solving the inverse problem for soil moisture and temperature profiles by sequential assimilation of multifrequency remotely sensed observations', *IEEE Trans. Geosci. Remote Sensing*, **32**, 438–448.
Famiglietti, J. S. and Wood, E. F. 1991. 'Evapotranspiration and runoff for large land areas: land surface hydrology for atmospheric general circulation models', *Surv. Geophys.*, **12**, 179–204.
Famiglietti, J. S. and Wood, E. F. 1994. 'Multi-scale modelling of spatially variable water and energy balance processes', *Wat. Resour. Res.*, **30**, 3061–3078.
Field, C. B., Holbrook, N. M. and Benning, T. L. 1994. 'Why do we need soil moisture data? An ecological perspective', paper presented at the *NASA Soil Moisture Workshop, Tiburon, CA, 25–27 January 1994*.
Freeze, R. A. 1980. 'A stochastic–conceptual analysis of rainfall runoff processes on a hillslope', *Wat. Resour. Res.*, **16**, 391–408.
Georgakakos, K. P. and Kavvas, M. L. 1987. 'Precipitation analysis, modelling and prediction in hydrology', *Rev. Geophys.*, **25**, 163–178.
Goodrich, D. C. and Woolhiser, D. A. 1994. 'Comment on "Physically based hydrologic modelling, 1, A terrain based model for investigative purposes" by R. B. Grayson, I. D. Moore and T. A. MacMahon', *Wat. Resour. Res.*, **30**, 845–847.
Goward, S. N., Waring, R. H., Dye, D. G. and Yang, J. 1994. 'Ecological remote sensing at OTTER: satellite macroscale observations', *Ecol. Appl.*, **4**, 322–343.
Grayson, R. B., Moore, I. D., and MacMahon, T. A. 1992a. 'Physically based hydrologic modelling, 1. A terrain-based model for investigative purposes', *Wat. Resour. Res.*, **28**, 2639–2658.
Grayson, R. B., Moore, I. D., and MacMahon, T. A. 1992b. 'Physically based hydrologic modelling, 2. Is the concept realistic?' *Wat. Resour. Res.*, **28**, 2659–2666.
Grayson, R. B., Bloschl, G., Barling, R. D. and Moore, I. D. 1993. 'Process, scale and constraints to hydrologic modelling in GIS', *HydroGIS Conference, IAHS Publ.*, **211**, 83–92.
Hay, L. E., Battaglin, W. A., Parker, R. S. and Leavesley, G. H. 1993. 'Modeling the effects of climate change on water resources in the Gunnison River Basin, Colorado', in Goodchild, M. F., Parks, B. O. and Steyaert, L. T. (Eds), *Environmental Modeling with GIS*. Oxford University Press, Oxford. pp. 173–181.
Helmlinger, K. R., Kumar, P. and Foufoula-Georgiou, E. 1993. 'On the use of digital elevation model data for Hortonian and fractal analyses of channel networks', *Wat. Resour. Res.*, **29**, 2599–2613.
Jenson, S. K. and Domingue, J. O. 1988. 'Extraction of topographic structure from digital elevation data for Geographic Information System analysis', *Photogr. Engin. Remote Sensing*, **54**, 1593–1600.
LaBarbera, P. and Rosso, R. 1989. 'On the fractal dimension of stream networks', *Wat. Resour. Res.*, **25**, 735–741.
Lammers, R. B. and Band, L. E. 1990. 'Automating object descriptions of drainage basins', *Comput. Geosci.*, **16**, 787–810.
Leavesley, G. H. and Hay, L. E. 1993. 'A nested model approach for investigating snow accumulation and melt processes in mountainous regions', *EOS*, **74**, 237.
L'Homme, J. P. 1992. 'Energy balance of heterogeneous terrain: averaging the controlling parameters', *Agric. Forest Meteorol.*, **61**, 11–21.

Loague, K. and Freeze, R. A. 1985. 'A comparison of rainfall runoff modelling techniques on small upland catchments', *Wat. Resour. Res.*, **21**, 229–248.

Mark, D. M. and Csillag, F. 1990. 'The nature of boundaries on 'area-class' maps', *Cartographica*, **27**, 65–78.

Marks, D., Dozier, J. and Frew, J. 1984. 'Automated basin delineation from digital elevation data', *GeoProcessing*, **2**, 299–311.

Maxwell, J. C. 1960. 'Quantitative geomorphology of the San Dimas National Forest, California', *Tech. Rep. No. 19*, Dept. Geology, Columbia University, Project NR389–042, 95 pp.

Montgomery, D. R. and Dietrich, W. E. 1988. 'Where do channels begin?' *Nature*, **336**, 232–234.

Moore, I. D. and Grayson, R. B. 1991. 'Terrain based prediction of runoff with vector elevation data', *Wat. Resour. Res.*, **27**, 1177–1191.

Moore, I. D., O'Loughlin E. M., and Burch, G. J. 1988. 'A contour-based topographic model for hydrological and ecological applications', *Earth Surf. Process. Landforms*, **13**, 305–320.

Moore, I. D., Grayson, R. B. and Ladson, A. R. 1991. Digital terrain modelling: a review of hydrological, geomorphological and biological applications', *Hydrol. Process.*, **5**, 3–30.

Moore, I. D., Gessler, P. E., Nielsen, G. A. and Peterson, G. A. 1993a. 'Soil attribute prediction using terrain analysis', *Soil Sci. Soc. Am. J.*, **57**, 443–452.

Moore, I. D., Turner, A. K., Wilson, J. P., Jenson, S. K. and Band, L. E. 1993b. 'GIS and land surface-subsurface modelling', in *Geographic Information Systems and Environmental Modeling* (Eds Goodchild, M. F., Parks, B. O. and Stayaert, L. T.), Oxford University Press, Oxford. pp. 196–230.

Moore, I. D., Lewis, A. and Gallant, J. C. 1993c. 'Terrain attributes: estimation methods and scale effects', in *Modelling Change in Environmental Systems* (Eds Jakeman, A. J., Beck, M. B. and McAleer, A.), J. Wiley, Chichester. pp. 189–214.

Moore, I. D., Norton, T. W. and Williams, J. E. 1993d. 'Modelling environmental heterogeneity in forested landscapes', *J. Hydrol.*, **150**, 717–747.

Nielsen, D. R. and Bouma, J. 1985. 'Soil spatial variability' in *Proceedings of a Workshop of the International Soil Science Society and the Soil Science Society of America, Pudoc, Wageningen*, 243pp.

Nemani, R. and Running, S. W. 1989. 'Estimating regional surface resistance to evapotranspiration from NDVI and thermal IR AVHRR data', *J. Climate Appl. Meteorol.*, **28**, 276–294.

Nemani, R., Pierce, L. L., Running, S. W. and Band, L. E. 1993. 'Ecological processes at the watershed scale: sensitivity to remotely sensed leaf area index estimates', *Int. J. Remote Sensing.*, **14**, 2519–2534.

O'Callaghan, J. F. and Mark, D. M. 1984. 'The extraction of drainage networks from digital elevation data', *Comput. Vision, Graphics Image Process.*, **28**, 323–344.

O'Loughlin, E. M. 1981. 'Saturation regions in catchments and their relations to soil and topographic properites', *J. Hydrol.*, **53**, 229–246.

O'Loughlin, E. M. 1986. 'Prediction of surface saturation zones in natural catchments by topographic analysis', *Wat. Resour. Res.*, **22**, 794–804.

O'Loughlin, E. M. 1990. 'Modelling soil water status in complex terrain', *Agric. Forest Meteorol.*, **50**, 23–38.

Quinn, P., Beven, K., Chevallier, P. and Planchon, O. 1991. 'The prediction of hillslope flow paths for distributed hydrological modelling using digital terrain models', *Hydrol. Process.*, **5**, 59–79.

Rodriguez-Iturbe, I. 1993. 'The geomorphological unit hydrograph' in *Channel Network Hydrology.* (Eds Beven, K. and Kirkby, M. J.), Wiley, Chichester. pp. 43–68.

Running, S. W. and Nemani, R. 1988. 'Relating seasonal patterns of the AVHRR vegetation index to simulated photosynthesis and transpiration of forests in different climates', *Remote Sensing Environ.*, **24**, 3477–3467.

Running, S. W., Nemani, R. R., and Hungerford, R. D. 1987. 'Extrapolation of synoptic meteorological data in mountainous terrain and its use for simulating forest evapotranspiration and photosynthesis', *Can. J. Forest Res.*, **17**, 472–483.

Sharma, M. L., Luxmore, R. L., DeAngelis, R., Ward, R. C. and Yeh, G. T. 1987. 'Subsurface water flow simulated for hillslope with spatially dependent soil hydraulic characteristics', *Wat. Resour. Res.*, **23**, 472–483.

Shreve, R. L. 1966. 'Statistical law of stream numbers', *J. Geol.*, **74**, 17–37.

Sivapalan, M., Wood, E. F. and Beven, K. J. 1987. 'On hydrologic similarity, 2, a scaled model of storm runoff production', *Wat. Resour. Res.*, **23**, 2266–2278.

Skidmore, A. K., Ryan, P. J., Dawes, W., Short, D. and O'Loughlin, E. O. 1991. 'Use of an expert system to map forest soils form a geographical information system', *Int. J. Geogr. Information Syst.*, **5**, 431–445.

Spanner, M. A., Pierce, L. L., Peterson, D. L. and Running, S. W. 1990. 'Remote sensing of temperate forest leaf area index: influence of canopy closure, understory vegetation and background reflectance', *Int. J. Remote Sensing*, **11**, 96–111.

Spanner, M. A., Johnson, L., Miller, J., McCreight, R., Freemantle, J., Runyon, J. and Gong, P. 1994. 'Remote sensing of seasonal leaf area index across the Oregon transect', *Ecol. Appl.*, **4**, 258–271.

Tribe, A. 1992. 'Automated recognition of valley lines and drainage networks from grid digital elevation models: a review and a new method', *J. Hydrol.*, **137**, 263–293.

USACERL 1991. *GRASS4.0 User's Manual*. United States Army Corps of Engineering, Construction Research and Engineering Laboratory, Champagne, Urbana.

Vertessey, R. A., Hatton, T. J., O'Shaughnessy, P. J. and Jayasuriya, M. D. A. 1993. 'Predicting water yield from a mountain ash forest catchment using a terrain analysis based catchment model', *J. Hydrol.*, **150**, 665–700.

Waring, R. F. and Peterson, D. L. 1994. 'Overview of the Oregon Transect Ecosystem Research Project', *Ecol. Appl.*, **4**, 211–225.

Wigmosta, M. S., Vail, L. W. and Lettenmaier, D. P. 1994. 'A distributed hydrology–vegetation model for complex terrain', *Wat. Resour. Res.*, **30**, 1665–1680.

Wolock, D. M. and Price, C. V. 'Effects of digital elevation model map scale and data resolution on a topography based watershed model', *Water Resources Research*, **30**, 3041–3051.

Wood, E. F., Sivapalan, M., Beven, K. and Band, L. 1988. 'Effects of spatial variability and scale with implications to hydrologic modelling', *J. Hydrol.*, **102**, 29–47.

Wood, E. F., Sivapalan, M. and Beven, K. 1990. 'Similarity and scale in catchment storm response', *Rev. Geophys.*, **28**, 1–18.

Wood, E. F., Lettenmaier, D. P. and Zartarian, V. G. 1992. 'A land-surface hydrology parameterization with subgrid variability for general circulation models', *J. Geophys.*, **97**(D3), 2717–2728.

Woolhiser, D. A. and Goodrich, D. C. 1988. 'Effect of storm rainfall intensity patterns on surface runoff', *J. Hydrol.*, **102**, 335–354.

Woolhiser, D. A., Smith, R. E. and Goodrich, D. C. 1990. 'KINEROS, A kinematic runoff and erosion model: documentation and user manual', *Rep. ARS-77*, Agricultural Research Service USDA, Tucson.

Zhu, A. 1994. Soil pattern inference using GIS under fuzzy logic, *PhD Dissertation*, Department of Geography, University of Toronto.

Zhu, A. and Band, L. E. A knowledge based approach to data integration for soil mapping. *Canad. J. Remote Sensing*, **20**, 408–417.

10

DELINEATING HYDROLOGICAL RESPONSE UNITS BY GEOGRAPHICAL INFORMATION SYSTEM ANALYSES FOR REGIONAL HYDROLOGICAL MODELLING USING PRMS/MMS IN THE DRAINAGE BASIN OF THE RIVER BRÖL, GERMANY

WOLFGANG-ALBERT FLÜGEL

Geographisches Institut, Universität Jena, Löbdergraben 32, D-07743 Jena, Germany

ABSTRACT

A modified concept of hydrological response units (HRUs) for regional modelling of river basins using the PRMS/MMS model is presented. The HRUs are delineated by geographical information system (GIS) analysis from physiographic basin properties such as topography, soils, geology, rainfall and land use using a thorough hydrological systems analysis. The HRUs, once classified by GIS analysis, preserve the three-dimensional heterogeneity of the drainage basin. The River Bröl basin ($A = 216\,\mathrm{km}^2$), Rheinisches Schiefergebirge, Germany was selected to apply the concept. In total, 23 HRUs were delineated and tested with the PRMS/MMS model using a 20-year hydrometeorological daily database. The hydrological systems analysis revealed that interflow is the dominant flow process through the basin's slopes and the major contribution to groundwater recharge and river runoff. This was accounted for by parameterizing the HRUs in the model control file to drain their surplus water not used for satisfying the demand of evapotranspiration to a common conceptual subsurface storage. This storage was simulated by interflow drainage to the groundwater aquifer in the valley floor, which in turn drained to the channel network. The PRMS/MMS model simulated the observed daily discharge very well and the fit was described by a daily correlation coefficient of $r = 0.91$. The NASIM and HSPF models using different means to represent the basin's physiographic heterogeneity were applied to the Bröl basin as well, but did not achieve this correlation. The HRU concept was found to be a reliable method for regional hydrological basin modelling and allows spatial up- and downscaling. Future research on this concept will focus on incorporating the variable precipitation distribution into the classification of HRUs and on the hydrodynamic routing of the modelled discharge. Additionally, satellite imagery must be used for classifying land use in macroscale drainage basins.

INTRODUCTION

During the last few decades, various hydrological models have been developed and extensive published work about this subject is available. Publications by Horton (1945), Penman (1948), Philip (1954), Richards *et al.* (1956), Nash (1957), Crawford and Linsley (1966), Kirkby (1978), Flügel (1979), Leavesley *et al.* (1983), Anderson and Burt (1985), Johanson (1984), Schulze (1989), Flügel (1990), Flügel and Lüllwitz (1993) are examples of the research into developing the conceptual and physical understanding of river basin hydrology. Through time, hydrological basin models have become more distributed and physically based to reflect heterogeneous and complex basin structures, and the various interactive processes controlling basin response. To some extent they also account for spatial physiographic heterogeneity to simulate the hydrological processes at different basin scales (Leavesley *et al.*, 1983; Johan son, 1984; Schulze, 1989).

Transfer between scales is of paramount importance if empirical, physically based process models from microscale field studies are to be regionalized to meso- or even macroscale drainage basins. Three basic questions concerning the up- and downscaling for regionalization are:

1. Is the basin's spatial physiographic heterogeneity in terms of topography, soils, geology, rainfall and land use controlling its hydrology represented in the distributive simulation done by the model?
2. Are the process dynamics described in the model only of local importance, or do they contribute to the runoff generation in the entire basin?
3. Is the transfer between scales intended only spatially or is a combined scale transfer in space and time required?

Work presented by Flügel (1993) addressed these questions by developing a method to regionalize the results of a six year interflow study from a microscale test slope to a mesoscale, loess-covered drainage basin in Germany. The modularly structured hydrological basin model PRMS/MMS (Leavesley *et al.*, 1983) accounts for the heterogeneity of hydrological systems, thereby allowing model adaptations to various hydrological regimes. Detailed process studies (Flügel, 1979; 1990; Flügel and Schwarz, 1983) are essential to develop and validate modules for this model. Geographical information systems (GISs) (Maidment, 1991; Moore, 1993), satellite image processing (SIP) (Engman and Gurney, 1991) combined with the PRMS/MMS model (Leavesley *et al.*, 1983) present an appropriate tool set to deal with the issue of a basin's heterogeneity. The challenge for regional hydrological modelling is to combine these techniques and integrate them into a hydrological systems analysis. Work presented by Flügel and Lüllwitz (1993) from comparative modelling exercises in drainage basins of different scales and climate encourage further development to meet this challenge.

OBJECTIVES

The objectives of the research study are six-fold: (i) based on a hydrological systems approach to further develop the hydrological response unit (HRU) concept as a tool for scale- related regional hydrologic modelling; (ii) to describe the GIS analysis needed to delineate HRUs by preserving the three-dimensional physiographic heterogeneity of a drainage basin; (iii) to apply the modified HRU concept and the GIS method to the basin of the River Bröl, Rheinisches Schiefergebirge, Germany; (iv) to use the delineated HRUs with the PRMS/MMS model to simulate the hydrology of the Bröl basin using a 21 year data set; (v) to evaluate the model's performance using different evapotranspiration models and compare the simulation from PRMS/MMS with those obtained by the NASIM and HSPF models applied to the Bröl basin as well; and (vi) to discuss the modified HRU concept used in the PRMS/MMS model with respect to scale transfer in space and time.

HYDROLOGICAL SYSTEMS ANALYSIS AND THE HRU CONCEPT

A systems approach is essential if the hydrology of a drainage basin is to be analysed and modelled. The process dynamics in the heterogeneously structured soil–vegetation–atmosphere (SVAT) interface is of paramount importance for basin hydrology as it affects evapotranspiration, infiltration, surface runoff and runoff generation. Consequently, regional hydrological models must account for these processes by preserving the three-dimensional physiographic heterogeneity of the basin's 'real world'. This important condition is not met if small-scale general circulation model (GCM) pixels (the terms 'large scale' and 'small scale' are used in this paper with respect to mapping scales) or tributaries delineated from small-scale maps are associated with an 'average' appearance of the physiographic basin's properties.

Figure 1 shows a schematic three-dimensional cross-section of such a 'real world' basin segment, with special emphasis on the presentation of the SVAT interface. The topography can be differentiated into high plains, slopes of varying gradients and the valley floor. Land use differs in space and also in time due to seasonal vegetation changes and economic influences. With respect to the water fluxes through the SVAT interface shown in Figure 1, some basic conclusions regarding the HRU concept can be drawn:

1. Topography and soil types are closely associated with each other due to the processes of weathering and erosion. This interdependency is described for various landscapes by different soil catenas. For

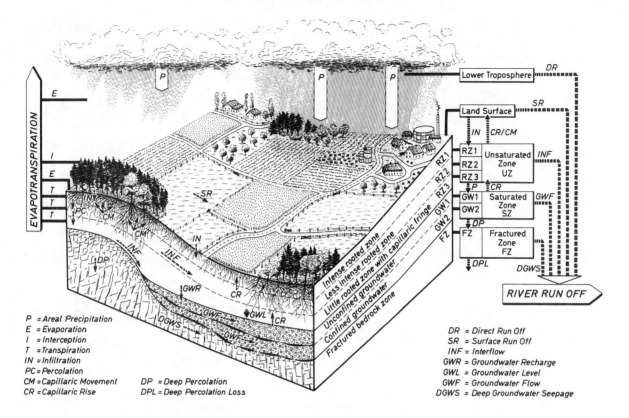

Figure 1. Schematic presentation of basin storages and their interlinked water fluxes

example, a gley soil can almost exclusively be found at the valley floor with shallow groundwater, but rarely on the plains, and certainly not on the slopes.

2. Soils and topographic sequences can therefore be grouped together into pedo-topographical association. In Figure 1, for example, they consist of gley soils at the valley floor with shallow groundwater, slopes with soils of dominant interflow dynamics and high plains with hydromorphic soils.

3. If the underlying geology is not impervious the pedo-topographical associations must be differentiated even further to form pedo-topo-geological associations.

4. Land use can generally be classified into agriculture, rangeland, forest and impervious areas (settlements, streets, etc.). Agriculture will be found on fertile, deep developed soils on the valley floor and on gentle hills. Rangeland and forests are restricted to less productive soils on steeper slopes and on the plains. The development of impervious areas is mainly controlled by socio-economic factors not included in this discussion.

5. Each land-use class is located on a specific pedo-topo-geological association. They contribute different amounts of the rainfall input to evapotranspiration, and to river runoff by surface and subsurface flow.

Based on this brief analysis of the SVAT interface the concept of HRUs uses the following two basic assumptions: (i) each land-use class on a specific pedo-topo-geological association has a homogeneous set of hydrological process dynamics; and (ii) these dynamics are controlled by the land-use management (type of vegetation) and the physical properties of the respective pedo-topo-geological association.

The unit entities consisting of land-use class and its underlying pedo-topo- geological association form HRUs and are defined as follows:

Hydrological response units are distributed, heterogeneously structured entities having a

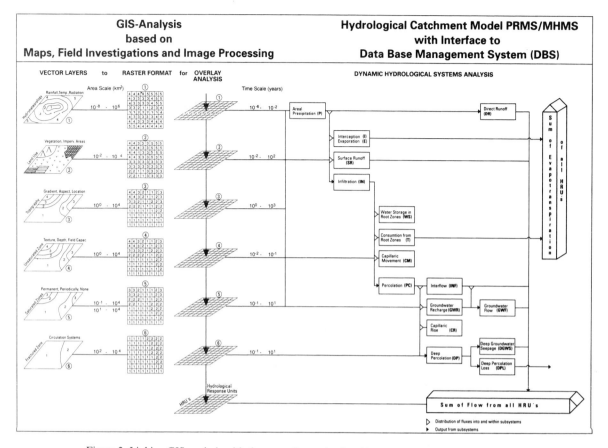

Figure 2. Linking GIS analysis with the water fluxes simulated by the basin model PRMS/MMS

common climate, land use and underlying pedo-topo-geological associations controlling their hydrological dynamics

The crucial assumption for each HRU is that the variation of the hydrological process dynamics within the HRU must be small compared with the dynamics in a different HRU. To evaluate this assumption a prerequisite thorough hydrological systems analysis including field investigations, is required before HRUs can be delineated at the computer desk.

GIS ANALYSIS AND THE PRMS/MMS MODEL

As shown in Figure 1, the SVAT interface has a three-dimensional physiographic structure and is heterogeneously distributed within the basin. Both conditions must be preserved by the delineated HRUs. Additionally, the delineation method used must account for the water fluxes occurring within the SVAT interface as simulated by the PRMS/MMS model (Leavesley *et al.*, 1983). This can be done by disaggregating the three-dimensional heterogeneity into two- dimensional layers, corresponding to the respective conceptual storages simulated in the PRMS/MMS model. The appropriate tool for this task is a GIS and the GIS methods used are digitizing, classification, reclassification and overlay analyses.

The linkage of GIS analysis with the PRMS/MMS model simulation is schematically presented in Figure 2, and can be performed in two different ways:

1. A conceptual link is established, as for this research study, by carrying out the GIS analysis according to the concept of the PRMS/MMS model to simulate the different water fluxes in each HRU.

2. The linkage is interactive, and model parameter files are generated during the delineation of the HRUs in the GIS.

The GIS analysis presented in Figure 2 reveals the different steps needed to delineate HRUs:

1. Each physiographical property of the SVAT interface in Figure 1 is represented by a separate two-dimensional polygon coverage: (i) the atmosphere by the rainfall distribution; (ii) the topography by slope and aspect; (iii) the vegetation by the land use; (iv) the soil by the unsaturated zone; and (v) the geology by the fractured bedrock zone.
2. The polygon coverages are then transformed into a raster format (some GIS use quadtree format instead) for use in the overlay analyses.
3. The HRUs are delineated by screening through the various raster files and sorting out those grid cells which fulfill common physiographic criteria defined by the analyst.
4. After each overlay, the generated smaller classes have to be merged with similar larger classes to keep the number of HRUs manageable.

From this discussion it is evident that the delineation of HRUs, especially the merging of classes, has to be based on the insight gained from the hydrological systems analysis. Only then can the criteria defined as common for the respective HRUs be identified according to the hydrology in the 'real world' basin. Such common criteria can be: *Rangeland* on *gley soil* at the *valley floor with shallow groundwater* over *impervious bedrock*.

The GIS analysis in Figure 2 also shows that the delineated HRUs are distributed over the basin depending on the scatter of land use which is superimposed on the usually less distributed pedo-topo-geological associations. For the location on the valley floor the HRU *rangeland on gley soil etc.* can be just next to the HRU *forest on gley soil etc.*, which in turn may be a neighbour to the HRU *agriculture on gley soil*.

APPLICATION TO THE BRÖL DRAINAGE BASIN

The GIS analysis discussed in the preceding section was used to delineate HRUs in the mesoscale basin of the River Bröl ($A = 216 \, \text{km}^2$). The basin is located in the Rheinische Schiefergebirge, a middle mountain range with an oceanic-type climate, north-east of Bonn, Germany. The mean annual rainfall varies with elevation between 900 and 1200 mm, averaging about 1070 mm. The River Bröl is a tributary of the River Sieg ($A = 2832 \, \text{km}^2$) flowing into the River Rhine north of Bonn.

Hydrological system analysis

The hydrology of the basin was studied by intensive field investigations. Soil profiles were dug in the valley floors, the slopes and the plains, and groundwater and interflow measurements were made. Additionally, discharge hydrographs and the spatial rainfall distribution were analysed from the hydrometeorological database (Dornberg, 1992). From these combined field and computer data analyses the hydrology of the Bröl basin was characterized as follows:

1. The underlying geology from Devonian shale is almost impermeable, and deep percolation in the underlying fractured bedrock zone can be considered as negligible.
2. Surface runoff occurs only during extreme winter snowmelt events with steep temperature rises in the presence of frozen soil, or when extreme summer storms with high rainfall intensities hit the dry soil surface.
3. Rainfall intensities are normally buffered by the forest and grassland vegetation and the soil infiltration capacity can account for the throughfall.
4. The infiltrating water is stored in the unsaturated root zone for consumption by evapotranspiration. If the field capacity is exceeded, water percolates further downwards, creating a wetted zone apparent in soil pits by its greyish mottles above the underlying impervious bedrock.
5. If coarse pores from earthworms or former root channels are common, this slower percolation process is enhanced by the fast downwards flow of draining rainfall from the soil surface towards this zone.

6. Interflow through the shallow soil of the slopes is the dominant mechanism of water fluxes in the basin and was often observed during storms. It is not only restricted to the wetted zone, but during winter storms can flow through the entire soil profile, as was observed in January 1994.

7. Groundwater is restricted to the valley floors and is recharged by direct percolation from the unsaturated soil zone above. The majority of its recharge, however, is received by interflow from the adjacent slopes. During storms the aquifer buffers the storm interflow, causing an additional runoff delay of this discharge component. However, the buffering capacity of the Bröl aquifer is small as the valley floor seldom exceeds 100 m in width, and its deposits vary only between 3 and 5 m in depth. Additionally, these sediments consist mainly of coarse gravel with high permeabilities between 10^{-3} and 10^{-2} m/s, allowing fast groundwater flow.

HRU delineation by GIS analysis

The GIS database used to delineate the HRUs for the Bröl basin was generated from the following data sources:

1. Daily rainfall from six stations located within and in the vicinity of the basin.
2. Land use mapped in the field on a 1 : 5000 scale was generalized to the 1 : 25 000 scale of the GIS base map.
3. A DEM with 50×50 m grid point resolution.
4. The soils map at the 1 : 50 000 scale from the Geological Survey, Nordrhein- Westfalen was photographically enlarged to the 1 : 25 000 scale of the GIS base map.

All maps were digitized and imported into the GIS SPANS, together with the DEM and the rainfall data to create the polygon coverages shown in Figure 2.

Precipitation and land use. The Thiessen polygons of the six rainfall stations were generated by the GIS, and the mean daily rainfall of the Bröl basin was calculated. The average daily rainfall was assigned uniformly for each HRU and this coverage was therefore excluded from the overlay analyses.

Altogether 17 land-use classes were mapped during the field campaign but for the modelling exercise these were reclassified into forest (deciduous + coniferous, $A = 33\cdot8\%$), rangeland (grazing + meadows, $A = 52\cdot6\%$), agriculture (all crops, $A = 2\cdot7\%$) and impervious areas (settlements, streets, $A = 10\cdot9\%$).

Topography and aspect. The drainage basin of the River Bröl has a middle mountain topography, which is typical of the Rheinische Schiefergebirge. The former Permian peneplain was geotectonically uplifted during the Tertiary and was cut through and segmented by the river network. Using the DEM the slopes were grouped into six gradient classes: 0–2% ($A = 2\cdot2\%$); 2–5% ($A = 10\cdot6\%$); 5–10% ($A = 36\cdot4\%$), 10–20% ($A = 42\cdot9\%$); and >20% ($A = 8\cdot0\%$).

The slope aspect is an important factor for the energy budget calculated in PRMS/MMS for each HRU. Using the DEM the four aspect classes: N–E (0–$90°$, $A = 20\cdot2\%$), E–S (90–$180°$, $A = 27\cdot3\%$), S–W (180–$270°$, $A = 24\cdot9\%$) and W–N (270–$360°$, $A = 27\cdot6\%$) were defined.

Soils and geology. Six soil classes were digitized from the soil map, but after comparing their grain size distributions, porosities, water-holding capacities and hydromorphic characteristics they were merged and reclassified as gley soil ($A = 14\cdot0\%$), brown soil ($A = 78\cdot8\%$) and hydromorphic brown soil ($A = 7\cdot2\%$).

The underlying bedrock was considered to be uniform impervious for each HRU, and this coverage therefore was excluded from the overlay analysis

Overlay analyses. As rainfall and geology had properties uniform for all HRUs in the basin, only the coverages of topography, soils, land use and aspect in the listed order were used by the overlay analyses.

The basin's topography was classified according to slope gradients. Therefore it was impossible to distinguish between the flat valley floor and the flat high plains (Figure 1) as they were both grouped together in the same gradient class of 0–10%. The concept of pedo-topo-geological associations described earlier solved this problem. When linking slope classes with their corresponding soil catenas, valley floors and plains could be classified differently.

In the next step the land use was overlayed and finally the coverage of the aspects was used. After each

Table I. Area and physiographic properties of the delineated HRUs for the Bröl catchment

HRU No.	Area (km^2)	Area (%)	Average elevation (m aSL)	Topo-sequence	Slope (%)	Soil types	Land use	Aspect (N, E, S, W)
1	23·4	10·9	213·1	All	All	All	Imperv.	All
2	3·2	1·5	235·4	Plains	0–10	Hy_brs	Forest	N–E
3	4·0	1·9	230·5	Plains	0–10	Hy_brs	Forest	E–S
4	5·6	2·6	231·9	Plains	0–10	Hy_brs	Forest	S–W
5	5·4	2·5	237·7	Plains	0–10	Hy_brs	Forest	W–N
6	3·3	1·5	239·0	Slope	10–20	Brs	Forest	N–E
7	5·2	2·4	214·6	Slope	10–20	Brs	Forest	E–S
8	5·2	2·4	197·2	Slope	10–20	Brs	Forest	S–W
9	7·6	3·5	190·7	Slope	10–20	Brs	Forest	W–N
10	13·8	6·4	211·5	Slope	>20	Brs	Forest	N–E
11	10·7	4·9	213·8	Slope	>20	Brs	Forest	E–S
12	8·8	4·1	153·8	Slope	>20	Brs	Forest	S–W
13	5·0	2·3	219·0	Slope	>20	Brs	Forest	W–N
14	13·1	6·1	213·0	Valley	0–10	Gley	Forest	All
15	15·8	7·3	217·8	Plains	0–10	Hy_brs	Rangel.	N–E
16	9·9	4·6	223·9	Plains	0–10	Hy_brs	Rangel.	E–S
17	4·4	2·1	227·5	Plains	0–10	Hy_brs	Rangel.	S–W
18	6·4	3·0	208·9	Plains	0–10	Hy_brs	Rangel.	W–N
19	10·0	4·7	204·7	Slope	10–20	Brs	Rangel.	N–E
20	18·0	8·4	207·1	Slope	10–20	Brs	Rangel.	E–S
21	9·8	4·5	216·2	Slope	10–20	Brs	Rangel.	S–W
22	8·5	3·9	214·6	Slope	10–20	Brs	Rangel.	W–N
23	18·2	8·5	174·9	Valley	0–10	Gley	Rangel.	All

Abbreviations: m aSL = metres above sea level; imperv. = impervious' rangel; −range land; hy_brs = hydromorphic brown soil; and brs = brown soil.

overlay, the generated smaller classes were merged with similar larger classes, and the coverage was reclassified. During this procedure agricultural areas were merged with the rangeland as their subclasses had areas less then 0·2%, and in each HRU slopes with less then 10% gradient were grouped together into the gradient class 0–10%.

After working through the GIS database 23 the HRUs listed in Table I were delineated. Each HRU was defined by a common set of three- dimensionally composed physiographical properties, which were used to set their respective model parameters in the PRMS/MMS control file.

TESTING THE HRU CONCEPT USING PRMS/MMS

The PRMS/MMS model developed by the US Geological Survey is a modularly structured, deterministic, distributed physical hydrological basin model (Leavesley *et al.*, 1983). The model is process-orientated and uses physical laws and empirical equations to describe the flux of water within the HRUs as shown in Figure 3.

A discussion of the model design is beyond the scope of this paper, but is described in the user manual (Leavesley *et al.*, 1983) and published applications (Flügel and Lüllwitz, 1993).

Parameterizing HRUs

The parameter control file for PRMS/MMS contains basin-related and HRU-specific parameters. Both were set according to the physiographical properties of the basin and the different HRUs.

Land use. The 23 HRUs had three land-use classes: forest, rangeland and impervious settlements. Coniferous and deciduous forest were evenly mixed at a stand scale and each forest type formed about 50% of the forested areas. By setting the seasonal coverage density parameters accordingly, this mixture of forest types was accounted for. The different management practices of hay harvesting and grazing could not be

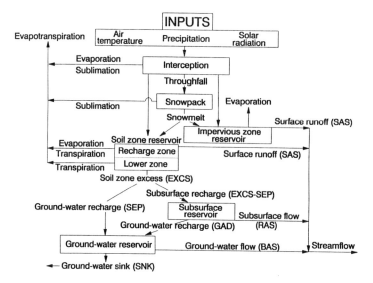

Figure 3. Process-controlled water fluxes between conceptual storages simulated by the PRMS/MMS model for each HRU

parameterized, and therefore all rangeland HRUs were given the same vegetation parameter values describing seasonal changes in interception and coverage density. The percentage of impervious surface for the settlements was obtained from the field survey and by setting this parameter accordingly the various densities of urban and rural settlements were accounted for.

Interflow dynamics in the conceptual subsurface storages. In Table I the HRUs are characterized by their land use, soils and location in different topo-sequences. They infiltrate rainfall into their common unsaturated storage covering the plains, slopes and the valley floor (Figure 1). This storage is drained by interflow to the stream or to the groundwater aquifer in the valley floor. The transfer of the interflow dynamics from the 'real world' (Figure 1) into a conceptual flow chart is shown in Figure 4.

Interflow from each HRU can be represented in PRMS/MMS by parameteri zing the exponential drainage functions applied to the unsaturated storages on the slopes and the plains. Thereby interflow was routed through the soils to the groundwater aquifer in the valley floor and to the river channel. The unsaturated storage of all HRUs on the valley floor was parameterized to drain only to the underlying shallow groundwater.

Modelling strategy

For the modelling exercise, a 21 year data set (1970–1990) of discharge, rainfall and climatological data was available. The modelling was performed in the following way:

1. The first two years were used as an initialization period.
2. The driest and wettest years were selected for calibrating the parameters according to the two criteria: (i) simulated flow should fit best with observed data; and (ii) model parameters should be reasonable with respect to the physical properties of each HRU.
3. Parameter optimization and sensitivity analysis were performed using the 1980–1985 data period.
4. Parameter validation was carried out by running the model with the complete 21 year data set.

To evaluate the model's performance with respect to land-use sensitivity, five different evapotranspiration models were tested with the data set using monthly adaptation factors to account for seasonal changes of the vegetation cover density.

The results obtained from the PRMS/MMS model combined with the HRU concept were compared with those of the NASIM and HSPF models, also applied in the Bröl basin (Mülders, 1992; Daamen, 1993).

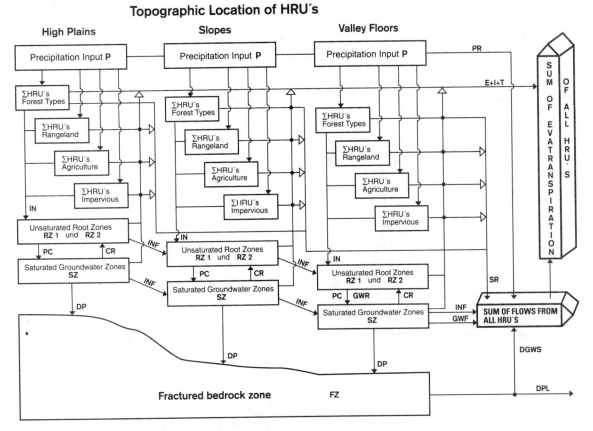

Figure 4. Interflow in the conceptual subsurface storages between HRUs located on different topo-sequences

Model results

The PRMS/MMS model offers many outputs to evaluate the model's performance, and each of its conceptual storages (Figure 3) can be evaluated separately. Such a intensive discussion, however, is beyond the scope of this paper and the following sections therefore will focus only on the HRU-related issues of the model's performance.

Fit between observed and simulated data. PRMS/MMS produced a good fit between simulated and observed discharge, as shown for the water year 1985 in Figure 5, and thereby proved the reliability of the HRU concept combined with the PRMS/MMS model. The model, however, tended to underestimate winter storm hydrographs and base flow during dry weather periods.

Two plausible reasons for this underestimation can be given:

1. Precipitation data were not corrected for systematic errors caused by wind drift and splash, which can cause up to 10% error (Sevruk, 1982). Separate snow measurements were not available and during snowfall only the snow catch of the raingauge ($A = 200\,\text{cm}^2$) was recorded.
2. The model did not simulate deep percolation from the fractured zone into the river channel as the underlying bedrock was considered to be impervious. In reality, however, there might be some inflow from the fractured zone into the river, supporting higher baseflow than was simulated by the model.

The good fit between observed and simulated discharge during the 21 year time period is described by a high correlation coefficient of $r = 0.91$. The discharge variability not explained by the model is only 18% and can probably be reduced if improved data for rainfall and the fractured zone are available.

Sensitivity analysis. The PRMS/MMS model was moderately sensitive to the parameters describing

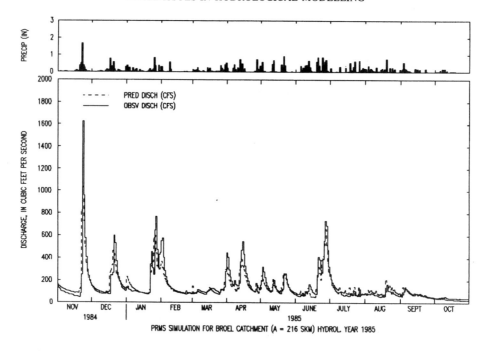

Figure 5. Modelled and observed discharge of the River Bröl in the water year 1985 (1 inch = 25·4 mm; 200 cfs = 5·664 m³/s)

interception storage capacity. High sensitivity was found for parameters describing the water holding capacity of the unsaturated storages. The latter were defined by the rooting depth of the vegetation in each HRU. These parameters are directly related to the vegetation cover and reflect the importance of accounting for the land-use heterogeneity in regional hydrological basin models.

EVALUATION OF MODEL PERFORMANCE

The performance of the PRMS/MMS model combined with the HRU concept was also evaluated by using different submodels for evapotranspiration. Additionally, the simulation obtained from PRMS/MMS was compared with those obtained by two other physically based, distributed basin models (NASIM and HSPF), both of which use different methods to represent the physiographic heterogeneity of the drainage basin. These comparisons provided additional insight into the application of the HRU concept for regional hydrological modelling.

Table II. Results of discharge simulation using different evapotranspiration models

Criteria of fit	Jensen and Haise	Haude model	Penman–Monteith	Kimberly Penman	Fao–Penman
r	0·8846	0·8502	0·9087	0·8757	0·8698
r^2	0·7812	0·7228	0·8257	0·7669	0·7565
$\Sigma Q_{sim}(m^3/s)$ ($\Sigma Q_{obs} = 30\,144$)	31 822	35 792	29 614	29 605	26 116
$\Sigma Q_{sim} - \Sigma Q_{obs}$	1678	5648	−530	−539	−4028
$Q_{min}(m^3/s)$ ($Q_{min\,obs} = 0·317$)	0·054	0·055	0·035	0·037	0·032
$Q_{max}(m^3/s)$ ($Q_{max\,obs} = 70·125$)	56·287	71·699	71·065	71·291	70·428

Evapotranspiration models

To evaluate the influence of land use on the simulated hydrology, five recognized evapotranspiration models were tested. Of these models only the Penman–Monteith model accounts for the vegetation cover in the different HRUs. All five models modified the flow simulation performed by PRMS/MMS, and different qualities of fit between observed and simulated discharge were obtained, as listed in Table II.

The correlation coefficients varied between 0·85 for the Haude model and 0·91 for the Penman–Monteith model. This figure is confirmed if the mean annual discharge and the minimum and maximum values are compared. The result of this model exercise indicates the importance of representing the land- use cover in the distributed modelling of basin hydrology.

Comparison with other physically based basin models

To evaluate the simulation obtained with the PRMS/MMS model in comparison with other modelling approaches, the two hydrological basin models NASIM and HSPF were applied to the Bröl basin by co-workers of the author (Mülders, 1992; Daamen, 1993). Both modelling exercises made use of the same hydrometeorological and GIS database.

The two models differ in respect of their ability to represent the three- dimensional heterogeneity of the basin. The NASIM model uses distributed subbasins derived from topographical maps, which were grouped together in a hierarchical structure. The topography is simplified by an 'open book' type of scheme with uniform slopes left and right of the channel and a headwater area having the shape of a circle segment. The HSPF model uses representative hydrological units (RHUs) delineated along homogeneous stream segments. The delineation of these RHUs is subjective and linked with the distribution of the rainfall stations in the basin they received their input from. However, soils and land use can be parameterized for each RHU in a similar manner to the HRUs in PRMS/MMS. NASIM and HSPF use a modified Penman model to simulate evapotranspiration from the modelled entities.

As a result of the model comparison, the PRMS/MMS model combined with the HRU concept was found to give a better fit between observed and simulated discharge than the simulations performed with NASIM and HSPF. After the parameter optimizations were completed for all models, the NASIM simulation achieved a correlation coefficient of $r = 0·88$ (Mülders, 1992) and the HSPF simulation a coefficient of $r = 0·86$. Both coefficients are below the correlation obtained by the PRMS/MMS model using the Penman–Monteith evapotranspiration model for the different HRUs.

HRUs and PRMS/MMS APPLIED FOR TRANSFER BETWEEN SCALES

The discussion about the use of the HRU concept combined with the PRMS/MMS model for transfer between scales refers back to the three questions put forward in the introduction of this paper. They were, in brief: (i) Is the basin's physiographic heterogeneity represented in the model? (ii) Are the processes simulated by the model subject for regionalization? (iii) Is scale transfer required in time and space? Associated problems such as pixel accuracy, advantages of quadtree structured GIS and the accuracy of DEMs are not covered herein, and the reader is referred to GIS textbooks.

Representing spatial heterogeneity

The GIS method to delineate HRUs *per se* is not restricted to a particular spatial scale and can be applied for basins ranging in size from field plots to macroscale drainage basins, providing the computer hardware is capable of carrying out the required calculations. The limiting factor for the spatial resolution of the HRU concept is the GIS coverage with the smallest scale in the GIS database. In the Bröl basin this was the DEM with a raster resolution of 50×50 m. The smallest spatial unit entity which can be derived from this DEM has an area of 2500 m^2. Although the land use mapping was performed more accurately on a larger scale, this had no effect on the spatial resolution of the HRUs derived from overlay analyses using this DEM. Three important conclusions in respect to spatial scale transfer can be derived from this: (i) the basic assumption for each HRU having a unique hydrological dynamics is always referring to the smallest unit entity that can be delineated by the GIS according to the limit of spatial resolution of its database; (ii) the scale limit for preserving the basin's spatial physiographic heterogeneity is determined by the GIS coverage of the smallest scale; and (iii) by using the HRU concept these unit entities can be assembled in larger HRUs according to their common land use and physiographic properties, allowing upscaling and downscaling in the entire basin. However, due to the merging of small subclasses with similar larger classes a certain degree of heterogeneity is lost. As a rule of thumb, merging should always be performed in such a way that the characteristics of the basin are preserved. This can be achieved by creating single HRU maps in the GIS and evaluating them together with the topography displayed from the DEM.

Scaling of simulated process dynamics

All physically based basin models, to different degrees, use parameterized process equations derived from empirical field studies. These process models, in principle, are restricted in their application to the study area they were developed from. However, if they use parameters accounting for the physical principles underlying the simulated processes, they can be regionalized to a certain extent providing these parameters can also be reasonably well estimated for other regions.

Limitations on the degree of scale transfer are imposed by the time resolution of the hydrological database. In the Bröl basin, the PRMS/MMS model was used at the daily time-scale mode and the process modules in the model were developed for daily data. Keeping this time-scale, upscaling of the simulated process dynamics in space therefore is possible as the different HRUs only become larger, but the process dynamics of their unit entities stays the same. Downscaling in time requires a different hydrometeorological database with a higher time resolution and the storm modelling mode provided by PRMS/MMS. However, research about the application of the HRU concept at this time-scale is still needed.

Scale transfer in time and space

For hydrological regionalization, intending to combine scale transfer in time and space, the smallest time resolution in the hydrometeorological database and the smallest spatial resolution in the GIS database set the limitations for up- and downscaling. Additionally, the model must provide modules to simulate the hydrological process dynamics at the smallest resolution in time and space. For example, a DEM having a 1×1 m spatial resolution is of little use if only daily data are available for PRMS/MMS in the daily mode. On the other hand using 15-minute observed data with PRMS/MMS in the storm mode having a land-use coverage of a $1 : 500\,000$ resolution does not make much sense either.

The HRU concept combined with the PRMS/MMS model, however, is an appropriate tool set for spatial up- and downscaling on a daily time-scale according to the spatial limitation set by the coverage having the smallest scale in the GIS database.

CONCLUSIONS AND FUTURE RESEARCH NEEDS

Summarizing the discussions the following conclusions can be drawn:

1. The concept of HRUs is capable of preserving the heterogeneity of the three-dimensional physiographic properties of the drainage basin. It therefore is not only applicable to such basins as the Bröl, but can also be used in regional hydrological modelling in other climatic regions.
2. The delineation of HRUs must be based on a thorough and detailed hydrological systems analysis. The insight gained from this analysis is prerequisite to defining the criteria to be common for each HRU, such as land use and the underlying pedo-topo-geological association.
3. GIS analyses combined with the physically based, distributed hydrological basin model PRMS/MMS is an appropriate tool set to delineate HRUs, and to simulate the water fluxes between them.
4. Model parameters can either be derived during the GIS analyses or set afterwards according the physiographic properties of each HRU. By parameterizing the conceptual unsaturated storages interflow dynamics and groundwater recharge can be accounted for.
5. The HRU concept and their GIS delineation was successfully applied to the basin of the River Bröl in the Rheinische Schiefergebirge, Germany. Using the generated GIS database generated for this basin 23 HRUs were delineated.

6. By testing of the HRU concept combined with the PRMS/MMS model using a 21 year daily hydro-meteorological data set for the basin, a good fit between observed and simulated flow was achieved described by a correlation coefficient of $r = 0.91$.

7. The PRMS/MMS simulation was sensitive to model parameters directly related to the land use, indicating its importance for the hydrological dynamics of the drainage basin.

8. This conclusion was also supported by results obtained from two comparative modelling exercises. (i) By testing five evapotranspiration models, the Penman–Monteith method which accounts for the vegetation cover obtained the simulation with the best fit. (ii) By applying the two comparative models NASIM and HSPF, which use different means to represent the basin's physiographic heterogeneity, the HRU concept combined with the PRMS/MMS model again obtained a better simulation fit.

9. If scale transfer is required the scale limitations of the hydrometeorological data and the GIS database must be considered. They restrict the extent of up- and downscaling possible in time and space. The HRU concept, combined with the PRMS/MMS model in the daily mode, however, can be used for spatial scale transfer in regional hydrological modelling.

Based on the results presented herein, the following future research needs can be identified:

1. For large-scale basins, spatially distributed physiographical properties have to be derived from satellite imagery and combined with the GIS–PRMS/MMS tool set.

2. The spatial heterogeneity of the precipitation input has to be incorporated into the GIS delineation of HRUs.

3. The presented application did not include hydrodynamic discharge routing, and further research is needed to determine the basin size and stream length for which such flood routing must be included.

4. The modular structure of PRMS/MMS allows extension of the model towards more flexibility and process orientation, and additional process modules must be developed to improve the existing model.

ACKNOWLEDGEMENTS

The author is pleased to acknowledge the continuous scientific co-operation and intensive discussions with Dr. George Leavesley, US Geological Survey, Denver, USA. Acknowledgement is also given to the US Geological Survey, Denver, the Deutsche Wetterdienst, Offenbach, the Landesamt für Wasser- und Abfallwirtschaft, Düsseldorf and the Landes Vermessungs Amt Nordrhein-Westfalen, Bonn, Germany, for supply of the model, the hydrometeorological database, and the DEM. Grateful acknowledgement is given to the Deutsche Forschungsgemeinschaft, Bonn, Germany, which financed the study between 1991 and 1994.

REFERENCES

Anderson, M. G. and Burt, T. P. 1985. *Hydrological Forecasting*. John Wiley & Sons, Chichester, 604 pp.

Crawford, N.H. and Linsley, R. 1966. 'Digital simulation in hydrology: Stanford watershed model IV', *Tech. Rep. No. 39*, Stanford Univ., Dept. of Civil Engineering.

Daamen, K.-H. 1993. 'Das hydrologische Flußeinzugsgebiets-modell HSPF und seine Anwendung im Einzugsgebiet der Bröl', *Dipl. Arb.*, Geogr. Inst., Univ. Bonn, 138pp.

Dornberg, P. 1992. 'Ermittlung des Gebietsniederschlags im Einzugsgebiet der Sieg', *Dipl. Arb.*, Geogr. Inst., Univ. Bonn, 298pp.

Engman, E.T. and Gurney, R.J. 1991. *Remote Sensing in Hydrology*. Chapman & Hall, London. 225pp.

Flügel, W.-A. 1979. 'Untersuchungen zum Problem des Interflow', *Heidelberger Geogr. Arb*, H. 56, 176.

Flügel, W.-A. 1990. 'Water balance and discharge simulation of an oceanic antarctic catchment on King George Island, Antarctic Peninsula', *Beitr. Hydrol.*, 11.2, 29–52.

Flügel, W.-A. 1993. 'Hierarchically structured hydrological process studies to regionalize interflow in a loess covered catchment near Heidelberg', *IAHS Publ.*, 212, 215–223.

Flügel, W.-A. and Lüllwitz, Th. 1993. 'Using a distributed model with GIS for comparative hydrological modelling of micro- and meso-scale catchments in the USA and in Germany', *IAHS Publ.*, 214, 59–66.

Flügel, W.-A. and Schwarz, O. 1983. 'Oberflächenabfluß und Interflow auf einem Braunerde-Pelosol Standort im Schönbuch; Ergebnisse eines Beregnungsversuchs', *Allg. Forst- u. Jagdzeitung*, 154, 59–63.

Horton, R.E. 1945. 'Erosional development of streams and their drainage basins: hydrophysical approach to quantitative morphology', *Bull. Geol. Soc. Am.*, 56, 275–370.

Johanson, R.C., Imhoff, J.C., Kittle, J.L., Donigian, A.S., Anderson-Nichols & Co. 1984. '*Hydrological simulation program-FOR-TRAN (HSPF): users manual for release 8.0*', EPA-600/3-84-066, 767pp.

Kirkby, M.J. 1978. *Hillslope Hydrology*. Wiley, Chichester. 389pp.

Leavesley, G.H., Lichty, B.M., Troutman, L.G., and Saindon, L.G. 1983. 'Precipitation–runoff modeling system: user's manual, *USGS Water-Resources Investigations Rep. 83–4238*, USGS, Denver.

Maidment, D.R. 1991. 'GIS and hydrologic modeling' in *Proceedings of the First International Symposium/Workshop on GIS and Environmental Modelling, 15–19 September 1991, Boulder, Colorado, USA*. Goodchild, M. D., Parks, B. O. and Steyaert, L. J. (Eds). Oxford University Press, New York.

Moore, I. 1993. 'Hydrologic modelling and GIS', in *2nd International Conference on Integrating Geographic Information Systems and Environmental Modeling, 26–30 September 1993, Breckenridge, Colorado, USA*. NCGIA, Washington, DC.

Mülders, R. 1992. 'Anwendung des NASIM-Modells zur hydrologischen Modellierung im Bröleinzugsgebiet', *Dipl. Arb.*, Geogr. Inst., Univ. Bonn, 90pp.

Nash, J.E. 1957. 'The form of the instantaneous unit hydrograph', *IAHS Publ.*, **45**, 114–121.

Penman, H.L. 1948. 'Natural evaporation from open water, bare soil and grass', *Proc. Roy. Soc. A*, **193**, 120–146

Philip, J.R. 1954. 'An infiltration equation with physical significance', *Soil Sci.*, **77**, 153–157.

Richards, L.A., Gardner, W.R. and Ogata, G. 1956. 'Physical processes determining water losses from soil', *Soil Sci. Soc. Am. Proc.*, **20**, 310–314.

Schulze, R. 1989. 'ACRU: background, concepts and theory', *Agric. Catchments Research Unit, Rep. No. 35*, Univ. Pietermaritzburg.

Sevruk, B. 1982. 'Methods of correction for systematic error in point precipitation measurement for operational use', WMO, *Operational Hydrology Rep. No. 21, WMO-No. 589*, Secreteriat of the World Meteorological Organisation, Geneva, 91pp.

11

PROCESS CONTROLS AND SIMILARITY IN THE US CONTINENTAL-SCALE HYDROLOGICAL CYCLE FROM EOF ANALYSIS OF REGIONAL CLIMATE MODEL SIMULATIONS

J. S. FAMIGLIETTI

Department of Geological Sciences, University of Texas at Austin, Austin, TX 78712, USA

B. H. BRASWELL

University of New Hampshire, Durham, NH, USA

AND

F. GIORGI

National Center for Atmospheric Research, Boulder, CO, USA

ABSTRACT

The surface hydrological output of a regional climate model is investigated with implications for process controls on the spatial-temporal variability of the water cycle over the continental USA. Principal component analysis was performed on the seasonal and annual hydrological cycles to determine their dominant modes of spatial variability. At both seasonal and annual time-scales, the first principal component is dominated by precipitation, which controls seasonal wetness and evaporation and accounts for only 52 to 58% of the variability in the continental-scale hydrological cycle. The second principal component is related to both snowmelt runoff and the time variability of weather (via its influence on the residence time of soil moisture near the land surface) and explains another 22% to 34% of the variability in the hydrological cycle. Based on these findings, a classification of hydroclimatological similarity is proposed in which two areas are similar in their hydroclimatology if their first and second principal components are similar. The classification scheme differs from classical approaches because it is based on dominant modes of variability rather than specific indices such as vegetation or seasonal wetness.

INTRODUCTION

A major component of climate dynamics involves the interactions between global climate and the hydrological cycle at the continental scale. However, significant uncertainties exist in our understanding of the water and energy cycles over the continents. This is largely a result of imperfect observations and a lack of appropriate technology. Observations tend to be limited in their spatial coverage (e.g. clustered around more populated regions), vary in quality (e.g. between local, state and federal agencies) and be limited in type (e.g. evapotranspiration, groundwater, soil moisture and atmospheric water balance components are traditionally neglected). A key component of the GEWEX Continental-Scale International Project (GCIP) is targeted at alleviating this data shortage in the USA so that fundamental questions regarding the continental-scale hydrological cycle can be addressed. How does the hydrological cycle vary in space and time within the continental USA? What are the dominant process controls on this space–time variability? How do these controls change with geographical location and increasing spatial scale? The answers to these questions will help further our understanding of the role of land–atmosphere interaction in driving climate. From a modelling perspective, they will help to improve land surface parameterizations for hydro-meteorological and hydroclimatological studies.

The purpose of this paper is to report recent research into process controls on the space–time variability of the USA continental-scale hydrological cycle. This work is part of a longer term effort directed at understanding the underlying mechanisms which determine how and why the water and energy balance varies across the continental USA. This first and most basic level of analysis seeks to develop a framework for understanding the relative roles of soil moisture, precipitation, evapotranspiration, runoff and snow depth in controlling spatial variability in the water balance over the continent. This analysis will also provide insight into the role of subcontinental-scale variability in driving the larger, continental-scale hydrological cycle. We introduce the concept of *hydroclimatological similarity* in which subcontinental-scale regions are classified as having similar hydroclimatology if their process controls are similar. In the remainder of this paper we discuss our methodology, results and implications of these results with regards to GCIP and coupled hydrological–atmospheric modelling.

METHODS

Regional climate model and simulation

To understand better the dominant modes of variability in the continental-scale hydrological cycle we performed principal component analysis (PCA) on the surface hydrology output of the National Center for Atmospheric Research (NCAR) nested regional climate model (RegCM) (Giorgi et al., 1993a; 1993b). The regional climate model is a limited area meteorological model (LAM) nested in either global climate model (GCM) output or large-scale observations. The LAM is based on the NCAR/Penn State mesoscale model (version MM4) with certain physics parameterizations suitably modified for long-term simulations (e.g. radiative transfer, land surface processes). The driving GCM or observations provides the nested model with large-scale meteorological fields and the LAM converts the fields into high-resolution regional climate variables by physically representing subGCM grid scale forcings such as topography, surface hydrology and vegetation. A number of application and validation studies conducted with the regional climate model have demonstrated its feasibility for regional climate simulation [see Giorgi et al., (1993a) for a review of these studies].

The land surface parameterization incorporated in RegCM is the Biosphere-Atmosphere Transfer Scheme (BATS) described by Dickinson et al. (1986). BATS is interactively coupled with RegCM so that land surface–atmosphere interaction is appropriately represented. Although from a hydrological point of view BATS is a rather crude, lumped hydrological model, from an atmospheric modelling perspective it represents a state of the art parameterization of the key biophysical processes that control the exchange of momentum, energy and water at the land–atmosphere interface.

At the grid scale of RegCM (60 km), BATS receives inputs of solar and infra-red radiation, precipitation, wind speed and humidity from the lowest atmospheric model level. Based on the properties and hydrological states of the soil and vegetation in a model grid, it converts the meteorological variables into the surface fluxes of energy, moisture and momentum. These fluxes are then returned to the atmospheric model and are used as flux boundary conditions at the lowest model level. The BATS structure includes a one-layer canopy with fractional surface cover, a surface snow layer, a 10 cm surface soil layer and a deeper soil layer which extends from the surface to a depth of 1–2 m depending on the type of vegetation or land cover within the model grid. The hydrological processes represented include wet canopy evaporation, dry canopy transpiration and bare soil evaporation; runoff and infiltration at the soil surface; drainage from the surface soil layer into the deeper soil layer; and drainage from the deep layer to groundwater runoff.

The output to be analysed in this work resulted from a recent multi-year simulation of the regional climate model nested in NCAR's community climate model (CCM) output over the continental USA. This simulation is fully described by Giorgi et al. (1994) and is only briefly reviewed here. The driving GCM is a version of the CCM known as the Global ENvironmental and Ecological Simulation of Interactive Systems (GENESIS) (Pollard and Thompson, 1993; in press). GENESIS truncates atmospheric and topographic components rhomboidally at wave number 15 (R15), which corresponds to a horizontal latitude × longitude grid of $4.5° × 7.5°$. Surface components are run on a $2.0° × 2.0°$ grid.

GENESIS was first run for 20 years to allow the coupled ocean–atmosphere system to equilibrate, and then run for an additional four years to produce driving fields for RegCM nesting. The RegCM simulation was initiated mid-way through the first year of the four-year period and was run continuously for the remaining 3·5 years. The surface hydrological output represents averages over the last three years of the RegCM simulation. Previous analysis of this output (Giorgi et al., 1994) has shown that the regional climate model reproduces the basic seasonal and spatial hydrological characteristics of the major drainage basins of the USA. In the next section we describe how this analysis was extended to identify dominant modes of spatial variability in the hydrological cycle and the processes responsible for them within the regional climate model.

Principal component analysis

Empirical orthogonal function (EOF), or principal component analysis (PCA), provides a convenient methodology for analysing spatial variability within the simulated hydrological cycle of the conterminous USA. (See Bretherton et al. (1992) and Peixoto and Oort (1992) for recent reviews of the mathematical derivation of EOF analysis and its application in meteorology, oceanography, and hydrology; see Guetter and Georgakakos (1993) and Lins (1985a,b) for recent applications to continental-scale hydrologic variability.) The approach identifies a set of orthogonal spatial modes (eigenvectors) such that, when ordered, each successive mode explains the maximum amount possible of remaining variance in the data. The eigenvectors are arranged so that each explains less variance than its predecessor, and the variability associated with a given eigenvector is independent of the other eigenvectors (modes). Spatial patterns associated with the eigenvectors which explain the most variance are often representative of physically realistic and identifiable spatial features. Since PCA has the important feature that any two modes are uncorrelated, modes can be interpreted as being representative of physically independent mechanisms. One advantage of the PCA approach to analysing spatial variability is that much of the information regarding process controls can be compressed into relatively few, uncorrelated dominant modes.

In this paper we conducted PCA on 60 km model output fields of surface soil moisture content (upper 10 cm), total soil moisture content (upper 1–2 m), precipitation, evapotranspiration, runoff and snow depth. The analysis was performed on both seasonal and annual time-scales with the intention of understanding the relative roles of these hydrological variables in controlling the variability in water balance within the continent. (see Figure 1 which shows the annually averaged model output fields.) PCA defines a new coordinate system for the six-dimensional water balance space in which, as described earlier, the new axes (eigenvectors), when ordered, explain the maximum possible amount of remaining variance in the data. The eigenvectors, or principal components (PCs), are defined as linear combinations of the weighted water balance variables so that the PCs y_i, are given by

$$y_i = \sum_{j=1}^{6} \alpha_{ij} x_j \quad (i = 1, \ldots, 6)$$

where x_1 = moisture content in the upper soil layer (mm); x_2 = moisture content in the total soil layer (mm); x_3 = precipitation (mm/h); x_4 = evapotranspiration (mm/h); x_5 = runoff (mm/h); x_6 = snow depth (mm); and α_{ij} = weighting coefficient for eigenvector i on water balance variable j. Interpretation of the relative magnitudes and signs of the weights, α_{ij}, indicates which variable or group of variables dominates the mode of variability of successive principal components.

Note that our intention here was not to substitute modelled data for observations. Rather, it was our goal to demonstrate a methodology for understanding process controls and spatial variability using currently available data which represent the fully coupled land–atmosphere system. We intend to repeat this work in more detail using high-resolution observations as GCIP products become available, and with coupled hydrological–atmospheric models as land parameterizations develop to the point where subgrid-scale spatial variability is more realistically represented (see Famiglietti and Wood, 1994).

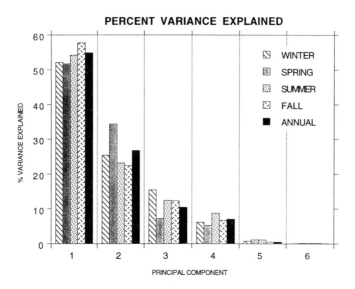

Figure 3. Percentage variance in the continental-scale hydrological cycle explained by successive eigenvectors

RESULTS

The percentage of variance in the seasonally and annually averaged hydrological cycles explained by successive eigenvectors is summarized in Figure 3 and Table I. Two features of the histogram shown in Figure 1 are immediately evident. Firstly, the percentage of variance explained by successive spatial modes is similar at both seasonal and annual time-scales. Secondly, most of the variability in the continental-scale hydrological cycle is explained by the first two or three principal components. The first eigenvector explains 52–58% of the variance, the second another 22–34% and the third an additional 6–15%. In fact, the first and second principal components alone account for roughly 77–86% of the variability.

To understand the process controls associated with the first two principal components, the coefficients or weights on the water balance variables must be examined. These are summarized in Table I for the seasonally and annually averaged hydrological cycles. Analysis of the relative magnitudes and signs of the coefficients of the first eigenvector indicates that this dominant mode of variability is highly correlated with the spatial pattern of seasonally/annually averaged precipitation rates. At both seasonal and annual time-scales, the weighting coefficient on the average precipitation rate, α_{13}, is consistently highest, indicating its strong influence on the first spatial mode of variability. All α_{1j}'s are positive, indicating that the remaining water balance components vary in synchronization with precipitation. This result is not surprising in that seasonal/annual precipitation controls seasonal/annual wetness and thus evapotranspiration and runoff. This fact is confirmed by the high correlation of the patterns of seasonally/annually averaged precipitation rates with the patterns of total soil moisture content, evapotranspiration and runoff rates. However, the percentage variance explained by this first eigenvector was much lower than anticipated based on our intuition alone. The spatial pattern of the first principal component is shown in the top of Figure 2 for the continental USA.

Figure 2 also shows the spatial pattern of the second principal component.

To understand the process controls associated with this spatial mode we again rely on a physical interpretation of the relative magnitudes and signs of the α_{ij} listed in Table I. In developing a physical interpretation of successive modes of variability, it is important to keep in mind that the modes are uncorrelated and thus representative of mechanisms controlled by physically independent factors. Therefore our interpretation must be unrelated to seasonal/annual precipitation and its control on season/annual wetness, evapotranspiration and runoff.

Three key observations lead to our interpretation of the process controls associated with the second principal component. The first observation was that the coefficients weighting the soil moisture content

Table I. Eigenvectors (y_i) and their coefficients (α_{ij}) for the seasonally and annually averaged hydrological cycles

Y_i	α_{i1}	α_{i2}	α_{i3}	α_{i4}	α_{i5}	α_{i6}	% var
Annual							
$i=1$	0·24	0·50	0·54	0·49	0·38	0·12	54·9
2	0·59	−0·25	−0·10	−0·31	0·33	0·61	26·8
3	−0·13	0·27	−0·12	0·24	−0·62	0·67	10·5
4	0·75	0·13	−0·08	0·06	−0·49	−0·39	7·1
5	0·03	−0·72	0·57	0·29	−0·26	0·06	0·6
6	0·07	−0·27	−0·60	0·72	0·23	−0·03	0·1
Winter							
$i=1$	0·15	0·52	0·55	0·51	0·38	0·01	52·1
2	0·70	−0·13	0·01	−0·26	0·22	0·62	25·4
3	0·14	−0·31	0·08	−0·30	0·67	−0·59	15·4
4	0·67	0·24	−0·21	0·09	−0·42	−0·52	6·2
5	0·14	−0·73	0·45	0·41	−0·27	−0·05	0·8
6	−0·04	0·17	0·66	−0·64	−0·34	−0·07	0·1
Spring							
$i=1$	0·32	0·49	0·53	0·47	0·33	0·21	51·7
2	0·46	−0·31	−0·18	−0·37	0·49	0·53	34·4
3	−0·59	0·17	−0·10	0·09	−0·15	0·77	7·2
4	0·58	0·23	−0·31	0·09	−0·68	0·22	5·3
5	0·06	−0·73	0·49	0·30	−0·32	0·17	1·2
6	0·00	−0·22	−0·59	0·73	0·27	−0·06	0·2
Summer							
$i=1$	0·34	0·47	0·53	0·48	0·38	0·19	54·2
2	0·30	−0·47	0·04	−0·41	0·43	0·58	23·2
3	−0·72	0·26	−0·14	0·12	0·08	0·62	12·4
4	0·44	0·10	−0·17	0·14	−0·70	0·50	8·8
5	−0·29	−0·52	0·70	0·16	−0·35	0·06	1·2
6	0·04	−0·49	−0·41	0·74	0·20	−0·02	0·2
Autumn							
$i=1$	0·22	0·50	0·53	0·48	0·42	0·07	57·8
2	0·64	−0·19	−0·07	−0·29	0·19	0·66	22·4
3	0·46	−0·25	0·07	−0·24	0·36	−0·73	12·3
4	0·56	0·25	−0·02	0·19	−0·74	−0·17	6·7
5	0·06	0·60	−0·74	0·08	0·28	−0·08	0·6
6	0·11	−0·47	−0·41	0·76	0·13	0·01	0·2

in the upper 10 cm soil layer (α_{21}) and snow depth (α_{26}) are generally higher than the others. The second observation was that the signs of α_{21}, α_{25} (the coefficient weighting the average runoff rate) and α_{26} were generally opposite those of α_{22}, α_{23} and α_{24} (the coefficients weighting moisture content in the total soil layer, average precipitation rate and average evapotranspiration rate, respectively). The third observation was that α_{23} was close to zero for most of the year.

The reason for the high weighting coefficient on snow depth is clear. It is a major control on snowmelt and consequently runoff, particularly in the spring. However, this is not a mechanism which acts over the entire USA or over the entire year, so it is only a partial explanation of the physical mechanisms associated with the second principal component.

The fact that the signs on α_{21}, α_{25} and α_{26} are apposite those of α_{22} α_{23} and α_{24} suggests a mode of variability in which the moisture content in the upper 10 cm soil layer, runoff rates and snow depth are high, whereas the moisture content in the total soil layer and evapotranspiration rates are low (and vice versa). The near-zero values of while $_{23}$ indicate that the physical mechanism responsible for this mode of variability is relatively independent of precipitation rates. A mechanism consistent with the situation described here results from the repeated time sequence of storm–interstorm events. During a storm, the surface soil moisture content increases rapidly, as does runoff or snow depth. However, the total soil moisture content shows less response and evapotranspiration is minimal. This mechanism is therefore related to the face that a storm has *occurred* rather than the *rate* of rainfall during the storm. During inter-storm periods, the opposite signs result because evapotranspiration occurs at a higher rate while the upper soil zone dries down rapidly. Runoff and snow accumulation during these periods are zero.

Consequently, our interpretation of this mode of variability is that it is also highly correlated with the time variability of weather events which manifests itself in the rapid wetting and drying (i.e. in the turnover time or residence time of soil water) in the surface 10 cm soil layer, explaining the relatively high values of α_{21} in Table I. This mechanism is also consistent with the near-zero values of α_{23} since, as mentioned above, it is related to the time sequence of precipitation events, not the long-term average of the rate itself. Since both of the mechanisms identified — snow depth control of runoff and the time variability of weather events — operate within the continent, both exert a control over the second spatial mode of variability. It is the nature of snowpack that makes the continental-scale hydrological cycle a mixed system with respect to process controls on the second spatial mode. PCA was performed on the entire continent, yet at both seasonal and annual time-scales, significant subcontinental-scale regions remained snow-free.

IMPLICATIONS

Based on these results we introduce the concept of hydroclimatological similarity — that two locations with the same first and second principal components have the same hydroclimatology. At present, we limit this definition to the first and second principal components because we were able to develop physical interpretations for these dominant modes of variability, and because higher order components often represent system noise. The bottom panel of Figure 2 shows a classification of the continental USA into regions of similar hydroclimatology based on clustering analysis in principal component space.

Clusters were identified via the K-means algorithm (Kaufmann and Rousseeuw (1990)), which is particularly well suited for analysis of large data sets (relative to computationally-intensive distance matrix-based methodologies). In the K-means method (also known as the migrating means method), given data with N degrees of freedom and K desired classes, K initial points (means) are injected into the space. The distance between each point and each mean is then calculated. Points are grouped according to which mean is closest, and a new mean for each group is calculated, replacing the original set. This procedure iterates to a convergent solution of K groups, or classes. Since the number of classes, K, is prescribed, we varied K over two orders of magnitude and plotted K versus the average distance between cluster means. A distinct discontinuity was observed in this relationship at K = 7 and K = 11, and K = 7 was retained in this work.

Such a classification scheme differs from more standard climate classification schemes because it is based on dominant modes of variability rather than specific indices such as vegetation or seasonal wetness. It is interesting to note, however, that the classification scheme proposed here actually incorporates elements of those more classical schemes. For example, the first principal component is dominated by the pattern of precipitation, and thus seasonal wetness and evapotranspiration, not unlike the Thornthwaite (1948) scheme. The second principal component reflects the time variability of weather events, as in air mass frequency schemes (Hidore, 1966). Finally, patterns of vegetation, which form the backbone of Köppen-type scheme (Köppen, 1931), are implicit in the system described here as they strongly influence the hydrological cycle of the regional climate mode.

The general features of the similar regions shown in Figure 2 are consistent with recognizable spatial patterns related to actual physiographic provinces. For example, the hydroclimatology of the Pacific Northwest differs from that of the Great Basin, the Rockies, the Mississippi Basin, the Gulf Coast Basin, the Appalachians, etc. However, close inspection of Figure 2 reveals some inconsistencies in the regions classified as similar — e.g. it is not realistic that the northern Great Plains are hydrologically similar to northern Mexico. These inconsistencies are likely the result of a number of interacting causes: the coarse resolution of the vegetation classification used in the model; model sensitivity to the soil moisture content initialization scheme (based on the input pattern of vegetation); the relative crudeness of BATS compared to more 'hydrologically-realistic' land surface models, and edge effects which result when the regional climate model output is dominated by the boundary conditions imposed by the driving GCM. They could also be caused by the restricted set of variables chosen for the EOF analysis. Although these variables are comprehensive in their representation of the water cycle, energy balance variables such as solar radiation and surface temperature were omitted from this round of analysis. As mentioned earlier, the concept of hydroclimatological similarity will be developed further using more comprehensive sets and as observations and models improve in the future.

At present, the work described in this paper may have the following implications for GCIP and the development of coupled hydrological–atmospheric models. Firstly, such a classification can provide a framework for understanding how hydrology–climate interaction is the same within regions and different between regions. With regards to the GCIP, it may suggest that some effort be spent understanding differences between the Mississippi Basin and other regions shown in the bottom of Figure 2, rather than concentrating entirely on the Mississippi (since, in this initial classification, much of the basin falls within a single class). Secondly, the classification provides a framework for improving both macroscale hydrological models and regional climate models. As our understanding of the spatial variation in process controls improves, land parameterizations can be modified on a region- specific basis so that the inconsistencies described here can be minimized.

SUMMARY

The purpose of this study was to analyse the surface hydrological output from a nested regional climate model to provide insight into process controls on the space–time variability of the hydrological cycle over the continental USA. Empirical orthogonal function (principal component) analysis was performed on model-simulated 60 km output fields of soil moisture content in the upper 10 cm soil layer, soil moisture content in the total soil layer (1–2 m deep), precipitation rates, evapotranspiration rates, runoff rates and snow depth.

The results indicated that 77–86% of the variance in the hydrological cycle is explained by the first two principal components. The first principal component explains 52–58% of the spatial variability in the hydrological cycle at seasonal and annual time-scales. Analysis of this eigenvector showed that its mode of variability is dominated by seasonal/annual precipitation rates, which control seasonal/annual wetness, evapotranspiration and runoff. The second principal component explains another 22–34% of the variance. Our analysis showed that the dominant mode of variability associated with this eigenvector is highly correlated with both snow depth and soil moisture content in the upper 10 cm soil layer. Snow depth controls the runoff resulting from snowmelt. Our interpretation of the correlation with soil moisture content in the upper soil layer (and anti-correlation with total soil moisture) is that this mode of variability is a reflection of the frequency of weather events, which manifests itself in the time scale of wetting and drying at the soil surface or the residence time of soil water in the upper soil layer.

Based on the EOF analysis we propose the concept of hydroclimatological similarity—that two locations with the same first and second principal components have the same process controls and thus similar hydroclimatology. A classification of the continental USA into regions of similar hydroclimatology based on clustering analysis in principal component space is presented. The classification scheme differs from classical approaches because it is based on dominant modes of variability rather than specific indices such as vegetation or seasonal wetness. Implications for large-scale observational programmes, regional climate modelling and the study of hydrology–climate interaction are described.

REFERENCES

Bretherton, C. S., Smith, C., and Wallace, J. M. 1992. 'An intercomparison of methods for finding copled patterns in climate data', *J. Climate*, **5**, 541–560.

Dickinson, R. E., Henderson-Sellers, A., Kennedy, P. J., and Wilson, M. F. 1986. Biosphere–atmosphere transfer scheme (BATS) for the community climate model, NCAR Technical Note TN-275 + STR.

Famiglietti, J. S. and Wood, E. F. 1994. 'Multi-scale modeling of spatially-variable water and energy balance processes', *Wat. Resour. Res.*, **30**, 3061–3078.

Guetter, A. K. and Georgakakos, K. P. 1993. 'River outflow of the conterminous United States, 1939–1988', *Bull. Amer. Meteor. Soc.*, **74**, 1873–1891.

Giorgi, F., Marinucci, M. R., and Bates, G. T. 1993a. 'Development of a second generation regional climate model (RegCM2) I: Boundary layer and radiative transfer processes', *Monthly Weather Rev.*, **121**, 2794–2813.

Giorgi, F., Marinucci, M. R., Bates, G. T., and DeCanio, G. 1993b. 'Development of a second generation regional climate model (RegCM2) II: Convective processes and assimilation of lateral boundary conditions', *Monthly Weather Rev.*, **121**, 2814–2832.

Giorgi, F., Hostetler, S. W., and Shields-Brodeur, C. 1994. 'Analyis of the surface hydrology in a regional climate model', *Q. J. Roy. Meterol. Soc.*, **120**, 161–183.

Ilidore, J. J. 1966. 'An introduction to the classification of climate', *J. Geogr.*, **65**, 2–57.

Köppen, W. 1931. *Grundiss der Klimkunde*. Walter de Gruyter, Berlin.

Lins, H. F. 1985a. 'Interannual streamflow variability in the United States based on principal components', *Wat. Resour. Res.*, **21**, 691–701.

Lins, H. F. 1985b. 'Streamflow variability in the United States: 1931–78', *J. Climate Applied Meteor.*, **24**, 463–471.

Peixoto, J. P. and Oort, A. H. 1992. *Physics of Climate*. American Institute of Physics, 520 pp.

Pollard, D. and Thompson S. L. 1993a. 'Sea-ice and CO_2 sensitivity in a global climate model', *Atmosphere–Ocean*, **32**, 449–467

Pollard, D. and Thompson, S. L. 1995. 'Use of a Land Surface Transfer Scheme (LSX) in a global climate model: The response to doubling stomatal resistance', *Global and Planetary Change* (in press).

Thornthwaite, C. W. 1948. 'An approach toward a rational classification of climate', *Geogr. Rev.*, **38**, 55–94.

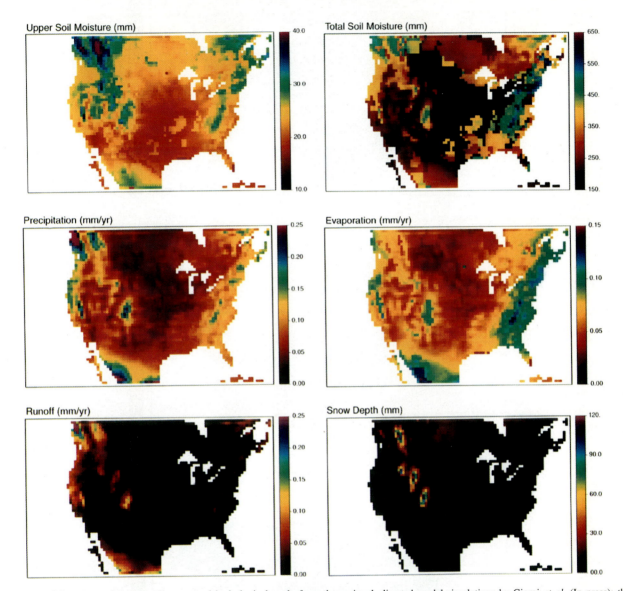

Figure 1. Spatial patterns of the annually averaged hydrological cycle from the regional climated model simulations by Giorgi *et al.* (In press); their previous analysis has shown that the regional climate model reproduces the basic seasonal and spatial hydrological characteristics of the major drainage basins in the USA. Note that the scale on the pattern of runoff is distorted by the extremely high values of snowmelt runoff in the Pacific Northwest and the Rocky Mountains

First PC

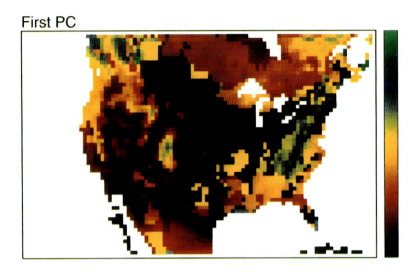

Second PC

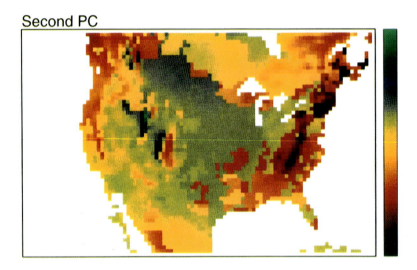

Classification

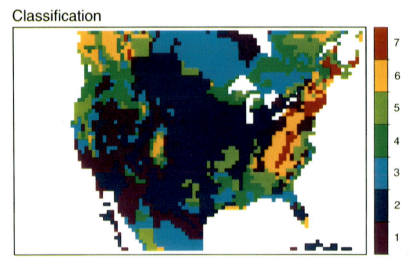

Figure 2. Upper panel: Spatial pattern of the first principal component of the continental-scale hydrological cycle. Middle panel: Spatial pattern of the second principal component. Lower panel: Classification of continental-scale hydroclimatological similarity based on PCR of the hydrological cycle simulated by NCAR's regional climate model

12

PREDICTING CATCHMENT-SCALE SOIL MOISTURE STATUS WITH LIMITED FIELD MEASUREMENTS

J. D. KALMA*

CSIRO, Division of Water Resources, Canberra Laboratory, GPO Box 1666, Canberra, ACT 2601, Australia

B. C. BATES

CSIRO, Division of Water Resources, Perth Laboratory, Private Bag, PO Wembley, WA 6014, Australia

AND

R. A. WOODS

Centre for Water Research, The University of Western Australia, Nedlands, WA 6009, Australia

ABSTRACT

A catchment-wide soil moisture index based on spatially distributed point measurements of soil moisture is used to describe the temporal trend in regional soil moisture status in a $26 \, \text{km}^2$ catchment in south-eastern Australia. The temporal variation in runoff, evaporation and soil moisture storage is simulated with a modification of the lumped SFB water balance model of Boughton (1984), which assumes a fixed bucket size, and with the variable infiltration capacity (VIC) model of Wood *et al.* (1992), which assumes a variable bucket representation. Comparison of simulated catchment soil moisture storage and the soil moisture index based on measurements indicates that both models can make useful predictions of soil moisture status at the catchment scale, with the VIC model performing slightly better than the SFB model. It is also shown that the quasi-distributed VIC model can predict the relative wetness at individual locations, given their relative frequency of occurrence, thus allowing the disaggregation of catchment-scale storage values to point-scale soil moisture values.

INTRODUCTION

Lumped water balance models use simple one-dimensional representations of hydrological processes to provide catchment-wide estimates of the water balance components. Such models use spatial averages of precipitation, soils, geology or topography. The spatial averaging implies that the whole catchment can be represented mathematically using only the dimensions of time and depth (Eeles *et al.*, 1990).

Many regional and macro-scale studies of the hydrosphere still rely on simple 'bucket' models, although there is now increasing effort towards accounting for spatial variability in land surface parameters. This may be achieved by assuming statistical distributions for key (sub)surface parameters (e.g. Entekhabi and Eagleson, 1989; Wood *et al.*, 1992) or by disaggregating regions, basins and global circulation model (GCM) grid squares into more homogeneous subregions to reduce spatial variability in land surface parameters. For example, Jolley and Wheater (1993) used the MORECS system (Thompson *et al.*, 1981) to make water balance calculations for each major land-use class present in $40 \times 40 \, \text{km}$ grid squares covering the $10\,000 \, \text{km}^2$ Severn River catchment. Fluxes and soil moisture deficit values are averaged according to land-use fractional areas, similar to the method proposed by Avissar and Pielke (1989).

Lumped rainfall–runoff models are usually highly parameterized. They require calibration using

---

* Present address: Department of Civil Engineering and Surveying, University of Newcastle, Callaghan, NSW 2308, Australia.

streamflow and rainfall data because estimates of the other model parameters cannot be determined by physical measurement owing to their conceptual nature, or because of the lack of appropriate field techniques and data for determining the parameter values at the scale of interest. However, lumped models are widely used in many hydrological applications in preference to spatially distributed models, which (i) have data requirements that are often difficult to meet; (ii) are often at least as heavily parameterized as lumped models in spite of claims to the contrary; and (iii) have a high computational cost, which may makes them impractical for some applications (e.g. in GCMs).

Alley (1984) described comparisons between several two- to six-parameter regional water balance models with 50 year records of monthly streamflow. He observed that simulated values of state variables such as soil moisture storage differ substantially between the models. A similar finding has been reported by Chiew *et al.* (1993). The results of Alley's comparison of simulated and monitored groundwater levels suggest that caution should be used in attaching physical significance to model parameters and in using state variables of models in indices of drought and basin productivity.

Published work on hydrology contains few examples of catchment studies where distributed measurements of soil moisture storage values have been compared with values simulated by lumped conceptual water balance models. Johnston and Pilgrim (1976) showed a comparison between soil moisture modelled with an earlier version of the Boughton SFB model and adjusted soil moisture data based on field measurements made to an arbitrary depth. They observed good agreement between model results and measurements, providing an independent assessment of the lumped model performance. Kuczera (1983) used soil moisture and throughfall measurements with a lumped rainfall–runoff model. He noted that the use of data on runoff, soil moisture and interception with catchment models can yield substantial reductions in the uncertainty of model parameters.

In this paper we explore how 3·5 years of soil moisture observations made to different depths at 41 measuring points in a 26 km^2 catchment in south-eastern Australia can describe the temporal trend in regional soil moisture status, i.e. in catchment wetness. Our objective is to relate these individual point measurements of soil moisture to catchment-scale soil moisture status. Two approaches are used in this paper.

In the first approach we have adopted a simple bucket representation of the surface hydrology. Measurements over the top 50 cm of the soil profile at each measuring point yield an average soil moisture index for the whole network. The observed temporal changes in average soil moisture index are then compared with estimates obtained with a modified version of the lumped SFB water balance model of Boughton (1984), which, assuming an *invariable* bucket size, predicts soil moisture storage, runoff and evaporation from rainfall measurements and estimates of potential evaporation.

The second approach is based on a *variable* bucket representation. It uses all soil measurements at each measuring point (i.e. over the full depth of measurements) and views the difference between the highest and lowest soil moisture measurements at each point as representing the soil moisture storage capacity at that point. These soil moisture observations are compared with results obtained with a modified version of the so-called variable infiltration capacity (VIC) model first described by Wood *et al.* (1992). This simple quasi-distributed hydrology model uses a statistical distribution of soil depths (or storage capacities) across the catchment. The model has recently been used by Sivapalan and Woods (1995) in a study aimed at evaluating the effects of spatial variability of rainfall intensities and the resulting soil moisture in a hypothetical GCM grid square.

The present study was carried out to investigate the concept of a catchment-wide soil moisture index, based on spatially distributed point measurements of soil moisture. Both the SFB and VIC models predict the temporal variation of runoff, evaporation and total soil moisture storage in the catchment. Thus simulated storage values can be compared with the measured data to test whether they make useful predictions of catchment-wide soil moisture status. In addition, the VIC model predicts the statistical distribution of relative wetness across the catchment so it can be used to predict the relative wetness at individual measurement points, given their relative frequency of occurrence. If the predictions are useful then the statistical distribution approach of the VIC model is a useful method for scaling up from point measurements to catchment-wide values.

STUDY AREA

The field measurements described in this paper were made in the 26.1 km^2 catchment area of Lockyersleigh Creek (Figure 1), located in the Goulburn-Marulan region in New South Wales, Australia. The terrain is undulating and the primal tree and shrub vegetation has been largely cleared. Elevations range from 600 to 762 m above mean sea level.

The catchment is used for cattle and sheep grazing. The pasture is a mixture of native and introduced grasses, with annual and perennial species. Tussock grass is present in most creek depressions and native rushes are widespread, which is indicative of local areas of sustained high water content. The higher eastern and south-eastern parts of the catchment are covered by eucalypt-dominated woodlands. The duplex soils in the area have a sandy/silty A horizon which changes abruptly, generally at a depth of 40–60 cm below the surface, to a heavy clay B horizon. Heavier clay soils occur in the major creek beds.

MEASUREMENTS

Soil moisture data were obtained with a Campbell Pacific neutron moisture meter (NMM) on 88 days (i.e. at approximately two-weekly intervals) between 7 January 1987 and 17 July 1990 along transects A, B and C (Figure 1). Table I gives details of transect length, elevation range and the number of access tubes for each transect.

The length of the 44 access tubes ranged from 60 to 150 cm. Tube length was determined by the soil depth which could be penetrated without damaging the aluminium access tubes: 16 tubes were between 60 and 80 cm; five tubes were in the 80–100 cm class; five tubes in the 100–120 cm class; nine tubes in the 120–140 cm class; and the remaining nine tubes were between 140 and 160 cm. Depending on tube length, NMM count measurements were taken at 10, 20, 30, 40, 60, 80, 100, 120 and 150 cm below the soil surface.

The NMM counts were converted to volumetric water content (cm^3/cm^3) and to depth of water (mm) for each soil layer using two separate regression equations: one for the NMM readings at 10 cm below the surface and one for all other depths of measurements. The regression equations relating NMM counts and volumetric soil moisture content are based on calibrations carried out by two methods: (i) excavation on three separate occasions of three NMM tubes in each of the three transects; and (ii) near-surface measurements at six locations on three separate days. The first three tubes excavated early in the measuring programme have not been used in the analysis described here. Further details on the monitoring programme and calibration procedures are provided by Alksnis *et al.* (1990).

The elevation ranges of the three NMM transects shown in Table I indicate that all soil moisture measurements have been made at elevations between 604 and 656 m. Figure 1 shows that about 65% of the catchment is below 660 m. This part of the catchment is mostly occupied by cleared grasslands, whereas the part above 660 m is largely occupied by open woodlands. The soil moisture data presented in this paper have all been obtained in cleared grasslands similar to those which exist in the two- thirds of the catchment which are in the lowest one-third of its elevation range.

The spatial representativeness of the soil moisture measurements has been investigated by comparing the distribution of several topographic attributes for the NMM network and for the catchment as a whole. The spatial distribution of soil water content can be characterized by a wetness index χ of the form (Sivapalan *et al.*, 1987; Moore *et al.*, 1988)

$$\chi = \ln[A_s/(\tan \beta_i)] + [\ln(T_e) - \ln(T_i)] \tag{1}$$

where A_s is the specific catchment area (i.e. the contributing upslope area draining across a contour segment of width b orthogonal to the flow, m^2/m), β_i is the slope angle (degrees), T_i is the local soil transmissivity (depth-integrated hydraulic conductivity, m^2/day) in the ith element and $\ln(T_e)$ is the areal average value of $\ln(T_i)$. The wetness index given by Equation (1) considers only topographic (first term) and soil factors (second term). It was developed initially for predicting saturated source areas and depth to water-tables. Wood *et al.* (1990) have demonstrated for the Kings Creek catchment that the variation in the topographic variable $\ln[A_s/(\tan \beta_i)]$ is far greater than the local variation in transmissivity $\ln(T_i)$. This indicates that the topographic variable alone may be a useful approximation of the wetness index.

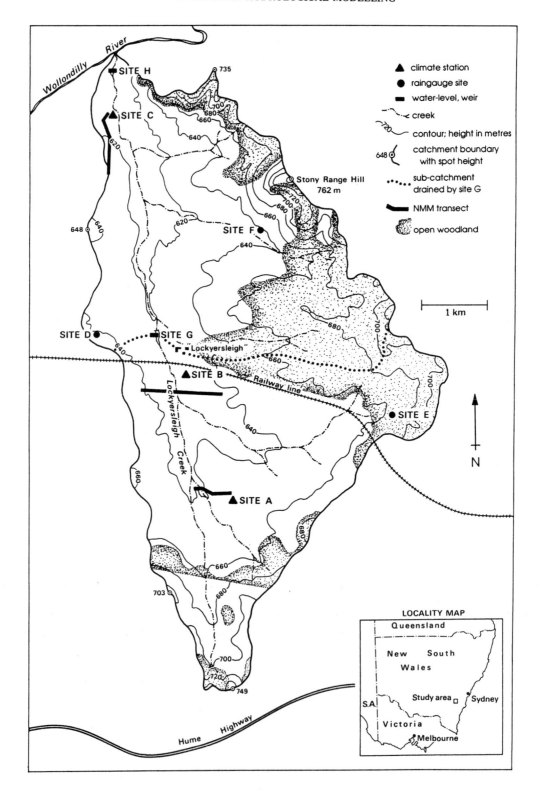

Figure 1. Lockyersleigh catchment, showing sites of measurement and locality map

Table I. Transect length, elevation range and number of access tubes

Transect	Length (m)	Elevation range (m + MSL)	Number of access tubes
A	600	636–656	13
B	1200	626–637	16
C	900	604–631	15

A grid-based method of terrain analysis (TAPES-G) described by Moore *et al.* (1991; 1993) has been used with a 20 m grid digital elevation model (DEM) of the Lockyersleigh catchment to calculate a number of terrain attributes at each node. These terrain attributes are: slope (β_i, degrees), upslope area (catchment area A_s above a short length of contour, ha) and the wetness index (approximated by the topographic variable $\ln[A_s/(\tan\beta_i)]$). Values of these attributes were also obtained for each of the NMM measuring points by interpolation from nearby grid nodes.

Figure 2 compares the cumulative frequency distributions of β_i, A_s and $\ln[A_s/(\tan\beta_i)]$ for the network of 44 NMM tubes and for all 65 469 nodal points of the 20 m DEM of the Lockyersleigh catchment. It is shown that for the entire cumulative frequency range, slopes for the NMM network are always slightly less than those for the catchment as a whole. The frequency distributions for upslope catchment area of the NMM network and the whole catchment are similar. The frequency distributions of $\ln[A_s/(\tan\beta_i)]$ for the NMM measuring points and for all 65 469 nodal points indicate that the NMM network perhaps under-represents the drier, higher areas in the catchment with wetness index values of less than about 8.

Climate data used in this study are based on continuous records of 10-minute weather data obtained between 10 January 1987 and 1 July 1993 with automatic climate stations at sites A, B and C (Figure 1). All observations are aggregated to daily (midnight to midnight) data. They comprise daily totals of rainfall, windrun, incoming global radiation net all-wave radiation, and mean daily values of air temperature and vapour pressure (see Daniel *et al.*, 1994). Streamflow in the Lockyersleigh Creek is measured by a broad-crested weir at site G (Figure 1). The size of the subcatchment above G is 14.7 km². Full details of the weir calibration have been provided by Daniel *et al.* (1994).

DESCRIPTION OF THE LUMPED PARAMETER MODELS

SFB lumped parameter model

The SFB model was developed by Boughton (1984) for use in the analysis of daily climate and streamflow data for small ephemeral catchments where only surface runoff occurs and for those catchments where baseflow is significant. The model transforms daily rainfall and potential evaporation estimates into daily streamflow with surface runoff and baseflow contributions separated and actual evaporation estimated. The model has been used most often to simulate monthly runoff (Boughton, 1984; Nathan and McMahon, 1990a; 1990b; Chiew *et al.*, 1993; Bates *et al.*, 1994).

The original model structure is indicated in the upper part of Figure 3. It has three parameters to be optimized: the total capacity of the (invariable) surface store (S); a daily infiltration capacity (F), which controls the movement of water from the drainable portion of the surface store to a lower store; and the baseflow factor (B), which determines the partitioning of the daily depletion of the lower store between baseflow and deep percolation.

The surface store is partitioned by a factor (NDC) into a drainable section on which parameter F operates and a non-drainable lower section which is only depleted by evaporation. The lower store is drainable by deep percolation and has a baseflow threshold (SDR_{max}) at which baseflow generation ceases. Evaporation from the upper store is defined by

$$E_a = \min\{E_{max} \cdot US/(NDC \cdot S); E_p\} \tag{3}$$

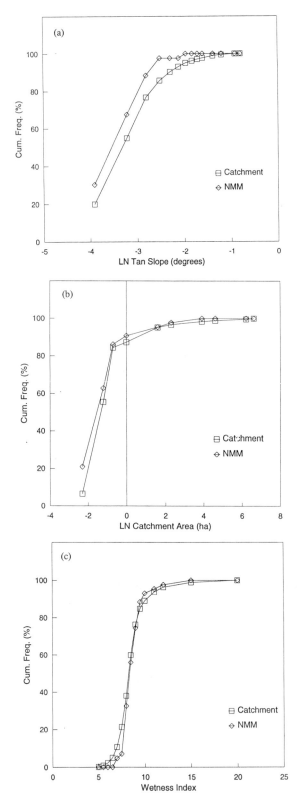

Figure 2. Frequency distributions of slope [ln(tan β_i)], upslope catchment area [lnA_s], and wetness index {ln [A_s/(tanβ_i)]} for the network of 44 NMM tubes and for all 65 469 nodal points of the Lockyersleigh digital elevation model

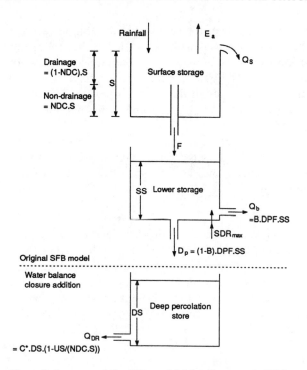

Figure 3. Structure of the SFB model (after Bates *et al.*, 1994)

where E_{max} denotes the maximum limiting rate of evapotranspiration, US the depth of water in the non-drainable component of the surface store and E_p the potential evapotranspiration. No evaporation occurs from the lower store or from deep percolation so that the water balance is not closed if $B < 1$. The daily contents of the lower store (SS) are depleted at the rate $DPF \cdot SS$ where DPF denotes a depletion factor.

Boughton (1984) showed that, for his data sets, a number of model parameters could be fixed. The fixed values were $NDC = 0.5$; $E_{max} = 8.9$ mm/day; $DPF = 0.005$; and $SDR_{max} = 25$ mm.

The outflow components of the model are derived as follows. Surface runoff, Q_s, occurs when a rainfall excess P^* is generated by daily rainfall exceeding the storage deficit in the upper store at the end of the previous day, but modified to account for daily losses due to the infiltration factor F

$$Q_s = P^* - F \tanh(P^*/F) \tag{4}$$

The depletion of the lower store is partitioned between baseflow generation

$$Q_b = B \cdot DPF \cdot SS \qquad \text{if } SS > SDR_{max} \tag{5}$$

and deep percolation

$$D_p = (1 - B) \cdot DPF \cdot SS \tag{6}$$

In this study a closure variant of *SFB* was used which has been described by Bates *et al.* (1994). The modification of the model structure is illustrated in the lower part of Figure 3. Deep percolation accumulates in a further store from which moisture is admitted to the non-drainable component of the upper store at a rate

$$Q_{DR} = C^* \cdot DS \cdot [1 - US/(NDC \cdot S)] \tag{7}$$

when that store is depleted by evaporation. Here C^* is a drainage coefficient and DS is the depth of water in the deep percolation store. This process may be rationalized as the regional groundwater supplying down-slope and riparian vegetation during drier periods.

In this study we have increased the number of parameters to be fitted by the optimization algorithm as set out in Bates *et al.* (1994). It is a variant of the simulated annealing algorithm of Kirkpatrick *et al.* (1983) and Press *et al.* (1992). The model generates daily runoff totals, but following the practice of Nathan and McMahon (1990a; 1990b) and Chiew *et al.* (1993) we used an objective function based on monthly totals of simulated and observed discharges. The objective function is therefore calculated using monthly totals of simulated and observed discharges. Although the program has the capacity to fit all of the model parameters, preliminary sensitivity tests showed that the following parameters could be fixed: $NDC = 0.8$, $SDR_{max} = 0$ and $E_{max} = 8.9$ mm/day.

The data inputs of the model are weighted daily catchment precipitation P, an estimate of potential evaporation based on the Priestley–Taylor algorithm using measured daily net radiation and temperature at site B, and monthly observed streamflow. The outputs of the model are daily values of streamflow, soil moisture storage and evaporation.

Variable infiltration capacity model

The VIC model is a statistical water balance model described by Wood *et al.* (1992). In the terminology of Wood *et al.* (1992) infiltration capacity means the total volumetric capacity of a soil column to hold water. This variable bucket model assumes that such storage capacities and therefore runoff generation and evaporation will vary within a region with topography, soils and vegetation. It allows for the components of the water balance to be predicted within 'the statistical framework provided by the concept of a distribution of storage elements of various capacities' (Wood *et al.*, 1991).

Wood *et al.*(1992) have described a comparison in the 767 km^2 French Broad River catchment of streamflow simulations with the VIC model and the GFDL hydrology model, which uses a Budyko simple bucket parameterization. The VIC model has been used by Sivapalan and Woods (1995) in simulation experiments designed to assess the biases in modelled water balance fluxes within a hypothetical GCM square if spatial heterogeneity were to be neglected.

The VIC model of Wood *et al.* (1992) assumes that scaled infiltration (i.e. storage) capacity, S, is a random variable with cumulative distribution function given by the Xinanjiang distribution

$$F_S(s) = 1 - (1 - s)^\beta \qquad (8)$$

where β is an empirical parameter of the model. In this paper, an additional parameter s_{min} is added to Equation (8) to allow more flexibility in modelling of runoff generation

$$F_S(s) = 1 - [(1 - s)](1 - s_{min})]^\beta \qquad (9)$$

The infiltration (storage) capacity at any point in a catchment is defined as the maximum depth of rainfall which can infiltrate at that point (i.e. under the driest possible initial conditions). Scaled infiltration capacity, s, is the local infiltration capacity divided by the largest infiltration capacity for any point in the catchment. If z is the soil depth (to bedrock) at any point, with a maximum value z_{max} and if the soil porosity, $\Delta\Theta$, is assumed constant throughout the catchment, then $s = (z\Delta\Theta)/(z_{max}\Delta\Theta)$.

The soil moisture status for the entire catchment at a particular time can be described by the scaled soil moisture variable, v, which represents the scaled soil moisture in storage at every point in the catchment. Antecedent soil moisture status is indicated by v_0 (see Figure 4). Under the assumption of uniform recharge to the soil throughout the catchment, v_0 is constant throughout the catchment. Those points on the land surface with $s < v_0$ are considered to be saturated before any rainfall begins. If all soil water in the catchment is assumed to be held in saturated soil, then the scaled soil moisture can be written as $v_0 = y_0/z_{max}$, where y_0 is the height of the water-table above bedrock (y_0 is assumed constant throughout the catchment). If $v_0 < s_{min}$, then no part of the catchment is saturated.

For a given v, the fraction of the land surface which is saturated is denoted α, and the total soil moisture held in the catchment is denoted by w (see Figure 4). From Equation (9) it follows that the relationships between v, w and α are

$$w(v) = v \qquad\qquad\qquad\qquad\qquad\qquad\qquad\qquad v < s_{min} \qquad (10a)$$

$$w(v) = \{1 - [(1 - v)/(1 - s_{min})]^{(\beta+1)}\} * [(1 - s_{min})/(\beta + 1)] + s_{min} \qquad v > s_{min} \qquad (10b)$$

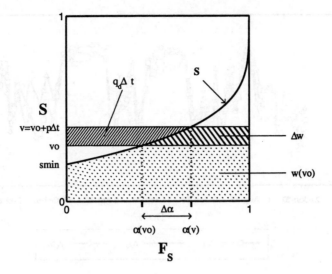

Figure 4. Schematic diagram of variable infiltration capacity model (after Sivapalan and Woods, 1995)

$$\alpha(v) = 0 \qquad\qquad\qquad v < s_{min} \qquad (11a)$$

$$\alpha(v) = 1 - [(1 - v)/(1 - s_{min})]^\beta \qquad\qquad\qquad v > s_{min} \qquad (11b)$$

$$w(\alpha) = [1 - (1 - \alpha)^{(\beta+1)/\beta}] * (1 - s_{min})/(\beta + 1) + s_{min} \qquad (12)$$

$$v(\alpha) = 1 - (1 - \alpha)^{(1/\beta)} * (1 - s_{min}) \qquad (13)$$

$$v(w) = w \qquad\qquad\qquad w < s_{min} \qquad (14a)$$

$$v(w) = 1 - (1 - s_{min}) * \{1 - [(w - s_{min}) * (\beta + 1)/(1 - s_{min})]^{1/(\beta+1)} \qquad w > s_{min} \qquad (14b)$$

$$\alpha(w) = 0 \qquad\qquad\qquad w < s_{min} \qquad (15a)$$

$$\alpha(w) = 1 - [1 - (w - s_{min}) * (\beta + 1)/(1 - s_{min})]^{\beta/(\beta+1)} \qquad w > s_{min} \qquad (15b)$$

Given values of β and s_{min}, any one of v, w or α is sufficient to define the moisture status of the entire catchment. When uniform rainfall $p\Delta t$, scaled by $z_m\Delta\Theta$, is applied to the VIC model, the soil moisture variable v

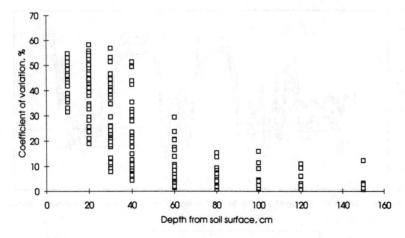

Figure 5. Coefficient of variation (%) of measurements of volumetric soil moisture content in each individual NMM plotted against the depth of measurement for the entire data set

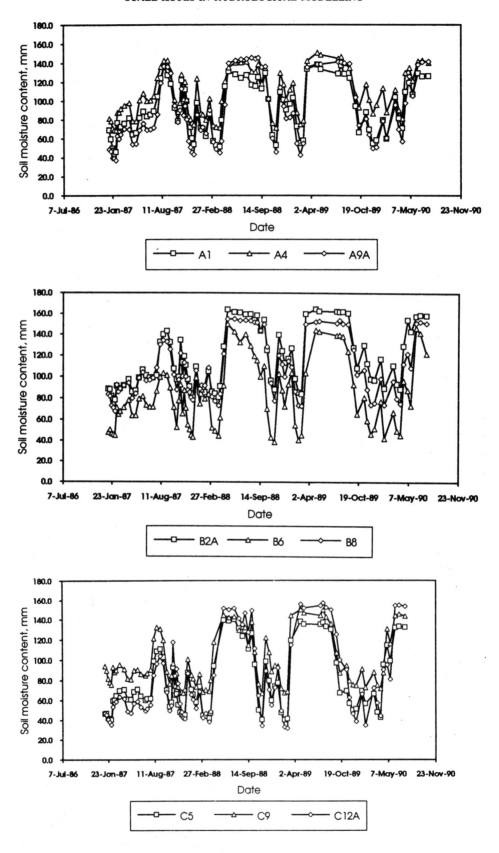

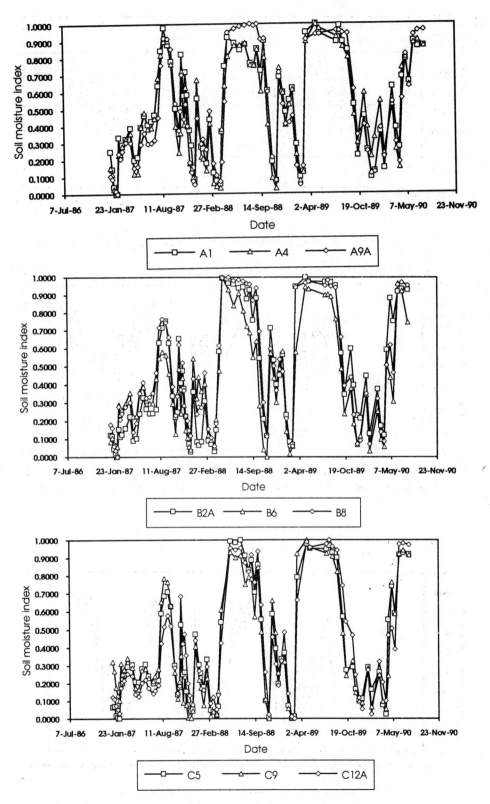

Figure 6. Plots of (a) volumetric soil moisture content (SM_{50}) and (b) soil moisture index (SMI_{50}) over the entire period of measurements for three selected positions along each of the three transects

Table II. Highest, lowest and average values of the maximum soil moisture content (SM_{max}) and the minimum soil moisture content (SM_{min}) in the top 50 cm as observed in the three NMM transects

Transect	SM_{max} (mm)			SM_{min} (mm)		
	Highest	Lowest	Average	Highest	Lowest	Average
A	168	93	142	100	27	54
B	181	102	153	107	31	62
C	168	109	146	96	28	47

increases (see Figure 4), with

$$v = v_0 + p\Delta t \tag{16}$$

This leads to an increase $\Delta\alpha$ in the fraction of the land surface which is saturated and an increase Δw in the total soil moisture storage.

Evaporation for the VIC model is calculated using the method of Sivapalan and Woods (1995), in which a point-scale model of evaporation (which depends on local soil moisture conditions via a two-parameter model) is integrated over the distribution of soil moisture conditions for the whole catchment, giving a catchment- scale evaporation model. Subsurface flow is calculated as a linear function of average soil moisture storage, but is assumed to be negligible for this application. Surface runoff (q_d) is then calculated by a simple water balance.

RESULTS

Variation of soil moisture content with depth

Comparison between tubes will be facilitated if soil moisture data are calculated for a standard soil depth. This common soil depth should preferably encompass most of the seasonal variation in soil moisture content. In Figure 5 the coefficient of variation (CV) of volumetric soil moisture content is shown for each depth of measurement based on all 88 days of measurement and all NMM tubes. It is shown that the CV drops rapidly to about 20–25% at a depth of 60 cm and remains in the 10–20% range for greater depths.

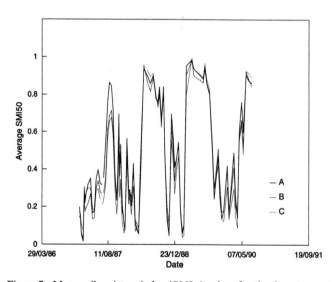

Figure 7. Mean soil moisture index (SMI_{50}) values for the three transects

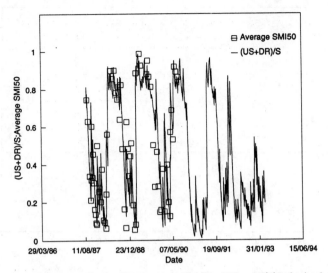

Figure 8. Time plot of daily values of the normalized soil moisture availability $(US + DR)/S$ obtained with the SFB model and average SMI_{50} values obtained on 88 days of soil moisture measurements

Figure 5 implies that the bulk of the seasonal variation is captured by the soil moisture content of the top 50 cm (SM_{50}, mm). Changes between any two successive days of measurements will almost entirely be restricted to that top layer. The top 50 cm is represented by measurements at the first four depths which are available for all tubes, i.e. at 10, 20, 30 and 40 cm.

In this paper we will first report a comparison of SM_{50} measurements with predictions obtained with the SFB model, which assumes a fixed depth of bucket across the catchment. Later in the paper we will consider the soil moisture content over the entire depth of the NMM tubes as used for comparisons with the results obtained with the VIC model, which assumes variable depths.

Spatial variability in soil moisture content of the top 50 cm

Figure 6a shows plots of SM_{50} over the entire period of measurements for three selected positions along

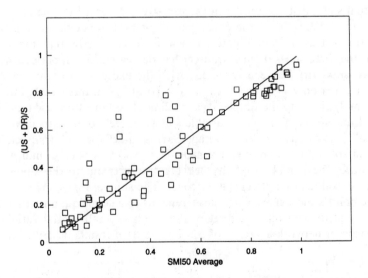

Figure 9. Normalized soil moisture availability $(US + DR)/S$ obtained with the SFB model plotted against average SMI_{50} values for 88 days of soil moisture measurements

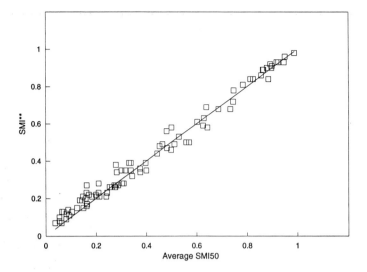

Figure 10. Comparison between SMI^{**} and $(SMI_{50})_{av}$ for all days of soil moisture measurement

each of the three transects. The nine plots show that significant similarity exists between the three transects in the general trends observed at the selected locations. The spatial variability in SM_{50} values *within* each of the three pasture transects was considerable, as illustrated in Table II, which shows the highest, lowest and average values of maximum and minimum SM_{50} values observed in each transect throughout the entire period of measurements.

The spatial differences in maximum and minimum SM_{50} values (and hence water holding capacity) shown in Table II appear to be caused by soil physical differences between measuring points which are linked to topographical position in the landscape. To make comparisons between locations with different water-holding capacity, the following procedure was used. For each NMM tube we determined the highest $[(SM_{50})_{max}]$ and lowest $[(SM_{50})_{min}]$ soil moisture values observed throughout the entire period of measurements and obtained for each day of measurement a soil moisture index (SMI_{50}) value from

$$SMI_{50} = [SM_{50} - (SM_{50})_{min}]/[(SM_{50})_{max} - (SM_{50})_{min}]$$

where SM_{50} is the actual soil moisture content in the top 50 cm observed on the day.

Figure 6b shows plots of SMI_{50} over the entire period of measurements for the same three selected positions along each transect as used in Figure 6a. The nine plots show that the spatial variability in SMI_{50} values between the three tubes in each transect has decreased in comparison with the SM_{50} data. There is also improved similarity between transects, with the exception of transect A in the winter of 1987. Rainfall during that period was sufficient to saturate locations in transect A, but not those in transects B and C. This anomaly is due to the simple linear scaling of soil moisture used in SMI_{50}, which is not sufficient to capture all the effects of spatially variable soil moisture storage capacity.

Figure 7 shows for all days of measurement average SMI_{50} values for each transect. The small range in SMI_{50} values on any day of soil moisture measurements between transects also indicates that the use of a soil moisture index reduces the spatial variability between transects due to the topographic position and differences in soil texture, soil depth and vegetation cover. The simple SMI_{50} index approach thus makes it possible to aggregate point-scale soil moisture measurements to a single quantity that is representative of soil moisture availability at the catchment- or regional-scale. The small spatial variability in SMI_{50} also allows for the comparison of network averages of SMI_{50} and SFB results described in the following.

Application of the SFB model

The lumped water balance model (SFB) has been used with rainfall and other climate data (temperature, humidity, wind and radiation) obtained at site B between 1988 and 1993. In the case of missing values, the

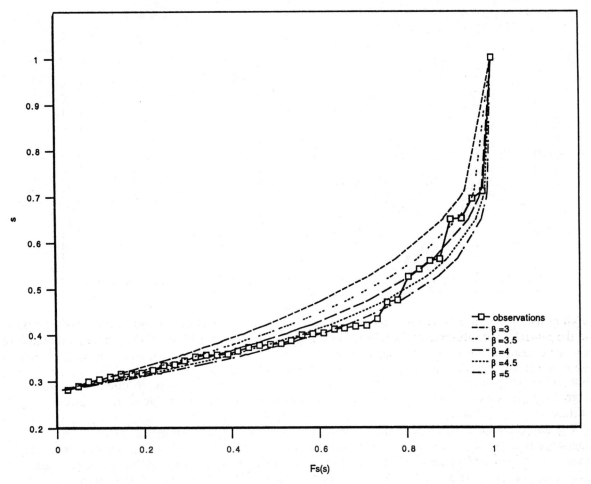

Figure 11. Cumulative frequency distribution of the scaled total storage capacity calculated from $s = (SM^*_{max} - SM^*_{min})/$ $(SM^*_{max} - SM^*_{min})_{max}$ based on 41 NMM tubes, showing that $s_{min} = 0.28$. Also shown are curves representing (9) with selected values of β

data have been taken from site A. The streamflow data used were observed at site G, which gauges the $14.7\,\text{km}^2$ subcatchment upstream from the causeway, which is 73% cleared pasture and 27% native pastures with open woodland.

Boughton (1984) lists the recommended values for the model parameters S, F and B for use in ungauged catchments. These values are based on catchment characteristics and a semi-quantitative knowledge of flow characteristics. Thus, whereas the calibration of the SFB model was not originally intended and is not strictly necessary, it can be beneficial if concurrent rainfall, streamflow and climate data are available.

The SFB model has been calibrated using an objective function based on total monthly discharge. The first simulated annealing runs were carried out with a predefined feasible parameter space and used $SDR_{max} = 0$ as the lower bound on that parameter. During preliminary optimization runs, the SDR_{max} values converged to this lower value in every instance. It was therefore decided to fix $SDR_{max} = 0$ and to repeat the estimation process. During early runs NDC had been fixed at 0·5. Changing to $NDC = 0·8$ and incorporating the deep percolation store led to a noticeable reduction in the residual sum of squares without compromising the realism of the parameter estimates.

The above calibration procedure resulted in the following parameter estimates: $S = 145\,\text{mm}$; $F = 3·0\,\text{mm/day}$; $B = 0·23$; $E_{max} = 8·9\,\text{mm/day}$; $NDC = 0·8$; $DPF = 0·010$; and $C^* = 0·33$. Note that $S = 145\,\text{mm}$ is close to the average SM_{max} measured with the NMM measurements in the top 50 cm

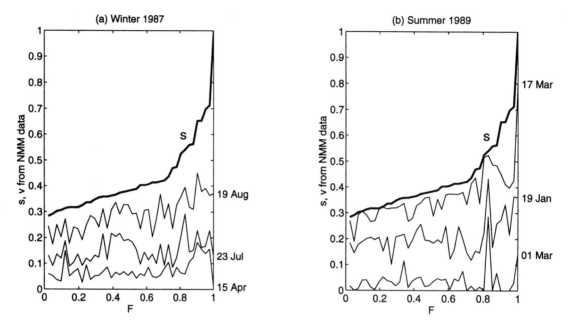

Figure 12. Distribution of scaled storage capacity (*s*) and scaled soil moisture content (*v*) for selected measurement dates in (a) 1987 and (b) 1989

(see Table II). Following calibration, the model was run for daily time steps to obtain daily values of evaporation, streamflow and estimates of the drainable (*DR*) and non-drainable (*US*) components of the surface store.

The results obtained with the invariable bucket representation of the SFB model may be compared with soil moisture index values based on soil moisture measurements in the top 50 cm of soil. Figure 8 shows a time plot of simulated values of the contents of the SFB surface store expressed as the (normalized) soil moisture availability $(US + DR)/S$ and average SMI_{50} values on the days of soil moisture measurements. In Figure 9 the two variables have been plotted against each other for the 88 days of soil moisture

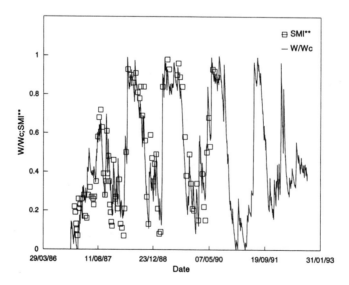

Figure 13. Time plot of simulated daily (w/w_c) values and SMI^{**} values computed from soil moisture measurements in all NMM tubes

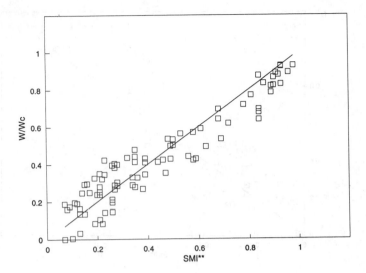

Figure 14. Values of the catchment scale wetness index w/w_c obtained with the VIC model plotted against SMI^{**} values for 88 days of soil moisture measurements

measurements between 7 January 1987 and 17 July 1990. Linear regression resulted in

$$(US + DR)/S = 0.879 SMI_{50} + 0.065 \qquad [r^2 = 0.869; \text{s.e.}(y) = 0.102]$$

The general agreement between the normalized modelled soil moisture values and the measured SMI_{50} values as shown in Figures 8 and 9 is encouraging in the light of the observation by Alley (1984) that simulated values of state variables such as soil moisture storage differed substantially in his comparisons between several two- to six-parameter regional water balance models. The present results seem to indicate that state variables such as soil moisture storage do have physical significance for the study catchment. They confirm the view expressed by Johnston and Pilgrim (1976) that such agreement will provide greater confidence in the physical realism of lumped models which average data in area and in time. The present results also imply that these lumped models therefore have a part to play in other applications such as drought assessment and crop-yield prediction.

Analysis of soil moisture data over the entire depth of measurement

The length of the NMM tubes varied between 60 and 160 cm. This length is closely related to the depth of soil which could be penetrated without damaging the aluminium tubes. For the purpose of this present study it is assumed that these depths correspond to soil depths and thus represent 'buckets' of different sizes.

For each NMM tube and for each day of measurement, the local value of the volumetric moisture content of the total soil profile (SM^*, mm) is obtained as the sum of the corresponding values for each layer of measurement. For each NMM tube the highest and lowest SM^* values were determined over the entire 3·5 years of measurements. The soil moisture index SMI^* for the tube is then defined as

$$SMI^* = (SM^* - SM^*_{min})/(SM^*_{max} - SM^*_{min}) = A/B$$

where $A = (SM^* - SM^*_{min})$ is the removable component of total storage (mm) in the local 'bucket' and $B = (SM^*_{max} - SM^*_{min})$ represents the maximum storage capacity (mm). SMI^* values thus range between 0 and 1. They have been calculated for each tube and for each day of measurement.

Finally, it is assumed that for each day of measurement representative network values denoted by SMI^{**} may be estimated from $SMI^{**} = (\Sigma A/\Sigma B)$ based on all measurements on that day. In Figure 10 these SMI^{**} values are compared with the average SMI_{50} values denoted by $(SMI_{50})_{av}$. Linear regression

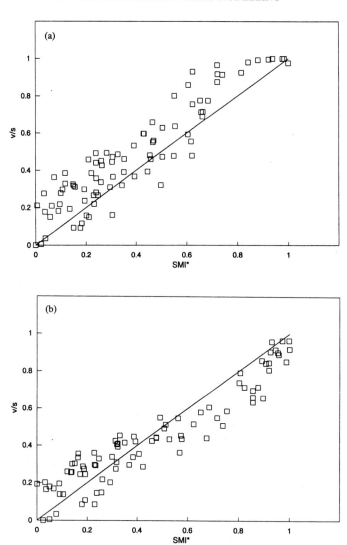

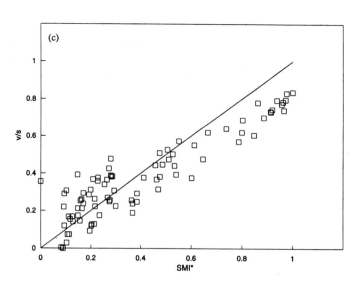

resulted in

$$SMI^{**} = 0{\cdot}962(SMI_{50})_{av} + 0{\cdot}028 \qquad [r^2 = 0{\cdot}984; \text{s.e.}(y) = 0{\cdot}035]$$

It is shown that there is very good agreement between the two NMM network values.

Application of the VIC model

In addition to requiring input rainfall and potential evaporation data, the VIC model requires one initial condition, w_0, the total soil moisture stored at the start of a simulation, and six parameters: β and s_{min}, which define the distribution of infiltration capacity; z_m and $\Delta\Theta$, which define the scaling of soil moisture (only the product of z_m and $\Delta\Theta$ is needed); and β_e and Ψ_c, which define the evaporation sub- model. Note that the subsurface flow parameter k_c of Sivapalan and Woods (1995) was fixed at zero.

To obtain an approximate value of s_{min} and β, we have assumed that the NMM network provides a set of total storage capacity values which is representative for the catchment as a whole. We have obtained the storage capacity values for all 41 NMM tubes from the difference between the highest and lowest soil moisture measurements over the entire period of monitoring, $(SM^*_{max} - SM^*_{min})$ and we have scaled these values by the highest value of $(SM^*_{max} - SM^*_{min}) = 268$ mm which was observed for tube B_0 and which, for an average porosity of $0{\cdot}3$, corresponds to a z_m value of approximately 900 mm (i.e. the hydrologically active soil depth).

Figure 11 shows the cumulative frequency distribution of the scaled total storage capacity s, calculated from

$$s = (SM^*_{max} - SM^*_{min})/(SM^*_{max} - SM^*_{min})_{max}$$

based on the 41 NMM tubes. Note that the cumulative frequency values $F_S(s)$ may be obtained by ranking all tubes according to $(SM^*_{max} - SM^*_{min})$ and dividing the rank for each tube by the total number of tubes (this assumes that the tubes are a representative sample from the catchment). Note that s_{min} represents the lowest value of s at which saturation will occur anywhere in the catchment. The data indicate that $s_{min} = 0{\cdot}28$, the s value for tube B3. The figure also shows for $s_{min} = 0{\cdot}28$, curves representing Equation (9) with selected values of β. By trial and error it was estimated that the empirical data points were best represented with $\beta = 4$.

The VIC model assumes that soil moisture status can be approximated by a single variable v, the uniform scaled soil moisture storage across the catchment. Figure 12 investigates this assumption for several measurement days: separate estimates of v for all 41 NMM tubes are plotted against the cumulative frequency $F_S(s)$, where

$$v = (SM^* - SM^*_{min})/(SM^*_{max} - SM^*_{min})_{max}$$

If the VIC model is applicable, data points should lie either on the catchment- specific $s - F_S(s)$ curve (for saturated locations) or on a horizontal line which rises and falls as catchment wetness changes over time (for unsaturated locations). This assumption is partially supported by the data shown in Figure 12. During dry periods the v-curve appears to be horizontal (15 April 1987), but in wetter periods the v-curve can sometimes slope upward and to the right, parallel to the s-curve, but below it (19 August 1987). This suggests that either the estimates of storage capacity (s) may be inaccurate (they do depend on accurately measuring extremes) or that the soil moisture is not in equilibrium across the catchment during relatively wet periods.

The rainfall, climate and discharge data used with the SFB model were also used with the VIC model. Preliminary sensitivity tests showed that the VIC model outputs were sensitive to s_{min}, β_e and Ψ_c. The following starting values were used in model simulations: $w_0 = 0{\cdot}0$ (the simulation began in mid-summer); β, s_{min}, z_m and $\Delta\Theta$ as above, $\beta_e = 4$, and $\Psi_c = 0{\cdot}2$. Automatic optimization with a simplex technique (Nelder and Mead, 1965) was used to match the VIC modelled runoff to measured daily flow data, using a sum of squared errors objective function.

Figure 15. Comparisons of (v/s) and SMI^* for (a) NMM tube B4 $[F_S(s) = 0{\cdot}098; r^2 = 0{\cdot}85]$; (b) NMM tube A6 $[F_S(s) = 0{\cdot}488; r^2 = 0{\cdot}89]$; and (c) NMM tube C1 $[F_S(s) = 0{\cdot}978; r^2 = 0{\cdot}84]$

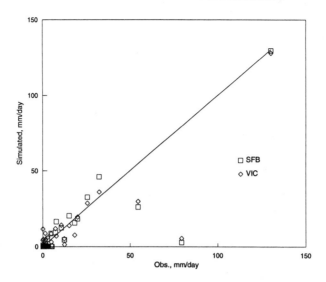

Figure 16. Comparison of daily streamflow observations and streamflow simulated with the SFB and VIC models

The resulting optimized parameter values were $s_{min} = 0.599$, $\beta_e = 1.54$ and $\psi_c = 0.186$. With the exception of one rainstorm when the rain clearly fell within the day before the runoff was recorded, the VIC model was able to usefully simulate the highly intermittent measured flow data (see also Figure 16).

Optimization of the VIC model parameters using monthly runoff data was also carried out, but the final model output was not significantly affected. The parameter values found by monthly optimization were $s_{min} = 0.734$, $\beta_e = 0.686$ and $\Psi_c = -0.078$.

The differences between parameter values found by optimization on daily and monthly data are not surprising, given that the runoff generation process has a threshold type of behaviour. The entire catchment remains unsaturated for much of the summer and the optimization process mainly involves ensuring that the minimum soil moisture deficit ($s_{min} - v$) at the beginning of winter will generate the correct amount of runoff. This can be achieved by adjusting either s_{min} or the evaporation parameters (which will then change v at the end of summer). The treatment of thresholds is easier when data such as soil moisture are available: however, the point of this exercise was to make predictions using only the runoff data.

The VIC model produces a time series of w, the total soil moisture storage for the entire catchment. If the w values are divided by w_c, the maximum possible value of w when all the soil is saturated, then the ratio w/w_c is a catchment-scale wetness index. For the modified Xinanjiang distribution Equation (9)

$$w_c = s_{min} + (1 - s_{min})/(\beta + 1) \tag{17}$$

In Figure 13, a time plot is shown of w/w_c and SMI^{**} values based on measurements at all depths for the entire NMM network calculated from

$$SMI^{**} = \Sigma(SM^* - \Sigma SM^*_{min})/\Sigma(SM^*_{max} - SM^*_{min})$$

In Figure 14 the two variables have been plotted against each other. Linear regression resulted in

$$w/w_c = 0.845 SMI^{**} + 0.061 \qquad [r^2 = 0.898; \text{s.e.}(y) = 0.079]$$

It is shown that values of catchment-wide soil moisture storage cluster tightly about the 1:1 line, with the model explaining 95% of the variance in the measured data. This was achieved without using the measured soil moisture data in the calibration process. However, there is a tendency for the VIC model to overpredict during dry periods and underpredict in wet periods.

The w values from the VIC model can be transformed to v values, representing the uniform height above bedrock to which the soil is saturated across the whole catchment.

A point in the catchment is saturated if $v > s$, where s is the local scaled infiltration capacity. If $v \leqslant s, v/s$ is a wetness index at that point. To compare individual NMM tubes with the model output, a particular value of s must be associated with each tube. This was achieved by again assuming that the tubes are a representative sample of the catchment, and using the 41 sample values of $F_S(s)$ (Figure 11) in Equation (9) to obtain s values (with $\beta = 4$ and $s_{min} = 0.599$).

Thus simulated values of v and s values computed from Equation (11) for individual NMM tubes may be used to compare v/s and $SMI^* = (SM^* - SM^*_{min})/(SM^*_{max} - SM^*_{min})$ throughout the entire period of measurements. This is shown in Figure 15, where v/s is compared with SMI^* for a selection of three tubes (B4, A6 and C1), which can be described as shallow [$F_S(s) = 0.098$], medium [$F_S(s) = 0.488$] and deep [$F_S(s) = 0.878$]. In cases where $v/s > 1$, the v/s ratio has been set to unity. The model does not predict individual tubes as well as it did the catchment-wide soil moisture, but for $F_S(s)$ between 0.1 and 0.9, the observed r^2 values are encouraging.

For deep tubes with $F_S(s) > 0.9$, the model underpredicts during wet periods. This is most likely caused by underestimating the SM^*_{max} value for this tube. As soil moisture measurements were invariably taken on days when it was not raining, the soil would have drained below its maximum capacity. Some of the effect could also be caused by underestimating the maximum soil depth in the catchment, as the tubes do not represent an exhaustive sample of soil depths across the catchment.

For shallow tubes with $F_S(s) < 0.1$ the VIC model appears to yield v/s values which are higher than the corresponding SMI^{**} values across the entire range of values. Figure 2 shows that the NMM network does not fully represent the drier parts of the catchment where the wetness index is less than about 8: the $F_S(s)$ values used for the shallow tubes are too low. However, a small increase in $F_S(s)$ in Equation (11) will only cause a very small increase in s and a corresponding decrease in v/s.

DISCUSSION AND CONCLUSIONS

It is generally agreed that lumped water balance models can play an important part in the aggregation of catchment hydrological processes and the scaling up to catchment and regional scales. However, such models are highly parameterized, they require calibration and physical measurement of the model parameters is difficult. The utility of many hydrological models is limited by difficulties in estimating actual evaporation and by uncertainty about the physical meaning of the various model parameters and about the realism of simulated values of state variables such as soil moisture status. It is against that background that this paper addresses the physical realism of the two lumped water balance models.

We have simulated the temporal variation in runoff, evaporation and soil moisture storage in the 26 km^2 Lockyersleigh catchment with a modification of the lumped SFB water balance model, which assumes a fixed bucket size, and with a variable bucket representation as originally expressed in the VIC model based on a statistical distribution of soil depths. The simple (invariable) bucket representation of the SFB model assumes that spatial variability in terrain, soils and vegetation may be ignored. The variable bucket VIC model makes the same assumption with one major difference: the model introduces a statistical distribution of storage capacities across the catchment. The degree of fit between observed and modelled streamflow (Figure 16) suggests that SFB and VIC are reasonable models to apply to this study area. This view is also partially supported by the results of Figure 12, where scaled soil moisture is shown to be consistent with some of the assumptions behind the VIC model.

Comparison of simulated catchment soil moisture storage and the soil moisture index based on measurements indicates that both models can make useful predictions of soil moisture status at the catchment scale. The SFB model results were compared with measurements in the top 50 cm, whereas the VIC model results were compared with measurements over the entire depth. We found good agreement between observed and simulated soil moisture values, with the VIC model performing slightly better than the SFB model. These results indicate that state variables such as soil moisture in the SFB and VIC models do have physical significance for the study catchment.

The availability of spatially distributed measurements of soil moisture was a key factor in defining a meaningful catchment-scale soil moisture index. It seems likely that previous modelling studies were unsuccessful in predicting soil moisture because of insufficient data to adequately sample the considerable spatial variability of soil moisture. In choosing a catchment suitable for analysis by a lumped model, a careful balance needs to be reached between the major competing sources of spatial variability in soil moisture. It is conceivable that larger catchments, if well- instrumented, will have a better defined soil moisture status than smaller catchments, but eventually the effects of large-scale rainfall variability can exceed the limitations of most lumped models and most simple soil moisture indices. At this larger scale, explicit treatment of large-scale spatial variability may be necessary.

This study has shown that the quasi-distributed VIC model can predict the relative wetness at individual locations, given their relative frequency of occurrence, thus providing a scaling function for the disaggregation of catchment-scale storage values to point-scale soil moisture values.

The paper has also highlighted several limitations of the study. Firstly, it has been shown that the NMM network is not fully representative of the catchment as a whole and therefore does not provide an exhaustive sample of soil depths across the catchment. Secondly, the soil moisture observations also do not necessarily provide reliable estimates of total storage capacity. Thirdly, catchment-scale soil moisture status is estimated from an incomplete coverage of the catchment with actual measurements, rather than a complete coverage with surrogate measurements, as would have been the case with remote sensing data. Fourthly, both models use very simple evapotranspiration models which do not adequately consider the role of vegetation. These limitations must be overcome in future experiments which address scaling questions.

The results of this study are encouraging in that a useful connection has been made between point- and catchment-scale soil moisture status. The VIC model exploits this relationship successfully, while still maintaining the simplicity of a lumped model. The results also provide important evidence that lumped parameter models can predict catchment soil moisture status. The successful use of distributed soil moisture measurements in validating such lumped models provides confidence in their use for catchment management studies and other applications, including drought studies.

ACKNOWLEDGEMENTS

We acknowledge the assistance of H. Alksnis, S. P. Charles, P. Daniel, P. M. Fleming, L. Guerra, M. Sivapalan and N. R. Sumner with the preparation of this paper. We also thank the anonymous referees for their helpful comments on an earlier draft of this paper.

REFERENCES

Alksnis, H., Daniel, P., and Kalma, J. D. 1990. 'The regional evaporation project: soil moisture and rainfall data', *CSIRO Div. of Wat. Resour. Tech. Memo. 90/1.*

Alley, W. M. 1984. 'On the treatment of evapotranspiration, soil moisture accounting and aquifer recharge in monthly water balance models', *Wat. Resour. Res.*, **20**, 1137–1149.

Avissar, R. and Pielke, R. A. 1989. 'A parameterization of heterogeneous land surfaces for atmospheric numeric models and its impact on regional meteorology', *Monthly Weather Rev.*, **117**, 2113–2136.

Bates, B. C., Charles, S. P., Sumner, N. R., and Fleming, P. M. 1994. 'Climate change and its hydrological implications for South Australia', *Trans. R. Soc. S. Aust.*, **118**, 35–43.

Boughton, W. C. 1984. 'A simple model for estimating the water yield of ungauged catchments', *Civ. Eng. Trans., I. E. Aust.*, **CE26**, 83–88.

Chiew, F. H. S. , Stewardson, M. J., and McMahon, T. A. 1993. 'Comparison of six rainfall-runoff modelling approaches', *J. Hydrol.*, **147**, 1–36.

Daniel, P., Alksnis, H., and Kalma, J. D. 1994. 'The regional evaporation project: climate and streamflow data', *CSIRO Div. Wat. Resour. Tech. Memo. 94/14.*

Eeles, C. W. O. , Robinson, M., and Ward, R. C. 1990. 'Experimental basins and environmental models' in Hooghart, J. C., Posthumus, C. W. S. and Warmerdam P. M. M. (Eds), *Hydrological Research Basins and the Environment.* TNO, Verslagen en Mededelingen, **44**, 3–12.

Entekhabi, D. and Eagleson, P. S. 1989. 'Land surface hydrology parameterization for atmospheric general circulation models including subgrid scale spatial variability', *J. Climate*, **2**, 816–831.

Johnston, P. R. and Pilgrim, D. H. 1976. 'Parameter optimization for watershed models', *Wat. Resour. Res.*, **12**, 477–486.

Jolley, T. J. and Wheater, H. S. 1993. 'Macromodelling of the River Severn' in *Macro-scale Modelling of the Hydrosphere* (Ed. (Wilkinson, W. B.), *Proc. Int. Symp., Yokohama, July 1993, IAHS Publ.*, **214**, 91–100.

Kirkpatrick, S., Gelatt, C. D. Jr, and Vecchi, M. P. 1983. 'Optimization by simulated annealing', *Science*, **220**, 671–680.

Kuczera, G. 1983.' Improved parameter inference in catchment models. 2. Combining different kinds of hydrologic data and testing their compatibility', *Wat. Resour. Res.*, **19**, 1163–1172.

Moore, I. D., Burch, G. J., and Mackenzie, D. H. 1988. 'Topographic effects on the distribution of surface soil water and the location of ephemeral gullies', *Trans. Am. Soc. Agric. Engin.*, **31**, 1098–1107.

Moore, I. D., Grayson, R. B., and Ladson, A. R. 1991. 'Digital terrain modelling: a review of hydrological, geomorphological and biological applications', *Hydrol. Process.*, **5**, 3–30.

Moore, I. D., Gallant, J. C., Guerra, L., and Kalma, J. D. 1993. 'Modelling the spatial variability of hydrologic processes using GIS' in *HydroGIS 93, International Conference on Application of Geographic Information Systems in Hydrology and Water Resources Management, Baden (Vienna), 19–22 April 1993* (Eds Kovar, K. and Nachtnebel, H.P.), *IAHS Publ.*, **211**, 161–169.

Nathan, R. J. and McMahon, T. A. 1990a. 'The SFB model Part 1 — validation of fixed model parameters', *Civ. Engin. Trans . I. E. Aust.*, **CE32**, 157–161.

Nathan, R. J. and McMahon, T. A. 1990b. 'The SFB model Part 2 — operational considerations', *Civ. Engin. Trans . I. E. Aust.*, **CE32,**, 162–166.

Nelder, J.A. and Mead, R. 1965. 'A simplex method for function minimization', *Comput. J.*, **7**, 308–313.

Press, W. H., Teukolsky, S. A., Vetterling, W. T., and Flannery, B. P. 1992. *Numerical recipes in FORTRAN: the Art of Scientific Computing*. 2nd edn. Cambridge Univ. Press, Cambridge.

Sivapalan, M. and Woods , R. A. 1995. 'Evaluation of the effects of GCM subgrid variability and patchiness of rainfall and soil moisture on land surface water balance fluxes', *Hydrol. Process.*, **9**, 697–717.

Sivapalan, M., Beven, K. J., and Wood, E. F. 1987. 'On hydrologic similarity. 2. A scaled model of storm runoff production', *Wat. Resour. Res.*, **23**, 2266–2278.

Thompson, N., Barrie, I. A., and Ayles, M. 1981. 'The Meteorological Office rainfall and evaporation calculation system: MORECS', *Hydrol. Memo. No. 45*, UK Meteorological Office, Bracknell.

Wood, E. F., Sivapalan, M., and Beven, K. 1990. 'Similarity and scale in catchment storm response'. *Rev. Geophys.*, **28**, 1–18.

Wood, E. F., Lettenmaier, D. P., and Wallis, J. R. 1991. 'Comparison of an alternative land surface parameterization with the GFDL high resolution climate model' in *Hydrological Interactions between Atmosphere, Soil and Vegetation* (Eds Kienitz, G., Milly, P. C. D., Van Genuchten, M. Th., Rosbjerg, D., and Shuttleworth, W. J.), *Proc. Int. Symp., Vienna, August 1991, IAHS Publ.*, **204**, 53–64.

Wood, E. F., Lettenmaier, D. P., and Zartarian, V. G. 1992. 'A land-surface hydrology parameterization with subgrid variability for general circulation models'. *J. Geophys. Res.*, **97**, 2717–2728.

13

ESTIMATION OF SUBGRID SCALE KINEMATIC WAVE PARAMETERS FOR HILLSLOPES

GARRY WILLGOOSE AND GEORGE KUCZERA

Department of Civil Engineering and Surveying, University of Newcastle, NSW, 2308, Australia

ABSTRACT

The kinematic wave approximation is commonly used in the new generation of so-called physically based, distributed rainfall–runoff models. However, although the kinematic wave approximation is commonly accepted for channel and experimental flows, its applicability to actual hillslopes remains unvalidated. Because it is not possible to measure, nor model, all of the details of the flow on any realistic surface, we use subgrid approximations to provide an effective parameterization of the processes that occur on scales smaller than those that can be modelled. This paper explores different effective parameterizations, the data required to identify the correct parameterization, and the implications of not being able to identify all of the parameters on the scale dependence of flood hydrology. Data from small-scale plot experiments ($100 \, \text{m}^2$) and large-scale catchments ($1 \, \text{km}^2$) are used to explore these issues. It has been found that infiltration parameters can be adequately calibrated from small-scale plots. However, it is more difficult to calibrate the kinematic wave parameters using small-scale data alone. The conveyance properties of the hillslope cross-sections are parameterized by two kinematic wave parameters, c_r and e_m, to yield the discharge $Q = c_r A_{cs}^{e_m} S^{0.5}$ with S being the slope and A_{cs} the cross-sectional area. It is shown that these two parameters are highly correlated, particularly when inferred from small-scale data. The surface roughness, amount of rilling and undulations of the surface all influence the kinematic wave parameters. The runoff response at large scales is very sensitive to changes in c_r and e_m, yet is not readily apparent in small-scale data. Unfortunately, using small-scale data c_r and e_m cannot be estimated with acceptable precision to reliably extrapolate to larger scales. The significance of this behaviour is demonstrated and some possible solution strategies are discussed.

INTRODUCTION

The scale or area dependence of runoff is of crucial importance in the development of accurate predictive models of the surface hydrology response of catchments. Indeed, '. . . the issue of the linkage and integration of formulations at different scales has not been addressed adequately. Doing so remains one of the outstanding challenges in the field of surficial processes' (National Research Council, 1991).

Linear runoff-routing models have been used in the past, and sometimes in the present, the most common being the deterministic rational method and the unit hydrograph. It has, however, long been recognized that the runoff response of catchments is non-linear, as reflected in currently adopted subcatchment based computer methods such as RORB (Mein *et al.*, 1974) and Field–Williams (Field and Williams, 1987).

Generally, these models are calibrated to site rainfall and runoff data and then used for the prediction of design hydrographs, either at the same site or at some site slightly upstream or downstream. How good are these models for predicting the runoff response for much larger catchments in which the calibration site is just a small subcatchment?

With the appearance of digital terrain map (DTM) based rainfall–runoff models (Moore *et al.*, 1988; Moore and Grayson, 1991; Huang and Willgoose, 1992; 1993) this question has much greater significance. These models can, in principle, predict the discharge/unit width for any point in the catchment down to the resolution of the digital terrain map (as low as every 6.25 m for the gridded DTM data in

New South Wales, but more typically around 30 m as in the USGS data). Subcatchment based models predict flows at the outlets of each subcatchment, and these subcatchments are rarely less than a hectare in size. If the model is calibrated to small-scale data, such as from runoff plots, then it is tempting to use the runoff-routing components of these models to scale up the predicted runoff response to larger design catchments. Similarly, if the DTM model is calibrated to data from a large catchment, then it is tempting to believe that, if the catchment properties are uniform, the flows within the catchment are also accurate. These extrapolations are replete with uncertainty.

This paper focuses on the kinematic wave approximation for the hillslope surface runoff model commonly used in DTM based rainfall–runoff models. One of the major uncertainties in their application is with the subgrid parameterization used for representation of the kinematic wave routing. At length scales less than the DTM resolution there is no possibility of a model exactly representing the surface roughness, the flow paths and the routing of runoff. Some effective or average representation is required of the routing processes over the grid element that captures the average effect of the microscale roughness and flow concentrations. At length scales greater than the DTM resolution the surface roughness, the flow paths and the kinematic routing can, in principle, be modelled adequately by the surface representation of the hydrological model.

This paper investigates the identifiability of the effective parameters in a hillslope kinematic wave model. In this way we explore what type of data are required to identify the kinematic wave subgrid parameterization. This paper will demonstrate that the subgrid parameterization adopted has an important effect on the scaling behaviour of a peak discharge predicted by the kinematic wave based rainfall–runoff model. We will demonstrate that it does not appear possible to calibrate all the parameters in the kinematic wave parameterization with data collected from either a small runoff plot or a larger catchment. It will be shown that the only effective means of identifying these parameters is by using runoff data collected from both small and large catchment areas.

In the calibration of a rainfall–runoff model for a rehabilitated mine site, rainfall–runoff data are often non-existent, particularly if the model is required as part of the design process. Alternative means of model calibration must be found. The normal method is to use small runoff plots to calibrate the parameters of the model. These plots are constructed on box-cut spoils or other waste rock from the early stages of the mining. This paper addresses these issues of scale with a case study, the runoff from a proposed waste rock dump at the Ranger Uranium Mine, in the Northern Territory, Australia. We show that: (1) reliably scaling up a calibrated runoff model is difficult and (2) the parameters of the hillslope runoff-routing component of the DTM based model cannot be reliably estimated.

DTM RAINFALL–RUNOFF MODEL

A DTM based rainfall–runoff model called DISTFW is used in this study. It is based on the subcatchment based Field–Williams generalized kinematic wave model (Field and Williams, 1987). The key enhancement of the original model is the ability to use a DTM on a square grid to define the catchment's hydrological flowlines. Each grid point is assumed to have a subcatchment. The area of each grid point is normally assumed equal, though the area of a grid point can be varied if desired. Flow into a grid point from upstream is allowed. The drainage connectivity of the grid points is determined by the algorithm used by the SIBERIA catchment evolution model (Willgoose et al., 1991), where, to first order, drainage flows into one of the adjacent eight points on the square grid in the steepest downslope direction.

The model parameters are varied depending on whether the point is a channel or hillslope. Channels are normally identified in one of two ways. The first is that a channel occurs when a channel threshold function is exceeded (Montgomery and Dietrich, 1989; Willgoose et al., 1991; Williams and Riley, 1992)

$$\frac{\beta_5 q^{m_5} S^{n_5}}{a_t} \begin{array}{l} > 1 \quad \text{channel} \\ < 1 \quad \text{hillslope} \end{array} \tag{1}$$

where β_5 is a rate constant and a_t is the channel initiation threshold, m_5 and n_5 are parameters, q is the mean peak discharge/unit width (from flood frequency analysis) and S is the slope. The second is that the coordinates of all the channel heads can be identified (from, for instance, the blue lines on maps or aerial photography), with channels being defined for all downstream points.

The model simulates Hortonian runoff with Philip infiltration. The infiltration rate is determined using the time compression algorithm (TCA) with the instantaneous infiltration rate, i, a function of cumulative infiltration, I

$$i = \phi + \frac{S_\phi^2}{4I} \left\{ 1 + \left[1 + \left(\frac{4I\phi}{S_\phi^2} \right) \right]^{1/2} \right\} \tag{2}$$

where S_ϕ is the sorptivity and ϕ is the long-term infiltration rate. Infiltration excess supplies a non-linear surface storage that can be used to model surface depressions (e.g. deep rip patterns on mine rehabilitation). The discharge/unit area of storage, s_s, is

$$s_s = \left(\frac{h_s}{cB^\gamma} \right)^{\frac{1}{\gamma}} \tag{3}$$

where h_s is the depth of storage, B the width of the grid element and c_s and γ are parameters. The non-linear surface store supplies water to the routing component, a kinematic wave. Normally only one of the kinematic wave or non-linear stores is activated during simulations because of the difficulty in identifying parameters. The original Field–Williams model was designed as an event model, but has been extended to model continuous flow series by the addition of an evaporation and infiltration recovery module that modifies the infiltrated volume in the TCA algorithm (Willgoose and Riley, 1993b).

The kinematic wave routing is the component of DISTFW that most influences the scale dependent properties of the runoff hydrograph discussed in this paper (Huang and Willgoose, 1992; 1993), whereas infiltration rates mainly influence the volume of runoff. The conveyance properties in the model of the hillslope and channels are different and are allowed to change with discharge, modelling, for instance, overbank flow for channels or flooded rills on the hillslope. The kinematic assumption that the friction slope equals the bed slope is used and discharge is determined from Manning's equation

$$q = \frac{R^{5/3} S^{1/2} P}{n} = \frac{A_{cs}^{5/3} P^{-2/3} S^{1/2}}{n} \tag{4}$$

where q is the discharge/unit width, R the hydraulic radius, P the wetted perimeter/unit width, n the Manning roughness, A_{cs} is the flow cross-sectional area/unit width and S the slope. This can be formulated as

$$q = KS^{1/2} \tag{5}$$

where K is the channel conveyance.

The channel conveyance is approximated by a power law function involving the flow cross-sectional area

$$K = c_r A_{cs}^{e_m} \tag{6}$$

where c_r and e_m are coefficients that are defined by the flow geometry and surface roughness. The parameters c_r and e_m are the two parameters that determine the kinematic wave component of the model. These parameters are allowed to change with discharge and from element to element (Field and Williams, 1987). Each grid element thus models a one-dimensional kinematic wave discharging into the next downstream element. The two-dimensional drainage convergence is modelled by the SIBERIA drainage algorithm, which allows many elements to drain into one.

To ease the task of identification of these parameters from measured data DISTFW has been integrated with a non-linear regression package NLFIT designed for parameter estimation (Kuczera, 1989; 1994). This package allows the user to fit (1) parameters to a single storm for a single site or (2) best-fit parameters for a number of storms at a single site. The diagnostic output of NLFIT includes error bounds on parameter estimates, error bounds on both calibrated and predicted hydrographs and output indicating parameter interactions. The catchment parameters fitted are (1) infiltration S_ϕ, ϕ, (2) surface storage c_s, γ and (3) kinematic wave c_r, e_m, whereas the parameters for each runoff event fitted are (4) the initial wetness, V, defined as the cumulative infiltration at the beginning of the event in the TCA formulation of infiltration and (5) the timing error between the observed runoff and rainfall.

The sensitivity of the model to changes in these parameters has been studied in detail elsewhere (Huang and Willgoose, 1992; 1993). This paper concentrates on the identification of the kinematic wave parameters and their influence on the scale dependence of the runoff hydrology.

KINEMATIC FLOW PARAMETERS AND SURFACE ROUGHNESS

The normal assumption when calibrating the overland flow kinematic wave of both DTM-based and sub-catchment-based models is uniform overland sheetflow (Moore and Grayson, 1989; Woolhiser, et al., 1990; Goodrich, et al., 1991)—that is, flow that has constant or irregular depth over the entire width of the hillslope and that does not concentrate into rills (i.e. Figure 1a and 1b). This traditional view of flow over hillslope is often adopted despite the fact it is widely discredited (Parsons et al., 1990), except in laboratory controlled circumstances. More commonly, flow concentrates into rivulets so that only a portion of the hillslope contributes to the downslope flow (Figure 1c and 1d).

The typical resolution of DTM models is too coarse to model the hydraulics of all but the largest of the rills; a subgrid scale representation for the kinematic wave which describes the subgrid scale surface roughness is needed. In this paper we examine the effect of this surface roughness on the scaling properties of the runoff-routing, and what type of runoff data are required to calibrate a subgrid scale kinematic wave model without direct information on the surface roughness. Three subgrid conceptualisations of surface roughness are considered.

Sheetflow

This case is illustrated in Figure 1a and 1b. The parameters of Equation (6) can be easily determined from Equation (4). If the surface roughness is such that the wetted perimeter/unit width is constant with increasing discharge, then

$$A_{cs} \propto P$$

$$c_r = \frac{1}{P^{2/3}n} \tag{7}$$

$$e_m = 5/3$$

It may be noted in passing that $A_{cs} \propto P$ has the minimum exponent on the wetted parameter possible so that $e_m = 5/3$ is an upper bound on the value of this parameter.

Triangular rillflow

This case is illustrated in Figure 1c. The parameters of Equation (6) can be derived from the cross-sectional geometry. For a channel with side slope of 1:a, then

$$A_{cs} \propto P^2$$

$$c_r = N \frac{\left[\dfrac{a}{4(1 + a^2)} \right]^{1/3}}{n} \tag{8}$$

$$e_m = 4/3$$

where N is the number of rills/unit width. The parameters are those governing the propagation of the kinematic wave when the triangular rills are partially flooded. If the rills are flooded, the wetted perimeter is

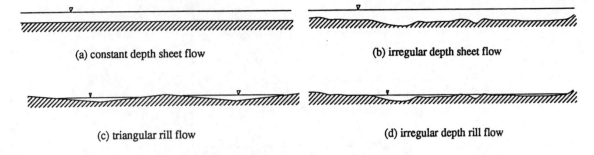

(a) constant depth sheet flow   (b) irregular depth sheet flow

(c) triangular rill flow   (d) irregular depth rill flow

Figure 1. Different overland sheetflow and rillflow geometries

independent of flow depth and the parameters are given by Equation (7). As the number of rills/unit width increases the threshold discharge at which this flooding occurs will decrease.

Natural surfaces

This case is illustrated in Figure 1d. The parameters for the kinematic equation can be determined from analysis of cross-sections. Parsons *et al.* (1990) provide detailed contour data of a 18 m by 35 m plot from the Walnut Gulch experimental catchment in Arizona (their Figure 1). Five cross-sections were taken at right angles to flow (Figure 2), the relationship between the wetted perimeter and area was determined (Figure 3) and input into Equation (6) to yield

$$A = 0 \cdot 0076 P^{1 \cdot 49} \qquad R^2 = 0 \cdot 88$$

$$c_{\mathrm{r}} = \frac{0 \cdot 113}{n} \qquad\qquad\qquad (9)$$

$$e_{\mathrm{m}} = 1 \cdot 21$$

These results for the natural surface assume: (1) uniform flow occurs within each cross-section, which means that the cross-sections vary gradually downstream; (2) cross-sections fill with water from the bottom up so that perched channels at higher elevations do not carry significant water; (3) Manning's n roughness does not vary with flow depth; and (4) the rills may be totally flooded above some threshold discharge so that the sheetflow results apply for these higher discharges. Abrahams (pers. comm.) notes that condition (2) can be violated. A perched channel may contribute flow from an upstream area not drained by the deeper part of the cross-section. He also noted variations in the elevation of the free surface across the cross-section during some storm events.

Two points should be noted about these analyses in general. Firstly, other workers have suggested higher values for e_{m} of up to three (Julien and Simons, 1985). These higher values assume laminar flow, which is unlikely to apply other than near catchment boundaries. Secondly, these analyses indicate that the presence

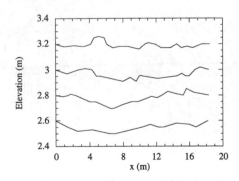

Figure 2 Hillslope cross-sections from the Parsons *et al.* (1990) plot

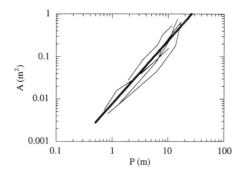

Figure 3. Area versus wetted perimeter for the cross-sections in Figure 2. The heavy line is the fitted relationship

of rills or surface irregularities substantially decreases the value of e_m compared with the value for sheet-flow. The exact value of e_m for a given catchment will need to be calibrated and what type of data is required to do this successfully is the main subject of this paper. Despite this it is believed that surface ir-regularities will result in a kinematic wave response that is closer to linearity ($e_m = 1$) than the sheetflow assumption normally used in distributed kinematic wave models ($e_m = 1.67$) would suggest.

We do not claim that these cases are exhaustive of the possible parameter values, only indicative of various physical configurations. However, these simple results provide a starting point for an analysis of the cross-sectional flow model and the effect of surface roughness on design flood predictions.

PARAMETER ESTIMATION

This section examines the ability to identify the parameters of the kinematic wave from rainfall–runoff data and the ability to scale up the calibrated model to larger catchments. Four case studies based on data from Ranger Uranium Mine are used to examine the ability to estimate parameters for the kinematic routing model. In the first two cases data for a $102 \, m^2$ plot is used to calibrate the model and error bounds for a design catchment of about $1 \, km^2$ (Figure 4) are examined: this is our small-scale data. The runoff data were collected from natural rainfall events on an unvegetated plot of gneiss derived mine spoil—this plot is designated as CWT2 (Willgoose and Riley, 1993a). The hydrographs and hyetographs are presented in Figure 5. Measures of the scale dependence of the runoff model are examined. In the third and fourth case studies simulated runoff data from a larger catchment (part of the proposed Ranger Uranium Mine waste rock dump) is used to augment the plot data during the calibration of the model: this is our large-scale data.

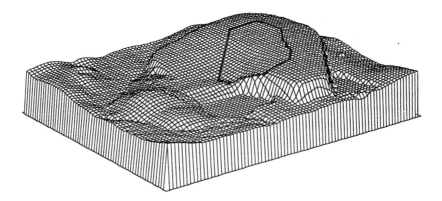

Figure 4. Design catchment digital terrain map (after Willgoose and Riley, 1993). The heavy outline indicates the extent of the catch-ment used in this study. DTM resolution is 30 m

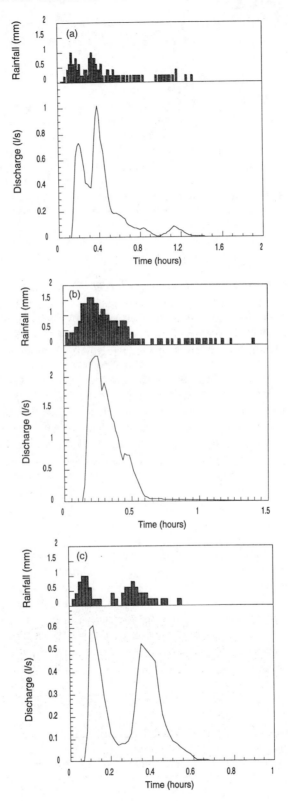

Figure 5. Observed runoff hydrographs and rainfall hyetographs for plot CWT2 for (a) 10 January 1991, (b) 21 January 1991 and (c) 7 January 1991

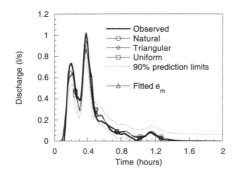

Figure 6. Fitted hydrographs for (a) calibration of c_r with e_m fixed and (b) calibration of both c_r and e_m simultaneously. The 90% prediction limits are for the sheetflow example and are similar to the other results. The rainfall and runoff data for 10 January have been used in all instances

CWT2 catchment calibrated to a single storm event

Rainfall–runoff data for plot CWT2 for 10 January 1991 were used to calibrate DISTFW using NLFIT (Figure 5a).

In the first stage, three parameters were fitted—c_r, S_ϕ and ϕ—for the three values of e_m corresponding to the three flow cross-sections discussed earlier (Figure 6). The other components of the model were turned off in this paper. It was found that the sorptivity was effectively zero for the material and therefore it was decided to fix $S_\phi = 0$ for subsequent calibrations. Final parameters and their errors of estimation (standard deviation) are listed in Table I. The coefficients of variation for c_r are about 5%. The data are equally well fitted by values of e_m in the range 1–2·5. The hydrographs for the design catchment (using the rainfall from 10 January 1991) and corresponding prediction limits are illustrated in Figure 7.

In the second stage, the parameters c_r, e_m, and ϕ (Table I) were fitted with the value of S_ϕ fixed at 0. The fit to the data is unchanged (Figure 6), the error of estimation of c_r is significantly increased (coefficients of variation for $c_r = 50\%$ and $e_m = 6\%$), whereas the prediction limits for the design catchment are significantly larger than those for the case of e_m fixed. This increase in the prediction limits and errors of estimation indicates that the latter model is overparameterized; i.e. the model parameters of the subgrid representation of the kinematic wave cannot be estimated with adequate precision from the available plot runoff data.

CWT2 catchment jointly calibrated to multiple storm events

The three parameters of the model c_r, e_m and ϕ were jointly calibrated to three storm events measured on the CWT2 plot on 7 January, 10 January and 21 January 1991 (Figure 8). The final parameter fits are listed in Table II. Note that the parameters and their errors of estimation are not substantially less than those for the single hydrograph fit. This suggests that the information in the extra two storm events is of only marginal use for improving the parameter estimates. The marginal tightening of the prediction limits for the design catchment compared with the single hydrograph estimates (Figures 7 and 9) confirms this conclusion. Again it appears that e_m cannot be reliably estimated from the data.

Table I. Calibration results for the CWT2 single event (mean ± standard deviation)

Parameter	Sheet flow assumption	Triangular rills assumption	Natural section assumption	Fitted e_m, single event
c_r (mm/hr$^{0.5}$)	$38\cdot1 \pm 2\cdot6$	$5\cdot63 \pm 0\cdot26$	$2\cdot87 \pm 0\cdot11$	$9\cdot90 \pm 4\cdot97$
e_m	$1\cdot67$	$1\cdot33$	$1\cdot21$	$1\cdot401 \pm 0\cdot088$
ϕ (mm/hr)	$7\cdot77 \pm 0\cdot23$	$7\cdot87 \pm 0\cdot23$	$7\cdot90 \pm 0\cdot22$	$7\cdot58 \pm 0\cdot20$

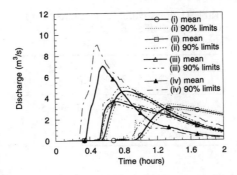

Figure 7. Predicted hydrographs and their prediction limits for the design catchment using the calibration results of Figure 6: (i) Parsons' natural cross-section; (ii) triangular rills; (iii) sheetflow; and (iv) fitted e_m. The rainfall data for 10 January have been used in all instances

Calibration of design catchment to a single storm event

The ability to estimate the parameters using storm event data from a larger catchment was tested. Synthetic data were generated for the design catchment (area 1 km^2) with the rainfall data of 10 January 1991 and used together with the plot data for the storm of CWT2 on the same day for the parameter estimation. The synthetic data were generated by using the parameters from the single site, single storm fit with e_m equal to 1·21, corresponding to a surface with triangular rills. These data were then corrupted with errors with a relative error variance equal that of the single site–single event prediction limits (Figure 6) and with a lag-one correlation of 0·7. This was an attempt to approximate the systematic errors observed in the plot data, the lag-one correlation approximating the error structure of the plot data residuals.

Using only the design catchment data, a reasonable fit to the data could be obtained. However, extrapolating that fit to the plot scale yielded unsatisfactory prediction limits for the plot simulations, with more

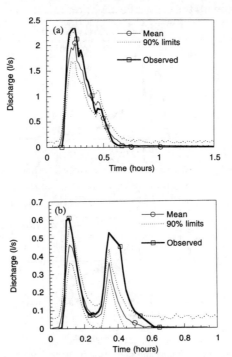

Figure 8. Fits of the calibration of c_r and e_m for (a) 21 January, 1991 and (b) 7 January, 1991 using all three runoff hydrographs measured on plot CWT2 in the calibration in a joint fit

Table II. Joint calibration results for the CWT2 3 runoff events (mean ± standard deviation)

Parameter	Fitted e_m, three events
c_r (mm/hr$^{0.5}$)	11.6 ± 5.0
e_m	1.438 ± 0.077
ϕ (mm/hr)	4.02 ± 0.17

than 10% of the observed data falling outside the 90% prediction limits (Figure 9a). For the case where the fitting was to the design catchment alone, the upper prediction limits for the flows on the plot are very high, approximately 15 times the predicted hydrograph, whereas the lower limits are effectively zero, indicating that very low confidence can be placed in the small-scale predictions based on fitting to the large-scale data.

Joint calibration of CWT2 and design catchment to a single storm event

Using both the design catchment data and the plot data better parameter estimates are obtained (Table III). The prediction limits for design catchment are larger than for the fit on the design catchment, reflecting that the fit is a compromise between that for the small and large catchment.

Notably the prediction limits for the joint fit of both c_r and e_m are significantly improved over that for the fits using the plot data alone (Figure 9a). The use of the design catchment data alone significantly tightens the prediction limits at the larger scale, whereas the addition of the plot data reduces the errors of estimation further. Perhaps surprisingly, the scaling down of the fit for the catchment data yields a poor prediction

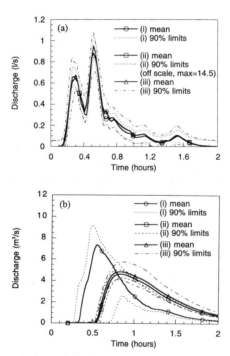

Figure 9. (a) Plot and (b) design catchment hydrographs and prediction limits for the cases of the calibrations using (i) the three plot runoff events alone, (ii) the large catchment data alone and (iii) both the plot and the large catchment data. Note that for downscaling the catchment calibration [i.e. (a)(ii)] the upper prediction limit is offscale, lying in the range 10–14·5. The rainfall data for 10 January have been used in all instances

Table III. Calibration results for the design catchment (mean ± standard deviation)

Parameter	Fitted e_m, single event, multiple sites
c_r (mm/hr$^{0.5}$)	4.86 ± 0.18
e_m	1.342 ± 0.051
ϕ (mm/hr)	6.18 ± 0.14

of the plot data, indicating that using the data from the large catchment alone is insufficient to identify both c_r and e_m (Figure 9a).

DISCUSSION

The scale dependence of the runoff-routing model can be further examined by looking at the trend of mean peak discharge with increasing area: the $Q–A$ curve. To derive the $Q–A$ curve rainfall events for a range of durations, and two year return period, are determined from intensity–frequency–duration data and applied to the design catchment. At each DTM grid point the peak discharge for each of the storms is determined and the peak of the storm that is critical at that point is correlated with the contributing area at that point. The $Q–A$ graph in Figure 10 demonstrates that the scaling behaviour is strongly influenced by the subgrid scale model used for kinematic wave runoff routing. Intuitively, we argue that one of the parameters, c_r if e_m is fixed, determines the ordinate of the $Q–A$ curve for small areas, whereas the other parameter determines the slope of the $Q–A$ curve.

This dependence of the $Q–A$ curve on both kinematic parameters is very important, and rather disturbing if only one of them can be reliably calibrated. It indicates that the scale dependence of the flood hydrology cannot be scaled up from small-scale data, nor scaled down from large-scale data. For erosion modelling it is important to know the variation of discharge with area if the areas of erosion and deposition, and the degree of concentrated versus sheet erosion is to occur (Willgoose and Riley, 1994).

It is asserted that the result reported here is the best that is likely to be observed. The use of field data for the design catchment will degrade these results for a number of reasons, including: (1) the signal to noise ratio will decline because of unknown error structure in the residuals; (2) systematic variation in c_r and e_m as rills and channels develop downstream or as rills drown will create systematic errors in the model; (3) variations in rainfall and infiltration rates will be significant at large scales; and (4) the observation errors of hydrographs inferred from rating curves normally used for large streams are larger than those observed for the flumes normally used in plot studies. Despite this, and perhaps because of point (2), it is believed that the use of two different scale catchments is the best means of estimating c_r and e_m.

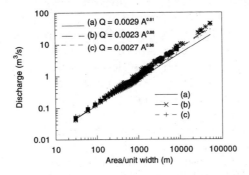

Figure 10. Discharge–area relationship for the calibrated c_r in Figure 6, assuming (a) Parsons' natural surface ($e_m = 1.21$), (b) sheetflow ($e_m = 1.67$) and (c) triangular rills ($e_m = 1.33$)

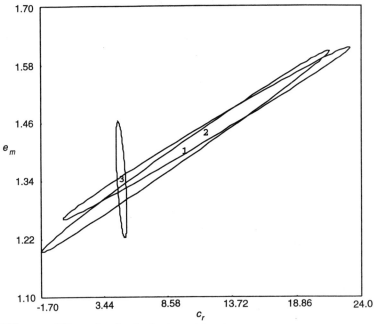

Figure 11. Approximate 95% compatibility regions for the fit of c_r and e_m using (1) plot data for 10 January, (2) plot runoff data for all three events and (3) plot and large catchment data for the runoff event of 10 January. The numbers in the centre of the ellipses on the graph correspond to these instances

In kinematic wave conceptualizations of overland and channel flow the conveyance of a cross-section ($K = c_r A_{cs}^{e_m}$) is parameterized by two parameters c_r and e_m. We have demonstrated that both of these parameters are important in characterizing the rate at which the peak discharge of a design hydrograph changes with area as parameterized by the relationship between peak discharge and area. We believe that this result also has implications for many of the subcatchment based non-linear and linear storage models commonly used in practice. We cannot reliably scale up or down these models, nor can we rely on predictions of discharges for internal catchments. Linear storage models (e.g. unit hydrographs) do not eliminate this problem; they effectively assume a value for unity for e_m to make the response linear.

It is important to recognize that, though our models are nominally physically based, if we cannot determine parameters from observed data, or fix them *a priori* with an acceptable degree of certainty, then we cannot rely on the internal dynamics of these models (Grayson *et al.*, 1992). For small-scale hydraulics the kinematic wave is an established approximation of open channel flow for high slope, zero backwater domains. On the larger scale, for rough or rilled hillslopes, the accuracy of this approximation has never been established. With branching and braiding of overland flow, hydraulic jumps behind obstructions, backwater from debris dams and subcritical/supercritical flow transitions, it is likely that the kinematic approximation used for the subgrid scale averaged flow at the resolution of the digital terrain map is a poor approximation of the actual physics. Its one saving grace from the practitioner's viewpoint is that it has few enough parameters that it can be calibrated with observed data. This fitting should not be confused with validation of the physics of the subgrid scale approximation. Moreover, inference of Manning's n from, for instance, Equation (7), should be viewed with the appropriate degree of scepticism.

The key reason for the difficulty of calibration of c_r and e_m is the strong interaction between the parameters at the small-scale. One means of demonstrating this interaction is to show the 95% probability region on these parameters, so-called compatibility regions (Kuczera, 1990). Although these regions are based on the multi-normal approximation and are therefore, at best, approximations to the actual region, they provide a useful guide to the compatibility of parameters derived from different data sources. Figure 11 shows the compatibility regions for three cases: (1) calibration with one event: (2) calibration with three events; and (3) calibration with a single event for the plot and larger catchment. The long diagonal regions are indicative of strong correlation between the parameters. The similarity between the regions is confirmation that the extra runoff events do not significantly reduce the parameter interaction.

In contrast with the small-scale plot results, we find for the joint large- and small-scale data that the ellipse is shorter and parallel to the axes. The uncertainty in both c_r and e_m has been dramatically reduced.

CONCLUSIONS

The identifiability of the two parameters in the kinematic wave approximation of runoff routing $K = c_r A_{cs}^{e_m}$ has been explored in this paper. It was shown that calibrating both parameters of a kinematic wave runoff-routing model to a single runoff event on a plot yielded unsatisfactory estimation errors. Joint calibration to a number of runoff events from the same runoff plots yielded only marginally reduced estimation errors, suggesting that little extra information can be extracted from extra storms. The prediction limits at the larger design catchment scale for multiple calibration events were only marginally tighter than those for a single storm calibration. Moreover, it was found that calibration of the model to runoff data from a large-scale catchment also yielded unsatisfactory estimation errors on the parameters. To obtain satisfactory estimation errors and prediction limits it was found to be necessary to calibrate the model to runoff data from both small-scale plots and larger scale design catchments.

The normal assumption in DTM models that kinematic wave routing occurs over a flat plane ($e_m = 1.67$) is shown to be one extreme of a continuum of acceptable physically based models of the hydraulics of overland flow, with e_m as low as 1.2. Although one of the kinematic wave parameters can be fitted to data obtained for plots, it is not possible to reliably fit both the parameters. A corollary is that equally good fits to the plot data can be obtained for values of e_m ranging from 1.2 to 1.7. By generating prediction limits for design storms on larger catchments it can be clearly shown that it is necessary to identify both parameters to have acceptably small limits of prediction at larger scales. Moreover, data from a large catchment also only yields limited information about both parameters. In this latter case prediction limits for plots are unacceptably large.

The only conclusive means of calibrating both the parameters of kinematic wave model was by using jointly data for plots and for larger catchments. In this instance the prediction limits were considerably smaller. This joint fit works because the model is being forced to operate over a large range of A_{cs}, sufficient to expose the power law dependence. The question of what is the minimum ratio of large to small catchment areas necessary to identify c_r and e_m remains unanswered.

Finally, this study does not answer the question of the validity of the kinematic wave approximation for representing hillslope runoff. However, we have defined one means to accurately estimate the parameters of the kinematic wave representation so that the prediction limits are small. In this way we can design an experimental arrangement that is capable of testing the validity of the one-dimensional kinematic wave approximation used for the representation of hillslope routing processes.

ACKNOWLEDGEMENTS

The runoff and rainfall plot data and the digital terrain map were supplied by workers in the Geomorphology Branch of Environmental Research Institute of the Supervising Scientist, Jabiru. Discussions with S. Riley are acknowledged, as are the comments of the three anonymous reviewers.

REFERENCES

Field, W. G. and Williams, B. J. 1987. 'A generalized kinematic catchment model', *Wat Resour. Res.*, **23**, 1693–1696.
Goodrich, D. C., Woolhiser, D. A., and Keefer, T. O. 1991. 'Kinematic routing using finite elements on a triangular irregular network', *Wat. Resour. Res.*, **27**, 995–1004.
Grayson, R. B., Moore, I. D., and McMahon, T. A. 1992. 'Physically based hydrologic modeling, 2, is the concept realistic', *Wat. Resour. Res.*, **28**, 2659–2666.
Huang, H. Q. and Willgoose, G. R. 1992. 'Numerical analyses of relations between basin hydrology, geomorphology and scale', *Res. Rep. No 075.04.1992*, The University of Newcastle, Department of Civil Engineering and Surveying.

Huang, H. Q. and Willgoose, G. R. 1993. 'Flood frequency relationships dependent on catchment area: an investigation of causal relationships', in *Towards the 21st Century, Hydrology and Water Resources Symposium, Newcastle, 1993.*

Julien, P. Y. and Simon, D. B. 1985. 'Sediment transport capacity of overland flows', *Trans. Am. Soc. Agric. Engin.*, **28**, 755–762.

Kuczera, G. 1989. 'An application of Bayesian nonlinear regression to hydrologic models', *Adv. Engin. Software*, **11**, 149–155.

Kuczera, G. 1990. 'Estimation of runoff-routing model parameters using incompatible storm data', *J. Hydrol.*, **114**, 47–60.

Kuczera, G. 1994. 'NLFIT: a Bayesian nonlinear regression program suite', *Res. Rep.*, The University of Newcastle, Department of Civil Engineering and Surveying.

Mein, R. G., Laurenson, E. M., and McMahon, T. A. 1974. 'Simple nonlinear model for flood estimation', *J. Hydr. Div. ASCE*, **100** (HY11), 1507–1518.

Montgomery, D. R. and Dietrich, W. E. 1989. 'Source areas, drainage density and channel initiation', *Wat. Resour Res.*, **25**, 1907–1918.

Moore, I. D. and Grayson, R. B. 1991. 'Terrain-based catchment partitioning and runoff prediction using vector elevation data', *Wat. Resour. Res.*, **27**, 1177–1192.

Moore, I. D., O'Loughlin, E. M., and Burch, G. J. 1988. 'A contour based topographic model for hydrological and ecological applications', *Earth Surf. Process Landforms*, **13**, 305–320.

National Research Council 1991. *Opportunities in the Hydrologic Sciences.* National Academy Press, Washington DC.

Parsons, A. J., Abrahams, A. D., and Luk, S. H. 1990. 'Hydraulics of interrill overland flow on a semi-arid hillslope, southern Arizona', *J. Hydrol*, **117**, 255–273.

Willgoose, G. R. and Riley, S. J. 1993a. 'Application of a catchment evolution model to the prediction of long term erosion on the spoil heap at Ranger Uranium Mine', *Open File Rep. 107*, Supervising Scientist for the Alligator Rivers Region, Canberra.

Willgoose, G. R. and Riley, S. J. 1993b. 'The assessment of the long-term erosional stability of engineered structures of a proposed mine rehabilitation', in Chowdhury, R. N. and Sivakumar, M. (Ed.), *Environmental Management: Geowater and Engineering Aspects.* Balkema, Rotterdam.

Willgoose, G. R. and Riley, S. J. 1994. 'Long term erosional stability of mine spoils', in *Australian Institute of Mining and Metallurgy Conference, Darwin, 1994.*

Willgoose, G. R., Bras, R. L. and Rodriguez-Iturbe, I. 1991. 'A physically based coupled network growth and hillslope evolution model: 1 theory', *Wat. Resour. Res.*, **27**, 1671–1684.

Williams, D. K. and Riley, S. J. 1992. *Some Geomorphic Thresholds Related to Gullying, Tin Camp Creek. Arnhem Land Northern Territory Australia, IR 59.* Supervising Scientist for the Alligator Rivers Region, Canberra.

Woolhiser, D. A., Smith, R. E., and Goodrich, D. C. 1990. *KINEROS, A Kinematic Runoff and Erosion Model: Documentation and User Manual, ARS 77*, US Department of Agriculture, Agricultural Research Service.

14

APPLICATION OF THE META-CHANNEL CONCEPT: CONSTRUCTION OF THE META-CHANNEL HYDRAULIC GEOMETRY FOR A NATURAL CATCHMENT

JOHN SNELL AND MURUGESU SIVAPALAN

Centre for Water Research, Department of Environmental Engineering, University of Western Australia, Nedlands, 6009, Australia

ABSTRACT

This is the second in a series of three papers about the meta-channel concept which illustrates the derivation of the principles behind the concept, the construction of the hydraulic geometry and the application of the concept to flood routing, respectively. It was shown in the first of these that a channel network in a catchment can be conceptualized into a single 'effective channel' representation: a meta-channel. This study uses this conceptualization to show how such a meta-channel can be constructed. The techniques derived are applied to one catchment in New Zealand. We derive hydraulic geometries expressed as functions of flow distance throughout this catchment based on the Leopold and Maddock power laws. This derivation uses classical published values for hydraulic geometry coefficients and exponents, regional parameterization of the index flood relationship for New Zealand as a whole, together with local knowledge regarding the order of magnitude of the channel roughness. Conservation principles derived from the continuity and mechanical energy balance equations are used to construct the hydraulic geometry of the meta-channel of this catchment. A meta-channel long profile is established and compared against the mainstream long profile. The effectiveness of the Leopold and Maddock power law assumptions is tested by comparing the derived hydraulic geometry against available field cross-sectional data for the gauging site at the outlet of the catchment.

INTRODUCTION

Since the original work of Sherman (1932), the concept of a unit hydrogaph has underlain much of flood routing work in surface hydrology. Implicit within this usage is an assumption that the instantaneous unit hydrograph (IUH) provides sufficient approximation to the instantaneous response function for a catchment. Instantaneous response functions (Wang *et al.*, 1981), derived for hillslopes and small catchments, show considerable non-linearity and are strongly dependent on antecedent conditions and rainfall intensity (Robinson and Sivapalan, submitted). In contrast, channel response functions tend to be weakly non-linear (Beven and Wood, 1993); consequently, routing of floods is often performed by solving the convective-diffusion equation, Equation (1), in which a diffusion coefficient, D, and celerity, c, are considered independent of discharge, Q

$$\frac{\partial Q}{\partial t} + c\frac{\partial Q}{\partial x} = D\frac{\partial^2 Q}{\partial x^2} \tag{1}$$

One solution to this equation is the inverse Gaussian function (Troutman and Karlinger, 1985). It is this function which is often taken to be the IUH of individual channels, (Mesa and Mifflin, 1986). Suitably weighted by a geomorphological function such as the width function, it produces a geomorphological IUH (GIUH), or channel network response function (Troutman and Karlinger, 1985; 1986; Mesa and Mifflin, 1986; Naden, 1992). It should be stressed that this approach to the catchment response problem is scale dependent in that the linear or weakly non-linear behaviour of the channels only dominates the non-linearity of the hillslope response functions for sufficiently large catchments.

The St Venant equations for single channel flow, which can be derived from the Navier–Stokes equations for incompressible fluid flow (Strelkoff, 1969), are non-linear and scale independent. Manipulation of these equations into more generalized forms (Sivapalan and Larsen, submitted) still maintains the non-linearities and scaling properties inherent in the original equations. Although it is often convenient to make assumptions about the extent of linearity within specific physical processes such as the routing of floods through networks, non-linear functions can be integrated using numerical techniques. The main practical reasons for using linearization is in the convenience of the mathematical techniques that are allowed and the ease of the resulting computations. To date, there are few tools which provide insight into either the issues of network response as against single channel response, or the effects of increasing catchment size on the non-linearities within the flood routing process.

The advent of modern computing techniques and digitized elevation models (DEMs) has made it possible to develop tools for the extraction and manipulation of components of network systems (O'Callaghan and Mark, 1984; Band, 1986; Jenson and Domingue, 1988). Snell and Sivapalan (1994a) have extended the work on the extraction of the catchment width function, the number of channels at a given distance from the outlet, to an automated extraction of the catchment area–distance function, the catchment area convergent to a specific flow distance from the outlet. Differences between the width and area–distance functions for a catchment can be thought of as providing a measure of the underlying variability in the constant of channel maintenance throughout a catchment, which in turn is expressing an underlying spatial variability of the catchment's geology and pedology. However, a catchment functions as an integrator in which the underlying spatial variability is effectively overshadowed or dispersed by variability in flow paths to the catchment outlet. Current models tend to be either conceptually lumped or physically distributed. We feel that, intrinsically, there is a place for physically based lumped models in which the certainty of distributed knowledge is replaced by the probability of lumped statistical distributions of physical parameters without loss of confidence in the end-product — a hydrograph. Motivated primarily towards the development of techniques of representing channel networks which preclude assumptions regarding linearities and independence of coefficients, we have defined a *meta-channel*, representing a collapsing of the channel network structure onto the single dimension of flow distance. This collapsing or lumping process is based on principles designed to conserve both mass and essential elements of the mechanical energy balance equation as originally derived by Strelkoff (1969). The 'effective' channel or meta-channel concept is justified on the basis of an inability of a catchment to maintain a history of the origins of variations induced in travel times to its outlet. Our long-term goal is to provide a mechanism of routing flow which is independent of linearity assumptions and which addresses a fundamental scaling issue in catchment response in that it provides more accurate estimations of discharge over a larger range of catchment scales than is currently possible with an IUH approach.

Before being able to route flow through this meta-channel we have to be able to derive the 'effective' physical characteristics of this channel. What does this channel look like? This paper briefly examines the principles behind the collapsing process [see Snell and Sivapalan (submitted) for details] and shows an application of the principles in the construction of a meta-channel for one catchment in New Zealand. We demonstrate in detail how the single channel representation of the catchment is achieved, paying particular attention to the derivation of geometrical form of this meta-channel. We show the effect of this geometry on an eventual single 'effective' channel representation of a network.

It should be noted that, in principle, the meta-channel we derive is purely concerned with the routing of flow which reaches the network. We take the pragmatic approach that networks only extend out to some support area threshold. We are cognizant that this support threshold area is highly dependent on geology, pedology and antecedent conditions, implying in turn that this support is both spatially and temporally variable. Important questions can therefore be asked regarding the determination of this support area. It is not our intention to answer this question, nor is it relevant to the meta-channel what the short-term spatial variations are — these factors being absorbed by some form of loss modelling. Routing by the meta-

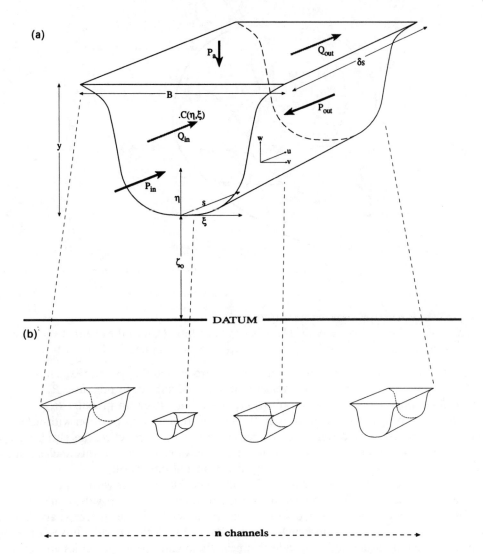

Figure 1. (a) Single channel control volume; (b) meta-channel control volume

channel model is potentially adaptable to handle long-term variability in rain fields and known spatial variability of pedology through modulations of the area–distance function.

PRINCIPLES GOVERNING THE COLLAPSE OF THE NETWORK

Control volume

Figure 1a depicts a control volume associated with one channel. Such a control volume can be extended across a set of $n(s)$ such channels (Figures 1b and 2). The extended version of the single channel control volume delimits the control volume of a meta-channel. Each channel segment building the meta-channel control volume extends between s and $s + \delta s$, where s is the flow distance between the outflow of the catchment and the outflow of each channel segment. To specifically remove all junctions and junction effects from subsequent analysis, we assume that the number of channels within this distance is invariant. Fluid travelling through a channel has local velocity $\vec{v}(\vec{x})$. The control volume is deformable such that its volume, $\nu(t)$, is a function of time, t, only, and each point of its surface has a local velocity of $\vec{w}$, a different

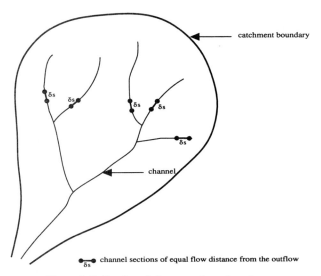

Figure 2. Collapsing of the network: a plan view

physical parameter to the fluid velocity, $\vec{v}$. There is a volume flux, Q, into the channel entrance and a lateral influx of q_l per unit length of channel. The control volume is contained by four surfaces:

- *Channel entrance* — normal to the predominant flow direction at that point. Fixed in space with respect to the flow distance, s, with no surface velocity, $\vec{w} = 0$, and subject to pressure, P_{in}.
- *Channel exit* — with similar properties to the entrance, but subject to pressure P_{out}.
- *Wetted surface* — primary source of drag in the system. It is fixed in position with respect to the flow distance. Both flow and surface velocities are zero ($\vec{v} = \vec{w} = 0$). In practice, this surface acts as a source and/or sink of fluid mass by interaction with the groundwater system. In this analysis, all such inputs are part of the lateral flows and this boundary is considered impervious.
- *Free surface* — fixed in space relative to the flow distance, but variable with respect to the bed normal. This surface expresses the deformable nature of the control volume. It provides a mass source by means of rainfall. This source is also lumped with the lateral flows. The free surface is also a kinetic energy source either from (i) rainfall — in which case it is lumped with the lateral influxes, or (ii) wind effects working on large free surface areas of wide channels. Fluid and surface velocities are generally different ($\vec{v} \neq \vec{w}$) but components of these velocities normal to the surface are equal, ($\vec{v} \cdot \vec{n} = \vec{w} \cdot \vec{n}$). The free surface is subject to atmospheric pressure, P_a.

Channel entrances and exits have maximum fluid depth, $y(s)$, top width, $B(s)$, and bed height above datum of $\zeta_0(s)$. Any point $C(\eta, \xi)$ on this surface is spatially characterized by its η and ξ coordinates as shown in Figure 1.

Equation of continuity

For continuity through a meta-channel control volume as described above (Snell and Sivapalan, submitted)

$$\frac{\partial}{\partial t}\sum_{i=1}^{n(s)} A_i + \frac{\partial}{\partial s}\sum_{i=1}^{n(s)} Q_i = \sum_{i=1}^{n(s)} q_i \qquad (2)$$

where A_i is the flow cross-sectional area of an individual channel, also known as the channel capacity of that channel, Q_i is the discharge for that channel and q_i is the lateral flow of that channel at that point; $n(s)$ is the width function at that point.

The first term on the left is the rate of change of total channel capacity with time at flow distance, s, where the total channel capacity is the aggregate of the individual channel capacities. The second term represents the net change of total mass flux through an infinitesimally thin section stretching across all the channels constructing the meta-channel at flow distance, s. The right-hand side represents aggregation of all lateral inflows into all channels. In applying the continuity principle to the meta-channel, the individual fluxes, channel capacities and lateral inflows for each channel have to be conserved across the set of channels making up the meta-channel at flow distance, s, from the catchment outflow.

Mechanical energy balance

Subject to a set of assumptions, the mechanical energy balance equations for the control volume consisting of a set of channel segments, is (Snell and Sivapalan submitted)

$$\frac{\partial}{\partial t}\sum_{i=1}^{n(s)}\left(\frac{\alpha_{1_i}Q_i\bar{u}_i}{2g}\right) + \frac{\partial}{\partial s}\sum_{i=1}^{n(s)}(Q_iH_i) = \sum_{i=1}^{n(s)}\left(\frac{Q_i\bar{\tau}_{0i}P_i}{\gamma A_i}\right) + \sum_{i=1}^{n(s)}(H_{l_i}q_i) \tag{3}$$

where $\bar{u}_i$ is the cross-sectionally averaged flow velocity for channel i, H_i is a total energy head for that channel consisting of a velocity head, a potential head due to the position of the channel above datum, a potential head due to the mass of water above the bed of the channel and a pressure head due to the imbalance of pressure force at the entrance and exit of the control volume. τ_{0_i} is the average stress induced at the wall of the channel, P_i is the wetted perimeter of the channel and H_{l_i} is a total head term contributed by the lateral inflows.

The four terms contributing to this equation are, respectively, from left to right: (1) Kinetic energy storage. (2) Kinetic energy flux and work done by pressure/gravity. (3) Energy losses. Currently, these energy losses are considered to be a function only of the frictional effects of the wetted surface. Energy losses due to transverse flows as caused by meandering, or three-dimensional turbulence effects of mixing at junctions or of hydraulic jumps are explicitly ignored. (4) Energy head due to lateral inflows. Contains similar terms to the total energy head of the main flow.

For conservation of the mechanical energy balance for a meta-channel, we preserve the same energy balance for the individual channels across the set of channels composing the meta-channel. Specifically we preserve the terms, $\alpha_{1_i}Q_i\bar{u}_i, Q_iH_i, H_{l_i}q_i, (Q_i\bar{\tau}_{0_i}P_o)/(\gamma A_i)$ in Equation (3).

APPLICATION OF THE PRINCIPLES IN CONSTRUCTING A META-CHANNEL

The meta-channel concept is applicable to any network — either natural, as extracted from DEMs, or artificially derived, as for example, Peano basins (Rinaldo et al., 1991), and optimum channel networks (Rigon et al., 1993). In this work, we limit ourselves to one naturally occurring network in New Zealand.

Assumptions

The generalized conservation equations become more amenable for deriving the geometry of a catchment meta-channel on making the following assumptions:

- *Steady state*. We assume a steady state system when determining hydraulic geometry of the meta-channel from hydraulic geometries of constituent channels. Consequently, discharge is related linearly to the area contributing to that discharge through a recharge parameter. This recharge parameter, R, contains the non-linearity and spatial variability inherent in both the rainfall and the soil characteristics within the catchment. The relationship is expressed as

$$Q(s) = R(\vec{x})M(s) \tag{4}$$

where $M(s)$ is the area of the catchment convergent at flow distance s. It forms a cumulative distribution function.

Initially, we consider the recharge to be uniform throughout the catchment — in effect, we ignore the spatial variability inherent within it. We reiterate that steady state is only assumed in forming the hydraulic

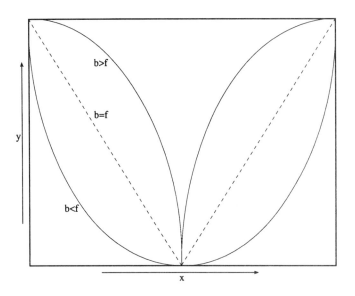

Figure 3. Effect of relative width and depth exponents on the shape of channels

geometry of the meta-channel; however, the meta-channel can still be used to model unsteady flow situations as in runoff-routing.

- *Gradually varied flow.* The gradually varied flow assumption neglects the effects of sharp changes in the channel geometry which bring about sharp changes in the flow height, e.g. hydraulic jumps.
- *No sediment transport.*
- *Negligible variation in velocity over the cross-section.* This sets all shape factors to ~ 1. This is in reasonable agreement with Van Driest (1946), but Henderson (1966) warns that these parameters can reach values of ~ 2 in meandering rivers.
- *Incompressibility*
- *One-dimensional flow.*
- *Resistance coefficients.* We assume resistance coefficients determined for steady, uniform and turbulent flow, hold for unsteady, non-uniform flow situations as in the case of flood routing.

Hydraulic geometry

In routing flow through the meta-channel, we need to derive the hydraulic geometry of the single channel representation. These hydraulic geometry relationships determine both celerity, $c(Q)$, and diffusion coefficient, $D(Q)$ (Henderson, 1966)

$$c(Q) = \frac{dQ}{dA}$$

$$D(Q) = \frac{Q}{2BS_0} \tag{5}$$

We usually refer to two distinct types of hydraulic geometry — *at a station* — identified with flow geometry and *downstream* — identified with channel geometry. This hydraulic geometry expressed in the form of power laws (Leopold and Maddock, 1953) is examined in depth in Appendix A. It should be noted that at-a-station hydraulic geometry characteristically shows a depth exponent, f_1, slightly greater than width exponent, b_1 (Leopold and Maddock, 1953), hence providing concave cross-sections with increasing discharge. In contrast, downstream hydraulic geometry characteristically shows the reverse effect (Leopold and Maddock, 1953), producing convex cross-sections and implying a flattening of downstream cross-section relative to upstream. Figure 3 illustrates these effects.

Hydraulic geometries as defined by Leopold and Maddock (1953) are smoothly varying power functions of discharge. For at a station hydraulic geometry this produces difficulties as the cross-section at any site normally shows one or more discontinuities as channel capacity is exceeded and discharge starts to enter the floodplain. Downstream geometries are affected by constrictions and expansions in the stream produced by the underlying spatial heterogeneity in both geological and pedological structures of the catchment. No attempt is made to handle these latter problems — they are ignored under the assumption of gradually varied flow and by treating the hydraulic geometry as stochastic rather than deterministic functions.

For the meta-channel, ideally, we would consider the downstream hydraulic geometry of the system. However, these are typically established at mean annual flood or bankfull discharge. Such relatively high discharge display a non-linear relationship with contributing area, principally through heterogeneity in rainfall across the catchment, leading to power law relationship of the form

$$Q = \psi M^{\theta} \tag{6}$$

where M is the catchment area contributing discharge to the network. In such situations, bankfull discharge across the network becomes disconnected in terms of the recurrence period required to produce the bank-full discharge. This leads to the paradox that conservation of mass no longer appears to hold.

As continuity is one of the fundamental principles of the meta-channel, we make the steady-state assumption mentioned previously. We establish continuity by maintaining a linear relationship between contributing area and discharge, the spatial heterogeneity mentioned earlier being absorbed by the recharge term. For the initial investigation, we assume a uniform recharge rate throughout the catchment. Future investigations will relax this constraint.

We consider the hydraulic geometry of individual channels as being of at a station type, with sites of equal flow distance from the outlet being aggregated such that the total channel capacity and total discharge weighted flow velocities and depths are conserved. Although at a station scaling exponents can remain effectively constant as we travel downstream, there can be large variations in the scaling coefficients. We show in Appendix A that at a station hydraulic geometry scaling coefficients are determined from knowing the hydraulic geometry exponents typical of at a station geometry in a region together, with hydraulic geometry exponents for downstream hydraulic geometry; these latter exponents are again taken to be characteristic of a region in which a catchment occurs. Thus, for each position on the network, we determine the width, mean and maximum depths, and velocity.

Hydraulic geometry can only provide the mean flow depth. In determining the hydraulic geometry parameters for a meta-channel, we need also to know the maximum flow depth in a channel and the wetted perimeter of that channel. We show (Appendix A) that the maximum fluid depth in a stream for any recharge R and at any flow distance s, $y = y(R, s)$, for each position is related to mean flow depth as follows

$$y = \bar{y}(1 + \phi^{-1}) = \frac{c_1}{a_1^{\phi}}(1 + \phi^{-1})B^{\phi} \tag{7}$$

where $\bar{y}$ is mean depth, $\phi = f_1/b_1$ is the ratio of depth to width exponent, a_1 is width scaling constant, c_1 is depth scaling constant and B is top width.

It is trivial to derive (Appendix A) the wetted perimeter by line integration across the top width of the channel as follows

$$P = 2\int_0^y \left[1 + \frac{1}{4}\left(\frac{d\xi}{d\eta}\right)^2\right]^{\frac{1}{2}} d\eta \tag{8}$$

where ξ and η are dummy variables expressing the width and depth of a channel, respectively.

Aggregation across the width function

By using the conservation equations described earlier, we collapse the geometries based on the conservation of discharges, cross-sectional areas, flow depths, heights above datum and energy losses due to stresses induced in the fluid by the solid boundary. We now require one further assumption in preserving the wetted perimeter across a set of channels.

The mechanical energy balance equation provides both a kinetic energy storage, $Q\bar{u}$, and a kinetic energy flux term, $Q\bar{u}^2$, by which we can preserve discharge velocity — see Appendix B. The flux conserved velocity, u_f, is used in all subsequent calculations in this application as, in the light of our steady-state assumption, the storage term is neglected. However, in routing problems we will be dealing with non-steady situations. To meet this problem, we need to establish a storage conserved velocity, u_s, from which we derive, λ, a parameter used for converting between the two velocities. This parameter becomes in effect a shape factor and can be considered to maintain all the variability usually conveyed by individual channel shape factors but neglected in our aggregation process

$$\lambda = \frac{u_s}{u_f} \tag{9}$$

Assuming all shape factors ≈ 1, we formulate the aggregation of hydraulic parameters as

$$A = \sum_{i=1}^{n(s)} A_i \quad Q\bar{u}_s = \sum_{i=1}^{n(s)} Q_i\bar{u}_i \quad Q\bar{u}_f^2 = \sum_{i=1}^{n(s)} Q_i\bar{u}_i^2 \quad Qy = \sum_{i=1}^{n(s)} Q_iy_i \tag{10}$$

and

$$P = \sum_{i=1}^{n(s)} P_i \quad \bar{u}_f^3 fP = \sum_{i=1}^{n(s)} \bar{u}_i^3 fP_i \tag{11}$$

Determination of the friction factor of each stream in the network

The shear stress, τ_0, at the bed of the channel is related to the Darcy–Weisbach friction factor, f, through

$$\tau_0 = \rho u_*^2 = 0{\cdot}125\rho f\bar{u}^2 \tag{12}$$

where u_* is shear velocity.

For fully turbulent flow this friction factor is related to roughness height, ϵ, as established by Keulegan (1938) and extended by Bray (1979)

$$f^{-\frac{1}{2}} \approx 0{\cdot}248 + 2{\cdot}28\log\left(\frac{d}{\epsilon}\right) \tag{13}$$

The variable, d, is some depth scale. Keulegan (1938) used hydraulic radius, Bray (1979), mean depth; we use maximum depth, y.

We consider ϵ to be an *effective* roughness height of the size of the 50th percentile in the sediment size distribution. We further assume roughness height to be uniform throughout the catchment.

Determination of slopes

It is feasible to establish local slopes from pixel heights in a DEM. However, we found this produced errors due to random fluctuations in the digitizing process. The diffusion coefficient is particularly sensitive to the slope parameter — see Equation (5). In this work we apply a polynomial fitting to the aggregated elevation data. The degree of the polynomial is estimated as the minimum value providing a reasonable fit by eye. We consider this point again in the discussion.

Determination of meta-channel parameters

Appendix B outlines the algorithm used to determine each of the meta-channel parameters from the hydraulic geometries of the individual streams. As shown there, discharge, cross-sectional area and wetted perimeters are conserved by direct aggregation across the width function. Discharge velocity is established from aggregating a discharge weighted kinetic energy flux term. Maximum flow depth and bed height are established in analogous procedures. The friction factor is established from the aggregated energy loss term, together with the meta-channel parameters of discharge velocity and wetted perimeter. The top width is determined as the derivative of the power law function between cross-sectional area and maximum

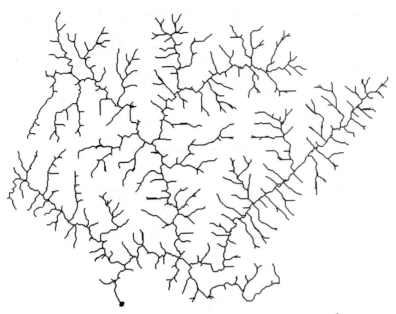

Figure 4. Network extracted from the Hutt catchment DEM at $46\,875\,\mathrm{m}^2$ resolution

depth. Cross-sectional area and top width then provide mean depth. Celerity, $c(Q)$, is determined by differentiating the power law function determined from the regression between discharge and cross-sectional area. $D(Q)$ is determined from the slope curve obtained by polynomial fitting the meta-channel long profile and discharge and top width according to Equation (5). Finally, hydraulic radius, obtained from cross-sectional area and wetted perimeter, is used with the friction factor to obtain Manning's n.

RESULTS

In this section, we provide an example of how the concept was used for a naturally occurring catchment in the North Island of New Zealand. The Hutt catchment occurs on the southern edge of the island and has a gauging station at the Kaitoke site. This site drains an area of approximately $90\,\mathrm{km}^2$ and shows considerable relief, with a height drop of over 1 km in a maximum flow distance of 17·5 km. The water course is boulder-strewn in the upper reaches and gravel-bedded throughout the channels above the gauging station. According to Hicks and Mason (1991), the sediment size profile shows a 50 centile value of 90 mm. At the Taita Gorge site on the same river, but draining an area of $555\,\mathrm{km}^2$, the 50th percentile value is 86 mm, illustrating uniformity in the ability of the system to maintain its sediment load throughout a major part of its length.

Figure 4 shows the channel network extracted using a steepest gradient approach of determining flow paths (Band, 1993) from the DEM of the catchment. A threshold area of $0·31\,\mathrm{km}^2$ was chosen for the initiation of a channel. This network was used as a basis for constructing the width function (Figure 5a) and the area–distance function (Figure 5b) for the catchment and accumulated area functions for the total network and for each individual link in that network. From this network of area functions, meta-channel parameters were derived according to the algorithm outlined in Appendix B. Eight recharge values were used ranging uniformly between 1·8 and $7·2\,\mathrm{mm\,h}^{-1}$.

At a station and downstream hydraulic geometry exponents were taken from classical published work on hydraulic geometry. Regional downstream hydraulic geometry scaling coefficients and the coefficient and exponent of the discharge–catchment area power law, all corresponding to mean annual flood, were obtained from Mosley (1992). Roughness height was obtained from Hicks and Mason (1991). Parameters are shown in Table I.

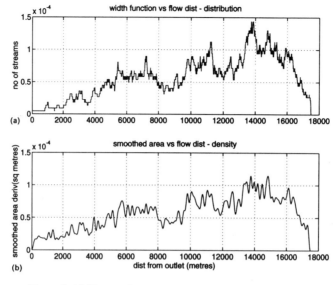

Figure 5. (a) Hutt catchment width function; (b) area function

In Figures 6–9, we present each parameter determined for the meta-channel as a function of flow distance, but defer discussion of these results to the next section. Polynomial fitting of the meta-channel bed elevations was performed to the sixth order with a coefficient of determination of 0·995. The fit obtained is shown in Figure 10a. For interest, we compare meta-channel bed height with main stream bed height in Figure 10b. Table II shows meta-channel parameters derived from the model compared with published values as provided by Hicks and Mason (1991) for Kaitoke gauging station.

DISCUSSION

Figures 6–9 imply that although hydraulic geometries of individual channels making up a meta-channel follow the Leopold–Maddock power law formulation, the hydraulic geometry behaviour of the meta-channel itself is not necessarily that described by those same power laws. Instead of smoothly increasing functions expressed by those laws, we find that all the derived hydraulic geometry relationships display a different form, which can be traced back to a distinct dependency on the network width function. This

Table I. Hydraulic geometry parameters for the Hutt catchment

Parameter	Value
At a site velocity exponent	0·34
At a site width exponent	0·33
At a site depth exponent	0·33
Downstream velocity exponent	0·1
Downstream width exponent	0·5
Downstream depth exponent	0·4
Downstream velocity scaling coefficient	0·61
Downstream width scaling coefficient	7·09
Downstream depth scaling coefficient	0·23
Regional discharge–area exponent	0·8
Regional discharge–area scaling coefficient	2
Roughness height	90 mm

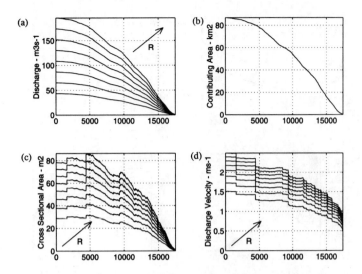

Figure 6. (a) Discharges against flow distance for a set of recharges. (b) contributing area against flow distance. (c) cross-sectional area against flow distance for a set of recharges. (d) discharge velocity against flow distance for a set of recharges

dependency itself can be attributed to the non-linearity inherent in these laws being expressed through discontinuities in the relevant meta-channel parameters. The diffusion coefficient particularly shows a marked impact from discontinuities in the top width. Both the top width and wetted perimeter are particularly affected to the point where they exhibit an underlying envelope characteristic of the width function, $n(s)$. To some extent this is to be anticipated in that, for the chosen downstream hydraulic geometry exponents, the top width exponent, b_2, was slightly greater than the depth exponent, f_2.

Both the celerity and the diffusion coefficients for this catchment show marked non-linearity throughout the flow distance and discharge range. This introduces reasonable doubt as to whether flood routing models involving constant parameters would perform adequately in such a catchment. The lack of requirement of the meta-channel concept to be tied down to basic assumptions of linearity allows this inherent non-linearity to express itself freely through the $c(Q)$ and $D(Q)$ relationships.

Friction factors and Manning's n show considerable instability at low discharge, but become smoother

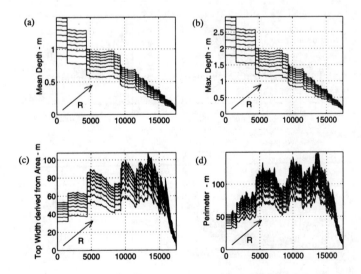

Figure 7. (a) Mean flow depth against flow distance for a set of recharges. (b) maximum flow depth against flow distance for a set of recharges. (c) top width against flow distance for a set of recharges. (d) wetted perimeter against flow distance for a set of recharges

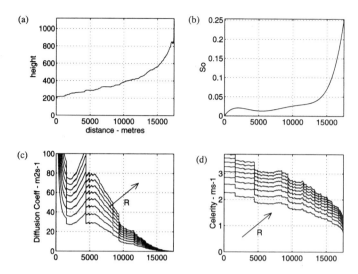

Figure 8. (a) Meta-channel bed elevation against flow distance for a set of recharges. (b) meta-channel bed slope against flow distance for a set of recharges. (c) diffusion coefficient against flow distance for a set of recharges. (d) celerity against flow distance for a set of recharges

(apart from junction effects) at higher discharges. This is a result of the relationship between roughness height and depth of flow and the relative values of these two parameters, especially for links draining small areas. The assumption of uniform roughness height is open to question, although as indicated from the 50th percentiles of the two different sites widely separated on the same river, there is some observational basis for this assumption. However, roughness height will obviously vary and Hack (1957) provides an empirical relationship of the bed slope with contributing area and roughness height from which this value may be estimated.

Similarly, the resistance law used in the determination of friction factors as based on the work of Keulegan (1938) and Bray (1979) has underlying problems. The law used, Equation (13), explains only about 39% of the variation is the friction factor. In fact, better correlations were obtained by Bray (1979) between the friction factor and bed slope (r^2 of 0·53). We prefer the relationship with the roughness height as being more consistent with the physical bases underlying the conservation equations.

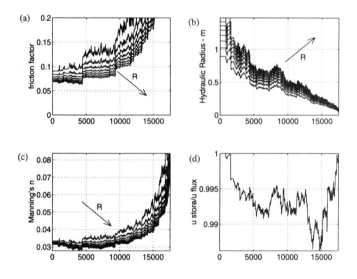

Figure 9. (a) Friction factor against flow distance for a set of recharges. (b) Hydraulic radius against flow distance for a set of recharges. (c) Manning's n against flow distance for a set of recharges. (d) velocity correction parameter against flow distance

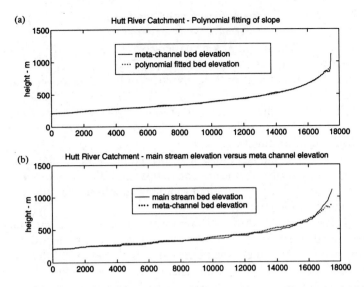

Figure 10. (a) Polynomial fit of meta-channel long profile; (b) Meta-channel long profile compared with mainstream long profile

Stress at the solid boundary

$$\tau_w = \gamma A S_0 \qquad (14)$$

implies from Equation (12) that the friction factor is related to bed slope

$$f = \frac{8gAS_0}{\bar{u}^2} \qquad (15)$$

This is a strong indication, as mentioned previously, that the estimation of friction effects will be facilitated with more satisfactory techniques of determining local bed slopes.

We have found the use of local slopes as calculated from DEMs not to be advisable due to the extreme variability from inaccuracies inherent in the digitization process. Rather, a local slope determined from a power law relationship with catchment area as derived by Tarboton *et al.* (1989) or a multiscaling slope exponent approach (Gupta *et al.* submitted) would be preferable. This is a natural extension to work performed in this study. The approximation of bed slope by a polynomial function shows a good fit in the lower reaches, but a poor fit in the highest reaches. This could possibly be improved by, for instance, piecewise exponential fitting, as indicated by Richards (1982). This would also provide a more realistic basis to this process. The polynomial fit produces a false value of zero slope at the outflow, which in turn produces a spuriously high value of the diffusion coefficient. For the Hutt catchment, the upper reaches are extremely

Table II. Cross-sectional characteristics at the Kaitoke gauging site

Characteristic	Units	Calculated value	Published value
Discharge	$m^3 s^{-1}$	108	104
Discharge velocity	$m s^{-1}$	2·05	2·08
Cross-sectional area	m^2	52·9	50·5
Top width	m	43·4	37*
Mean depth	m	1·22	1·36*
Hydraulic radius	m	1·21	1·45
Friction factor		0·076	N/A
Manning's n		0·032	0·047

* Estimated from diagram and supplied cross-sectional area

steep, falling well outside basic assumptions underlying the analysis leading to the generalized conservation equations as applied to the meta-channel. Results show very small diffusion coefficients ($\sim 1\,\text{m}^2\,\text{s}^{-1}$) in the upper part of the meta-channel, indicating that kinematic wave solutions of the routing problem would be adequate within these reaches. The lower reaches show appreciable increases both in diffusion coefficients and celerities, which indicates that solutions of the convective–diffusion equation with constant coefficients are potentially open to doubt.

The conversion parameter between flux-derived and storage-derived velocity, λ, shows maximum deviation from a value of 1·0 of $\sim 1·5\%$ and can safely be ignored in further work on this catchment. Its magnitude is dependent on the width function.

The mainstream long profile and meta-channel long profile are very similar. Once again, the largest deviations appear at larger values of the width function. Apparent differences in meta-channel long profile between Figures 10a and 10b result from the different sampling lengths used in these two plots — every 1 m in Figure 10a and every 100 m in Figure 10b.

We concede that the geometries selected to illustrate this construction of a catchment meta-channel are simplistic in nature based, as they are, purely on Leopold and Maddock type power laws. No attempt has been made to include floodplains within these geometries. It would be expected that at very high discharges, discrepancies would occur by flow maintained in the channel that would naturally have entered the floodplain. Our assumption of Leopold and Maddock power laws is to a certain extent vindicated by the good agreement between the predicted and actual hydraulic geometries for the Kaitoke site (see Table II). This is especially so, considering that the only point knowledge of the site required was the roughness height and the area of catchment draining to that point — all other parameters used are either classical published generalizations for hydraulic geometry exponents (Leopold and Maddock, 1953), or regional parameters for the whole of New Zealand. Even for roughness height, the estimation only needs to be order of magnitude rather than exact as the relationship with the friction factor is logarithmic in nature. Generalization of this geometry to a series of power laws using different scaling coefficients and exponents, switchable on recharge value, or indeed to situations where series of cross-sections have been established through field study are trivial extensions to techniques already described, being easily incorporated within the model.

In this work several assumptions have been made. We examine the effects of the more important of these as follows:

1. *Steady-state assumption* — made to obtain meta-channel downstream hydraulic geometry. The assumption being made here is that a linear relationship exists between the discharge at a point and the catchment area converging to that point. There is empirical evidence that the main variability expressed in the discharge—catchment power law relationship is governed by rainfall processes (Gupta *et al.*, submitted). Ibbitt (1994) has found a linear relationship between discharge and contributing area under normal flow, steady-state conditions for the Ashley catchment in New Zealand. Our assumption implies that, ignoring rainfall variability, discharges of channel-forming significance are linearly related to contributing area.

2. *Backwater effects* — neglect of these effects becomes apparent in discontinuities evident in meta-channel hydraulic geometry parameters. Dynamic storage implicit in these back water controls is expected to induce smoothing in transitions across junctions.

3. *No sediment transport* — this impacts to some extent on the hydraulic geometry that forms. Certainly, Richards (1982) indicates that both at a station and downstream geometries are affected by the sedimentology of stream banks. This effect can be seen in power law regression plots whose slopes are not constant with discharge. Our assumption of a linear relationship for the power law regression leads to inaccuracy in hydraulic geometries, especially at higher flows.

4, *Uniform roughness height* — as roughness height is used only within a logarithmic relationship, it would not be expected to markedly affect the hydraulic geometry demonstrated by a meta-channel.

CONCLUSIONS

Starting with generalized equations of continuity and mechanical energy balance as established for a meta-channel control volume (Snell and Sivapalan, submitted), a technique for the construction of such a meta-channel involving DEMs, regional parameterizations of hydraulic geometry parameters and a working knowledge of local roughness heights has been developed. The results of the prediction of the hydraulic geometry at a single channel outflow from a catchment channel network system are encouraging. This single channel representation of a catchment network provides a powerful tool for the subsequent routing of flows through that network. It allows any single channel routing method to be applied to the channel network as a whole.

The meta-channel concept addresses scaling in two regards. Firstly, it demonstrates the technique of upward scaling in which spatially distributed information is lumped into a statistically distributed system. Secondly, it allows the routing of flows to be investigated through a range of catchment sizes. The advantages of representing the network in this way lies in the absence of linearity assumptions implicit in unit hydrograph approaches. As non-linearity is fundamentally handled by this technique, we have a more powerful tool available for examining the scaling behaviour of catchments. Furthermore, results to date, based on the Leopold and Maddock power law relationships for hydraulic geometries of individual channels, indicate that some of the work performed with an assumed linearity of response for larger catchments may be open to reasonable doubt. The technique has the potential for modelling at large catchment scales in that it provides a natural linkage between hydrological and general circulation models. A set of meta-channels representing primary drainage basins can be used to route water between general circulation model grid cells and hence improve water balance 'accounting'.

Many questions still remain unresolved concerning this meta-channel concept — not the least of which are how well it deals with field-measurement hydraulic geometries compared with the power law distributions currently modelled, and how well does the routing of flow through a meta-channel compare against routing through distributed network systems. Both junction effects and meandering, important agents of energy dissipation in river systems, are neglected. These need to be addressed in future.

We are currently assessing the meta-channel construct for routing of flow through a catchment using variable parameter Muskingum–Cunge techniques.

ACKNOWLEDGEMENTS

Thanks are expressed to the officers of National Institute of Water and Atmospheric Research Ltd of New Zealand for providing a suitable analysis of DEM data for the Hutt River catchment. This research was supported in part by a special Environmental Fluid Dynamics grant awarded by the University of Western Australia to the second author. J. D. Snell was supported by a University of Western Australia Research Scholarship, a Centre for Water Research Scholarship and an Australian Research Council Grant (Small Grants Scheme, Grant Ref. 04/15/031/254). Centre for Water Research Reference No. ED 886 JS.

REFERENCES

Band, L. E. 1986. 'Topographic partition of watersheds with digital elevation models', *Wat. Resour. Res.*, **22**, 15–24.

Band, L. E. 1993. 'Extraction of channel networks and topographic parameters from digital elevation data' in Beven, K. and Kirkby, M. J. (Eds), *Channel Network Hydrology*, Wiley, Chichester. pp. 13–42.

Beven, K., and Wood, E. F. 1993. 'Flow routing and the hydrological response of channel networks' in Beven, K. and Kirkby, M. J. (Eds), *Channel Network Hydrology*, Wiley, Chichester. pp. 99–128.

Bray, D. I. 1979. 'Estimating average velocity in gravel-bed rivers', *J. Hydr. Div. ASCE*, **105**(HY9) 1103–1122.

Gupta, V. J., Mesa, O. J., and Dawdy, D. R. 'Multiscaling theory of flood peaks: regional quantile analysis', *Wat. Resour. Res.*, submitted.

Hack, J. T. 1957. 'Studies of longitudinal stream profiles in Virgina and Maryland', *Prof. US Geol. Surv. Pap. 294B*.

Henderson, F. M. 1966. *Open Channel, Flow*. MacMillan, New York.

Hicks, D. M., and Mason, P. D. 1991. *Roughness Characteristics of New Zealand Rivers. Water Resources Survey*. DSIR Marine and Freshwater, Wellington.

Ibbitt, R. P. 1994. 'Optimal channel networks: validation in a small catchment', *NIWA Misc. Rep. No. 197*, 32 pp.

Jenson, S. K., and Dominigue, J. O. 1988. 'Extracting topographic structure from digital elevation data for geographic information system analysis', *Photogr. Engin. Remore. Sensing*, **54**, 1593–1600.

Keulegan, G. H. 1938. 'Laws of turbulent flow in open channels', *J. Res. Nat. Bur. Standards, Res. Pap. RP1151*, **21**, 707–741.

Leopold, L. B., and Maddock, T. 1953. 'The hydraulic geometry of stream channels and some physiographic implications', *U.S. Geol. Surv. Prof. Pap. 252*, 9–16.

Mesa, O. J., and Muffins, E. R. 1986. 'On the relative role of hillslope and network geometry in hydrologic response' in Gupta, V. K., Rodriguez-Iturbe, I., and Wood, E. F. (Eds), *Scale Problems in Hydrology*, D. Reidel, Dordrecht. pp. 1–17.

Mosley, M. P. 1992. 'River morphology' in *Waters of New Zealand*. New Zealand Hydrological Society. pp. 285–304.

Naden, P. S. 1992. 'Spatial variability in flood estimation for large catchments: the exploitation of channel network structure', *Hydrol. Sci.* **37**, 53–71.

O'Callaghan, J. F., and Mark, D. M. 1984. 'The extraction of drainage networks from digital elevation data', *Comp. Vision Graphics Image Process.* **28**, 323–344.

Richards, K. 1982. *Rivers, Form and Process in Alluvial Channels*. Methuen, London.

Rigon, R., Rinaldo, A., Rodriguez-Iturbe, I., Bras, R. L., and Ijjasz-Vasquex, E. 1993. 'Optimal channel networks: a framework for the study of river basin morphology', *Wat. Resour. Res.*, **29**, 1635–1646.

Rinaldo, A. Marani, A., and Rigon, R. 1991. 'Geomorphological dispersion', *Wat. Resour. Res.*, **27**, 513–525.

Robinson, J. S., and Sivapalan, M. 'Instantaneous response functions of overland flow and subsurface storm flow for catchment models', *Hydrol. Process.*, submitted.

Sherman, L. K. 1932. 'Streamflow from rainfall by the unit-graph method', *Eng. News Rec.* **108**, 501–505.

Sivapalan, M., and Larsen, J. E. 'A generalized non-linear diffusive wave equation', *Wat. Resour. Res.*, submitted.

Snell, J. D., and Sivapalan, M. 1994. 'On geomorphological dispersion in natural catchments and the geomorphological unit hydrograph', *Wat. Resour. Res.*, **30**, 2311–2323.

Snell, J. D., and Sivapalan, M. 'The meta-channel — a basis for the modelling of rainfall–runoff at the drainage basin scale', *Wat. Resour. Res.*, submitted.

Strelkoff, T. 1969. 'One-dimensional equations of open-channel flow', *J. Hydr. Div. ASCE*, **95**(HY3), 861–876.

Tarboton, D. G., Bass, R. L., and Rodriguez-Iturbe, L. 1989. 'Scaling and elevation in river networks', *Wat. Resour. Res.*, **25**, 2037–2051.

Troutman, B. M., and Karlinger, M. R. 1985. 'Unit hydrograph approximations assuming linear flow through topologically random networks', *Wat. Resour. Res.*, **21**, 743–754.

Troutman, B. M., and Karlinger, M. R. 1986. 'Averaging properties of channel networks using methods in stochastic branching theory' in Gupta, V. K., Rodriguez-Iturbe, I., and Wood, E. F. (Eds), *Scale Problems in Hydrology*. D. Reidel, Dordrecht. pp. 185–216.

Van Driest, E. R. 1946. 'Steady turbulent-flow equations of continuity, momentum, and energy for finite systems', *J. App. Mech. ASME*, **3**, A-231–A-238.

Wang, C. T., Gupta, V. K., and Waymire, E. 1981. 'A geomorphologic synthesis of non-linearity in surface runoff', *Wat. Resour. Res.*, **17**, 545–554.

APPENDIX A:
HYDRAULIC GEOMETRY

Reconciliation of at a station and downstream hydraulic geometries

At a station hydraulic geometry expresses the temporal variation in geometrical parameters at a fixed point in space. For any site at a given flow distance from the outlet, this hydraulic geometry can be represented by

$$\text{Top width, } B(t) = a_1 Q_A^{b_1}(t)$$

$$\text{Mean flow depth, } \bar{y}(t) = c_1 Q_A^{f_1}(t) \tag{16}$$

$$\text{Mean flow velocity, } \bar{u}(t) = k_1 Q_A^{m_1}(t)$$

From Leopold and Maddock (1953) it can be seen that the scaling coefficients, a_1, c_1, k_1, by implication, are functions of space only and not of time, whereas the scaling exponents, b_1, f_1, m_1, are independent of both time and space.

Similarly, downstream hydraulic geometry expresses the spatial variation for the same parameters at a fixed return period

$$B(s) = a_2 Q_D^{b_2}(s)$$

$$\bar{y}(s) = c_2 Q_D^{f_2}(s) \tag{17}$$

$$\bar{u}(s) = k_2 Q_D^{m_2}(s)$$

In this instance, the scaling coefficients, a_2, c_2, k_2, by implication, are functions of time expressed through the return period only and not of space. Once again, the scaling exponents, b_2, f_2, m_2, are independent of both time and space.

For a fixed return period at a fixed site, then the two hydraulic geometries can be equated. Considering the top width as being typical of the hydraulic geometry parameters, then

$$a_1 = a_2 Q^{b_2 - b_1} \tag{18}$$

where $Q = Q_A = Q_D$

Furthermore, the regional variation of discharge with contributing area for a constant return period, either mean annual flood or bankfull discharge, is

$$Q(s) = \psi M(s)^\theta \tag{19}$$

This equation is only meaningful in the sense of downstream hydraulic geometry. The coefficient ψ is a function of return period, but not space.

From Equations (18) and (19)

$$a_1 = a_2 \psi^{(b_2 - b_1)} M(s)^{\theta(b_2 - b_1)} \tag{20}$$

Analogous equations are derived for both mean flow depth and mean flow velocity. As a_1 and M are both independent of the return period, then, by implication, so is the product term $a_2 \psi^{b_2 - b_1}$, even though individually a_2 and $\psi^{b_2 - b_1}$ are functions of the return period. By estimating this product term for a given return period, such as the mean annual flow, this term is established for all sites and return periods. As typically $b_2 > b_1$, the implication is that the downstream scaling coefficient for top width, a_2, should decrease monotonically with return period. For the mean flow velocity, then typically $m_2 < m_1$, implying that as ψ increases, $\psi^{m_2 - m_1}$ decreases montonically with the return period. This in turn implies that the downstream velocity scaling coefficient, k_2, will increase montonically with the return period. These conclusions are demonstrated by Leopold and Maddock (1953).

We are interested in deriving the set of at a station hydraulic geometries throughout the flowpath domain, s. For reasons previously stated, we have made a steady-state approximation, such that

$$Q_A(t) = MR(t) \tag{21}$$

Consequently, by suitably rearranging Equations (16), (17), (19) and (21), we obtain hydraulic geometry relationships reconciled in time and space

$$
\begin{aligned}
B(s,t) &= a_2 \psi(t)^{(b_2 - b_1)} M(s)^{\theta(b_2 - b_1)} Q_A(t)^{b_1} = K_B M(s)^{b_1 + \theta(b_2 - b_1)} R(t)^{b_1} \\
\bar{y}(s,t) &= c_2 \psi(t)^{(f_2 - f_1)} M(s)^{\theta(f_2 - f_1)} Q_A(t)^{f_1} = K_y M(s)^{f_1 + \theta(f_2 - f_1)} R(t)^{f_1} \\
\bar{u}(s,t) &= k_2 \psi(t)^{(m_2 - m_1)} M(s)^{\theta(m_2 - m_1)} Q_A(t)^{m_1} = K_u M(s)^{m_1 + \theta(m_2 - m_1)} R(t)^{m_1}
\end{aligned}
\tag{22}
$$

where s refers to any spatial point on a network and t refers to any generalized return period. Note constants $K_B = a_2 \psi(t)^{(b_2 - b_1)}, K_y = c_2 \psi^{(f_2 - f_1)}$, and $K_u = k_2 \psi^{(m_2 - m_1)}$.

Determination of maximum depth of flow from mean depth of flow

Hydraulic geometry considerations only supply mean flow depths. Stage–discharge relationships require maximum flow depths. For hydraulic geometries expressed in terms of Leopold and Maddock power laws, we determine maximum depth of flow, y, from mean flow depth, $\bar{y}$, as follows

$$\text{Cross-sectional area}, A = B\bar{y} \tag{23}$$

Eliminating, discharge, Q_A, between the top width and mean flow depth expressions from Equation (16), we obtain

$$\bar{y} = \frac{c_1}{a^{\frac{f_1}{b_1}}} B^{\frac{f_1}{b_1}} \tag{24}$$

Letting $\phi = f_1/b_1$, we obtain for cross-sectional area

$$A = \frac{c_1}{a_1^{\phi}} B^{\phi+1} \tag{25}$$

Differentiating with respect to top width, this relationship now becomes

$$\frac{dA}{dB} = \frac{c_1}{a_1^{\phi}}(\phi+1)B^{\phi} = \bar{y}(\phi+1) \tag{26}$$

We can relate the derivative dA/dB to $\bar{y}$, mean flow depth, through

$$B = \frac{dA}{dy} = \frac{dA}{dB} \cdot \frac{dB}{dy} \tag{27}$$

$$= \bar{y}(\phi+1)\frac{dB}{dy} \tag{28}$$

Substituting and rearranging

$$dy = \frac{c_1}{a_1^{\phi}}(\phi+1)B^{\phi-1}dB \tag{29}$$

Integrating flow depth over η, a dummy variable, between 0 and the maximum flow depth, y, and width over ξ, a dummy variable, between 0 and top width, B, we obtain

$$y = \frac{c_1}{a_1^{\phi}}(\phi+1)\phi^{-1}B^{\phi} = \bar{y}(\phi^{-1}+1) = \bar{y}\left(\frac{b_1}{f_1}+1\right) \tag{30}$$

Thus we form a simple relationship between mean and maximum flow depths based only on width and depth hydraulic geometry exponents.

Determination of wetted perimeter

The wetted perimeter is required for the determination of energy losses associated with stress induced by the bed of each stream. It is derived from a line integration of the top width across the depth domain. Let the domain of possible depths be represented by η and the domain possible top widths associated with those depths by ξ. An infinitesimal elemental increase in wetted perimeter, dP, is engendered by an incremental elemental increase in flow depth, $d\eta$. For a symmetrical cross-section

$$dP = \left[(d\eta)^2 + \left\{d\left(\frac{\xi}{2}\right)\right\}^2\right]^{\frac{1}{2}} \tag{31}$$

We use $d\left(\frac{\xi}{2}\right)$ because of symmetry in the cross-section.

Integrating this expression across total flow depth in the channel and multiplying by two, because of symmetry, we obtain

$$P = 2\int_0^y \left[1 + \frac{1}{4}\left(\frac{d\xi}{d\eta}\right)^2\right]^{\frac{1}{2}} d\eta \tag{32}$$

Using Equations (29) and (22) and recalling that $\phi = f_1/b_1$, this integration becomes

$$P = 2 \int_0^y \left[1 + \frac{1}{4} \left\{ \frac{a_1^{\frac{f_1}{b_1}}}{c_1 \left(\frac{f_1}{b_1} + 1 \right)} [K_B M(s)^{b_1 + \theta(b_2 - b_1)} R(t)^{b_1}]^{1 - \frac{f_1}{b_1}} \right\} \right]^{\frac{1}{2}} \mathrm{d}\eta \tag{33}$$

APPENDIX B:
ALGORITHM FOR THE DETERMINATION OF META-CHANNEL HYDRAULIC GEOMETRY, CELERITY AND DIFFUSION COEFFICIENT

$\forall R\{$

    $\forall s\{$

        $\forall$ streams at flow distance $s\{$

            $a_1 \leftarrow a_2 \psi^{(b_2 - b_1)} M_i^{\theta(b_2 - b_1)}$ $c_1 \leftarrow c_2 \psi^{(f_2 - f_1)} M_i^{\theta(f_2 - f_1)}$

            $k_1 \leftarrow k_2 \psi^{(m_2 - m_1)} M^{\theta(m_2 - m_1)}$

            $\bar{y}_i \leftarrow c_1 M_i^{f_1} R^{f_1}$ $B_i \leftarrow a_1 M_i^{b_1} R^{b_1}$

            $\bar{u}_i \leftarrow k_1 M_i^{m_1} R^{m_1}$

            $A_i \leftarrow \bar{y}_i B_i \Rightarrow \Sigma_{i=1}^{n(s)} A_i$ $Q_i \leftarrow M_i R \Rightarrow \Sigma_{i=1}^{n(s)} Q_i$

            $y_i \leftarrow \bar{y}_i (1 + \phi_{1i}^{-1})$ $f_i^{-\frac{1}{2}} = 0 \cdot 248 + 2 \cdot 36 \log_{10}(\frac{y_i}{\epsilon})$

            $P_i \leftarrow 2 \int_0^{y_i} \left[1 + \frac{1}{4} \left(\frac{\mathrm{d}\xi}{\mathrm{d}\eta} \right)_i^2 \right]^{\frac{1}{2}} \mathrm{d}\eta$

            Aggregation

            $Q_i y_i \Rightarrow \Sigma_{i=1}^{n(s)} Q_i y_i$ $Q_i \zeta_{0i} \Rightarrow \Sigma_{i=1}^{n(s)} Q_i 0_i$

            $Q_i \bar{u}_i \Rightarrow \Sigma_{i=1}^{n(s)} Q_i \bar{u}_i$ $Q_i \bar{U}_i^2 \Rightarrow \Sigma_{i=1}^{n(s)} Q_i \bar{u}_i^2$

            $\bar{u}_i^3 f_i P_i \Rightarrow \Sigma_{i=1}^{n(s)} \bar{u}_i^3 f_i P_i$

        $\}$

        $Q \leftarrow \Sigma_{i=1}^{n(s)} Q_i$ $A \leftarrow \Sigma_{i=1}^{n(s)} A_i$

        $\bar{u}_f \leftarrow \frac{1}{Q} \Sigma_{i=1}^{n(s)} Q_i \bar{u}_i$ $\bar{u}_s \leftarrow \left[\frac{1}{Q} \Sigma_{i-1}^{n(s)} Q_i \bar{u}_i^2 \right]^{\frac{1}{2}}$

        $P \leftarrow \Sigma_{i=1}^{n(s)} P_i$ $f \leftarrow \frac{1}{\bar{u}^3 P} \Sigma_{i=1}^{n(s)} \bar{u}_i^3 f_i P_i$

        $y \leftarrow \frac{1}{Q} \Sigma_{i=1}^{n(s)} Q_i y_i$ $\zeta_0 \leftarrow \frac{1}{Q} \Sigma_{i=1}^{n(s)} Q_i \zeta_{0i}$

        $S_0 \leftarrow \frac{\partial \zeta}{\partial s}$

    $\}$

$\}$

$A \leftarrow A(y)$ $Q \leftarrow Q(A)$ (see note 1)

Determination of hydraulic geometry

$\forall R\{$

    $\forall s\{$

        $B \leftarrow \frac{\mathrm{d}A}{\mathrm{d}y}$

$$\bar{y} \leftarrow \frac{A}{B} \qquad\qquad c \leftarrow \frac{\mathrm{d}Q}{\mathrm{d}A} \qquad\qquad D \leftarrow \frac{Q}{2BS_0}$$

$$\}$$

$$\}$$

Note: 1. Functional form is obtained by a power law fitting of the values obtained over the set of recharge values.

LIST OF SYMBOLS

A	Cross-sectional area
a	Width scaling constant in hydraulic geometry power law functions
B	Top width of the channel
b	Width scaling exponent in hydraulic geometry power law functions
c	Celerity in routing equations: depth scaling constant in hydraulic geometry
D	Diffusion coefficient
d	A representative depth scale
f	Darcy–Weisbach friction factor; depth scaling exponent in hydraulic geometry power law functions
g	Acceleration due to gravity
H	Head
H_l	Head introduced by lateral inflow
i	The ith component in a set of channels
K_B	Regional coefficient scaling top width to contributing area and recharge
K_y	Regional coefficient scaling mean flow depth to contributing area and recharge
K_u	Regional coefficient scaling mean flow velocity to contributing area and recharge
k	Discharge velocity scaling constant in hydraulic geometry power law functions
M	Contributing area
m	Discharge velocity scaling exponent in hydraulic geometry power law functions
$\vec{n}$	A vector component normal to a surface
$n(s)$	The number of channels at flow distance s from the outflow of the network
n	Manning's n
P	Wetted perimeter
P_a	Atmospheric pressure
Q	Volume flux $= \int_A u\,\mathrm{d}A$ or discharge
q_i	Lateral inflow
R	Recharge coefficient
R_h	Hydraulic radius $= A/P$
S_0	Bed slope
s	Flow direction
t	Dimension of time
u_*	Shear velocity
u_i	ith component of the velocity vector $\vec{u}$
$\bar{u}$	Discharge velocity
$\bar{u}_f$	Discharge velocity from conservation of kinetic energy flux
$\bar{u}_s$	Discharge velocity from conservation of kinetic energy storage
$\vec{v}$	Velocity vector
$\vec{w}$	Local velocity at any point on the surface of the control volume
$\vec{x}$	Position vector
y	Depth of water in the channel
$\bar{y}$	Mean depth of water in the channel
ν	Control volume
α_1	Energy storage shape factor

ψ Regional flood index scaling coefficient
δ_s Length of control volume
ϵ Effective roughness height of channel
η Height of an arbitrary point in the flow, normal to the bed
γ Specific weight
λ Conversion parameter between storage and flux determined velocities
ρ Density of the flow
τ_0, τ_w Shear stress at the solid boundary
θ Regional flood index scaling exponent
ξ Lateral position of a point P across the width of a channel
ζ_0 Height of the bed of the channel above a datum point
ϕ Ratio of depth exponent to width exponent

15

LINKING PARAMETERS ACROSS SCALES: SUBGRID PARAMETERIZATIONS AND SCALE DEPENDENT HYDROLOGICAL MODELS

KEITH BEVEN

Centre for Research on Environmental Systems and Statistics, Institute of Environmental and Biological Sciences, Lancaster University, Lancaster LA1 4YQ, UK

ABSTRACT

It is argued that the aggregation approach towards macroscale hydrological modelling, in which it is assumed that a model applicable at small scales can be applied at larger scales using 'effective' parameter values, is an inadequate approach to the scale problem. It is also unlikely that any general scaling theory can be developed due to the dependence of hydrological systems on historical and geological perturbations. Thus a disaggregation approach to developing scale-dependent models is advocated in which a representation of the distribution of hydrological responses is used to reflect hydrological heterogeneity. An appropriate form of distribution may vary with both scale and environment. Such an approach is dependent on the data available to define and calibrate the chosen subgrid parameterization. A parameterization based on a minimum patch representation is suggested and the problems of identification at the larger scale discussed.

INTRODUCTION

One of the driving forces behind the original symposium on scale problems in hydrology in 1982 was the search for hydrological laws at the catchment scale, instigated by the enthusiasm of Ignacio Rodriguez-Iturbe. This stemmed, at least in part, from the progress that had been made at that time in the theory of the geomorphological unit hydrograph (GUH), which could be made to reflect both the non-linearity and scaling of flow routing in catchments of different orders. The GUH appeared to work fairly well (after calibration of a single parameter), but at that first symposium it was pointed out that the problem is not so much how to route, but how much water to route, i.e. the partitioning of the rainfall into discharge, evapotranspiration (or other losses) and the intermediate step of storage. In 1982, most of the available models of runoff production were lumped explicit soil moisture accounting (ESMA) models (see O'Connell, 1991) and were not amenable to addressing the problem of scale effects in catchment response in other than an empirical way.

Since that time, this part of the scale problem has been given added importance by the search for an appropriate hydrological model at the general circulation model (GCM) grid scale. Early parameterizations of land surface hydrology used in GCMs were (to hydrologists) appallingly simple. Since then, they have become more sophisticated, but primarily in the *vertical* representation of the soil–plant–boundary layer system. Models such as BATs (Dickinson and Kennedy, 1991) or SiB (Sellers *et al.*, 1986) are essentially local or point-scale models that are used at the GCM grid scale with little or no account taken of the changed scale.

This simplicity has been primarily due to the very real computing constraints on the operation of GCMs, but it has been worrying to many hydrologists. Such an approach assumes that effective parameters can be used at any scale without any change in model structure. It has been shown that for runoff production processes the use of effective parameters with even the most 'physically based' distributed models cannot be expected to produce accurate predictions of discharges at the hillslope and small catchment scale (e.g. Wood *et al.*, 1988; Binley *et al.*, 1989), or even at the grid scale of such models (Beven, 1989a). This is basically because of the non-linearity of the processes involved, together with the heterogeneity of the natural system. The processes governing land surface–atmosphere interactions are also non-linear and heterogeneous. Thus there is no reason to suppose that the use of effective parameter values should be any more successful in reproducing the areal average of the subgrid fluxes, especially in cases where the water availability starts to exert an influence over vapour flux in a spatially heterogeneous way.

Thus in both small-scale and large-scale hydrology there is a need for a theory that will allow prediction of the correct partitioning of rainfall at any scale of interest. This I will call here the *scale problem*. There is also a need for a theory that will allow the use of information gained at one scale in making predictions at either smaller or larger scales. This can be called the *scaling problem*, or, alternatively, the aggregation/ disaggregation problem. Near-surface hydrology currently lacks a coherent theory for either problem, although some progress has been made in the more linear problem of groundwater flows. At present, a solution to the scaling problem appears to be far off, if not impossible, but it may be possible to address the scale problem in a coherent way. In what follows, some fundamental principles that will need to be considered in the development of such a theory will be discussed.

SCALING AND SCALE DEPENDENCE IN HYDROLOGICAL MODELS

The dendritic organization of catchments, as reflected in the stream order ratios of Horton (1945), suggests a certain similarity in structure across a range of scales. Numerical experiments have shown how this similarity of form can result from long periods of geomorphological development (Willgoose *et al.*, 1991); it has also been revealed in analyses of digitally encoded networks (e.g. Tarboton *et al.*, 1989; 1992; La Barbera and Rosso, 1989) and has been exploited in theories of the GUH (e.g. Valdes *et al.*, 1979; Wang *et al.*, 1981; Rosso, 1984; Rodriguez-Iturbe, 1993).

As a consequence, there might be a similar structure in catchment responses to storm rainfalls. Certainly, analyses of hydrographs suggest a certain consistency in behaviour with increasing catchment scale. It has been suggested, for example, that the catchment response becomes more linear (Wang *et al.*, 1981) and that the runoff per unit area becomes less variable between areas of similar size (Wood *et al.*, 1988; 1990), whereas peak flows per unit area tend to decrease with increasing area (Klein, 1976).

There are, however, good reasons to suggest that it may be difficult to find a scaling invariance in such behaviour. One reason is that runoff production is strongly affected by heterogeneity in geology, soils and vegetation characteristics, and the spatial variability of precipitation as well as the geomorphological form of the catchment. Such heterogeneity may also arise due to the history of the catchment, in particular in areas that have been subjected to different geomorphological processes in the past, such as glaciation or laterization, and to the more recent effects of humans on land use and vegetation type. These effects may be a significant constraint on any tendency towards a scaling invariance. It may be argued that each catchment is essentially unique and that the search for any scaling laws in the partitioning of discharge and evapotranspiration may be a transcientific problem. Consequently, in what follows, we shall concentrate on the principles that might be relevant to the definition of a scale-dependent model of catchment hydrology.

In expressing these principles, we will draw an analogy with those that underlie atmospheric models. These models are expressed in the form of general conservation laws of mass, energy and momentum. The resulting partial differential equations must be solved numerically and, consequently, the scaling problem enters through the problem of parameterizing the effects of processes operating at the subgrid scale of the numerical model. Such parameterizations can be highly complex and include components for the scaling of turbulent eddies, radiation, cloud formation, rainfall and the simple land surface components referred to above (see, for example, Hansen *et al.*, 1983; Dickinson and Henderson-Sellers, 1988; Slingo *et al.*, 1989; Smith, 1989; Henderson-Sellers and McGuffie, 1991; Dolman and Gregory, 1992;

Lean and Roundtree, 1993; Watterson, 1993). This can be viewed as a disaggregation strategy for defining a scale-dependent model (for a previous discussion of hydrological parameterizations, see Dooge, 1982).

The problem of defining a hydrological model at an appropriate scale has been approached in a very different way. This is primarily because there are not the same universal conservation laws for large-scale hydrological systems. The principle of mass conservation still holds (but may be difficult to demonstrate in practice) and water will always flow down-gradient from a high potential to low potential. However, any description of the flow processes will be dependent on the specification of empirical parameters (such as hydraulic conductivities and roughness coefficients) that are themselves dependent on the structure and geometry of the flow pathways. These flow pathways, particularly for subsurface flow, will be impossible to know in any geometrical or statistical detail given current technologies.

Given this dependence on structure and heterogeneity, it is consequently surprising that, in general, 'physically based' hydrological models have taken an aggregation strategy towards the problem of scaling water fluxes at the grid scale. In fact, they have tended to do so in a very unsophisticated way. It is accepted that hydrological theory, such as Darcy's law, valid at the scale of the representative elementary volume (the REV, at which it is assumed that a continuum description can be applied with local equilibration of potentials), can produce reasonably accurate predictions at the scale of laboratory experiments (say 0.01 to 0.1 m). Extension to larger scales has generally been a matter of assuming that the same equation holds and that effective parameters can be found appropriate to the scale required. This requires (a) that the flow is laminar and (b) that realistically representative pressure gradients can be defined at the grid scale. The utility of this approach for saturated groundwater systems is without doubt. However, it appears to be very simple minded given the non-linearities and structural heterogeneities of unsaturated systems where non-equilibrium behaviour and preferential flow at different scales will make any definition of effective parameters difficult (as discussed, for example, by Beven and Germann, 1982; Beven, 1989a; 1989b).

Even for saturated systems it appears that effective parameter values may be scale-dependent. This has been investigated for hydraulic conductivity (see, for example, Neuman, 1990; Desbarats, 1992; Durlofsky, 1992) and for the dispersion coefficient (see Gelhar, 1986; Wheatcraft and Tyler, 1988; Dagan, 1989; Neuman, 1990). For unsaturated flow, structural effects may mean that it is impossible to define an appropriate potential gradient at scales much larger than the REV (Beven and Germann, 1982). Thus it might be better to accept that effective values at different scales are not governed by any simple scaling law and might in fact be incommensurate, being incompatible with any consistent experimental methodology. Similar considerations will apply to roughness parameters for flow over heterogeneous surfaces or in heterogeneous channels.

The aggregation approach to hydrological scaling therefore seems to be ultimately doomed to failure. Even if it was possible, given unlimited computer power, to reduce the grid elements of a physically based model towards the scale of the REV, it would then be simply impossible to know the proper structural geometry of the system and assemble an appropriate set of parameter values for that scale. All such models must therefore use, to a greater or lesser extent, effective parameters at a grid scale where the detailed subgrid scale geometry is ignored. In the terminology of atmospheric models, this simple lumping is one form of subgrid scale parameterization.

If this argument is accepted, then the development of hydrological scaling theory will be precluded except for some special cases of known (or assumed) simple structures. It follows that the hydrological modeller should seek a more realistic, and possibly more parsimonious, subgrid scale parameterization. In doing so, it might prove better to approach the problem from a disaggregation viewpoint such that the development of a scale-dependent model is viewed explicitly as a problem in subgrid scale parameterization. A similar argument has been made by Shuttleworth (1988) in the context of evaporation modelling; he notes that an important need in taking such an approach is to identify the *minimum* complexity required for an adequate parameterization. Shuttleworth suggests that the way to proceed is then to redefine the parameters of a point model, perhaps empirically, with a consequent loss of physical significance of the large-scale parameter values. This has the advantage that the parameterization will have the correct small-scale limit, but the disadvantage that the model structure may not have the proper form to represent large-scale averaged fluxes *unless* the system is strictly linear. In what follows, we take a wider view of the factors involved.

TOWARDS A SCALE-DEPENDENT SUBGRID PARAMETERIZATION

Working then within a disaggregation framework, some basic objectives of developing a scale-dependent subgrid parameterization can be summarized fairly simply.

Objective 1.   Get the input fluxes and other boundary conditions right
Objective 2.   Get the partitioning between discharges, storage and latent heat fluxes right
Objective 3.   Get the timing of discharges and evapotranspiration fluxes right at an appropriate time-scale
Objective 4.   Get the partitioning between different flow pathways right (if necessary for objectives 2 and 3, or for calculations of water quality or sediment transport)

In that this disaggregation approach to subgrid parameterization must recognize the approximate nature of any subgrid model, a further objective should be added as:

Objective 5.   Get an estimate of the uncertainty in the subgrid scale predictions right

These five objectives provide, in a schematic fashion, the basic requirements of a subgrid model at any scale. It is much more difficult, of course, to produce an appropriate model definition to satisfy those objectives. There appears to be very few principles or constraints that can be used in developing such a model. Two, albeit both approximate, will be:

Principle 1.   The net fluxes must satisfy the conservation of mass
Principle 2.   Water (mostly) flows downhill, i.e. the dominant hydraulic gradient for discharges will be that due to gravity

The first principle will only be satisfied approximately as the boundary conditions can only be known approximately. The second will only be approximate in deeper subsurface flow systems.

It is much more difficult to invoke a conservation of energy principle in hydrological systems as, although the inputs to a system with known topography can be characterized in terms of potential energy, the resistance losses in the system are essentially unknowable, even in many river channels. However, for channel flows some attempts have been made to use a principle of minimum energy dissipation as a way of refining models of the GUH (see Rodriguez-Iturbe *et al.*, 1992).

Further principles may need to be defined and, indeed, may be contentious. The following are suggested to promote discussion of the issues involved:

Principle 3.   The equations of a subgrid parameterization appropriate at one scale may not be the same as at smaller scales
Principle 4.   Subgrid parameterizations may depend on the data available to support them, and the level of geometric and process complexity required may vary in different environments and for different purposes
Principle 5.   Preferential flow may exist at all scales
Principle 6.   The most important non-linearities in hydrological systems arise from the coupling of different processes, such as surface and subsurface runoff and soil water flows and transpiration
Principle 7.   At any grid scale, the extremes of the distribution of subgrid responses may be important in the partitioning of discharge and evapotranspiration
Principle 8.   Hydrological systems are not solely self-organizing and the response may be conditioned on the structural history of the grid element

There is one consideration that, at certain scales, might be important in shaping a subgrid parameterization. For example, at the hillslope scale the pattern of hillslope forms and related soil characteristics might be important in controlling the variability in responses of hillslopes and small catchments. At a certain scale a catchment or grid element might contain a sufficient sample of the hillslope and soil characteristics of the region. It is then no longer necessary to take account of the pattern of those characteristics, but only of their distribution function. The underlying variability may still be important in controlling both discharges and evapotranspiration fluxes, but the pattern is less important. The scale at which this happens has been called the representative elementary area (REA) by Wood *et al.* (1988; 1990). It may also be that a similar effect occurs at much larger scales. At the mesoscale, the pattern of land surface variability may be important in controlling mesoscale circulations, whereas at larger GCM scales, that variability may not be important provided that the distribution of fluxes of latent heat, sensible heat and momentum are represented adequately.

Thus, in approaching the definition of a subgrid parameterization based on distributions, the critical sources of variability to be represented must be identified. These will differ between environments and consequently the appropriate *type* of distribution function variables, as well as appropriate parameter values, may vary between grid elements at the macroscale. The type of distribution function may also differ across scales in line with the principles outlined above.

PARTITIONING FLUXES AT THE REA SCALE

The REA concept is not a direct analogy with the REV of soil physics. In the latter, the REV denotes a scale at which average quantities of potential and moisture content can be used in a continuum description of the fluxes. In the former the *distribution* of characteristics may still be important in determining the fluxes, but the REA denotes a scale at which the *pattern* of those fluxes is no longer important. Wood *et al.* (1988) made a hypothetical study of the effects of variable topography, soils and rainfall and, at least for short rainfall correlation lengths, showed that the REA for runoff generation predicted by their particular model and catchment characteristics was of the order of 1 km^2. This was the point at which the variance in storm runoff production between areas of similar size reached a minimum.

This can be interpreted as a point at which the local patterns in runoff production are sufficiently well integrated to produce a similarity in response, before non-stationarity in catchment characteristics (or hydrological processes) starts to increase the variances at larger scales. Such non-stationarity might be due to changes in geology, physiography or land use, changes in the scale of rainfall variability or changes in the nature of the hydrological processes. It may be, for some conditions, that there is no scale at which the variance reaches a minimum, whereas in general it should be expected that if an REA scale exists, it might vary between environments and processes.

The REA concept does, however, allow for the fact that within the REA area, heterogeneity may be important in predicting the hydrological responses and even if it is difficult to define an REA scale unequivocally, it may still be possible to use an approach based on the distribution functions of parameters (or variables) to provide realistic predictions of discharge and evapotranspiration fluxes within heterogeneous terrain. Actually, what is required is the distribution of hydrological *responses* in the landscape, however that might be achieved.

The model used by Wood *et al.* (1988), TOPMODEL, allows for heterogeneity by making calculations on the basis of the distribution function of an index of hydrological similarity, $\ln(a/T_0 \tan \beta)$, where a is the area draining through a point, $\tan \beta$ is the local slope angle at that point and T_0 is the local downslope transmissivity at soil saturation (Beven and Kirkby, 1979; Quinn and Beven, 1993; Beven *et al.*, in press). The index is derived from certain simplifying assumptions within a quasi-steady-state description of lateral saturated flows on hillslopes. It is not applicable everywhere as the assumptions will hold best for slopes of moderate relief with relatively shallow soils over an impermeable bedrock.

TOPMODEL is usually used for the prediction of discharges, but Quinn *et al.* (1995) have shown how the approach can be extended to take account of the effects of downslope flows on water availability in predicting evapotranspiration fluxes. Multiple distribution functions may now be involved, but the form of the model appears to be a good first approximation to that required for a scale-dependent model for partitioning rainfall inputs into discharges and latent heat fluxes. A distribution function approach allows for the fact that different parts of the catchment (or grid element) may have different significance for different hydrological processes. It can allow for the fact that the relationship between different catchment areas may change with wetting and drying. The problem is how to define an appropriate distribution or distributions to reflect, in a realistic way, the hydrological responses at a particular scale.

AVAILABLE DATA

Hydrological science is constrained by the measurement techniques that are available at the present time. Hydrological theory reflects the scale at which measurements are relatively easy to make. Approaching the subgrid parameterization problem as a disaggregation process, however, makes the question of the scale of available data of crucial importance. The utility of data will depend very much on the scale at which they are made available relative to the scale of the parameterization.

Consider the problem of defining a hydrological parameterization for the GCM grid scale. There are no hydrological measurements available directly at that scale. Indeed, hydrological measurements in catchments within any particular GCM grid square will suggest that the subgrid scale hydrology is fairly variable. The GCM itself can provide data at that scale on meteorological variables (which can be used in calculations of evapotranspiration) and rainfalls (at average intensities over the grid square, the so-called GCM 'drizzle'), remembering, of course, that those predicted variables are, at least in part, dependent on the particular land surface hydrology parameterization used in the GCM. Thus it would appear that any subgrid scale parameterization of the land surface hydrology should be developed from data available at the next (catchment) scale down and should reflect the variability inherent at that scale. The important question is then how that data can be used in developing a parameterization at the GCM scale.

This process can be continued, of course, to smaller (subcatchment) scales, and, if necessary, to smaller (hillslope, profile) scales still. Ideally, there will be a future scaling theory in which parameterizations at one scale will reflect the variability and data available to parameterize the responses at smaller scales. This may not, however, be the best way to proceed in practice and it might be better, when working at scales greater than that of gauged catchments, to make use of the available discharge measurements in calibrating the required parameterization.

Consider a mesoscale grid square of 40×40 km. This is the scale of the UK Meteorological Office MORECS predictions of potential and actual evapotranspiration based on simple soil moisture deficit accounting (Thompson *et al.*, 1981). This model is fairly similar to the early soil–vegetation–atmosphere transfer models used in GCMs (Manabe, 1969). It has been demonstrated that soil moisture deficit models of this type can reproduce the observed soil moisture deficits at the profile scale reasonably well (Calder *et al.*, 1983), despite making crude assumptions about the role of the vegetation and its interaction with soil moisture and the way in which 'runoff' is produced. What then can be proposed as an improved partitioning model at the 40 km grid scale? From the disaggregation viewpoint this will depend, in part, on the data available about the nature and processes of any particular grid square. In the UK, the type of data available might include 50 m elevation data (see Figure 1), 25 m remote sensing derived land-use data (see Figure 3), radar rainfall at a 2 km grid scale (at best), point rain gauge measurements and a smaller number of point meteorological measurements, point discharge measurements in one or more rivers within the grid square and other remote sensing images such as surface temperatures (but only at large time steps). In the future, other remote sensing data such as AVHRR, ERS1 radar, etc. may also become more easily available and will serve to reinforce the impression of heterogeneity given by the spatial patterns now available (Figures 1–3). At some particular grid squares other data, such as radiosonde data or some soil moisture data, might also be available locally.

Of these data, which are going to be useful in building a subgrid scale parameterization and which are not? To my mind, the important aspects are going to be:

- characterizing the rainfall inputs in terms of some spatial distribution derived from radar and rainfall measurements
- characterizing the vegetation in terms of some joint spatial distribution of vegetation type and associated parameters with available soil water status (with a possible dependence on geology, soil type, topography and potential evapotranspiration)
- characterizing runoff production in a way that makes use of any discharge measurements in conjunction with the distributions of rainfalls, vegetation and soil water.

The result will be a distribution of runoff production and evapotranspiration responses. The subgrid

Figure 1. Digital elevation data (50 m resolution) for a 25 × 20 km area of north-west England, including the catchment of the river Hodder, based on interpolation from digitized Ordnance Survey contor maps carried out at the Institute of Hydrology, Wallingford

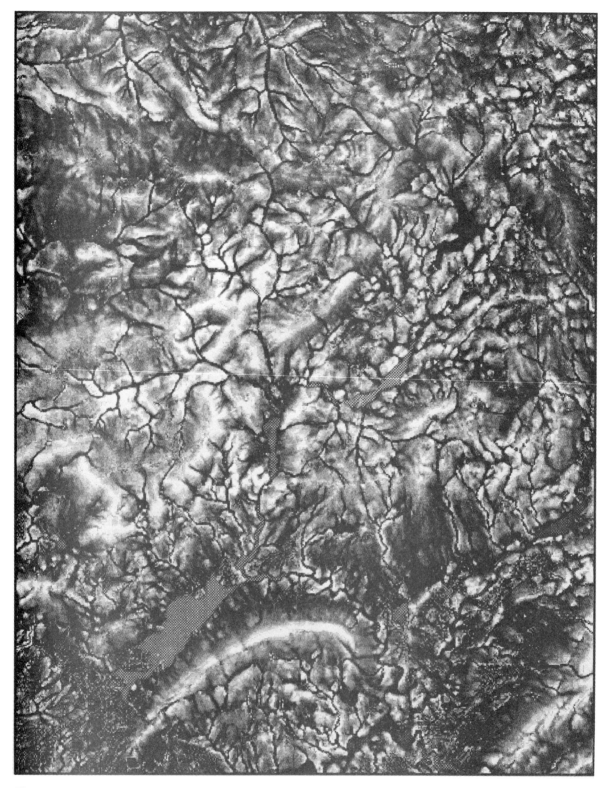

Figure 2. Map of the topographic index $\ln(a/\tan \beta)$ derived from the data of Figure 1 at Lancaster University using the multiple direction algorithm of Quinn *et al.* (1991)

Figure 3. Map of vegetation classes, at 25 m resolution, for the same area as Figures 1 and 2 derived at the Institute of Terrestrial Ecology, Monks Wood

parameterization needs to be only a minimal description of the functional that underlies this distribution, but which reflects adequately the range of responses within the grid square or landscape. It may be possible to achieve this using some type of landscape patch model in which this range of (non-linear) responses is represented as the integration of a distribution of patch models. Each patch model may be non-linear in its operation, but the integration operator might be constrained to be linear, i.e. an areal weighting associated with each patch.

One existing example of such a patch model is the topographic indexing system used in TOPMODEL (see Beven *et al.*, in press). This can be extended to take account of vegetation and soil type (e.g. Famiglietti and Wood, 1991). The patches in this instance are defined by values of the topographic index $\ln(a/\tan \beta)$ and are interlinked through a mean areal storage deficit. The areal weights for each patch come from the analysis of a digital terrain map (such as Figure 1) to derive values of the topographic index. The hydrological response depends in a non-linear way on the distribution of the topographic index. Knowledge of the areal pattern of the topographic index (Figure 2) does allow the hydrological predictions to be mapped back into space (Quinn and Beven, 1993).

It is possible to extend this approach to a combined soil–topographic index $\ln(a/T_0 \tan \beta)$ where T_0 is an effective lateral downslope transmissivity. Regardless of the validity of the simple theory on which TOPMODEL is based, T_0 is a fundamental parameter of hillslope hydrology that is essential in the prediction of surface and subsurface saturated contributing areas, but is not available from normal measurements. Fitting a catchment-average value from stream discharges (within some model structure) is possible, but this will then reduce the effects of soil variability in the grid-scale parameterization. Similar problems will arise in the subgrid characterization of the effect of vegetation on latent heat fluxes given only vegetation maps.

USE OF SPATIAL DISTRIBUTION FUNCTIONS IN SUBGRID SCALE MODELS

Spatial distribution functions have long been used in hydrological models at the catchment scale. The Stanford watershed model of Crawford and Linsley (1966) has an infiltration function that can be interpreted as a uniform spatial distribution over a certain range of infiltration capacities. There is a similar, but non-linear, function in the Xinanjiang model (Zhao, 1992), which has also been used in the Arno model (Dümenil and Todini, 1992) and in the macroscale hydrological model of Wood *et al.* (1992). Explicit distribution functions of storage elements have been used in the model of Moore and Clarke (1981; Moore, 1985). In addition, attempts to improve macroscale hydrological predictions have started to use simplified distribution functions of land surface parameters (Entekhabi and Eagleson, 1989; Famiglietti and Wood, 1991; Avissar, 1991; 1992; Dolman, 1992; Henderson- Sellers and Pitman, 1992; Koster and Suarez, 1992; Wood *et al.*, 1992; Blyth *et al.*, 1993; Bonan *et al.*, 1993; Johnson *et al.*, 1993; Jolly and Wheater, 1993).

On the other hand, Beven (1989b; in press) has shown how distribution functions can be used to reflect the full complexity of the local scale processes governing runoff production. This conceptualization involves distribution functions for the incoming rainfall intensities, effects of vegetation on throughfall and stemflow, soil characteristics including preferential flow pathways whether due to macroporosity or spatial heterogeneity, antecedent moisture in affecting depth to the saturated zone and the factors affecting the heterogeneity of downslope wave velocities. The analysis suggests the impossibility of identifying all the distribution functions involved without measurements of the associated internal states. An equivalent formulation of the factors controlling latent heat fluxes is even more complex because of the interactions between different spatially distributed variables.

TOPMODEL is an attractive candidate for a distribution function model at the landscape scale. It makes use of one set of data that can be made available over large areas of land, the digital elevation map (remembering that this should be of adequate resolution to define the inferred hillslope flow pathways). There are, however, some problems in using the TOPMODEL approach as a general macroscale model. Studies have shown that the values of $\ln(a/\tan \beta)$ derived from an analysis of digital terrain model (DTM) data depend on the grid resolution of the elevations, while the downslope transmissivities will not generally be known. In addition, for deeper subsurface systems, the soil surface slope may not be a good approximation to the local hydraulic gradient and in systems that dry out seasonally the effective upslope contributing area, a, may not extend all the way to the divide and may vary as the hillslopes wet and dry. The logarithmic form of the index is also dependent on the specific exponential form of the change in transmissivity with depth assumed in the original version of TOPMODEL (Beven and Kirkby, 1979; Beven et al., in press).

Most of these restrictions can be relaxed while retaining the relative simplicity of the TOPMODEL index approach. For example, in deeper subsurface systems the 'reference level' approach suggested by Quinn et al. (1991) could be used to define an effective local hydraulic gradient. There is a cost, however, that more data would be required. It is much easier to postulate the use of a reference level than to define a reference shape for a water-table in practice, especially over large areas.

In terms of the principles for a subgrid scale parameterization defined above, a TOPMODEL approach will satisfy principles 1 and 2, while recognizing that it may be difficult to specify an appropriate reference level of potentials for principle 2. It also allows that the form of the appropriate distribution functions may change with scale (principle 3). It is an attempt to make the best use of the available topographic data and can, in principle, be extended to use spatial soil, land-use and rainfall data (principle 4). There is, however, often little or no data actually available on the lateral transmissivity and factors controlling water use by different soil/vegetation communities that would be required (a problem common to all SVAT models). Thus some functional parameterization may be necessary, with no clear way of determining the parameters at large scales. A similar problem extends to preferential flow in the unsaturated zone, saturated zone and minor channels on the hillslopes (principle 5). Recognizing the significance of preferential flow may be of greater importance in understanding flow pathways for transport calculations than in calculating volumes of runoff and evapotranspiration. However, macropore flows through the unsaturated zone may be important in producing rapid subsurface discharges to stream channels and minor channels, and drains may be effective in reducing the effective contributing area to a point on a hillslope while not normally being included in an analysis of DTM data, certainly for large-scale areas (but see the original small catchment manual analyses of Beven and Kirkby, 1979).

In its contributing area calculations the TOPMODEL approach attempts to account for the important non-linear interaction between surface and subsurface runoff generation (principle 6). It can also be used to model interactions between downslope subsurface flows and water availability for calculations of evapotranspiration (Quinn et al., 1995). In both instances, catchment-average fluxes may be controlled by the extremes of the distribution of hydrological states (principle 7). Finally, the approach treats every area as it is and with its actual structure (in so far as is possible with a simplified theory and available data), rather than speculating about some underlying theory of organization (principle 8).

The TOPMODEL concepts provide a distribution function approach that is one basis for defining a methodology for scale-dependent modelling. Such an approach is, however, lacking in a number of important respects. What is required at a given scale are distribution functions that, when integrated, will give effective partitioning into the runoff and evapotranspiration fluxes of interest, including the effects of extreme values of the distribution of responses where these are important (such as small contributing areas in runoff production or deep-rooting trees tapping groundwater, thus maintaining latent heat fluxes in a dry environment). In what follows, a more flexible patch model approach is described that overcomes some of these limitations.

SIMPLE PATCH MODEL FOR SCALE-DEPENDENT MODELLING

Let a patch be defined as any area of the landscape that has broadly similar hydrological responses in terms of the quantities of interest, in this instances say, runoff production and evapotranspiration. Note that this definition does not thus far refer to similarity in characteristics, only in function. Similarity in hydrologically relevant characteristics should, of course, imply similarity in function but, as discussed earlier, it is not yet possible to characterize the landscape in such a way. If it was, then we would be much nearer to understanding scale dependence in modelling and a theory of scaling.

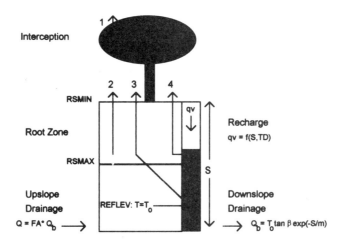

1. Evaporation from the interception store (canopy resistance = 0)
2. Evapotranspiration from the root zone store   (RSMIN<canopy resistance<RSMAX)
3. Evapotranspiration from the water table when in root zone
4. Evapotranspiration supplied by capillary rise from the water table

Figure 4. Schematic representation of a simple patch model

A patch defined in this way may have an arbitrary shape and need not be a contiguous area [just as in TOPMODEL points with the same $\ln(a/T_0 \tan \beta)$ index value are treated as hydrologically similar regardless of location within a catchment area]. In landscape-scale applications, the extremes of size for a patch would be the area of a geographical information system (GIS) pixel, for which there is no smaller scale information, and the area of a GCM grid square for a patch model in which it is assumed that effective parameter values can be defined to represent the hydrological responses (as with the SVAT models in current GCM applications). It should be possible to improve the hydrological representation at large scales by using more than one patch, representing different types of *functional* behaviour. The requirements of defining and calibrating an adequate patch model, however, remain a research question.

Figure 4 shows one conceptualization of a patch model that combines elements from current one-dimensional SVAT models with elements of an extended set of TOPMODEL concepts. It is similar to some of the bucket-type SVAT models, but with one critical difference: that it allows the possibility that water will enter the patch by subsurface flow from upslope and that water may be lost from the patch by drainage downslope or to a stream channel. This allows for more realistic modelling of the patch in the landscape: in riparian areas the effects of downslope flow in maintaining the water available for evapotranspiration is an important control on both runoff production and landscape–atmosphere fluxes in some environments (see, for example, Quinn and Beven, 1993; Quinn *et al.*, 1995).

In this proposal for a patch model these subsurface fluxes are simplified using a set of assumptions similar to those used in TOPMODEL, in which downslope flows are assumed to be in quasi- equilibrium with some recharge rate averaged over the upslope contributing area, and subject to a constant effective hydraulic gradient at each point (in TOPMODEL, this is assumed to be the surface slope $\tan \beta$, but this is not a necessary condition). This leads to a single-valued relationship between discharge and storage at any point, or for incremental downslope areas, both upslope and downslope discharges and storage. This is treated simply in this patch model by assuming that upslope discharge is a simple linear function of the downslope discharge controlled by the parameter FA (Figure 4), where FA is the ratio of the effective area draining into the patch to that draining out of the patch. For the simplest case of a patch that is a unit square pixel with an upslope contributing area of a, then $FA = a/(a + 1)$. In the general case, a will not necessarily be the whole of the upslope area (as normally assumed in digital terrain analysis)

Table I. Parameters of the patch model

RA (s/m)	Aerodynamic resistance
$RSMIN$ (s/m)	Minimum dry canopy resistance
$RSMAX$ (s/m)	Maximum dry canopy resistance at wilting point
$MAXINT$ (m)	Interception storage capacity
FA	Fractional upslope area
$REFLEV$ (m)	Reference level for soil transmissivity
$T_0 \tan\beta$ (m^2)	Product of saturated transmissivity at reference level and effective downslope hydraulic gradient
m (m)	Transmissivity profile (and recession curve) parameter
TD(h/m)	Effective wave speed per unit of deficit for recharge
$SOIL$	Soil type for capillary rise calculations

and might be expected to vary over time, at least in seasonally dry environments (see, for example, Barling et al., 1994). As this would introduce further unknown parameters, in this exploratory exercise FA has been assumed constant.

Evapotranspiration from the patch is calculated using a Penman–Monteith equation, which requires the specification of the meteorological forcing variables (which may vary in space as well as time), aerodynamic resistance and canopy resistance. Canopy resistance is here made a function of the water storage on the vegetation surface, or, when the canopy is dry, of the storage in the root zone. Where the predicted water-table is high enough, as maintained by the downslope flows, evapotranspiration rates may be maintained by capillary rise, as predicted by the quasi-steady-state approach of Eagleson (1978).

The prediction of downslope flows also requires the specification of a transmissivity function. In this application, following most applications of TOPMODEL, an exponential transmissivity profile with water-table storage deficit (S) has been used, characterized by the parameters T_0 and m. In addition, a parameter has been introduced to control the reference level for the exponential function (REFLEV). Whereas this is normally taken to be the soil surface, in the patch model this can be a specified depth allowing for the treatment of much deeper water-tables than in previous TOPMODEL applications. In fact, the T_0 parameter only ever enters the calculations in combination with the fixed hydraulic gradient $\tan\beta$, so that these can be treated as a single scaling parameter for the downslope flow.

A summary of the parameters of the patch model are given in Table I. Although simple and with a minimum parameterization, the patch model incorporates many of the elements required for predicting the variability of runoff production, storage and evapotranspiration in the landscape as the linear combination of patches of different characteristics of vegetation, soil and geology, topographic position and meteorological variables. Clearly some elements could be made more sophisticated, such as the partitioning of sensible heat inputs and the treatment of multi-layered canopies, but at the expense of adding more parameters.

IDENTIFICATION OF THE LANDSCAPE PATCH MODEL

The identification of current SVAT models has proceeded on the basis that the effective parameter values would be related directly to vegetation and soil types. Identification of one-dimensional models by calibration to data collected during detailed experiments on landscape–atmosphere fluxes at some locations in a landscape would then allow predictions at other places and other times. This approach has been a pragmatic response to the practical limitations of data and understanding, but is difficult to justify scientifically in that it ignores much of what is known about the effects of spatial variability in the landscape on hydrology. In that evapotranspiration rates in some environments are relatively conservative in space, such an approach might be reasonable; for runoff production it is unlikely to be reasonable.

In fact, given a characterization of the landscape, say, within a GIS system with digital maps of soil, vegetation and topography, it is clear that even the simple patch characterization postulated earlier is far more sophisticated than the data available to support it. In the past, this has commonly been considered to be a problem of model calibration, in which parameter values that cannot be identified directly by measurement are calibrated by comparing model predictions with those observations that are available. The limitations of this strategy, in producing uniquely defined parameter values, have been discussed by Beven (1993), who suggests that hydrological modellers must accept an equifinality principle, with the consequence that there may be many models and parameter sets that are equivalently consistent with the data available for calibration/validation and that any predictions should be associated with an estimate of the resulting uncertainty. Models can then be compared in terms of the predictive uncertainty and different types and sources of data can be evaluated in terms of constraining model uncertainty.

If the equifinality principle is accepted (recognizing that this remains contentious in many spheres of environmental modelling for both practical and philosophical reasons), then there may be many combinations or distributions of patches that may be contenders as landscape models in a particular environment in terms of being consistent with any available data on runoff production and evapotranspiration fluxes. Beven and Quinn (1994), for example, have shown how the described patch model can be used to produce a wide range of predicted cumulative evapotranspiration rates. Figure 5 shows the sensitivity of cumulative evapotranspiration to several of the patch model parameters, following many runs of the model with randomly chosen values of the parameters. All runs use the same input record of rainfalls and meteorological forcing variables from a site in the south of France. In each diagram the simulations have been grouped together into 10 sets, each representing a subdivision of the range of predicted cumulative evapotranspiration. The distribution of parameter values is then plotted for each evapotranspiration increment, with the scale of the parameter value being normalized from the sampled range to a range of unity (the x-axis). Thus a wide spread of distributions indicates the sensitivity of predicted evapotranspiration to that parameter (e.g. SRMAX); a narrow spread little sensitivity (e.g. *FA* over most of the range). SRMAX, the available water capacity for root extraction is, for these conditions, the dominant parameter.

Two things are of note. For most parameters it is possible to span the complete range of predicted cumulative evapotranspirations at any value of that parameter. Also, the important sensitivity to the parameters *FA* and REFLEV over part of their range, reflecting the role of the upslope inputs of subsurface flow when the soil starts to dry. The parameter showing the greatest sensitivity is the available water capacity of the soil. Similar types of analyses can be used to look at the sensitivity of predicted runoff production.

There will be one patch model that will perform best in predicting the data available on hydrological responses (normally, of course, primarily discharge responses). There will, almost certainly, be many linear combinations of two patches that will perform better. Combinations of many patches, in more realistically reflecting the pattern of variability in the landscape, should in principle perform better still until the limits of accuracy caused by errors in the input data and model structure are reached. As, with a model as simple as this, there is no real computational problem in running the whole gamut of feasible parameter sets (at least as an offline simulation), the problem becomes one of how to construct one or more combinations of patches that are acceptable simulators of the landscape hydrological system. Note that what is sought is a *linear* combination of patches to comprise the distribution function appropriate at a given scale. Thus we seek a solution to the problem

$$Y(t) = \sum W_i \Phi(\Theta_i, U, t) \qquad i = 1, N$$

where the Y values are some predicted variables, the W_i are the linear weights chosen such that $\Sigma W_i = 1, \Phi(\Theta_i, U, t)$ represents the response of the ith patch with parameters Θ_i and input variables U at time t and N is an appropriate number of patches to reflect the essential variability of responses in the landscape. The problem thus posed is similar to that of the combination of forecasts from different models studied extensively in other fields (see, for example, Bunn, 1989, and other papers in the same special issue of the *Journal of Forecasting*). In these instances, however, the number of contending models is usually small and defined *a priori*.

Here, the identification of the W_i weights is a problem with a large number of competing solutions and not many strong constraints: a problem that is decidedly ill-posed unless some constraints are imposed *a priori*. These *a priori* constraints will be more or less artificial. In past distribution function approaches

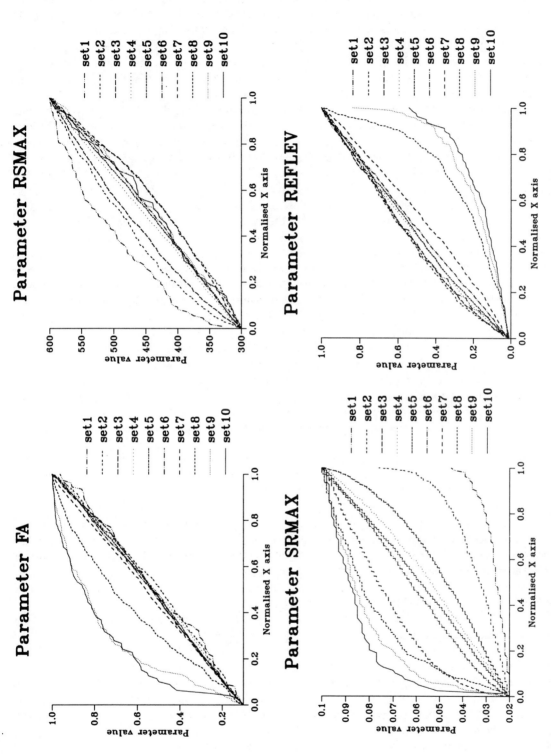

Figure 5. Sensitivity of patch model predictions of cumulative evapotranspiration derived from 10 000 randmomly chosen parameter sets for four of the patch model parameters. Each curve represents the distribution of parameter values (normalized across the range over which they were chosen) in the set associated with an increment of predicted evapotranspiration. The 10 sets span the full range of predicted evapotranspiration

to the landscape-scale modelling problem it has been assumed that these weights are known *a priori* from the joint distribution of soils and vegetation over the area (e.g. Entekhabi and Eagleson 1989; Jolley and Wheater, 1993). The effects of topography on both downslope fluxes and meteorological forcing variables could also be specified *a priori* in this way following, for example, a topographic domain approach. Such approaches may, or may not, be successful in predicting the observed hydrology, after perhaps some calibration of some model parameters. It is at least debatable, however, whether there will ever be enough information available to allow the correct specification of these weights from the data available. Some additional calibration of these models will inevitably be required.

Thus an alternative approach is worthy of consideration in situations where there are some observed data with which to compare model performance. Consider the patch model behaviours available from simulation in a purely functional sense. Assume that the parameter sets on which they are based have been chosen to be consistent with the available knowledge on catchment characteristics. Some parameters (aerodynamic resistance? available water capacity? minimum canopy resistance?) might be more closely constrained in this way than others (T_0? pattern of reference levels?). The identification problem could then be posed as the minimum number of patches required to obtain an adequate simulation of the observed discharge and estimates of actual evapotranspiration, subject to the constraint $\Sigma W_i = 1$. In some environments, and for small areas, it is possible that one patch might be enough to yield an acceptable simulation (the patch model described here is, after all, not very different from the lumped ESMA models used as functionally acceptable models at the catchment scale in the past). In other environments or for larger areas, two or more patches might be required. The choice of patches and their associated weights will require the combination of prior constraints and calibration, recognizing the inherent uncertainties involved. This approach to the macroscale modelling problem is being actively pursued within a Bayesian framework and results will be presented in a future paper.

Such an approach does pose the question as to what observables may be most effective in defining the patch weights or in constraining the uncertainty in predictions. It seems that the primary data at an appropriate scale that could be used to constrain the model predictions are, at present, catchment discharge data from current and historical observations in the area. This will be an inadequate constraint, particularly for predictions under changing conditions with poorly estimated inputs and boundary conditions, but the value of other types of data remains the subject for further research. Significant progress may, in fact, await the development of new measurement techniques providing useful data at larger scales, or new ways of using time sequences of remote sensing data to assess hydrological variablility.

CONCLUSIONS

These discussions have stressed the limitations of the aggregation strategy that has been taken in the past in attempting to scale point-scale hydrological processes and parameters to larger scales. In particular, it has been stressed that the search for scaling laws and scale invariance in hydrological behaviour may not be successful in many environments where the continuing dependence of hydrological responses on environmental history over time-scales of climate change means that the hydrological system is greatly influenced by external historical and geological forcings. This greatly limits the potential for scale-invariant behaviour and it is suggested that it may be better to recognize explicitly that scale-dependent model structures are required in which the characterization of the heterogeneity of responses is posed as a problem of subgrid- scale parameterization.

Thus a disaggregation approach to the development of scale-dependent models has been advocated where, at a given scale of interest, heterogeneity in hydrological response and boundary conditions can be represented in terms of joint distribution functions which must retain the essential non-linearities of the system while using the linear operation of integration to predict the areal fluxes at the scale defined. It is stressed that the structure of an appropriate distributional model may vary with the environment or hydrological regime, and that a full distributional model, taking account of all relevant processes and variables, is not identifiable. The TOPMODEL structure at scales greater than the REA is used as an example of a distributional structure. The limitations of the TOPMODEL assumptions and REA concepts are discussed. A more general patch model, defined to address some of these limitations, is suggested as one form of distribution function approach to the disaggregation or subgrid parameterization problem.

It is possible to run simulations for many different combinations of parameters within a patch model structure giving a range of functional behaviours for any set of forcing data. The identification problem then becomes one of defining the weights associated with a certain number of patches that best reproduces the landscape-scale responses. This is an ill-posed problem, currently solved by an *a priori* choice of weights. An alternative approach, more closely linked to observable variables, is suggested, in which a minimum patch representation is sought, consistent with the observed data, which at large scales will be predominantly the measured discharges. This alone will be an inadequate constraint and further progress in this area may well be dependent on the development of new large-scale measurement techniques or new methods of using time sequences of remote sensing data. A methodology for defining a multiple patch model in this way is the subject of current research.

ACKNOWLEDGEMENTS

Thanks are due to Renata Romanowicz for the preparation of Figures 1,2 and 3 and to Paul Quinn for the basis of Figures 4 and 5. The thought process leading to this paper has been stimulated by support under the UK NERC TIGER programme, grant GST/02/569, and clarified by the discussions that took place at the Robertson Scale Problems workshop. Thanks particularly to Ross Woods and other anonymous reviewers who did not like much of this discussion, but made some valuable suggestions for clarification. I am grateful to the organizers for enabling me to attend the workshop and for the chance to dedicate this small contribution to the memory to Ian Moore. His energy and enthusiastic leadership will be sorely missed in hydrology.

REFERENCES

Avissar, R. 1991. 'A statistical–dynamical approach to parameterise subgrid-scale land surface heterogeneity in climate models', in Wood, E. F. (Ed.), *Land Surface–Atmospheric Interactions for Climate Modeling*. Kluwer, Dordrecht. pp. 155–178.

Avissar, R. 1992. 'Conceptual aspects of a statistical–dynamical approach to represent landscape subgrid scale heterogeneities in atmospheric models', *J. Geophys. Res.*, **97**(D3), 2729–2742.

Barling, R. D., Moore, I. D., and Grayson, R. B. 1994. 'A quasi-dynamic wetness index for characterising the spatial distribution of zones of surface saturation and soil water content', *Wat. Resour. Res.*, **30**, 1029–1044.

Beven, K. J. 1989a. 'Changing ideas in hydrology: the case of physically-based models', *J. Hydrol.*, **105**, 157–172.

Beven, K. J. 1989b. 'Interflow' in Morel-Seytoux, H. J. (Ed.), *Unsaturated Flow in Hydrologic Modelling, Proc. NATO ARW, Arles.* Reidel, Dordrecht. pp. 191–219

Beven, K. J. 1993. 'Prophecy, reality and uncertainty in distributed hydrological modelling', *Adv. Wat. Resour.* **16**, 41–51.

Beven, K. J. 1994. 'Process, heterogeneity and scale in modelling soil moisture fluxes' in Sorooshian, S. and Gupta, V. K. (Eds), *Global Environmental Change and Land Surface Process in Hydrology: The Trials and Tribulations of Modelling and Measuring. Proc. NATO ARW, Tucson, Arizona.*, Springer-Verlag, New York, in press.

Beven, K. and Binley, A. M. 1992. 'The future of distributed models: model calibration and uncertainty prediction', *Hydrol. Process.*, **6**, 279–298.

Beven, K. J. and Germann, P. 1982. 'Macropores and water flow in soils', *Wat. Resour. Res.*, **18**, 1311–1325.

Beven, K.J. and Kirkby, M. J. 1979. 'A physically based variable contributing area model of basin hydrology', *Hydrol. Sci. Bull.*, **24**, 43–69.

Beven, K. J. and Quinn, P. F. 1994. 'Similarity and scale effects in the water balance of heterogeneous areas,' T. Keane and E. Doly (Eds) in *Proc. AGMET Conference on the Balance of Water–Present and Future.* AGMET, Dublin.

Beven, K. J., Lamb, R., Quinn, P. F., Romanowicz, R., and Freer, J. 'TOPMODEL' in Singh, V. P. (Ed.), *Computer Models of Watershed Hydrology.* Water Resource Publications, in press.

Binley, A. M., Beven, K. J., and Elgy, J. 1989. 'A physically-based model of heterogeneous hillslopes, 2. effective hydraulic conductivities', *Wat. Resour. Res.*, **25**, 1227–1233.

Blyth, E. M., Dolman, A. J., and Wood, N. 1993. 'Effective resistance of sensible- and latent-heat flux in heterogeneous terrain', *Q. J. Roy. Meteorol. Soc.*, **119**, 423–442.

Bonan, G. B., Pollard, D., and Thompson S. L. 1993. 'Influence of subgrid scale heterogeneity in leaf area index, surface resistance and soil moisture on grid scale land-atmosphere interactions', *J. Climate*, **6**, 1882–1897.

Bunn, D. 1989. 'Forecasting with more than one model', *J. Forecast.*, **8**, 161–167.

Calder, I. R., Harding, R. J., and Ropsier, P. T. 1983. 'An objective assessment of soil moisture deficit models', *J. Hydrol.*, **60**, 329–355.

Crawford, N. H. and Linsley, R. K. 1966. *'Digital simulation in hydrology: Stanford watershed model IV, Rep. No. 39*, Department of Civil Engineering, Stanford University.

Dagan, G. 1989. *Flow and Transport in Porous Media.* Springer-Verlag, New York. 465pp.

Desbarats, A. J. 1992. 'Spatial averaging of hydraulic conductivity in three-dimensional heterogeneous porous media', *Math. Geol.*, **24**, 249–267.

Dickinson, R. E. and Kennedy, P. J. 1991. 'Land surface hydrology in a general circulation model–global and regional fields needed for validation' in Wood, E. F. (Ed.), *Land Surface–Atmospheric Interactions for Climate Modeling*. Kluwer, Dordrecht. pp. 115–126.

Dickinson, R. E. and Henderson-Sellers, A. 1988. 'Modelling tropical deforestation: a study of GCM land surface parameterizations', *Q. J. Roy. Meteorol. Soc.*, **114**, 439–462.

Dolman, A. J. 1992. 'A note on areally averaged evaporation and the value of effective surface conductance', *J. Hydrol.*, **138**, 583–589.

Dolman, A. J. and Gregory, D. 1992. 'The parameterization of rainfall interception in GCMs', *Q. J. Roy. Meteorol. Soc.*, **118**, 455–467.

Dooge, J. C. I. 1982. 'The parameterization of hydrologic processes' in Eagleson, P. S. (Ed.), *Land Surface Processes in Atmospheric Global Circulation Models*, Cambridge University Press, Cambridge. 243–288.

Dümenil, L. and Todini, E. 1992. 'A rainfall–runoff scheme for use in the Hamburg climate model', in O'Kane, J. P. (Ed.), *Advances in Theoretical Hydrology. A Tribute to James Dooge*. Elsevier, Amdsterdam. pp. 129–158.

Durlofsky, L. J. 1991. 'Numerical calculation of equivalent grid block permeability tensors for heterogeneous porous media', *Wat. Resour. Res.*, **27**, 699–708.

Eagleson, P. S. 1978. 'Climate, soil and vegetation. 3. A simplified model of soil moisture movement in the liquid phase', *Wat. Resour. Res.*, **14**, 722–799.

Entekhabi, D. and Eagleson, P. S. 1989. 'Land surface hydrology parameterization for atmospheric general circulation models including subgrid scale spatial variability', *J. Climate*, **2**, 816–831.

Famiglietti, J. S. and Wood, E. F. 1991. 'Evapotranspiration and runoff from large land areas: land surface hydrology for atmospheric general circulation models', *Surv. Geophys.*, **12**, 179–204.

Gelhar, L. W. 1986. 'Stochastic subsurface hydrology from theory to applications', *Wat. Resour. Res.*, **22**, 135S–145S.

Hansen, J., Russell, G., Ring, D., Stone, P., Lacis, A., Lebedeff, S., Ruedy, R, and Travis, L. 1983. 'Efficient three-dimensional global models for climate studies. Models I and II', *Monthly Weather Rev.*, **111**, 609–662.

Henderson-Sellers, A. and McGuffie, K. 1991. *A Climate Modelling Primer*. Wiley, Chichester. 217pp.

Henderson-Sellers, A. and Pitman, A. J. 1992. 'Land-surface schemes for future climate models: specification, aggregation and heterogeneity', *J. Geophys. Res.*, **97**(D3), 2687–2696.

Horton, R. E. 1945. 'Erosional development of streams and their drainage basins: hydrophysical approach to quantitative morphology', *Geol. Soc. Am. Bull.*, **56**, 275–370.

Johnson, K. D., Entekhabi, D., and Eagleson, P. S. 1993. 'The implementation and validation of improved land surface hydrology in an atmospheric general circulation model', *J. Climate*, **6**, 1009–1026.

Jolley, T. J. and Wheater, H. S. 1993. 'Macromodelling of the River Severn' in Wilkinson, W. B. (Ed.), *Macroscale Modelling of the Hydrosphere. IAHS Publ.*, **214**, 91–100.

Klein, M. 1976. 'Hydrograph peakedness and basin area', *Earth Surf Process.*, **1**, 27–30.

Koster, R. D. and Suarez, M. J. 1992. 'Modelling the land surface boundary in climate models as a composite of independent vegetation stands', *J. Geophys. Res.*, **97**(D3), 2697–2715.

La Barbera, P. and Rosso, R. 1989. 'On the fractal dimension of stream networks', *Wat. Resourc. Res.*, **25**, 735–741.

Lean, J. and Rowntree, P. J. 1993. 'A GCM simulation of the impact of Amazonian deforestation using an improved canopy representation', *Q. J. Roy. Meteorol. Soc.*, **119**, 509–530.

Manabe, S. 1969. 'Climate and the ocean circulation 1. The atmospheric circulation and the hydrology of the earth's surface', *Monthly Weather Rev.*, **97**, 739–774.

Moore, R. J. 1985. 'The probability distributed principle and runoff production at point and basin scales', *Hydrol. Sci. J.*, **30**, 263–297.

Moore, R. J. and Clarke, R. T. 1981. 'A distribution function approach to rainfall-runoff modelling', *Wat. Resour. Res.*, **17**, 1367–1382.

Neuman, S. P. 1990. 'Universal scaling of hydraulic conductivities and dispersivities in geologic media', *Wat. Resour. Res.*, **26**, 1749–1758.

O'Connell, P. E. 1991. 'A historical perspective' in Bowles, D. S. and O'Connell, P. E. (Eds), *Recent Advances in the Modeling of Hydrologic Systems. Nato ASI Ser. C345*. Kluwer, Dordrecht. pp. 3–30.

Quinn, P. F. and Beven, K. J. 1993. 'Spatial and temporal predictions of soil moisture dynamics, runoff, variable source areas and evapotranspiration for Plynlimon, mid-Wales', *Hydrol. Process.*, **7**, 425–448.

Quinn, P. F., Beven, K., Chevallier, P. and Planchon, O. 1991. 'the prediction of hillslope flow paths for distributed hydrological modelling using digital terrain models', *Hydrol. Process.*, **5**, 59–79.

Quinn, P. F., Beven, K. J., and Culf, A. 1995. 'The introduction of macroscale hydrological complexity into land surface–atmosphere transfer function models and the effect on planetary boundary layer development', *J. Hydrol.*, **166**, 421–445.

Rodriguez-Iturbe, I. 1993. 'The geomorphological unit hydrograph', in Beven, K. J. and Kirkby, M. J. (Eds), *Channel Network Hydrology*. wiley, Chichester, pp. 43–68.

Rodriguez-Iturbe, I., Rinaldo, A., Rigon, R., Bras, R. L., Marani, A., and Ijasz-Vasquez, E. 1992. 'Energy dissipation, runoff production and the three-dimensional structure of river basins', *Wat. Resour. Res.*, **28**, 1095–1103.

Rosso, R. 1984. 'Nash model relation to Horton ratios', *Wat. Resour. Res.*, **20**, 914–920

Sellers, P. J., Mintz, Y., Sud, Y. C., and Dalcher, A. 1986. 'A simplified biosphere model (SiB) for use within general circulation models', *J. Atmos. Sci.*, **43**, 305–331.

Shuttleworth, W. J. 1988. 'Macrohydrology—the new challenge for process hydrology', *J. Hydrol*, **100**, 31–56.

Slingo, A., Widerspin, R. C., and Smith, R. N. B. 1989. 'The effect of improved physical parametrizations on simulations of cloudiness and the earth's radiation budget in the tropics', *J. Geophys. Res.*, **94**, 2281–2291.

Smith, R. N. B. 1989. 'A scheme for prediction of layer clouds and their water content in a Global Climate Model', *Q. J. Roy. Meteorol. Soc.*, **116**, 435–460.

Tarboton, D. G., Bras, R. L., and Rodriguez-Iturbe, I. 1989. 'Scaling and elevation in river networks', *Wat. Resour. Res.*, **25**, 2037–2051.

Tarboton, D. G., Bras, R. L., and Rodriguez-Iturbe, I. 1992. 'On the extraction of channel networks from digital elevation data', in Beven, K. J. and Moore, I. D. (Eds), *Terrain Analysis and Distributed Modelling in Hydrology*. pp. 85–106, Wiley, Chichester.

Thompson, N., Barrie, I. A., and Ayles, M. 1981. 'The Meteorological Office rainfall and evaporation calculation system, MORECS', *Hydrol. Memo. 45*, Mterological Office, Bracknell.

Valdes, J. B., Fialto, Y., and Rodriguez-Iturbe, I. 1979. 'A rainfall–runoff analysis of the geomorphological IUH', *Wat. Resourc. Res.*, **15**, 1421–1434.

Wang, C. T., Gupta, V. K., and Waymire, E. 1981. 'A geomorphic synthesis of nonlinearity in surface runoff', *Wat. Resour. Res.*, **17**, 545–554.

Watterson, I. G. 1993. 'Global climate modelling' in Jakeman, A. J., Beck, M. B. and McAleer, M. J. (Eds), *Modelling Change in Environmental Systems*. Wiley, Chichester. pp. 343–366.

Wheatcraft, S. W. and Tyler, S. W. 1988. 'An explanation of scale-dependent dispersivity in heterogeneous aquifers using concepts of fractal geometry', *Wat. Resour. Res.*, **24**, 566–578.

Willgoose, G., Bras, R. L. and Rodriguez-Iturbe, I. 1991. 'A coupled channel network growth and hillslope evolution model. 1. Theory', *Wat. Resour. Res.*, **27**, 1671–1684.

Wood, E. F., Sivapalan, M., Beven, K. J., and Band, L. 1988. 'Effects of spatial variability and scale with implications to hydrological modelling', *J. Hydrol.*, **102**, 29–47.

Wood, E. F., Sivapalan, M., and Beven, K. J. 1990. 'similarity and scale in catchment storm response', *Rev. Geophys.*, **28**, 1–18.

Wood, E. F., Lettenmaier, D. P., and Zartarian, V. G. 1992. 'A land surface hydrology parameterization with sub-grid variability for general circulation models', *J. Geophys. Res.*, **97**(D3), 2727–2728.

Zhao, R.-J. 1992. 'The Xinanjiang model applied in China', *J. Hydrol.*, **134**, 371–381.

16

SCALING THEORY TO EXTRAPOLATE INDIVIDUAL TREE WATER USE TO STAND WATER USE

THOMAS J. HATTON

CSIRO Division of Water Resources, GPO Box 1666, Canberra 2601, Australia

AND

HSIN-I WU

Center for Biosystems Modelling, Department of Industrial Engineering, Texas A&M University, College Station, TX 77843-3131, USA

ABSTRACT

Extrapolation of measurements of water use by individual trees to that for a stand of trees is a critical step in linking plant physiology and hydrology. Limitations in sampling resources and variation in tree sizes within a stand necessitate the use of some scaling relationship. Further, to scale tree water use in space as well as time, the relationship must reflect the changing availabilities of energy and water supply. It is argued here that tree leaf area is the most appropriate covariate of water use to achieve this aim. However, empirical results show that the relationship is not always linear. A theory is developed, based on the concepts of hydrological equilibrium (*sensu* Nemani and Running, 1989) and ecological field theory (Walker *et al.*, 1989) which accounts for (occasional) non-linear behaviour of the flux/leaf area relationship in evergreen trees. A key feature of this theory is the notion of a non-linear, quasi-equilibrium reflecting plant water stress. An equation is derived from these concepts and a standard, explicit treatment of tree water use (Landsberg and McMurtrie, 1984), which is used to characterise this relationship. This equation has the form $Q = aIA + b\Psi_s A^f$. The theory is tested against field data and published reports on *Eucalyptus* tree water use.

INTRODUCTION

Over the past 50 years, techniques to measure tree water use by individual stems have developed to such an extent that monitoring water use in a sample of stems (trees) taken from a population (stand) is now routine. Regardless of the specific technique used, a key question is how to extrapolate information from a limited number of trees to water use by the entire stand ('telling the forest from the trees', Denmead, 1984). This step is especially crucial to studies in which the objective is to estimate transpiration as part of the overall water balance of a site or region. For this type of study, it is also necessary to convert flux estimates expressed as volume per unit time to more useful hydrological expressions such as mm day^{-1}. Implicit in this conversion is an expression of the effective area occupied by the sampled trees. The most direct solution to this scaling problem is to measure the flux through every tree in a plot of known area (e.g. Doley and Grieve, 1966; Kelliher *et al.*, 1992). The limitations of this approach are obvious: to census water use by the population of stems in a plot of sufficient size to minimize edge effects is logistically and practically difficult.

In tree plantations or orchards with trees of the same size and regular spacing, or in closed forests with very few gaps in the canopies, the average area occupied by a given tree with respect to its contribution to the total transpired flux from the site may be reasonably approximated. However, in stands with gaps in the canopy and trees ranging in size (i.e. woodlands and open forests), the scaling of water use by an individual tree to total stand transpiration is not so readily resolved.

In stands with open or irregularly spaced canopies, an alternative is to scale tree water use by some expression of tree size. Ladefoged (1963) attempted to scale tree flux measurements to a mixed forest stand on the basis of (a) a relationship between sapflow and crown size; and (b) the area occupied by each tree in the stand. The linear correlation between tree flux and size was poor. Cermak and Kucera (1987) estimated stand transpiration from sapflow measurements based on relationships between tree flux and basal area, but reported no indication of the strength of nature of the relationship. Werk *et al.* (1988) extrapolated flux measurements to the stand level by means of leaf area estimates, but again without any indication of the nature or strength of the relationship. In an even-aged plantation, Hatton and Vertessy (1990) scaled flux measurements from a sample of *Pinus radiata* trees to the stand level on the basis of ground area occupied by each tree; the resulting values compared well with transpiration estimates based on micrometeorological methods. Thorburn *et al.* (1993) examined the relationships between sap flux density in *Eucalyptus largiflorens* and *E. camaldulensis* trees with sapwood area; in winter and autumn the relationship was good ($r^2 > 0.8$; $p < 0.01$), but in summer the slope of the relationship was not significantly different from zero.

These relationships enable the extrapolation of flux to other trees growing at a homogeneous site, and at a particular instant in time. Often this is the only objective. What is of more general use, however, is a scaling relationship that can be applied to a forest or woodland with spatial and temporal variation in the supply of energy and soil water. With such a relationship and a knowledge of the distribution of these parameters at a given time, and the spatial pattern in the local tree leaf area distribution, transpiration flux could be distributed over a region.

Cermak and Kucera (1987) suggested that the use of some expression of leaf area should lead to more precise extrapolation. The notion that the leaf area of a single plant (or stand) should be strongly related to the flux of water through the system is reflected in the dominant paradigm associated with scaling leaf resistance to water vapour flux to surface resistance (e.g. Running and Coughlan, 1988; McMurtrie *et al.*, 1990; Hatton *et al.*, 1992). The assumption is that leaves act in parallel (*sensu* the electrical analogue) such that there is some direct proportionality between leaf area index and conductance (Szeicz and Long, 1969; Tan and Black, 1976). However, there are examples which suggest that leaf efficiency (flux per unit leaf area) is not always constant, especially in open forests and woodlands. Greenwood and Beresford (1979) found that tree transpiration (as estimated from ventilated chambers) was not linearly related to tree leaf area in various species of *Eucalyptus*. Greenwood *et al.* (1982) found a curvilinear relationship between leaf area (or canopy area) and water use in regenerating *Eucalyptus wandoo* trees at a site with some restrictions to soil water availability; as the soil dried through summer, trees with larger leaf area used proportionately less water per unit leaf area. Greenwood *et al.* (1985) reported a linear relationship between leaf area and annual evaporation which accounted for 84% of the variation in the data, with little difference in the slope of the relationship among three species of *Eucalyptus* growing in a site well upslope from a saline seep, but with access to shallow groundwater. However, the senior author reported that large variations in leaf efficiency existed among species in a plantation near a saline seep. Thorburn *et al.* (1993) found a very poor linear relationship between the tree water use and leaf area in *Eucalyptus largiflorens* at the end of a dry summer and autumn on a semi-arid floodplain. Thus, a wide variety of relationships between leaf area and flux are reported. The question is, can the variation in the flux/leaf area relationship be encompassed within a general theory for scaling individual tree water use measurements to stand transpiration?

Implicit in this objective is the assumption that the transpiration–leaf area relationship is scale-dependent. It is important to recognize that there is a large body of published work based on the transpiration–LAI (leaf area index) relationship; by indexing leaf area to ground area, the scale dependence is eliminated. In this paper, however, we focus on the relationship of an individual plant's leaf area to water use. This is clearly a scale-dependent phenomenon, and the inferences about the form of the LAI–transpiration relationship (Grier and Running, 1977; Gholz, 1982) based on LAI will not necessarily hold.

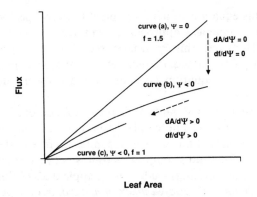

Figure 1. Forms of the theoretical scaling relationship between tree leaf area and flux. With unlimited soil water, the relationship is linear and the slope is limited only by irradiance (curve **a**). As the soil dries, the retention of leaves creates a quasi-equilibrium in which the leaf efficiency of the larger trees declines (curves **b**). If a drought is of sufficient extent and duration, the trees will drop leaves such that the relationship is in a new, lower hydrological equilibrium, and the curve may once again be linear (curve **c**)

This paper presents a scaling theory which predicts the nature of the flux/leaf area relationship by combining elements of ecological field theory (Walker *et al.*, 1989), the hydrological equilibrium theory (Eagleson 1982, Nemani and Running, 1989) and a standard treatment of the soil–plant–atmosphere continuum (Landsberg and McMurtrie, 1984). We derive a relationship of the form

$$Q = aIA + b\Psi_s A^f \tag{1}$$

where Q is the flux, I is the canopy average radiation (*sensu* Running and Coughlan, 1988; i.e. the radiation intercepted by the canopy), Ψ_s is the soil water potential, A is the leaf area, f is a scaling exponent and a and b are lumped parameter coefficients. In this formulation, when water is freely available, the slope of the relationship is linear and controlled by irradiance and the constant a. We further hypothesize that at hydrological equilibrium, this relationship will be linear. For instance, under prolonged, favourable soil water status, the relationship will be as in curve **a** Figure 1. Under extended periods of water stress, the relationship will take on a curvilinear form due to the episodic nature of leaf area adjustment in evergreen tree species (curve **b** in Figure 1). Following sufficient leaf area reduction, we predict that the new hydrological equilibrium will again be linear (curve **c** in Figure 1). With the return of more favourable soil water conditions, the relationship will be linear while leaf area recovers back to curve **a**.

Tests of the scaling theory are based on comparisons of the expected form of the flux/leaf area relationship against field data.

SINGLE TREE TO STAND FLUX SCALING THEORY

The flux of water Q through any stem in a stand of trees can be characterized by a gradient and associated resistances between the soil and atmosphere (Philip, 1957). An example of a detailed and explicit treatment of plant water relations is given by Landsberg and McMurtrie (1984). They proposed a model for water use by an *isolated* tree as follows:

$$\frac{dQ}{dt} = L_V(z,r)\delta V \kappa_r \theta(z,r,t)[\psi_s(z,r,t) - \psi_r(t)] \tag{2}$$

where dQ/dt = rate of water extraction by roots; $L_V(z,r)$ = rooting density at any point at radius r and depth z; δV = volume of toroid of radius r and depth z; κ_r = root conductance per unit root length; $\theta(z,r,t)$ = volumetric water content at (z,r) and time t; $\Psi_s(z,r,t)$ = soil water potential at z, r and time t; and $\Psi_r(t)$ = root water potential at time t.

These workers extended this equation to a system of N trees by specifying the rooting density of each tree at position x, y, z. This numerical approach provides an explicit, dynamic framework to model water at the stand level. However, this treatment requires an immense amount of *a priori* data characterizing the site at all points (x, y, z, t), making this approach impractical for application to the scaling question identified earlier. Equation (1) represents a simpler, more practical way to scale tree water use to the stand level.

Hydrological equilibrium theory (Grier and Running, 1977; Eagleson, 1982; Nemani and Running, 1989; Pierce *et al.*, 1993) suggests that in water-limited plant communities, a dynamic balance exists between leaf area, soil and climate. Empirical tests of this theory on coniferous forests (Grier and Running, 1977; Gholz, 1982; Nemani and Running, 1989) found strong relationships between stand leaf area (LAI) and site water balance. Results from competition (spacing) studies using apple trees (Atkinson *et al.*, 1976) indicated that *this equilibrium is valid not only for the leaf area index of a stand, but for the leaf areas of individual trees as well. This constitutes our first principal assumption.* For these apple trees, as competition for space increased, the individual root:shoot ratios remained constant, reflecting the fact that the above-ground biomass is constrained by the capacity of the root system to supply below-ground resources (cf. Landsberg and McMurtrie, 1984). This equilibrium is what Eagleson (1982) defined as short-term optimality and operates on the time-scale of one or more seasons, as opposed to diurnal time-scales (e.g. the time-scale of stomatal regulation) or long-term optimality (e.g. the time-scale of system evolution).

Ecological field theory (Walker *et al.*, 1989) states that the resources available to an individual plant are a function of the size of the domain D and the intensity F of the resource(s) within D. In well-stocked stands, the extent of D is dependent on the size of the plant as characterized by leaf area. Where domains of adjacent plants overlap, competition for resources exists and the local intensity within this overlap will tend to set a limit on the realized domain of the plant. Intensity is a dynamic quantity, which changes as a function of the inputs and outputs of some resource within the domain (Penridge *et al.*, 1987). We postulate that under the condition of hydrological equilibrium, a stand homogeneous with respect to climate, soils and topography will develop such that the spatial variability in F is minimized over time. *That is, each tree tends toward an equilibrium between its size (leaf area) and realized domain such that at no time resources available in local abundance within the stand remain untapped (e.g. the soil volume is equally accessed by roots). This constitutes our second principal assumption.* At equilibrium, the soil water potential experienced by each tree is identical and can be characterized by a mean value for the plot. This does not contradict observations by Ziemer (1968) or Eastham and Rose (1988) that soil water declines with distance from tree stems; our scaling theory is based on how the plants *perceive* moisture availability, not just the actual spatial distribution of soil water. We assume that soil evaporation operates uniformly in space from the plants' perspective. Further, we assume that understorey transpiration is governed by the same principals as tree water use, and its spatial distribution operates to minimize the spatial variance in available soil water.

Therefore, we follow this argument and characterize each domain on the basis of the volume of soil available to each tree. Equation (2) is a derivative expression, which needs integration over the entire rooting volume of a tree and over an extended period of time to obtain a flux equivalent to Q in Equation (1). We replace the toroidal volume by a total rooting volume V for simplicity and to facilitate the extension of a single tree to a stand. By this replacement, the mean rooting density for the stand is then the ratio of the total fine root biomass over V. For a given individual, V can be considered proportional to D^3. For the entire stand, the total volume of soil occupied by roots (V_T) is then proportional to the third power of the length scale L of the stand; we can express this relationship as

$$V_T = \lambda_V L^{f_V} \tag{3}$$

where λ_V is a proportionality constant with units $m^{(3-f_V)}$, where $f_V = 3$ for a cube and $f_V < 3$ for other solids. Likewise, total stand leaf area (A_T) is proportional to stand area

$$A_T = \lambda_A L^{f_A} \tag{4}$$

where $f_A = 2$ for a square and $f_A < 2$ for other two-dimensional shapes. Expressing V_T in terms of A_T by lumping the proportionality constants into λ, we have

$$V_T = \lambda A_T^f \tag{5}$$

where $f = f_V / f_A$ and $\lambda = 1 \, m^{(3-2f)}$.

Under prolonged water stress, adjustments to leaf area are paralleled by reductions in root biomass (Running and Gower, 1991). In the above formulation, we can characterize this behaviour as a change in rooting density or rooting volume. Treating the relative dimensions of leaf area and rooting volume as a fractal property, the exponent f in Equation (5) has an upper limit of 3/2; as root biomass decreases, the rooting volume can be conceived as approaching L^2, and thus the exponent in Equation (5) approaches unity.

At steady-state (i.e. the flux through the stem equals the flux through the roots), root water potential Ψ_R can be expressed in terms of leaf water potential Ψ_L by

$$\Psi_R = \kappa \Psi_L \frac{A}{\mu L_V V} \tag{6}$$

where κ is a dimensionless proportionality constant reflecting a normalized conductivity, μ is finite root surface area per unit root mass and Ψ_L is a function of the average energy available (intercepted) per unit leaf area (I)

$$\Psi_L = -k_L I \tag{7}$$

where k_L is a proportionality constant representing leaf (i.e. stomatal) conductivity; the negative sign arises from the water potential convention in which matrix water potentials less than saturated are expressed as less than zero.

The pressure gradient (Δp) between the soil and the roots, i.e. $\Delta P = (\Psi_S - \Psi_R)$, can therefore be expressed as

$$\Delta p = \Psi_s + kI \frac{A}{\mu L_V V} \tag{8}$$

where $k = \kappa k_L$. Flux can be expressed as

$$Q = \sigma \mu L_V \Delta p \tag{9}$$

where σ is flow rate per unit pressure difference and unit area. Substituting Equation (8) into Equation (9) we obtain

$$Q = \sigma [\mu L_v V \Psi_s + kIA] \tag{10}$$

As argued earlier, under the assumptions of hydrological equilibrium and ecological field theories, we can replace the local values L_V and Ψ_S by mean values for the entire stand.

Substituting V in Equation (10) by Equation (5) and lumping parameters into $a = \sigma k$ and $b = \sigma \mu \lambda L_V$ (units $m^3 \, MJ^{-1}$ and $m^{-(3-2f)} \, MPa^{-1} \, d^{-1}$, respectively), Equation (10) becomes

$$Q = aIA + b\Psi_s A^f$$

This equation expresses the expected relationship, at equilibrium, between the size (leaf area) of a tree and its expected water use, dependent on the average radiation and soil moisture availabilities. This equation holds for each tree in the stand and can be summed over all individuals to obtain stand water use. The development and nature of non-equilibrium conditions is presented in Appendix 1. A procedure to estimate a and b for a stand is presented in Appendix 2.

SUMMARY OF ASSUMPTIONS AND INTERPRETATIONS

The above derivation for a general relationship between tree water use and leaf area is based on a number of explicit and implicit assumptions. The scaling theory strictly holds only over sites which are homogeneous with respect to soil and climate. The site is idealized such that it is fully occupied (*sensu* field theory) and that it is always tending toward some hydrological equilibrium between available moisture and available energy, but that the time-scale for the realization of this equilibrium in terms of changes in leaf area is seasonal at best. Further, each individual is reacting to conditions in parallel with all other individuals in the stand; Hatton and Vertessy (1989) presented evidence to support this assertion in *Pinus radiata* plantations, as did Olbrich *et al.* (1993) in *E. grandis*, and we present here additional evidence for *Eucalyptus*. The implications of departures from these conditions on sampling variability are discussed in Appendix 2.

The simple demand function based on irradiance ignores advective contributions to evaporative demand. Thus, we assume equilibrium conditions (*sensu* McIlroy, 1984). More complex treatments of the canopy energy balance in non-equilibrium conditions may be substituted, but are not crucial to the derivation of our flux/leaf area relationship nor to its underlying assumptions or implications. The lumped parameter formulation presented here reduces the number of variables involved in scaling to allow a practical sampling framework.

In summary, the theory has a number of implications. Firstly, at any realized hydrological equilibrium, the relationship between tree water use and leaf area is linear. In a stand with unlimited soil water, the slope of this linear relationship will only be limited by intercepted irradiance (or total available energy) and the characteristic value of a for a given species. As soil water becomes limiting, the inability of trees to instantaneously adjust their leaf areas will result in a curvilinear relationship between flux and leaf area (see Appendix 1). Such a curve reflects hydrological disequilibrium, and cannot be sustained indefinitely. The degree to which a tree will hold onto its leaves under these conditions is species-dependent.

TESTS OF THE THEORY AGAINST FIELD DATA

Data from two widely separated sites were used to test model predictions regarding the form of the flux/leaf area relationship:

(1) A mixed species woodland (*E. polyanthemos*, *E. macrorhynca* and *E. blakelyi*) site at Lockyersleigh, New South Wales, as described by Kalma and Jupp (1990).
(2) A popular box (*E. populnea*) woodland site at Wycanna, Queensland, as described in Ross (1989).

Mixed species woodland data and results

Tree water use, meteorological data, soil moisture data and leaf area data were collected for three *Eucalyptus* species growing in the same paddock near Goulburn, New South Wales. The species were *E. blakelyi* (largest), *E. macrorhynca* and *E. polyanthemos* (smallest). Tree water use was measured by Sapflow Logging Systems (Greenspan Technology, Warwick, Queensland) and the data were collected over two-week periods in December 1992 and March 1993. Throughout the spring and summer of that year, soil moisture at the site was at or near field capacity. During this season, the trees of all the *Eucalyptus* species measured produced new leaves. Under these favourable conditions, the theory proposed in this paper predicts that the relationship between flux and leaf area should be linear at both these times.

Results shown in Figure 2 for both the sampling periods were linear, as predicted. Of additional interest is that different species of *Eucalyptus* fall on the same line, at least under these conditions. Leaf water

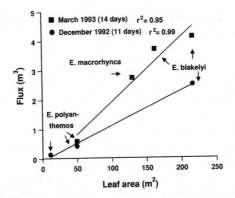

Figure 2. Flux/leaf area curves for December 1992 and March 1993 for a mixed woodland site near Goulburn, NSW. Soil moisture was non-limiting at both of these times. The three species of *Eucalyptus* fall on the same curve at both sampling times

efficiency was thus similar among *Eucalyptus* species, as for the species in the upslope plantings reported in Greenwood *et al.* (1985).

To test the implicit assumption that individual trees react in parallel to changes in the supplies of water and energy, the covariance among the 11 daily totals for tree water use for the three trees in the December 1992 sample was calculated. Daily flux ranged over a factor of three for all trees during this period. The linear coefficients of variation exceeded 0·94 for all of the three pairwise combinations of trees. Figure 3 shows the linear correlation in daily flux over two-week period between two trees (*E. blakelyi* and *E. polyanthemos*). This result is similar to that of Hatton and Vertessy (1989) and Olbrich *et al.* (1991), and indicates that trees at the same site do use water in proportion to one another over widely varying meteorological conditions. Data from these two sites and evergreen genera support one of the principal assumptions of our scaling theory: that sites can be approximated as spatially homogeneous with respect to radiation and soil water potential, and that the trees react in parallel to changes in ambient conditions.

Following the procedure outlined in Appendix 2, values for a were calculated for each sampling period for those days without rain. Values of $9{\cdot}7 \times 10^{-5}$ $(s = 1{\cdot}22 \times 10^{-5})$ and $9{\cdot}6 \times 10^{-5}$ $(s = 0{\cdot}71 \times 10^{-5})$ $m^3 MJ^{-1}$ were obtained for December and March data, respectively.

Poplar box data and results

Tree water use, meteorological data, soil moisture data and leaf area data were collected in April and October 1988; the former data were collected using Custom Heat Pulse Loggers (Aokoutere). The first

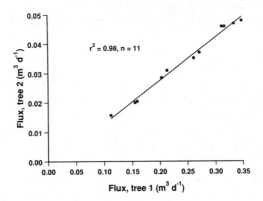

Figure 3. Correlation in daily water use by two species of *Eucalyptus* (*E. blakelyi* and *E. polyanthemos*) growing at the same site. Despite large differences in the leaf areas of the trees (216 versus 50 m², respectively), and the large range in the daily fluxes, the correlation is high. The same degree of correlation held over all pairwise linear correlations among the sample trees

of these periods was in April ($n = 12$ trees), at the end of a (southern hemisphere) summer of average rainfall. This resulted in a soil which was relatively moist at the surface (0–25 cm) from rain the preceding week, but dry at depth at the time of measurement; based on the soil moisture characteristic curve for this site reported in Hatton *et al.* (1993), the mean soil water potential to 3-m depth was estimated as -0.2 MPa. The second measurement period was in October (spring) of that year ($n =$ eight trees). At this time, the surface soil layer had dried, but there was substantial moisture at depth (mean $\Psi_S = -0.05$ MPa).

In April 1993, after an extended period of drought, the above measurements were repeated (using Greenspan Sapflow Logging Systems) for two plots of 10 and nine trees taken from the same stand measured in 1988 (as described in Moore *et al.*, 1993). The soil was dry throughout the profile (mean $\Psi_S = -1.0$ MPa). By this time, the canopy had dropped large numbers of leaves. Monitoring of the leaf area index of this site since 1982 indicated that a peak in LAI was reached in early 1988 (0.77), declining to 0.57 by October of that year and to 0.40 by April 1993.

These three sampling times should be characterized by predictable differences in the flux/leaf area relationship. The drying soil conditions in April 1988, in combination with a peak leaf area in the previous spring, should result in a state of hydrological disequilibrium (stress) and the form of the relationship should be as in curve **b**, Figure 1. Under the more favourable soil water conditions in October 1988, we would expect the form of the flux/leaf area relationship to be almost linear, as in curve **a**, Figure 1. Finally, soil moisture conditions in April 1993 were harsh. The trees had reacted to these conditions by massive leaf drop. By re-establishing an equilibrium between supply and demand in this fashion, the form of the flux/ leaf area curve should be more linear, as in curve **c**, Figure 1.

Parameters a and b in Equation (1) were fitted by a least-squares criterion to the data from October 1988, using values of $\Psi_S = -0.005$ MPa, $I = 7.0$ MJ m^{-2} day^{-1} and $f = 1.5$; the *a priori* value for f is consistent with the above argument based on peak leaf area index at this time. Values for a and b at this time were 8.7×10^{-5} m^3 MJ^{-1} and 3.2×10^{-4} MPa^{-1} day^{-1}, respectively. The value of a compares well with that found for the two sampling periods in the mixed woodland.

Using these values for a and b under the basic assertion that these constants are fixed for a given species, f was solved for using the April 1988 data and values of $I = 8.3$ MJ m^{-2} day^{-1} and $\Psi_S - 0.2$ MPa. The results (1.52) is consistent with theory, as is the increased curvature in the leaf efficiency relationship (Figure 4). Fixing $f = 1.5$ and solving for b resulted in a value $b = 3.2 \times 10^{-4}$ MPa^{-1} day^{-1}, which is close to the value estimated from the October data.

Using the October values for a and b and values $I = 12.0$ MJ m^{-2} day^{-1} and $\Psi_S = -1.0$ MPa, solving for f using the 1993 data resulted in $f = 1.21$ and 1.18 for plots 1 and 2, respectively. That this scaling factor should tend towards unity following considerable loss of leaf area under water stress is consistent with theory.

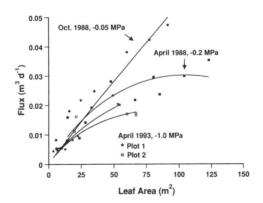

Figure 4. Flux/leaf area curves for April 1988, October 1988 and April 1993 for a poplar box woodland site

DISCUSSION

The theoretical scaling relationship derived in this paper from the principles of hydrological equilibrium, ecological fields and plant water relations is consistent with the measured flux/leaf area relationships reported here. Further, the relationships seem consistent with others reported elsewhere. The non-linear response for *E. wandoo* trees reported in Greenwood *et al.* (1982) for dry, summer conditions contrasts appropriately with the linear response for *Eucalyptus* species growing on a hillslope with access to shallow groundwater (Greenwood *et al.*, 1985). In the latter paper, the upslope trees were permanently phreatophytic; the variation in leaf efficiencies among species in a plantation near a saline seep was later ascribed to heterogeneity in soil water supply (Greenwood *et al.*, 1992).

The broad implication from this study is that estimates of tree water use extrapolated on the basis of leaf area can be affected by temporally variable, non-linear stress responses, especially water stress. In evergreen systems that are at least seasonally water-limited (semi-arid, Mediterranean, dry tropics, savanna systems), the duration of this water-stressed period may extend over much of the year. Thus any assumption of linearity (equivalency in leaf efficiency among individuals differing in size) will not generally hold for these systems.

The reason that these systems become non-linear in the flux/leaf area response is due to the fact that the required change in leaf area under increasing water stress is not instantaneous; there is a time lag. Responses to water stress in trees at the gross level involve losses in leaf water efficiency (conductance) or the leaves themselves. Over short time-scales (hours), changes in leaf conductance are largely sufficient to maintain the balance between transpiration, assimilation, respiration and root water uptake. To continue this strategy over longer time-scales (weeks or months) comes at the cost of reduced leaf efficiency with respect to net carbon assimilation (canopy respiration may exceed assimilation), or even leaf fatality due to high leaf temperatures. Following a prolonged period of stress, evergreen trees adjust leaf area to better match the available soil water with available energy (the new hydrological equilibrium). Such behaviour is well documented in *Eucalyptus* (Pook, 1985) and in *Quercus* (Griffin, 1973). The amount of leaf drop and the time lag at a given level of water stress varies among plant growth forms and species.

As a consequence of this argument, the degree of curvature in the flux/leaf area relationship is a direct measure of the quasi-equilibrium (stress) that the stand is under at its present leaf area. If this is true, then perhaps this is a better measure of the apparent stress of a stand than an arbitrary measure of soil water deficit. This is because it is stress as perceived by the trees.

Alternative explanations for observed non-linear relationships between leaf area and flux were considered. The most obvious alternative is that as trees develop increasing leaf areas, self-shading of leaves reduces the mean leaf efficiencies relative to smaller trees. Two lines of logic argue against this interpretation. The first of these is that, at least for *Eucalyptus*, tree architecture and leaf angles minimize self-shading. Secondly, the results presented in this paper show that the relationship changes from linear to curvilinear for the same stand of trees under varying conditions, but with similar leaf area indices. Clearly, the theory presented in this paper is limited to stands without above-ground interference; dominant/sub-dominant competition for light violates the theory's conceptual basis.

The documented dynamics of leaf areas of evergreen trees, especially under seasonal drought stress, has implications for extrapolating tree water use by other scalars. Good correlations between sapwood area and leaf area were reported by a number of investigators (Grier and Waring, 1974; Rogers and Hinckley, 1979; Kauffmann and Troendle, 1981; Brack *et al.*, 1985). However, relationships of this nature cannot be static if leaf areas are adjusting to stress at rates different to those of sapwood adjustments. For instance, if a stand has been under a favourable, extended hydrological equilibrium going into a drought, the time lag to leaf drop will be much shorter than any adjustment in sapwood area. Use of a relationship between leaf and sapwood areas to scale tree water use is likely to be unreliable over changing conditions. Thorburn *et al.* (1993) indicated the non-unique sapwood–sapflow velocity relationship over the three seasons reported above; in summer 1991, there was a poor correlation between sap flux density and sapwood area, but there was a good linear relationship in winter 1991 and autumn 1992.

If a particular application is limited to extrapolating tree water use to stand water use at a single time and over a homogeneous site, then scalars other than leaf area may be as accurate and are likely to be easier to measure. For instance, Moore *et al.* (1993) found that tree diameter was an accurate, linear scalar of flux for a stand of *Eucalyptus populnea*, ad was of comparable accuracy to leaf area or conducting wood area. The limitation to this relationship, however, is the lack of any predictive framework for accounting for changes in soil water availability or climate.

The properties of the scaling equation allow the practical estimation of parameters. The lumped values *a* and *b* can be estimated by simple graphical techniques, as explained in Appendix 2. Under wet soil conditions, the parameter *a* establishes the slope of the equilibrium curve (leaf efficiency) at a given intercepted irradiance. We speculate that *a* may vary little among species within a genus, as indicated in Figure 4, where the *Eucalyptus* species investigated conformed to the same equilibrium curve under similar ambient conditions. The results in Greenwood *et al.* (1985) provide similar evidence for the generality of *a* among species, at least those growing at a similar site. If this is true, then the *potential* water use by a stand of trees is limited much more by their leaf area than by species differences in leaf efficiency (but see Honeysett *et al.*, 1992). Actual water use may vary due to differences in temperature optima, leaf angle, the sensitivity of conductance to the vapour pressure deficit and how water potential is perceived (e.g. rooting depth, sensitivity to salinity, etc.), especially under stress. A constant value for *b* was sufficient for characterizing the leaf efficiency behaviour of poplar box under varying conditions.

After a prolonged drought, evidence was found to confirm the theorized dynamics in *f*; the scaling factor that relates rooting volume and leaf area was reduced from 3/2 towards unity following large reductions in the leaf areas of the trees. The theory and its associated equations and application could be substantially simplified were we not to account for the effects of catastrophic leaf drop following extended water stress. In this instance, *f* would remain constant (3/2) and the dynamic fractal relationship involving a constant rooting density, but a variable volume space, need not be invoked. The theory would remain largely sufficient for accounting for tree water use behaviour under normal seasonal variation.

Departures from the assumptions of plot homogeneity and equilibrium will tend to increase the sampling variability in parameters *a* and *b*. Spatial variability in soil water and irradiance can be accounted for at the landscape scale by characterizing the site into spatial elements homogeneous with respect to water availability and radiation; catchment models such as TAPES-G (Moore and Grayson, 1991) or TOPOG (O'Loughlin, 1990; Hatton *et al.*, 1992; Dawes and Hatton, 1993) have potential in this regard. Local disturbance such as thinning, defoliation by insects, or fire can impose local disequilibria and increase the variance in the flux/leaf area relationship as described in Appendix 1.

The increasing interest in the role of trees in the surface water balance, and the proliferation of measurements of tree water use, raise the need for an interpretive framework for flux measurements and their extrapolation in space which is both consistent with theory and practical. Strong physical and physiological reasonings suggest that leaf area must ultimate scale plant water use, and argue for a scaling framework based directly on this factor. Under low plant water stress, it appears that the relationship between flux and leaf area is linear. Scaling to stand water use in this instance is trivial. However, empirical evidence presented in this paper and elsewhere suggests that under water stress, linearity cannot be assumed. The theory presented here provides a predictive framework which accounts for variations in the flux/leaf area relationship as a function of irradiance, soil water potential, two lumped parameters and a scaling factor. We provide a simple sampling protocol to determine these latter parameters, which we consider as species characteristics. With values for these coefficients, extrapolation of tree water use to stand water use is straightforward. With a knowledge of the spatial distributions of soil water potential and irradiance across heterogeneous landscapes, and local leaf area distributions, scaling of tree water use measurements to the catchment scale is also possible.

ACKNOWLEDGEMENTS

The authors thank J. Walker, B. Li and W. Dawes for assistance in the development of the theoretical arguments. E. Greenwood and G. Bartle provided the 1988 sapflow data and S. Moore and P. Reece assisted with the 1993 data collection. The authors also thank H. Alksnis and P. Daniel for meteorological data from the mixed woodland site. The second author acknowledges the support of NSF Grant BSR-9109240, the US/Australian Fullbright Foundation, and the Texas A&M University Faculty Development Leave Program.

REFERENCES

Atkinson, D., Naylor, D., and Coldrick, G. A. 1976. 'The effect of tree spacing on the apple root system', *Hort. Res.* **16**: 89–105.
Atkinson, D. 1980. 'The distribution and effectiveness of the roots of tree crops', *Horticult. Rev.*, **2**, 425–490.
Brack, C. L., Dawson, M. P., and Gill, A. M. 1985. 'Bark, leaf and sapwood dimensions in Eucalyptus', *Aust. Forest Res.*, **15**, 1–7.
Cermak, J., and Kucera, J. 1987. 'Transpiration of mature stands of spruce (*Picea abies* (L.) Karst.) as estimated by the tree-trunk heat balance method'. In *Forest Hydrology and Watershed Management. IAHS–AISH Publ.*, **167**, 311–317.
Dawes, W. R., and Hatton, T. J. 1943. 'TOPOG_IRM. Model description', *CSIRO Tech. Memo.*, **93/5**, Canberra, CSIRO, Australia.
Denmead, O. T. 1984. 'Plant physiological methods for studying evapotranspiration: problems of telling the forest from the trees', *Agric. Wat. Manage.*, **8**, 167–190.
Doley, D., and Grieve, B. J. 1966. 'Measurement of sapflow in *Eucalyptus* by thermoelectric methods', *Aust. Forest Res.*, **2**, 3–27.
Eagleson, P. S. 1982. 'Ecological optimality in water limited natural soil–vegetation systems. 1. Theory and hypothesis', *Wat. Resour. Res.*, **18**, 325–340.
Eastham, J., and Rose, C. W. 1988. 'The effect of tree spacing on evaporation from an agroforestry experiment', *Agric. Forest Meteorol.*, **42**, 355–368.
Gholz, H. L. 1982. 'Environmental limits on aboveground net primary production, leaf area, and biomass in vegetation zones of the Pacific Northwest', *Ecology*, **63**, 469–481.
Greenwood, E. A. N., and Beresford, J. D. 1979. 'Evaporation from vegetation in landscapes developing secondary salinity using the ventilated-chamber technique. I. Comparative transpiration from juvenile *Eucalyptus* above saline groundwater seeps', *J. Hydrol.*, **42**, 369–382.
Greenwood, E. A. N., Beresford, J. D., Bartle, J. R., and Barron, R. J. W. 1982. 'Evaporation from vegetation in landscapes developing secondary salinity using the ventilated-chamber technique. II. Evaporation from a regenerating forest of *Eucalyptus wandoo* on land formerly cleared for agriculture', *J. Hydrol.*, **58**, 357–366.
Greenwood, E. A. N., Klein, L., Beresford, J. D., and Watson, G. D. 1985. 'Differences in annual evaporation between grazed pasture and *Eucalyptus* species in plantations on a saline farm catchment', *J. Hydrol.*, **78**, 261–278.
Greenwood, E. A. N., Milligan, A., Biddiscombe, E. F., Rogers, A. L., Beresford, J. D., Watson, G. D., and Wright, K. D. 1992. 'Hydrologic and salinity changes associated with tree plantations in a saline agricultural catchment in southwestern Australia', *Agric. Wat. Manag.*, **22**, 307–323.
Grier, C. C., and Running, S. W. 1977. 'Leaf area of mature conifrous forests: relation to site water balance', *Ecology*, **58**, 893–899.
Grier, C. C., and Waring, R. H. 1974. 'Conifer foliage mass related to sapwood area', *Forest Sci.*, **20**, 205–206.
Griffin, J. R. 1973. 'Xylem sap tension in three woodland oaks of central California', *Ecology*, **54**, 152–159.
Hatton, T. J., and Vertessy, R. A. 1989. 'Variability in sapflow in a *Pinus radiata* plantation and the robust estimation of transpiration' In *Hydrology and Water Resources Symposium, Christchurch*. pp. 6–10. Institution of Engineers, Australia No 89/19.
Hatton, T. J., and Vertessy, R. A. 1990. 'Transpiration of plantation *Pinus radiata* estimated by the heat pulse method and the Bowen ratio', *Hydrol. Process.*, **4**, 289–298.
Hatton, T. J., Walker, J., Dawes, W. R., and Dunin, F. X. 1992. 'Simulations of hydroecological responses to elevated CO_2 at the catchment scale', *Aust. J. Bot.*, **40**, 679–696.
Hatton, T. J., Pierce, L. L., and Walker, J. 1993. 'Ecohydrological changes in the Murray–Darling Basin. II. Development and tests of a water balance model', *J. Appl. Ecol.*, **30**, 274–282.
Honeysett, J. L., Beadle, C. L., and Turnbull, C. R. A. 1992. 'Evapotranspiration and growth of two contrasting species of eucalypts under non-limiting and limiting water availability', *Forest Ecol. Manage.*, **50**, 203–216.
Kalma, J. D., and Jupp, D. L. B. 1990. 'Estimating evaporation from pasture using infrared thermometry: evaluation of a one-layer resistance model', *Agric. Forest Meteorol.*, **51**, 223–246.
Kauffmann, M. R., and Troendle, C. A. 1981. 'The relationship of leaf area and foliage biomass to sapwood conducting area in four subalpine forest tree species', *Forest Sci.*, **27**, 477–482.
Kelliher, F. M., Kostner, B. M. M., Hollinger, D. Y., Byers, J. N., Hunt, J. E., McSeveny, T. M., Meserth, R., Weir, P. L., and Schulze, E. D. 1992. 'Evaporation, xylem sap flow, and tree transpiration in a New Zealand broad-leaved forest', *Agric. Forest Meteorol.*, **62**, 53–73.
Ladefoged, K. 1963. 'Transpiration of forest trees in closed stands', *Physiologia Plantarum*, **16**, 378–414.
Landsberg, J. J., and McMurtrie, R. 1984. 'Water use by isolated trees', *Agric. Wat. Manage.*, **8**, 223–242.
McIlroy, I. C. 1984. 'Terminology and concepts in natural evaporation', *Agric. Wat. Manage.*, **8**, 77–98.
McMurtrie, R. E., Rook, D. A., and Kelliher, F. M., 1990 'Modelling the yield of *Pinus radiata* on a site limited by water and nitrogen', *Forest Ecol. Manage.*, **30**, 381–413.
Moore, I. D., and Grayson, R. B. 1991. 'Terrain-based catchment partitioning and runoff prediction using vector elevation data', *Wat. Resour. Res.*, **27**, 1177–1191.
Moore, S., Hatton, T. J., and Reece, P. R. 1993. 'Estimating stand transpiration via the heat pulse method', In *Water Issues in Forests Today, Poster Abstracts. International Forest Hydrology Symposium, Canberra*. pp. 45–46.

Nemani, R. R., and Running, S. W. 1989. 'Testing a theoretical climate–soil–leaf area hydrologic equilibrium of forests using satellite data and ecosystem simulation', *Agric. Forest Meteorol.*, **44**, 245–260.

Olbrich, B. W., Le Roux, D., Poulter, A. G., Bond, W. J., and Stock, W. D. 1993. 'Variation in water use efficiency and ^{13}C levels in *Eucalyptus grandis* clones', *J. Hydrol.*, **150**, 615–633.

O'Loughlin, E. M. 1990. 'Modelling soil water status in complex terrain', *Agric. Forest Meteorol.*, **50**, 23–28.

Penridge, L. K., Walker, J., Sharpe, P. J. H., Spence, R. D., Wu, H., and Zou, G. 1987. 'RESCOMP: A resource competition model to simulate the dynamics of vegetation cover', *CSIRO Div. Wat. Land Resour. Tech. Memor.*, **87/5**, CSIRO, Canberra.

Philip, J. R. 1957. 'Evaporation, moisture and heat fields in the soil', *J. Meteorol.*, **14**, 354–366.

Pierce, L. L., Walker, J., Dowling, T. I., McVicar, T. R., Hatton, T. J., Running, S. W., and Coughlan, J. C. 1993. 'Ecohydrological changes in the Murray–Darling Basin III. A simulation of regional hydrological changes', *J. Appl. Ecol.*, **30**, 283–294.

Pook, E. W. 1985. 'Canopy dynamics of *Eucalyptus maculata* Hook. III. Effects of drought', *Aust. J. Bot.*, **33** 65–79.

Rogers, R., and Hinckley, T. M. 1979. 'Foliar weight and area related to current sapwood in oak', *Forest Sci.*, **25**, 298–303.

Ross, D. J. 1989. 'Soils of the Wycanna Woodlan Experiment Centre, Talwood, South Maranoa Region, Queensland', *CSIRO Div. Soils Rep. No. 83.*

Running, S. W. and Coughlan, J. C. 1988. 'A general mode of forest ecosystem processes for regional applications', *Ecol. Modelling*, **42**, 125–154.

Running, S. W., and Gower, S. T. 1991. 'FOREST-BGC, a general model of forest ecosystem processes for regional applications. II. Dynamic carbon allocation and nitrogen budgets', *Tree Physiol.*, **9**, 147–160.

Sziecz, G., and Long, I. F. 1969. 'Surface resistance of crop canopies', *Wat. Resour. Res.*, **5**, 622–633.

Tan, C. S., and Black, T. A. 1976. 'Factors affecting the canopy resistance of a Douglas-fir forest', *Boundary-Layer Meteorol.*, **10**, 475–488.

Thorburn, P. J., Hatton, T. J., and Walker, G. R. 1993. 'Combining measurements of transpiration and stable isotopes of water to determine groundwater discharge from forests', *J. Hydrol.*, **150**, 589–614.

Walker, J., Sharpe, P. J. H., Penridge, L. K., and Wu, H. 1989. 'Ecological field theory: the concept and tests', *Vegetatio* **83**, 81–95.

Werk, K. S., Oren, R., Schulze, E.-D., Zimmermann, R., and Meyer, J. 1988. 'Performance of two *Picea abies* (L.) Karst. stands at different stages of decline. III. Canopy transpiration of green trees', *Oecologia*, **76**, 519–524.

Ziemer, R. R. 1968. 'Soil moisture depletion patterns around scattered trees', *USDA Forest Serv. Res. Note PSW-166*, Pacific Southwest Forest and Range Experiment Station, 13 pp.

APPENDIX 1: LEAF STRESS AND CHANGES IN LEAF AREA

The instantaneous flux through a stem can be expressed as

$$Q_P = \frac{\Psi_r - \Psi_L}{R_P}$$

where the subscripts P, r and L refer to stem, root and leaves, respectively, and R is resistance. This equation is analogous to Ohm's law. Likewise, the instantaneous flux across the fine root surface can be written as

$$Q_r = \frac{\Psi_S - \Psi_r}{R_r}$$

where Ψ_S is soil water potential. At steady-state, $Q_P = Q_r$, and thus

$$\Psi_r = \frac{R_P \Psi_S + R_r \Psi_L}{R_P + R_r}$$

Assuming that Ψ_L will not be able to respond instantaneously, a slight change in Ψ_S will result in a change in Ψ_r as

$$\Delta\Psi_r = \frac{R_P}{R_P + R_r} \Delta\Psi_S$$

At steady state, the water potentials at the leaf and root levels are tightly related by a relationship analogous to a hydrostatic equilibrium, i.e.

$$\kappa \Psi_L A = \mu L_\nu V \Psi_r$$

where κ is a dimensionless proportionality constant reflecting a normalized stem conductivity and μ is the fine root surface area per unit root mass. The right-hand side of the above equation can be considered as the total pressure experienced at the root level. Thus a drop in Ψ_r will yield the following response:

$$\kappa(A\Delta\Psi_L + \Psi_L\Delta A) = \mu(L_V V\Delta\Psi_r + V\Psi_r\Delta L_V)$$

This equation reflects that a drop in Ψ_r (associated with a decrease in Ψ_S) can only be countered by an increase in fine root density L_V or a decrease in leaf area A (long-term response), or a change in leaf water potential by closing stomates (short-term response). While closing stomates is a short-term solution to this problem, it is at the cost of lower carbon assimilation, and (potentially fatal) higher leaf temperatures. Similarly, under stress, fine roots may increase to some degree, but we expect this response to be small. Thus, under extended periods of soil water stress, under the hydrological equilibrium theory, we expect changes in leaf area to dominate this reponse.

Plants cannot respond to these changes in (soil) water potential instantaneously. Eucalyptus trees, for example, tend to exhibit large and spontaneous leaf reductions after prolonged stress in an effort to readjust the equilibrium between supply of water to the roots and demand at the leaves (Pook, 1985). The above equation must therefore be viewed as a cumulative stress until such a time as the trees shed leaves.

Differentiating the flux Equation (1) of the text, we have

$$dQ = aI dA + bA^f d\Psi_s + b\Psi_s A^f \ln A df + bf A^{f-1}\Psi_s dA$$

This equation reflects the fact that a drop in Ψ_S, accompanied by an instantaneous decrease in leaf area, can retain the linear characteristics of the flux/leaf area relationship. As plants cannot generally respond with instantaneous changes in leaf area, the build-up of stress relative to the hydrological equilibrium will result in a curvilinear relationship between flux and leaf area. Without relief in the soil water deficit, we postulate that this stress will ultimately result in (perhaps catastrophic) leaf shedding. The degree of which a particular individual can maintain this stressed status will vary among species.

APPENDIX 2: ESTIMATING LUMPED PARAMETERS a AND b

The derivative of the flux equation [Equation (1) in text] with respect to A gives

$$\frac{dQ}{dA} = aI + f b\Psi_s A^{f-1}$$

At a time of no moisture stress (i.e. $\Psi_S = 0$), coefficient a can be obtained as the slope of a linear regression between Q and A at a given temperature and irradiance. Once a is determined, the parameter b can be estimated by the slope, dQ/dA, intercepted irradiance, soil water potential and leaf area according to the above equation; by sampling at a time of high leaf area index, f may be assumed to be 1·5. Note that irradiance I need not be the same between data sets collected to estimate a and b. Knowing a and b, the scaling factor f may be solved for in subsequent data sets either to test its expected value of 1·5 under prolonged favourable conditions, or its degree of reduction under prolonged stress.

Under strict spatial uniformity in radiation and soil moisture, few data points are necessary to estimate a. Departures from the latter assumption may increase under drying conditions, and thus the number of samples associated with estimating b may need to be greater. Note that the factor $\lambda_i\Psi_S$ can be interpreted as the local soil moisture field intensity (*sensu* Walker *et al.*, 1989), i.e.

replacing λ_i by $A_i^{f_i - f}$

$$\Psi_{s(i)} = \frac{A_i^{f_i}}{A_i^{f}} \Psi_s$$

where A_i is the leaf area of the ith tree and f_i is the corresponding local exponent. Thus, at equilibrium, all $f_i = f$ and there is no spatial variability in the availability of water to the trees. Where the stand has been locally disturbed, the local size/domain relationship will not be in equilibrium and the local f_i will vary from f and the local $\Psi_{S(i)}$ will depart from the mean stand Ψ_S.

17

TOWARDS A CATCHMENT-SCALE MODEL OF SUBSURFACE RUNOFF GENERATION BASED ON SYNTHESIS OF SMALL-SCALE PROCESS-BASED MODELLING AND FIELD STUDIES

JUMPEI KUBOTA

Department of Environmental Science and Natural Resources, Tokyo University of Agriculture and Technology, Fuchu, Tokyo 183 Japan

AND

MURUGESU SIVAPALAN

Centre for Water Research, University of Western Australia, Nedlands, Western Australia 6009, Australia

ABSTRACT

Methodologies for developing a macro-scale model of subsurface stormflow generation on a steep forested catchment in Japan are addressed. Field studies on this catchment have indicated that subsurface flow, consisting mainly of 'old' water displaced by 'new' rain water, dominates the storm response. Detailed field measurements on the catchment allowed simple, catchment-scale relationships to be developed between the volume of saturated groundwater, on the one hand, and discharge, and surface and subsurface saturated areas, on the other. Attempts are made to find linkages between these empirical relationships and physically-based descriptions of hydrological processes operating at smaller scales. A distributed model based on the saturated–unsaturated groundwater flow on steep catchments was developed and tested with field data collected on this catchment. Derivation of catchment-scale relationships can be carried out by a straightforward integration of the distributed model output. An alternative disaggregation–aggregation approach is presented, whereby the catchment is divided into a number of hillslope flow strips. By applying the distributed model on each flow strip, under steady-state conditions, it is possible to infer the spatial variability of groundwater volume in the various flow strips. This variability is related to a measure of hillslope topography and geometry. Knowing the catchment topography and geometry for any catchment, it is then possible to derive useful catchment-scale relationships which are applicable under quasi-steady-state conditions. Hydrological relationships derived in this manner are compared with corresponding empirical relationships. The methodologies presented are effective for understanding the linkages between catchment scale response and small-scale flow processes.

INTRODUCTION

The dominant mechanisms of runoff generation operating at a given location depend strongly on environmental conditions such as climate, soils and topography. Dunne (1983) presented a diagram which expressed the three main runoff generation mechanisms and their controlling factors, i.e. climate, topography, soil type, vegetation and land use. Surface runoff consisting of the infiltration excess (Horton-type) and saturation excess (Dunne-type) mechanisms tends to dominate in flat to moderately steep terrain and relatively arid to semi-arid climates. On the other hand, forest hydrologists have emphasized the importance of the subsurface runoff mechanism, especially in regions which have well vegetated and steep hillslopes, experiencing heavy rainfall. Forested catchments in Japan are typical of the regions where the subsurface runoff mechanism dominates storm runoff. In this study, we will focus on the subsurface runoff mechanism, from the point of view of developing macro- or catchment-scale models of runoff generation.

Indeed, we seek an understanding of catchment response, with respect to subsurface stormflow, so that we can build simple models at the catchment scale. However, we like to do this without losing important hydrological information about processes occurring within the catchment. In other words, we want to be able to integrate our knowledge of hydrological processes operating at the small scale to develop models at catchment scale. In this paper, we limit ourselves to the study of the links between catchment-scale hydrological response functions, for e.g. discharge versus volume of saturated groundwater, and small-scale flow process descriptions based, for example, on Darcy's law. The question of using these catchment-scale relations to build catchment-scale rainfall–runoff models is not addressed in this paper.

PREVIOUS WORK ON SUBSURFACE RUNOFF IN JAPANESE CATCHMENTS

The geography of Japan is characterized by narrow islands and steep land surfaces, with over 70% of the mountainous area being forested. The steep topography, combined with heavy rainfall from typhoons and other frontal systems, produces considerable erosion, often leading to landslides. The flow of debris can be so great as to lead to major disasters. As landslides are closely related to water movement within soils, Japanese forest hydrologists have been studying subsurface runoff processes by both field observations and theoretical analyses.

Modeling of subsurface flow on forested hillslopes

Since Richards (1931) proposed a description of water flow in soils as a combination of an equation of continuity with Darcy's law, the Richards' equation has been extended to unsaturated flows as well, and remains useful for the theoretical analysis of flows within hillslopes. To clarify the properties of base flow recessions on forested, mountainous catchments, Suzuki (1984) proposed a non-dimensional form of a fundamental equation by combining Richards' equation with functional relationships describing soil–water properties. On a sloping, orthogonal soil layer, as an approximation to steep, highly permeable topsoil layers on forested slopes illustrated in Figure 1, the two-dimensional saturated–unsaturated flow within soils is described by

$$C\frac{\partial \psi}{\partial t} = \frac{\partial}{\partial x}\left\{K(\psi)\left(\frac{\partial \psi}{\partial x} + \sin \omega\right)\right\} + \frac{\partial}{\partial z}\left\{K(\psi)\left(\frac{\partial \psi}{\partial z} + \cos \omega\right)\right\} \tag{1}$$

where x is the upslope axis, z is the axis normal to x, ψ is the soil water pressure head, $K(\psi)$ is the hydraulic conductivity; ω is the slope angle, C is the specific water capacity, and t is time.

In the case of the soil-water property relating the water content and pressure head, the following

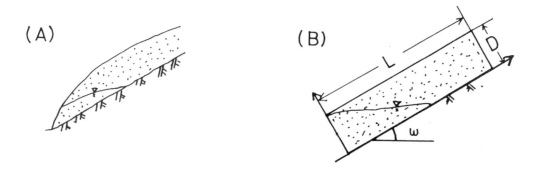

Figure 1. (A) Soil profile on a slope and (B) definition diagram for subsurface flow model on a hillslope (after Suzuki (1984), modified)

functional relationship proposed by Tani (1982) is used

$$C = \frac{\partial \theta}{\partial \psi} = -\frac{w\psi'}{\psi_0^2} \exp\left(-\frac{\psi'}{\psi_0}\right) \tag{2}$$

$$\theta = w\left(\frac{\psi'}{\psi_0} + 1\right) \exp\left(-\frac{\psi'}{\psi_0}\right) + \theta_r \tag{3}$$

$$S = \frac{\theta - \theta_r}{w} = \left(\frac{\psi'}{\psi_0} + 1\right) \exp\left(-\frac{\psi'}{\psi_0}\right) + \theta_r \tag{4}$$

$$w = \theta_s - \theta_r \tag{5}$$

where

$$\psi' = \begin{cases} \psi & (\psi \leqslant 0) \\ 0 & (\psi > 0) \end{cases},$$

ψ' is the pressured head, and θ is the volumetric water content, θ_s is the saturated water content; θ_r = residual water content, ψ_0 is the bubbling pressure head at air entry, w is the effective porosity, and S is the relative saturation. For the relationship between hydraulic conductivity and pressure head, the following functional form proposed by Brutsaert (1966) is used

$$K(\psi) = K_s S^\beta \tag{6}$$

where K_s is the saturated hydraulic conductivity at ψ and β is the decay parameter of hydraulic conductivity with relative saturation.

Equation (1) is transformed to non-dimensional form by using the non-dimensional variables

$$-\psi_*' \exp(-\psi_*')\frac{\partial \psi_*}{\partial t_*} = \frac{\partial}{\partial x_*}\left\{(\psi_*' + 1)^\beta \exp(-\psi_*' \cdot \beta)\left(\alpha\frac{\partial \psi_*}{\partial x_*} + 1\right)\right\}$$
$$+ \frac{1}{\gamma^2}\frac{\partial}{\partial z_*}\left\{(\psi_*' + 1)^\beta \exp(-\psi_*' \cdot \beta)\left(\alpha\frac{\partial \psi_*}{\partial z_*} + \delta\right)\right\} \tag{7}$$

$$x_* = x/L \tag{8}$$

$$z_* = z/D \tag{9}$$

$$t_* = t \cdot K_S \cdot \sin\omega/wL \tag{10}$$

$$\psi_* = \psi/\psi_0 \quad (\psi_*' = \psi'/\psi_0) \tag{11}$$

where D is the soil depth and L is the slope length.

The physical conditions of the catchment are characterized by the four non-dimensional parameters

$$\alpha = \psi_0/(L \cdot \sin\omega) \tag{12}$$

$$\gamma = D/L \tag{13}$$

$$\delta = D/L \cdot \sin\omega \tag{14}$$

α and β are the parameters that describe the characteristics of saturated–unsaturated flow and γ and δ are topographic parameters related to the slope length, soil depth and slope gradient. Suzuki (1984) examined the sensitivity of non-dimensional hydrographs obtained by numerical solutions of the governing Equation (1) to these non-dimensional parameters and suggested that the fundamental Equation (1) can be simplified to a form having only the parameter β for the the usual topographies and soil–water properties typical of steep forested catchments in Japan. This simplified form is given by

$$S = kq^p \tag{15}$$

$$wD \frac{\partial S}{\partial t} = \frac{\partial q}{\partial x} \tag{16}$$

where q is the flux per unit width and k and p are constant coefficients

$$k = \left(\frac{1}{D \cdot K_s \cdot \sin \omega}\right)^p, \qquad p = \frac{1}{\beta} \tag{17}$$

The simplified Equations (15), (16) and (17) indicate that capillary forces are negligible in comparison with gravitational forces. Suzuki (1984) showed, using non-dimensional simulations, that this assumption is justified for steep-forested slopes with relatively thin and highly permeable topsoil layers. Kubota *et al.* (1987) and Kubota and Suzuki (1988) supported this theoretical result by field observations in small forested catchments in western Japan. The intensive observations were made of the soil water potential field within the forest soils during several storms. The flux calaculations on steep forested slopes using the observed pressure head fields suggest that the downward flux along the slopes are approximated fairly well by the simplified form of the governing equation given above, which ignores capillary forces. This was attributed to two key factors: the steepness of the hillslope and the rapid decay of the soil hydraulic

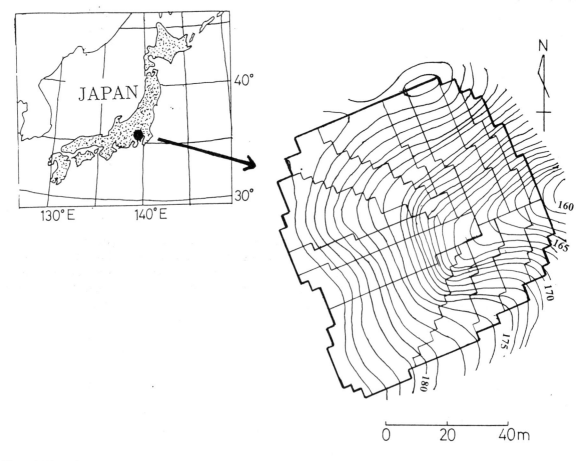

Figure 2. Map of the study area. Outer bold line is the boundary of the main catchment and inner thin lines are the boundaries of the flow strips

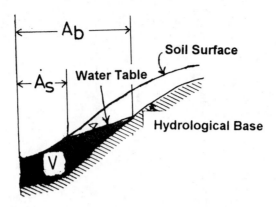

Figure 3. Schematic diagram describing source area (surface saturated area) A_s, subsurface saturated area A_b and saturated volume V

conductivity with soil moisture under unsaturated conditions. On the basis of these field observations and theoretical studies, Kubota *et al.* (1987; 1988) proposed a physically-based long-term water balance model by including a simple evapotranspiration sub-model. Their model is able to predict not only the flow hydrographs, but also the change in soil moisture storage everywhere within a catchment. The model considers the two-dimensional saturated–unsaturated flow on a grid-based topographic map by using the simplified form of the governing equation [Equations (15) and (16)].

These combined research efforts on subsurface runoff generation represent an example where process-based numerical modelling and field studies are used interactively to progress towards a simple, distributed model of runoff generation at catchment scale. This is the model that will be used in the simulations described in the following. The objective then is to develop an understanding of the causes of heterogeneity of runoff generation within a catchment and to quantify it in such way that a catchment-scale model can then be derived by aggregation.

Field observations on Hakyuchi catchment

Intensive field observations using about 100 tensiometers and observation wells were carried out on the Hakyuchi catchment, a small forested catchment in the tertiary area of Tama-hill near Tokyo, to investigate catchment-scale subsurface runoff response. The catchment area is 5200 m^2 and covered with deciduous trees. Figure 2 shows the location and topography of the Hakyuchi catchment. Figure 3 is a schematic diagram of a typical hillslope profile assumed for this catchment, providing definitions for source area, A_s (surface saturated area), subsurface saturated area, A_b, and saturated volume, V.

Ohta (1990) measured discharge and saturated volume, i.e. the volume of saturated groundwater above a 'hydrological base', on the Hakyuchi catchment during many storm events. He surveyed the structure of the soil mantle of the catchment by the cone penetration test proposed by the Public Works Research Institute, the Japanese Ministry of Construction. The N_c value is the number of hits on the cone to penetrate 10 cm of the soil mantle. He defined the hydrological base as that having $N_c = 30$. He found that the discharge of the catchment depends more strongly on the saturated groundwater volume than on source area. Figure 4(A) shows the relationship between discharge and saturated volume during four storm events, demonstrating the dependence of discharge on saturated volume. On the basis of this observational result, Ohta (1990) proposed that the 'variable source volume concept' is more appropriate than the 'variable source area concept' to describe runoff generation on steep forested catchments. During this study, we found further evidence which indicates that, in addition to discharge, both source area, A_s, (surface saturated area) and subsurface saturated area, A_b, also depend strongly on the saturated groundwater volume. This is presented in Figure 4(B); note, however, that the data for Figure 4(B) is available for only one storm event, that of 24 September 1988.

Kubota (in prep.) also carried out hydrograph separation work by using the stable isotope, oxygen-18. The isotope concentrations of rainfall, soil water, river water and the water draining from soil pipes were

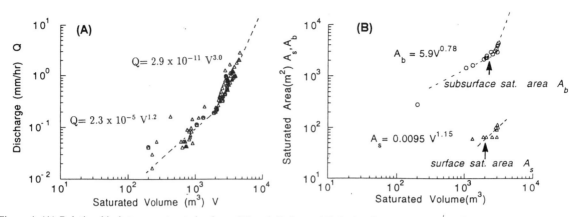

Figure 4. (A) Relationship between saturated volume (V) and discharge (Q) during four storm events (after Ohta (1990), modified). ($\triangle$) 10 August 1994; ($\blacktriangle$) 24 September 1988; ($\bigcirc$) 27 September 1988; and ($\bullet$) 5 October 1988. (B) Relationship between saturated volume (V) and source area (A_s) and subsurface saturated area (A_b) for the storm event on 24 September 1988

measured during storm events. These concentration measurements were then used for solving the water balance equations for each event, thus separating 'new event water' from 'pre-event water'. Figure 5 shows the result of hydrograph separation for the storm event of September 1992, and suggests that most of the subsurface runoff comes from 'displaced water' or 'pre-event water'. Unfortunately, field measurements of the saturated volume of groundwater and source areas, and the isotope studies, could not be conducted on the same events. In hindsight, a good opportunity was lost for carrying out such parallel field investigations alongside modelling studies. This remains the subject of ongoing research.

The observation results of Ohta (1990) and Kubota (in prep.) indicate strongly the dominant control of subsurface runoff by the saturated volume of groundwater. The heterogeneity of subsurface runoff generation is thus governed by the heterogeneity of the saturated volume of groundwater. The saturated volume of groundwater is a state variable; the critical question from the viewpoint of modelling is the dominant control of this state variable by catchment morphology and hydrogeology. This is the main subject of the remainder of the paper, where we follow the quasi-steady-state formulation adopted by Beven and Kirkby (1979) and by Sivapalan *et al.* (1987) that led to the topography index $\ln(a/\tan\omega_*)$, where a is the area drained per unit contour length and $\tan\omega_*$ is the slope of the ground surface at the location. We thus seek a similar topography index governing the spatial variability of the saturated volume of groundwater among the hillslopes.

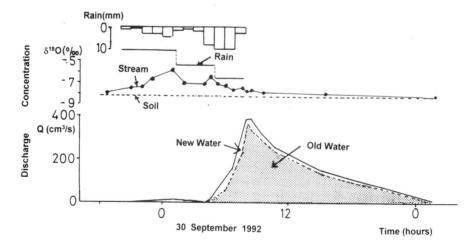

Figure 5. Hydrograph separation using stable isotope oxygen-18

MOTIVATION AND PROBLEMS TO BE SOLVED

The field observations carried out on the Hakyuchi catchment have suggested two important points. Firstly, the results of the hydrograph separation using stable isotopes have indicated that subsurface flow consisting mainly of old pre-event water displaced by new rain water dominates the storm response. Secondly, the observed catchment discharge Q and the volume of saturated groundwater V exhibit a simple relationship for a variety of storms, suggesting a power function relationship. Similar relationships were obtained for both surface and subsurface saturated areas. Two questions then arise:

1. In what way are these power function relationships related to small-scale process-physics, hydrogeology and hillslope topography?
2. Can such functions be used to build a catchment-scale subsurface stormflow model by integrating the small-scale heterogeneity in the topography, hydrogeology and soil properties?

In this paper we address the first of these questions by three different approaches (the second question is the subject of future work). These are:

1. *Observation*. This approach does not provide the 'solution', but rather the motivation. Investigations using detailed and carefully-planned field measurements can provide useful suggestions for possible conceptualisations. The purpose of this study is to provide additional support for the conceptual model proposed by Ohta (1990).
2. *Integration approach using distributed model*. We apply the simplified, distributed model proposed by Kubota *et al.* (1987; 1988) to a number of 'flow strips' identified within the catchment area. The discharge, volume of saturated groundwater and the source areas estimated for each flow strip by the model are then combined arithmetically to yield the corresponding catchment-scale quantities. Relationships are then formed between these catchment-scale quantities. We call this the 'integration approach'. This procedure is preceded by calibration of the distributed model using measured flow data for a number of storm events so that realistic estimates of a number of hydrogeological parameters can be estimated.
3. *Disaggregation–aggregation approach*. This approach uses the calibrated small-scale model used in the integration approach described above. The distributed model is applied to each flow strip with hypothetical, steady-state rainfall intensities. The resulting heterogeneity of the steady-state values of the volume of saturated groundwater among the flow strips is analysed. A topography-based index is developed which explains the variability of the antecedent state variable (saturated volume) and this then forms the basis of a disaggregation equation for handling small-scale variability within the catchment. It is then used in an aggregation procedure to develop catchment-scale functions or constitutive relations that will later be used in catchment-scale runoff generation models. This approach is especially useful for the development of regional relationships for catchments other than those on which the model was first implemented.

INTEGRATION APPROACH

We use here the distributed model proposed by Kubota *et al.* (1987; 1988). The model uses a simplified form proposed by Suzuki (1984) of the saturated–unsaturated Richards equation governing Darcian flow, predicting both flow hydrograph and soil moisture conditions at every point in a catchment.

The fundamental equations are as follows:

$$q_{x,y} = \int_0^D K(z) \sin \omega \, dz \tag{18}$$

$$w D \frac{\partial S_*}{\partial t} = \frac{\partial q_x}{\partial x} + \frac{\partial q_y}{\partial y} \tag{19}$$

where q_x and q_y are the fluxes of the directions x and y, respectively, $K(z)$ is the hydraulic conductivity at z and $S_* =$ averaged relative saturation in soil

$$S_* = \frac{1}{D} \int_0^D S \mathrm{d}z$$

Soil property variations are expressed in terms of the functions

$$K(z) = K_s(z) S^\beta \tag{20}$$

$$K_s(z) = K_0 \left(\frac{z}{D}\right)^f (z > 0), \qquad K_s(z) = 0 (z = 0) \tag{21}$$

where K_0 is the saturated conductivity at $z = D$ (soil surface), $K_s(z)$ is the saturated hydraulic conductivity at z; and f is the decay parameter of saturated hydraulic conductivity with soil depth.

In the original version of the model, hydraulic conductivity was assumed to be constant with soil depth. However, it is more general and effective to consider that the hydraulic conductivity decreases with soil depth. Following Beven (1981), the variation of saturated hydraulic conductivity with depth is assumed to be given by Equation (20). The model requires grid-based maps of topography and soil depth. The actual maps prepared by Ohta (1988) are available for the Hakyuchi catchment and are digitized on a 2·5 m square grid. The model input is hourly precipitation. We divide the catchment into 14 flow strips.

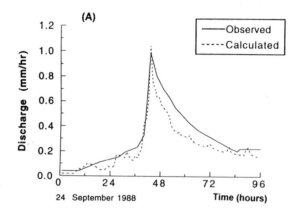

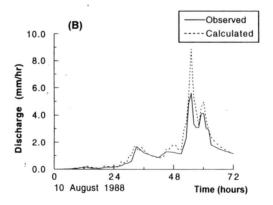

Figure 6. Validation of the distributed model using observed runoff data from two storm events on (A) 24 September 1988 and (B) 10 August 1994

The catchment discharge is the arithmetic sum of the discharges from the individual flow strips. The same is true of the other key hydrological quantities (i.e. volume of saturated groundwater, surface and subsurface saturated areas, etc.).

The model validation is carried out by applying the model to predict the catchment responses throughout the year in 1988, including the four storm events, which are the events presented in Figure 4. The model parameters are estimated by manual calibration of the model to these events, by visual comparison of the observed and predicted flow hydrographs. Figure 6 shows the comparison of the observed and predicted hydrographs for two events. In this study, we focus on the catchment-scale responses mainly during storm events; this is the reason we present the storm hydrographs. Parameters used here are: $K_0 = 1 \cdot 0 \times 10^{-3}$ (m/s), $f = 2 \cdot 0$, $\psi_0 = 0 \cdot 4$ (m), $\beta = 4 \cdot 5$.

It is clear that there is much room for improvement in the parameter calibration, but as our main interest is in the development of the methodology for building catchment-scale models, we did not pursue this here. During the simulations, the model produces not only the flow hydrographs, but also the changes in soil moisture conditions, so that we can obtain discharge, saturated volume of groundwater, source area and subsurface saturated area on each flow strip, as well as the corresponding values for of the whole catchment. The results will be presented later.

DISAGGREGATION–AGGREGATION APPROACH.

We next present a disaggregation–aggregation approach to derive the catchment-scale responses. The disaggregation is carried out in the same manner as in Sivapalan (1993). Firstly, the catchment is divided into 14 flow strips as in the distributed model. Next we run the distributed model on each flow strip with different rainfall intensities under steady-state conditions. For each flow strip, we form simple relationships between saturated volume and discharge on each strip under the steady-state conditions. In this study, a simple power function expresses the relationship between discharge and saturated volume on each flow strip. This can be expressed as

$$q_i = g(v_i) \tag{22}$$

where q_i is the discharge from the ith flow strip; v_i is the saturated volume of ith flow strip; and g is a functional expression between q_i and v_i.

The functional relationship between discharge q_i and volume v_i is governed by Darcy's law and depends on the saurated hydraulic conductivity at the surface K_0, the decay parameter f, the slope of the flow strip and the geometry of the hillslope cross-section (e.g. depth D). In idealised conditions, this relationship can be derived analytically. However, in this study we estimated it by simulation with the distributed model. Similar relationships have also been derived by simulation for each flow strip, between the surface and subsurface saturated areas and the volume of saturated groundwater, i.e.

$$a_{si} = h_s(v_i) \tag{23}$$

and

$$a_{bi} = h_b(v_i) \tag{24}$$

where a_{si} is the saturated area (source area) of ith strip, a_{bi} is the subsurface saturated area of ith flow strip, h_s is the functional relationship between a_{si} and v_i and h_b is the functional relationship between a_{bi} and v_i.

The next objective is to determine the topographic parameter that explains much of the variability in steady-state saturated volumes among the 14 flow strips. Motivated by Beven and Kirkby (1979) and Sivapalan (1993), two empirical topographic indices, $1/\tan \omega_i$ and $a_i/\tan \omega_i$ were tried where a_i is the area and ω_i is the mean slope angle of the ith flow strip. Figure 7 shows the result of this procedure. The lines in Figure 7 indicate the best-fit linear regressions for each rainfall intensity. According to Figure 7, $1/\tan \omega_i$ explains the variability of the steady-state saturated volumes slightly better than $a_i/\tan \omega_i$. However, this is not conclusive, and the analysis of many more, of the order of hundreds, of hillslopes, with widely varying properties are needed to help select a suitable topographic index. In this study, following

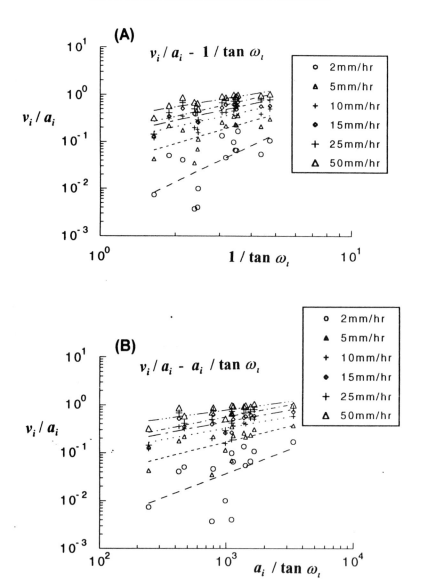

Figure 7. Tests of two alternative topographic indices, (A) $1/\tan\omega_i$ and (B) $a_i/\tan\omega_i$ for characterising the distribution of saturated ground volume between flow strips in terms of topography. Lines in the figure are best-fit regressions for six different rainfall intensities

Sivapalan (1993), we selected $1/\tan\omega_i$. The best-fitting relationship between saturated volume and $1/\tan\omega_i$ is

$$\frac{v_i}{a_i} = f_1 \left(\frac{1}{\tan\omega_i}\right)^\epsilon \tag{25}$$

$$V = \sum_{alli} v_i = f_1 \sum_{alli} a_i \left(\frac{1}{\tan\omega_i}\right)^\epsilon \tag{26}$$

where f_i and ϵ are the constants expressing the topographic effect.

Combining Equations (25) and (26) we can obtain the disaggregation equation for the state variable v_i

$$\frac{v_i}{a_i} = \frac{V}{A} + f_1 \left\{ \left(\frac{1}{\tan\omega_i}\right)^\epsilon - \mu \right\} \tag{27}$$

where A is the catchment area; $A = \sum_{alli} a_i$ and μ is the catchment average of topographic index $(1/\tan \omega_i)^\epsilon$. The constants f_1 and ϵ are dependent on rainfall intensity. f_1 varies 0·0014 to 0·09 and ϵ varies from 1·46 to 3·15.

Given the catchment-scale state variable (saturated volume of groundwater) V, and the topographic index $(1/\tan \omega_i)^\epsilon$ and its catchment average μ, we can then estimate the local state variable v_i in a straightforward manner. Equation (27) is thus a scaling relationship for the state variable representing the state of the saturated volume for an individual flow strip, v_i, linking it to the corresponding catchment value V. Combining Equations (22) and (27), it is also straightforward to derive a catchment-scale relationship between the saturated volume V and discharge Q. Similar procedures have been adopted, with Equations (23), (24) and (27), to derive the saturated area A_b versus saturated volume V and source area A_s versus saturated volume V relationships at the catchment scale.

DISCUSSION

In this section we compare the catchment-scale relationships, Q versus V, A_s versus V and A_b versus V obtained by the two model-based approaches presented above with the observational results shown in Figure 4. The results of model-based approaches are presented in Figure 8. Note that the observational results shown in Figure 4(A) are based on field measurements of Q versus V taken at two or three hour

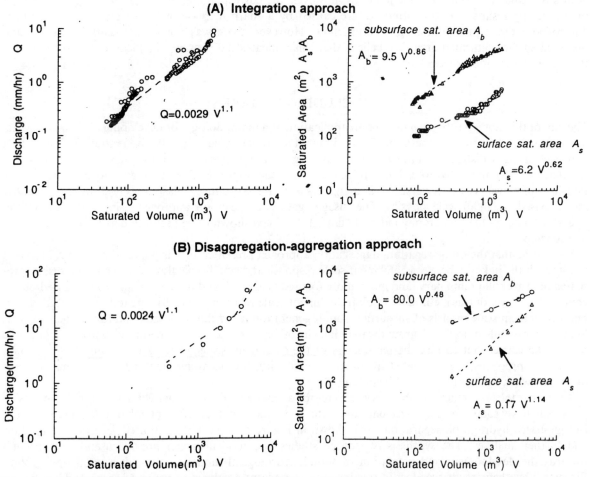

Figure 8. Comparisons of catchment-scale relationships between the two model-based approaches. (A) Integration based on the distributed storage model and (B) the steady-state disaggregation–aggregation approach using 14 flow strips

intervals during four storm events. However, the data set of measurements of A_s and A_b are available only for the event of 24 September 1988. Similarly, the results obtained by the integration approach, using the distributed storage model, includes hourly values for Q, V, A_s and A_b for these storm events. On the other hand, for the disaggregation–aggregation approach, as we carried out these simulations under steady-state conditions for only six different rainfall intensities, only six data points are presented in Figure 8 for this case.

On the relationship between saturated groundwater volume and discharge at the catchment-scale, the observed relationships differ considerably from those obtained by model-based approaches. Note that the observed saturated volume was measured above the hydrological base defined by Ohta (1990). On the other hand, the digitized grid map used in the distributed model was rather different. Ohta defined the hydrological base as that having $N_c = 30$. We believe that this is too deep for modelling subsurface generation. The differences in the saturated volume are probably caused by this discrepancy about the level of the impermeable base. Despite these differences, the basic tendency of the observed relationship is reproduced by the distributed model and by the disaggregation–aggregation approach. Importantly, the relationships obtained by the latter two approaches are almost identical, demonstrating that the steady-state disaggregation approach is a viable method for the estimation of catchment-scale relationships. Similar results were obtained in the case of both the A_s versus V and A_b versus V relationships.

Another feature that can be noted in Figures 4 and 8 is the discontinuities in slope in all of the relationships, especially in the relationship between Q and V. This is especially clear in the observational results, more so than in the modelling results. The critical value of saturated volume at which the discontinuity occurs is similar between the three methods; however, the corresponding discharge rates are different. In the modelling results, the discontinuities are caused by a sharp increase in the surface flow component due to both direct precipitation and return flow. However, there was not a significant increase in the observed surface saturated area. We believe that this indicates the possible significance of return flow in the runoff generation process.

CONCLUDING REMARKS

The aim of this study has been to present an application of a methodology for developing a catchment-scale model of subsurface runoff generation. Three different approaches, namely field observation, an integration approach using a distributed model and a disaggregation–aggregation procedure based on steady-state simulations, were presented to estimate the catchment-scale response. The observed catchment-scale relationships of Q versus V, A_s versus V and A_b versus V are reproduced reasonably well by a distributed model based on small-scale physics. The disaggregation–aggregation approach using a disaggregation operator based on the topography index $(1/\tan \omega_i)^\epsilon$ has been shown to be useful to derive catchment-scale relationships.

We believe that the disaggregation–aggregation approach presented in this paper is an effective method to establish the linkages between catchment-scale response and small-scale flow processes. The results presented here are still preliminary, and much more focused field work and modelling need to be carried out to confirm the power of this method to develop catchment-scale relations. The ultimate test of the method is to build catchment scale models of subsurface runoff generation using this approach and to test these against long-term rainfall–runoff and groundwater data in other catchments. Furthermore, sensitivity analyses need to be carried out to link the parameters of the catchment-scale relations to parameters describing small-scale processes, e.g. saturated hydraulic conductivity, the decay parameter f, etc., along the lines of similar work by Robinson and Sivapalan (1995).

As the complete discharge data for 1992 were not available and the model validation for the events in 1988 still have room for improvement, the data from the event of 30 September 1992, used for the isotope-based hydrograph separation work shown in Figure 5, could not be utilised for model validation and further analyses. The synthesis of isotope studies and modelling can provide useful new insights towards the development and validation of sound hydrological models. Although a good opportunity for such a synthetic study was lost in this instance, it remains a subject of active research and results are expected in the near future.

ACKNOWLEDGEMENTS

This study was carried out while the senior author was a visiting fellow at the Centre for Water Research (CWR), the University of Western Australia, with the support of a Grant-in-Aid from the Japanese Ministry of Education. He is pleased to acknowledge the kind support provided by the staff and students of the CWR, especially by Professor Jorg Imberger. We thank Dr Takehiko Ohta for providing us with his precious data sets and for giving many useful suggestions. We also record with thanks the many valuable comments provided by Dr Yoshinori Tsukamoto and Dr Masakazu Suzuki.

LIST OF SYMBOLS

A	Catchment area
A_b	Saturated area (subsurface saturated area)
A_s	Source area (surface saturated area)
a	Area drained per unit contour length
a_i	Area of the ith flow strip
a_{bi}	Subsurface saturated area of ith flow strip
a_{si}	Saturated area (source area) of ith strip
C	Specific water capacity
D	Soil depth
f	Decay parameter of saturated hydraulic conductivity with soil depth
f_1, ϵ	Constants expressing topographic effect
g	Functional relationship between q_i and v_i
h_s	Functional relationship between a_{si} and v_i
h_b	Functional relationship between a_{bi} and v_i
K_s	Saturated hydraulic conductivity
$K(z)$	Hydraulic conductivity at z
$K(\psi)$	Hydraulic conductivity at ψ
K_0	Saturated conductivity at $z = D$ (soil surface)
$K_s(z)$	Saturated hydraulic conductivity at z
k, p	Constant coefficients
L	Slope length
N_c	Number of hits on a simple penetrometer
Q	Catchment discharge
q	Flux per unit width
q_i	Discharge from ith flow strip
q_x, q_y	Fluxes of the directions x and y, respectively
S	Relative saturation
S_*	Averaged relative saturation in soil, $S_* = \dfrac{1}{D} \displaystyle\int_0^D S \, \mathrm{d}z$
t	Time
t_*	Non-dimensional form of t
V	Volume of saturated groundwater
v_i	Saturated volume of ith flow strip
w	Effective porosity
x	Upslope axis
x_*	Non-dimensional form of x
z	Axis normal to x
z_*	Non-dimensional form of z
α	Parameter describing characteristics of saturated–unsaturated flow
β	Decay parameter of hydraulic conductivity with relative saturation
γ, δ	Topographic parameters

μ Catchment average of topographic index $(1/\tan\omega_i)^\epsilon$

ω Slope angle

ω_* Slope of the ground surface

ω_i Mean slope angle of ith flow strip

θ Volumetric water content

θ_s Saturated water content

θ_r Residual water content

ψ Soil water pressure head

ψ' Pressured head, $\psi' = \begin{cases} \psi & (\psi \leqslant 0) \\ 0 & (\psi > 0) \end{cases}$

ψ_0 Pressure head at air entry

ψ_* Non-dimensional form of ψ

ψ'_* Non-dimensional form of ψ'

REFERENCES

Beven, K. J. 1981. 'On subsurface stormflow: predictions with simple kinematic theory for saturated and unsaturated flows', *Wat. Resour. Res.*, **18**, 1627–1633.

Beven, K. J. and Kirkby, M. J. 1979. 'A physically-based variable contributing area model of basin hydrology', *Hydrol. Sci. Bull.*, **24**, 43–69.

Brutsaert, W. 1968. 'The permeability of a porous medium determined from certain probability laws for pore-size distribution', *Wat. Resour. Res.* **4**, 425–434.

Dunne, T. 1983. 'Relation of field studies and modeling in the prediction of storm runoff', *J. of Hydrology*, **65**, 25–48.

Kubota, J. 'Hydrograph separation using stable isotope oxygen-18 in a forested catchment', (in prep.)

Kubota, J. and Suzuki, M. 1988. 'Two-dimensional distribution of soil water potential and flux on a hillslope', *J. Jpn Assoc. Hydrol. Sci.*, **18**, 62–73 [in Japanese with English abstract].

Kubota, J., Fukushima, Y., and Suzuki, M. 1987. 'Observation and modeling of the runoff process on a hillslope', *J. Jpn Forestry Soc.*, **69**, 258–269 [in Japanese with English abstract].

Kubota, J., Fukushima, Y., and Suzuki, M. 1988. 'Observation and modeling of the runoff process on a hillslope (II) water budget and location of the groundwater table and its rise', *J. Jpn Forestry Soc.*, **70**, 381–389 [in Japanese with English abstract].

Ohta, T. 1988. 'Storm runoff mechanism on forested slopes', *J. Jpn Soc. Hydrol. Wat. Resour.*, **1**, 31–38 [in Japanese with English abstract].

Ohta, T. 1990. 'A conceptual model of storm runoff on steep forested slopes', *J. Jpn Forestry Soc.*, **72**, 201–207 [in Japanese with English abstract].

Richards, L. A. 1931. 'Capillary conduction of liquids through porous mediums', *Physics*, **1**, 318–333.

Robinson, J. S. and Sivapalan, M. 1995. 'Catchment-scale runoff prediction equation based on aggregation and similarity analyses', *Hydrol. Process.*, **9**, 555–574.

Sivapalan, M. 1993. 'Linking hydrologic parameterizations across a range of scale: hillslope to catchment to region' in *Exchange Processes at the Land Surface for a Range of Space and Time Scales. IAHS Publ*, **212**, 115–123.

Sivapalan, M., Beven, K. J., and Wood, E. F. 1987. 'On similarity. 2. A scaled model of storm runoff production', *Wat. Resour. Res.*, **23**, 2266–2278.

Suzuki, M. 1984. 'The properties of a base-flow recession on small mountainous watersheds (I). Numerical analysis using the saturated-unsaturated flow model', *J. of Jpn Forestry Soc.*, **66**, 172–182 [in Japanese with English abstract].

Tani, M. 1982. 'The properties of a water-table rise produced by a one-dimensional, vertical, unsaturated flow', *J. of Jpn Forestry Soc.* **64**, 409–418 [in Japanese with English abstract].

18

CATCHMENT-SCALE RUNOFF GENERATION MODEL BY AGGREGATION AND SIMILARITY ANALYSES

JUSTIN S. ROBINSON AND MURUGESU SIVAPALAN

Centre for Water Research, Department of Environmental Engineering University of Western Australia, Nedlands, WA 6009, Australia

ABSTRACT

Runoff generation in natural catchments due to storm rainfall is highly complex and spatially and temporally heterogeneous. In recent work on seven small experimental catchments Larsen *et al.* (1994) showed that underlying the heterogeneity of runoff generation *within* the catchments, there is a degree of regularity *between* the catchments that could be quantified in terms of two dimensionless similarity parameters K_0^* and f^*. These two parameters, constants for a catchment, were able to characterize the relative dominance of the saturation excess (Dunne-type) and infiltration excess (Horton-type) mechanisms of runoff generation. Given that K_0^* and f^* can characterize the type of runoff generation on any catchment, it may follow that they can be used to define a catchment-scale runoff generation model. This idea is pursued in this paper. For the same catchments as studied by Larsen *et al.* (1994), a lumped, physically based model is developed that describes both the extent of saturated areas and the average infiltration capacity of the unsaturated areas during a storm. This is achieved by utilizing the distributed model used by Larsen *et al.* (1994) to aggregate the point-scale runoff generation responses, up to the catchment scale, from which the functional form and the parameters of the catchment-scale runoff generation model are inferred. The parameters of this lumped model are defined entirely in terms of the underlying distribution of topography, three similarity parameters K_0^*, f^* and B^*, the normalized average water-table depth, z^*, and the normalized cumulative volume of infiltration, G^*.

INTRODUCTION

The modelling of infiltration and runoff generation is exceedingly complex and difficult in natural catchments because soil properties are variable in space and initial soil moisture and rainfall are variable in both space and time. These heterogeneities can have a significant impact on runoff generation at the field and catchment scales (Smith and Hebbert, 1979; Sharma *et al.*, 1980; Sivapalan and Wood, 1986). A number of field and modelling studies have demonstrated the considerable spatial and temporal heterogeneity of runoff processes within a catchment, involving interacting surface and subsurface flow mechanisms (Dunne, 1978; Freeze, 1974). Surface runoff is generated by two different mechanisms: *infiltration excess* (Horton-type), when the rate of rainfall is greater than the rate of infiltration at any point, and *saturation excess* (Dunne-type), when the entire vertical soil profile is saturated, either before or during the storm event, and the infiltration rate is reduced to zero. The spatial variability of infiltration excess runoff can be related to patterns of rainfall intensity and soil hydraulic conductivity, with the runoff being generated mostly from areas where the rainfall intensity is high and soil hydraulic conductivity is low (Wood *et al.*, 1990). In the case of saturation excess runoff, links have been found between the locations of saturated areas producing such runoff and topography, with the saturated areas tending to develop at the bottom of hillslopes and near stream banks (Beven *et al.*, 1988).

Previous efforts made towards modelling runoff generation at the catchment scale have tended to concentrate on two distinct approaches. The first approach involved the use of point-scale, process-based equations in complex distributed numerical models, e.g. the SHE model (Abbott *et al.*, 1986), in an attempt to model the detailed, small-scale heterogeneity of the catchment response. The second approach captures the lumped catchment response using empirical, black box models or conceptual models, e.g. runoff coefficients (Chow *et al.*, 1988) and the Soil Conservation Service (SCS) method of abstractions (SCS, 1972). Models developed using the second approach are only marginally, if at all, physically based and the model parameters cannot easily be related to the physical factors which affect runoff generation, such as soil properties and topography. However, these simple models, some with only one or two parameters, have been shown to give an adequate empirical fit to the observed responses of particular catchments (Beven and Kirkby, 1979) and are used widely for design purposes. Process-based distributed models such as SHE, on the other hand, tend to be expensive in terms of data and computational requirements. Furthermore, the physical basis of these process-based models is also being questioned because of the spatial lumping inherent within the model grid cells (Beven and Kirkby, 1979; Beven, 1988).

Work towards finding connections between the physically based and conceptual models has been inadequate, although this is precisely what is required for the advancement of hydrological modelling for predictive purposes (O'Connell, 1991). This paper represents a preliminary attempt to do this. It presents a lumped, catchment-scale model of surface runoff generation by the saturation excess and infiltration excess mechanisms. This model is developed by aggregating the heterogeneous, point-scale responses predicted by a distributed model of runoff generation, based on TOPMODEL concepts (Beven and Kirkby, 1979). This lumped model is an advance over a similar, catchment-scale runoff generation model presented by Sivapalan *et al.* (1987). It is more general and can be applied to predict the runoff due to arbitrary, time-variable rainfall. The component models of saturation excess and infiltration excess runoff generation are both simple, involving a one-step calculation in each instance.

The resulting model has been tested on a number of small agricultural catchments in Western Australia and, following regionalization of its parameters, could be used for predicting runoff on a regional basis. By recasting the lumped model equations into dimensionless forms, within the similarity framework of Sivapalan *et al.* (1987) and Larsen *et al.* (1994), it has been possible to express the parameters of the lumped model in terms of three dimensionless similarity parameters. This presents a natural and physically consistent framework for the regionalizations of the model parameters and for linking parameterizations across scales.

SIMILARITY OF RUNOFF GENERATION IN EASTERN WHEAT BELT CATCHMENTS

The work presented in this paper was carried out on seven small agricultural catchments in the eastern wheat belt of Western Australia. The catchments are Moolanooka (MO), Minjin (MI), Nungarin North (NN), Nungarin South (NS), Kunjin North (KN), Holland (HO) and Jackitup Creek (JC). Storm runoff on the catchments is generated by infiltration excess and saturation excess overland flow. Subsurface flow does not contribute significantly to storm runoff (Coles, 1994). Some properties of each of the seven catchments are given in Table I.

To examine the patterns of runoff generation within the catchments, Larsen *et al.* (1993) and Coles (1994) developed a distributed version of TOPMODEL. The model is able to simulate runoff generation by both infiltration and saturation excess. It incorporates, in a distributed manner, the soil hydraulic properties and the topography based wetness index of Beven and Kirkby (1979), estimated from the field data. The model was applied to the most severe storm events recorded on the catchments. With minimal calibration, involving a single antecedent wetness parameter and two routing parameters, they achieved satisfactory predictions of storm runoff for 58 storm events (Larsen *et al.*, 1993; 1994). Simulation results for all of the catchments are presented in Larsen *et al.* (1993).

Using the distributed model of Larsen *et al.* (1993) and Coles (1994), Larsen *et al.* (1994) showed that there is an underlying similarity between the simulated runoff generation responses of catchments in the region and that this similarity could be quantified by two of the five dimensionless similarity parameters originally presented by Sivapalan *et al.* (1987). The definitions of these similarity parameters were slightly

Table I. Some catchment characteristics [adapted from Larsen et al. (1994) and Larsen et al. (1993)]

	A (km^2)	$\bar{K}_0$ (mm/hr)	σ_{K_0} (mm/hr)	$\bar{\psi}_c$ (mm)	$\bar{\theta}_s$ (10^{-2})	$\bar{\theta}_r$ (10^{-2})	λ (ln[m])	$\sigma_{\ln(a/\tan\beta)}$ (ln[m])	$\bar{f}$ (m^{-1})	$\bar{c}_0$ (m/h)	L_{max} (m)
Jackitup Creek (JC)	0·38	19·39	8·83	43·2	24·1	1·76	8·66	1·23	13·77	306	990
Holland (HO)	0·39	34·72	16·10	134·7	25·2	2·04	8·84	1·13	9·61	1125	1026
Kunjin North (KN)	0·18	9·47	5·68	60·2	29·3	1·73	9·06	0·95	3·24	914	674
Minjin (MI)	0·41	10·36	4·52	82·2	25·5	1·85	9·33	1·33	3·00	700	664
Moolanooka (MO)	0·24	10·04	6·16	74·7	33·0	2·99	8·42	1·20	3·52	950	673
Nungarin North (NN)	0·15	15·14	8·18	123·9	24·4	1·68	8·47	1·12	2·90	1833	502
Nungarin South (NS)	0·18	10·77	8·40	46·6	32·8	1·58	8·85	1·36	2·23	1400	656

modified by Larsen et al. (1994), and are given by:

$$C_{V,K_0} = \frac{\sigma_{K_0}}{\bar{K}_0} \tag{1}$$

$$C_{v,\ln[a/\tan\beta]} = \frac{\sigma_{\ln[a/\tan\beta]}}{\lambda^{-1}} \tag{2}$$

$$B^* = \bar{B} \tag{3}$$

$$f^* = \frac{\bar{\psi}_c \bar{f}}{\lambda} \tag{4}$$

$$K_0^* = \frac{L_{max}\bar{K}_0}{\bar{c}_0\bar{\psi}_c(\bar{\theta}_s - \bar{\theta}_r)} \tag{5}$$

C_{V,K_0} and $C_{V,\ln[a/\tan\beta]}$ are the coefficients of variation of the surface saturated hydraulic conductivity, K_0, and the topographic index, $\ln[a/\tan\beta]$, respectively. B^* is the areal average of the pore size distribution index, B (Brooks and Corey, 1964). f^* is the normalized, areal average of the parameter f, (i.e. $\bar{f}$), which describes the assumed exponential decrease of saturated hydraulic conductivity with depth. Finally, K_0^* is the normalized, areal average of the saturated hydraulic conductivity at the soil surface (i.e. $\bar{K}_0$). The definitions of the catchment properties used to construct the similarity parameters are given in the notation list at the end of the paper. Throughout this paper we set $\exp[\lambda^1]$ equal to one metre which results in λ^1 being equal to zero. The values of the similarity parameters estimated using Equations (1) to (5) and the catchment properties in Table I are presented in Table II.

For the wheat-belt catchments, Larsen et al. (1994) found that of the five similarity parameters, only f^* and K_0^* are necessary to characterize the dominant mechanism of runoff generation. By artificially varying soil properties over a wide range of values they were able to examine the range of variability of the runoff generation response and to quantify it in terms of the two similarity parameters f^* and K_0^*. To quantify the differences in runoff generation responses between catchments, Larsen et al. (1994) defined a dimensionless ratio, R, as the ratio of the volume of runoff generated by the saturation excess mechanism to the total

Table II. Dimensionless similarity parameters (reproduced from Larsen et al., 1994)

	C_{V,K_0}	$C_{V,\ln(a/\tan\beta)}$	B^*	f^*	K_0^*
Jackitup Creek (JC)	0·455	0·141	0·738	0·0687	6·500
Holland (HO)	0·464	0·128	0·264	0·1464	1·014
Kunjin North (KN)	0·599	0·105	0·597	0·0215	0·421
Minjin (MI)	0·425	0·142	0·416	0·0265	0·519
Moolanooka (MO)	0·614	0·143	0·657	0·0312	0·317
Nungarin North (NN)	0·541	0·132	0·389	0·0425	0·148
Nungarin South (NS)	0·780	0·153	1·005	0·0117	0·347

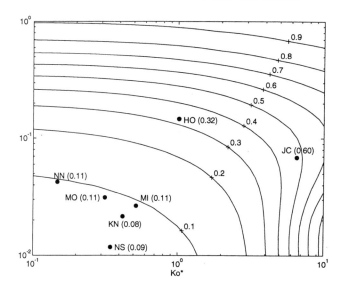

Figure 1. Contours of constant R values as a function of the two similarity parameters K_0^* and f^*. The closed circles (•) show the location of the wheat belt catchments on the K_0^*–f^* diagram, the letters and numbers, respectively, indicate the catchment and its corresponding R value (reproduced from Larsen et al., 1994)

volume of runoff. The effect of antecedent conditions is removed by calculating R as an average value based on a number of storms, over a range of antecedent soil moisture conditions. An R value equal to unity means that all of the runoff, in a hypothetical year, is generated by saturation excess, whereas an R value of zero corresponds to all runoff being of the infiltration excess type.

The simulation results of Larsen et al. (1994) are presented in Figure 1 in the form of a contour plot of the R values in K_0^*–f^* space. The results show that the similarity parameters f^* and K_0^* are able to quantitatively describe the type of runoff generation on all seven catchments. The results also show that the essential characteristics of the runoff generation responses, in this instance quantified by R, can be 'scaled' from one catchment to another, provided that the values of the dimensionless similarity parameters are also scaled, but *regardless* of the actual magnitudes of the individual variables that constitute the two dimensionless parameters.

The similarity results of Larsen et al. (1994) provided the primary motivation for the research presented in this paper. If the similarity parameters are able to quantify the dominant mechanism of runoff generation at the catchment scale, then it follows that they might also be used to characterize a catchment-scale runoff prediction equation.

CATCHMENT-SCALE RESPONSE

The catchment-scale runoff prediction equation developed in this paper consists of two components. The first describes the extent of the saturated areas producing saturation excess runoff, $A_c[t]$, and the second describes the average infiltration capacity, $\bar{g}[t]$, over the unsaturated parts of the catchment. Note that infiltration capacity, whether at the point scale, or at catchment scale, is defined as the maximum possible rate of infiltration at any given time, t. The rate of runoff generation on the catchment can then be expressed as follows:

$$\bar{p}_{\text{eff}}[t] = \bar{p}[t]\left(\frac{A_c[t]}{A}\right) + \left(1 - \frac{A_c[t]}{A}\right)\max[(\bar{p}[t] - \bar{g}[t]), 0] \tag{6}$$

where $\bar{p}_{\text{eff}}[t]$ is the catchment-average effective rainfall intensity (rate of runoff generation) and $\bar{p}[t]$ is the catchment-average rainfall intensity, assumed known, and A is the total area of the catchment. The first term on the right-hand side of Equation (6) represents saturation excess runoff and the last term represents infiltration excess runoff. Equation (6) ignores the spatial variability of infiltration capacity and possible reinfiltration. The effects of ignoring these two components are discussed later in the paper.

The lumped, catchment-scale equations that describe the temporal variation in A_c/A and $\bar{g}$, during a storm, are developed by first aggregating the point-scale responses predicted by the distributed model of Larsen *et al.* (1994). The aggregation is carried out using the following equations:

$$\frac{A_c[t]}{A} = \frac{1}{A} \sum_0^A a_c[t]_i \Delta A \tag{7}$$

$$\bar{g}[t] = \frac{1}{A(1 - A_c[t]/A)} \sum_0^A g[t]_i \Delta A \tag{8}$$

$$\bar{G}[t] = \frac{1}{A} \sum_0^A G[t]_i \Delta A \tag{9}$$

where $g[t]_i$ is the infiltration capacity at location i, $a_c[t]_i$ is given a value of *one* if the vertical soil profile at location i is saturated, otherwise it is given a value of *zero*, $G[t]_i$ is the cumulative volume (in depth units) of infiltration at location i, $\bar{G}[t]$ is the catchment-average cumulative volume (in depth units) of infiltration and ΔA is the size of the rectangular grid element. The summations are taken over all the grid elements within the catchment.

The infiltration capacity, $g[t]_i$, in the distributed model of Larsen *et al.* (1994), is estimated using the Green and Ampt (1911) equation extended to account for a finite water-table depth and an assumed exponential decrease of saturated hydraulic conductivity with depth. To model temporally variable rainfall, the time condensation approximation (Reeves and Miller, 1975) is used. The initial water-table depths are determined using a distributed version of TOPMODEL (Beven and Kirkby, 1979) and the initial moisture distribution above the water-table is determined by assuming full gravity drainage and the Brooks and Corey (1964) soil moisture characteristic relationships. During a storm the downslope redistribution of moisture is ignored. Details of the distributed model are not presented here as they are well covered elsewhere (Sivapalan *et al.*, 1987; Wood *et al.*, 1988; Larsen *et al.* 1994).

Simulations with the distributed model are then repeated for a set of artificial storms of varying intensity and temporal pattern, over a wide range of antecedent conditions. The objective is to seek simple relationships for the values of A_c/A and $\bar{g}$, obtained from the simulations, in terms of measurable catchment parameters. An important requirement of the relationships sought is universality, so that the models can be applied under a variety of catchment and climatic conditions, and for temporally variable rainfall.

TIME CONDENSATION APPROXIMATION AT THE CATCHMENT SCALE

The approach adopted to achieve universality of the catchment-scale runoff prediction equation has been motivated by the success of the time condensation approximation, or TCA (Reeves and Miller, 1975; Milly, 1986; Sivapalan and Milly, 1989) for the estimation of rainfall infiltration at a point. Coles (1994) and Larsen *et al.* (1994) used it to model infiltration capacity, and the runoff generation rate, at each grid point in their distributed model. The main idea behind the TCA is that when the infiltration capacity at any time is plotted as a function of the cumulative volume of infiltration, rather than as a function of time, the resulting relationship is relatively independent of the rainfall or infiltration history (Reeves and Miller, 1975). In this paper, we attempt to determine whether this idea has applicability at the catchment scale, in spite of the spatial variability, not only for the infiltration capacity, but also for the saturated area fraction.

To examine the validity of the TCA at the catchment scale, we assessed the effects of the rainfall intensity and temporal patterns on the average infiltration capacity, $\bar{g}$, and on the saturated contributing areas A_c/A. The analysis used design storms of different average recurrence intervals (ARI), durations and intensities, with realistic temporal patterns, constructed using the guidelines given in *Australian Rainfall and Runoff*

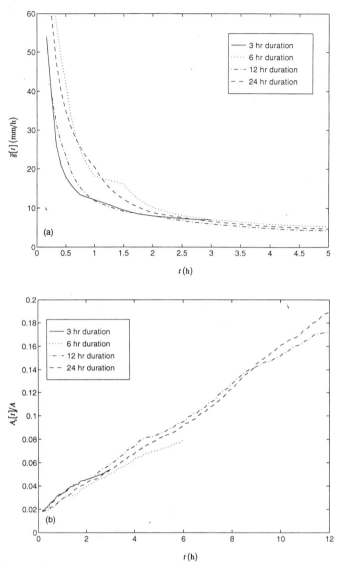

Figure 2. (a) Average infiltration capacity as a function of time for 10 year average recurrence interval design storms of different durations for Jackitup Creek catchment. (b) Growth of saturated contributing areas as a function of time for 10 year average recurrence interval design storms of different durations for Jackitup Creek catchment

(Institution of Engineers Australia, 1987). In view of the size of the catchments the rainfall intensity was assumed constant in space. Results of the analysis for Jackitup Creek catchment for four 10 year ARI storms of 3, 6, 12 and 24 hour duration are presented in Figure 2 ($\bar{g}$ versus t and A_c/A versus t) and Figure 3 ($\bar{g}$ versus $\bar{G}$ and A_c/A versus $\bar{G}$). The figures clearly support the adequacy of the TCA for runoff predictions at the catchment scale, in that when A_c/A and $\bar{g}$ are plotted against $\bar{G}$, rather than time t, curves obtained for the different design storms of varying duration collapse into just one curve for each case. Simulations on the other catchments yielded similar results. The catchment-scale runoff prediction equation, (6), can therefore be rewritten, approximately, as follows:

$$\bar{p}_{\text{eff}}[t] = \bar{p}[t]\left(\frac{A_c[\bar{G}]}{A}\right) + \left(1 - \frac{A_c[\bar{G}]}{A}\right)\max[(\bar{p}[t] - \bar{g}[\bar{G}]), 0] \tag{10}$$

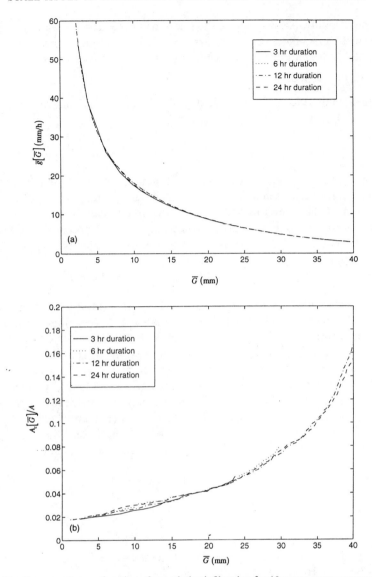

Figure 3. (a) Average infiltration capacity as a function of cumulative infiltration for 10 year average recurrence interval design storms of different durations for Jackitup Creek catchment. (b) Growth of saturated contributing areas as a function of cumulative infiltration for 10 year average recurrence interval design storms of different durations for Jackitup Creek catchment

The two components of Equation (10), A_c/A and $\bar{g}$ are no longer explicitly dependent on the pattern and intensity of rainfall, but depend only on the cumulative volume of infiltration and antecedent conditions.

The derivation of the runoff prediction equation given by Equation (10) involves the parameterization of three components: (i) antecedent conditions, (ii) saturation excess runoff, and (iii) infiltration excess runoff. In the following sections we present the development of these three components of the lumped catchment-scale model and comparisons against predictions by the distributed model of Larsen *et al.* (1994). In all three instances the parameterizations are recast into dimensionless forms within the framework of the similarity theory of Larsen *et al.* (1994) to allow the examination of questions of catchment similarity and of the potential for a physically based regionalization of the resulting runoff prediction equation.

ANTECEDENT SATURATED AREA AND UNSATURATED-ZONE SOIL MOISTURE CONTENT

A point in a catchment becomes saturated either as a result of the antecedent soil moisture conditions or as a result of the volume of infiltrated water accumulated during the storm equalling or exceeding the initial moisture deficit. The distributed model we use (Larsen *et al.*, 1994) predicts both the initial extent of saturation area and the antecedent soil moisture deficit using TOPMODEL concepts.

Prediction of antecedent saturation area

The relation between the initial local water-table depth at location i, z_i, and the initial catchment average water-table depth, $\bar{z}$, is given by (Sivapalan *et al.*, 1987)

$$z_i = \bar{z} - \frac{1}{f}[\ln[a/\tan\beta]_i - \lambda - \ln[K_0/f]_i + \gamma] \tag{11}$$

where $\ln[a/\tan\beta]_i$ and $\ln[K_0/f]_i$ are, respectively, a topographic index and a hydrogeological index, whereas λ and γ are, respectively, the corresponding catchment averages.

The areas of a catchment that are saturated before a storm are given by locations that have water-table depths less than or equal to the capillary fringe thickness, ψ_{c_i} (i.e. $z_i \leqslant \psi_{c_i}$). ψ_{c_i} is also known as the air entry pressure head. From Equation (11) this can be expressed by the inequality

$$\bar{z} \leqslant \psi_{c_i} + \frac{1}{f_i}[\ln[a/\tan\beta]_i - \lambda - \ln[K_0/f]_i + \gamma] \tag{12}$$

Larsen *et al.*'s (1994) distributed model uses Equation (12) to determine whether a point in the catchment is saturated. However, because the right-hand side of Equation (12) combines a number of soil and topographic parameters, all of which can vary in space, it is difficult to form a catchment-scale equation which is mathematically tractable. For simplicity, we have chosen to treat the soil parameters in Equation (12), ψ_{ci}, f_i and K_{0i}, as constants equal to their estimated catchment-average values $\overline{\psi_c}$, $\bar{f}$ and $\overline{K_0}$. This assumption is supported by the work of Wood *et al.* (1990) and Famiglietti *et al.* (1992), who showed that the spatial variation of the topographic index, $\ln[a/\tan\beta]$, is large compared with that of the hydrogeological index, $\ln[K_0/f]$. This is also the case in the wheatbelt catchments (Larsen *et al.*, 1993). Larsen *et al.* have also shown that runoff predictions by the distributed model are unaffected when the f parameter is treated as a constant, equal to its catchment average, $\bar{f}$. With these assumptions, Equations (11) and (12) simplify to

$$z_i = \bar{z} - \frac{1}{\bar{f}}(\ln[a/\tan\beta]_i - \lambda) \tag{13}$$

and

$$\bar{f}(\bar{z} - \bar{\psi}) + \lambda \leqslant \ln[a/\tan\beta]_i \tag{14}$$

If the areal distribution of $\ln[a/\tan\beta]_i$ is described by the cumulative distribution function $F_T[\ln(a/\tan\beta)]$ then the area that is initially saturated, A_{c0}, is given by

$$A_{c0}/A = 1 - F_T[\lambda + \bar{f}(\bar{z} - \bar{\psi}_c)] \tag{15}$$

Equation (15) is a catchment-scale equation that describes the initial extent of the saturated area as a function of the catchment-scale parameters ψ, $\bar{f}$ and λ, the underlying distribution of topography expressed through the topography index, $\ln[a/\tan\beta]$, and a single catchment-scale antecedent wetness parameter, $\bar{z}$.

It is straightforward to transform Equation (15) into dimensionless form, consistent with the similarity formulation of Larsen *et al.* (1994). This yields

$$A_{c0}^* = 1 - F_T^*[1 + (z^* - f^*)] \tag{16}$$

where $A_{c0}^* = A_{c0}/A$ is the normalized initial saturation area, $F_T^*[x]$ is the cumulative distribution function of the normalized topographic index, $x = \ln[a/\tan\beta]/\lambda$, and $z^* = \bar{f}\bar{z}/\lambda$ is the normalised initial depth to the water-table.

Prediction of antecedent unsaturated zone soil moisture content

The models to be developed in the next two sections require the estimation of the average soil moisture

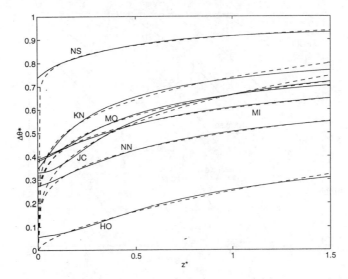

Figure 4. Normalized available water content as a function of the normalized depth to the water-table; a comparison between the distributed model (solid line) and the approximation by Equation (19) (broken line)

content deficit, $\overline{\Delta\theta} = \overline{\theta}_s - \overline{\theta}_0$, over the unsaturated part of the catchment. Here, $\overline{\theta}_s$ is the average saturation moisture content and $\overline{\theta}_0$ denotes the initial soil moisture content, averaged over the entire unsaturated zone above the water-table, before to the storm. This section derives a simple parameterization of $\overline{\Delta\theta}$ in terms of $\bar{z}$ to reduce the number of parameters required to specify the initial soil moisture state of the catchment.

The antecedent soil moisture content, θ_0, varies considerably in space and time. Thus, in reality, two separate parameters are required to characterize the antecedent wetness: $\overline{\Delta\theta}$ and $\bar{z}$. This is impractical and we sought to reduce the number of antecedent wetness parameters by using the full gravity drainage assumption. If full gravity drainage is assumed, then the soil moisture distribution above the water-table is uniquely related to the water-table depth (Sivapalan *et al.*, 1987) and $\overline{\Delta\theta}$ is uniquely related to $\bar{z}$. It should be noted that if additional information about antecedent wetness information is available, then it can be used to specify $\overline{\Delta\theta}$ and $\bar{z}$ directly.

By definition, $\overline{\Delta\theta}$ is given by:

$$\overline{\Delta\theta} = \frac{\sum_{A_{c0}}^{A} M_i \Delta A}{\sum_{A_{c0}}^{A} (z_i - \psi_{ci}) \Delta A} \tag{17}$$

where M_i is the initial soil moisture deficit at location i. The summations in Equation (17) are carried out over the unsaturated elements of the catchment only. At full gravity drainage M_i is related to z_i by (Sivapalan *et al.*, 1987)

$$M_i = (\theta_s - \theta_r)\left\{ z_i - \psi_{ci} - \frac{1}{1 - B_i}\left[\left(\frac{\psi_{ci}}{z_i}\right)^{B_i} z_i - \psi_{ci} \right] \right\} \tag{18}$$

We computed $\overline{\Delta\theta}$ for a wide range of values of $\bar{z}$ by combining Equations (11) and (18), then carrying out the summations in Equation (17) using the distributed model. In this way it has been possible to develop relationships between $\bar{z}$ and $\overline{\Delta\theta}$ for each catchment. The results are presented in dimensionless form by normalizing the variables $\bar{z}$ and $\overline{\Delta\theta}$ using the dimensionless formulation presented by Larsen *et al.* (1994). The normalized variables are $z^* = \bar{f}\bar{z}/\lambda$ and $\Delta\theta^* = \overline{\Delta\theta}/(\overline{\theta}_s - \overline{\theta}_r)$. The relationships between z^* and $\Delta\theta^*$ are shown in Figure 4 for the seven wheat belt catchments. Figure 4 also includes approximations to these relationships of the form

$$\Delta\theta^* = \mu(z^*)^\eta \tag{19}$$

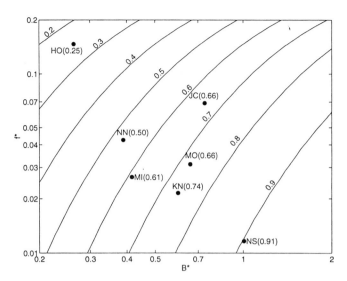

Figure 5. Contours of coefficient μ as a function of the two similarity parameters f^* and B^* obtained by varying B and $\bar{f}$ within the soil property fields for the Holland Catchment

The coefficients μ and η were evaluated for each catchment by fitting the equation by least squares in the range $0.1 < z^* < 1.5$. This range corresponds to the wettest and driest conditions expected on the wheat-belt catchments (Larsen *et al.*, 1994).

We next developed equations for μ and η in terms of the similarity parameters defined by Equations (1) to (5). For this we used a similar methodology to that adopted by Larsen *et al.* (1994). As both B_i and ψ_{c_i} appear in the point-scale Equation (18) for soil moisture deficit, M_i, it was expected that μ and η would primarily be governed by B^* and f^*. This was tested by creating a number of artificial catchments covering a wide range of B^* and f^* values, and evaluating the coefficients μ and η for these using the same procedure adopted for the natural catchments.

From the μ and η values obtained for these artificial catchments, we generated contour plots of μ and η in $B^* - f^*$ space. For example, the contour plots shown in Figure 5 (μ) and Figure 6 (η) were obtained by

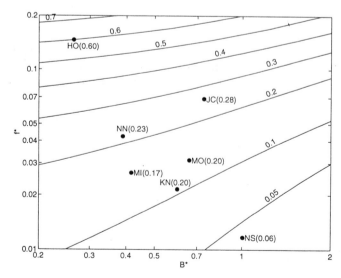

Figure 6. Contours of coefficient (η) as a function of the two similarity parameters f^* and B^* obtained by varying B and $\bar{f}$ within the soil property fields for the Holland catchment

varying $\bar{B}$ and $\bar{f}$, while maintaining the other soil and topographical property fields of the Holland catchment. Similar contour plots were derived by varying the soil property fields of the other catchments and are not presented here.

Figures 5 and 6 also show the positions of the seven experimental catchments in $B^* - f^*$ space, with their corresponding values of μ and η presented in brackets. Note that in most instances, the similarity theory has been able to accurately predict the μ and η values obtained for the natural catchments when the similarity parameters B^* and f^* were identical. Early results suggest that the differences between μ and η for the actual catchments and those predicted using the similarity theory are a result of actual spatial patterns of soil variability which are not accounted for in the similarity theory. Given B^* and f^*, μ and η can be obtained from Figures 5 and 6. These and z^* are then used with Equation (19) to estimate $\Delta\theta^*$.

GROWTH OF SATURATED AREA DURING STORM EVENT

By definition, saturation excess runoff arises due to surface saturation by water-table rise (Dunne, 1978). We estimated the average water-table rise, $\overline{\Delta z}$, due to a specified catchment-averaged cumulative volume (in depth units) of infiltration, $\bar{G}$. In view of the considerable spatial and temporal variability of soil moisture in the unsaturated zone above the water-table, and due to the expansion of the saturation areas, we decided to do this in a rather simple, but intuitive manner, similar to the approach adopted by Sivapalan et al. (1987). We assumed that $\overline{\Delta z}$ can be approximated by

$$\overline{\Delta z} = \frac{\bar{G}}{\epsilon(1 - A_{c0}/A)\overline{\Delta\theta}} \tag{20}$$

where coefficient ϵ is a parameter that takes into account some of the effects of spatial variation in antecedent soil moisture and the increase in saturated areas during the storm.

The average water-table rise given by Equation (20) can be combined with Equation (15) to yield a dynamic extension of Equation (15) to enable it to predict both the initial saturated area and its growth during a storm. This is given by

$$\frac{A_c[\bar{G}]}{A} = 1 - F_T\left[\lambda + \bar{f}\left(\bar{z} - \frac{\bar{G}}{\epsilon(1 - A_{c0}/A)\overline{\Delta\theta}} - \bar{\psi}_c\right)\right] \tag{21}$$

The appropriate value for ϵ was found by comparing the saturated area fraction predicted by Equation (21) with that predicted by the distributed model of Larsen et al. (1994). It was found that ϵ varies in a small range (generally between 0·7 and 0·85). Given its small variation it was not parameterized. However, it is expected that ϵ may vary outside this small range for catchments other than those in the wheat belt region and additional work will be required if the model is to be applied to such catchments.

As before, Equation (21) can be transformed into a dimensionless form using the definitions of dimensionless variables adopted by Sivapalan et al. (1987) and Larsen et al. (1994). This yields the dynamic extension of Equation (16) as follows:

$$A_c^*[G^*] = 1 - F_T^*\left[1 + \left(z^* - f^* - \frac{f^*G^*}{\epsilon(1 - A_{c0}^*)\Delta\theta^*}\right)\right] \tag{22}$$

where $A_c^* = A_c[G^*]/A$ is the normalized dynamic saturated area and $G^* = \bar{G}/\bar{\psi}_c(\bar{\theta}_s - \bar{\theta}_r)$ is the normalized cumulative depth of infiltration. Note that $\Delta\theta^*$ has already been parameterized in terms of the similarity parameters f^* and B^*.

Comparisons of the $A_c^* - G^*$ relation with the catchment-scale model Equation (22) and the distributed model of Larsen et al. (1994) for different z^* values are shown for Jackitup Creek and Minjin catchments in Figures 7 and 8, respectively. Similar results were also obtained for the other five catchments, but are not presented here. These results show that the catchment-scale model is a reasonable approximation to the distributed model. The lumped model is, of course, much more efficient in terms of computational effort. In view of the similarity framework within which it is presented, the lumped model can be scaled

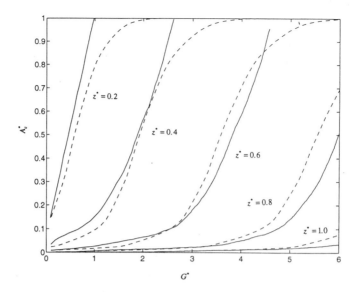

Figure 7. Growth of saturated contributing areas as a function of cumulative infiltration (Jackitup Creek catchment). Distributed model (solid line) and catchment-scale model (broken line) ($\epsilon = 0.8$)

to predict the responses of other catchments in the region, provided the similarity parameters f^* and B^* are identical. Note that the other three similarity parameters either do not play any part in the parameterization of dynamic saturated area (i.e. K_0^*), or have been determined to be relatively invariant within the wheat belt region (i.e. C_{V,K_0} and $C_{V,\ln[\alpha/\tan\beta]}$). These results are consistent with those of Larsen $et\ al.$ (1994) (Figure 1), which show that the dominance of saturation excess is primarily determined by f^*. It is interesting to note that Larsen $et\ al.$ (1994) show that R is independent of B^*, which might seem inconsistent with the catchment-scale model derived here. The difference is due to the fact that R in Larsen $et\ al.$ (1994) is calculated as an average over a uniform distribution of antecedent water-table depths, which removes the effects of B^*.

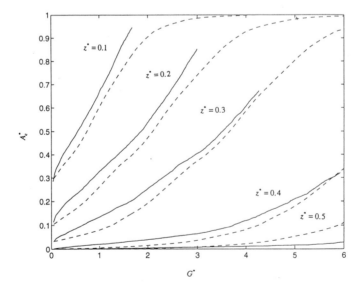

Figure 8. Growth of saturated contributing areas as a function of cumulative infiltration (Minjin catchment). Distributed model (solid line) and catchment-scale model (broken line) ($\epsilon = 0.8$)

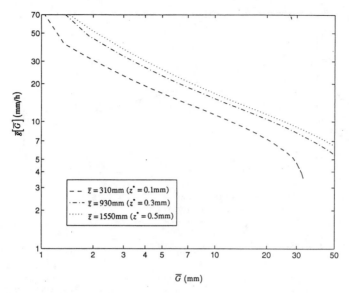

Figure 9. Average infiltration capacity curves as a function of cumulative infiltration (Minjin catchment) on log–log scale

CATCHMENT–AVERAGE INFILTRATION EXCESS RUNOFF

In this section we develop a catchment-scale model that describes the average infiltration capacity, $\bar{g}$, of the unsaturated part of the catchment in terms of the catchment-average cumulative volume of infiltration, $\bar{G}$. The difficulty of obtaining analytical relations for infiltration capacity at the point scale is well known in published work. At the catchment scale this difficulty becomes insurmountable. For this reason, a simulation approach is adopted using the distributed model to infer a simple, parsimonious catchment-scale relationship between $\bar{g}$ and $\bar{G}$.

Figure 9 presents the simulation results, plotted on log–log paper, obtained by applying the distributed model to Minjin catchment for a number of hypothetical storm events with different values of $\bar{z}$. It relates $\bar{g}$ and $\bar{G}$ estimated using Equations (7) and (8). The curves plot as straight lines, indicating a possible power function relationship, although the curves do depart from this linear relationship at very low and high values of $\bar{G}$. This was not considered to be significant as small errors in the predicted value $\bar{g}$ at either the beginning or end of a storm event are not important in the predictive sense. Therefore, the following functional form was deemed sufficient to describe the relationship

$$\bar{g} = \exp(a')\bar{G}^{b'} \tag{23}$$

where a' and b' are constants.

For the power function relation to be meaningful, the two variables a' and b' need to be parameterized to take into account the effects of variability in antecedent soil moisture, soil properties and topography. In the Green and Ampt (1911) formulation, which assumes a sharp wetting front, the infiltration capacity is expressed as a function of the depth to the wetting front. It is therefore intuitively appealing to recast Equation (23) in terms of an effective wetting front depth to remove the effects of antecedent conditions. This is achieved using the same simple approach adopted to describe the average water-table rise, $\overline{\Delta z}$. Recasting Equation (23) in this way yields

$$\bar{g} = \exp(a'') \left(\frac{\bar{G}}{(1 - A/A_{c0})\overline{\Delta\theta}} \right)^{b''} \tag{24}$$

In the case of the effective wetting front depth, above, the parameter, ϵ, is included in parameter a''. When the infiltration capacity $\bar{g}$ is plotted against $\bar{G}/(1 - A_c/A)\overline{\Delta\theta}$, instead of against $\bar{G}$, all the infiltration

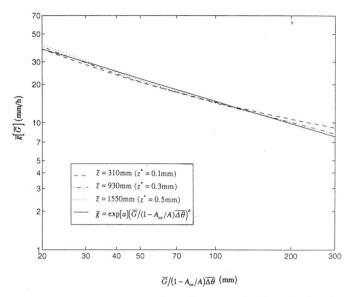

Figure 10. Average infiltration capacity curves as a function of an 'effective' wetting front depth and a comparison between the distributed model and the catchment scale model of Equation (24) (Minjin catchment)

capacity curves collapse into a single curve, as shown in Figure 10 for Minjin catchment. Figure 10 also shows that Equation (24) provides a good approximation of the $\bar{g} - \bar{G}$ relationship. Similar results were found for all of the other catchments.

To develop consistent, regional parameterizations of the infiltration capacity, Equation (24) is recast into a dimensionless form using the similarity formulation of Larsen *et al.* (1994) in the following way

$$g^* = \exp(\phi) \left(\frac{G^*}{(1 - A^*)\Delta\theta^*} \right)^\chi \tag{25}$$

where $g^* = L_{max}\bar{g}/\bar{c}_0\bar{\psi}_c(\bar{\theta}_s - \bar{\theta}_r)$ is the normalized average infiltration capacity. L_{max} and $\bar{c}_0$ are, respectively, a characteristic length and velocity scale that together determine the characteristic time-scale used by Larsen *et al.* (1994).

The parameters ϕ and χ are obtained empirically for each of the seven catchments by estimating the slopes and intercepts of the $\ln[g^*]$ versus $\ln[G^*/(1 - A^*)\Delta\theta^*]$ curves. As Equation (25) is analogous to an early time approximation of the Green and Ampt (1911) equation if χ is equal to -1, it was considered likely that f^* and K_0^* would be the primary determinants of the coefficients ϕ and χ. To test this hypothesis, several artificial catchments were created covering a wide range of f^* and K_0^* values. The values of the two coefficients ϕ and χ were then estimated for these catchments, using an identical procedure to that used previously for the natural catchments. Examples of these are shown in Figures 11 and 12. Figure 11 shows contours of the coefficient ϕ as a function of the two similarity parameters f^* and K_0^* and was obtained by artificially varying the values of $\bar{f}$ and $\bar{K}_0$ while keeping the other soil and topographic properties fixed corresponding to the Holland catchment. Similarly, Figure 12 presents contours for the coefficient χ. The positions of the individual wheatbelt catchments in the $f^*-K_0^*$ diagram are included in Figsures 11 and 12, with their corresponding values of ϕ and χ being presented alongside them in brackets.

In the case of ϕ, all of the seven catchments appear to be located at or near the locations predicted by the similarity theory. Thus the differences between catchments are captured accurately by the two similarity parameters, f^* and K_0^*. In the case of χ there are some differences between the χ values of the actual catchments and the values predicted by the similarity theory. Early results suggest that the difference is a result of actual spatial patterns of soil variability which are not accounted for in the similarity theory.

The two contour plots for ϕ and χ (Figures 11 and 12) also indicate that the coefficient ϕ is dependent mainly on K_0^*, except for large values of f^* where it is dependent on both. The coefficient χ is almost

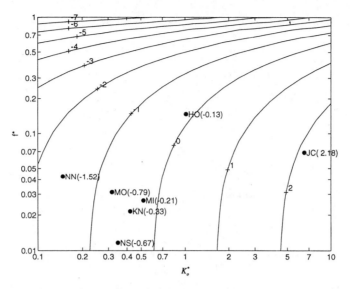

Figure 11. Contours of coefficient ϕ as a function of the two similarity parameters f^* and K_0^* obtained by varying $\bar{K}_0$ and $\bar{f}$ within the soil property fields for the Holland catchment

entirely dependent on f^*. Combined together, the contour plots presented in Figures 11 and 12 indicate that the linkages between point- and catchment-scale parameterizations of infiltration capacity are complex and non-linear. The similarity results are also consistent with those of Larsen *et al.* (1994) (Figure 1), which show that the dominance of infiltration excess mechanism is primarily determined by K_0^*.

MODEL APPLICATION

To illustrate the ability of the catchment-scale model [consisting of Equations (10), (19), (22) and (25)] to predict the responses of the wheat belt catchments to actual rainfall events, we compared the observed

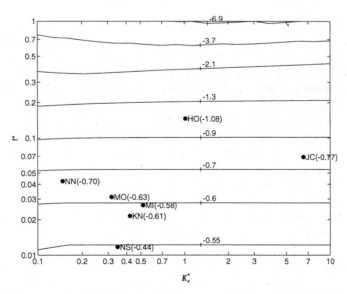

Figure 12. Contours of coefficient χ as a function of the two similarity parameters f^* and K_0^* obtained by varying $\bar{K}_0$ and $\bar{f}$ within the soil property fields for the Holland catchment

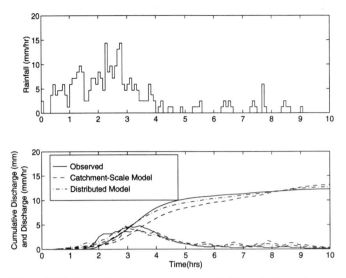

Figure 13. Hydrographs for a storm event on Jackitup Creek catchment

runoff hydrographs and the hydrographs predicted by the catchment-scale model and the distributed model. The simulations were carried out for two storm events, one on Jackitup Creek catchment, and the other on Minjin. According to the similarity study of Larsen *et al.* (1994), the dominant mechanisms of runoff generation on these two catchments are saturation excess (JC) and infiltration excess (MI), respectively.

Note that the catchment-scale infiltration capacity equation, Equation (25), is an effective equation and assumes that infiltration excess runoff is generated only when the entire catchment is saturated 'from above'. In reality, runoff could be generated earlier in the less permeable parts of the catchment due to the spatial variability of soil properties and antecedent soil moisture. By using the lumped equation we are effectively assuming that all such runoff is re-infiltrated back until the entire catchment becomes saturated. The distributed model, on the other hand, assumes no re-infiltration. This difference between the two models needs to be kept in mind when comparing the two hydrographs.

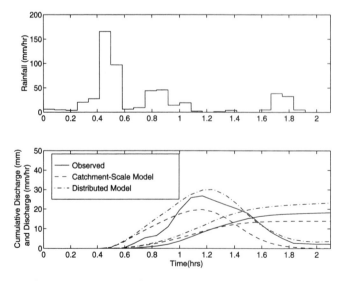

Figure 14. Hydrographs for a storm event on Minjin catchment

Table III. Calibrated parameters and error analysis for the two simulated events

	$\bar{z}$ (mm)	D (m²/h)	c_0 (m/h)	Error catchment-scale model* (mm/hr)	Error distributed model* (mm/hr)
Jackitup Creek (JC)	180	$1 \cdot 12 \times 10^{11}$	500	0·71	0·44
Minjin (MI)	1000	$4 \cdot 12 \times 10^{9}$	650	4·14	3·57

Error is the square root of the mean squared difference between the observed and predicted discharge hydrographs

To compute the outflow hydrograph at the outlet for the catchment-scale model, it is necessary to couple it to a routing scheme. This is achieved by using the approach of Mesa and Mifflin (1986), also adopted by Larsen *et al.* (1994). The discharge at the catchment outlet, $Q(t)$, at time, t, is given by

$$Q(t) = \int_0^t \int_0^{L_{max}} w(x) p_{eff}(t) h(x, t - \tau) \, dx \, dt \qquad (26)$$

where

$$h(x, t) = \frac{x}{(4\pi D t^3)^{1/2}} \exp\left[\frac{-(x - c_0 t)^2}{4Dt}\right] \qquad (27)$$

The function $w(x)$ is an area-weighting function, D is a diffusion coefficient and c_0 is the wave celerity (average overland flow velocity). Given the lack of clearly defined flow paths within the catchment, the function $w(x)$ is taken to be the distribution of catchment area at a radial distance, x from the catchment outlet. In this work, both c_0 and D are used as calibration parameters. Note that the antecedent soil moisture parameter, $\bar{z}$, and routing parameters, c_0 and D, estimated with the distributed model are kept the same for the two models.

The resulting hydrographs for the selected storm events on the Jackitup Creek and Minjin catchments are illustrated in Figures 13 and 14, respectively. The calibration parameters, together with the square root of the average squared error between the measured and predicted hydrographs, are given in Table III. Figures 13 and 14 and the error estimates in Table III demonstrate that the lumped model does almost as well as the distributed model. Note, however, that the catchment-scale model tends to predict less runoff, especially for the Minjin catchment where runoff is dominated by infiltration excess runoff. This is a consequence of the differences in the way re-infiltration is handled in the two models, as mentioned earlier. These results confirm that it is possible to develop a lumped model of the catchment response, of a complexity comparable with simple conceptual models, while retaining the physical basis of a distributed model.

DISCUSSION AND CONCLUSIONS

The aggregation approach presented in this paper represents one possible methodology for the development of physically based, spatially lumped models of runoff generation at the catchment scale. When the actual processes are complex and spatially heterogeneous, it appears that a combined approach using simulation and semi-analytical arguments, together with intuitive reasoning, will yield satisfying results. Analytical approaches are extremely difficult under these circumstances.

At the catchment scale there appears to be a certain invariance in the runoff generation response when the response function is expressed in terms of cumulative infiltration, rather than in terms of time. This idea is an extension of the time condensation approximation to the catchment scale, for the prediction of not just the infiltration capacity, but also the growth of the saturated areas by saturation excess during a storm event.

The resulting lumped, physically based model, being a simple one-step equation, is much less complex than the distributed model from which it was derived. The distributed model requires a large amount of parameter information for every grid point within the catchment. In contrast, the lumped model is defined entirely in terms of the underlying distribution of topography, three similarity parameters K_0^*, f^* and B^*, the normalized average water-table depth, z^*, and the normalized cumulative volume of infiltration, G^*. This reduced parameter set should be commensurate with the information content of the hydrological record.

The aggregation approach adopted in this paper for the development of the catchment-scale model could be classified as belonging to the 'upwards approach' (Klemes, 1983). Its validity is closely tied to that of the distributed model. The approach will obviously fail if the small-scale descriptions of runoff generation used in the distributed model are not truly representative of the actual processes taking place on the catchment.

The paper has also illustrated the power of the similarity theory of Sivapalan et al. (1987) and Larsen et al. (1994). In the case of the eastern wheat belt catchments, three similarity parameters, namely f^*, B^* and K_0^* have been found to describe catchment storm responses adequately. Regionalization of the run-off prediction in the eastern wheat belt region, ideally, should be based on these parameters.

In conclusion, this paper presents a methodology for the development of simple, process-based, catchment-scale models. The methodology presented in this paper has been motivated by the representative elementary area (REA) concept (Wood et al., 1988), which has been suggested as the scale at which only the underlying statistical distributions of parameters, such as soil properties, topography and rainfall, are sufficient to model the catchment response. We have not been able to address the applicability of the REA concept for hydrological parameterizations in these catchments because of their relatively small size. However, some of the discrepancies found in the results presented here could be attributed to the fact that actual patterns of variability may have been important in these small catchments. This needs to be examined in future studies.

ACKNOWLEDGEMENTS

This research was supported by a Special Environmental Research Grant awarded to the second author by the University of Western Australia. The authors are grateful to Neil Coles for making available the data from the eastern wheat belt catchments, and to Jens Larsen and Ross Woods for their critical review of the paper. Report ED 845 JR.

REFERENCES

Abbott, M. B., Bathurst, J. C., Cunge, J. A., O'Connell, P. E. and Rasmussen, J. 1986. 'An introduction to the European Hydrological System—Systeme Hydrologique Europeen, "SHE", 2. Structure of a physically-based, distributed modelling system', J. Hydrol., **87**, 61–77.

Beven, K. J. 1988 'Changing ideas in hydrology—the case of physically based models ', J. Hydrol., **105**, 61–77.

Beven, K. J. and Kirkby, M. J. 1979. 'A physically-based variable contributing area model of basin hydrology', Hydrol. Sci. Bull., **24**, 43–69.

Beven, K. J., Wood, E. F., and Sivapalan, M. 1988. 'On hydrological heterogeneity—catchment morphology and catchment response', J. Hydrol., **100**, 353–375.

Brooks, R. H. and Corey, A. T. 1964. 'Hydraulic properties of porous media', Hydrol. Sci. J., **27**, 505–521.

Chow, V. T., Maidment, D. R., and Mays, L. W. 1988. Applied Hydrology. McGraw-Hill, New York. 572pp.

Coles, N. 1994. 'Soil factors affecting runoff in the south-west of Western Australia', Unpublished PhD dissertation, Department of Soil Science, University of West Australia, Nedlands.

Dunne, T. 1978. 'Field studies of hillslope processes' in Kirkby, M. J. (Ed) Hillslope Hydrology. McGraw- Hill, New York, pp. 145–175.

Famiglietti, J.S., Wood, E.F., Sivapalan, M., and Thongs, D.J. 1992. 'A catchment scale water balance model for FIFE', J. Geophys. Res., **97**(D19), 18 977–19 077.

Freeze., R. A. 1974. 'Streamflow generation', Rev. Geophys. Space Res., **12**, 627–647.

Green, W. H. and Ampt, G. A. 1911. 'Studies in soil physics, 1: the flow of air and water through soils', J. Agric. Sci., **4**, 1–24

Institution of Engineers Australia (IEA) 1987 Australian Rainfall and Runoff: a Guide To Flood Estimation. Vol.1. 3rd edn. IEA, Canberra.

Klemes, V. 1983. 'Conceptulization and scale in hydrology', J. Hydrol., **65**, 1–23.

Larsen, J. E., Linnet, P. E., Coles, N. A., and Sivapalan, M. 1993. 'Heterogeneity and similarity of catchment responses in small agricultural catchments in Western Australia', Rep. WP 503 MS, Centre for Water Research, University of Western Australia, Nedlands.

Larsen, J. E., Sivapalan, M., Coles, N. A., and Linnet, P. E. 1994. 'Similarity analysis of runoff generation in small agricultural catchments', Wat. Resour. Res., **30**, 1641–1652.

Mesa, O. J. and Mifflin, E. R. 1986. 'On the relative roles of hillslope and network geometry in hydrologic response' in Gupta, V. K., Rodriguez-Iturbe, I., and Wood, E. F., (Eds), *Scale Problems in Hydrology*. D. Reidel, Dordrecht. pp. 1–16.

Milly, P. C. D. 1986. 'An event-based simulation model of moisture and energy fluxes at a bare soil surface', *Wat Resour. Res.*, **22**, 1680–1692.

O'Connell, P. E. 1991. 'A historical perspective' in Bowles, D. S. and O'Connell, P. E. (Eds), *Recent Advances in the Modeling of Hydrologic Systems. Proceedings of the NATO Advanced Study Institute.* Kluwer Academic, Dordrecht. pp. 3–30.

Reeves, M. and Miller, E. E. 1975. 'Estimating infiltration for erratic rainfall', *Wat. Resour. Res.*, **11**, 102–110.

Sharma, M. L., Gander, G. A., and Hunt, C. G. 1980. 'Spatial variability of infiltration in a watershed', *J. Hydrol.*, **45**, 101–122.

Sivapalan, M. and Milly, P. C. D. 1989. 'On the relationship between the time condensation approximation and the flux-concentration relation', *J. Hydrol.*, **105**, 357–367.

Sivapalan, M. and Wood, E. F. 1986. 'Effect of scale on the infiltration response in spatially variable soils', in Gupta, V. K., Rodriguez-Iturbe, I., and Wood, E. F. (Eds), *Scale Problems in Hydrology*. D. Reidel, Dordrecht. pp. 81–106.

Sivapalan, M., Beven, K., and Wood, E. F. 1987. 'On similarity, 2. A scaled model of storm runoff production', *Wat Resour. Res.*, **23**, 2266–2278.

Smith, R. E. and Hebbert, R. H. B. 1979. 'A Monte Carlo analysis of the hydrologic effects of spatial variability of infiltration, *Wat. Resour. Res.*, **15**, 419–429.

Soil Conservation Service 1972. *National Engineering Handbook, Section 4: Hydrology.* US Department of Agriculture, available from US Government Printing Office, Washington, DC.

Wood, E. F., Sivapalan, M., Beven, K. J., and Band, L. 1988. 'Effects of spatial variability and scale with implications to hydrologic modeling', *J. Hydrol.*, **102**, 29–47.

Wood, E. F., Sivapalan, M., and Beven, K. J. 1990. 'Similarity and scale in catchment storm response', *Rev. of Geophys.*, **28**, 1–18.

NOTATION

a_{ci}	Indicator of saturation at location i
A	Catchment area
A_c	Saturated contributing area
A_{c0}	Antecedent saturated contributing area
A_c^*	Normalized A_c
A_{c0}^*	Normalized A_{c0}
B	Pore size distribution index
B_i	Pore size distribution index at location i
$\bar{B}$	Areal (catchment) average of B
B^*	Equal to $\bar{B}$ (dimensionless similarity parameter)
c_0	Storm average wave celerity (characteristic flow velocity)
$\bar{c}_0$	Catchment-storm average wave celerity (characteristic flow velocity)
C_{V,K_0}	Coefficient of variation of K_0 (dimensionless similarity parameter)
$C_{V,\ln[a/\tan\beta]}$	Coefficient of variation of $\ln[a/\tan\beta]$ (dimensionless similarity parameter)
D	Hydrodynamic diffusion coefficient for runoff routing
f	Exponential decrease (with depth) parameter for saturated hydraulic conductivity
$\bar{f}$	Areal (catchment) average of f
f^*	Normalized f (dimensionless similarity parameter)
f_i	f at location i
$\bar{g}$	Average (unsaturated part of the catchment) infiltration capacity
$F_T[x]$	Cumulative distribution function of $x = \ln[a/\tan\beta]$
$F_T^*[x]$	Cummulative distribution function of $x = \ln[a/\tan\beta]/\lambda$
g^*	Normalised $\bar{g}$
$\bar{G}$	Average (catchment) cumulative infiltration
G^*	Normalised $\bar{G}$
K_0	Saturated hydraulic conductivity at the surface
$\bar{K}_0$	Areal (catchment) average of K_0
K_0^*	Normalized $\bar{K}_0$ (dimensionless similarity parameter)
L_{max}	Flow distance from catchment outlet to the furthest point on the catchment
M_i	Initial soil moisture deficit at location i

$\ln[a/\tan\beta]$	Topographic index
$\ln[a/\tan\beta]_i$	Topographic index at location i
$\ln[K_0/f]$	Hydrogeological index
$\ln[K_0/f]_i$	Hydrogeological index at location i
$\bar{p}$	Areal (catchment) average of rainfall intensity
$\bar{p}_{\text{eff}}$	Areal (catchment) average of effective $\bar{p}$ (the rate of runoff generation)
R	Larsen et $al.$'s (1994) ratio volume of runoff generated by saturation excess to the total volume of runoff generated in a typical year
$w[x]$	Area weighting function
z_i	Initial water table depth at location i
$\bar{z}$	Areal (catchment) average of the initial water-table depths
z^*	Normalized dimensionless $\bar{z}$
ΔA	Size of rectangular grid element
$\overline{\Delta z}$	Average (catchment) water-table rise
$\overline{\Delta\theta}$	Average (unsaturated part of the catchment) antecedent soil moisture deficit
$\Delta\theta^*$	Normalized $\overline{\Delta\theta}$
χ	Parameter of lumped average infiltration capacity model
ϵ	Parameter of lumped saturation area model
ϕ	Parameter of lumped average infiltration capacity model
γ	Average (catchment) of $\ln[K_0/f]$
η	Parameter of lumped average antecedent soil moisture deficit model
λ	Average (catchment) of $\ln[a/\tan\beta]$
μ	Parameter of lumped average antecedent soil moisture deficit model
θ_0	Initial water content (unsaturated zone)
$\bar{\theta}_0$	Catchment average initial water content (unsaturated zone)
θ_r	Residual water content
$\bar{\theta}_r$	Catchment average of the residual water content
θ_s	Saturated water content
$\bar{\theta}_s$	Catchment average of saturated water content
σ_{K_0}	Standard deviation of K_0
$\sigma_{\ln[a/\tan\beta]}$	Standard deviation of $\ln[a/\tan\beta]$
ψ_{c_i}	Thickness of the capillary fringe at location i
$\bar{\psi}_c$	Average (catchment) ψ_{c_i}

19

STREAM NETWORK MORPHOLOGY AND STORM RESPONSE IN HUMID CATCHMENTS

PETER A. TROCH AND FRANCOIS P. DE TROCH

Laboratory of Hydrology and Water Management, University of Ghent, Ghent, Belgium

MARCO MANCINI

DIIAR, Politecnico di Milano, Milan, Italy

AND

ERIC F. WOOD

Water Resources Program, Princeton University, Princeton, NJ, USA

ABSTRACT

Addressing scaling issues in hydrological modelling involves, among other things, the study of problems related to hydrological similarity between catchments of different scales. Recent research about catchment similarity relationships is based on distributed conceptual models of surface runoff production. In this type of hydrological modelling both infiltration excess and saturation excess runoff production mechanisms are considered. In many humid lowland areas overland flow is a rare phenomenon because of the specific conditions that prevail: moderate rainfall, high infiltration capacity and low relief. The complete drainage system in these regions consists of surface and subsurface components which have organized themselves in a given geological, geomorphological and climatic situation. A surface drainage network has developed through sapping erosion at the zone of groundwater exfiltration. The resulting hierarchical stream network is in equilibrium with large time-scale conditions and adjusts itself dynamically to the inter-year and seasonal meteorological fluctuations. Greater understanding of the interrelationships that underlie the storm response of catchments in humid lowland regions can be expected by focusing on stream network morphology as a function of topography, geology and climate. This paper applies the physically based mathematical model of stream network morphology, developed by De Vries (1977), to the Zwalmbeek catchment, Belgium. Based on this model and for different climatic conditions (expressed in terms of rainfall characteristics) the first-order stream spacing versus average water-table depth relationship is calculated. From field observations, digital elevation model derived channel network drainage densities and flood event analysis it is concluded that the 1% exceedance probability rainfall can be suggested as representative for the shaping climatic conditions in the catchment under study. The corresponding curves relating channel network characteristics, such as stream spacing, drainage density and channel geometry, to average water-table depth are basin descriptors and could be used for comparative studies (e.g. regional flood frequency analysis). The model further allows for the prediction of the expansion and shrinkage of the first-order channel network as a function of catchment wetness expressed in terms of the effective water-table depth.

INTRODUCTION

In the past two decades important contributions to the understanding of hydrological similarity have been possible through the use of catchment-scale physically based models of runoff generation (Beven, 1986; Sivapalan *et al.*, 1987; 1990). Sivapalan *et al.* (1987) investigated catchment similarity using a physically based model which incorporated both the infiltration excess and saturation excess mechanisms of runoff generation. They were able to define five dimensionless similarity parameters as potential measures of hydrological response. According to their theory, two catchments can be considered hydrologically similar, in terms of their runoff generation response, if they have similar values of all five similarity parameters. One of the shortcomings of these similarity parameters is that they are based mainly on point-scale soil and topography characteristics. Obviously, these characteristics are important to identify runoff production, a process which typically exists at that scale. Although useful to classify dominant runoff processes in drainage basins as either Hortonian or Dunnian, these similarity parameters contain little information about subsurface flow processes. Therefore, it is doubtful that they are able to describe fully the hydrological response at the catchment scale. Hydrogeomorphic features, relevant at the catchment scale (such as the channel network), need to be taken into consideration in this type of study. Drainage network evolution and the resulting stream network morphology is important in understanding surface and subsurface hydrological response at the catchment (and regional) scale.

The study of drainage network evolution in areas of uniform lithology and negligible structural control has been based on both deterministic and stochastic modelling. Stochastic models of stream network development are based on random topology (Shreve, 1966; 1967). An example of a deterministic approach to stream network development is the well-known Horton model (Horton, 1945), based on surface erosion. De Vries (1974) developed another deterministic model of stream network development applicable to humid lowland areas. Because of the moderate rainfall intensities, the high infiltration capacity and the low relief, unchannelled overland flow (Hortonian overland flow) is a rare phenomenon in these catchments and therefore almost all precipitation surplus percolates to the subsurface to become part of a groundwater drainage system. This groundwater flow eventually re-appears at the surface as soon as the subsurface discharge capacity is exceeded by precipitation surplus. At this zone of groundwater exfiltration, a surface drainage network may develop through sapping erosion. Such a channel system represents an interface between subsurface and surface drainage components. The channel system that develops in this way is a function of subsurface permeability, climate and topography.

The groundwater drainage capacity is primarily a function of climate and storage capacity. The latter is associated with soil properties and average groundwater depth at the beginning of a wet period. In contrast with the average groundwater depth, the seasonal fluctuation of the groundwater table is a variable within an existing drainage system. It regulates the imbalance between recharge and discharge by the combination of storage and hydraulic gradients. This feedback mechanism includes the changing number of drainage channels that participate in the drainage process (De Vries, 1993).

A drainage system consists of surface and subsurface components which have organized themselves in such a way as to satisfy the drainage requirements in a given geological, geomorphological and climatic situation. This self-regulating process results in the development of a hierarchical stream network which is in equilibrium with conditions on a longer time-scale, and which adjusts itself dynamically to the inter-year and seasonal meteorological fluctuations by shifts in the number and character of streams that interface with the groundwater discharge. Therefore, channel network morphology is important to understand the hydrological response in humid lowland areas. Carlston (1963) and Gregory and Walling (1968) were about the first to relate the variation of drainage density within a catchment to streamflow. Carlston (1963) showed, based on the groundwater flow model developed by Jacob (1943), that baseflow Q is related to D_d^{-2}, with D_d drainage density. This paper expands these earlier ideas and presents a physically based model of stream network morphology to predict the fluctuation of drainage basin characteristics during wet and dry periods. The mathematical model is based on the groundwater outcrop erosion model developed by De Vries (1977). The model is applied to the catchment of the Zwalmbeek, a tributary of the Scheldt river in Belgium. For different rainfall characteristics we obtain different curves relating first-order stream spacing and drainage density to average water-table depth. Predictions about stream spacing and drainage density in function of initial storage capacity and rainfall characteristics are compared with map derived and digital elevation model (DEM) derived channel networks. From this analysis it is hypothesized that the 1% exceedance probability rainfall intensity and duration can be regarded as representing the shaping climatic conditions in the catchment under study. It is shown that the corresponding drainage density curves can be used to explain the expansion and shrinkage of the drainage network during storm and interstorm periods. The results of this work are important in extending research

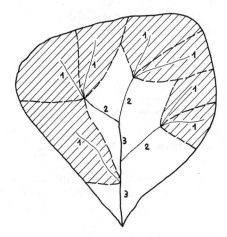

Figure 1. Typical channel network of a third-order (Strahler) drainage basin

on hydrological similarity between catchments. The drainage density curves derived here take into account information on the basin topography, geology and climate and, as such, can be interpreted as basin descriptors of the surface and subsurface components of the drainage system and therefore could be used for comparative studies (e.g. regional flood frequency analysis).

STREAM NETWORK GROWTH MODEL

Stream network morphology

The mathematical model for stream network morphology in humid regions presented here is based on the groundwater outcrop erosion model developed by De Vries (1977).

Let us consider a humid drainage basin of given Strahler order Ω (Figure 1; $\Omega = 3$). The stream network morphology of the basin is characterized by the bifurcation ratio, R_B, the length ratio, R_L, and the drainage area ratio, R_A. Let us isolate two adjacent first-order drainage basins with corresponding drainage area A' and A'' (Figure 2). The groundwater flow system towards the first-order streams with given length l' and l'' is shown in cross-section OO' in Figure 2. For the given cross-section, groundwater flow to the streams can

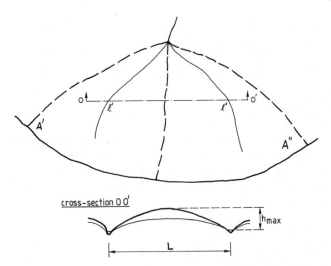

Figure 2. Definition sketch of the phreatic aquifer draining towards a set of parallel drains

be schematized as flow towards a set of more or less parallel drains. For a given steady recharge rate, R, this flow can be described by the formula given by Ernst (1956)

$$\frac{h}{R} = \frac{L^2}{8T} + \frac{L}{\pi K_r} \ln\left(\frac{aD_r}{P}\right) \tag{1}$$

where h = difference in hydraulic head between the divide and the stream level; L = stream spacing; T = horizontal transmissivity of the aquifer; K_r = radial hydraulic conductivity near the stream; D_r = depth of radial flow; P = wetted perimeter of stream; and a = shape factor; for a homogeneous aquifer $a = 1$.

The first term on the right-hand side of Equation (1) is the horizontal flow resistance, whereas the second term represents the radial flow resistance near the drain. For an aquifer which is completely saturated, $h = h_{max}$, and introducing the expression $s^* = 2h_{max}/L$, where s^* is the average transversal slope of the terrain, Equation (1) yields

$$R_{max} = \frac{0.5s^*}{\dfrac{L}{8T} + \dfrac{1}{\pi K_r}\ln\left(\dfrac{aD_r}{P}\right)} \tag{2}$$

where R_{max} represents the maximum recharge rate for given topographic and geological conditions.

Under steady-state conditions, the discharge in the first-order channels midway between the source and the junction is given approximately by

$$Q = 0.5 R \bar{A}_1 \tag{3}$$

where $\bar{A}_1$ is the average first-order drainage area. For half-circular channels with radius r and bankful uniform flow conditions, the maximum discharge Q_{max} is given by the Strickler–Manning equation

$$Q_{max} = k_M r^{2.67} s^{0.5} \tag{4}$$

where s is the average slope of the first-order streams and k_M Strickler–Manning coefficient. For drainage basins where overland flow is a rare phenomenon, the maximum recharge rate can be expressed as

$$R_{max} = \frac{r^{2.67} s^{0.5} k_M}{0.5 \bar{A}_1} \tag{5}$$

To avoid ponding, the following condition has to be satisfied

$$t\bar{R} \geqslant ti - S \tag{6}$$

where t = length of rainfall period; i = average rainfall rate for the given period t; $\bar{R}$ = average recharge rate during t; and S = available initial storage in the unsaturated zone.

Rainfall characteristics corresponding to the climatic conditions in a given area can be expressed in terms of the intensity–duration–frequency (IDF) relationship (Demarée, 1985; De Vries, 1974)

$$i = ct^{-m} \tag{7}$$

where c and $0 < m < 1$ are local climatic parameters which depend on the exceedance probability p. Substitution of the IDF relationship in the water balance Equation (6) and taking the first derivative with respect to t yields the following expression for $\bar{R}_{max}$

$$\bar{R}_{max} = c(1-m)\left(\frac{S}{mc}\right)^{m/m-1} \tag{8}$$

$\bar{R}_{max}$ is the maximum average groundwater discharge corresponding to a given exceedance probability of rainfall intensity and duration. As a first approximation we can assume that $R_{max} = 2\bar{R}_{max}$. The available initial storage capacity, S, is a function of the average depth to the water-table, $\bar{z}$, and soil hydraulic characteristics. If we assume that before storm rainfall the pressure distribution in the unsaturated zone is hydrostatic and we use the Brooks–Corey soil water retention characteristic (Brooks and Corey, 1964), S is given by (Sivapalan et al., 1987)

$$S = (\theta_s - \theta_r)\left\{\bar{z} - \psi_c - \frac{1}{1-B}\left[\left(\frac{\psi_c}{\bar{z}}\right)^B \bar{z} - \psi_c\right]\right\} \tag{9}$$

where θ_s = saturated volumetric moisture content; θ_r = residual volumetric moisture content; ψ_c = height of capillary fringe; and B = Brooks–Corey pore size distribution index.

Catchment wetness index

The average depth to the water-table, $\bar{z}$, as an index of catchment wetness, can be related to the observed baseflow Q. Troch et al. (1993) developed a method to determine the effective or average depth to the water-table, $\bar{z}$, as a measure of the initial storage capacity before storm rainfall. This method is valid for humid regions characterized by a relatively shallow groundwater table. The method is based on Boussinesq's standard hydraulic groundwater theory for horizontal aquifers. Troch et al. (1993) derived the following expression that relates $\bar{z}$ to the ratio Q/Q_0

$$\frac{Q}{Q_0} = \frac{1 \cdot 673(D - \bar{z})^2}{D^2} \tag{10}$$

where D is the effective depth to the water-table and Q_0 is a critical value of the baseflow corresponding to a situation where the aquifers start to behave in accordance with the large time solution of Boussinesq's equation (Boussinesq, 1904). This value is given by (Troch et al., 1993)

$$Q_0 = 3 \cdot 450kD^2 D_d L_t \tag{11}$$

where k can be considered as catchment-scale effective hydraulic conductivity, D_d represents drainage density and L_t is total length of the channels. An estimate of the effective hydraulic conductivity is based on the baseflow analysis as suggested by Brutsaert and Nieber (1977)

$$\frac{dQ}{dt} = -\alpha Q^\beta \tag{12}$$

For the large time solution of Boussinesq's equation the parameters α and β are given by

$$\alpha = \frac{4 \cdot 804k^{1/2}L_t}{fA^{3/2}}; \qquad \beta = 3/2 \tag{13}$$

where f is the drainable porosity. For the catchment under study (the Zwalmbeek, Belgium), Troch et al. (1993) showed that the observed baseflow recession curves behaves as predicted by Equations (12) and (13).

Stream network growth

In humid catchments, baseflow fluctuates seasonally and within storms. The method described in the preceding section relates baseflow observations to the average depth to the water-table in the catchment. For a given range of $\bar{z}$ values the available initial storage capacity, S, can be computed from Equation (9). Then from Equation (8) the maximum average groundwater discharge corresponding to a given exceedance probability of rainfall intensity and duration is calculated. Using the approximation $R_{max} = 2\bar{R}_{max}$ and simple expressions for steady-state groundwater and channel flow [Equations (2) and (5), respectively], we have a way to predict analytically the expansion and shrinkage of first-order channels (expressed in terms of hydraulic geometry and stream spacing or drainage density) during wet and dry periods. The shape of the resulting functions relating first-order channel radius and stream spacing with depth to the water-table can then be used to characterize subsurface flow processes in humid catchments, as these relationships include information about topography, climate and geology. As such, these relationships can be used in hydrological similarity studies. Ongoing research applying this method to different catchment addresses this problem. To illustrate the above-developed model we now present in the following section the results from one humid catchment in Belgium.

APPLICATION TO THE ZWALMBEEK

Physical characteristics of the catchment

The geological formulations relevant for understanding soil conditions in the catchment consist of Tertiary and Quartenary sediments (Table I).

The lithology of the Tertiary rocks in this region is composed of several alternated subhorizontal layers of clay, sand and mixtures of both (Eocene and Miocene). The structure presents a gentle inclination to the NNE. During the last glacial period (Wurm glacial) the eroded Tertiary surface was covered by niveo-aeolic sandy loam and loam (loess) soils. The depth of the Quartenary cover depends on the geomorphology of the region. On the gentle slopes, oriented to the east, the depth of the Pleistocene cover ranges between 5 and 15 m, whereas on the steep slopes, oriented to the south-west, this cover is shallow or not present. In general, over the whole region the loess cover is almost 3–7 m thick. During the Holocene the valleys were partially covered with alluvial clayey and loamy material.

Of particular interest to the objectives of this study is the genesis of the valleys in the region. Detailed information about this and other geomorphological aspects of the Zwalmbeek catchment are found in Vanmaercke-Gottigny (1967). Vanmaercke-Gottigny (1967) determined the genesis of valleys by observing the spreading and orientation of the valley heads on the topographic maps. She distinguished three main orientations. The SSW–NNE valley heads all have springs or fossil spring basins. Their orientation is roughly the same as the main inclination of the geological structure. These valleys are labeled spring valleys. The west–east valley heads are numerous and situated on the more gentle valley sides which are

Table I. Summary of geological formations in the Zwalmbeek catchment

Quartenary		
Holocene:	recent colluvium from hillslopes	
	recent alluvium from Zwalmbeek and tributaries	
Pleistocene:	niveo-aeolic sandy loam and loam, mixed with	
	Tertiary sediments	
Tertiary		
Eocene:	Bartoon:	heavy clay rich in glauconite
	Ledian:	fine sand and sandstone
	Paniselian:	sand and clay rich in glauconite, on heavy clay

covered with loess. They are sedimentary dells which principly rose at the same time of the gentle loess slopes. The north–south to east–west valley heads are not so frequent and their valleys are shorter. Mostly they are erosive dells on the steep, loessless northern and eastern valley sides. Many valleys are asymmetrical. One of the valley sides is often gentle, loess-covered and always oriented on the west or the east. It has a characteristic declivity of 3%. This value corresponds to an equilibrium value, resulting from niveoaeolian loess deposit and periglacial wash and mass movement on this sediment. On the base of such slopes, the concavity of loess deluvium presents a declivity of 1–2%. On the other, steep side of the valley, where the Eocene sediments appear, no characteristic slope value could be observed (Vanmaercke-Gottigny, 1967).

The main valleys are about south–north. Possibly they are consequent rivers arisen with Diestian regression (Miocene). About each kilometre these valleys have bifurcations. Such a bifurcated structure of valleys seems to be caused by springs and their regressive erosion. The main valleys in the region are the Scheldt valley, the Zwalm valley and the Dender valley. These structures define the primary topography of the area. The secondary topography is determined by the frequent west–east valleys on the gentle, loess-covered slopes and the less frequent north–south to east–west short valleys.

De Troch (1977) determined, based on topographic maps of scale 1/10 000, geomorphologically characteristic parameter values for the stream network, such as the bifurcation ratio, R_B, the length ratio, R_L, the drainage area ratio, R_A, and the slope ratio, R_S. These values are based on Strahler's stream network ordering system. Table II summarizes the main geomorphological characteristics of the drainage network. The total number of first-order streams on these topographic maps is 219, the average length $\bar{l}_1 = 230\,\text{m}$ and the average drainage area $\bar{A}_1 = 111\,900\,\text{m}^2$. The average slope of first-order streams, $\bar{s}_1$, is estimated as 3·67%. The first-order basin drainage density equals $2 \times 10^{-3}\,\text{l/m}$. In this study, the average transversal slope for first-order basins, $\bar{s}_1^*$ is estimated as equal to the average slope of first-order streams. This estimate is motivated by the observations of Vanmaercke-Gottigny (1967).

Climatic conditions can be described as humid temperature. The yearly mean air temperature is 10°C, the average of the coldest month (January) being 3°C and the average of the warmest month (July) being 18°C.

Table II. Stream network morphology model parameters for the Zwalmbeek

Soil hydraulic characteristic parameters (Troch, 1993)

$\theta_s = 0\cdot40$	$\psi_c = 0\cdot10$–$1\cdot0\,\text{m}$
$\theta_r = 0\cdot14$	$B = 0\cdot934$

Aquifer characteristic parameters (Troch et al., 1993)

$K = 20\,\text{m/day}$	$T = 42\,\text{m}^2/\text{day}$
$D = 2\cdot1\,\text{m}$	$K_r = 20\,\text{m/day}$
$f = 0\cdot05$	

Stream network geomorphologic characteristics (De Troch, 1977)

$R_B = 3\cdot89$	$N_1 = 219$
$R_L = 2\cdot35$	$\bar{A}_1 = 111\,900\,\text{m}^2$
$R_A = 5\cdot49$	$\bar{l}_1 = 230\,\text{m}$
$R_S = 0\cdot46$	$\bar{s}_1 = 3\cdot67\%$
$\bar{D}_{d,1} = 2 \times 10^{-3}\,\text{l/m}$	$\bar{s}_1^* = 3\cdot67\%$

Intensity–duration — frequency relationship parameters (Demaree, 1985)

Exceedance probability	m	c (mm day$^{-0\cdot2318}$)
1	0·7682	80·25
5	0·7682	61·22
10	0·7682	54·02
20	0·7682	46·91

The mean yearly rainfall is 775 mm and is distributed almost uniformly over the year. The mean yearly evaporation is about 450–500 mm (De Troch, 1977). This indicates that, on average, a yearly rainfall surplus of about 275–325 mm exists. For a rain gauge station nearby, viz. Uccle, Demarée (1985) derived the intensity–duration–frequency relationships based on a reference period 1934–1983. Demarée (1985) used a five-parameter model to fit the observations. From the fitted relationships it can be noted that for durations ranging from one hour to several days these relationships are linear on log–log paper. Therefore expression (7) used by De Vries (1974) becomes applicable for durations beyond one hour. The exponent m is independent of exceedance probability and equals 0.7682, whereas the coefficient c depends on the chosen frequency.

The average characteristics for the three different soil types in the catchment (sandy loam, loam and clay) are given in Troch (1993). In this study we will use the values for the loamy soils as they represent about 80% of the soils present in the basin. The saturated volumetric soil moisture content, θ_s, is equal to 0.42; the residual volumetric soil moisture content, θ_r, is 0.14; the capillary fringe, ψ_c, is less well defined and is estimated to range between 0.10 and 1.0 m; and the pore size distribution index B in the Brooks–Corey soil water retention characteristic (Brooks and Corey, 1964) equals 0.934. These values are determined by laboratory procedures. Troch *et al.* (1993) estimated the hydraulic characteristics of the aquifers for the Zwalm basin based on the Brutsaert–Nieber drought flow analysis (Brutsaert and Nieber, 1977). They found that, for a typical value of drainable porosity of 5%, the horizontal hydraulic conductivity is of the order of 20 m/day and the average depth to the impervious layer, D, is of the order of 2.1 m. This results in a transmissivity value of $42\,\mathrm{m}^2/\mathrm{day}$ (Table II).

DEM-derived channel networks

With the advent of DEMs it is possible to carry out detailed geomorphological analysis automatically. Automatic techniques to extract and calculate geomorphological characteristics of a catchment and its subcatchments from DEMs is gaining increasing popularity. During the last decade, hydrologists have used DEMs to extract drainage networks and related information, useful for the development of hydrological models. A key issue in applying this technique for hydrological purposes is the choice of the correct threshold area value needed to define channels.

Tarboton *et al.* (1989) proposed a method of predicting the channel network based on the stability threshold of Smith and Bretherton (1972), which defines the transition from stable diffusive processes to unstable channel-forming processes. The different scaling behaviour in the stable and unstable regimes was shown to be equivalent to a change in the sign of the slope–area scaling function gradient. Thus the area at which a break of slope in a slope–area plot occurred was interpreted as the scale at which stability changes and was then used as the theoretical threshold area to define channel initiation. Based on the above concepts, Gyasi-Agyei *et al.* (in press). developed a method to estimate the 'correct' threshold area for the extraction of a channel network from digital elevation data. The method is based on a scaling analysis of link slope versus link drainage accumulation area. It is found that the slope–area scaling exponent, which reflects the concavity of a catchment, decreases with increasing threshold area, and stabilizes beyond a certain threshold area value. The threshold area at which the stabilization of the slope–area scaling exponent begins defines an approximate drainage density for hydrological modelling. Estimates about hydrogeomorphic characteristics following this approach can now be compared with predictions from the stream network morphology model and with estimates from topographic maps. We will use the method developed by Gyasi-Agyei *et al.* (in press) to define the possible range of first-order drainage density values for the Zwalmbeek catchment. Figure 3 shows the evolution of drainage density of first-order basins with drainage accumulation threshold area, expressed in terms of number of pixels in the DEM (one pixel has an area of $30 \times 30\,\mathrm{m}^2$). This figure indicates that the first-order drainage density, as derived from topographic features of the terrain, ranges from about 2.5×10^{-3} to about 0.7×10^{-3} l/m. Gyasi-Agyei *et al.* (in press) report that the threshold area at which a break in scaling exponent is observed corresponds to about 300 pixels and results in a drainage density

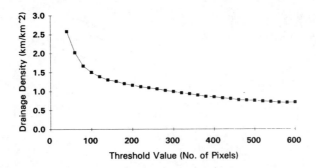

Figure 3. Evolution of drainage density of first-order basins with drainage accumulation threshold area (Zwalmbeek)

value of $1\cdot0 \times 10^{-3}$ l/m. From the topographic maps, scale 1/10 000, De Troch (1977) estimated the drainage density of first-order basins equal to 2×10^{-3} l/m. It is noted that this value corresponds to a threshold area of 60 pixels (Figure 3).

Results and discussion

Figure 4 shows the IDF relationships for Uccle, Belgium. For durations ranging from one to several days these relationships can be expressed in terms of simple power laws of the form of Equation (7). The parameter values of expression (7) for four different frequencies are given in Table II. Equation (9) allows the computation of the available storage capacity for a range of $\bar{z}$ values. The maximal average groundwater discharge, $\bar{R}_{max}$, can now be computed using Equation (8) with the parameter values corresponding to a given exceedance probability (Table II). Equation (5) yields the characteristic channel radius of first-order streams and Equation (2) allows the computation of the characteristic stream spacing. Results of these computations for the Zwalmbeek catchment are given in Figures 5 and 6. Figure 5 shows the variation of first-order stream spacing with average depth to the water-table. From this figure we can see, for example, that for an average depth to the water-table less than $0\cdot40$ m and for the 1% exceedance probability case the catchment becomes close to complete saturation (average stream spacing <10 m). Figure 6 gives the evolution of first-order drainage density with average depth to the water-table. The range of possible drainage densities for the catchment as derived from DEMs and defined in Figure 3 corresponds to a $\bar{z}$ value between $0\cdot60$ and $0\cdot90$ m for the 1% exceedance probability case. Troch *et al.* (1993) used 17 rainfall–runoff events observed between 1974 and 1987 to test the applicability of the technique, developed in that paper, to determine the initial storage capacity before a given flood event.

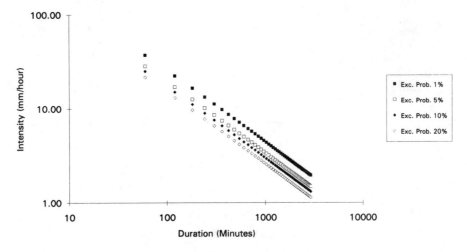

Figure 4. Intensity–duration–frequency curves for Uccle, reference period 1934–1983

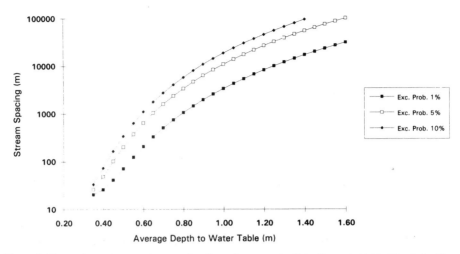

Figure 5. First-order stream spacing as a function of average depth to the water-table (Zwalmbeek)

Figure 7 shows the relationship between the estimated initial depth to the water-table, $\bar{z}$, and the runoff coefficient for these 17 flood events. For more detailed information about the 17 flood events, we refer to Troch *et al.* (1993). The runoff coefficient given in Figure 7 is defined as the ratio between direct runoff and total areal rainfall. An almost linear relationship between $\bar{z}$ and the runoff coefficient is observed. Important to note is that $\bar{z}$ values in the range 0·60–0·90 m correspond to relatively high runoff coefficients (>0·20). Therefore it is reasonable to assume that these initial conditions, in combination with the 1% probability rainfall conditions, result in extreme flood events with sufficient periodicity to be considered as landscape — shaping fluvial processes. These results indicate that the 1% exceedance probability can be proposed as representing the 'shaping' climatic conditions which have determined stream network morphology. Figure 8 further confirms these findings: the characteristic channel radius corresponding to the range of these $\bar{z}$ values is 0·10 m (Figure 8). This value agrees fairly well with observed channel geometry in the south part of the catchment. This part of the catchment underwent little modifications to its surface and subsurface drainage system. On the contrary, in the north part of the catchment, intensive agricultural activities have altered the discharge

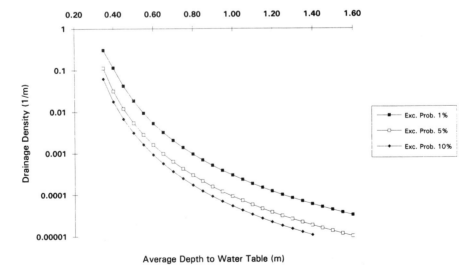

Figure 6. First-order drainage density as a function of average depth to the water-table (Zwalmbeek)

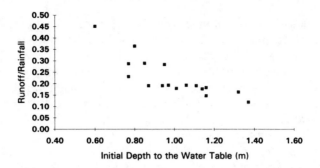

Figure 7. Relationship between runoff coefficient and initial average depth to the water-table (Zwalmbeek)

capacity of the stream system to improve agricultural productivity. As a consequence, our estimated characteristic channel radius represents an underestimation of the observed characteristic channel radius of first-order streams for this area. Another observation that could explain the discrepancy between observed and calculated radius is the fact that the slopes in the north part are more gentle than in the south. Our estimate is based on an average parameter value for transversal hillslopes.

From the results obtained it is advocated that the model presented here yields realistic predictions of first-order channel network characteristics for the catchment under study. There is also some evidence that the 1% relationship represents the climatic conditions that governed the erosion processes that took place in the catchment. It is therefore reasonable to assume that the corresponding curves relating channel network characteristics, such as stream spacing, drainage density and channel geometry, to average water-table depth are useful to describe the hydrological response of the drainage basin to storm rainfall. As such they form basin descriptors and could be used for comparative purposes (e.g. regional flood frequency analysis). The major benefits to be expected from this approach is that it is based on geological, climatic and topographic information about the catchment. Results from this work are important to extend recent research on hydrological similarity for humid catchments. The characteristic channel network curves described here will help to classify catchments with respect to their hydrological response. This can be demonstrated by means of Monte Carlo simulation with physically based subsurface stormflow models. Troch *et al.* (submitted) developed a conceptual framework for modelling stormflow in areas where the hydrological response is dominated by subsurface flow. This subsurface stormflow model for

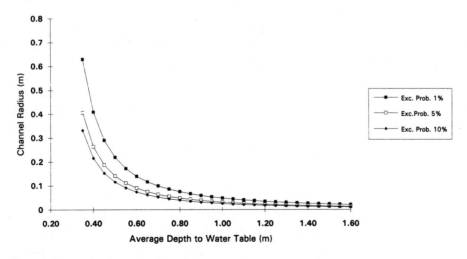

Figure 8. First-order channel radius as a function of average depth to the water-table (Zwalmbeek)

humid basins is based on Boussinesq's hydraulic groundwater theory for sloping aquifers. The catchment-scale aquifer parameters are estimated based on a drought flow analysis which is consistent with the groundwater theory used to develop the model. A technique based on groundwater table dynamics is developed to estimate the initial contributing area to subsurface stormflow, such that the model is applicable for event-based rainfall–runoff simulation at the catchment scale. This model allows us to study the controlling effect of different basin characteristics on flood frequency characteristics for catchments of different scales. This approach to regional flood frequency analysis is similar to that adopted by many different workers (Hebson and Wood, 1982; Cordova and Rodriguez-Iturbe, 1983; Diaz-Granados et al., 1984; Beven, 1986; Sivapalan et al., 1987; 1990; Troch et al., 1994). Future research will apply this modelling methodology to catchments under different climatic and geomorphological conditions.

Although the above-presented model of stream network evolution applies to the first-order drainage basins, it should be noted that the characteristics curves are representative of a given region. Indeed, seasonal and interstorm variations of network density is expected to take place in these drainage basins and as catchment size (or order) increases the drainage density will stabilize. Therefore the relationship between network density and catchment scale is dominated by first-order basin characteristics. Of course, upscaling from first-order basins to the regional scale needs to consider variability in climatic conditions. In this study and during the derivation of the algorithms it is assumed that the rainfall characteristics can be described by the IDF relationships observed in a point in the region. Dupriez and Demarée (1988) analysed data from 11 rain gauge stations in Belgium where more than 80 years of observations are available. They identified the IDF relationships for these stations and concluded that the exceedance probabilities for rainfall totals aggregated over periods ranging from one to 30 days is considerably variable between regions. Based on 143 rain gauge stations distributed over the country where more than 30 years of observations were available, Dupriez and Demarée (1989) constructed maps of maximum probable rainfall totals for given aggregation periods. From these maps it is clear that these statistics show little variation in the region of interest in this study, viz East Flanders. It is therefore reasonable to accept the initial assumption (rainfall characteristics described from point observation in the region). Further research is needed to investigate the sensitivity of the stream network evolution model to spatially variable rainfall characteristics.

SUMMARY

This paper relates geomorphological characteristics observed in humid lowland areas with the catchment storm response. We applied a simple physically based mathematical model of stream network morphology, developed by De Vries (1977), to the Zwalmbeek catchment, Belgium. The model uses information about the geology, climate and topography of the area to predict stream spacing and drainage density as a function of the initial storage capacity of the basin. These predictions were then compared with field observations and DEM derived channel networks. It is found that the 1% exceedance probability rainfall can be suggested as the shaping climatic conditions for the catchment under study. Coupled with the method advocated by Troch et al. (1993) to estimate the effective depth to the water-table as a function of observed baseflow, the model proposed by De Vries (1977) allows for the prediction of the expansion and shrinkage of the channel network during storm and interstorm periods. As such, the methodology used in this paper relates channel network morphology and storm response in humid catchments.

ACKNOWLEDGEMENTS

The first author thanks the Belgian Fund for Scientific Research (NFWO) for supporting this research. The authors express their gratitude towards two anonymous reviewers and Dr Sivapalan for the useful comments on this paper.

REFERENCES

Beven, K. J. 1986. 'Runoff production and flood frequency in catchments of order n: an alternative approach' in Gupta, V. K., Rodriguez-Iturbe, I., and Wood, E. F. (Eds), *Scale Problems in Hydrology*, Reidel, Dordrecht. pp. 107–131.

Boussinesq, J. 1904. 'Récherches théoriques sur l'écoulement des nappes d'eau infiltrées dans le sol et sur le débit des sources', *J. Math. Pure Appl., 5th Ser.*, **10**, 5–78.

Brooks, R. H., and Corey, A. T. 1964. 'Hydraulic properties of porous media', *Hydrol. Pap. No. 3*, Colorado State University, Fort Collins.

Brutsaert, W., and Nieber, J. L. 1977. 'Regionalized drought flow hydrographs from a mature glaciated plateau', *Wat. Resour. Res.*, **13**, 637–643.

Carlston, C. W. 1963. 'Drainage density and streamflow', *US Geol. Surv. Prof. Pap. 422–C*.

Cordova, J. R., and Rodrigez-Iturbe, I. 1983. 'Geomorphologic estimation of extreme flow probabilities', *J. Hydrol.*, **65**, 159–173.

Demarée, G. 1985. *Intensity–Duration–Frequency Relationship of Point Precipitation at Uccle, Reference Period 1934–1983*. Royal Meteorological Institute, Brussels.

De Troch, F. P. 1977. 'Studie van de oppervlaktewaterhydrologie van een Oost-Vlaams stroombekken, De Zwalmbeek te Nederzwalm', *PhD Dissertation* Ghent University, Ghent, 188 pp [in Dutch].

De Vries, J. J. 1974. 'Groundwater flow systems and stream nets in The Netherlands', *PhD Thesis*, Vrije Universiteit Amsterdam, Editions Rodopi, Amsterdam.

De Vries, J. J. 1976. 'The groundwater outcrop–erosion mode, evolution of the stream network in The Netherlands', *J. Hydrol.*, **29**, 43–50.

De Vries, J. J. 1977. 'The stream network in The Netherlands as a groundwater discharge phenomenon', *Geo. Mijnb.*, **56**, 103–122.

De Vries, J. J. 1993. 'Stream network morphology as a function of topography, geology and climate in sandy lowland areas' In *concepts and Methodlogy in Hydrogeomorphology, Workshop held at IAMAP/IAHS Meeting, Yokohama, July 1993*.

Diaz-Granados, M. A., Valdes, J. B., and Bras, R. L. 1984. 'A physically-based flood frequency distribution', *Wat. Resour. Res.*, **20**, 995–1002.

Dunne, T. 1969. 'Runoff production in a humid area', *PhD Thesis*, Johns Hopkins University, Baltimore.

Dupriez, G. L., and Demarée, G. 1988. 'Contribution à l'étude des relations intensité-durée-fréquence des précipitations: totaux pluviométriques sur des périodes continues de 1 à 30 jours, 1. Analyse de 11 series pluviométriques de plus de 80 ans', *Koninklijk Meteorologisch Instituut, Miscelanea Series A*, **8**, 154 pp.

Dupriez, G. L., and Demarée, G. 1989. 'Contribution à l'étude des relations intensité-durée-fréquence des précipitations: Totaux pluviométriques sur des périodes continues de 1 à 30 jours, 2. Analyse des séries pluviométriques d'au moins 30 ans', *Koninklijk Meteorologisch Instituut, Miscelanea Series A*, **9**, 53 pp.

Ernst, L. F. 1956. 'Calculations of the steady flow of groundwater in vertical sections', *Neth. J. Agric. Sci.*, **4**, 126–131.

Gregory, K. J., and Walling, D. E. 1968. 'The variation of drainage density within a catchment', *Bull. Int. Assoc. Sci. Hydrol.*, **13**, 61–68.

Gyasi-Ageyi, Y., De Troch, F. P., and Troch, P. A. 'A dynamic hillslope response model in a geomorphology based rainfall-runoff model, 1, model development', *J. Hydrol.*, in press.

Horton, R. E. 1945. 'Erosional development of streams and their drainage basins, hydrophysical approach to quantitative morphology', *Geol. Soc. Am. Bull.*, **56**, 257–370.

Jacob, C. E. 1943. 'Correlation of groundwater levels and precipitation on Long Island, N.Y.: Theory', *Am. Geophys. Union Trans.*, 564–573.

Shreve, R. L. 1966. 'Statistical laws of stream numbers', *J. Geol.*, **74**, 17–37.

Shreve, R. L. 1967. 'Infinite topologically random channel networks', *J. Geol.*, **75**, 178–186.

Sivapalan, M., Beven, K. J., and Wood, E. F. 1987. 'On hydrologic similarity, 2, a scaled model of storm runoff production', *Wat. Resour. Res.*, **23**, 2266–2278.

Sivapalan, M., Wood, E. F., and Beven, K. J. 1990. 'On hydrologic similarity, 3, a dimensionless flood frequency model using a generalized geomorphologic unit hydrograph and partial area runoff generation', *Wat. Resour. Res.*, **26**, 43–58.

Smith, T. R., and Bretherton, F. P. 1972. 'Stability and the conservation of mass in drainage basin evolution', *Wat. Resour. Res.*, **8**, 1506–1529.

Tarboton, D. G., Bras, R. L., and Rodriguez-Iturbe, I. 1989. 'Scaling and elevation in river networks', *Wat. Resour. Res.*, **25**, 2037–2051.

Troch, P. A. 1993. 'Conceptual basin-scale runoff process models for humid catchments, analysis, synthesis and applications', *PhD Dissertation*, University of Ghent, Ghent, 247 pp.

Troch, P. A., De Troch, F. P., and Brutsaert, W. 1993. 'Effective water table depth to describe initial conditions prior to storm rainfall in humid catchments', *Wat. Resour. Res.*, **29**, 427–434.

Troch, P. A., Smith, J. A., Wood, E. F., and De Troch, F. P. 1994. 'Hydrologic controls of large floods in a small basin: central Appalachian case study', *J. Hydrol.*, **156**, 285–309.

Troch, P. A., Pauwels, V., and De Troch, F. P. 'A catchment-scale subsurface stormflow model for humid catchments', *Wat. Resour. Res.*, submitted.

Vanmaercke-Gottigny, M. C. 1967. 'De geomorfologische kaart van het Zwalmbekken', *Verhandelingen van de Koninklijke Vlaamse Academie voor Wetenschappen, Letteren en Schone Kunsten van Belgie*, **99**, 93 pp. [in Dutch with an English summary].

Wood, E. F., Sivapalan, M., and Beven, K. J. 1990. 'Similarity and scale in catchment storm response', *Rev. Geophys.*, **28**, 1–18.

20

SCALE ISSUES IN BOUNDARY-LAYER METEOROLOGY: SURFACE ENERGY BALANCES IN HETEROGENEOUS TERRAIN

M. R. RAUPACH AND J. J. FINNIGAN

CSIRO Centre for Environmental Mechanics, GPO Box 821, Canberra, ACT 2601, Australia

ABSTRACT

This paper, part review and part new work, falls into three main sections. The first is a review of scale issues in both hydrology and meteorology, focusing on their origins in the water and energy conservation equations, integrated over control volumes of different scales. Several guidelines for scale translations are identified.

The second section reviews the upscaling problem in boundary-layer meteorology, setting out two 'flux-matching' criteria for upscaling models of land–air fluxes: the conservation requirement that surface fluxes average linearly and the practical requirement that model form be preserved between scales. By considering the effects of boundary conditions, it is shown that the combination or Penman–Monteith equation is a model for elemental energy fluxes which leads to physically consistent flux-matching rules for upscaling surface descriptors (resistances). These rules are tested, along with some other possibilities, and found to perform well.

The third section tests the hypothesis that regionally averaged energy balances over land surfaces are insensitive to the scale of heterogeneity, X. Heterogeneity is classified as microscale when $X \leqslant U_m T_*$, mesoscale when $U_m T_* \leqslant X \leqslant U_m T_e$, and macroscale when $U_m T_e \leqslant X$ [where U_m is the mean wind speed in the convective boundary layer (CBL) and T_* and T_e the convective and entrainment time scales, respectively]. A CBL slab model is used to show that regionally averaged energy fluxes are remarkably insensitive to X in both the microscale and macroscale ranges. Other reviewed evidence suggests that the mesoscale range behaves similarly in dry conditions. Questions remain about the consequences of clouds and precipitation for regionally averaged surface energy fluxes.

SCALE ISSUES IN HYDROMETEOROLOGY

Preamble

This paper has three main aims. The first is to compare and contrast scale issues in boundary-layer meteorology and hydrology, with the intention of constructing a framework and some general guidelines for the systematic analysis of scale problems. The second is to consider scale translations in boundary-layer meteorology specifically, by outlining and evaluating methods for 'upscaling', or aggregation, of land-surface descriptors and fluxes. The third is to apply the upscaling considerations to the question 'how does the surface energy balance of a heterogeneous region change as the scale of heterogeneity changes?' To answer this question, we use both a simple convective boundary layer (CBL) slab model (for which new results are reported here) and a review of other approaches. These three broad aims are treated in three separate sections. The present paper is a companion to that by Raupach and Finnigan (in press). This paper deals with scale issues, particularly for evaporation and the surface energy balance, whereas Raupach and Finnigan (in press) discuss flow and scalar transfer over topography, including the application of scaling concepts presented here.

Scale issues and conservation principles

The 'scaling problem', if such a thing exists, clearly means different things to different people. Some of the differences in perception arise from the differing aims and scale of interest of the practitioners — for instance, hydrologists and meteorologists. Ecologists also have strong interests in scale issues (Wessman, 1992), and we hope that some of the following discussion will be relevant in general terms to them (if only for purposes of contrast). However, we concentrate here on the 'fluid' disciplines of hydrology and meteorology.

Hydrologists are interested primarily in water balances, a key question being the lateral movement of liquid water for the purposes of calculating catchment runoff or recharge. Their models have typical time steps of days or weeks, whereas horizontal areal scales range from subcatchments of a few hectares to hundreds of square kilometres. On the other hand, meteorologists concerned with the atmospheric part of the hydrological cycle concentrate on modelling the partition of radiant energy into sensible and latent heat — that is, the surface energy balance. Their motivations include the modelling of atmospheric motions and cloud processes, which depend critically on the fluxes of sensible and latent heat into the air from land (and sea) surfaces. The lateral redistribution of liquid water following rain is of interest primarily because it leads to the temporary removal of some precipitation from the short-term hydrological cycle through runoff, and produces spatial heterogeneity in the availability of water for evaporation. Typical model time-scales are necessarily fine (a few minutes to less than an hour) because of the need to resolve the diurnal cycle of the energy balance; horizontal areal scales range from individual leaves up to regions of thousands of square kilometres, in which the near-surface meteorology is largely self-organizing on diurnal time-scales, as argued later. [Coincidentally, this regional scale is also approximately the area of a grid cell in current global climate models (GCMs). However, it is better to choose length and area scales on physical grounds than on the basis of current computing practices and capabilities.]

Despite these differences in aims and scales, the disciplines of hydrology and meteorology share several common aspects of the 'scaling problem'. Both disciplines aim to predict the fluxes of conserved entities across explicitly or implicitly specified surfaces: hydrologists seek liquid water fluxes across catchment boundaries, whereas meteorologists seek the sensible and latent heat fluxes across the land–air interface. Both disciplines do this by writing 'balance equations' which are actually conservation equations for the appropriate scalars, integrated over volumes bounded partly by the surfaces over which the fluxes are sought. Hydrologists use the water balance

$$[\text{runoff}] = [\text{precipitation}] - [\text{deep drainage}] - \frac{d[\text{storage}]}{dt} - [\text{evaporation}] \tag{1}$$

where the conserved scalar is water (phase changes such as snowmelt here being part of the storage term) and the integration volume is the hydrologically active soil and vegetation layer across a catchment, bounded above by the atmosphere. Meteorologists use the surface energy balance

$$F_A = F_H + F_E \tag{2}$$

where F_H and F_E are the land–air fluxes of sensible and latent heat and F_A is the available energy flux (net irradiance less heat flux into storage). In this instance the conserved scalar is energy and the integration volume is a thin layer enveloping the land–air interface. To partition the sum $F_H + F_E$ into separate sensible and latent heat terms, extra information must be supplied: this is often done by means of a surface resistance. Of course, the evaporative water flux is the link between the water and energy balance equations.

To find the runoff from Equation (1), or F_H and F_E from Equation (2) with ancillary partitioning equations, we have to do two things: choose the horizontal extent of the integration volume, thereby setting the averaging area for the fluxes; and measure or model all the terms on the right-hand side of Equation (1) or (2) (plus ancillary partitioning equations) at this areal scale. However, the process is beset by several difficulties, which include:

1. On all but idealized land surfaces, the terms to be measured or modelled are actually strongly spatially heterogeneous. This fact leads to the other difficulties.
2. When terms have to be modelled, or measurements interpreted with the aid of models, it is necessary to supply descriptors of the land surface. Various terms in Equations (1) or (2) require descriptors such as

land slope, soil hydraulic properties, albedo, aerodynamic resistance (or surface roughness) and surface resistance (or parameters which determine it). These descriptors may be available across the heterogeneous landscape at a fine scale, requiring amalgamation in some way for use at larger scales (slope being a prime example); or they may be available only at large scales, requiring disaggregation to allow flux calculations over smaller areas (for instance, lumped catchment hydraulic properties).

3. Not only the descriptors, but also the models themselves are often scale-specific. An appropriate model may not be available at the scale of interest, forcing the use of a model designed for another scale. For example, surface resistance is a parameter originally intended to describe the physiological properties of individual leaves, and its use at larger scales involves several assumptions (Raupach and Finnigan, 1988); this is true at canopy scales (Kelliher *et al.*, 1995) and even more at regional scales. Likewise, one-dimensional (vertical) versions of Darcy's law and the Richards equation are often used to model soil water movement in the unsaturated zone and thence the storage and drainage terms in Equation (1). However, these theories describe water movement only when the soil is sufficiently uniform in the horizontal, raising difficulties for their application at larger scales where macropores and preferred pathways are significant hydraulic features.

4. The aggregation or disaggregation of both descriptors and models is often complicated by strong non-linearities. For example, the averaged roughness length of a region is not a linear average of the roughness lengths of individual patches in the region (Mason, 1988; Wood and Mason, 1991); and a linear average of hydraulic conductivities does not produce a useful hydraulic parameter to describe an inhomogenous soil (Binley and Beven, 1989).

All these troubles, and more, contribute to what is loosely called the 'scaling problem'. Efforts to solve them have proceeded apace in both hydrology and meteorology, as demonstrated at this workshop. Such efforts generally distinguish the problem of aggregating models or measurements (or 'upscaling') from that of disaggregating (or 'downscaling'). From our viewpoint as meteorologists, these terms can be defined as follows:

1. *Upscaling* is the derivation of areally averaged land–air fluxes across a heterogenous landscape, using spatially aggregated meteorological conditions (such as solar irradiance, wind speed, air temperature and air humidity) and suitable aggregated land surface descriptors (such as albedo, roughness length and surface resistance).

2. *Downscaling* is the inference of the spatially resolved pattern of land–air fluxes across a heterogeneous landscape, from aggregated meteorological conditions and spatially resolved surface descriptors.

Equivalent definitions exist for hydrology. Let us discuss the meteorological case, beginning with upscaling. Meteorological conditions determining surface fluxes are usually available only at coarse spatial scales, from weather data or GCM output. Information on the land surface descriptors is available at fine scales (from land-use surveys, soil maps or remote sensing), either as probability distributions across the heterogeneous landscape or in the form of deterministic position attributes, as in a geographical information system. However, it is often impractical or impossible to use this fine-scale information to calculate fluxes from each small homogeneous element (patch) of the heterogeneous surface. Therefore, the upscaling problem amounts to finding ways of averaging the land surface descriptors for each element to produce a set of bulk descriptors which (when combined with large-scale meteorological information) yields the areally averaged land–air fluxes directly. Among the applications for upscaling methods are the specification of land–air fluxes in GCMs, with averaging over grid cells; determination of the overall evaporation across heterogeneous catchments; and determination of the sensitivity of fluxes at a large scale to the details of the distribution of descriptors at a finer scale.

Downscaling methods are techniques for inferring patch-scale values for land–air fluxes across an entire heterogeneous landscape, rather than measuring them directly. In principle, this is the inverse of upscaling; however, in practice the data requirements are more intensive, because disaggregation amplifies uncertainties, whereas aggregation attenuates them. Meteorological downscaling methods have significant practical applications, one being the estimation of evaporation for use in small-scale hydrological models (recalling that evaporation is common to both the energy and water balances). A second application is the use of near-surface flux data in the validation of large-scale models such as GCMs. This arises because the outputs of large-scale models are spatially aggregated and are therefore not observable. For example, at its finest resolution the evaporation calculated by a GCM is an average value over at least one grid cell (actually more, for numerical reasons), which does not necessarily apply at any given point in the grid cell. For validation, the model output must be disaggregated until it can be compared with a field observation. Even if the field observations are themselves averaged over the grid cell, as may occur with satellite measurements of surface temperature or sampling of a large area by aircraft transects, the averaging operations applied to the observations and implicit in the models are generally different, and have to be reduced to the common denominator of their distributions at finer scales to enable model validation.

Guidelines for scale translations

Scale translations (including those discussed later in this paper) often implicitly follow two broad guidelines, which it is worthwhile to state explicitly. The first can be called the 'conservation of complication'. As the scale of interest increases, thereby encompassing more processes, the detail with which each can be described reduces to preserve the total level of complication embraced by the model. To describe larger scale features, we must (for purely practical reasons) relinquish smaller scale detail. This notion has two important consequences: (1) as scale increases, 'sub-scale' processes must be parameterized in such a way that the essential features of their interaction with the larger picture are preserved; and (2) to preserve these essential features, each parameterization must incorporate a sufficient degree of mechanistic understanding of both the individual process in question and also its part in the whole system.

The second guideline is that 'fluids flow and mix, but solid boundaries stay where they are'. This truism also has some significant consequences. The critical transformations of energy and matter at land surfaces, primarily the conversion of radiant energy to sensible and latent heat, take place at points on solid soil and plant surfaces that are reasonably immobile (plant waving notwithstanding). The fluxes resulting from these transformations, F_H and F_E, are controlled partly by surface properties and partly by the properties of the surrounding fluids — the temperature, humidity and velocity of the air, and the soil moisture distribution in the root zone. At local scales, those aspects of the transformation processes that are controlled by surface properties reflect every detail of the land-surface heterogeneity, but the aspects which are controlled by fluid properties are advected and naturally smoothed by flow and mixing in the fluid. The mixing is strong in the atmosphere and far weaker (though still often significant) in the terrestrial hydrosphere, the water moving across and below the land surface.

Fluid flow and mixing impose an influence from distant events on local energy balances, and therefore naturally integrate events at one scale into larger scales, and vice versa. As the scale increases, the range

Table I. Atmospheric mixing processes

Length scale (m)	Time-scale (s)	Process	Typical situations
10^{-5}	10^{-5}	Molecular diffusion in static fluid	Substomatal cavity
10^{-3}	10^{-1}	Advection–diffusion in laminar flow	Leaf boundary layer
10^{-1}–10^2	10^{-1}–10^2	Shear-driven turbulence	Canopy layer and surface layer
10^2–10^3	10^2–10^3	Buoyancy-driven turbulence	Convective mixed layer
10^3–10^5	10^3–10^5	Buoyancy-driven mean flows	Mesoscale circulations; thermally driven hill and valley flows; mountain waves, bores, hydraulic jumps (flow systems locked into topography); frontal systems; convective storms
10^5–10^7	10^5–10^6	Two-dimensional turbulence, gyres	Weather systems on continent scale

Table II. Mixing by motion of liquid water in and above the ground

Length scale (m)	Time-scale (s)	Process	Typical situation
10^{-1}	10^5	Darcy flow in unsaturated zone	Redistribution after rainfall
10^0	10^3	Water movement from roots to leaves	Extraction of soil water by plant roots
10^0–10^1	10^5–10^6	Darcy flow in saturated zone	Downslope or lateral movement in hilly or texture contrast soils after rain; discharge zones in hilly terrain; river margins
10^2–10^3	10^4–10^5	Overland flow	Heavy rains, floods
10^3–10^5	10^4–10^6	River flow	Supply of river margins, wetlands, floods
10^3–10^6	10^6–	Groundwater/aquifer flows	Recharge/discharge regions, major aquifers

and extent of fluid mixing also increases through a hierarchy of physical processes encompassing many orders of magnitude in both space and time-scales. For the atmosphere, this hierarchy of processes is shown in Table I. At the largest scale shown, and even larger scales, the time-scale for mixing is significantly longer than that of the energy source that drives the hydrological and climate systems (diurnal radiative forcing), so that on a diurnal time-scale, little mixing occurs. It is this property which allows us to define a regional scale (from the atmospheric viewpoint) as the area over which the atmospheric boundary layer is self-organizing on diurnal time-scales, and largely uninfluenced by events in neighbouring regions.

A similar hierarchy (Table II) exists for the terrestrial hydrosphere. There are several major differences between this and the hierarchy in the atmosphere: the typical velocity and diffusivity scales are far smaller than in the atmosphere; water movement is influenced to a far greater extent by the heterogeneity of the soil medium through which it travels; and water flows downhill. This last truism also has the important consequence that the heterogeneity of the water distribution across the land surface increases with time (following a saturating rain event, say) and with spatial scale (through river networks, for instance), in contrast with the atmosphere. A final truism with significant consequences is that (most) plants have roots, which explore the soil to substantial depths and are able to conduct water to the surface from throughout the root zone during transpiration. Therefore, plant roots act to homogenize water extraction from the soil column.

UPSCALING IN BOUNDARY-LAYER METEOROLOGY

We now focus on scale translations in boundary-layer meteorology. Our ultimate goal is to consider (in the next section) the effect of scale on regionally averaged surface energy balances in heterogeneous terrain. However, our starting point, and the subject of this section, is a discussion of upscaling — that is, methods for averaging land-surface descriptors to produce spatially aggregated models for land–atmosphere fluxes.

The need for upscaling

There is some debate about whether upscaling methods are necessary at all. For instance, Koster and Suarez (1992) distinguished two ways of calculating land–atmosphere fluxes over heterogeneous terrain for GCM applications: a 'mosaic' approach, in which fluxes are calculated independently for each patch and the fluxes then summed with area weighting, thus avoiding the need for any upscaling procedure; and 'mixture' approach, in which upscaling is used to define aggregated parameters representing a uniform, composite surface, for which the area-averaged fluxes are calculated. They found by numerical comparisons that these two approaches gave very similar results, and commented that the 'mixture' approach was more complicated to implement. This illustrates that for some practical problems, upscaling methods and 'mixture' approaches may not be optimal. However, there are at least two reasons why a study of upscaling methods is useful. Firstly, land surfaces are heterogeneous on a very wide range of scales (stomata to regions), so horizontal averaging and upscaling is implicit at any practical scale, no matter how small; a study of the upscaling operation helps to make explicit the assumptions involved. Secondly, the treatment of some kinds of heterogeneity is made far simpler by reducing the surface to an averaged composite through upscaling: for instance, small-scale heterogeneity with a blending height within the surface layer, analysed in the next section.

Three scales of dominant practical importance in land–atmosphere exchanges are 'leaf' scale (that of the single energy-exchanging element, often, but not always, an actual leaf), 'canopy' scale (that of the homogeneous land-surface patch, often, but not always, covered by a vegetation canopy) and 'regional' scale (that of the heterogeneous landscape or region, on which the atmospheric boundary layer is approximately self-organizing on a diurnal time-scale). Between these scales, there are two significant upscaling translations: leaf–canopy and canopy–region. This section treats both within the same framework, referring to arbitrary 'small' and 'large' scales which can be either leaf–canopy or canopy–region.

Constraining the upscaling problem

A primary constraint on areal averaging is provided by conservation requirements, which demand that land–atmosphere fluxes average linearly with area weighting; see Raupach (1995) for a formal demonstration. Thus

$$F = \sum a_i f_i \tag{3}$$

where F is the areally averaged scalar flux density across a large area of the land–atmosphere interface, f_i is the flux density across element i of a set of surface elements or patches which together make up the large area and a_i is the normalized area (area per unit ground area of the averaged surface) of element i. For leaf–canopy translations $\sum a_i$ is the leaf area index; for canopy–region translations, $\sum a_i$ is unity for level terrain and greater than unity for non-level terrain (it is the average of the secant of the slope). Equation (3) introduces a convention to be used henceforth: areally averaged (large-scale) quantities are denoted by capital letters and elemental (small-scale) quantities by the corresponding small letters. Also, from now on we refer to flux densities simply as fluxes.

Suppose that the small-scale flux f is a specified function $h(x, y, \ldots)$ of the small-scale independent variables $x, y, \ldots$ (henceforth we will explicitly show only two of these variables). The function $h(x, y)$ constitutes a small-scale model for f; we seek a corresponding large-scale model $F = H(X, Y)$ for the averaged flux. The upscaling problem is basically to find how the large-scale model function H and variables X and Y are related to their small-scale counterparts. To constrain the problem, two requirements were imposed by McNaughton (1994) and Raupach (1995): (A) the flux f averages linearly according to Equation (3); and (B) the model form is the same at both scales, so that $h = H$. Requirement (A) is a consequence of scalar conservation, so any upscaling procedure which does not satisfy (A) has the highly undesirable property of violating scalar conservation. Requirement (B) has three practical (rather than fundamental) motivations: firstly, $h = H$ is necessary if H is a large-scale analogue to a model h which is more rigorously justified at a small scale. Secondly, when $h = H$, parameters in the model (for instance, resistances) can be interpreted in complementary ways at small and large scales. Thirdly, $h = H$ is a strong form of the 'conservation of complication' guideline outlined above, as the model itself is identical at both scales. An upscaling scheme satisfying requirements (A) and (B) will be called a 'flux-matching' scheme (Raupach, 1995).

If h is linear in x and y ($f = c_1 x + c_2 y$), then both requirements can be satisfied easily by defining the large-scale independent variables X and Y as linear averages of their small-scale counterparts: $X = \sum a_i x_i$ and likewise for Y. However, in the more usual case that h is non-linear, we cannot simultaneously satisfy both of the above requirements and retain linear averages for X and Y. Alternative, non-linear definitions of X and Y are therefore required. This problem has long been recognized for leaf–canopy translations, especially the definition of the bulk canopy resistance (Philip, 1966; Monteith, 1973; Tan and Black, 1976; Finnigan and Raupach, 1987; Raupach and Finnigan, 1988; Baldocchi *et al.*, 1991). Definitions of X and Y require two independent relationships (if there are m independent variables $X, Y, Z, \ldots$, then m relationships are needed). One of these is supplied by the flux-matching relationship following from requirements (A) and (B)

$$F = H(X, Y) = h(X, Y) = \sum a_i h(x_i, y_i) \qquad (4)$$

The other can be supplied in several ways, including *ad-hoc* assumption and term by term matching. This latter option means that, if $h(x, y)$ can be split into additive terms $h_1(x, y) + h_2(x, y)$, then the terms can be matched individually to give $H_1(X, Y) = \sum a_i h_1(x_i, y_i)$ and likewise for H_2, thus yielding the two relationships needed to define both X and Y. The assumption that terms can be matched individually amounts to requiring that the ratios among h_1, h_2 and h be the same for the large-scale model as for the summed small-scale models, which is usually reasonable.

A simple example will illustrate the process and some of the difficulties. Consider scaler (say heat) transfer obeying the elemental model $f = (c_0 - C_b)/r_{aC}$, where c_0 is the elemental surface concentration, C_b is the concentration at a height in the atmosphere where the scalar is 'blended' (horizontally uniform) and r_{aC} is the scalar aerodynamic resistance between the surface and this 'blending height'. Because C_b is uniform for all elements, it is therefore a bulk, capitalized quantity, unlike c_0 which can vary from element to element. Matching the flux at smaller and larger scales, we obtain

$$F = (C_0 - C_b)/R_{aC} = \sum a_i (c_{0i} - C_b)/r_{aCi} \qquad (5)$$

where R_{aC} and C_0 are large-scale, averaged quantities. Equation (5) is one of the two relationships needed to define R_{aC} and C_0; let us examine two alternatives for obtaining the other. Firstly, term by term matching of Equation (5) gives

$$\frac{1}{R_{aC}} = \sum \frac{a_i}{r_{aCi}}, \qquad C_0 = R_{aC} \sum \frac{a_i c_{0i}}{r_{aCi}} \qquad (6)$$

so the bulk resistance is a simple parallel sum of the elemental resistances, but the bulk surface concentration C_0 is not the linear average $\sum a_i c_{0i}$; rather, it is an average weighted with the conductance $1/r_{aCi}$, so that C_0 is weighted to the parts of the surface on which transfer is most effective. Secondly, we can require that C_0 average linearly. This leads to

$$C_0 = \sum a_i c_{0i}$$

$$R_{aC} = \frac{\sum a_i f_i r_{aCi}}{\sum a_i f_i} = \frac{C_0 - C_b}{\sum a_i (c_{0i} - C_b)/r_{aCi}} \qquad (7)$$

in which the two apparently dissimilar definitions of R_{aC} are actually equivalent if C_0 is a linear average. All three definitions of R_{aC} and C_0 in Equations (6) and (7) satisfy requirements (A) and (B) above, as all are consistent with Equation (5). However, all carry the severe disadvantage that knowledge of the elemental surface concentrations or fluxes is required to determine the large-scale parameters. In other words, to parameterize the large-scale transfer, we have to solve the small-scale problem in detail, thus making irrelevant the need for parameterizing the large-scale problem in the first place. Equations (6) and (7) are therefore impractical as upscaling equations. Blyth *et al.* (1993) pointed out that this awkward situation is avoided if the boundary condition on the small-scale problem is sufficiently simple (either constant-flux or constant-concentration): if the surface flux is constant, then R_{aC} is the series sum $\sum a_i r_{aCi}$ [from the first definition of R_{aC} in Equation (7)], and if the surface concentration is constant, then R_{aC} is the parallel sum $\left(\sum a_i r_{aCi}^{-1}\right)^{-1}$ [from Equation (6) or the second definition of R_{aC} in Equation (7)]. An important case in which the constant-concentration condition applies is momentum transfer, as the elemental surface velocity is always zero by the no-slip condition. Therefore, the average kinematic momentum flux T (capital τ, the square of the area-average friction velocity U_*) can be written

$$T = U_*^2 = \sum a_i \tau_i = \sum a_i u_{*i}^2 = \sum \frac{a_i U_b}{R_{aMi}} = \frac{U_b}{R_{aM}}; \qquad \frac{1}{R_{aM}} = \sum \frac{a_i}{r_{aMi}} \qquad (8)$$

where τ_i are the patch kinematic momentum fluxes (each the square of a patch friction velocity u_{*i}) and U_b

is the uniform velocity at the blending height. Flux-matching the momentum fluxes (not friction velocities!) requires that the bulk aerodynamic resistance R_{aM} is defined by parallel summation. Equation (8) was used by Wieringa (1986) and Mason (1988) to define area-averaged momentum roughness lengths.

Unfortunately, one of the most important problems in practice — that of coupled sensible and latent heat — has boundary conditions which cannot in general be reduced to simple constant-flux or constant-concentration conditions (Philip, 1959; 1987; McNaughton, 1976). Therefore, neither the series-sum nor the parallel-sum definitions of R_{aC} are tenable in this case. Blyth *et al.* (1993) suggested the *ad hoc* approach of using the average of the two; however, a better alternative is available. This is the subject of the next sub-section.

Upscaling coupled sensible and latent heat transfer

For coupled sensible and latent heat transfer between a vegetated land surface and the atmosphere, neither the surface temperature nor the surface humidity is an independent boundary condition; rather, both depend on the partition of energy into sensible and latent heat. This partition is controlled by both meteorological and surface conditions, the latter usually expressed in terms of a surface resistance. Hence, for an upscaling analysis, an appropriate small-scale model for the sensible and latent heat fluxes is the Penman–Monteith or combination equation (henceforth the CE). This equation eliminates the surface temperature and humidity through an approximate linearisation of $q_{sat}(T)$ (defined below) to express the fluxes in terms of meteorological variables and resistances which, for the present purpose, can be taken as independent and externally prescribed.

The CE is usually written to describe the latent heat flux, but a complementary form exists for the sensible heat flux. These two equations, and the associated energy balance equation, are (in the notation used by Raupach, 1995)

$$f_E = \frac{\epsilon r_{tH}\phi_A + \rho\lambda d}{r_d} \tag{9}$$

$$f_H = p\left[\frac{r_{tE}\phi_A - \rho\lambda d}{r_d}\right] \tag{10}$$

$$f_A = f_E + f_H, \qquad \phi_A = f_E + p^{-1}f_H \tag{11}$$

with

$$r_{tH} = \left(r_{aH}^{-1} + r_r^{-1}\right)^{-1}, \qquad r_{tE} = r_{aE} + r_s \tag{12}$$

$$r_d = \epsilon r_{tH} + r_{tE}, \qquad p = r_{tH}/r_{aH} \tag{13}$$

Here f_E and f_H are the latent and sensible heat fluxes; f_A is the available energy flux and ϕ_A the isothermal available energy flux, equal to f_A with the outward longwave flux evaluated at air temperature (note that isothermal fluxes are distinguished by using the Greek ϕ in place of f). Also, $d = q_{sat}(\theta) - q$ is the potential saturation deficit of the ambient air, with θ potential temperature and q specific humidity; $\epsilon = (\lambda/c_p)dq_{sat}/dT$ is the dimensionless slope of the saturation specific humidity $q_{sat}(T)$ as a function of temperature T; ρ is air density; λ is the latent heat of vaporisation of water; and c_p is the isobaric specific heat of air. The total latent heat resistance $r_{tE} = r_{aE} + r_s$ is the series sum of the aerodynamic resistance r_{aE} for latent heat and the surface resistance r_s, whereas the total sensible heat resistance r_{tH} is the parallel sum of the aerodynamic resistance r_{aH} for sensible heat and the radiative resistance $r_r = \rho c_p/(4\epsilon_s\sigma T^3) \approx 230\,\mathrm{s\,m^{-1}}$ (with ϵ_s the surface emissivity and σ the Stefan–Boltzmann constant). The 'deficit resistance' r_d defined by Equation (13) is the denominator in the CE, and is closed related to the resistance controlling the flux of saturation deficit [see Equation (B14)].

The replacement of the available energy f_A with its isothermal counterpart ϕ_A, and the inclusion of r_r in the total thermal resistance in parallel with r_{aH}, account for 'radiative coupling'. This is the dependence of the outward longwave part of the net irradiance on surface temperature, which is eliminated in the CE. The difference between ϕ_A and f_A is the difference in the outward longwave flux $f_{L\uparrow}$ between bodies at the surface temperature T_0 and the air temperature T. The ratio $p = r_{tH}/r_{aH}$ also expresses the significance of radiative coupling, as $p = 1$ when radiative coupling is absent ($r_{tH} = r_{aH}$) and $p < 1$ when radiative coupling is included ($r_{tH}/r_{aH} < 1$). Typically, p is in the range 0·5–0·9 (lower as surface roughness or wind speed decreases so that r_{aH} increases relative to r_r). Many workers ignore radiative coupling and possible differences between r_{aH} and r_{aE}. This will be called the 'simple case' henceforth; it is obtained by setting $p = 1$ and $r_{aH} = r_{aE} = r_a$, so that $\phi_A = f_A$ and $r_d = (\epsilon + 1)r_a + r_s$.

When applied to the CE for latent heat flux, Equation (9), the flux-matching requirements (A) and (B) demand that

$$F_E = \frac{\epsilon R_{tH}\Phi_A + \rho\lambda D_b}{R_d} = \sum_i a_i \left(\frac{\epsilon r_{tHi}\phi_{Ai} + \rho\lambda D_b}{r_{di}} \right) \tag{14}$$

Here all surface elements are assumed to be exposed (via their respective aerodynamic resistances) to the same uniform, 'blended' saturation deficit D_b. Also, ϵ takes a uniform value, representative of all patches, for simplicity. The assumption of uniform D_b is good for sufficiently small land surface patches in the daytime atmospheric boundary layer (later defined as 'microscale' patches). It is also good for leaves in a plant canopy if the deficit changes across individual leaf boundary layers are large compared with the changes in the canopy air.

We now carry out a term by term match on Equation (14), first making the extra simplification that Equation (3) can be used to average the isothermal available energy flux: $\Phi_A = \sum a_i \phi_{Ai}$. This is a good approximation, rather than an exact equality, because ϕ_A is an isothermal flux rather than a net scalar flux (Raupach, 1995). Matching the second terms in Equation (14), we obtain

$$\frac{1}{R_d} = \sum \frac{a_i}{r_{di}} \tag{15}$$

so the bulk deficit resistance R_d is defined by parallel summation of the corresponding elemental resistances. Matching the first term, we obtain

$$R_x = \frac{R_d}{\Phi_A} \sum \frac{a_i \phi_{Ai} r_{xi}}{r_{di}} \tag{16}$$

with $x = tH$. Thus, R_{tH} is defined by averaging the elemental r_{tH} values, weighted not only with area fraction but also with radiation (ϕ_A) and r_d. It can be shown that this same weighting factor defines R_{tE} in the general case, and also defines R_{aH}, R_{aE} and R_s in the simple case ($p = 1$, $r_{aH} = r_{aE} = r_a$). It is reasonable to adopt this as a constraint in the general case also, so that Equation (16) applies with $x = tH$, tE, aH, aE and s.

Equations (15) and (16) define spatially averaged resistances R_d, R_{tH} and R_{tE}, and their close relatives R_{aH}, R_{aE} and R_s, in terms of small-scale quantities. When used in the large-scale CE for latent heat flux [the first part of Equation (14)], these large-scale resistances yield the correct large-scale F_E, defined by requirement (A). When used in the CE for sensible heat flux, Equation (10), they yield nearly, but not exactly, the correct large-scale F_H in the general case; the outcome for F_H is exact only in the simple case (when $p = 1$). The reason for the inexactness in the general case is the non-linear appearance in Equation (10) of p (alternatively, r_r), which must also be averaged to produce an equivalent large-scale parameter P. No choice of P is available which exactly satisfies the flux-matching requirement for F_H, Equation (3). However, the simple linear average

$$P = \sum a_i P_i \tag{17}$$

provides an adequate approximation, as confirmed in the following.

The upscaling scheme for coupled sensible and latent heat transfer defined by Equations (9) to (17) is a flux-matching scheme satisfying requirements (A) and (B). It was derived by Raupach (1991, 1995) and McNaughton (1994). It is not the only possible flux-matching scheme for F_E and F_H which can be derived from the CE, because the inclusion of radiative coupling leaves some room for choice, for instance in the definition of p. Several possibilities were compared by Raupach (1995), the one above being scheme M_2, p_1 in the notation of that paper. In the general case, all schemes considered were equivalent and exact for F_E and nearly equivalent for F_H, whereas in the simple case ($p = 1$), all schemes were equivalent and exact for both F_E and F_H.

Comparison of alternative upscaling schemes

Staying with the case of coupled sensible and latent heat transfer, let us compare the averaged fluxes F_E and F_H over a heterogeneous land surface which are produced by several averaging schemes mentioned already. The six schemes to be compared are designated L, M, N, S, P and Q.

1. L: Scheme L is the linear average, Equation (3), which gives the true average fluxes F_E and F_H. This is not an upscaling scheme but an average over elemental fluxes. It provides the 'truth' against which the following upscaling schemes are judged.

2. M: Scheme M is the flux-matching scheme given by Equations (9) to (17). The elemental deficit resistances r_d (linear combinations of aerodynamic and surface resistances) average in parallel, but the elemental aerodynamic (r_{aH}, r_{aE}) and surface (r_s) resistances average with weighting from both r_d and radiation (ϕ_A). This reflects the combined roles of aerodynamic, surface and radiative factors in determining the energy partition from each element.

3. N: There may be occasions when it is inconvenient to carry a radiative weighting in the definition of the bulk resistances. In some circumstances, such as canopy–region translations in level terrain, ϕ_A does not vary greatly from patch to patch (but this is clearly not always a tenable approximation, for instance in leaf–canopy translations). Therefore, scheme N tests the utility of the bulk resistances

$$R_x = R_d \sum \frac{a_i r_{xi}}{r_{di}} \qquad (18)$$

(with $x = tH$, tE), equivalent to Equation (16), but omitting the radiative weighting. These bulk resistances still depend on a combination of surface and aerodynamic factors at elemental scale, as in scheme M.

4. S, P and Q: The last three options test the notion that upscaling schemes can be obtained from series resistances (scheme S), parallel resistances (scheme P) or the mid-point of the two (scheme Q, suggested by Blyth et al., 1993). These rules were used to calculate average values of R_{tH} and R_{tE}. Many other ad hoc combinations along the same lines could be tested.

We tested these schemes by applying them to a series of hypothetical heterogeneous surfaces consisting of just two surface types, chosen from four possibilities: crop, desert, forest and lake. Table III shows the assumed elemental albedos (α_s), emissivities (ϵ_s), roughness lengths (z_0) and surface resistances (r_s), and Table IV the imposed meteorological conditions (all identical to assumptions in Raupach, 1995). The elemental fluxes f_E and f_H are completely determined by these specifications. For each pairing of surface types (say crop–desert), the average fluxes F_E and F_H under each scheme were calculated for the complete range

Table III. Properties of surface types used for (1) tests of averaging schemes L, M, N, S, P and Q (Figure 1 and Table V, with r_s held constant at value shown); and (2) calculations with CBL slab model and SCAM-B [Figures 2 and 3, with r_s responsive to light and saturation deficit according to Equation (B7), where $r_{s(min)}$ takes the value shown, and $f_{S\downarrow}^* = 200\,\mathrm{W\,m^{-2}}$, $d_0^* = 20\,\mathrm{g\,kg^{-1}}$ for all surface types]

	Desert	Crop	Lake	Forest
Albedo α_s	0·3	0·15	0·05	0·10
Emissivity ϵ_s	0·99	0·99	0·99	0·99
Roughness length z_0 (m)	0·01	0·05	0·0001	2·0
Surface resistance r_s (s m^{-1})	2000	50	0·1	200

Table IV. Meteorological conditions used in tests of averaging schemes L, M, N, S, P and Q (Figure 1 and Table V). Downward irradiances were held uniform for all patches, and patch isothermal available energy ϕ_{Ai} calculated using Equation (B4)

Blended potential temperature Θ_b	25°C
Blended specific humidity Q_b	$10\,\mathrm{g\,kg^{-1}}$
Downward shortwave irradiance $f_S\downarrow$	$500\,\mathrm{W\,m^{-2}}$
Downward longwave irradiance $f_L\downarrow$	$374\,\mathrm{W\,m^{-2}}$
Aerodynamic resistances r_{aH}, r_{aE}	$\ln(Z_b/z_0)\ln(5Z_b/z_0)/(\kappa^2 U_b)$
Wind speed U_b	$5\,\mathrm{m\,s^{-1}}$
Blending height Z_b	50 m

of surface fractions of desert (fraction a_1) and crop (fraction $1 - a_1$). For scheme L this involved a linear average of the elemental fluxes using Equation (3), whereas for the upscaling schemes (M, N, S, P and Q), F_E and F_H were obtained from appropriate values of R_{tE} and R_{tH} in the large-scale versions of Equations (9) and (10), with P defined from Equation (17).

Figure 1 shows the results for the crop–desert pairing, by plotting F_E and F_H from each averaging scheme against the desert fraction a_1. Scheme L (the 'truth') appears as a straight line on this plot, so the error introduced by an averaging scheme can be judged by its curvature. Scheme M (the full flux-matching scheme) is exact for F_E, coinciding with the straight line, and shows a very slight departure for F_H, for reasons given above. Scheme N leads to slight departures from the straight line for both F_E and F_H. The *ad hoc* schemes S, P and Q all lead to very large errors, reflected by high curvatures. Similar plots can be produced for all six possible pairings between the four assumed surface types. Table V summarizes these results by showing the largest percentage errors (at any area fraction a_1) in F_E and F_H from schemes M, N, S, P and Q, relative to scheme L. The findings from Figure 1 are confirmed. In summary, scheme M is exact

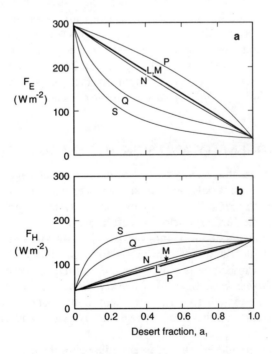

Figure 1. Predictions of averaging schemes L, M, N, S, P and Q for the areally averaged (a) latent heat flux F_E, and (b) sensible heat flux F_H, plotted against desert fraction a_1, for the crop–desert paired surface. Assumed surface and meteorological conditions given in Tables III and IV. Scheme L (the 'truth') forms a straight line; high curvature implies high error

Table V. Largest percentage errors in F_E and F_H (at any area fraction) for averaging schemes M, N, S, P and Q, relative to the linear average, scheme L, for six surface pairings. Assumed surface and meteorological conditions given in Tables III and IV

		M	N	S	P	Q
Crop–desert	F_E	0·0	−6·5	−61·1	36·7	−41·3
	F_H	4·2	11·0	143·2	−21·8	90·3
Lake–desert	F_E	0·0	−11·6	−49·4	5·6	−32·2
	F_H	−5·6	2·4	45·3	−9·1	25·6
Forest–desert	F_E	0·0	−2·8	−49·6	13·8	−34·2
	F_H	1·4	2·6	31·2	−3·0	21·4
Crop–forest	F_E	0·0	1·0	1·3	1·1	1·2
	F_H	−6·6	−8·4	−9·0	−8·9	−8·9
Lake–forest	F_E	0·0	−1·1	6·6	−13·9	−3·5
	F_H	−8·0	−6·7	−15·3	12·3	−4·4
Lake–crop	F_E	0·0	0·0	−2·8	−0·2	−1·6
	F_H	9·9	10·0	20·2	10·6	15·6

for F_E and produces small errors (typically a few per cent) for F_H. Scheme N is usually a reasonable approximation under the assumed conditions, leading to typical errors of around 10% in both F_E and F_H. The *ad hoc* schemes are always inferior (falling roughly in the order P, Q, S) and often lead to very large errors.

Flux-matching schemes similar to M have several applications for both leaf–canopy and canopy–region translations. For leaf–canopy translations, they enable the canopy resistance (pertaining to vegetation only) to be related to the distribution of leaf stomatal resistance, a photosynthetic parameter which is highly variable through the canopy. Also, they quantify the separate vegetation and soil contributions to the overall surface resistance. For canopy–region translations, they enable regional–scale resistances and fluxes to be obtained from the distributions of resistances (and also radiation) across individual patches.

We conclude this section by restating that upscaling schemes are dependent on the model being scaled, or on the function h in Equation (4). For instance, different large-scale aerodynamic resistances are required for momentum, Equation (8), and latent and sensible heat, Equation (16). McNaughton (1994) examined the upscaling of semi-mechanistic models for CO_2 exchange, obtaining different large-scale resistances again. He also examined the consequences of linearly averaging surface temperature, following L'Homme (1992), rather than latent heat flux as in Equation (14); such a scheme does not satisfy the flux-matching requirement (A).

HETEROGENEITY AND SURFACE ENERGY BALANCES

We have emphasized several times that on any particular patch of a heterogeneous land surface, the sensible and latent heat exchanges depend not only on local surface properties, but also on the transfer of heat and matter from elsewhere by fluid mixing. Transfers can occur between neighbouring patches by horizontal advection and also from much further afield by entrainment from the tropospheric flow above the atmospheric boundary layer (McNaughton and Jarvis, 1983; Jarvis and McNaughton, 1986). In this section, we analyse the consequences of these transfers for regionally averaged surface energy balances. The problem being considered can be put as a hypothesis: *for practical purposes, regionally averaged surface energy balances over heterogeneous landscapes are insensitive to the scale of heterogeneity.* It is implicit that the average is taken over a large enough area, or a sufficient number of patches, that it is statistically meaningful. Also, the statement clearly applies only to regional averages, and emphatically not to individual patches.

There are a number of ways to address this problem, one being through mesoscale numerical modelling of atmospheric motions (e.g. Pielke *et al.*, 1992; 1993; Avissar and Chen, 1993; Avissar, 1995). Another, more analytical approach is to use a simple 'slab' or 'mixed-layer' model for the development of the daytime CBL. In the highly idealized case of an enclosed, zero-entrainment CBL with a heterogeneous surface, the surface energy fluxes depend only on patch available energies and resistances and are independent of patch geometry and scale (Raupach, 1991). More realistically, the scale-insensitivity hypothesis is supported by simple slab CBL model calculations, including entrainment (Raupach, 1993). This section presents new, more detailed calculations with the CBL slab model, incorporating improved descriptions of surface fluxes and blending heights. We also review the simplifications of this approach and compare it with other approaches.

Convective boundary layer

McNaughton and Spriggs (1986), McNaughton (1989) and Cleugh and Grimmond (1993) argued that a slab CBL model provides a rational link between the surface energy balance and larger scale (tropospheric) atmospheric motions. The dynamic simplifications of the slab CBL model are appropriate for studies of land–air energy exchanges because the key property of the CBL which influences the surface energy balance is the rate of entrainment of overlying (generally dry) air into the CBL. With an appropriate entrainment parameterization to describe the CBL growth rate, this property can be described fairly realistically in a slab model. However, there are clear restrictions to the applicability of CBL slab models: they are inappropriate over terrain with large topographic variations, or in disturbed meteorological conditions such as fronts. These cases are excluded from the following discussion. Also, CBL slab models cannot resolve mesoscale circulations — a topic to which we return.

The slab model assumes that the CBL is well mixed by convective turbulence (except possibly within a relatively thin surface layer) and is bounded above by a capping inversion at height $Z_i(t)$. The scaler conservation equation, height-integrated from $z = 0$ to $z = Z_i$ in an air column moving with the mean wind field, is then

$$\frac{dC_m}{dt} = \frac{F_C}{Z_i} + \left(\frac{C_+(Z_i) - C_m}{Z_i}\right)\left(\frac{dZ_i}{dt} - W_+\right) \tag{19}$$

where $C_m(t)$ is the mixed-layer scalar concentration (uniform in the CBL), W is the mean vertical velocity, the + subscript denotes conditions in the troposphere just above $z = Z_i$, and F_C is the scalar flux at the surface. The slab model can be solved for the daytime evolution with time t of five variables, all of which are areally averaged over small-scale heterogeneities (therefore capitalized): $\Theta_m(t)$ and $Q_m(t)$ (the mixed-layer potential temperature and specific humidity), $F_H(t)$ and $F_E(t)$ (the land–air fluxes of sensible and latent heat) and $Z_i(t)$. Of the five equations needed to do this, two are supplied by writing Equation (19) for potential temperature $[C = \Theta, F_C = F_\Theta = F_H/(\rho c_p)]$ and specific humidity $[C = Q, F_C = F_Q = F_E/(\rho\lambda)]$. Two are supplied by Equations (9) and (10) in large-scale form, which yield F_E and F_H from given (generally time-dependent) radiation and resistances, and a mixed-layer deficit D_m determined by Θ_m and Q_m. The fifth equation is an entrainment parameterization specifying dZ_i/dt, for instance

$$\frac{dZ_i}{dt} = \frac{C_K W_*^3}{C_T W_*^2 + gZ_i(\Theta_{v+} - \Theta_{vm})/T_0} \tag{20}$$

where g is gravitational acceleration, $\Theta_v = \Theta(1 + 0.61Q)$ is potential virtual temperature, $W_* = (hgF_{\Theta v}/T_0)^{1/3}$ is the convective velocity scale, $F_{\Theta v}$ is the land–air flux of Θ_v, and $C_K = 0.18$ and $C_T = 0.8$ are empirical coefficients. This form was recommended by Rayner and Watson (1991) for low-shear conditions ($U_* \ll W_*$, where U_* is the areally averaged friction velocity). Many alternative, conceptually similar forms for dZ_i/dt are available. The integration of the differential equations for $\Theta_m(t)$, $Q_m(t)$ and $Z_i(t)$, and the algebraic equations for $F_H(t)$ and $F_E(t)$, proceeds through the day from dawn ($t = 0$). Initial conditions $\Theta_m(0)$, $Q_m(0)$ and $Z_i(0)$ are given, together with the tropospheric temperature and humidity profiles $\Theta_+(z)$ and $Q_+(z)$ above the CBL, the downward shortwave and longwave irradiances, and the surface albedo, emissivity and bulk resistance properties.

Scales of heterogeneity and blending heights

Length scales of heterogeneity have been classified in several ways in the past; see, for instance, Avissar (1995). Here we use a process-oriented classification based on two dynamic time-scales for CBL processes, the mixing time-scale $T_* = Z_i/W_*$ (the time-scale for a scalar to mix fully through the CBL) and the entrainment time scale $T_e = Z_i/(\mathrm{d}Z_i/\mathrm{d}t - W_+)$ (the time-scale for renewal of the air in the CBL by entrainment from above). Typically, T_* is of order of 10^2 s in the early morning, quickly rising to around 10^3 s later in the day as Z_i increases; and T_e is of the order of the time since dawn. The corresponding length scales are $U_m T_*$ (typically 1–5 km) and $U_m T_e$ (typically 100 km or more), where U_m is the mean wind speed in the well-mixed bulk of the CBL. Over *microscale heterogeneity* with patch length scale $X \leqslant U_m T_*$, air temperature and humidity respond to surface conditions only in a thin surface layer of depth $\ll Z_i$, above which they take the uniform, average values Θ_m and Q_m. Microscale advection (the interaction between vertical and horizontal gradients in the surface layer) is significant in these conditions. For *macroscale heterogeneity* with patch scale $X \geqslant U_m T_e$, the CBLs over adjacent patches are energetically independent, because air cannot move from patch to patch during the time it takes the daytime CBL to evolve. The CBL evolution is controlled by the surface fluxes and the tropospheric conditions above the CBL ($\mathrm{d}\Theta_+/\mathrm{d}z$, $\mathrm{d}Q_+/\mathrm{d}z$). This is the scale at which the CBL can be regarded as self-organizing, subject to these external conditions. Between these extremes, *mesoscale heterogeneity* occurs when $U_m T_* \leqslant X \leqslant U_m T_e$. Patches of this scale are not necessarily energetically independent, because Θ and Q (although well-mixed vertically through the bulk of the CBL) continue to adjust to changes in surface state over distances much larger than $U_m T_*$. This is advection at the CBL scale, rather than at the surface-layer scale as for microscale heterogeneity. In some instances, contrasts in surface energy balances may be large enough to generate mesoscale circulations akin to sea breezes (Avissar and Chen, 1993; Avissar, 1995; Pielke *et al.*, 1993), which alter the nature of the entrainment processes in the boundary layer and make the use of a slab model questionable for mesoscale heterogeneity.

These heterogeneity definitions, based on the dividing length scales $U_m T_*$ and $U_m T_e$, follow Raupach (1993). They coincide roughly with the broad Orlanski classification (Avissar, 1995) in which the mesoscale extends from 2 to 2000 km, though our upper mesoscale limit is less — typically a few hundred kilometres. The 'disordered' heterogeneity of Shuttleworth (1988) coincides with our microscale heterogeneity, and his 'ordered' heterogeneity embraces our two larger classes.

A key property of the scale of heterogeneity is the blending height Z_b above the surface, at which atmospheric properties such as wind or scalars (Θ, Q, D) become well mixed (uniform) horizontally. For both mesoscale and macroscale heterogeneity, this height is in the troposphere above the CBL, so $Z_b = Z_i$. (The distinction between mesoscale and macroscale is that, on diurnal time-scales, mesoscale patches are connected by CBL circulations, whereas macroscale patches are energetically independent.)

In microscale heterogeneity, Z_b has been studied both for wind speed (Wieringa, 1986; Mason, 1988) and scalars (Claussen, 1991; Wood and Mason, 1991; Blyth *et al.*, 1993). One order of magnitude estimate for Z_b for wind speed, in thermally neutral conditions, was proposed by Mason (1988)

$$Z_b/X = 2[U_*/U(Z_b)]^2; \qquad Z_b[\ln(Z_b/Z_0)]^2 = 2\kappa^2 X \tag{21}$$

where U_* and Z_0 are the area-averaged friction velocity and roughness length, respectively; $\kappa = 0.4$ is the von Karman constant; and X is a length scale of differentiation in the streamwise direction (e.g. $X = \Lambda/2\pi$ for sinusoidal heterogeneity with wavelength Λ). This can be compared with the 'inner layer depth' in the theory of flow over low hills, extended to roughness changes on flat terrain by Belcher *et al.* (1990), their Equation (3.7)

$$Z_b/X = 2\kappa U_*/U(Z_b); \qquad Z_b \ln(Z_b/Z_0) = 2\kappa^2 X \tag{22}$$

The scaling arguments leading to Equations (21) and (22), summarized in Appendix A, indicate why these two estimates are so different. Equation (21) estimates the depth of the equilibrated layer over the down-wind surface following a transition in surface properties; this is the layer within which the velocity and scalar concentration profiles have approximately readjusted to logarithmic forms (in neutral conditions) appropriate to the downwind surface. Its depth is $O[X(U_*/U_b)^2]$, where U_b is a mean velocity scale for the layer. In contrast, Equation (22) estimates the depth of the 'influenced' layer, or the height to which perturbations can diffuse by turbulent transport. Within a neutral surface layer, its depth is $O[X(U_*/U_b)]$. Finnigan *et al.* (1990) showed experimentally that this is a good estimate for the height to which mean vorticity perturbations diffuse in the flow over low hills. Many workers have used similar advection–diffusion principles to suggest equations for the depth of a developing internal boundary layer consistent with a depth of $O[X(U_*/U_b)]$; see the comprehensive reviews of Garratt (1990) and Kaimal and Finnigan (1994).

To summarize, three layers appear in the flow downwind of a surface transition: an upper layer uninfluenced by the new surface (U), a lower layer in which the flow is fully equilibrated to the new surface (L) and a transition layer (T) between layers U and L. The slopes of $U–T$ and $T–L$ interfaces are $O(U_*/U_b)$ (typically 1:10) and $O[(U_*/U_b)^2]$ (typically 1:100), respectively. These numerical ratios are familiar from micrometeorological folklore as the height to fetch ratios required for the initiation and completion (respectively) of flow adjustment to a new surface.

In the present application, we are interested in the height above which atmospheric conditions can be assumed uniform. This is given by Equation (22), rather than Equation (21). Such an estimate, of $O(XU_*/U_b)$, is also consistent with our earlier estimate for the upper length scale of microscale heterogeneity: if the characteristic diffusion velocity in the bulk of the CBL is W_* (rather than U_*) and the mean velocity is U_m, we obtain $Z_i U_m/W_* = U_m T_*$ as the order of magnitude of the fetch X at which Z_b equals the inversion height Z_i.

Regionally averaged surface energy balances

We now apply the CBL slab model to test the scale-insensitivity hypothesis stated above, by calculating F_E and F_H over surfaces with microscale and macroscale heterogeneity. The intermediate, mesoscale case cannot be treated formally with the CBL slab model, because mesoscale circulations may lead to violation of the model assumptions; instead, we argue that in most situations, the mesoscale case is constrained between the other two. We study hypothetical surfaces consisting of pairs of surface types, chosen from the same four possibilities (crop, desert, forest and lake) used in the last section.

Within the CBL slab model, the surface fluxes are calculated using a 'simplified canopy–atmosphere model' for heterogeneous surfaces, which we will call SCAM-B. Described fully in Appendix B, SCAM-B is a development of the very simple SCAM outlined in Raupach (1991; 1993), but does not claim to be a complete biosphere model for (say) GCM application. Rather, it is a means of calculating surface fluxes to test hypotheses about the properties of heterogeneous surfaces. On a patch scale, SCAM-B accounts for: (1) radiative coupling; (2) stability effects on patch aerodynamic resistances; and (3) the response of patch surface resistances to radiation and surface saturation deficit (and also surface temperature and water stress, though these sensitivities are not included in the present calculations).

The main novel aspect of SCAM-B is the incorporation of microscale advection between patches, when the scale of heterogeneity (X) is small enough. Instead of specifying the scale X directly, it is convenient to characterize the heterogeneity by specifying the blending height Z_b, which is related to X through Equation (22) (typically $Z_b/X \approx 0.05$ to 0.1 in vegetated terrain). Microscale advection is significant when $Z_b < Z_m$, where Z_m is the lower height limit of the mixed layer or upper height limit of the surface layer, typically around $0.1Z_i$. The range of X for which this occurs can be found from Equation (22) by setting $Z_b < 0.1Z_i$, which implies $X < 0.1(2\kappa)^{-1}Z_i(U_b/U_*)$, or typically $X \lesssim 1$ km. The blending-height deficit D_b then differs from the mixed-layer deficit D_m (and likewise for other air properties) by $R_a(b,m)F_D$, where F_D is the deficit flux and $R_a(b,m)$ the aerodynamic resistance for scalar entities (including Θ, Q and the deficit D) across the upper part of the surface layer, between Z_b and Z_m. This layer is horizontally uniform, so both F_D and $R_a(b,m)$ depend on the properties of all patches, rather than any particular patch. The inclusion of a blending height $Z_b < Z_m$ (when X is small enough) allows the local energy balance of any one patch to be influenced by neighbouring patches, through fluid mixing in the surface layer. Accounting properly for these influences, SCAM-B calculates surface fluxes from meteorological conditions specified in the mixed layer (at Z_m), rather than at the blending height (Z_b).

To quantify the aerodynamic resistances r_{ai} (for individual patches, from the surface to Z_b) and $R_a(b,m)$

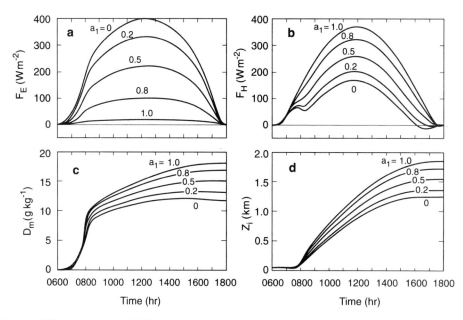

Figure 2. Prediction of CBL slab model and SCAM-B for diurnal variation of (a) areally averaged latent heat flux F_E, (b) areally averaged sensible heat flux F_H, (c) mixed-layer deficit D_m and (d) CBL depth Z_i, for microscale-heterogeneous crop–desert surfaces with desert fraction $a_1 = 0$, 0·2, 0·5, 0·8, 1·0. Assumed surface and meteorological conditions given in Tables III and VI; blending height $Z_b = Z_m$

(from Z_b to Z_m), SCAM-B uses conventional Monin–Obukhov similarity theory for the surface layer. This formulation clearly cannot incorporate details of the microscale advection process such as the undershoot and overshoot in *local* (point by point) aerodynamic transfers caused by changes in roughness or moisture availability (e.g. Kaimal and Finnigan, 1994). However, it produces reasonable estimates for the *patch-averaged* resistances (Wood and Mason, 1991), provided Z_b is not so small as to be comparable with the heights of roughness elements.

Predictions of regionally averaged surface energy balances have been made using the CBL slab model, coupled to SCAM-B, for heterogeneous surfaces consisting of pairs of surface types chosen from the hypothetical crop, desert, forest and lake surfaces specified in Table III. Figure 2 shows the diurnal course of the areally averaged surface energy fluxes F_E and F_H, the mixed-layer deficit D_m and the CBL depth Z_i, for microscale crop–desert heterogeneity with $Z_b = Z_m$ (so X is large enough for the blending height to be above the surface layer, and $D_b = D_m$). The meteorological conditions and computational parameters are

Table VI. Meteorological conditions and computational parameters for calculations with CBL slab model and SCAM-B (Figures 2 and 3)

Downward shortwave irradiance $F_{S\downarrow}$ (uniform over all patches)	$F_{S\,max} \sin(\pi 2t/T_{day})$ with $F_{S\,max} = 800\,\mathrm{W\,m^{-2}}$, $T_{day} = 86400\,\mathrm{s}$
Downward longwave irradiance $F_{L\downarrow}$ (uniform over all patches)	Swinbank (1963) formula: $F_{L\downarrow}(\mathrm{W\,m^{-2}}) = 5\cdot31 \times 10^{-14} T^6$ (T in K)
Conditions above CBL	$\Theta_+(z) = 15 + 0\cdot005z\,(C)$ $Q_+(z) = 0\,(\mathrm{g\,kg^{-1}})$
Initial conditions	$\Theta(0) = 10^\circ\mathrm{C}$ $Q(0) = Q_{sat}$ $Z_i(0) = 50\,\mathrm{m}$
Depth of surface layer Z_m	50 m
Integration period	12 hours (0600 to 1800 local time)
Integration time step	15 minutes

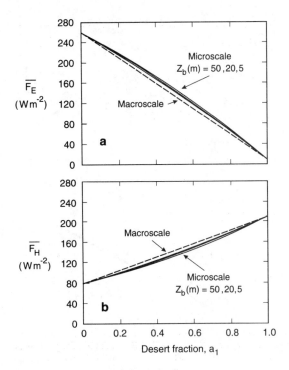

Figure 3. Prediction of CBL slab model and SCAM-B for the daytime and areally averaged latent and sensible heat fluxes $\bar{F}_E$ and $\bar{F}_H$, for macroscale and microscale crop–desert surfaces, plotted against desert fraction a_1. Assumed surface and meteorological conditions are given in Tables III and VI. For microscale heterogeneity, predictions are shown for blending heights $Z_b = 50$ m ($= Z_m$), 20 m and 5 m

given in Table VI (in particular, the surface-layer depth Z_m is set at 50 m). Five choices of the desert fraction a_1 (0, 0·2, 0·5, 0·8 and 1) are shown, to span the range from pure crop to pure desert. Not surprisingly, increasing the desert fraction decreases F_E and increases F_H, D_m and Z_i. The sum $F_E + F_H$ is not constant with changing desert fraction (despite the forcing downward shortwave and longwave irradiances being kept to the same form throughout) because of radiative coupling.

For the same crop–desert heterogeneous surface, Figure 3 shows the daytime averages of the areally averaged latent and sensible heat fluxes, $\bar{F}_E$ and $\bar{F}_H$, defined by

$$\bar{F}_E = \frac{1}{T} \int_0^T F_E \mathrm{d}t = \frac{1}{T} \int_0^T \sum a_i f_{Ei} \mathrm{d}t \tag{23}$$

and similarly for $\bar{F}_H$, where T is the integration period (12 hours from 0600 solar time). These averages provide a convenient means of comparing the daily energy fluxes over heterogeneous surfaces of different scales. They are plotted against the desert fraction a_1, for microscale heterogeneity with $Z_b = 5$, 20 and 50 m (the last representing the case $Z_b = Z_m$), and macroscale heterogeneity. In the macroscale case, the regionally averaged $\bar{F}_E$ and $\bar{F}_H$ values across several patches are simply the area-weighted linear averages of the corresponding fluxes for pure desert and pure crop surfaces

$$\bar{F}_E(a_1) = a_1 \bar{F}_E (\text{desert}) + (1 - a_1) \bar{F}_E (\text{crop}) \tag{24}$$

because the crop and desert patches act independently. This forms a straight line in Figure 3. The differences between the various scales of microscale heterogeneity (curves) and the macroscale case (straight line) are all quite small*, reaching about 10% at most for very small-scale heterogeneity with $Z_b = 5$ m.

---

* Similar calculations in Raupach (1993) omitted effects (1) to (4), accounted for in SCAM-B. They produced the same overall result, but with slightly higher differences between the microscale and macroscale averages, because they were inadvertently carried out with an unrealistically low $Z_m = 2$ m.

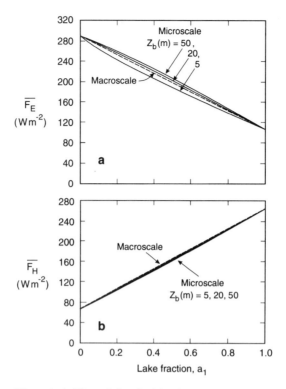

Figure 4. As Figure 3, but for lake–forest surface pairing

Similar results are obtained for the other surface pairings, with the differences between microscale and macroscale heterogeneity being even smaller than Figure 3 in most instances. Another example is shown in Figure 4, for the lake–forest pairing. The overall dependence on scale of heterogeneity is again small, but in this instance the effect of decreasing Z_b in the microscale case is to decrease $\bar{F}_E$, in contrast with the crop–desert pairing. This is probably associated with the distinction between a (moist, rough) and (dry, smooth) pairing such as crop–desert, and the converse (moist, smooth) and (dry, rough) pairing for lake–forest.

These results suggest that the scale-insensitivity hypothesis for regionally averaged surface energy balances is supported. The basic reason is the energetic constraint that regionally averaged latent and sensible heat fluxes are controlled by the net fluxes of radiant energy and saturation deficit into a well-mixed control volume (conveniently the CBL), and by the surface properties. Although all these quantities are strongly sensitive to the *nature* of the heterogeneous surface, none is strongly sensitive (in a regionally averaged sense) to the *scale* of heterogeneity.

Mesoscale heterogeneity cannot be treated by the above analysis if mesoscale circulations are significant. Several workers have studied mesoscale circulations numerically. Hadfield *et al.* (1991; 1992) used large-eddy simulation (LES) methods over level, patchy terrain with a variable heat flux, finding significant circulations in conditions of zero mean wind, which were enhanced as the terrain wavelength increased, but drastically disrupted by even a light wind. The *areally averaged* atmospheric properties, including fluxes, were not significantly influenced by the heterogeneity or the circulations — a similar result to ours. Walko *et al.* (1992) obtained similar results from LES simulations of the flow over low hills. Avissar and Chen (1993) and Avissar (1995) used the Colorado State University (CSU) regional atmospheric modelling system (RAMS) (Pielke *et al.*, 1992) to study idealized mesoscale deforestation in light $(0.5\,\mathrm{m\,s^{-1}})$ wind conditions. They found significant mesoscale circulations from which they calculated substantial 'mesoscale fluxes', vertical transports due to time-averaged flow fields with significant vertical wind components, locked into the terrain. At heights comparable with Z_i, the mesoscale fluxes exceeded the conventional turbulent fluxes. However, the mesoscale fluxes decay to zero at the ground (by continuity), and their main significance may be through the problem they raise for subgrid-scale parameterizations of atmospheric fluxes above the surface in GCMs. Pielke *et al.* (1993) studied the effects of real landscape heterogeneity, similarly finding significant mesoscale circulations and associated mesoscale fluxes. (It is interesting that analogous fluxes maintained by time-averaged vertical motions have long been studied in vegetation canopies (Kaimal and Finnigan, 1994) where they have been termed 'dispersive fluxes'.)

In summary, detailed numerical studies indicate that mesoscale circulations can generate substantial local effects, including mesoscale fluxes above the surface, in light wind conditions; but the available evidence suggests that these circulations alone do not lead to strong departures from the scale-insensitivity hypothesis for regionally averaged surface fluxes in a simple, dry CBL. The likely reason is that they do not significantly affect the regionally averaged saturation deficit in the CBL. It is possible that strong non-linearities could be induced in plots such as Figures 3 and 4 by 'wet' processes, including clouds and precipitation, which may affect the regionally averaged CBL saturation deficit. However, these are outside the scale of our simple CBL slab model. Avissar (1995) suggested that these 'wet' processes respond strongly to scale in idealized mesoscale deforestation experiments (see above), but the effects on regionally averaged surface energy balances need further investigation.

SUMMARY AND CONCLUSIONS

Our summary follows the three main parts of the paper. The first part discussed scale issues in hydrometeorology as problems in parameterizing terms in the equations for water and energy conservation, integrated over control volumes of different scales. Upscaling was defined (from a meteorological viewpoint) as the process of aggregating land-surface descriptors and fluxes, with a converse definition for downscaling. Several guidelines were identified which explicitly or implicitly underlie most upscaling and downscaling analyses: the 'conservation of complication' and the truism that 'fluids flow and mix'.

The second part of the paper dealt with upscaling in boundary-layer meteorology, first establishing two 'flux-matching' constraints on the upscaling models for surface exchange processes: the conservation requirement that net scalar fluxes at land surfaces average linearly, and the practical requirement that model form be preserved across scales. By considering a simple transfer process described by the elemental law $f = (c_0 - C_b)/r_{aC}$, we showed that the choice of boundary condition c_0 is critical in determining the upscaling relationships: unphysical boundary conditions produce unphysical results at large scales. This led to the use of the CE as a description of the small-scale fluxes, because it eliminates the surface condition c_0 in favour of externally prescribed properties. Rational upscaling relationships for latent and sensible heat exchange were derived in this way, tested against other possibilities, and shown to perform well in all situations explored.

The third part of the paper applied these considerations to the question of the sensitivity of regionally averaged energy balances over land surfaces to the scale of heterogeneity, X. Microscale, mesoscale and macroscale heterogeneity were defined on physical grounds. A CBL slab model was used to show that regionally averaged energy fluxes are remarkably insensitive to X, being very similar for X in the microscale range (accounting for blending height variations with X) and in the macroscale range. The mesoscale range could not be studied directly in this way, but evidence from other modelling work indicates that, despite the occurrence of mesoscale circulations, scale insensitivity of regionally averaged surface energy fluxes occurs in this scale range also.

This work suggests several areas for future research. One is the need to assess the response of regionally averaged energy fluxes to cloud and precipitation processes: does the inclusion of these processes lead to strong dependences of regionally averaged energy fluxes on the scale of heterogeneity? A second is downscaling, a topic which we have largely neglected here in favour of upscaling. A third is the interface between the meteorological models we have discussed and their hydrological (and ecological?) counterparts. The broad issue is the mutual interaction of two fluids, air and water, with the vegetation, soils and topography of heterogeneous landscapes, to determine the overall behaviour of the biosphere.

ACKNOWLEDGEMENTS

We are grateful to Drs H. A. Cleugh, J. D. Wilson and R. Leuning for comments on a draft of this paper, and to Ms Mona Loofs for editorial assistance. This work contributes to the CSIRO Climate Change Research Program, and is partly funded through the National Greenhouse Research Program.

REFERENCES

Avissar, R. 1995. 'Scaling of land–atmosphere interactions: an atmospheric modelling perspective', *Hydrol. Process.*, **9**, 679–695.

Avissar, R. and Chen, F. 1993. 'A method to bridge the gap between microscale land-surface processes and land–atmosphere interactions at regional and continental scales' in Bolle, H.-J., Feddes, R. A., and Kalma, J. D. (Eds), *Exchange Processes at the Land Surface for a Range of Space and Time Scales, Proc. Symp. J3.1, Joint Scientific Assembly of IAMAP and IAHS, Yokohama, Japan, 11–23 July 1993, IAHS Publ.*, **212**, 315–323.

Baldocchi, D. D., Luxmoore, R. J., and Hatfield, J. L. 1991. 'Discerning the forest from the trees: an essay in scaling canopy stomatal conductance', *Agric. Forest Meteorol.*, **54**, 197–226.

Belcher, S. E., Xu, D. P., and Hunt, J. C. R. 1990. 'The response of a turbulent boundary layer to arbitrarily distributed two-dimensional roughness changes', *Q. J. Roy. Meteorol. Soc.*, **116**, 611–635.

Binley, A. and Beven, K. 1989. 'A physically based model of heterogeneous hillslopes. 2. Effective hydraulic conductivities', *Wat. Resour. Res.*, **25**, 1227–1233.

Blyth, E. M., Dolman, A. J., and Wood, N. 1993. 'Effective resistance to sensible and latent heat flux in heterogeneous terrain', *Q. J. Roy. Meteorol. Soc.*, **119**, 423–442.

Claussen, M. 1991. 'Estimation of areally-averaged surface fluxes', *Boundary-Layer Meteorol.*, **54**, 387–410.

Cleugh, H. A. and Grimmond, C. S. B. 1993. 'A comparison between measured local-scale suburban and areally-averaged urban heat and water vapour fluxes' in Bolle, H.-J., Feddes, R. A., and Kalma, J. D. (Eds), *Exchange Processes at the Land Surface for a Range of Space and Time Scales, Proc. Symp. J3.1, Joint Scientific Assembly of IAMAP and IAHS, Yokohama, Japan, 11–23 July 1993, IAHS Publ.*, **212**, 155–163.

Finnigan, J. J. and Raupach, M. R. 1987. 'Transfer processes in plant canopies in relation to stomatal characteristics' in Zeiger, E., Farquhar, G. D., and Cowan, I. R. (Eds), *Stomatal Function*. Stanford University Press, Stanford. pp. 385–429.

Finnigan, J. J. and Raupach, M. R. 'The influence of topography on meteorological variables and surface–atmosphere interactions', *J. Hydrol.*, in press.

Finnigan, J. J., Raupach, M. R., Bradley, E. F., and Aldis, G. K. 1990. 'A wind tunnel study of turbulent flow over a two-dimensional ridge', *Boundary-Layer Meteorol.*, **50**, 277–317.

Garratt, J. R. 1990. 'The internal boundary layer — a review', *Boundary-Layer Meteorol.*, **50**, 171–203.

Garratt, J. R. 1992. *The Atmospheric Boundary Layer*. Cambridge University Press, Cambridge. 316 pp.

Hadfield, M. G., Cotton, W. R., and Pielke. R. A. 1991. 'Large-eddy simulations of thermally forced circulations in the convective boundary layer. Part I: a small-scale circulation with zero wind', *Boundary-Layer Meteorol.*, **57**, 79–114.

Hadfield, M. G., Cotton, W. R., and Pielke, R. A. 1992. 'Large-eddy simulations of thermally forced circulations in the convective boundary layer. Part II: the effect of changes in wavelength and wind speed', *Boundary-Layer Meteorol.*, **58**, 307–328.

Jackson, P. S. and Hunt, J. C. R. 1975. 'Turbulent wind flow over a low hill', *Q. J. Roy. Meteorol. Soc.*, **101**, 929–955.

Jarvis, P. G. and McNaughton, K. G. 1986. 'Stomatal control of transpiration: scaling up from leaf to region', *Adv. Ecol. Res.*, **15**, 1–47.

Kaimal, J. C. and Finnigan, J. J. 1994. *Atmospheric Boundary Layer Flows: Their Structure and Measurement*. Oxford University Press, Oxford. 289 pp.

Kelliher, F. M., Leuning, R., Raupach, M. R., and Schulze, E. D. 1995. 'Maximum conductances for evaporation from global vegetation types', *Agric. Forest Meteorol.*, **73**, 1–16.

Koster, R. D. and Suarez, M. J. 1992. 'A comparative analysis of two land surface heterogeneity representations', *J. Climate*, **5**, 1379–1390.

L'Homme, J. P. 1992. 'Energy balance of heterogeneous terrain: averaging the controlling parameters', *Agric. Forest Meteorol.*, **61**, 11–21.

McNaughton, K. G. 1976. 'Evaporation and advection I: evaporation from extensive homogeneous surfaces', *Q. J. Roy. Meteorol. Soc.*, **102**, 181–191.

McNaughton, K. G. 1989. 'Regional interactions between canopies and the atmosphere' in Russell, G., Marshall, B., and Jarvis, P. G. (Eds), *Plant Canopies: their Growth, Form and Function*. Cambridge University Press, Cambridge. pp. 63–81.

McNaughton, K. G. 1994. 'Effective stomatal and boundary-layer resistances of heterogeneous surfaces', *Plant, Cell Environ.*, **17**, 1061–1068.

McNaughton, K. G. and Jarvis, P. G. 1983. 'Predicting the effects of vegetation changes on transpiration and evaporation' in Kozlowski, T. T. (Ed.), *Water Deficits and Plant Growth*. Vol. VII. Academic Press, New York. pp. 1–47.

McNaughton, K. G. and Spriggs, T. W. 1986. 'A mixed layer model for regional evaporation', *Boundary-Layer Meteorol.*, **34**, 243–262.

Mason, P. J. 1988. 'The formation of areally-averaged roughness lengths', *Q. J. Roy. Meteorol. Soc.*, **114**, 399–420.

Monteith, J. L. 1973. *Principles of Environmental Physics*. Arnold, London. 241 pp.

Philip, J. R. 1959. 'Evaporation, moisture and heat fields in the soil', *J. Meteorol.*, **14**, 354–366.

Philip, J. R. 1966. 'Plant water relations: some physical aspects', *Ann. Rev. Plant Physiol.*, **17**, 245–268.

Philip, J. R. 1987. 'Advection, evaporation and surface resistance', *Irrig. Sci.*, **8**, 101–114.

Pielke, R. A., Cotton, W. R., Walko, R. L., Tremback, C. J., Lyons, W. A., Grasso, L. D., Nicholls, M. E., Moran, M. D., Wesley, D. A., Lee, T. J., and Copeland, J. H. 1992. 'A comprehensive meteorological modelling system — RAMS', *Meteorol. Atmos. Phys.*, **49**, 69–91.

Pielke, R. A., Dalu, G. A., Lee, T. J., Rodriguez, H., Eastman, J., and Kittel, T. G. F. 1993. 'Mesoscale parameterization of heat fluxes due to landscape variability for use in general circulation models' in Bolle, H.-J., Feddes, R. A., and Kalma, J. D. (Eds), *Exchange Processes at the Land Surface for a Range of Space and Time Scales. Proc. Symp. J3.1, Joint Scientific Assembly of IAMAP and IAHS, Yokohama, Japan, 11–23 July 1993, IAHS Publ.*, **212**, 331–342.

Raupach, M. R. 1991. 'Vegetation–atmosphere interaction in homogeneous and heterogeneous terrain: some implications of mixed-layer dynamics', *Vegetatio*, **91**, 105–120.

Raupach, M. R. 1993. 'The averaging of surface flux densities in heterogeneous landscapes' in Bolle, H.-J., Feddes, R. A., and Kalma, J. D. (Eds), *Exchange Processes at the Land Surface for a Range of Space and Time Scales, Proc. Symp. J3.1, Joint Scientific Assembly of IAMAP and IAHS, Yokohama, Japan, 11–23 July 1993, IAHS Publ.*, **212**, 343–355.

Raupach, M. R. 1995. 'Vegetation–atmosphere interaction and surface conductance at leaf, canopy and regional scales', *Agric. Forest Meteorol.*, **73**, 151–179.

Raupach, M. R. and Finnigan, J. J. 1988. 'Single-layer evapotranspiration models are incorrect but useful, whereas multilayer models are correct but useless: discuss', *Aust. J. Plant Physiol.*, **15**, 715–726.

Rayner, K. N. and Watson, I. D. 1991. 'Operational prediction of daytime mixed-layer heights for dispersion modelling', *Atmos. Environ.*, **25A**, 1427–36.

Shuttleworth, W. J. 1988. 'Macrohydrology — the new challenge for process hydrology', *J. Hydrol.*, **100**, 31–56.

Swinbank, W. C. 1963. 'Long-wave radiation from clear skies', *Q. J. Roy. Meteorol. Soc.*, **89**, 339–348.

Tan, C. S. and Black, T. A. 1976. 'Factors affecting the canopy resistance of a Douglas Fir forest', *Boundary-Layer Meteorol.*, **10**, 475–488.

Walko, R. L., Cotton, W. R., and Pielke, R. A. 1992. 'Large-eddy simulations of the effects of hilly terrain on the convective boundary layer', *Boundary-Layer Meteorol.*, **58**, 133–150.

Wessman, C. A. 1992. 'Spatial scales and global change: bridging the gap from plots to GCM grid cells', *Annu. Rev. Ecol. Syst.*, **23**, 175–200.

Wieringa, J. 1986. 'Roughness-dependent geographical interpolation of surface wind speed averages', *Q. J. Roy. Meteorol. Soc.*, **112**, 867–889.

Wood, N. and Mason, P. J. 1991. 'The influence of static stability on the effective roughness lengths for momentum and temperature', *Q. J. Roy. Meteorol. Soc.*, **117**, 1025–1056.

APPENDIX A: SCALE ANALYSIS OF BLENDING HEIGHT

For steady flow over a flat but heterogeneous surface in neutral conditions, the perturbation part of the linearized equation of motion is

$$U(z)\frac{\partial u'}{\partial x} = \frac{\partial \tau'}{\partial z} \tag{A1}$$

where $u(x, z) = U(z) + u'(x, z)$ is the time-mean velocity and $\tau(x, z) = T + \tau'(x, z)$ the time-mean stress. Capitals denote horizontal averages (assumed to represent the background state about which the perturbation is taken and primes the perturbation part of the mean field (*not* the turbulent fluctuation). The pressure term is ignored (Kaimal and Finnigan, 1994). Using a mixing-length model to estimate τ for neutral conditions, we can write

$$\tau = \kappa^2 z^2 \left(\frac{\partial u}{\partial z}\right)^2; \qquad \tau' = 2\kappa U_* z \frac{\partial u'}{\partial z} \tag{A2}$$

where $U_* = T^{1/2} = \kappa z(\partial U/\partial z)$ is the background friction velocity and the second equation is linearized by neglecting products of perturbation terms. We now equate order of magnitude estimates for the left- and right-hand sides of Equation (A1). The term on the left is of order $U_b u'/X$, where X is a length scale for streamwise differentiation and U_b a background advection velocity. The term on the right is estimated in different ways by different workers:

1. Mason (1988) assumed that u' 'adjusts to the new surface in the usual logarithmic way', implying

$$\partial u'/\partial z \sim [u'/U(z)](U_*/\kappa z); \qquad \tau' \sim 2U_*^2 u'/U(z) \tag{A3}$$

Then, to find the height Z_b at which the two sides of Equation (A1) are in balance, Mason assumed $\partial\tau'/\partial z \sim \tau'/Z_b$. This gives $Z_b/X \sim 2(U_*/U_b)^2$, Equation (21), where $U_b = U(Z_b)$. The resulting Z_b is an estimate of the height up to which equilibrium conditions prevail over the new surface.

2. Jackson and Hunt (1975) introduced an order of magnitude analysis for the inner layer depth in the

linear theory of flow over low hills, which has since become standard (see Kaimal and Finnigan, 1994 for a review) and has been adapted to roughness changes by Belcher *et al.* (1990). If Z_b is a height scale for the inner layer, then Equation (A2) gives

$$\frac{\partial \tau'}{\partial z} = 2\kappa U_* \frac{\partial}{\partial z}\left(z\frac{\partial u'}{\partial z}\right) \sim 2\kappa U_* \frac{u'}{Z_b} \tag{A4}$$

where u' and Z_b are here to be understood as velocity and height scales and the second relationship includes a dimensionless $O(1)$ function of position. Equating this with $U_b u'/X$, we obtain $Z_b/X \sim 2\kappa U_*/U_b$, Equation (22). The resulting Z_b is an estimate of the height of the transition zone in which stress divergence is large.

APPENDIX B: CALCULATING PATCH SURFACE FLUXES

For patch i, the latent heat, sensible heat and momentum fluxes are

$$f_{Ei} = \frac{\epsilon r_{tHi}\phi_{Ai} + \rho\lambda D_b}{r_{di}} \tag{B1}$$

$$f_{Hi} = p_i\left[\frac{r_{tEi}\phi_{Ai} - \rho\lambda D_b}{r_{di}}\right] \tag{B2}$$

$$u_{*i}^2 = U_b/r_{aMi} \tag{B3}$$

where D_b and U_b are the saturation deficit and wind speed at the blending height Z_b (hence constant across patches). The isothermal available energy ϕ_{Ai} is

$$\phi_{Ai} = (1-\alpha_{si})f_{S\downarrow i} + \epsilon_{si}(f_{L\downarrow i} - \sigma\Theta_b^4) \tag{B4}$$

where $f_{S\downarrow i}$ and $f_{L\downarrow i}$ are the downward shortwave and longwave irradiances, α_{si} and ϵ_{si} are the patch albedo and emissivity, and the outward longwave component is expressed in terms of the blending height potential temperature Θ_b instead of the patch surface temperature. Note that $\Theta_b = T_b + \Gamma Z_b$, where T_b is the blending height actual temperature and Γ the dry adiabatic lapse rate. In the calculations, we assume uniform downward irradiances ($f_{S\downarrow i} = F_{S\downarrow}, f_{L\downarrow i} = F_{L\downarrow}$) and use the Swinbank (1963) formula to estimate $F_{L\downarrow}$ from T_b.

The aerodynamic resistances for scalar (say heat) and momentum transfer between the surface and Z_b are

$$r_{aHi} = \frac{[\ln(Z_b/z_{0Hi}) - \Psi_H(Z_b/L_i)][\ln(Z_b/z_{0Mi}) - \Psi_M(Z_b/L_i)]}{\kappa^2 U_b} \tag{B5}$$

$$r_{aMi} = \frac{[\ln(Z_b/z_{0Mi}) - \Psi_M(Z_b/L_i)]^2}{\kappa^2 U_b} \tag{B6}$$

where z_{0Hi} and z_{0Mi} are the patch scalar and momentum roughness lengths ($z_{0Hi} = z_{0Mi}/5$ is assumed); $L_i = -(\rho c_p T_b u_{*i}^3)/[\kappa g(f_{Hi} + 0.07 f_{Ei})]$ is the patch Monin–Obukhov length; and Ψ_H and Ψ_M are the diabatic influence functions for scalar and momentum transfer, for which we use the common forms given by Garratt (1992). The patch surface resistance is assumed to respond to incoming solar radiation and saturation deficit at the surface, according to

$$r_{si} = \frac{r_{s(\min)i}}{s_1(f_{S\downarrow i})s_2(d_{0i})}$$

$$s_1(f_{s\downarrow i}) = 1 - \exp(f_{S\downarrow i}/f^*_{S\downarrow i}); \qquad s_2(d_{0i}) = \frac{1}{1 + d_{0i}/d^*_{0i}} \tag{B7}$$

where $r_{s(\min)i}$ is the minimum patch surface resistance for patch i; s_1 and s_2 are physiological stress functions describing the effects of low solar irradiance (f_{Si}) and high surface saturation deficit (d_{0i}), respectively; and f^*_{Si} and d^*_{0i} are empirical parameters. Stress functions for moisture and surface temperature can be included, but are omitted here.

Equations (B1) to (B7) form a closed but implicit set which can be solved for the patch fluxes f_{Ei}, f_{Hi} and u^2_{*i}, using specified patch properties $\alpha_{si}, \epsilon_{si}, z_{0Mi}, z_{0Hi}, r_{s(min)i}, f^*_{Si}$ and d^*_{0i}, and meteorological forcing variables $F_{S\downarrow}, F_{L\downarrow}, U_b, D_b$ and Θ_b. In this work, the irradiances $F_{S\downarrow}$ and $F_{L\downarrow}$ are supplied externally, and the blending-height properties U_b, D_b and Θ_b are calculated from the corresponding mixed-layer properties U_m, D_m and Θ_m (with U_m supplied externally, and D_m and Θ_m calculated from the CBL slab model). The solution of Equations (B1) to (B7) requires some care, as simple iteration often diverges. The following methods are used (1) to calculate the patch stability parameter Z_b/L_i; (2) to find U_b, D_b and Θ_b from U_m, D_m and Θ_m; and (3) to perform the overall calculation.

(1) *Stability parameter Z_b/L_i.* Iterative solutions for patch Z_b/L_i often diverge (from initial aerodynamic resistances, calculate fluxes, then L_i, then new aerodynamic resistances, then new fluxes ...). Instead, the following method was used. The stability parameter is written as an implicit function of itself

$$\zeta = Z_b/L_i = \frac{\kappa g Z_b(f_{Hi} + 0.07 f_{Ei})}{\rho c_p u^3_{*i} T_b} = y(\zeta) \tag{B8}$$

where the function y is determined by substituting Equations (B5) and (B6) into (B1) to (B3) and then (B8). Thus, y is a fully determined function for given $z_{0Mi}, z_{0Hi}, \phi_{Ai}, r_{si}, U_b, D_b$ and T_b. Equation (B8) can then be solved by standard methods for implicit equations (other than direct iteration, which diverges when $|dy/d\zeta| > 1$).

(2) *Blending height properties.* If the blending height $Z_b > Z_m$ (where $Z_m \approx 0.1 Z_i$ is the top of the surface layer or bottom of the mixed layer), then blending-height and mixed-layer properties are identical: $U_b = U_m, D_b = D_m, \Theta_b = \Theta_m$. The mixed-layer properties (externally specified or computed from the CBL slab model) can then be used directly in the patch surface flux calculations. On the other hand, if $Z_b < Z_m$, then

$$U_m - U_b = R_{aM}(b,m)T = R_{aM}(b,m)\sum a_i \tau_i \tag{B9}$$

$$D_b - D_m = R_{aH}(b,m)F_D = R_{aH}(b,m)\sum a_i f_{Di} \tag{B10}$$

where

$$f_{Di} = (\epsilon f_{Hi} - f_{Ei})/(\rho\lambda) \tag{B11}$$

is the patch flux of saturation deficit, and $R_{aM}(b,m)$ and $R_{aH}(b,m)$ are the momentum and scalar (heat and deficit) aerodynamic resistances between Z_b and Z_m. These resistances are

$$R_{aH}(b,m) = \frac{\ln(Z_m/Z_b) = \Psi_H(Z_m/L_m) + \Psi_H(Z_b/L_m)}{\kappa U_*} \tag{B12}$$

and similarly for $R_{aM}(b,m)$, where U_* and the Monin–Obukhov length $L_m = -(\rho c_p T_b U^3_*)/[\kappa g(F_H + 0.07 F_E)]$ are defined from the areally averaged fluxes. When $Z_b < Z_m$, the air properties at Z_b depend on the fluxes from all patches, not any single patch i; so to find the fluxes from patch i, we must find the fluxes from all patches first. Iteration is unworkable, diverging when $R_a(b,m)$ exceeds a typical patch aerodynamic resistance from the surface to Z_b. Fortunately, the problem is linear and iteration is unnecessary. For U, Equations (B9) and (8) give

$$U_b = \frac{U_m}{1 + R_{aM}(b,m)/R_{aM}(0,b)} \tag{B13}$$

with $[R_{aM}(0,b)]^{-1} = \sum a_i/r_{aMi}$, the bulk aerodynamic resistance for momentum from Equation (8). For

D, we express the patch deficit flux F_{Di} in terms of the equilibrium saturation deficit δ_{eq}, a function of ϕ_{Ai} and the patch resistances [see Raupach (1995), Equations (9) and (10)]

$$f_{Di} = \frac{(p_i\epsilon + 1)(\delta_{eqi} - D_b)}{r_{di}}$$

$$\delta_{eqi} = \frac{p_i\epsilon\phi_{Ai}(r_{si} + r_{aEi} - r_{aHi})}{(p_i\epsilon + 1)\rho\lambda} \tag{B14}$$

Combining this with Equation (B9), we obtain

$$D_b = \frac{D_m + \Delta'_{eq}R_{aH}(b,m)/R'_d}{1 + R_{aH}(b,m)/R'_d} \tag{B15}$$

where

$$\frac{1}{R'_d} = \sum \frac{a_i(p_i\epsilon + 1)}{r_{di}}; \qquad \Delta'_{eq} = R'_d \sum \frac{a_i(p_i\epsilon + 1)\delta_{eqi}}{r_{di}} \tag{B16}$$

are large-scale counterparts to the patch quantities $r_{di}/(p_i\epsilon + 1)$ and δ_{eqi}. They are obtained by applying the flux-matching requirements to the deficit flux $F_D = \sum a_i f_{di}$, and are slightly different from the equivalent large-scale quantities obtained from the CE for F_E, which omit the factor $p_i\epsilon + 1$; see Equation (15), and Raupach (1995). (The difference becomes a simple factor when $p_i = 1$.) Equations (B13) and (B15) express U_b and D_b in terms of weighted means of specifiable patch properties.

(3) *Overall calculation procedure.* The techniques given above for finding Z_b/L_i and the blending-height properties do not completely remove the need for iteration, but they do make the eventual iteration process strongly convergent and stable. The procedure used to calculate the patch surface fluxes, carried out at each time step of the integration of the CBL slab model, is:

1. From mixed-layer properties (U_m, Θ_m, Q_m), calculate fluxes using Equations (B1) to (B8), using the initial estimates $U_b = U_m$, $D_b = D_m$, $\Theta_b = \Theta_m$, and d_{0i} from the last time step (the initial step uses $d_{0i} = 0$).
2. Find improved estimates of U_b, D_b and Θ_b, using Equations (B13), (B15) and similar.
3. Find an improved estimate of $d_{0i}(= D_b + f_{Di}r_{aHi})$ and thence r_{si} from Equation (B7).
4. Repeat steps 1 to 3 to convergence: four iterations are ample, omitting steps 2 and 3 on the last one.

The fluxes from this procedure are used to integrate the CBL slab model, Equations (19) and (20), forward to the next time step.

21

ESTIMATION OF TERRESTRIAL WATER AND ENERGY BALANCES OVER HETEROGENEOUS CATCHMENTS

R. P. SILBERSTEIN AND M. SIVAPALAN

Centre for Water Research, Department of Environmental Engineering, The University of Western Australia, Nedlands, Western Australia, 6009, Australia

ABSTRACT

Modelling experiments have been undertaken to address the effects of land surface heterogeneity on the energy and water fluxes at catchment scales. The simulation results indicate that in the presence of strong contrasts (i.e. patchiness) in the land surface characteristics (for example, soil moisture, leaf area index or vegetation type) significant inter-patch advection can result. For example, small areas with high levels of soil moisture surrounded by drier areas have a disproportionately high latent heat flux. The sensible heat flux showed a complementary suppression. The response to changes in leaf area index, however, was found to be much more complex. At constant soil moisture levels, it was found that under some conditions the latent heat flux of patches of high leaf area index increased as its proportion of the surface increased as a result of the increase in net radiation and roughness. There was also an increase in sensible heat flux. This effect was also found on surfaces with low moisture levels and strong contrasts in surface vegetation. Although these results do depend on the initial boundary layer and terrestrial conditions, a consequence of this is that significant biases can be generated in modelling catchment output if the heterogeneity effects are not fully accounted for. The model simulations demonstrate that the fluxes from a 'homogenized' surface with catchment-average land surface properties (e.g. soil moisture) can be significantly higher than that for a heterogeneous surface with explicitly modelled inter-patch interactions. These results have particular implications for nested catchment models where the responses of individual subcatchments are as important as that of the total catchment. They are also significant in efforts towards developing lumped land surface parameterizations for use in atmospheric models such as general circulation models.

INTRODUCTION

Hydrologists are interested in modelling the responses of the land surface at a variety of time and space scales. Those scales are dictated not only by the characteristics of the surface itself and the atmospheric processes which occur above it, but also by the modifications of the surface resulting from anthropogenic activities.

The influence which surface conditions can have on the microclimate within the atmospheric boundary layer, has been demonstrated in a number experiments conducted with numerical atmospheric circulation models (e.g. Avissar, 1992; Collins and Avissar, 1994). Avissar applied such a model over different homogeneous surfaces (of infinite extent) and determined the differences in conditions of the resulting atmospheric boundary layer. Over bare ground, by the mid-afternoon, the boundary layer was shown to grow to be significantly higher, significantly warmer and drier than an equivalent boundary layer over wet or vegetated land. Collins and Avissar (1994) assessed the atmospheric response to 10 surface parameters. Soil moisture availability, surface roughness, albedo, leaf area index (LAI) and stomatal conductance were identified as the five characteristics of greatest significance for atmospheric response. Albedo, surface soil wetness (particularly where there is no vegetation) and LAI have previously been shown to have significant influence and feedback effects on climate at large scales (Charney *et al.*, 1977; Shukla and Mintz, 1982; Rowntree and Bolton, 1983).

The importance of such atmosphere–surface interactions for larger scale atmospheric modelling is well recognized (Wood, 1991), but they may also be significant for surface hydrological modelling at catchment scales. By modifying the boundary layer, surface conditions may affect the evaporative flux, thus having a feedback effect on soil moisture and hence catchment runoff and overall water balance.

Effects of surface heterogeneity

The question that next arises is what happens if the surface is not homogeneous? Shuttleworth (1988) suggested a length scale of the order of 10 km above which a heterogeneous surface makes a discernible impact on the atmospheric boundary layer (ABL), but below which acts as a homogeneous surface without an identifiable ABL response. Studies have demonstrated the effects of vegetation heterogeneities on meso-scale atmospheric circulation (André *et al.*, 1989), but is the possible enhancement or suppression of surface fluxes to the atmosphere at smaller scales important for hydrological models? Are there interactions between patches over heterogeneous landscapes important enough to be considered in catchment models?

Looking at the effects of heterogeneity and scale for runoff generation, Wood *et al.* (1988) found that small-scale variations need not be considered explicitly, but may be parameterized so long as they are at a scale less than a certain 'representative elementary area' (REA). Using TOPMODEL, a topographically based catchment model (Beven and Kirkby, 1979), Wood *et al.* found the REA to be ~ 1 km^2. According to Wood *et al.*, the implication of these results is that at scales below the REA probability distribution functions, rather than the actual surface patterns, are adequate to describe the characteristics and behaviour of the system. Processes can then be modelled at larger scales using lumped representations. Extending this work, Wood and Lakshmi (1993) found that the same REA seemed to apply to catchment-scale evaporation.

The interactions between the atmosphere and a heterogeneous surface at mesoscales were investigated by Avissar and Pielke (1989). Avissar and Pielke used the distribution of heterogeneous patches inside a grid cell, and treated each as an independent (and homogeneous) surface–atmospheric pathway. The grid-scale fluxes were obtained by averaging according to the distribution of the surface patch types. This was compared with 'bulk layer parameterization' (Deardorff, 1978) in which the entire grid cell was treated as homogeneous with characteristics representative of the mean of the internal patch characteristic. It was demonstrated that the micrometeorological conditions simulated by the two representations (that is, grid-scale mean and subgrid-scale weighted mean) were very different. They also demonstrated that when surface heterogeneities result in sharp contrasts in sensible heat flux strong local circulations can be generated.

Koster and Suarez (1992) examined ways for averaging surface conditions for use in general circulation models (GCMs), with surface heterogeneity described as a distribution of 'tiles' within a GCM grid square. The energy balance of each 'tile' was determined using a simplified version of the simple biosphere (SiB) model (Sellers *et al.*, 1986) and could be written in the form of the Penman–Monteith equation. The resultant grid-scale fluxes from the 'mosaic' of tiles were compared with that from a grid-scale SiB model with the same vegetation types within it. In the mosaic approach the individual tiles, with a single surface type, each interact with the atmosphere independently, whereas with the SiB model all types are included in a single multilayered model and interact with a common 'canopy air'. Koster and Suarez (1992) found that the multiple independent patch approach ('mosaic') resulted in similar grid-scale energy partitioning to the single cell multilayered SiB approach. This contrasted with the fairly strong differences found by Avissar and Pielke (1989) in their similar comparison.

Blyth *et al.* (1993) used a two-dimensional boundary layer model to examine averaging techniques in pursuit of an 'effective' surface aerodynamic resistance to characterize a heterogeneous surface. They found that the aerodynamic resistances for momentum, latent and sensible heat fluxes should be determined separately. More particularly, they found that for heterogeneous surfaces 'effective' resistances for these fluxes do not have the same relationship with each other as they do for homogeneous surfaces. The sensible heat flux was the most sensitive to errors in the averaging for the effective surface resistance.

Raupach (1991; 1993) derived length scales by which surface heterogeneity may be categorized by its impact on the ABL. He viewed the heterogeneous surface as a patchwork of different surfaces acting in parallel coupling with the atmosphere. He classified a surface as heterogeneous at the 'microscale' if the ABL could not fully adjust to a change in the surface conditions. Patches influenced only the near-surface layer, above which a well-mixed ABL takes uniform values across the entire region. Energy exchanges over patches are governed by the average ABL conditions, and inter-patch advection is at its maximum. This length scale is essentially determined by the time-scale for convective mixing within the ABL, and is typically ~1–10 km. Surface heterogeneity at length scales larger than this were classified by Raupach as 'mesoscale' heterogeneous. Neighbouring patches do not influence each other except as a boundary layer scale advection source to the downwind patch. Examining the responses of surfaces with his 'microscale' heterogeneity, Raupach used a simple 'slab' model of the daytime convective atmospheric boundary layer (CBL) (McNaughton and Spriggs, 1986), with the surface energy balance determined by the Penman–Monteith equation. Varying stomatal conductance and roughness height between patches, Raupach showed that a non-linear response occurs as patch fractions of surfaces are varied between the two extremes of all of one type to all of another. He found that patch-scale evaporation may be enhanced by up to 20% from high flux regions in an otherwise low flux surface.

Nested catchment approach to modelling

The present study is part of an ongoing programme of large-scale catchment modelling using a nested catchment approach. The primary interest is in understanding the effects of land-use changes on catchment response. The central idea behind 'nesting' catchments for modelling is to be able to describe not only the response of the catchment as a whole, but also of the subcatchments. The number and size of the subcatchments will depend primarily on the topography and on the natural variability of soil moisture related properties and vegetation systems within the catchment, and on land use. If some of the subcatchments are gauged this allows testing of the model at a range of scales, giving insight into the internal mechanisms of the catchment.

Simple conceptualisations are used to describe the key hydrological processes at the subcatchment scale, in terms of macro-scale state variables (for example, soil moisture stores). In this way we take advantage of the computational efficiencies available from conceptual lumped models, by limiting to some extent the number of parameterizations required, while preserving the physical basis and relevance to identifiable characteristics of the catchment. The estimation of parameters involves a calibration procedure which makes use of long-term rainfall and streamflow records.

Existing approaches to evaporation modelling

Hydrological models traditionally use very simplistic descriptions of evaporation, with a uniform evaporative demand assigned across the domain, regardless of internal heterogeneity. Limitations in the availability of data have resulted in the use of models based on evaporation pan measurements, or on atmospheric variables such as air temperature or daily radiation. These may be used alone or together with average wind speed and humidity, usually with a 'crop coefficient' to estimate the evaporation demand (i.e. potential evapotranspiration) over vegetated surfaces (TVA, 1972; Doorenbos and Pruitt, 1975). Actual evaporation is then estimated by multiplying the evaporation demand by some specified function of soil moisture to represent the uptake of water by vegetation.

At a more sophisticated level is the 'big leaf' approach, representing the vegetation system as a single big leaf which in some sense behaves like a real leaf, with stomatal response to some combination of atmospheric and soil moisture conditions. A simple example of this is the Penman–Monteith equation (Monteith, 1965) in which the soil moisture control on evapotranspiration is expressed in terms of a canopy-scale 'stomatal' resistance. Slightly more complex formulations based on the same principle, but involving multi-layer representations of the vegetation, have been developed to account for the presence of a mix of vegetated and bare soil surfaces (e.g. Sellers et al., 1986; Choudhury and Monteith, 1988). More detailed models with many layers attempting to simulate all the physical and physiological processes affecting evaporation are also available (Sinclair et al., 1976), but they are seldom used as the data required to run them are not available and they are computationally expensive.

Modelling spatial heterogeneity within catchments

The motivation for this work is to address the issue of spatial heterogeneity within catchments, and its effect on evaporation. In spite of their complexity, none of the models described in the previous section, neither the simple conceptual models based on potential evapotranspiration, nor the more sophisticated models based on the 'big leaf' approach, can explicitly incorporate the full complexity of the spatial heterogeneity of evaporation processes occurring within catchments.

In the past the main variability which has been incorporated in catchment models has been rainfall, with a heterogeneous distribution across a catchment. There are a variety of methods for dealing with this, such as the Thiessen polygon method, kriging and the isohyetal method, which try to account for systematic rainfall variability by explicitly including it in large-scale models. However, variations in meteorology have often not been included. This is clearly not always valid, especially if the catchment is large. Micrometeorological conditions will vary across the catchment as the characteristics referred to earlier vary (i.e. topography, albedo, surface roughness, soil moisture, vegetation, etc.). It is rare that a good meteorological record is available at one site within a catchment. It is extraordinary if there are more than one.

It is important to know if the 'patchiness' which characterizes real catchments, particularly those subject to land-use changes, has a hydrological influence in its own right. If the interpatch advection found by Raupach (1993) occurs at the subcatchment scale, this could be significant for hydrology, particularly on long-term water balance studies involving nested catchments. The most readily available data we have for a catchment, particularly in the long term, is streamflow. However, this flow makes up only 10–15% of the precipitation in regions with Mediterranean climates in general and in Western Australian catchments in particular. A 20% enhancement of evaporation from a subcatchment could be equivalent to 100% of the runoff total at that scale. It is thus an important question which needs to be resolved.

Models of hydrological processes at large scales (such as the scales of GCMs) should implicitly integrate the subgrid-scale processes which they cannot resolve. For this reason we need to understand the interactions resulting from subgrid-scale heterogeneity. How can we parameterize subcatchment-scale behaviour, and response to heterogeneity, in large-scale lumped models? We can resolve the large-scale influences, rainfall patterns, topographic effects and even meteorology, but as yet we do not know how to resolve the subcatchment-scale heterogeneity in an efficient and accurate way. We do not even know whether this is necessary.

In this paper we introduce an approach aimed at determining the significance of heterogeneity at the subcatchment scale. We wish to assess the biases introduced by assuming uniform conditions across patches and across catchments at these scales, and how useful average or 'effective' catchment-scale parameters, such as a bulk canopy resistance, are for describing and quantifying land surface evaporation at these scales. In particular, we consider variations in and distributions of vegetation and soil moisture. These questions are investigated using a subcatchment- or 'patch-' scale soil–vegetation–atmospheric transfer model which is coupled to a model of the ABL under convective conditions. We use this model to study the extent to which inter-patch interactions are important for their impact on the microclimate of subcatchments, thereby affecting the total catchment evaporative flux. The interactions between patches occur through the impact of the individual patches on the common ABL and the feedback influence the modified ABL then has on fluxes from the neighbouring patches. With this representation a catchment is considered as a set of nested subcatchments, each with its own interaction with the ABL, producing a pattern of atmospheric 'fingering' analogous to the process of water infiltrating into soil. At a patch level the vapour (or sensible heat) travels in an almost definable path, but reaches a level where the 'fingers' coalesce into a larger scale mean flux and the boundary layer conditions are dependent on all patches simultaneously. We interpret our results from a water balance modelling perspective.

A brief outline of the model, together with our calibration and validation procedures, is described in the next section. The modelling experiments undertaken are then described, and our results presented and discussed.

THE MODEL

We follow the lead of the natural hydrological systems and treat a subcatchment as the fundamental response unit, and a catchment as a set of such response units. The nominal scale of the model is $\sim 1\,km^2$, which corresponds to the REA of Wood *et al.* (1988), coincides roughly with the scale of Raupach's (1993) 'microscale' heterogeneity and is the area of the catchment pair used in our study.

The model (COUPLE) is composed of three modules

- a catchment-scale water balance model (Silberstein and Sivapalan, 1993; 1994)
- an energy balance model at a canopy surface and the soil surface (based on Choudhury and Monteith, 1988; Choudhury, 1989)
- a 'slab' CBL model (McNaughton and Spriggs, 1986; Raupach, 1993).

Figure 1 is a schematic diagram of the model, illustrating the connections between the soil moisture storages, soil water balance, surface energy balance and the ABL. The model has been developed to study the long-term impacts of vegetation changes and other management practices on water resources.

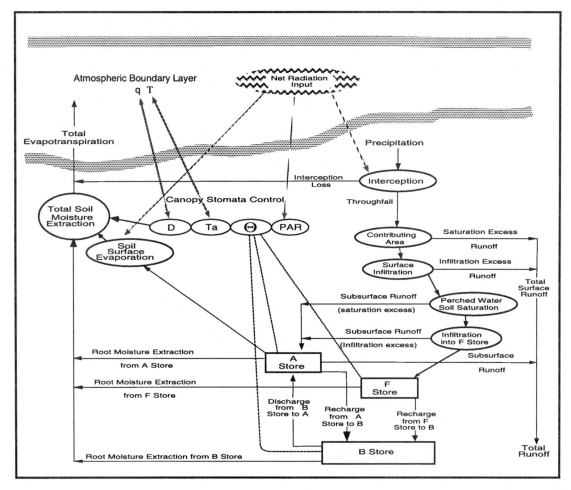

Figure 1. Schematic diagram of the model COUPLE. The coupling between the water balance and energy balances is indicated by lines connecting vapour flux pathways (bare soil and stomata) to moisture storages. D is the moisture deficit in the canopy air; T_a, the temperature of the canopy air; PAR, the photosynthetically active radiation; Θ, the soil moisture; q, the CBL moisture deficit; and T the temperature of the CBL. Broken lines indicate the soil moisture input to stomatal resistance from each of the three stores, dependent on root distribution. The feedback connection between canopy air humidity and temperature and those of the ABL is indicated with stippled arrows

Short-term (15 minute) atmospheric fluxes are calculated and the daily hydrological response and long-term water balance can also be modelled.

The water balance component of the model centres around three lumped conceptual soil moisture storages on a catchment scale, referred to as the A, B and F stores. The A store represents the shallow perched aquifer system responsible for most of the runoff; the B store is the deep permanent groundwater storage; and the F store is an intermediate unsaturated zone, which has a high degree of seasonality, especially under forest. The F store is the major moisture source for the forest during the summer drought (about six months), but is not accessible to shallow rooted annual pastures. The dynamics of the stores, through the distribution of water within the profile, controls the short-term runoff response and the long-term water balance, as well as having a major influence on the short-term energy balance. Runoff occurs as saturation excess, infiltration excess or subsurface flow, as dictated by soil moisture storage, rainfall rate and soil hydraulic characteristics. Soil moisture availability both to bare soil evaporation and transpiration is critical to the energy balance and is explicitly included in the model.

The energy balance is modelled using the well-known electrical resistance circuit analogy (Choudhury and Monteith, 1988; Choudhury, 1989), with bare soil evaporation acting in parallel with the transpiration, into a common canopy air layer. The canopy stomatal resistance is a function of atmospheric moisture deficit (D in Figure 1), ambient temperature (T_a), photosynthetically active radiation (PAR) and catchment average soil moisture saturation (Θ). There is a resistance to turbulent diffusion between the canopy air and the overlying ABL. The resistances for sensible and latent heat fluxes are assumed to be the same, and atmospheric stability adjustments are made using the scheme of Choudhury (1989). Soil thermal behaviour is modelled using the 'force–restore' approach (Deardorff, 1978). Canopy surface evaporation of intercepted rainfall (interception loss) is modelled using a modified Rutter scheme (Rutter *et al.*, 1971; 1975; Silberstein and Sivapalan, 1993; 1994).

The CBL is modelled as a well-mixed 'slab' (McNaughton and Spriggs, 1986; Raupach, 1993), with daily growth from an early morning level driven by convection. As the CBL grows, warm dry air is entrained from above a capping inversion. This air is instantaneously (within one time step) mixed into the entire CBL, along with the vapour and sensible heat transported from the surface. There is feedback between the mixed layer and the surface through the diurnal development of CBL temperature and humidity modifying the gradient-driven surface fluxes.

Sub-model linkages and inter-'patch' connections

The three submodels are linked as follows:

1. The water balance is coupled to the energy balance through: (a) soil moisture influence on stomatal control and transpiration; (b) soil surface evaporation; and (c) interception loss. Further details are given elsewhere (Silberstein and Sivapalan, 1994). As moisture is removed through vaporization (either evaporation or transpiration) the soil storages are adjusted and hence the runoff response is affected. Similarly, as runoff proceeds moisture is also removed from storages, thus influencing stomatal resistance (through root systems), and hence transpiration, as well as affecting soil surface vapour pressure and soil surface evaporation.

2. The soil and vegetation energy balance interacts with the CBL through energy partitioning. (a) As evapotranspiration proceeds, humidifying the atmosphere, so atmospheric demand will be modified, hence feeding back on energy partitioning in the next time step. In turn, the energy partitioning governs the buoyancy flux development (through the sensible heat flux) which drives the daily growth of the boundary layer. (b) Equally, as the boundary layer grows entraining warm dry air from aloft, the boundary layer conditions are also modified, resulting in changed atmospheric demand for vapour flux, and hence the CBL–surface coupling.

Model operation

The model requires vegetation height, LAI, leaf dimensions and throughfall parameters (that is, the proportion of rainfall reaching the ground for a given vegetation at a given LAI; usually estimated from field data). As configured for this study, the model requires wind speed and incoming shortwave and long-wave radiation at 15 minute intervals. The surface is initialized with soil moisture storages and soil temperature at the surface and at depth. The CBL is initialized with specific humidity, surface air temperature, boundary layer height and inversion conditions (specific humidity, potential temperature and jump change in these, and lapse rate aloft). We have used mean monthly data obtained from the Commonwealth of Australia Bureau of Meteorology for the inversion conditions, and used a local meteorology station for the surface conditions. The initial soil moisture storages and all soil hydraulic parameters were obtained in a previous calibration exercise, which is described briefly in the next section and more fully elsewhere (Silberstein and Sivapalan, 1994).

Model calibration and validation

The three sections of the model are validated separately. Only a brief discussion is given here, with more details given by Silberstein and Sivapalan (1993; 1994) and Silberstein et al. (1994a). The calibration exercise was undertaken in two parts. Firstly, the water balance submodel was tested and calibrated using an 18-year rainfall and runoff record from a catchment pair (Salmon Brook and Wights) in the south-west of Western Australia. One of the pair (Salmon Brook) retains native jarrah and marri forest (*Eucalyptus marginata* and *Eucalyptus calophylla*), and the other (Wights) has been cleared for grazing of annual (non-irrigated) pastures. Meteorological data are only available over the last five years of the record. To determine the conceptual model parameters, the water balance submodel was run independently of the energy balance with a synthetic potential evaporation applied to it (based on monthly evaporation pan measurements). The first 10 years of record were used to calibrate and the rest of the record was used as a test series. Daily stream flow and long-term water balance were used as the response variables and the parameter sets derived from this procedure were maintained for all subsequent simulations.

The model also requires calibration of the stomatal response to soil moisture storage. Five years of meteorological data are available from a site ~50 km from the catchment pair, and these were used as a forcing data set for the surface energy balance model coupled with the water balance model. The first year of these data was used to calibrate the stomatal resistance dependence on the soil moisture storages. The primary test variables were again daily flow and long-term water balance; in this instance for just the one year. Simulations were then run for the next four years and results reported elsewhere (Silberstein et al., 1994a) showed the model reproduced the recorded streamflow very well (summarized in Table I).

As part of the energy balance calculations, canopy and soil surface temperatures are calculated. An effective surface temperature can be defined (Choudhury, 1989), as

$$T_{\text{eff}} = T_v \times [1 - \exp(-\text{LAI}/2)] + T_{ss} \times \exp(-\text{LAI}/2)$$

Table I. Water balance statistics for four-year simulation with coupled energy and water balances model, without CBL model

Catchment	Totals				
	Measured rainfall (mm)	Measured runoff (mm)	Modelled runoff (mm)	RMS error (annual totals) (mm)	Modelled evaporation (mm)
Forest	5034	581	613	13·2	3279
Pasture	5034	2010	2022	26·9	2489
Coefficients of determination (r^2) for various runoff accumulation periods					
	Daily	2-Daily	Weekly	Monthly	Yearly
Forest	0·84	0·89	0·92	0·96	0·99
Pasture	0·80	0·86	0·92	0·95	0·97

where T_v is the canopy surface temperature, T_{ss} is the soil surface temperature and LAI is the leaf area index of the vegetation system. This gives an additional catchment-scale property which can be measured. We have compared T_{eff} with the catchment areal average surface temperature measured by satellite using 12 Landsat-TM images distributed over a four year period, also with very good results (Silberstein *et al.*, 1994a; 1994b). A field experiment aimed at measuring the complete energy balance over forest was undertaken late in spring 1993 and late summer 1994, thus giving a contrast in soil moisture conditions from maximum to minimum levels. Processed data from this experiment are not yet available, but will be used to validate the model when applied to forested catchments.

For the simulations just described, the ABL was not explicitly included in the model, but provided boundary conditions through the forcing meteorology. The model was then coupled with the CBL model. A lack of specific detailed boundary layer data has meant only qualitative validation of the CBL model growth has been achieved — through comparison with published field observations at other sites in the south-west of Western Australia (Rayner and Watson, 1991) and the diurnal development of temperature and humidity by comparison with the available meteorological data.

We have found our model to perform accurately over the long term, achieving the correct water balance while still maintaining short-term runoff response. This obviously requires correct evaporation determination in the long term. Additionally, we believe we model the energy balance well in the short term by virtue of the corroboration with Landsat-TM surface temperature measurements. The CBL growth compares well with previously published field observations recorded in this part of Western Australia.

INVESTIGATIONS OF THE SURFACE HETEROGENEITY PROBLEM

We have applied our model to investigate the concept of 'microscale' and 'mesoscale' heterogeneity (as defined by Raupach, 1993) with respect to a number of subcatchments. The numerical experiments were conducted in a manner that attempted to preserve as much of the full complexity of the surface–vegetation–soil moisture system as possible. We varied the surface and soil conditions appropriate to a change in vegetation or soil moisture and then observed the response of the whole system. The stomatal resistance and aerodynamic resistance will change as a result of our modifications. We do not prescribe them *a priori*. We feel this gives a better representation of the total response of the system and a closer representation of the environmental response to a change in conditions.

We have examined the response of a heterogeneous surface to atmospheric forcing and the affect a patch has on the response of its neighbours. This is done by examining the response of heterogeneous surfaces, made up of multiple patches of given types, and comparing the individual patch responses with those from homogeneous surfaces of the same type as each of the patches. The patches interact through their impact on ABL conditions. As all patches feed into a common ABL, the conditions in the ABL reflect the whole surface and in turn affect the surface through the common forcing to all patches.

We have also investigated the effects of heterogeneity in surface conditions as opposed to 'equivalent' homogeneous surfaces whose characteristics are equal to the areal averages of the heterogeneous surfaces. To some extent our work follows that by Avissar and Pielke (1989) and also that of Koster and Suarez (1992). We differ in our approach as we define our boundaries in terms of subcatchments, and in the parameterizations used. Each has its own soil moisture conditions which are a product of its rainfall, runoff and evapotranspiration history. The current state of a 'patch', or subcatchment, is due to its prior response to forcing and to changes in vegetation which affect that response. We are interested in determining the extent to which atmospheric coupling between patches (through modifications to the ABL and hence evaporative demand) can affect the catchment water balance and runoff.

Modelling experiments

Modelling experiments were undertaken to simulate the effects of heterogeneity in soil moisture, LAI and vegetation type (forest, pasture and bare ground), and varying the meteorology from a summer day to winter and spring. In each instance two 'patch' types were chosen, and their proportion of surface coverage varied from 0 to 100%. By varying the patch fraction, different heterogeneous surfaces were created. We then constructed 'equivalent homogenized' surfaces for each instance, from the two fractional patches. This homogenized surface had its LAI, soil moisture storage, or vegetation and soil parameters, as the case may be, determined as the areal average of the two patches in proportion. Thus we compared a surface consisting of two patches with such a single, 'equivalent', homogeneous surface. The surface characteristics are determined as follows.

We define Φ_A and Φ_B to be the fractions of surface area which are patch 'A' and patch 'B', respectively. For example, if we take a surface with two different forest patches, with one patch having LAI = 4 (patch 'A') and the other LAI = 0.5 ('B'), our surface would then vary between homogeneous with LAI = 4 ($\Phi_A = 1$, $\Phi_B = 0$) through incremental steps to homogeneous with LAI = 0.5 ($\Phi_A = 0$, $\Phi_B = 1$). For this example the 'equivalent homogenized' surface has an LAI

$$LAI_{hom} = \Phi_A \times LAI_A + \Phi_B \times LAI_B$$

where at all times

$$\Phi_A + \Phi_B = 1$$

and where LAI_{hom}, LAI_A and LAI_B are the leaf area indices of the homogenized patch and patches 'A' and 'B', respectively. If $\Phi_A = 0.2$ and $\Phi_B = 0.8$, then

$$LAI_{hom} = 0.2 \times 4 + 0.8 \times 0.5 = 0.84$$

Soil moisture heterogeneities were investigated by having two patch surfaces with a common vegetation system and the same density of cover (LAI), but with different soil moisture storage levels. The moisture storages under pastured surfaces differ from those under forest by virtue of the shallower and seasonal root regime, seasonal shading and summer senescence. Long-term simulation runs were used to select the moisture variability appropriate to the vegetation systems by choosing winter conditions for the wet patch and summer conditions for the dry. Extremes of moisture variability were then given by allowing

Table II. Summary of model parameter values used in this study

Moisture stores (mm)	Patch 'A'	Patch 'B'
Forest		
A	120	90
B	1700	1400
F	700	130
Pasture		
A	150	90
B	2910	2650
F	540	370
LAI		
Forest	4	0.5
Pasture	3	0.1

CBL initialization [values are for summer (winter)]:
Time of dawn: 5:15(7:15)
Specific humidity at dawn, $q(0)$: (g/kg) 5.9(3.7)
Mixed-layer temperature at dawn, $T(0)$: (K) 279(272)
Surface potential temperature from inversion
    at dawn, (Θ_0): (K) 283(276)
Height at dawn, $h(0)$: (m) 100(100)
Potential temperature lapse rate above
    inversion (γ_Θ)): (K/m) 0.0042(0.0048)
Specific humidity gradient above $-8 \times 10^{-7}(-8 \times 10^{-7})$
    inversion (γ_q): (g/g/m)

a 10% greater seasonal variation than seen during a typical annual cycle. This was done to maintain the surface and soil conditions within the range of validity of the model — that is, not too far beyond the range that the model has been calibrated for. We present the soil moisture conditions for the equivalent 'homogenized' surface as the spatial mean of each of the stores in the same way as for the LAI, but all stores are variety synchronously. Thus, for our equivalent example (but this time keeping LAI constant between patches)

$$F_{hom} = 0.2 \times F_A + 0.8 \times F_B$$

where F_{hom}, F_A and F_B are the F store values for the 'homogenized' surface, patch 'A' and patch 'B', respectively; and similarly for the A and B stores.

In most hydrological models mean values for catchments and subcatchments are prescribed because of the difficulty of collecting or deducing the spatially distributed properties. This lumping of characteristics is performed despite known non-linearities in the system and explicit non-linearities in the models because there appears to be no alternative. It is the result of this lumping of parameters that we wish to test. In making these experiments we are assuming that our model is equally valid at the patch scale as it is at a scale of up to 10 patches.

We used measured wind speed and solar radiation from dawn on the day of simulation. The ABL was initialized with the specific humidity and temperature of the actual atmosphere on the day of simulation and, following Raupach (1993), the dawn boundary layer height was initialized at 100 m. The conditions at the inversion were taken from mean monthly data supplied by the Bureau of Meteorology for Perth and extrapolated to our site. Details are given in Table II. All model runs described in this paper were from dawn to dusk for a single day.

RESULTS

Table II summarizes the model initialization details for the simulation runs presented here. Attempts were made to give as wide a representation of as many scenarios as possible, including some totally artificial ones, such as pasture with LAI = 2.4 in mid-summer. This particular example would only occur in south-west Western Australia under irrigated conditions, and then the soil moisture would perhaps be akin to our 'high moisture' patch, which has been taken from winter. Only a sample of results are shown here, as the general theme of the results is common to many of the simulations. In all figures patch 'A' is the patch with the higher latent heat flux — that is, having the higher LAI or the higher soil moisture availability.

Heterogeneity in soil moisture

For a surface with soil moisture heterogeneity, the diurnal growth of the CBL, CBL moisture deficit (D), surface sensible heat (H) and latent (LE) fluxes, are shown in Figure 2a–d. In this instance the vegetation was forest with a high LAI (LAI = 4), which corresponds to dense growth of either mine rehabilitation or a plantation of paper pulp species in this area. In each of the figures five fractional surface covers are shown from $\Phi_A = 0$ (all low moisture) to $\Phi_B = 1$ (all high moisture) and intermediate fractions $\Phi_A = 0.2$, 0.5 and 0.8 (and $\Phi_B = 0.8$, 0.5, 0.2).

Figures 3 and 4 summarize the results for heterogeneity in soil moisture for high LAI and low LAI, respectively. In each figure the sensible heat flux and latent heat flux for forested and grassed surfaces are shown. The four lines in each plot give the daily total flux for the heterogeneous and the equivalent 'homogenized' surfaces, and the flux density from the higher moisture content and lower moisture content patches ('A' and 'B', respectively).

The more moist patches have an enhanced latent heat flux, when part of a heterogeneous surface, over that which would obtain over a surface which was made up entirely of that type. This is indicated by the difference between the patch 'A' flux density at a given Φ, and that for the homogeneous surface made up entirely of that type ($\Phi_A = 1$). For example, in Figure 3a the latent heat flux from an infinitesimal patch 'A' ($\Phi_A \to 0$) (high moisture) surrounded by a surface of patch 'B' (low moisture) tends towards $\sim 19 \, \text{MJ} \, \text{m}^{-2}$

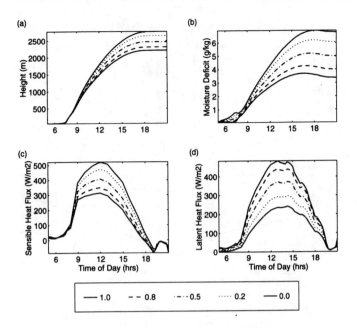

Figure 2. Diurnal variation of (a) height of the CBL, (b) moisture deficit of the CBL, (c) sensible heat flux and (d) latent flux from surfaces covered with forest having LAI = 4 and variations in soil moisture. The numbers in the legend refer to the proportion of surface covered by the high moisture patch

for the day. This is an enhancement of about 25% over the large-scale mean for this surface, which can be seen from the plot (at $\Phi_A = 1$) to be $\sim 15\,\mathrm{MJ\,m^{-2}}$. This is similar in magnitude to Raupach's (1993) observations of this effect. The complementary suppression effect on the low flux patch can be seen in this example to be significantly greater as a fraction of total patch flux, as the flux from an infinitesimal drier patch surrounded by moist areas tends towards $\sim 4\,\mathrm{MJ\,m^{-2}}$. An equivalent result can be seen in the case of

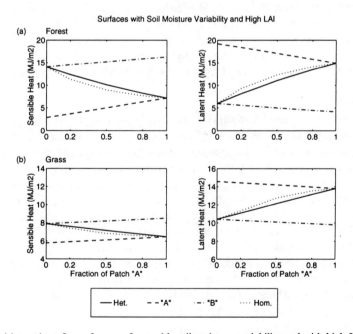

Figure 3. Total sensible and latent heat fluxes from surfaces with soil moisture variability and with high LAI (forest LAI = 4, grass LAI = 3). Patch 'A' has the high moisture content. The legend refers to the response of the heterogeneous (solid line) and 'equivalent' homogenized surfaces (dotted line) and of the patches 'A' (dashed line) and 'B' (dot-dashed line) individually

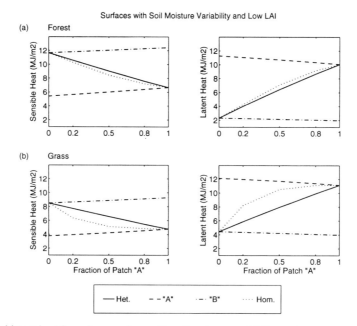

Figure 4. Total sensible and latent heat fluxes from surfaces with soil moisture variability and with low LAI (forest LAI = 0·5, grass LAI = 0·1). Patch 'A' has the high moisture content. The legend is the same as for Figure 3

sensible heat flux, but there is a depression in H from the wetter patch. This occurs because of the enhanced latent heat flux from this patch, which means that there is an effective latent heat flux from the moist areas to the dry areas and an effective sensible heat flux in the opposite direction. Thus the total net radiation for the surface remains about the same.

Figures 3 and 4 also show that for a surface with variability in soil moisture, the latent heat flux from the equivalent homogenized surface can be 10–15% more than that from the heterogeneous surface. There is a slight upward curvature to the lines for total flux from both surfaces, indicating that the surface total flux is higher than would be obtained from a straight average of the individual patch fluxes. There is a complementary reduction in sensible heat flux, with a slight increase in net radiation as the wet fraction is increased. This is enabled by lower surface temperatures, resulting in lower outgoing longwave radiation.

Heterogeneity in leaf area index

Figures 5 and 6 summarize the results of varying the LAI while maintaining the soil moisture availability constant across the surface for a high moisture and low moisture surface, respectively. The response to LAI variability is different to that shown by the moisture variations. In Figure 5 it can be seen that under conditions of readily available moisture both patches preferentially increase their latent heat flux slightly as the proportion of high LAI patch is increased. This apparently counter-intuitive result occurs because the reduction in albedo which accompanies the increase in LAI results in a significant increase in net radiation. There is also an increase in surface roughness, as we parameterize roughness to be dependent on LAI, and these two lead to greater sensible heat flux, driving increased CBL growth, which entrains more dry air from aloft. A relatively high moisture deficit is therefore maintained in the CBL. The effect is more pronounced for the forested surface (Figure 5a). The effect is similar, although much reduced, for the forested surface when soil moisture availability is low (Figure 6a). Note that in both Figures 5b and 6b the homogenized grass surface has a reduced latent heat flux at low LAI levels, becoming greater than that for the heterogeneous surface as LAI increases.

Heterogeneity in vegetation type

In these simulations, a grassed and forested patch were combined under low and high LAI and moisture

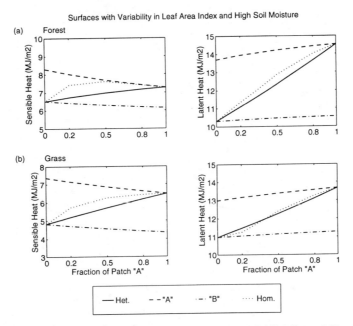

Figure 5. Total sensible and latent heat fluxes from surfaces with leaf area index variability (forest LAI = 0·5–4, grass LAI = 0·1–3) and with a high moisture availability. Patch 'A' has the higher LAI of the two (see Table II). The legend is the same as for Figure 3

conditions to produce a series of mixed vegetation surfaces. Figure 7 summarizes the surface flux results, and Figure 8 the maximum daily surface temperatures of a surface with a forested catchment under approximately natural conditions adjacent to a cleared catchment in summer and spring. In summer the grass has LAI = 0·1 as it has dried off by December, whereas the forest retains its leaves throughout the year, the LAI has been set at 2·4 for both simulations. In summer the forested catchment is supplying

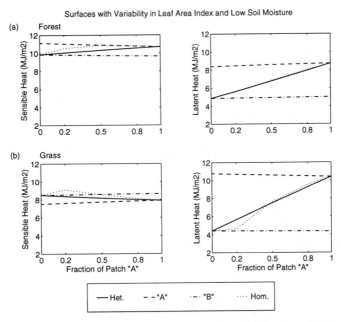

Figure 6. Total sensible and latent heat fluxes from surfaces with leaf area index variability (forest LAI = 0·5–4, grass LAI = 0·1–3) and with a low moisture availability. Patch 'A' has the higher LAI of the two (see Table II). The legend is the same as for Figure 3

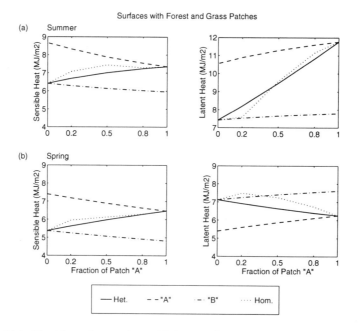

Figure 7. Total sensible and latent heat fluxes from surfaces with forest and grass patches for summer (minimum soil moisture, grass LAI = 0·1, forest LAI = 2·4) and spring (maximum soil moisture, grass LAI = 3, forest LAI = 2·4). Patch 'A' is forested. The legend is the same as for Figure 3

the bulk of the latent heat flux and the sensible heat flux due to its much higher net radiation (lower albedo) and greater roughness. In this example, as with the LAI variation, there is an enhancement of the latent heat flux from both patches as the forested proportion is increased.

In spring (Figure 7b) the LAI for the grass has been assumed to have increased to 3, and it has the dominant latent heat flux. The latent heat flux from the equivalent homogenized surface shows particularly

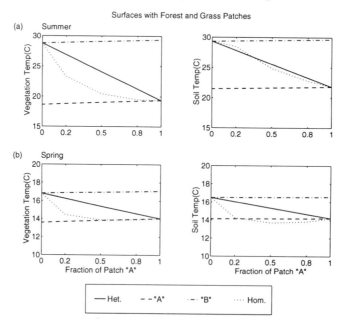

Figure 8. Maximum daily surface 'effective' temperatures and soil temperatures for surfaces with forest and grass patches for summer (minimum soil moisture, grass LAI = 0·1, forest LAI = 2·4) and spring (maximum soil moisture, grass LAI = 3, forest LAI = 2·4). Patch 'A' is forested. The legend is the same as for Figure 3

interesting behaviour at low fractions of forest cover. Under low moisture conditions (summer, Figure 7a) a little forest acts to reduce the latent heat flux and increase soil temperature, whereas with high soil moisture the latent heat is increased and soil temperatures are reduced. In both instances of summer and spring conditions, the mean catchment vegetation temperatures were reduced by adding a proportion of forest to the homogeneous catchment surface. These results are applicable for the particular conditions of these simulations, but must be taken as indicative of the complex responses which may be expected from catchments of this type.

DISCUSSION

Heterogeneous surfaces

Our results for soil moisture variability (Figures 3 and 4) show the same patch-scale enhancement found by Raupach (1993), who used a simple Penman–Monteith representation of the surface energy balance. We have extended that work to the case of a complete subcatchment model. Our modelling experiments have indicated that patch-scale enhancements in evaporative flux can be from 10 to 25% for high moisture patches surrounded by large areas of drier surface. There is a complementary suppression of the latent heat flux from drier patches.

When surfaces have heterogeneous LAI distributions the response is more complex. The result here contrasts with the moisture variability case, when flux from the higher flux patch decreased as its proportion increased. Figures 5 and 6 both showed that the latent heat flux from areas of high LAI increased slightly as their proportion increased. The result is possible as the reduced albedo, and hence increased net radiation, and increased roughness result in a significant increase in sensible heat flux. This drives an enhanced CBL growth entraining dry air from aloft, thus enhancing the latent heat flux.

When patches of very different characteristics are combined to make up a surface, the interactions can result in complex behaviour as in Figures 7 and 8. Under low moisture (summer) with nearly bare ground alongside forest it was found that the latent heat flux from the (high flux) forested patch was actually enhanced as its areal proportion was increased. This appeared to be counter-intuitive, as the higher total latent heat flux humidifies the boundary layer, normally leading to suppression of the patch-scale flux. However, in this instance as the proportion of forest increases, the decrease in total surface albedo leads to a total increase in net radiation, enhancing the sensible heat flux and boundary layer growth. This leads to an increase in atmospheric moisture deficit and a consequent increase in moisture flux.

At subcatchment scales these effects would have a considerable impact on the quantity of runoff, as the runoff from forested catchments in south-west Western Australia is typically less than 15% of rainfall. Thus modifications to the water balance may be of the same order as the entire annual runoff. We stress we have not tested this for a complete year of meteorology and these effects are not likely to persist for the whole year. However, the possible impact may still be substantial. The effects on sensible heat flux have implications for boundary layer development, as it is primarily the sensible heat flux which drives CBL growth.

There is also a potential impact on water quality. Salinization is a significant problem for streams in this area, due to rising groundwater in catchments which have been cleared for agriculture. Modifications to evaporative flux also affect the groundwater level. There is currently an active programme of replanting trees on farmland, particularly in water resource catchments, to restore the hydrological balance. The balance has been upset by the removal of native vegetation and its replacement with annual pastures, which have essentially no latent heat flux for at least four months of the year. Our work suggests that neighbouring catchments to pasture which are heavily forested will dominate the moisture flux, but there may be an increased moisture loss from both the forested catchment and cleared areas. Thus planting trees in a subcatchment can have an effect on neighbouring subcatchments through modifications of the CBL. The implication of this is that such tree planting efforts should be co-ordinated on a subcatchment basis.

The temperature responses shown in Figure 8 also show complex behaviour. The vegetation temperatures on the homogenized surfaces were considerably reduced relative to the heterogeneous surface in all instances. The situation for soil temperature was more complex. Increasing the forest fraction initially resulted in a slight increase in soil temperature in summer (Figure 8a), but as the forest fraction increased, the temperature reduced slightly, relative to the heterogeneous surface. A similar response was found by Avissar and Pielke (1989). In spring there was a significant decrease in temperature for all homogeneous surfaces relative to the heterogeneous surfaces.

Subcatchment averaging

As can be seen in Figures 3–8, the homogenized catchment does not respond in the same way as the heterogeneous surface it is meant to represent. In nearly every instance there is a significant enhancement of latent heat flux from the homogenized surface. The sensible heat flux is reduced as the moisture content increases (while the LAI is constant), but for a given moisture content, increasing the LAI can result in increasing the sensible heat flux and latent heat flux as the net radiation and effective roughness increase. It is likely that there will be some modification of this result if a better parameterization is used for roughness at very low LAI (Stannard, 1993; Shuttleworth and Wallace, 1985). The soil and vegetation temperatures are also generally reduced, although particular combinations of conditions can led to soil temperatures of homogenized surfaces being higher than the heterogeneous surfaces (Figure 8). These results are similar to those of Avissar and Pielke (1989) and the magnitude of effect of averaging is greater than that found by Koster and Suarez (1992). The implication of this is that, particularly for surfaces with relatively low latent heat flux, a more even distribution of moisture source results in a higher estimation of evaporation. Thus the practice of averaging a flux over an entire catchment can lead to significant bias in the evaporation estimate and that same bias will influence any water balance and runoff modelling which is carried out, particularly in long-term studies. At the same time however, this result suggests that a distributed planting will have a more significant impact on groundwater levels. Both these are important for water quality issues as discussed earlier.

CONCLUSIONS

We have conducted modelling experiments with a hydrological model which represents soil moisture dynamics and surface–atmosphere interactions at the scale of a small catchment ($\sim$1 km^2). This scale corresponds to the REA of Wood *et al.* (1988) and to Raupach's (1993) definition of 'microscale' heterogeneity. We have investigated the effects of heterogeneity in surface moisture availability and vegetation cover on surface fluxes to the atmosphere. Our results illustrate some of the complexities inherent in determining fluxes over heterogeneous surfaces.

Our results show similar patch-scale non-linearities to those found by Raupach (1993). If LAI, and hence roughness, remain constant, small areas of high flux density increase their local flux rate above their 'mesoscale' rate — that is, above the rate that they would have if the entire surface was covered by that patch type. High latent heat flux surfaces produce enhanced flux rates, when surrounded by low latent heat flux patches, and high sensible heat flux surfaces do the same. There is a complementary reduction in the fluxes from the low flux patches.

The situation is more complex for surfaces with varying LAI. The reduction in albedo and increase in roughness accompanying an increase in LAI can result in the patch-scale latent heat flux actually increasing as the high LAI fraction of a surface increases. This is essentially contrary to the result for variable moisture with constant LAI. Thus surfaces which include areas of forest surrounded by bare ground or pasture may have variable net responses in moisture flux as atmospheric conditions change.

Modelling experiments also found that averaging a surface characteristic may introduce significant bias as opposed to modelling the heterogeneous surface more explicitly. In nearly all instances averaging at the scale of our subcatchments produced a higher latent heat flux, although in some of the examples presented the effect was to reduce the surface total latent heat flux. The magnitude of the effect is similar to that found by Avissar and Pielke (1989) and is rather more than suggested by Koster and Suarez (1992). This would result in a consequent reduction in the estimation of groundwater storage and runoff response if this was used in a total water balance study.

Our results suggest that a knowledge of the distribution of patches within a 'grid element', at least on the 'micro-' to 'meso-' scale and a good description of those patches will produce a better representation of the surface, and better determination of the fluxes, than an attempt to 'homogenize' the surface into a lumped parameter set. Neglect of this effect could produce significant errors in long-term water balance studies, and may also have implications for water quality issues.

Work is in progress to assess the extent to which this result can be applied over larger catchments and over annual time periods. The model is to be applied to a catchment with multiple patch types using long time series (several years) of radiation and wind data to provide the forcing, with a simple scheme to simulate nocturnal ABL behaviour. Extension of the work also includes investigation of the effects of different vegetation types, and how they are distributed across a catchment, and how different ABL conditions give rise to the different surface responses reported here. Ultimately, accurate and efficient averaging schemes are sought which may be applied over large catchments. These schemes, possibly directed towards derivation of a simple catchment-scale 'stomatal resistance', will have to take account of the distribution of conditions within catchments so as not to produce the biases which linear averaging may have.

ACKNOWLEDGEMENTS

The work described herein was greatly enhanced by the assistance of Dr Michael Raupach, who allowed us to use his computer code of the CBL slab model, which we adapted to our purposes. Thanks are expressed to the officers of the Water Authority of Western Australia, Surface Water Branch, for data from the Salmon and Wights paired catchment study and the nearby climate station used in calibrating and validating the water balance and coupled models. This research was supported in part by a special Environmental Fluid Dynamics grant of the University of Western Australia, and a grant from the Australian Research Council Small Grants Scheme (Ref. 04/15/031/254), both awarded to the second author. R. P. Silberstein was supported by a University of Western Australia Research Scholarship, and a Centre for Water Research Scholarship.

REFERENCES

André, J.-C., Bougeault, P., Mahfouf, J.-F., Mascart, P., Noilhan, J., and Pinty, J.-P. 1989. 'Impact of forests on mesoscale meteorology', *Phil Trans. Roy. Soc. London*, **B324**, 407–422.

Avissar, R. 1992. 'Conceptual aspects of a statistical–dynamical approach to represent landscape subgrid-scale heterogeneities in atmospheric models', *J. Geophys. Res.*, **97**, 2729–2742.

Avisar, R. and Pielke, R. A. 1989. 'A parameterisation of heterogeneous land surfaces for atmospheric numerical models and its impact on regional meteorology', *Monthly Weather Rev.*, **117**, 2113–2136.

Beven, K. and Kirkby, M. J. 1979. 'A physically-based variable contributing area model of basin hydrology', *Hydrol. Sci. J.*, **29**, 425–434.

Blyth, E. M., Dolman, A. J., and Wood, N. 1993. 'Effective resistance to sensible- and latent-heat flux in heterogeneous terrain', *Q. J. Roy. Meteorol. Soc.*, **119**, 423–442.

Charney, J., Quirk, W., Chow, S., and Kornfield, J. 1977. 'A comparative study on the effects of albedo change on drought in semi-arid regions', *J. Atmos. Sci.*, **34**, 1366–1385.

Choudhury, B. J. 1989. 'Estimating evaporation and carbon assimilation using infrared temperature data: vistas in modelling' in Asrar, G. (Ed.), *Theory and Applications of Optical Remote Sensing*, Wiley, New York. pp. 628–690.

Choudhury, B. J. and Monteith, J. L. 1988. 'A four layer model for the heat budget of homogeneous land surfaces', *Q. J. Roy. Meteorol. Soc.*, **114**, 373–398.

Collins, D. and Avissar, R. 1994. 'An evaluation with the Fourier Amplitude Sensitivity Test (FAST) of which land-surface parameters are of greatest importance for atmospheric modelling', *J. Climate*, **7**, 681–703.

Deardorff, J. W. 1978. 'Efficient predictions of ground surface temperature and moisture with inclusion of a layer of vegetation', *J. Geophys. Res.*, **83**, 1889–1903.

Doorenbos, J. and Pruitt, W. O., 1995. 'Crop water requirements', *Irrigation and Drainage Paper 24*, FAO, Rome.

Koster, R. D. and Suarez, M. J. 1992. 'A comparative analysis of two land-surface heterogeneity representations', *J. Climate*, **5**, 1379–1390.

McNaughton, K. G. and Spriggs, T. W. 1986. 'A mixed-layer model for regional evaporation', *Boundary Layer Meteorol.*, **34**, 243–262.

Monteith, J. L., 1965. 'Evaporation and environment', in Fogg, G. E. (Ed.), *The State and Movement of Water in Living Organisms*, Symp. Soc. Exper. Biol. Academic Press, New York, U.S.A., **19**, pp. 205–234.

Raupach, M. R. 1991. 'Vegetation–atmosphere interaction in homogeneous and heterogeneous terrain: some implications of mixed-layer dynamics', *Vegetatio*, **91**, 105–120.

Raupach, M. R. 1993. 'The averaging of surface flux densities in homogeneous landscapes' in Bolle, H.-J., Feddes, R. A., and Kalma, J. D. (Eds), *Exchange Processes at the Land Surface for a Range of Space and Time Scales* (*Proceedings of the Yokohama Symposium, July, 1993*). *IAHS Publ.*, **212**, 343–355.

Rayner, K. N. and Watson, I. D. 1991. 'Operational prediction of daytime mixed layer heights for dispersion modelling', *Atmos. Environ.*, **25A**, 1427–1436.

Rowntree, P. R. and Bolton, J. R. 1983. 'Simulation of the atmospheric response to soil moisture anomalies over Europe', *Q. J. Roy. Meteorol. Soc.*, **109**, 501–526.

Rutter, A. J., Kershaw, K. A., Robins, P. C., and Morton, A. J. 1971. 'A predictive model of rainfall interception in forests. I: Derivation of the model from observations in a plantation of Corsican Pine', *Agric. Meteorol.*, **9**, 367–384.

Rutter, A. J., Morton, A. J., and Robins, P. C. 1975. 'A predictive model of rainfall interception in forests. II: Generalisation of the model and comparison with observations in some coniferous and hardwood stands', *J. Appl. Ecol.*, **12**, 367–380.

Sellers, P. J., Mintz, Y., Sud, Y. C., and Dalcher A. 1986. 'A simple biosphere model (SiB) for use within general circulation models', *J. Atmos. Sci.*, **43**, 505–531.

Shukla, J. and Mintz, Y. 1982. 'Influence of land-surface evapotranspiration on the earth's climate', *Science*, **215**, 1498–1501.

Shuttleworth, W. J. 1988. 'Macrohydrology — the new challenge for process hydrology', *J. Hydrol.*, **100**, 31–56.

Shuttleworth, W. J. and Wallace, J. S. 1985. 'Evaporation from sparse crops — an energy combination theory', *Q. J. Roy. Meteorol. Soc.*, **111**, 839–855.

Silberstein, R. P. and Sivapalan, M. 1993. 'Modelling terrestrial water and energy balances at local ($1 \, km^2$) scales' in Bolle, H.-J., Feddes, R. A., and Kalma, J. D. (Eds), *Exchange Processes at the Land Surface for a Range of Space and Time Scales* (*Proceedings of the Yokohama Symposium, July, 1993*). *IAHS Publ.*, **212**, pp. 105–114.

Silberstein, R. P. and Sivapalan, M. 1994. 'A coupled model of terrestrial water and energy balances under different vegetation systems', *Rep. ED 712 RS*, Centre for Water Research, Univ. of Western Australia.

Silberstein, R. P., Sivapalan, M., and Wyllie, A. 1994a. 'Coupling terrestrial water and energy balances at small catchment ($\sim 1 \, km^2$) scales', *Rep. ED 927 RS*, Centre for Water Research, Univ. of Western Australia.

Silberstein, R. P., Wyllie, A., Sivapalan, M., and Smith, R. C. G. 1994b. 'Application of remote sensing to modelling terrestrial water and energy balances', in *Proceedings of the 7th Australasian Remote Sensing Conference, Melbourne, 1–4 March 1994*. Remote Sensing and Photogrammetry Association of Australasia Ltd, Perth, Western Australia, 2, pp. 753–760.

Sinclair, T. R., Murphey, C. E., and Knoerr, K. R. 1976. 'Development and evaluation of simplified models simulating canopy photosynthesis and transpiration', *J. Appl. Ecol.*, **13**, 813.

Stannard, D. I. 1993. 'Comparison of Penman–Monteith, Shuttleworth–Wallace, and modified Priestley–Taylor evapotranspiration models for wildland vegetation in semiarid rangeland', *Wat. Resour. Res.*, **29**, 1379–1392.

Wood, E. F. (Ed.) 1991. *Land Surface–Atmosphere Interactions for Climate Modelling. Observations, Models and Analysis.* Kluwer Academic, Dordrecht. Reprinted from *Surv. Geophys.*, **12**(1–3).

Wood, E. F. and Lakshmi, V. 1993. 'Scaling water and energy fluxes in climate systems: three land–atmosphere modeling experiments', *J. Climate*, **6**, 839–857.

Wood, E. F., Sivapalan, M., Beven, K., and Band, L. 1988. 'Effects of spatial variability and scale with implications to hydrologic modelling', *J. Hydrol.*, **102**, 29–47.

22

ESTIMATION OF LAND SURFACE PARAMETERS USING SATELLITE DATA

HUANG XINMEI AND T. J. LYONS

Environmental Science, Murdoch University, Murdoch, WA 6150, Australia

R. C. G. SMITH

Leeuwin Centre for Earth Sensing Technologies, Floreat, WA 6019, Australia

AND

J. M. HACKER

Flinders University of South Australia, Adelaide, SA 5000, Australia

ABSTRACT

Surface observations and NOAA advanced very high resolution radiometer (AVHRR) satellite data are combined to provide area-averaged values of albedo, canopy resistance, leaf area index and fractional vegetation cover. Albedo, fractional vegetation cover and leaf area index are derived from the reflectance of the visible and infra-red NOAA AVHRR channels. Canopy resistance is estimated by closing the surface energy balance equation using the surface infra-red temperature and the normalized difference vegetation index. These land surface parameters are evaluated against independent measurements and used as input into a numerical model to simulate the energy exchange between the surface and the overlying atmosphere. Simulation results are validated against detailed aircraft observations undertaken in south-western Australia over both natural and agricultural vegetation.

INTRODUCTION

The exchange of energy between the atmosphere and the underlying surface strongly influences both climate and weather systems. The inadequate representation of these interactions is a major failure of current climate models. Andre *et al.* (1990) suggest that the atmosphere develops a coherent response for surfaces with characteristic length scales greater than 10 km. It is thus necessary for climate modelling to characterize the land surface at the appropriate scale, yet conventional techniques with point measurements are only representative of a local area. Satellite remote sensing methods, associated with surface observations, provide the possibility of estimating representative land surface parameters over reasonably large areas (Sellers *et al.*, 1990).

Land surface parameters obtained from remotely sensed data include surface albedo (Gutman *et al.*, 1989; Cess *et al.*, 1991), surface resistance (Nemani and Running, 1989) and vegetation amount (Price, 1992). Clevers (1989) used corrected near infra-red reflectance to estimate the leaf area index (LAI), whereas Carlson *et al.* (1990) combined a land surface–boundary layer model, surface temperature and reflectance measurements to estimate soil moisture availability and fraction of vegetation cover. The difficulty in applying such methods to satellite data is that the relatively low resolution may be unable to resolve a sufficiently representative distribution over inhomogeneous terrain.

Table I. Input land surface parameters required in the
model

	Parameter
z_0	Surface roughness (m)
d	Displacement (m)
A	Albedo
F_c	Fraction of vegetation cover
LAI	Leaf area index ($m^2\,m^{-2}$)
R_{min}	Minimum stomata resistance ($s\,m^{-1}$)

The primary objective of this work is to obtain average land surface parameters such as canopy resistance, albedo, LAI and fraction of vegetation cover from NOAA Advanced Very High Resolution Radiometer (AVHRR) data and surface-based observations. These land surface parameters will be used as input into a modified soil–vegetation–atmosphere model to study the daily surface energy budget. Model results are validated against aircraft and surface observations taken as part of the buFex experiment (Lyons *et al.*, 1993).

MODEL

A modified one dimensional soil–vegetation–boundary layer model based on Ek and Mahrt (1989) treats the soil heat and latent heat fluxes separately for vegetation and bare soil and averages the heat fluxes by the fraction of vegetation cover (Huang and Lyons, 1995). The land surface input parameters for this model are listed in Table I. Among these, canopy resistance, albedo and fraction of vegetation cover are difficult to obtain by routine observations over relatively large areas and their determination from satellite data is our primary focus.

Determination of the surface albedo from satellite observations requires a correction from narrow band reflectances to broad band reflectances as well as correcting for atmospheric effects (Briegleb *et al.*, 1986; Cess *et al.*, 1991). Considering the different reflectances of vegetation in the visible and near infra-red bands, the planetary broad band albedo, A_p, is calculated using a representative spectral reflectance of the red and near infra-red bands (Brest and Goward, 1987)

$$A_p = 0.526(B_4) + 0.362(B_7) + 0.112[0.5(B_7)] \qquad B_7/B_4 > 2.0 \ (\text{Veg})$$
$$A_p = 0.526(B_4) + 0.474(B_7) \qquad B_7/B_4 \leqslant 2.0 \ (\text{Non-Veg}) \tag{1}$$

where B_4 and B_7 are the reflectances of red and near infra-red bands ($0.5–0.6$, $0.8–1.1\,\mu m$) of the Landsat-MSS, respectively. As NOAA AVHRR channels 1, 2 cover similar parts of the spectra to B_4, B_7, the reflectances of channels 1, 2 (R_1, R_2) were substituted for B_4, B_7 in Equation (1).

Chen and Ohring (1984) found that surface albedo alone was able to explain 98% of the variance of planetary albedo and thus a one-predictor regression scheme could be used for the atmospheric correction. The relationship of planetary albedo, A_p, to surface albedo, A_g, is written as

$$A_p = a + bA_g \tag{2}$$

where a is the albedo of the atmosphere, describing the planetary albedo of a clear sky above a non-reflecting surface, and b is a coefficient to account for the fraction of solar radiation reflected back to space. The values of a and b, as a function of solar zenith angle, given by Chen and Ohring (1984) were adopted.

Thus, using the reflectances observed by the NOAA AVHRR satellite, the surface albedo is derived as

$$A_g = \frac{A_p - a}{b} \tag{3}$$

The normalized difference vegetation index (NDVI) is a measure of the fractional green vegetation cover

Table II. Ground observations used in estimating fractional vegetation cover and LAI

		NDV_{vg}	$NDVI_{so}$	$R_{s,ir}$
31 August	Agriculture	0·72	0·12	0·46
24 April	Agriculture	0·63	0·12	0·46
24 April	Native vegetation	0·63	0·13	0·29

and its surface value can be calculated from

$$NDVI_g = \frac{R_{2g} - R_{1g}}{R_{2g} + R_{1g}} \tag{4}$$

where R_{1g}, R_{2g} are the reflectances of channels 1, 2, respectively, incorporating an atmospheric correction following Partridge and Mitchell (1990) and assuming aerosol optical depths of 0·1 for channel 1 and 0·07 for channel 2. Fractional vegetation cover (F_c), is estimated by assuming that

$$F_c = \frac{NDVI_g - NDVI_{so}}{NDVI_{vg} - NDVI_{so}} \tag{5}$$

where $NDVI_g$, $NDVI_{so}$ and $NDVI_{vg}$ are the surface NDVI values incorporating atmospheric correction, the surface NDVI for soil and the surface NDVI for dense canopy, respectively. Estimates of $NDVI_{so}$, $NDVI_{vg}$ require on-site measurements (Table II).

Clevers (1989) suggested using a corrected near infra-red reflectance R_{ir} to estimate the LAI

$$LAI = -\frac{1}{\alpha} \ln\left(1 - \frac{R_{ir}}{R_{1,ir}}\right) \tag{6}$$

where α is a complex combination of extinction and scattering coefficients and $R_{1,ir}$ is the asymptotically limited value for the corrected near infra-red reflectance, R_{ir}, which can be expressed as

$$R_{ir} = R_{tir} - R_{s,ir}(1 - F_c) \tag{7}$$

where R_{tir} is the total measured near infra-red reflectance and $R_{s,ir}$ the near infra-red reflectance of soil, and values of $\alpha = 0.255$, $R_{1,ir} = 0.72$ were adopted from Clevers (1989), who regressed LAI with corrected near infra-red reflectance based on his experimental data. We have assumed that

$$R_{tir} = R_{2g} \tag{8}$$

whereas estimates of $R_{s,ir}$ require on-site observations (Table II).

Minimum stomatal resistance R_{min} is evaluated by closing the surface energy balance at the time of the afternoon satellite overpass. In particular, following Huang et al. (1993), the latent heat flux is estimated as the residual of the surface energy balance and potential evaporation (E_p) is calculated by setting the canopy and soil resistances to zero. Thus we solve

$$\begin{aligned} R_n &= G + \rho C_p C_h(\theta_p - \theta_a) + \rho C_h L[q_s(T_p) - q_a] \\ &= G + \rho C_p C_h(\theta_p - \theta_1) + LE_p \end{aligned} \tag{9}$$

where R_n is the net radiation, G the ground heat flux, ρ the air density, C_p the specific heat for air, L the latent heat associated with the phase change of water, q_s the saturated specific humidity, θ_p the apparent surface potential temperature if potential evaporation occurs, θ_a and q_a the air potential temperature and specific humidity, respectively, and C_h the heat exchange coefficient (Louis, 1979; Louis et al., 1982). Following Huang et al. (1993), R_n is calculated from Savijarvi (1990) and Satterlund (1979) and G is estimated as a function of net radiation and NDVI (Kustas and Daughtry, 1990).

By equating the expression for surface resistance in Monteith (1965), the latent heat flux, LE (W m^{-2}), can be related to potential evaporation as

$$LE = LE_p \left[\frac{(RR + \Delta)(1 - F_c)}{r_{so} C_h RR + RR + \Delta} + \frac{(RR + \Delta)F_c}{r_c C_h RR + RR + \Delta} \right] \tag{10}$$

with

$$RR = 4\sigma \frac{T_a^3}{\rho C_p C_h} + 1 \tag{11}$$

$$\Delta = \frac{L}{C_p} \frac{dq_s(T_a)}{dT}$$

where r_{so} and r_c are the bare soil and canopy resistances, respectively, T_a is the air temperature and σ the Stefan–Boltzmann constant.

If the bare soil resistance is estimated from the soil water content (Lee and Pielke, 1992), the canopy resistance can be evaluated, leading to R_{min} being estimated as (Noilhan and Planton, 1989)

$$R_{min} = r_c LAI F_1^{-1} F_2 F_3 F_4 \tag{12}$$

where F_1, F_3, F_4 are evaluated following Noilhan and Planton (1989) and F_2 is evaluated following Pan and Mahrt (1987).

COMPARISON WITH buFex OBSERVATIONS

Data from the buFex experiment conducted in the Lake King region (Figure 1) of south-western Australia (Lyons *et al.*, 1993) were used to evaluate the model. This district is characterized by areas of woodland association composed of *Eucalyptus* tree species with a mallee growth habit and with an understorey of sclerophyllous shrubs about 1 m high. The adjoining agricultural lands are devoted to raising crops of winter wheat and limited grazing. The Lake King district represents the eastern extremity of agricultural development in Western Australia and is bounded to the east by a vermin-proof fence, which forms a natural demarcation between agricultural and native vegetation. This demarcation is based purely on political rather than physical considerations.

Two cloud-free days, 31 August 1991 (during the crop growing season) and 24 April 1992 (before sowing), with post-flight calibrated data from NOAA 11 AVHRR corresponding to the buFex flight track (Figure 1), were used to evaluate the land surface parameters. The afternoon overpass NOAA AVHRR reflectances were used to estimate A_p, and the solar zenith angle was calculated from the satellite overpass time. Both $NDVI_{so}$ and $NDVI_{vg}$ for 24 April 1992 were based on surface on-site observations using a hand held radiometer (Smith *et al.*, 1992), whereas measurements of dense wheat NDVI during September 1993 were used as $NDVI_{vg}$ for agriculture on 31 August 1991 (Table II) as this corresponded to a similar period in the growing cycle.

During buFex, infra-red surface temperature, humidity, potential temperature and the three-dimensional wind components were measured by a GROB G109A single-engine motorglider (Hacker and Schwerdtfeger, 1988) at a sampling rate of 13 Hz (Lyons *et al.*, 1993). Sensible heat flux, latent heat flux and momentum were evaluated by eddy correlation methods. On 24 April 1992, the albedo and vegetation index were also measured by the aircraft. All aircraft and satellite observations were averaged over 20 km for both native and agricultural vegetation.

The mean albedoes calculated from NOAA 11 AVHRR and the aircraft are shown in Table III, whereas the albedo estimated along the buFex flight track from both aircraft and satellite on 24 April 1992 are shown in Figure 2. In this figure, the aircraft data have been averaged over 1 km to be consistent with the satellite footprint and are in close agreement with the satellite estimates.

The fraction of vegetation cover was estimated from the aircraft observations by evaluating the fraction of the aircraft traverse flown over vegetation in each of the adjoining areas. In particular, the aircraft measured outgoing radiance at wavelengths of 630 and 830 nm, with a half-width of 10 nm, corresponding

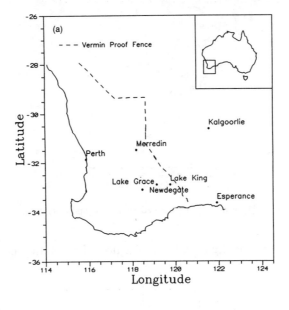

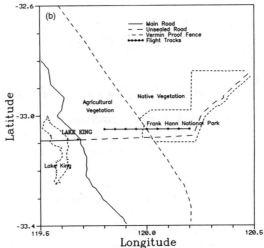

Figure 1. Lake King study area

to NOAA AVHRR channels 1, 2. As the aircraft did not measure reflectance, an equivalent vegetation index to NDVI was defined as

$$VI_a = \frac{I_2/I_{02} - I_1/I_{01}}{I_2/I_{02} + I_1/I_{01}} \tag{13}$$

where I is the irradiance and subscripts 1, 2, 0 correspond to the 630, 830 nm bands and the top of the atmosphere, respectively.

Assuming that the soil has a representative NDVI, $NDVI$so, as observed by Smith *et al.* (1992), the fraction of vegetation cover can be found by defining a critical VI_{ac} corresponding to vegetation. If an individual value of VI_a is greater than or equal to this critical value, it is assumed to correspond to vegetation, whereas if it is less than the critical value, it is assumed to correspond to soil. That is, by defining V_i as

$$\begin{aligned} V_i &= 1 \qquad VI_a \geqslant VI_{ac} \\ &= 0 \qquad VI_a < VI_{ac} \end{aligned} \tag{14}$$

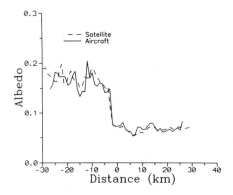

Figure 2. Comparison between albedo measured by aircraft and satellite on 24 April 1992

and summing V_i over the length of the flight path, the fraction of vegetation cover is defined as

$$F_c = \frac{\sum_{i=1}^{n} V_i}{n} \qquad (15)$$

where n is the total number of aircraft observations, sampled at 13 Hz, taken along the flight path.

The received irradiance can be related to the surface reflectance using the atmospheric correction suggested by Partridge and Mitchell (1990)

$$\frac{I_1/I_{01}}{I_2/I_{02}} = \frac{g_1(m)\exp(-m\tau_1)(1+m\tau_1)(1-M_2)}{g_2(m)\ exp(-m\tau_2)(1+m\tau_2)(1-M_1)}\frac{R_{1g}}{R_{2g}} \qquad (16)$$

where m is the air mass and M, $g(m)$ are atmospheric correction factors.

From the on-site red and near infra-red reflectance measurements for different canopy and soil conditions (Smith *et al.*, 1992), $R_{1g}/R_{2g} = 0.40$ was chosen as defining vegetation. Substituting this into the above equation leads to $I_1/I_2 = 0.43$ and consequently a value of $VI_{ac} = 0.40$ to distinguish vegetation. Comparisons between the fractional vegetation cover estimated using this aircraft-based index with that estimated from the satellite on 24 April 1992 are shown in Table IV and illustrate good agreement.

Surface observations of temperature, humidity and horizontal wind speed over the agricultural area were obtained from standard meteorological records at Newdegate (Dear *et al.*, 1990), whereas surface meteorological data over the native vegetation were inferred from the aircraft. These observations enabled the estimation of surface roughness.

In the surface layer under neutral conditions, wind speed is expressed as

$$u = \frac{u_*}{k}\ln\left(\frac{(z-d)}{z_o}\right) \qquad (17)$$

where u_* is the friction velocity, k the von Karmen constant assumed to be 0.4, z_0 the surface roughness, d

Table III. Surface albedo evaluated from NOAA AVHRR and measured from aircraft

	Agriculture	Native vegetation
NOVA 11 AVHRR		
31 August 1991	0.17 ± 0.03	0.08 ± 0.009
24 April 1992	0.17 ± 0.02	0.07 ± 0.005
Aircraft		
24 April 1992	0.16 ± 0.02	0.08 ± 0.01

Table IV. Fractional vegetation cover evaluated from NOAA AVHRR and aircraft on 24 April 1992

	AVHRR		Aircraft
	$NDVI_g$	F_c	F_c
Agriculture	0·40	0·55	0·58
Native vegetation	0·47	0·68	0·66

the displacement height and u the wind speed at height z. u_* can be estimated from the variation of wind direction (Beljaars and Holtslag, 1991), as

$$u_* \propto u\sigma_d \tag{18}$$

where σ_d is the standard deviation of wind direction.

Combining Equations (17) and (18) surface roughness can be estimated from the standard deviation of the wind direction as

$$\ln\left(\frac{(z-d)}{z_o}\right) = \frac{C_u k}{\sigma_d} \tag{19}$$

where C_u is a constant set to 3·1 by comparison with the aircraft observations over the agricultural area.

Aircraft observations were used to estimate the friction velocity and Monin–Obuhov length and, consequently, the surface roughness following Huang *et al.* (1993). A comparison of the surface roughness estimated from both methods is shown in Table V and is comparable.

The boundary layer model was initialized with atmospheric temperature and humidity profiles (Figure 3) obtained at 0700 WST (local standard time) from the nearest inland routine Bureau of Meteorology radiosonde released from Kalgoorlie (Figure 1). Geostrophic wind speed was estimated from the routine surface synoptic maps prepared by the Bureau of Meteorology.

As the soil water content profile was not observed on these days, the initial soil water content on 31 August 1991 was assigned by reference to the soil water content measured on September 1992 up to 30 cm, being at a similar time in the growing season. The 5 cm soil water potential measured at the agriculture site on 24 April 1992 was used for both agriculture and native vegetation sites and 0·2–1 m soil water contents were assigned as similar to 31 August 1991. The initial soil water contents used for the model are listed in Table VI. Initial soil temperature was set equal to the first model level air temperature.

All model input parameters, including estimated albedo, fractional vegetation cover, *LAI*, R_{min} and surface roughness are summarized in Table VII. Modelled daily surface energy components and aircraft

Table V. Surface roughness estimated from the Newdegate surface observations ($z = 3\,\text{m}$) and aircraft observations

Date	Station			Aircraft
	Time (WST)	σ_d	$z_0(\text{m})$	$z_0(\text{m})$
28 August 1991	0900	0·183	0·0035	
29 August 1991	1700	0·187	0·0039	
31 August 1991	0900	0·182	0·0033	
Mean			0·0036	0·0046
24 April 1992	1700	0·216	0·0096	
25 April 1992	1700	0·211	0·0084	
26 April 1992	1700	0·212	0·0086	
Mean			0·0088	0·0062

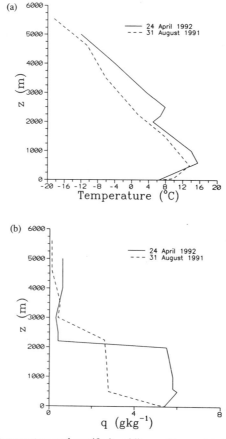

Figure 3. Initial temperature and specific humidity profiles used to initialize the model

measured sensible and latent heat fluxes for both native vegetation and agriculture are shown in Figures 4–6. These show that the model simulated surface heat fluxes are realistic compared with the aircraft observations over both native and agricultural vegetation. On both days, the model simulates a higher early afternoon sensible heat flux over the native vegetation, which is in agreement with the earlier observations of Lyons *et al.* (1993). During the wheat-growing season (Figure 4), the model simulates a dominance of the latent heat flux over agricultural lands.

DISCUSSION

Although we found good agreement between the surface albedo evaluated from NOAA AVHRR

Table VI. Initial soil water content ($m^3 m^{-3}$) used in model

z (m)	31 August 1991		24 April 1992	
	Agriculture	Nature	Agriculture	Native
−0·05	0·12	0·07	0·07	0·07
−0·10	0·12	0·09	0·09	0·09
−0·20	0·13	0·12	0·13	0·12
−0·50	0·14	0·17	0·14	0·17
−1·0	0·14	0·17	0·14	0·17

Table VII. Model input parameters

	31 August 1991		24 April 1992	
	Agriculture	Nature	Agriculture	Native
z_0 (m)	0·0036	0·15	0·0088	0·15
d (m)	0·02	0·8	0·05	0·8
Albedo	0·17	0·08	0·18	0·07
F_c	0·68	0·68	0·55	0·68
LAI	1·6	1·2	0·88	1·2
R_{min} (S m^{-1})	30	118	15	118
P (hpa)	966	966	977	977
Geostrophic wind (m s^{-1})	6·2	6·2	7·5	7·5

reflectances and simultaneous aircraft observations, Vulis and Cess (1989) have shown that estimates based on narrow band reflectances are dependent on the solar zenith angle. In particular, if the solar zenith angle was large at the satellite overpass time, this leads to errors in albedo estimation. Obviously, this limits the applicability of the above technique, but simultaneous surface observations throughout the year were not available to clarify the magnitude of this error.

Siddique *et al.* (1989) observed wheat LAIs at Merredin (Figure 1) of between 1 and 2·5 during August and September. This is comparable with our satellite-derived value of 1·6 for the agricultural area, given that it includes a mixture of wheat and pasture. Within the model, the LAI is used to adjust R_{min} and the simulated latent heat flux is relatively insensitive to changes in LAI.

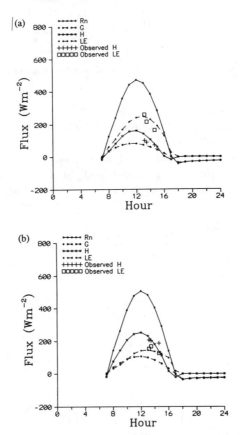

Figure 4. Simulated surface energy balance for 31 August 1991 over (a) agricultural and (b) native vegetation

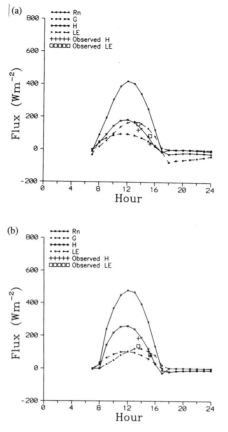

Figure 5. Simulated surface energy balance for 24 April 1992 over (a) agricultural and (b) native vegetation

The fractional vegetation cover influences the average ground heat flux and canopy resistance. As the soil heat flux is generally not the dominant heat flux component, errors in the fractional vegetation cover do not significantly influence the total available energy.

Canopy resistance is derived from the estimated latent heat flux, potential evaporation, soil resistance and fractional vegetation cover. As the average latent heat flux is of greater interest, the total evapotranspiration over the land surface is more important than the amount of transpiration from the canopy

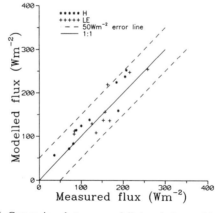

Figure 6. Comparison between modelled and observed heat fluxes

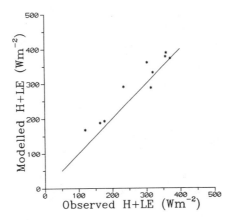

Figure 7. Comparison between modelled and observed available energy

or evaporation from the soil individually. The advantage of estimating the canopy resistance from closure of the energy balance is that even if the soil resistance and canopy resistance values are in error, as long as the estimated latent heat flux evaluated at the satellite overpass time is realistic, the modelled partitioning between sensible and latent heat flux should also be realistic. In other words, the partitioning of the fluxes throughout the day relies directly on the partitioning estimated through the closure of the surface energy balance at the time of the satellite overpass.

Soil water content and soil water capacity are the other important factors for surface heat fluxes. Unfortunately, detailed measurements of soil water content were not available during buFex, and we have inferred appropriate soil moisture profiles. Although the simulated average latent heat fluxes are realistic, evaporation from the soil and transpiration cannot be evaluated individually with this data set.

Average soil moisture is difficult to determined as soil moisture is highly variable both spatially and temporarily, as a result of the inhomogeneity of soil properties, topography, land use and the non-uniformity of rainfall and evapotranspiration. Although wet soil will generally have a lower albedo than dry soil, the albedo is also affected by other factors, such as vegetation cover and angle of incidence. Huete and Warrick (1990) found that due to the complex dynamics of soil surface drying and variability in soil properties, soil water content at the surface (0–5 cm) could not be determined with the Thematic Mapper moisture bands. Microwave remote sensing has been shown to provide a possible estimate of soil moisture (Engman, 1991), but in contrast with the reflectance of red and near infra-red irradiance, there is no existing microwave satellite system for soil moisture measurement.

The comparison between measured and estimated available energy is shown in Figure 7. Modelled available energy, H + LE, is generally higher than the corresponding aircraft measurements. Dugas *et al.* (1991) and Walker *et al.* (1989) also found that heat fluxes measured by eddy correlation, the technique used by the aircraft, were lower than those observed by the Bowen ratio. In addition, many large-scale experiments have noted a significant, systematic loss in aircraft measurements of energy fluxes (ie. Schuepp *et al.*, 1987), suggesting that some flux transfer may be at length scales much longer than those associated with the atmospheric boundary layer (Shuttleworth, 1991). Within these limitations, the estimated fluxes have a similar accuracy to those estimated by Huang *et al.* (1993) from radiant surface temperature and NOAA AVHRR reflectances.

CONCLUSION

By using reflectances and surface temperatures from NOAA AVHRR and ground observations, the average albedo, LAI, fractional vegetation cover and canopy resistance are estimated. These estimated land surface parameters are used as input into a modified one dimensional land surface–atmosphere model (Huang and Lyons, 1995) to simulate the surface energy components over both native and agricultural vegetation. Compared with the aircraft observations, the simulated sensible and latent heat fluxes are realistic. Combining NOAA AVHRR satellite data with the ground observations can provide realistic land surface parameters at an appropriate scale for modelling the surface energy balance.

ACKNOWLEDGEMENTS

This study forms part of the buFex programme, which is supported by the Australian Research Council. Throughout it, Huang Xinmei was in receipt of an Australian Overseas Postgraduate Research Award and associated stipend through Murdoch University. NOAA AVHRR data were kindly supplied by the Leeuwin Centre for Earth Sensing Technologies, whereas the Newdegate data set was provided by Dr Ian Foster of the West Australian Department of Agriculture and the radiosonde data were supplied by the Australian Bureau of Meteorology. All of this assistance is gratefully acknowledged.

REFERENCES

Andre, J-C., Bougeault, P., and Goutorbe, J-P. 1990. 'Regional estimates of heat and evaporation fluxes over non-homogeneous terrain. Examples from the HAPEX-MOBILHY programme', *Boundary Layer Meteorol.*, **50**, 77–108.
Beljaars, A. C. M. and Holtslag, A. A. M. 1991. 'Flux parameterization over land surfaces for atmospheric models', *J. Appl. Meteorol.*, **30**, 327–341.
Brest, C. L. and Goward, S. 1987. 'Deriving surface albedo measurements from narrow band satellite data', *J. Remote Sensing*, **8**, 351–367.
Briegleb, B. P., Minnies, P., Ramanmathan, V., and Harrison, E. 1986. 'Comparison of regional clear-sky albedo infra-red from satellite observation and model computations', *J. Climate Appl. Meteorol.*, **25**, 214–226.
Carlson, T. N. 1991. 'Modelling stomatal resistance: an overview of the 1989 workshop at the Pennsylvania State University', *Agric. Forest Meteorol.*, **54**, 103–106.
Carlson, T. N., Perry, E. M., and Schmugge, T. J. 1990. 'Remote estimation of soil moisture availability and fractional vegetation cover for agriculture fields', *Agric. Forest Meteorol.*, **52**, 45–69.
Cess, R. D., Dutton, E. G., Delush, J. J., and Jiang, Feng. 1991. 'Determining surface solar absorption from broadband satellite measurements for clear skies: comparison with surface measurements', *J. Climate*, **4**, 236–247.
Chen, T. S. and Ohring, G. 1984. 'On the relationship between clear sky planetary and surface albedos', *J. Atmos. Sci.*, **41**, 156–158.
Clevers, J. G. P. W. 1989. 'The application of a weighted infra-red-red vegetation index for estimating leaf area index by correcting for soil moisture', *Remote Sensing Environ.*, **29**, 25–37.
Dear, S. J., Bell, M. J., and Lyons, T. J. 1990. 'Western Australian wind atlas', *Rep. 64*, Minerals and Energy Research Institute of Western Australia, 229pp [Available from MERIWA, Mineral House, 100 Plain Street, East Perth, WA 6004, Australia].
Dugas, W. A., Fristchen, L. J., Gray, L. W., Held, A. A., Mathias, A. D., Reicosky, D. C., Steduto, P., and Steiner, J. L. 1991. 'Bowen ratio, eddy correlation and portable chamber measurement of sensible and latent heat fluxes over irrigated spring wheat', *Agric. Forest Meteorol.*, **58**, 1–20.
Ek, M. and Mahrt, L., 1989. *A User's Guide to OSU1DPBL*. Oregon State University, Corvallis. 106pp [Available from Department of Atmospheric Science, Oregon State University, Corvaillis, Oregon, USA].
Engman, E. T. 1991. 'Applications of microwave remote sensing of soil moisture for water resources and agriculture', *Remote Sensing Environ.*, **35**, 213–226.
Gutman, G., Ohring, G., Tarrley, D., and Ambroziak, R. 1989. 'Albedo of the U.S. great plains as determined from NOAA-9 AVHRR data', *J. Climate*, **2**, 608–617.
Hacker, J. M. and Schwerdtfeger, P. 1988. *The FIAMS Research Aircraft System Description*. 2nd edn. Flinders Institute for Atmospheric and Marine Sciences, Adelaide. 60 pp [Available from FIAMS, Flinders University of South Australia, GPO Box 2001, Adelaide, SA 5000, Australia].
Huang, X. and Lyons, T. J. 1995. 'The simulation of surface heat fluxes in a land surface-atmosphere model', *J. Appl. Meteorol.*, **34**, 1099–1111.
Huang, X., Lyons, T. J., Smith, R. C. G., Hacker, J. M., and Schwerdtfeger, P. 1993. ' Estimation of surface energy balance from radiant surface temperature and NOAA AVHRR sensor reflectances over agricultural and native vegetation', *J. Appl. Meteorol.*, **32**, 1441–1449.
Huete, A. R. and Warrick, A. W. 1990. 'Assessment of vegetation and soil water regimes in partial canopies with optical remotely sensed data', *Remote Sensing Environ.*, **32**, 155–167.
Kustas, W. P. and Daughtry, C. S. T. 1990. 'Estimation of soil heat flux/net radiation ratio from spectral data', *Agric. Forest Meteorol.*, **49**, 205–223.
Lee, T. J., and Pielke, R. A. 1992. 'Estimating the soil surface specific humidity', *J. Appl. Meteorol.*, **31,**, 480–484.
Louis, J. F. 1979. 'A parametric model of vertical eddy fluxes in the atmosphere', *Boundary Layer Meteorol.*, **17**, 187–202.
Louis, J. F., Tiedtke, M., and Geleyn, J. F. 1982. 'A short history of the operational PBL-parameterization of ECMWF' in *Workshop on Planetary Boundary-layer Parameterization*. European Centre for Medium Range Weather Forecasts, Reading, [Available from ECMWF, UK].
Lyons, T. J., Schwerdtfeger, P., Hacker, J. M., Foster, I. J., Smith, R. C. G., and Huang, X. 1993. Land–atmosphere interaction in a semiarid region: the bunny fence experiment', *Bull. Am. Meteorol. Soc.*, **74**, 1327–1334.
Monteith, J. L. 1965. 'Evaporation and environment', *Symp. Soc. Exp. Biol.*, **19**, 205–234.

Monteith, J. L. 1973. *Principles of Environmental Physics*. Arnold, London.

Nemani, R. R. and Running, S.W. 1989. 'Estimation of regional surface resistance to evapotranspiration from NDVI and thermal-IR AVHRR data', *J. Appl. Meteorol.*, **28**, 276–284.

Noilhan, J., and Planton, S. 1989. 'A simple parameterization of land surface processes for meteorological models', *Monthly Weather Rev.*, **117**, 536–549.

Pan, H, and Mahrt, L. 1987. 'Interaction between soil hydrology and boundary layer development', *Boundary Layer. Meteorol.*, **38**, 185–202.

Partridge, G. W., and Mitchell, R. M. 1990. 'Atmospheric and viewing angle correction of vegetation indices and grassland fuel moisture content derived from NOAA-AVHRR', *Remote Sensing Environ.*, **8**, 121–135.

Price, J. C. 1992. 'Estimating vegetation amount from visible and near infra-red reflectances', *Remote Sensing Environ.*, **41**, 29–34.

Satterlund, D. R. 1979. 'An improved equation for estimating longwave radiation from the atmosphere', *Wat Resour. Res.*, **15**, 1649–1650.

Savijarvi, H. 1990. 'Fast radiation parameterization schemes for mesoscale and short-range forecast model', *J. Appl. Meteorol.*, **29**, 437–447.

Schuepp, P. H., Desjardins, R. L., MacPherson, J. I., Boisvert, J., and Austin, L. B. 1987. 'Airborne determination of regional water use efficiency and evapotranspiration: present capabilities and initial field tests', *Agric. Forest Meteorol.*, **41**, 1–19.

Sellers, P. J., Rasool, S. I. and Bolle, H. J. 1990. 'A review of satellite data algorithms for studies of the land surface', *Bull. Am. Meteorol. Soc.*, **71**, 1429–1447.

Siddique, K. H. M., Belford, R. K., Perry, M. W., and Tennant, D. 1989. 'Growth, development and light interception of old and modern wheat cultivars in a Mediterranean-type environment', *Aust. J. Agric. Res.*, **40**, 473–487.

Smith, R. C. G., Huang, X., Lyons, T. J., Hacker, J. H. and Hick, P. T. 1992. 'Change in land surface albedo and temperature in southwestern Australia following the replacement of native perennial vegetation by agriculture: satellite observations' in *World Space Congress 1992—43rd Congress of the International Astronautical Federation, August 28–September 5, 1992, Washington, DC*. [Available from International Astronautical Federation, 3–5, Rue Mario-Nikis, 75015 Paris, France].

Shuttleworth, W. J. 1991. 'Insight from large-scale observational studies of land/atmosphere interactions', *Surv. Geophys.*, **12**, 3–30.

Vulis, I. L. and Cess, R. D. 1989. 'Interpretation of surface and planetary directional albedos for vegetated regions', *J. Climate*, **2**, 986–996.

Walker, C. D., Brunel, J. P., Dunin, F. X., Edwards, W. R. N., Hacker, J. M., Hartmann, J., Jupp, D. L., Reyenga, W., Schwerdtfeger, P., Shao, Y., and Williams, A. E. 1989. 'A joint project on water use of the Mallee community in late summer' in Noble, J. C., Joss, P. J., and Jones, G. K. (Eds), *The Mallee Lands—A Conservation Perspective. Proceedings of the National Mallee Conference*. CSIRO, Adelaide.

23

METEOROLOGICAL IMPACT OF REPLACING NATIVE PERENNIAL VEGETATION WITH ANNUAL AGRICULTURAL SPECIES

HUANG XINMEI AND T. J. LYONS

Environmental Science, Murdoch University, Murdoch, WA 6150, Australia

AND

R. C. G. SMITH

Leeuwin Centre for Earth Sensing Technologies, Floreat, WA 6019, Australia

ABSTRACT

Following the clearing of native perennial vegetation for agriculture based on winter growing annual species, the surface characteristics of south-western Australia have been significantly altered. Analysis of the annual variation of these characteristics based on satellite data and a one-dimensional boundary layer model suggests that convective mixing over the cleared land is no longer able to reach the lifting condensation level for a significant period of the year. This implies a decrease in convective cloud formation and a reduction in the convective enhancement of rainfall.

INTRODUCTION

In clearing native vegetation for agriculture, land surface characteristics, such as albedo, surface roughness and canopy resistance, are changed. Consequently, surface energy components are redistributed and changes in the partitioning between sensible and latent heat flux will affect boundary layer development and the vertical transport of heat and water vapour in the atmosphere. These effects could influence cloud formation and precipitation, as Rabin *et al.* (1990), and Lyons *et al.* (1993) suggest that clouds form earliest over regions characterized by high sensible heat flux.

Lake King (Figure 1), in the Lakes District of the Great Southern Region of Western Australia, is characterized by a Mediterranean climate with hot, dry summers and mild, wet winters. It is dominated by the subtropical ridge (Gentilli, 1971). During summer, heat troughs form along the western coast of Australia and move inland in accordance with a regular progression of anticyclones (Watson, 1980). Winter rainfall comes with the northward movement of the ridge and incursions of mid-latitude cyclones and cold fronts. Rainfall decreases towards the east and inland from the coast.

Native vegetation of the region is characteristically a woodland called mallee, with *Eucalyptus eremophila* the most dominant species. Patches of eucalypt woodland occur on the lower ground, and scrub heath and *Casuarina* thickets are found on residual plateau soils. The topography is gently undulating country of low relief with duplex mallee soils — that is, sand overlying clay (Beard, 1979). The vermin-proof fence (Figure 1) marks the eastern extremity of agricultural lands.

Since the beginning of this century approximately 13×10^6 ha of native perennial vegetation to the west of the fence have been cleared for agriculture, which is based on winter growing annual species, predominantly wheat. Analysis of long-term rainfall data indicates that following the extensive clearing between 1950 and 1980 of the native vegetation, which transpires year-round, and its replacement with annual vegetation that only transpires during the winter, winter rainfall has declined by about 20% (Pittock, 1983; Williams,

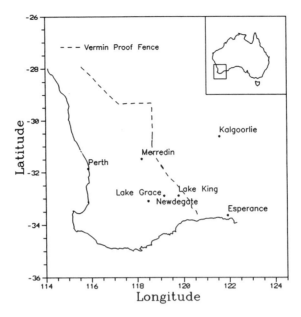

Figure 1. Study area centred on Lake King

1991). Pittock (1983) and Allan and Haylock (1993) have ascribed this decrease to large-scale circulation changes, but such a process of desertification initiated by deforestation is also in accord with observational and numerical studies (Anthes, 1984; Segal *et al.*, 1988; Otterman, 1974; Shukla and Mintz, 1982; Charney, 1975) and highlights the need to gain a greater insight into the climatic impact of the large-scale clearing of native perennial vegetation and its replacement by winter growing annual species. Such a change is known to cause a significant reduction in evapotranspiration, as evidenced by rising water-tables and increased salinity (Greenwood *et al.*, 1985). Its climatic impact is relevant to proposed plans to replant trees and clear further land for agriculture.

Thus we seek to characterize the impact of these land-use changes by using satellite data over both agricultural and native vegetation combined with long-term climatic data to initialize a simple boundary layer model. Such an approach will illustrate the potential impact of surface clearing and is a first step to understanding its longer term climatic effect.

DATA

Twelve NOAA advanced very high resolution radar (AVHRR) cloud-free afternoon overpasses (Table I) covering the area of native vegetation and agriculture for 100 km around the vermin-proof fence were

Table I. Summary of NOAA 11 satellite overpasses used between April 1990 and March 1991

Date	Time (WST)	Date	Time (WST)
8 Apr 1990	1421	26 Oct 1990	1436
24 May 1990	1424	5 Nov 1990	1426
21 Jun 1990	1421	10 Dec 1990	1441
9 Jul 1990	1426	6 Jan 1991	1442
31 Aug 1990	1444	10 Feb 1991	1442
28 Sep 1990	1444	9 Mar 1991	1453

Table II. Five areas of native vegetation selected for study from the NOAA satellite data. The five agricultural areas selected were 0·2–0·4 degrees to the west of each native vegetation site

Area	Latitude	Longitude
1. Lake Magenta Reserve	−33·54	119·00
2. Commander Rocks	−33·20	119·45
3. Dragon Rocks	−32·70	118·98
4. Vermin Proof Fence	−33·20	120·20
5. Vermin Proof Fence	−32·70	119·70

selected from the Leeuwin Remote Sensing Centre archive to cover a 12 month period from April 1990 to March 1991 (Smith *et al.*, 1992). Five matched sites (Table II), representative of land surface conditions before and after clearing, were sampled by displaying each overpass, identifying two 1 km^2 pixels of native vegetation in each area and recording their digital counts. A matched pair of agricultural pixels at the same latitude was selected to the west of each native vegetation area, giving 20 values each month made up of 5 sites × 2 vegetation types × 2 replicates. The mean digital counts for NOAA AVHRR channels 1, 2, 4 and 5 were converted to reflectance or surface temperature.

The normalized difference vegetation index (NDVI) is a measure of the fractional green vegetation cover and is defined as

$$NDVI = \frac{R_2 - R_1}{R_2 + R_1} \qquad (1)$$

where R_1, R_2 are the reflectances of channels 1, 2, respectively. The annual variation of NDVI for both agricultural and native vegetation is shown in Figure 2. This illustrates the seasonal variation observed over the agricultural area as the crop progresses through its life cycle compared with the relative constancy of the native vegetation. As a result of local cloud on 21 June (day 172), a slightly different area of native vegetation had to be selected, resulting in the anomalous peak NDVI observed.

Individual reflectances of channels 1 (red band) and 2 (near infra-red band) are shown in Figure 3. Both the reflectance of red and near infra-red irradiance were higher in the agricultural area than over the native vegetation throughout the year. Red reflectance declines during winter through absorption by the

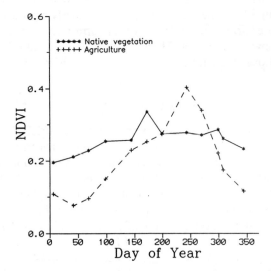

Figure 2. Annual variation of NDVI

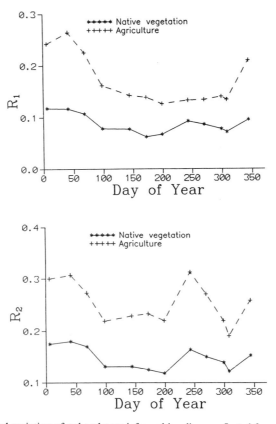

Figure 3. Annual variation of red and near infra-red irradiance reflected from the surface

chlorophyll and pigments of new vegetative growth and this is illustrated between May and September with the emergence and growth of the wheat crop. It increases from December to March after harvest as bare soil emerges.

Near infra-red reflectance over the agricultural area illustrates a more complex pattern, leading to higher values during August and September with new leaf growth, followed by a decrease with senescence of the crop. It increases again during November to January as the underlying soil is exposed after harvest.

Surface albedo was estimated from the visible and near infra-red bands of NOAA AVHRR following Huang *et al.* (1995) and is shown in Figure 4. Throughout the year, the albedo over the agricultural region is greater than that observed over the native vegetation. Over both, it is higher in summer than in winter.

BOUNDARY LAYER MODEL

Twelve days, in the middle of each month, were chosen to simulate surface heat fluxes and boundary layer development using a one-dimensional soil–vegetation–boundary layer model based on Ek and Mahrt (1989). This model was originally developed at Oregon State University and has been described by Mahrt and Pan (1984), Troen and Mahrt (1986), and Pan and Mahrt (1987). The latent heat flux is calculated separately for the soil and canopy and the total latent heat flux weighted by the fraction of vegetation cover. Huang and Lyons (1995) have shown that a modified version of this model is able to provide a realistic simulation of energy fluxes over a variety of surface conditions. No cloud was simulated in the model and all results were computed under cloud-free conditions. The model was initialized by climatological monthly average radiosonde profiles taken at 0700 WST (local standard time) for the nearest inland Bureau of Meteorology radiosonde station at Kalgoorlie (Figure 1) between 1957 and

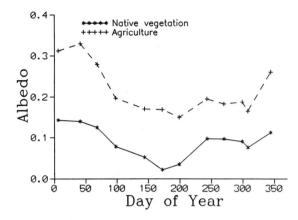

Figure 4. Annual variation of albedo

1975 (Maher and Lee, 1977). The mean wind speed observed at 700 hPa was assumed to be representative of the prevailing geostropic wind speed.

Loamy sandy soil and the initial soil water content was assumed to be the same for both agricultural and native vegetation. The typical driest and wettest vertical profiles of the soil water content observed for this soil in the Western Australian wheat belt (Tennant *et al.*, 1991) were used to estimate initial soil water content. The soil water content for each month was interpolated based on the weighted average monthly rainfall recorded by the Bureau of Meteorology at Lake Grace (Table III).

Following Huang *et al.* (1995), the annual variation of land surface parameters for both agricultural and native vegetation were estimated and are shown in Table IV. Leaf area index, LAI, was derived from near infra-red reflectance (Clevers, 1989) and fractional vegetation cover, F_c, estimated from the NDVI. The surface roughness, z_o, and displacement height, d, were estimated from measurements of surface wind speed. Minimum stomatal resistance, R_{min}, was estimated from the heat fluxes derived from satellite and ground meteorological measurements at selected times throughout the year (Huang *et al.*, 1993). Over the agricultural area, the land surface parameters were assumed to vary with the crop growing cycle and NDVI based on parameters estimated during the growing season and before sowing (Huang *et al.*, 1995). The albedos estimated for May, June and July (Figure 4) may have errors due to the low solar angle (Vulis and Cess, 1989) and hence the albedo has been assumed to be constant between April and September. Over the native vegetation all surface parameters except the albedo were assumed to be constant throughout the year given the relative constancy of the NDVI.

Table III. Monthly rainfall at Lake Grace and initial soil water content $(m^3 m^{-3})$ assumed in model

	Rainfall (mm)	z(m) −0·05	−0.10	−0·20	−0·50	−1·0
Jan	5	0·04	0·03	0·03	0·06	0·14
Feb	7	0·04	0·04	0·04	0·07	0·14
Mar	12	0·05	0·05	0·05	0·08	0·15
Apr	17	0·06	0·06	0·07	0·09	0·16
May	42	0·11	0·12	0·14	0·15	0·20
Jun	48	0·12	0·14	0·16	0·17	0·21
Jul	45	0·12	0·13	0·15	0·16	0·20
Aug	38	0·11	0·11	0·13	0·13	0·19
Sep	30	0·09	0·09	0·10	0·11	0·18
Oct	19	0·07	0·07	0·07	0·09	0·17
Nov	12	0·05	0·05	0·05	0·08	0·15
Dec	9	0·04	0·04	0·04	0·07	0·15

Table IV. Land surface parameters

	z_o(m)	d(m)	Albedo	F_c	LAI	R_{min} $(S\,m^{-1})$
Agricultural area						
Jan	0·006	0·03	0·32	0·0		
Feb	0·006	0·03	0·32	0·0		
Mar	0·006	0·03	0·27	0·0		
Apr	0·006	0·03	0·20	0·20	0·2	15
May	0·004	0·02	0·18	0·55	0·3	30
Jun	0·004	0·02	0·18	0·55	0·8	30
Jul	0·005	0·03	0·18	0·68	1·2	30
Aug	0·005	0·03	0·18	0·68	1·6	30
Sep	0·01	0·05	0·18	0·68	1·5	30
Oct	0·01	0·05	0·20	0·68	1·2	40
Nov	0·01	0·05	0·20	0·45	0·4	50
Dec	0·006	0·03	0·27	0·0		
Native vegetation						
Jan–Dec	0·15	0·8	—*	0·68	1·2	118

* Albedo: Jan, Feb 0·14; March 0·12; Apr–Nov 0·08; and Dec 0·12

A comparison of the accumulated sensible, H_{cu}, and latent heat, LE_{cu}, fluxes between 0800 and 1700 WST over agricultural and native vegetation are shown in Figure 5. Sensible heat flux over the native vegetation is greater than over the agricultural area throughout the year, whereas the latent flux is greater over the agricultural area during the winter growing season of the agricultural crops. Both vegetated surfaces illustrate a marked increase in evaporation throughout the winter as the supply of soil moisture increases and the agricultural crops develop. The significant difference in latent heat flux between these two areas occurs in August, September and October as the wheat crop matures. Total evapotranspiration over the agricultural area is greater than over the native vegetation for all days modelled.

Changing land surface characteristics result in a changing surface heat budget, which in turn affects the development of the planetary boundary layer (PBL). Figure 6 compares the modelled PBL height at 1400 WST over both agricultural and native vegetation with the estimated lifting condensation level. The lifting condensation level was calculated from the monthly mean air temperature and relative humidity observed at 1500 WST at Lake Grace (Bureau of Meteorology, 1988). As this was the only long-term surface meterological data available in the immediate vicinity, the estimated lifting condensation level is assumed to be the climatologically appropriate height of convective cloud formation over both native and agricultural vegetation. Clearly this approximation implies that large-scale climatological atmospheric properties have not adjusted to local surface conditions.

Throughout the year, the atmospheric boundary layer develops to a greater height over the native vegetation than is observed over the agricultural area. The lifting condesation level is higher than both PBLs during the summer, attesting to the high aridity of the environment and the observed lack of local cloud formation. During the rest of the year, the lifting condensation level is below the PBL over the native vegetation, but only below the PBL over the agricultural lands during August, September and October.

DISCUSSION

Satellite observations show that after the replacement of native perennial vegetation by winter growing agricultural species, the reflectance of red and near infra-red irradiance has been increased. The annual variation of NDVI also increases. Field observations show that the bare soil has a higher reflectance of both red and near infra-red irradiance than vegetation and the reflectance of agricultural crops is higher than native vegetation (Smith *et al.*, 1992). Stanhill (1970) found that the albedo was inversely related to

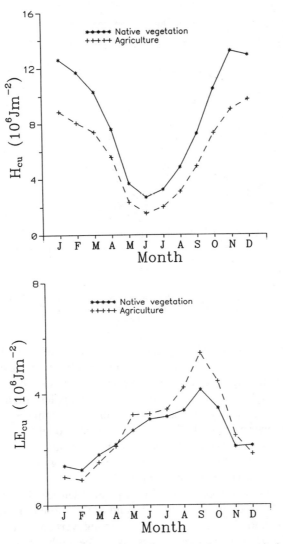

Figure 5. Annual variation of accumulated sensible and latent heat flux over agricultural and native vegetation

vegetation height due to multiple reflectance between adjacent leaves and stems (Monteith, 1979). Therefore increasing the exposure of soil in summer in the agricultural area results in increasing the reflectance and albedo. The replacement of high native vegetation by short annual agricultural species leads to an increase in albedo during the winter growing season compared with the native vegetation. Thus clearing native vegetation for agriculture in south-western Australia increases the surface albedo and reduces the absorption of solar radiation.

Replacing higher native vegetation with agriculture also reduces the surface roughness and canopy resistance, resulting in changes in the surface heat fluxes and boundary layer development. Tables V and VI illustrate the sensitivity of these parameters to changes in albedo, surface roughness, R_{min} and soil water content where the initial input and profiles are based on the simulation for native vegetation in September. They also show that increasing the albedo and decreasing the surface roughness leads to a decrease in daytime sensible and latent heat flux as well as boundary layer height. Reducing R_{min} and increasing the soil water content leads to an increased latent heat flux, while the sensible heat flux and boundary layer height decrease.

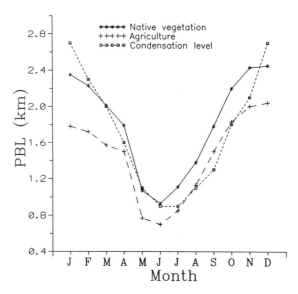

Figure 6. Annual variation of the height of the planetary boundary layer (PBL) over agricultural and native vegetation compared with the height of the lifting condensation level

Novak (1990) suggested that the sum of daily average sensible and latent heat flux decreases as the surface resistance and albedo increase and the surface roughness decrease. Such effects were apparent in these simulations in summer when agricultural areas appear as dry soil, but the native vegetation still has limited transpiration. Although the canopy resistance of crops is less than that of native vegetation, which is adapted to a semi-arid environment, the reduced surface roughness and increases albedo result in a decrease in the available energy over the agricultural area throughout the year. The difference in available energy between the two surfaces is greater in summer than in winter and spring as the latent heat fluxes are enhanced over the agricultural region during the growing season. Consequently, the sensible heat flux is much higher over the native vegetation. Mahrt and Ek (1993) also showed that low albedo and high surface roughness over a forest area correspond to a greater sensible heat flux. Greater sensible heat over the native vegetation results in stronger convection and higher boundary layer development.

Convective cloud formation and development require strong vertical convection and enough water vapour in the atmosphere. During the summer, attesting to be high aridity of the environment, dry conditions result in a lack of local cloud formation as the boundary layer height does not reach the lifting condensation level. On the other hand, although the maximum rainfall occurs in June, the convective enhancement is limited by weak vertical mixing.

Although from August to October conditions are favourable for convective cloud formation over both areas, comparison between modelled boundary layer heights and the lifting condensation level suggests that from July to November conditions are more favourable to cloud formation and development over native vegetation. Visible satellite imagery (Lyons et al., 1993) has also shown convective cumulus clouds associated with the native vegetation, but not with the agricultural regions. Rabin et al. (1990) suggested that clouds form earliest over regions characterized by high sensible heat flux and are suppressed over

Table V. Summary of sensitivity test

Test 1	Change albedo from 0·08 to 0·17
Test 2	Change z_o from 0·15 to 0·01 m
Test 3	Change R_{min} from 118 to 40 S m^{-1}
Test 4	Change soil water content profile from 0·15 to 0·25 m^3 m^{-3}

Table VI. Sensitivity of heat fluxes and PBL to surface parameters

	Test 1	Test 2	Test 3	Test 4
ΔH_{cu} (10^4 J m^{-2})	−97	−14	−94	−86
ΔLE_{cu} (10^4 J m^{-2})	−27	−8	123	69
ΔPBL (m)	−165	−54	−146	−108

regions characterized by high latent heat fluxes. Otterman (1989) and Otterman *et al.* (1990) have argued that increases in precipitation resulting from land-use are attributable to intensification of the dynamic processes of convection and advection resulting from plant-induced enhancement of daytime sensible heat flux from generally dry surface.

From analysing rainfall data in south-west Australia, Williams (1991) found that from 1945 to 1990 the rainfall decrease was statistically significant over the agricultural region from May to October, but between November to April it was not statistically significant.

Rainfall in this region is primarily associated with onshore flow or the passage of fronts from the Southern Ocean. Hence if decreased convective mixing over the agricultural region is associated with the observed decline in rainfall, we would need to show that this percentage of the rainfall has been locally generated by convective enhancement of precipitation. Such an analysis is beyond the scope of this study, but Brubaker *et al.* (1992) have suggested that the local contribution to total monthly precipitation generally lies between 10 and 30%, which is of similar magnitude to the observed winter rainfall decrease (Pittock, 1988).

Thus convective clouds are more likely to form over the native vegetation than over the agricultural area. Although there is no evidence as to the importance of convective processes in providing precipitation to the south-west of Australia, this analysis clearly shows that agricultural practices have resulted in a decrease in convective mixing during the time that rainfall has been recorded to decrease over the agricultural area.

ACKNOWLEDGEMENTS

buFex is a collaborative experiment between Murdoch University, Flinders University of South Australia, Western Australian Department of Agriculture, CSIRO Division of Exploration Geoscience and the Remote Sensing Application Centre of the Western Australian Department of Land Administration and is being supported by the Australian Research Council, Murdoch University's Special Research Grant, a Flinders University/CSIRO Collaborative Research Grant and the Western Australian Department of Agriculture. NOAA AVHRR data was kindly supplied by the Western Australia Satellite and Technology Applications Consortium (WASTAC) and radiosonde and rainfall data were provided by the Australian Bureau of Meterology. All of this associated is gratefully acknowledged.

REFERENCES

Allan, R. J. and Haylock, M. 1993. 'Circulation features associated with the winter rainfall decrease in southwestern Western Australia: implications for greenhouse and natural variability studies', *J. Climate*, **6**, 1356–1367.
Anthes, R. A. 1984. 'Enhancement of convective precipitation by mesoscale variations in vegetative cover in semi-arid regions', *J. Climate Appl. Meterol.*, **23**, 541–554.
Beard, J. S. 1979. 'Phytogeographic regions' in Gentilli, J. (Ed.), *Western Landscapes*. University of Western Australia Press. pp. 107–121.
Brubaker, K. L., Entekhabi, D., and Eagleson, P. S. 1992. 'Estimation of continental precipitation recycling', *J. Climate*, **6**, 1077–1089.
Bureau of Meterology 1988. *Climatic Averages Australia*. Australia Government Publishing Service, Canberra.
Charney, J. G. 1975. 'Dynamics of deserts and droughts in the Sahel', *Q. J. Roy. Meteorol. Soc.*, **101**, 193–202.
Clevers, J. G. P. W. 1989. 'The application of a weighted infrared-red vegetation index for estimating leaf area index by correcting for soil moisture', *Remote Sensing Environ.*, **29**, 25–37.
Ek, M. and Mahrt, L. 1989. *A One-dimensional Planetary Boundary Layer Model With Interactive Soil Layers and Plant Canopy*. Department of Atmospheric Sciences, Oregon State University, Corvallis. 105 pp.
Gentilli, J. (Ed.) 1971. *Climates of Australia and New Zealand*. Vol. 13. World Survey of Climatology, Elsevier, Amsterdam. 345 pp.

Greenwood, E. A. N., Klein, L., Beresford, J. D., and Watson, G. D. 1985. 'Differences in annual evaporationl between grazed pasture and Eucalypts species in plantations on a saline farm catchment', *J. Hydrol.*, **78**, 261–278.

Huang, X. and Lyons, T. J. 'The simulation of surface heat fluxes in a land surface–atmosphere model', *J. Appl. Meteorol.*, **34**, 1099–1111.

Huang, X., Lyons, T. J., Smith, R. C. G., Hacker, J. M., and Schwerdtfeger, P. 1993. 'Estimation of surface energy balance from radiant surface temperature and NOAA AVHRR sensor reflectances over agricultural and native vegetation', *J. Appl. Meteorol.*, **32**, 1441–1449.

Huang, X., Lyons, T. J., Smith, R. C. G., and Hacker, J. M., 1995. 'Estimation of land surface parameters using satellite data', *Hydrol. Process.*, **9**, 631–643.

Lyons, T. J., Schwerdtfeger, P., Hacker, J. M., Foster, I. J., Smith, R. C. G., and Huang, X. 1993. 'Land–atmosphere interaction in a semiarid region: the bunny fence experiment', *Bull. Am. Meteorol. Soc.*, **74**, 1327–1334.

Maher, J. V. and Lee, D. M. 1977. *Upper Air Statistics Australia*. Bureau of Meteorology, Australian Government Publishing Service, Canberra.

Mahrt, L. and Ek, M. 1993. 'Spatial variability of turbulent fluxes and roughness lengths in HAPEX-MOBILHY', *Boundary Layer Meteorol.*, **65**, 381–400.

Mahrt, L. and Pan, H. 1984. 'A two-layer model of soil hydrology', *Boundary Layer Meteorol.*, **29**, 1–20.

Monteith, J. L. 1979. *Principles of Environmental Physics*. Arnold, London.

Novak, M. D. 1990. 'Micrometeorological changes associated with vegetation removal and influencing desert formation', *Theor. Appl. Climatol.*, **42**, 19–25.

Otterman, J. 1974. 'Baring high-albedo soils by overgrazing: a hypothesized desertification mechanism', *Science*, **186**, 531–533.

Otterman, J. 1989. 'Enhancement of surface–atmosphere fluxes by desert fringe vegetation through reduction of surface albedo and of soil heat flux', *Theor. Appl. Climatol.*, **40**, 67–79.

Otterman, J., Manes, A., Rubin, S., Alpert, P., and O'C. Starr, D. 1990. 'An increase of early rains in southern Israel following land use change?', *Boundary Layer Meteorol.*, **53**, 333–351.

Pan, H. and Mahrt, L. 1987. 'Interaction between soil hydrology and boundary layer development', *Boundary Layer Meteorol.*, **38**, 185–202.

Pittock, A. B. 1983. 'Recent climate change in Australia: implications for a CO_2-warmed earth', *Climatic Change*, **5**, 321–340.

Pittock, A. B. 1988. 'Actual and anticipated change in Australia's climate' in Pearman, G. I. (Ed.), *Greenhouse: Planning for Climate Change*. CSIRO Australia, Melbourne. pp. 35–53.

Rabin, R. M., Stadler, S., Wetzel, P. J., Stensrud, D. J., and Gregory, M. 1990. 'Observed effects of landscape variability on convective clouds', *Bull. Am. Meteorol. Soc.*, **71**, 272–280.

Segal, M., Avissar, R., McCumber, M. C., and Pielke, R. A. 1988. 'Evaluation of vegetative effects on the generation and modification of mesoscale circulations', *J. Atmos. Sci.*, **45**, 2268–2292.

Shukla, J. and Mintz, Y. 1982. 'Influence of land-surface evapotranspiration on the earth's climate', *Science*, **215**, 1498–1500.

Smith, R. C. G., Huang, X., Lyons, T. J., Hacker, J. M., and Hick, P. T., 1992. 'Change in land surface albedo and temperature in southwestern Australia following the replacement of native perennial vegetation by agriculture: satellite observations' in *World Space Congress 1992 — 43rd Congress of the International Astronautical Federation, IAF-92-0117*. International Astronautical Federation, Paris.

Stanhill, G. 1970. 'Some results of helicopter measurements of albedo', *Solar Energy*, **13**, 59–66.

Tennant, D., Siddique, K. H. M., and Perry, M. W. 1991. 'Crop growth and water use' in Perry, M. W. and Hillman, B. (Eds), *The Wheat Book. A technical Manual for Wheat Procedures*', *Bull. No. 4196*, Western Australia Department of Agriculture.

Troen, I. and Mahrt, L. 1986. 'A simple model of the atmospheric boundary layer: sensitivity to surface evaporation', *Boundary Layer Meteorol.*, **37**, 129–148.

Vulis, I. L. and Cess, R. D. 1989. 'Interpretation of surface and planetary directional albedos for vegetated regions', *J. Climate*, **2**, 986–996.

Watson, I. D. 1980. 'A dynamic climatology of the Australian west coast trough', *Unpublished PhD Thesis*, Department of Geography, University of Western Australia, 245 pp.

Williams, A. 1991. 'Climate change in the southwest of Western Australia', *Unpublished BSc(Hons) Thesis*, Environmental Science, Murdoch University, 116 pp.

24

GLOBAL ATMOSPHERIC WATER BALANCE AND RUNOFF FROM LARGE RIVER BASINS

TAIKAN OKI* AND KATUMI MUSIAKE

Institute of Industrial Science, University of Tokyo, Tokyo, Japan

HIROSHI MATSUYAMA AND KOOITI MASUDA

Department of Geography, Tokyo Metropolitan University, Tokyo, Japan

ABSTRACT

Atmospheric vapour flux convergence is introduced for the estimation of the water balance in a river basin. The global distribution of vapour flux convergence, $-\nabla_H \cdot \vec{Q}$ is estimated using the European Centre for Medium-Range Weather Forecasts global analysis data for the period 1980–1988. From the atmospheric water balance, the annual mean $-\nabla_H \cdot \vec{Q}$ can be interpreted as the precipitation minus evaporation. The estimated $-\nabla_H \cdot \vec{Q}$ is compared with the observed discharge data in the Chao Phraya river basin, Thailand. The mean annual values are not identical, but their seasonal change corresponds very well. The four year mean $-\nabla_H \cdot \vec{Q}$ is also compared with the climatological runoff of nearly 70 large rivers. The multi-annual mean runoff is calculated from the Global Runoff Data Centre data set and used for the comparison. There is generally a good correspondence between the atmospheric water balance estimates and the runoff observations on the ground, especially in the mid- and high latitudes of the northern hemisphere. However, there are significant differences in many instances. The results emphasize the importance of accurate routine observations in both the atmosphere and river runoff. The global water balance of the zonal mean is compared with prior estimates, and the estimated value from this study is found to be smaller than previous estimates. The annual water balance in each ocean and each continent are also compared with previous estimates. Generally, the global runoff estimation using the conventional hydrological water balance is larger than the result by the atmospheric water balance method. Annual freshwater transport is estimated by atmospheric water balance combined with geographical information. The results show that the same order of freshwater is supplied to the ocean from both the atmosphere and the surrounding continents through rivers. The rivers also carry approximately 10% of the global annual freshwater transport in meridional directions as zonal means.

INTRODUCTION

Water balance is the most fundamental aspect of the hydrological cycle. Traditionally, water balance has been estimated using observational data at the ground surface. There are many studies in progress that seek to observe or to estimate evapotranspiration and precipitation over large spatial scales using radar or satellite remote sensing. However, it is still difficult to obtain reliable estimates. On the other hand, water

---

* Correspondence to: Dr Taikan Oki, Institute of Industrial Science, University of Tokyo, 7-22-1 Roppongi, Minato-ku, Tokyo 106, Japan.

balance estimation using atmospheric data, namely the atmospheric water balance method, is becoming easier to apply due to the availability of high-resolution atmospheric data.

Three important research areas that need to be investigated by hydrologists addressing the problems of global change are:

1. a better understanding of the mechanisms associated with global water circulation and balance
2. the development of climate models which can represent the regional-scale water circulation and balance, including precise hydrological surface models at general circulation model (GCM) grid scales
3. the interpretation of the model forecasts for societal benefit

In each of these research areas, the atmospheric water balance can play a significant part: it can help in the estimation of global water circulation and balance, in the validation of the GCM grid-scale hydrological models and in the interpretation of the model forecasts for water resources assessment.

This study attempted to estimate the water balance in both river basins and over the global domain using the atmospheric water balance method. The basic concepts of the atmospheric water balance method are presented in the next section. The atmospheric data and data handling algorithm are then described. The atmospheric water balance method was applied to the Chao Phraya river basin in Thailand. Part of this study was presented at the IUGG/IAHS conference in 1991, in Vienna (Oki *et al.*, 1991b). The estimated atmospheric water balance is validated by river discharge data and by prior estimates. Annual freshwater transport from the continents to the oceans and the meridional means are also estimated. Part of this section was presented at IAMAP/IAHS 1993 in Yokohama (Oki *et al.*, 1993).

ATMOSPHERIC AND RIVER BASIN WATER BALANCE

It is well known in the field of climatology that vapour flux convergence gives water balance information that can complement the traditional hydrological elements such as precipitation, evapotranspiration and discharge (Peixóto and Oort, 1983). The basic concept of using atmospheric data to estimate the terrestrial water balance was first presented by Starr and Peixóto (1958). The application of this concept to regional studies (Rasmusson, 1968) or the region and period with special observations (Peixóto, 1970), has been troublesome because there were only a few scattered observation stations of upper air soundings (Bryan and Oort, 1984). Brubaker *et al.* (1994) interpolated these upper air-sounding data onto a regular grid and analysed the atmospheric water vapour fluxes over North and South American. Rasmusson (1977) suggested that such a method to estimate the regional water balance using vapour flux convergence should be useful and accurate for climatological estimates over areas larger than $10^6 \, \text{km}^2$ and over monthly or longer periods, with the operational rawin sonde network and current observational schedules. Since 1980, 'objective analyses data' have been prepared in the context of prescribing initial values for daily numerical weather forecasting using GCMs and observational data (Daley, 1991). Such data sets are

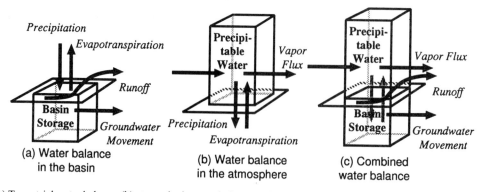

Figure 1. (a) Terrestrial water balance; (b) atmospheric water balance and (c) combined atmosphere–land surface water balance. Parts corresponds to Equations (8), (1) and (10), respectively

now available, and their spatial resolution is more improved than the operational rawinsonde network. Therefore, it is worth applying such atmospheric data to the estimation of water balance in river basins.

Water balance in the atmosphere

The atmospheric water balance is described by the equation

$$\frac{\partial W}{\partial t} + \frac{\partial W_\mathrm{c}}{\partial t} = -\nabla_H \cdot \vec{Q} - \nabla_H \cdot \vec{Q}_\mathrm{c} + (E - P) \tag{1}$$

where W, W_c, $\vec{Q}$, $\vec{Q}_\mathrm{c}$, E and P represent precipitable water (column storage of water vapour), column storage of liquid and solid water, vertically integrated two-dimensional vapour flux, vertically integrated two-dimensional water flux in the liquid and solid phases, evapotranspiration and precipitation, respectively. These terms are shown in Figure 1a. The term $\nabla_H \cdot$ represents the horizontal divergence. $\vec{Q} = (Q_\lambda, Q_\phi)$ is the vapour flux vector and the components are directed towards the east and north

$$Q_\lambda \equiv \int_0^{p_0} qu \frac{\mathrm{d}p}{g} \tag{2}$$

$$Q_\phi \equiv \int_0^{p_0} qv \frac{\mathrm{d}p}{g} \tag{3}$$

$$W \equiv \int_0^{p_0} q \frac{\mathrm{d}p}{g} \tag{4}$$

where W, q, u, v, g, p and p_0 represent the precipitable water, specific humidity, wind velocity E–W and N–S, gravitational acceleration, pressure at the point and pressure at the ground surface. Water vapour flux convergence is computed assuming that the earth is a sphere which has radius of R_e

$$\nabla_H \cdot \vec{Q} = \frac{1}{R_\mathrm{e} \cos \phi} \left(\frac{\partial Q_\lambda}{\partial \lambda} + \frac{\partial Q_\phi \cos \phi}{\partial \phi} \right) \tag{5}$$

Generally, the water content in the atmosphere in the solid and liquid phases is negligible, and Equation (1) is simplified as

$$\frac{\partial W}{\partial t} = -\nabla_H \cdot \vec{Q} + (E - P) \tag{6}$$

Water balance of river basins

The water balance of a river basin is described as

$$\frac{\partial S}{\partial t} = -\nabla_H \cdot \vec{R}_\mathrm{o} - \nabla_H \cdot \vec{R}_\mathrm{u} - (E - P) \tag{7}$$

where S represents the storage in the basin and $\vec{R}_\mathrm{o}$ and $\vec{R}_\mathrm{u}$ represent surface runoff and the ground water movement, respectively. S includes snow accumulation in addition to soil moisture and groundwater storage. If the area of water balance is set within an arbitrary boundary, $\vec{R}_\mathrm{o}$ represents the net outflow from the region of consideration (i.e. the outflow minus inflow from surrounding areas). In this study, all groundwater movement is considered to be observed at the gauging point of the river ($\nabla_H \cdot \vec{R}_\mathrm{u} = 0$) and Equation (7) becomes

$$\frac{\partial S}{\partial t} = -\nabla_H \cdot \vec{R}_\mathrm{o} - (E - P) \tag{8}$$

Combined atmospheric–river basin water balance

With respect to the term of $E - P$, Equations (6) and (8) are combined into

$$-\frac{\partial W}{\partial t} - \nabla_H \cdot \vec{Q} = (P - E) = \frac{\partial S}{\partial t} + \nabla_H \cdot \vec{R}_\mathrm{o} \tag{9}$$

The following further assumptions are used in the annual water balance computations

1. Annual change of atmospheric vapour storage is negligible ($\partial W / \partial t = 0$)
2. Annual change of basin water storage is negligible ($\partial S / \partial t = 0$)

With these assumptions, Equation (9) is simplified as

$$-\nabla_H \cdot \vec{Q} = (P - E) = \nabla_H \cdot \vec{R}_o \qquad (10)$$

In this simplified equation, the water vapour convergence (precipitation – evaporation) and runoff are equal over the annual period. If a river basin is selected as the water balance region, $\nabla_H \cdot \vec{R}_o$ is simply the discharge from the basin.

Estimation of large-scale evapotranspiration

The large-scale mean evapotranspiration can be estimated if the large-scale mean precipitation is available. The equation

$$E = \frac{\partial W}{\partial t} + \nabla_H \cdot \vec{Q} + P \qquad (11)$$

is obtained from Equation (6) without the annual mean assumptions and is then applicable over shorter periods. If atmospheric data and precipitation data are available over short time-scales such as months or days, evapotranspiration can be estimated at the corresponding time-scales. The region over which to estimate evapotranspiration is not limited to a river basin, but depends only on the scale of atmospheric and precipitation data.

The atmospheric water balance method to estimate large-scale evapotranspiration thus complements the traditional river basin water balance, which calculates the annual evapotranspiration as a loss within a river basin.

Estimation of total water storage in river basins

From the Equations (8) and (6), we obtain

$$\frac{\partial S}{\partial t} = -\frac{\partial W}{\partial t} - \nabla_H \cdot \vec{Q} - \nabla_H \cdot \vec{R}_o \qquad (12)$$

which indicates that the change in basin water storage can be estimated with atmospheric and runoff data. Even though an initial value is required to obtain the absolute value of storage, the atmospheric water balance can be useful in estimating the seasonal change in the total water storage in a river basin.

Application to a regional hydrological study

Accurate measurements can be carried out in comparatively small river basins to observe areal evapotranspiration. However, it is practically impossible to measure evapotranspiration over large continental-scale catchments. In such a case, the atmospheric water balance method using atmospheric data and precipitation data [Equation (11)] will offer another way to obtain the areal evapotranspiration.

Total water storage in river basins is one of the important state variables in hydrological processes, which is directly related to the discharge and surface soil moisture. Generally, it is also difficult to measure, except for a small experimental river basin with intensive observations. The atmospheric water balance method utilizing runoff data [Equation (12)] will provide this key information about hydrological processes in large river basins. Further investigation will be made in the near future to analyse the relationship between storage and runoff, and storage and evapotranspiration.

Many papers in this special issue are concerned with methods for estimating the macroscale hydrological processes from observations and by models. The atmospheric water balance method has the capability to provide the macroscale validation data for them. In the case of intensive hydrological field observations, it is desirable to design a rawinsonde observational network which encloses the region of interest. The atmospheric water balance will then estimate the areal mean evapotranspiration and the change in total water storage in the area. The required time intervals for the atmospheric observation depend on the spatial scale. The relationship $v\Delta t \leqslant \Delta x$ may be used to determine the intervals. If the wind velocity v is $10 \, \mathrm{m \, s^{-1}}$ and the horizontal distance Δx is $100 \, \mathrm{km}$, it will be better to make observations at intervals of $\Delta t \approx 3$ hour or less. On the contrary, twice-daily (i.e. $\Delta t = 12$ hour) data can provide a good estimation for horizontal scales of approximately 400–500 km or more.

Application to the global hydrological cycle

Another advantage of the atmospheric water balance method is the availability of data. It is not easy to collect discharge data and cover all the continents by observed runoff data. Even though the spatial density of the observational network varies among regions, the atmospheric data cover the whole world and there are fewer political problems to be faced when handling these data. In this paper, Equation (10) is basically used for the quantitative examination of both the atmospheric and discharge data. It can be also used to estimate the global distribution of discharge.

The meridional distribution of the zonally averaged annual energy transport by the atmosphere and the ocean is well known and has been discussed (Masuda, 1988a). However, a corresponding distribution of water transport has not often been studied, although the cycles of energy and water are closely related. Wijffels *et al.* (1992) used the $-\nabla_H \cdot \vec{Q}$ from Bryan and Oort (1984) and discharge data from Baumgartner and Reichel (1975) and estimated the transport of freshwater by the ocean and atmosphere, but not by rivers.

The annual water transport in the north–south direction over continents $R^L(\phi_0)$ and oceans $R^O(\phi_0)$ can be estimated from the vertically integrated vapour flux convergence $-\nabla_H \cdot \vec{Q}$ with geographical information and river discharge. The governing equations can be written as

$$R^L(\phi_0) = - \int_{-\frac{\pi}{2}}^{\phi_0} \oint_{\text{Land}} R_e^2 \cos\phi \nabla_H \cdot \vec{Q} \mathrm{d}\lambda \mathrm{d}\phi - \int_{-\frac{\pi}{2}}^{\phi_0} D(\phi)\mathrm{d}\phi \tag{13}$$

$$R^O(\phi_0) = - \int_{-\frac{\pi}{2}}^{\phi_0} \oint_{\text{Sea}} R_e^2 \cos\phi \nabla_H \cdot \vec{Q} \mathrm{d}\lambda \mathrm{d}\phi + \int_{-\frac{\pi}{2}}^{\phi_0} D(\phi)\mathrm{d}\phi \tag{14}$$

where $D(\phi)$ represents the total discharge from continents to the oceans at latitude ϕ, and R_e and λ represent the radius of the earth and longitude, respectively. $\oint_{\text{Land}} \mathrm{d}\lambda$ and $\oint_{\text{Sea}} \mathrm{d}\lambda$ mean the zonal integration only over land and sea, respectively.

ESTIMATION OF WATER VAPOUR FLUX CONVERGENCE

The European Centre for Medium Range Weather Forecasts (ECMWF) produces the 'objective analysis' data set, obtained through the 'four-dimensional data assimilation system' (Hoskins, 1989) and used as initial values in weather forecasting. This method is based on dynamically consistent temporal extrapolation using a GCM and statistical spatial interpolation of observed data using the 'optimum interpolation method' (Daley, 1991). Rawinsondes, satellite temperature and moisture, cloud track winds, surface observations by ships, ocean buoys, land stations, aircraft reports, etc. are used as observational values. If there are reliable observations, the first-guess value estimated by the GCM would be almost replaced by these observational data. However, these values are always dominated by GCM forecasts in regions with sparse observation.

The physical parameters of z (geopotential height), u (wind velocity E–W), v (wind velocity N–S), w (vertical pressure velocity), T (temperature) and R_h (relative humidity) are located at each 2·5° grid point; they cover the globe in a 144×73 matrix. There are seven layers at 1000, 850, 700, 500, 300, 200 and 100 hPa heights and values are defined on these levels. Data archives are available after the First GARP Global Experiment (FGGE) in 1979. FGGE was the first opportunity for such a numerical weather forecasting centre to develop the four-dimensional data assimilation system in practice. Twice daily data are used in this study because there are mainly twice daily observations by atmospheric soundings, even though ECMWF have made intermittent six-hour analysis–initialization–forecast cycles.

The algorithm used in this study to estimate the $-\nabla_H \cdot \vec{Q}$ by Equations (2), (3) and (5) from the ECMWF data set is described in the remaining part of this section. Different algorithms may be required for different data sources to implement the atmospheric water balance method.

Surface pressure

If the surface pressure is below 1000 hPa, and sometimes below 850/700 hPa in mountainous regions, these levels are below the ground surface in reality. However, extrapolated values are stored at these pressure levels in the data sets used in this study, and no information is provided for the values at the ground surface. Surface pressure is necessary for the vertical integration, and it requires a computation similar to that used for the reduction to mean sea level.

1. The tentative mean temperature between the lowest level 1000 [hPa] and ground surface is computed, using the temperature ($T_{1000}[°C]$) and pressure height (Z_{1000} [gpm]) at the lowest level of the data. The equation is

$$T_m = T_{1000} + \Gamma(Z_{1000} - Z_h) \tag{15}$$

where Z_h [m] is the altitude at that grid and the adiabatic temperature lapse rate Γ is set to $0.005[°C/m]$

2. Moisture effect is not taken into account in the estimation of T_m, therefore a temperature correction factor ϵ_m is computed as

$$\epsilon_m = \begin{cases} 0.0, & T_m \leqslant -20.0 \\ \frac{3}{55}(T_m + 20.0), & -20.0 \leqslant T_m \leqslant 35.0 \\ 3.0, & 35.0 \leqslant T_m \end{cases} \tag{16}$$

Then the corrected mean temperature T'_m [K] is computed from the equation $T'_m = 273 \cdot 15 + T_m + \epsilon_m$

3. Surface pressure p_s is given using the lowest pressure p_{1000}, gas constant R_d ($= 287 \cdot 05$ [m²/s² K]) and the gravity constant g ($= 9 \cdot 80$ [m/s²])

$$p_s = p_{1000}e^{\frac{g(Z_h - Z_{1000})}{R_d T'_m}} \tag{17}$$

Topography, the ground surface height information, on $2 \cdot 5° \times 2 \cdot 5°$ mesh was obtained from EPOPO5 (Edwards, 1986), the digital elevation map of five minutes spatial resolution.

Values at ground surface

Physical parameters at the ground surface A_s, such as pressure, temperature, etc., are interpolated using values A_k and A_{k+1} at the levels of p_k and p_{k+1} ($p_k \geqslant p_s \geqslant p_{k+1}$). The values $p_k = 1000$ [hPa] and $p_{k+1} = 850$ [hPa] are used for extrapolation in the case of $p_s \geqslant 1000$ [hPa]. Logarithmic values of pressure are used for interpolation (or extrapolation). The equation is

$$A_s = A_k + (A_{k+1} - A_k)\frac{\log p_s - \log p_k}{\log p_{k+1} - \log p_k} \tag{18}$$

Specific humidity

Specific humidity can be computed using the pressure p, temperature T [K] and relative humidity $R_h[\%]$, according to the equation

$$q = \frac{\epsilon \dfrac{R_h}{100} e_s(T)}{p - (1 - \epsilon)\dfrac{R_h}{100} e_s(T)} \tag{19}$$

where $\epsilon (\approx 0 \cdot 622)$ is the ratio of the density between dry air and the water vapour. Saturated vapor pressure $e_s(T)$ [hPa] at temperature T [K] is calculated using Goff–Gratch's equation (Goff and Gratch, 1946)

$$
\begin{aligned}
\log_{10} e_s = {} & 10{\cdot}79574\left(1 - \frac{T_1}{T}\right) - 5{\cdot}02800\log_{10}\left(\frac{T}{T_1}\right) \\
& + 1{\cdot}50475 \times 10^{-4}\left\{1 - 10^{-8{\cdot}2969\left(\frac{T}{T_1}-1\right)}\right\} \\
& + 0{\cdot}42873 \times 10^{-3}\left\{10^{4{\cdot}76955\left(1-\frac{T_1}{T}\right)} - 1\right\} + 0{\cdot}78614
\end{aligned}
\tag{20}
$$

$T_1(= 273{\cdot}16\,[\mathrm{K}])$ is the triple equilibrium temperature of water and the saturated vapour pressure for liquid water surface is also used in the case of $T < 0[^{\circ}\mathrm{C}] = 273{\cdot}15\,[\mathrm{K}]$.

Vertical integration and horizontal convergence

The vertical integration of $W = \int_0^{p_s} q\,dp/g$ is computed as $W = 1/g \sum_{i=1}^{7} q_i \delta p_i$. The thickness of the layer δp_i is given by

$$
\delta p_i = \begin{cases}
0{\cdot}0 & p_i > p_s \\[4pt]
\dfrac{p_i - p_{i+1}}{2} & p_i = p_s \\[6pt]
\dfrac{p_{i-1} - p_{i+1}}{2} & p_i < p_s, \quad 1 < i < 7 \\[6pt]
\dfrac{p_{i-1} - p_i}{2} & i = 7
\end{cases}
\tag{21}
$$

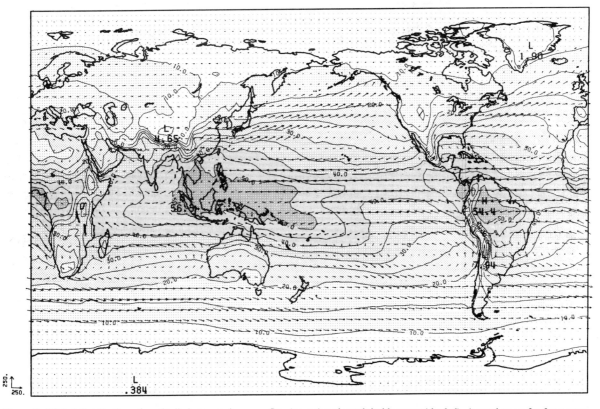

Figure 2. Global distribution of vertically integrated vapour flux (arrow) and precipitable water (shaded). Annual mean for four years from 1985 to 1988

The components of the two-dimensional vapour flux $\vec{Q} = (Q_\lambda, Q_\phi)$, $Q_\lambda \equiv \int_0^{p_s} qu\,dp/g$ and $Q_\phi \equiv \int_0^{p_s} qv\,dp/g$ are similarly computed. In the discrete equation, the convergence $-\nabla_H \cdot \vec{Q}[i, j]$ at latitude of $\lambda[i]$ and longitude of $\phi[j]$ is computed using the centred difference approximation

$$-\nabla_H \cdot \vec{Q}[i, j] = \frac{Q_\lambda[i+1, j] - Q_\lambda[i-1, j]}{R_e \cos(\phi[j])(\lambda[i+1] - \lambda[i-1])}$$
$$+ \frac{\cos(\phi[j+1])Q_\phi[i, j+1] - \cos(\phi[j-1])Q_\phi[i, j-1]}{R_e \cos(\phi[j])(\phi[j+1] - \phi[j-1])} \tag{22}$$

Vapour flux convergence is given with the dimensions of [kg/m² s] and is converted into [mm/s] by assuming the density of water to be 1·0, which simplifies the comparison with annual runoff.

These data processing steps are made for each twice daily data sets, and the monthly and annual mean value are integrated from these twice daily estimates of W, $\vec{Q}$ and $-\nabla_H \cdot \vec{Q}$. The sampling effect on these estimates is discussed by Phillips *et al.* (1992) and the diurnal variation, especially in tropical areas, is very large (Oki and Musiake, 1994). However no detailed discussion is made in this paper of these questions.

Four years mean of vapour flux and its convergence

The vertically integrated atmospheric water vapour flux and the precipitable water in an atmospheric column are shown in Figure 2 for annual mean, Figure 3 for January and Figure 4 for July. The arrows show the direction and the strength of vertically integrated vapour flux, and the shading indicates the precipitable water. For the sake of consistency, these figures are displayed as the four years mean from 1985 to 1988. Precipitable water W is essentially described by the surface temperature and its distribution is more or less symmetrical with respect to the equator; it moves north and south according to the seasonal change in

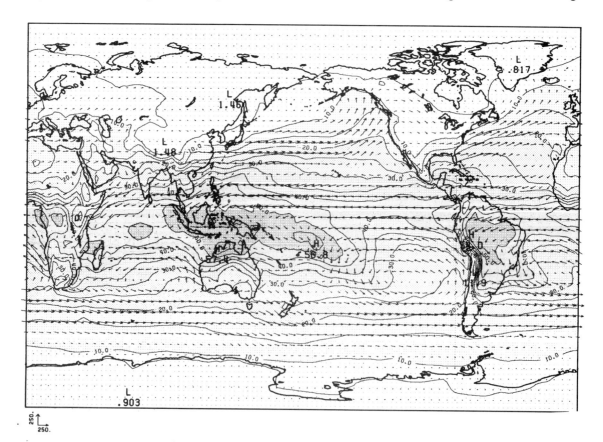

Figure 3. Same as Figure 2, but for monthly mean in January

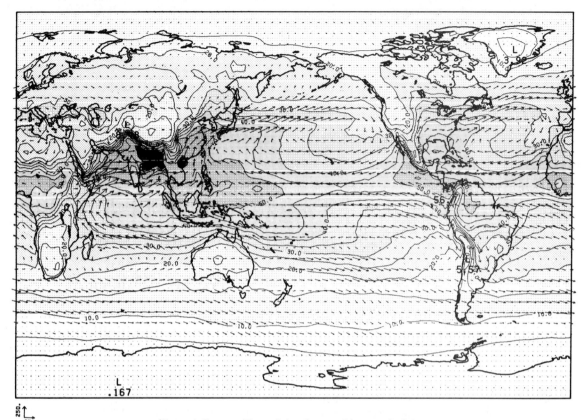

Figure 4. Same as Figure 2, but for monthly mean in July

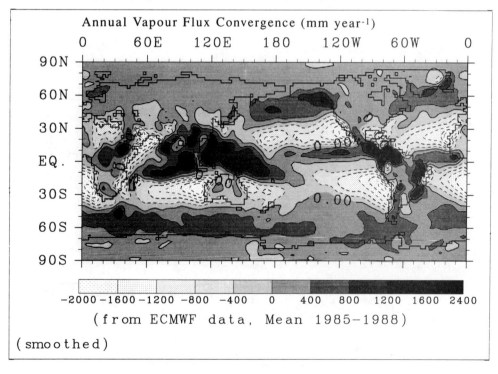

Figure 5. Vertically integrated annual water vapour convergence mean from 1985 to 1988

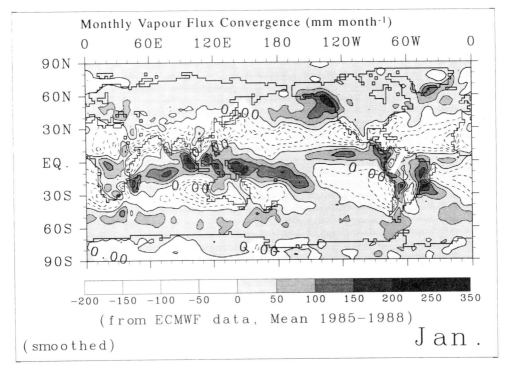

Figure 6. Same as Figure 5, but for January

temperature. The vertical contour lines, which can be seen in the southern Pacific Ocean for a year and in the northern Pacific and Atlantic oceans in July, indicate strong mixing or transport of water vapour in the north–south direction. The difference between the land and ocean is significant, and W is small over continents except for the Amazon River Basin and South-East Asia. The vapour transport occurs mainly over

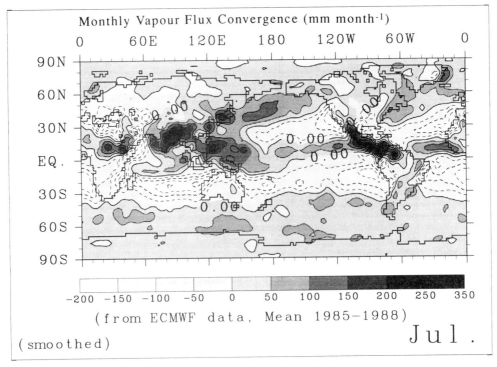

Figure 7. Same as Figure 5, but for July

the oceans, but the Amazon Basin in January, and the Asian monsoon system from Madagascar, the Somalian Peninsula, Arabian Sea, Bay of Bengal to East China Sea in July, carry significant vapour water fluxes. The effect of the monsoon is difficult to see on annual average plots because the summer (July) south-west monsoon and winter (January) north-east monsoon nearly cancel each other out.

Flux convergences are shown from Figures 5 to 7 for the annual mean, January and July, respectively. The mean $-\nabla_H \cdot \vec{Q}$ over the globe, which theoretically should be 0, was -0.07 mm year^{-1}. The negative $-\nabla_H \cdot \vec{Q}$ region indicates that the annual evaporation exceeds the annual precipitation [see Equation (10)]. Such a situation over land may occur in some part of an inland river basin or in the lower part of a large river. Almost every negative estimated $-\nabla_H \cdot \vec{Q}$ value falls in such regions. However, Masuda (1988b) pointed out that the observations at points surrounding arid areas may cause an increase in vapour divergence (negative convergence). The improvement in data analysis techniques may eliminate the negative annual convergence regions over land.

Tropical regions and mid- and high latitudes are convergence zones globally. The tropical convergence zone from the Indian Ocean to the west Pacific Ocean is wider than any other. The high convergence regions generally correspond to the region of high precipitable water; however, the convergence zone at the northern ends of the Pacific and Atlantic oceans do not have high precipitable water. The polar frontal low depression tends to grow in these areas. Subtropical oceans are generally regions of divergence, but parts of them represent negative divergence zones at the so-called South Pacific Convergence Zone, off-shore eastern Australia in January, and in the Asian monsoon convergence zone in the west Pacific Ocean in July.

ATMOSPHERIC AND BASIN WATER BALANCE IN CHAO PHRAYA RIVER

The GEWEX Asian Monsoon Experiment (GAME) has been proposed as a part of the Global Energy and Water Cycle Experiment (GEWEX) under the World Climate Research Programme (WCRP). Process studies are one component in GAME, and four regions are planned to be intensively investigated. They are Siberia (taiga/tundra), the Tibetan Plateau, the subtropical/temperate monsoon region (Yantze, Huai-He and Huang River, China) and the tropical monsoon region. In this regional study of GAME in the tropics the Chao Phraya river basin is selected for the water balance study and macroscale hydrological modelling.

The Chao Phraya River flow through the centre of Thailand and its catchment area is about 178 000 km^2. On a global scale, this size is not large (Figure 8).

Discharge data used here were provided by the Electricity Generating Authority of Thailand. The first data set is inflow to the Bhumibol dam at the Ping River, which is a tributary running through the north-western part of the basin. Another is the inflow to the Sirikit dam at the Nan River, which is also a tributary within the north-eastern portion. Discharge data observed by the Royal Irrigation Department were also used. Station P.1 is located upstream of the Ping River. As an index of the total discharge from the Chao Phraya River, observations at station C.2 were also used, which is located downstream of the confluence of four main tributaries, but before the Chao Phraya River divides into several channels running through its delta. The catchment area at each station point is given in Table I. Rainfall data were observed by the Thai Meteorological Department, and the arithmetic means of the data at each station in the catchment area were used.

Estimated annual atmospheric water vapour flux convergence (i.e. the arithmetic mean of surrounding 5×5 grid points, at $2.5°$ resolution of 95–105°E and 12.5–22.5°N), annual discharge and annual

Table I. Catchment area at each observation station

	Bhumibol dam	Sirikit dam	P.1	C.2
Catchment Area (km^2)	26 396	15 718	6355	110 569
Mean runoff (mm year^{-1})	201	335	210	212

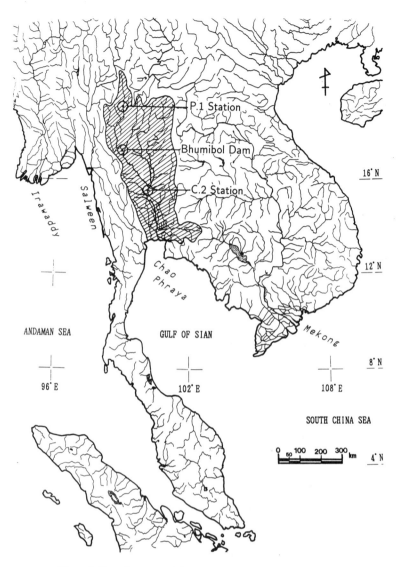

Figure 8. Location map of Chao Phraya River Basin, Thailand

precipitation are shown in Figure 9. The annual change in water storage both in the catchment ($\partial S/\partial t$) and in the atmosphere ($\partial W/\partial t$) are comparatively negligible. Therefore the annual vapour flux convergence should match the annual discharge from Equation (10). However, the quantitative agreement is not perfect. One main reason is that this basin is surrounded by mountains except in the southern direction. The basin is shielded from both the south-west and north-east monsoons and has less water vapour convergence than the surrounding regions, which are wind-ward of the monsoons. Regions on the wind-ward side are known to have more rainfall (Oki *et al.*, 1991a), but evapotranspiration may differ less. Consequently, the actual convergence value in the basin should be much less than the value estimated here, which is the mean of 25 grid points, $10 \times 10°$ degree area. The annual precipitation map clearly indicates such a distribution of moisture convergence.

An apparent gap can be seen in $-\nabla_H \cdot \vec{Q}$ between the years of 1984 and 1985 (Figure 9). There were some significant changes in the data analysis system/algorithm of ECMWF during this period, especially regarding the GCM spatial resolution, treatment of cloud and water vapour parameterizations (Hoskins, 1989).

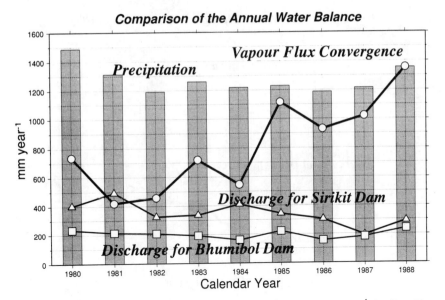

Figure 9. Estimated annual vapour flux convergence, precipitation and discharge (mm year^{-1}) in Chao Phraya River Basin

Because the products of later years are expected to be improved, the data after 1985 are used in this study.

Even though the absolute value does not give good results, the temporal variation is similar between vapour flux convergence and observed discharge. To clarify this point, a reduction factor ξ was applied to the mean $-\nabla_H \cdot \vec{Q}$ of every month during 1985–1988. A value of $\xi = 0.18$ was chosen such that the four-year total $-\nabla_H \cdot \vec{Q}$ will agree with the corresponding observational discharge at Bhumibol Dam. Rasmusson (1968) found a net increase in storage of 210 mm within five years over the continental USA, and he corrected $\nabla_H \cdot \vec{Q}$ by the uniform addition of 3·5 mm month^{-1}. The development of a coherent correction method is beyond the scope of this paper. In any case, this rough correction should be replaced in the future when more temporally and spatially resolved data sets could be obtained. Corrected monthly

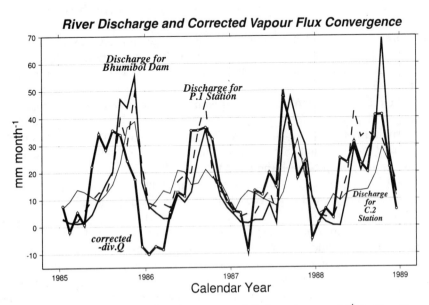

Figure 10. Corrected monthly vapour flux convergence and discharge (mm month^{-1}) in Chao Phraya River

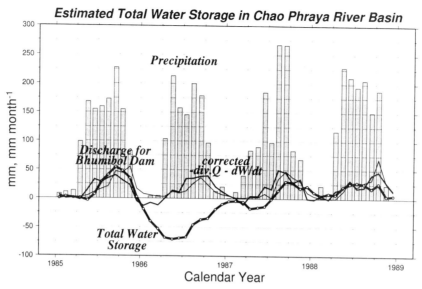

Figure 11. Relative basin water storage (mm) in Chao Phraya River estimated by the atmospheric water balance method using corrected vapour flux convergence (mm month^{-1}) and discharge (mm month^{-1}) for Bhumibol Dam

net water vapour convergence $[(-\partial W/\partial t) - \xi\nabla_H \cdot \vec{Q}]$ is calculated, and a comparison with the observed monthly runoff is shown in Figure 10. The time variations agree very well, and it is concluded that inclusion of a correction factor provides a better estimation for the case under study.

The temporal change of storage S is neglected in Equation (10) for the annual water balance, but it cannot be neglected for the monthly water budget and the storage may make the discharge appear delayed compared with $-\nabla_H \cdot \vec{Q}$. Approximately one month of delay can be seen in Figure 10 for this case.

By applying the atmospheric water balance method, the storage S can be estimated using observed discharge data [Equation (12)] and evapotranspiration can be estimated using observed precipitation data [Equation (11)]. Figure 11 shows the change in total water storage in the Chao Phraya River basin. Discharge and precipitation are from surface observations. Corrected $\nabla_H \cdot \vec{Q}$ is used and the initial storage is set to zero. In 1986, the precipitation was similar to the other years, but the basin storage S was estimated to be less than in other years. The discharge was also less than in other years and it certainly reflects the shortage of basin storage S. On the other hand, the basin storage and river discharge were comparatively high in 1985. Note that the seasonal change is much larger than the intra-annual variation. The storage value recovers close to zero at the beginning of each year. It supports the assumption of $(\partial S/\partial t) \approx 0$ for the annual water balance.

Figure 12 shows the estimated evapotranspiration values by atmospheric water balance, Penman equation and model output of the ECMWF-GCM given in the TOGA-COARE CD-ROM (Oki and Sumi, 1993). The original spatial resolution of the ECMWF data in TOGA-COARE CD-ROM was 2·5°, and it is integrated over the $10 \times 10°$ region corresponding to the estimated $-\nabla_H \cdot \vec{Q}$. Penman evapotranspiration has its maximum in March and April; the ECMWF model evapotranspiration has its peak in August; and the evapotranspiration from the atmospheric water balance has a similar seasonal change to precipitation. Penman evapotranspiration would only represent a potential value and it is not realistic because there should be limits to the water availability for evapotranspiration at the end of the dry season in March to April. The ECMWF model output has a small seasonal variation compared with the atmospheric water balance estimation; the reason could be that $10 \times 10°$ region contains the sea surface evaporation. Atmospheric water balance estimates, on the other hand, show negative evapotranspiration at the end of 1986; the true value for evapotranspiration is expected to be between the ECMWF model output and the atmospheric water balance estimation. For an improved understanding of the large-scale patterns of evapotranspiration, intensive observations are inevitably required, especially in the dry season. Therefore, the

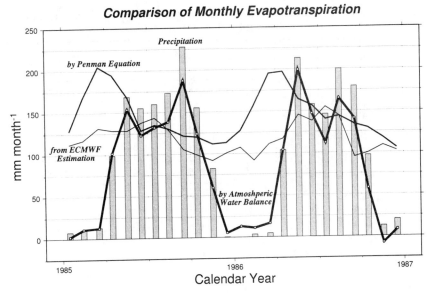

Figure 12. Large-scale monthly evapotranspiration (mm month^{-1}) in Chao Phraya River. Comparison among estimations by the atmospheric water balance method using corrected vapour flux convergence and precipitation, by ECMWF model forecast and by the Penman equation

main issues in the intensive field observation within the Chao Phraya River Basin under GAME are

1. to observe regional evapotranspiration more accurately by point measurement of the latent and sensible heat flux at the ground surface, and by the atmospheric water balance method with enforced rawinsonde networks
2. to evaluate the dry-up effect for the suppression of evapotranspiration compared with the potential evapotranspiration

Even though the quantitative corresponding between $-\nabla_H \cdot \vec{Q}$ and discharge is not perfect, the corrected $-\nabla_H \cdot \vec{Q}$ value seems adequate for understanding the basic hydrological cycle in the Chao Phraya River basin. The atmospheric water balance method gives valuable information about the water balance within a river basin, such as the large-scale monthly evaporation and the seasonal change of basin storage. This information is especially helpful in regions of sparse observation in the tropics (Matsuyama, 1992; Matsuyama *et al.*, 1994).

APPLICATION TO GLOBAL RUNOFF ESTIMATION

Comparison of vapour flux convergence and river runoff

The annual vapour flux convergence has been summarized in each of 70 river basins and compared with mean annual runoff (Table II). River runoff observed at gauging stations is from the Global Runoff Data Centre (GRDC, 1992) and also from Unesco (1969). Long-term mean values are calculated from these data by averaging over the available data period in each station. Seventy rivers which have basin areas greater than approximately 100 000 km^2 were selected. The continental land mass was manually divided into river basins using published world maps in 2·5° grids. In Table II, the atmospheric vapour convergence to the Chao Phraya River basin is estimated from only three grids instead of the surrounding 25 grids.

Normalizing for basin area, the mean $-\nabla_H \cdot \vec{Q}$ in these 70 rivers is approximately 220 mm year^{-1}. The size of the river basin is considered in the averaging procedure. This value is larger than that for whole continents (i.e. 165 mm year^{-1} in Table IV), but it is smaller than the mean observed runoff of these specific rivers, which is 314 [mm/year] according to GRDC and 365 [mm/year] according to Unesco. These 70 river

Table II. Four year mean annual discharge and vapour flux convergence in 70 large river basins in the world. D_r is the climatological mean discharge (mm year^{-1}) at the station shown in the table

River	Station	Years	D_r	$-\nabla_H \cdot \vec{Q}$	$-\nabla_H \cdot \vec{Q} - D_r$	$\dfrac{-\nabla_H \cdot \vec{Q} - D_r}{D_r}$
Amazon	Obidos	56	1053	153	−900	−0·855
Zaire	Kinshasa	81	365	245	−119	−0·328
Missouri	Hermann, MO	92	53	109	55	1·039
Blue Nile	Khartoum	71	149	254	105	0·704
Ob	Salekhard	55	134	64	−69	−0·520
Yenisei	Igarka	49	222	243	20	0·093
Lena	Kusur	50	215	204	−11	−0·053
Parana	Corrientes	80	267	460	192	−0·719
Niger	Malanville	29	35	−27	−62	−1·761
Amur	Komsomolsk	52	178	157	−20	−0·117
Changjiang	Datong	4	464	708	243	0·525
Mackenzie	Norman Wells	19	167	234	66	0·395
Volga	Volgograd Power Plant	106	183	172	−11	−0·064
Zambeze	Matundo-Cais	4	112	−254	−366	−3·267
St Lawrence	Cornwall (Ontario)	12	323	290	−33	−0·102
Ganges	Paksey	7	428	−56	−485	−1·132
Murray	Lock 9 Upper	20	8	−87	−95	−11·659
Nelson	Above Bladder Rapids	19	75	209	133	1·760
Oranje	Vioolsdrif	23	5	313	308	57·380
Indus	Kotri	7	85	−53	−139	−1·629
Orinoco	Puente Angostura	67	1176	1032	−144	−0·123
Yukon	Carmacks	19	35	593	558	15·926
Tocantins	Conceicao do Araguaia	4	604	1087	482	0·799
Chari	Ndjamena (Fort Lamy)	43	63	135	71	1·129
Danube	Ceatal Izmail	64	252	42	−210	−0·832
Mekong	Mukdahan	64	651	714	63	0·097
Tegwani	Tegwani Weir			−2		
Euphrates	Kadaheyeh	4	268	35	−233	−0·868
Huanghe	Sanmenxia	4	55	−143	−199	−3·587
Sao Francisco	Traipu	3	133	207	73	0·552
Brahmaputra	Pandu	9	1529	406	−1122	−0·734
Columbia	The Dalles, Oreg.	111	264	132	−132	−0·500
Syr-Darya	Tyumen-Aryk	55	78	−113	−191	−2·453
Kolyma	Sredne-Kolymsk	58	192	253	61	0·319
Colorado	Limite Norte	4	3	−41	−45	−12·429
Bravo	Matamoros	4	4	336	331	68·144
Dniepr	Dniepr Power Plant	33	100	80	−20	−0·199
Amu-Darya	Chatly	43	96	−183	−280	−2·899
Senegal	Bakel	80	103	−528	−632	−6·093
Limpopo	Chokwe	4	33	−845	−878	−26·224
Xijiang	Wuzhou 3	8	680	493	−186	−0·275
Irrawaddy	Sagaing	11	2187	390	−1796	−0·822
Don	Razdorskaya	94	66	199	133	2·021
Volta	Senchi (Halcrow)	44	88	−509	−597	−6·737
Northern Dvina	Ust-Pinega	103	298	290	−7	−0·025
Indigirka	Vorontsovo	48	163	141	−21	−0·134
Pechora	Ust-Tsilma	53	434	234	−199	−0·459
Godavari	Polavaram	79	322	249	−72	−0·226
Parnaiba	Porto Formoso	6	93	1956	1862	19·863
Neva	Novosaratovka	126	318	123	−195	−0·613
Magdalena	Calamar	9	863	4202	3339	3·869
Krishna	Vijayawada	79	207	970	762	3·669
Churchill	Above Granville Falls	19	119	260	141	1·191
Rhein	Rees	49	451	−192	−644	−1·426
Ural	Kushum	70	49	−77	−126	−2·561

Table II. Continued

River	Station	Years	D_r	$-\nabla_H \cdot \vec{Q}$	$-\nabla_H \cdot \vec{Q} - D_r$	$\dfrac{-\nabla_H \cdot \vec{Q} - D_r}{D_r}$
Fraser	Hope	73	396	571	174	0·440
Yana	Dzanghky			111		
Olenek	River Pur	20	158	199	41	0·259
Wisla	Tczew	88	171	466	294	1·717
Kura	Surra	55	97	322	225	2·309
Fitzroy	The Gap	12	54	−1495	−1549	−28·391
Burdekin	Clare	20	85	−547	−633	−7·366
Santiago	El Capomal	17	71	1374	1303	18·225
Elbe	Neu-Darchau	4	193	340	147	0·761
Huaihe	Bengbu	4	111	−236	−347	−3·128
Albany	Near Hat Island	19	252	312	60	0·237
Brazos	Richmond, Tex.	20	53	−312	−366	−6·793
Odra	Gozdowice	88	154	−3	−157	−1·022
Loire	Montjean	117	234	152	−82	−0·350
Chao Phraya	Nakhon Sawan	10	197	1370	1173	5·950

Vapour flux $-\nabla_H \cdot \vec{Q}$ is calculated from ECMWF objective analysis data 1985–1988.
Climatological mean discharge D_r is calculated from the data from the Global Runoff Data Center.

basins cover only 55% of the land surface of the earth, yet they hold 75% of the total $-\nabla_H \cdot \vec{Q}$ within the land and their mean runoff is larger than the global average, indicating that the areas where large rivers exist have comparatively high runoff rates. It suggests that the global water balance estimated by the extrapolation of the large river water balance may overestimate the true global value for the continents.

Comparison with observed river runoff is shown in Figure 13 based on the absolute difference between the two data sets. The tendency is not clear, but larger river basins have smaller differences between $-\nabla_H \cdot \vec{Q}$ and observed runoff. The $-\nabla_H \cdot \vec{Q}$ tends to be smaller than the corresponding river runoff (Table II) and it may be caused by the weak divergence of the wind in the ECMWF objective analysis data set (Masuda, 1988b).

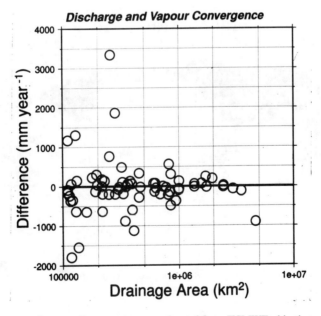

Figure 13. Difference between annual vapour flux convergence estimated from ECMWF objective analysis data mean from 1985 to 1988 and climatological annual runoff observed at gauging stations in large river basins

If the estimation error is evaluated by the ratio of predicted to observed values (as seen in the last column of Table II), the ratio of the error (expected to be zero) is within the range ± 1.0 in 34 rivers among the 68 rivers of comparison. Owing to the lack of discharge data, no comparison was made for Tegwani and Yana rivers. The major difficulties confronted in the comparison of atmospheric vapour convergence and river runoff here are:

1. Flux convergence is computed for whole basin in the model drainage basin, but runoff observation stations are not always representative of the whole basin runoff
2. There are obvious errors in a $2.5°$ template that divides the continents into river basins. In the case of small rivers, it is hard to define a river basins accurately using a $2.5°$ grid. It is also difficult to define basin boundaries in steep mountainous regions, such as the Andes of the Amazon or the Himalayas of the Ganges and Brahmaputra. $-\nabla_H \cdot \vec{Q}$ itself seems to be erroneous in these mountainous areas because of the representation of topography in the data assimilation system
3. The observation periods for atmospheric data and runoff are not congruent in time
4. A period of four years is not long enough for the assumption of $\partial S/\partial t = 0$. Inter-annual variation of the soil moisture storage will cause this difference
5. The quality of the runoff observations at each river gauging station is not known, and may vary over stations

Even though their data sources and time–space scales are completely different, the atmospheric vapour flux convergence and river runoff correspond well in many instances. These difficulties will be overcome by using the forthcoming 're-analysis data' (see later for explanation), by collecting the discharge data in the corresponding period and by using the higher resolution geographical information of basin boundaries with the atmospheric data of higher resolution.

The observed runoff may be influenced by human activities, such as the intake for an irrigation scheme or the storage in a reservoir. However, such an effect should appear in a similar manner in the results of the atmospheric water balance method. Therefore, anthropogenic effects will not be the cause of any error. In the case of hydrological simulations which do not incorporate human activities, numerical models may predict a different result from that of the atmospheric water balance method as well as the observed runoff.

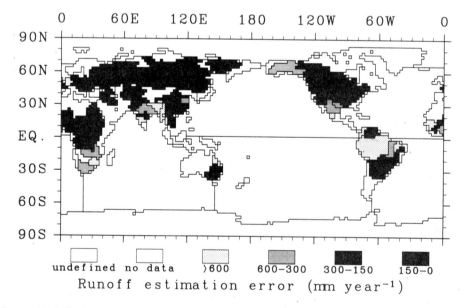

Figure 14. Geographical distribution of the absolute difference (mm year^{-1}) between annual mean vapour flux convergence and observed river runoff

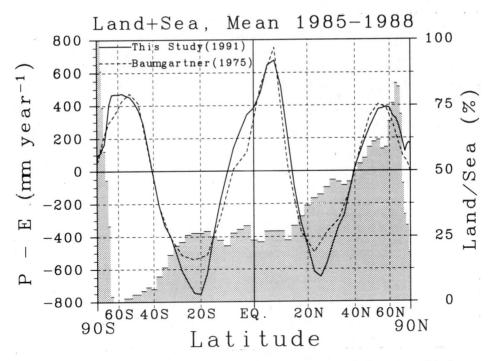

Figure 15. Zonal mean vapour flux convergence compared with latitudinal distribution of discharge (= precipitation − evaporation = vapour flux convergence) in annual mean basis. Bar indicates the fractional percentage of land in each latitude

The geographical distribution of the absolute differences between the predicted mean convergence and observed runoff are shown in Figure 14. The poor correspondence occurs in South Africa, central and monsoon Asia and the South America. These areas have very few observational points or a large amount of precipitation. On the contrary, there are many basins that show good correspondence (the absolute difference is less than 150 mm year^{-1}) in the mid- and high-latitude regions of the northern hemisphere. The error index ratio from Table II also has a similar geographical distribution. This distribution may reflect the high quality and high density of the observations, especially with respect to the atmosphere, which emphasizes the importance of actual observations in performing accurate data assimilations.

Latitudinal distribution of global runoff

Figure 15 shows the zonal (E–W direction) mean of vapour flux convergence and a comparison with previous results (Baumgartner and Reichel, 1975) of the latitudinal distribution of precipitation minus evaporation. The two methods generally show good agreement quantitatively. However, the results of this study show that the evaporation excess over precipitation is greater than the result obtained by Baumgartner and Reichel (1975) at subtropical latitudes.

Table III. Global water balance (precipitation − evaporation) estimations over oceans (mm year^{-1})

All oceans	Arctic	Indian	Pacific	Atlantic	Reference
−111	50	−250	91	−384	Baumgartner and Reichel (1975)
−132	263	−97	−56	−333	Korzun (1974)
−18	(Atlantic)	−53	20	−136	Bryan and Oort (1984)
−66	163	−113	12	−190	Masuda (1988b, FGGE-ECMWF)
−114	175	−147	14	−345	Masuda (1988b, FGGE-GFDL)
−78	185	−126	6·3	−236	This study

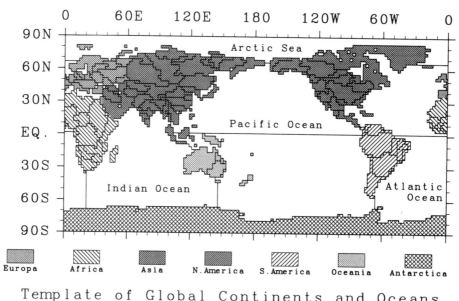

Template of Global Continents and Oceans

Figure 16. Land–sea template of 2·5° mesh, used in the regional integrations

Global annual runoff from whole continents

Global annual runoff from whole continents is calculated using the land-sea distribution (Figure 16). Tables III and Table IV compare the results with the available estimates. Baumgartner and Reichel (1975), Lvovitch (1973) and Korzun (1978) estimated runoff by hydrological methods using surface observations. Bryan and Oort (1984), Masuda (1988b) and this study used the atmospheric water balance method. Note that the Middle East is included in Europe, not in Asia, and the Arctic Ocean is included in the Atlantic Ocean in the Bryan and Oort (1984) estimates.

The results of the present study are comparable with those of FGGE-ECMWF (Masuda, 1988b), but appear to be relatively small compared with the results of basin water balance analyses and with the values presented by Bryan and Oort (1984). According to Korzun (1978), land cover on the earth is approximately 29%, but 31% was assumed in the topographical model used in this study and this may underestimate the mean convergence over land.

The signs of $-\nabla_H \cdot \vec{Q}$ in Table III are almost the same among the estimates, but the absolute values vary. In the Pacific Ocean, precipitation and evaporation approximately balance each other; evaporation exceeds precipitation in the Atlantic and the Indian oceans; and vice versa in the Arctic Ocean.

Table IV. Global water balance (precipitation – evaporation) estimations over continents (mm year^{-1}). Estimated for all continents (Asia, Europe, Africa, North America, South America, Australia and Antarctica)

All lands	Asia	Europe	Africa	North America	South America	Australia	Antarctica	Reference
256	260	255	113	223	611	267	143	Baumgartner and Reichel (1975)
269	281	273	140	258	578	222	157	Lovovitch (1973)
303	300	273	153	315	678	278	164	Korzun (1974)
42	32	−181	7	162	333	−400	43	Bryan and Oort (1984)
152	94	164	63	227	442	56	100	Masuda (1988b, FGGE-ECMWF)
260	100	91	333	223	850	211	107	Masuda (1988b, FGGE-GFDL)
165	235	136	−100	263	415	54	112	This study

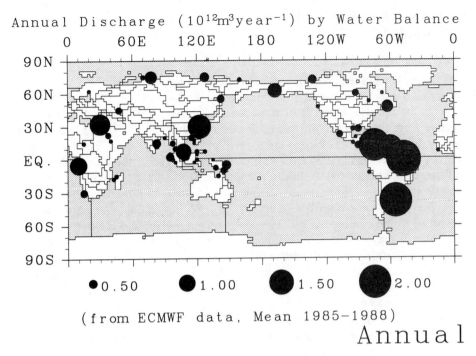

Figure 17. Mean annual discharge from selected rivers (10^{12} m^3 year^{-1})

In Table IV, the results from the hydrological method produce similar estimates, and the results from the atmospheric water balance method generate smaller values than the hydrological estimates, except for Africa and South America [i.e. Masuda's (1988b) estimate using GFDL-FGGE data]. Generally, the annual river runoff is approximately 200–300 mm year^{-1}, except for South America, where the estimated annual runoff is from 400 to 800 mm year^{-1}. Continents with high aridity (Africa) and with very cold regions (Antarctica) have a small runoff. Although the results seem to have smaller runoff values than previous estimates, the atmospheric water balance method can cover the whole globe simultaneously. Owing to the consistency of the data quality, it is difficult to see the inter-annual variation of global water balances using objective analysis data sets. However, the 're-analysis' project is now under way in ECMWF and also in Numerical Weather Center of the USA, which applies the four-dimensional assimilation again to the historical records (i.e. from 1980 to 1994 in the case of ECMWF) by fixing the latest algorithm

Table V. Annual freshwater transport from continents to each ocean (10^{15} kg year^{-1}). 'Inner' indicates the runoff to the inner basin in Asia and Africa. $-\nabla_H \cdot \vec{Q}$ indicates the direct freshwater supply from the atmosphere to the ocean

	North Pacific	South Pacific	North Atlantic	South Atlantic	Indian	Arctic	Inner	Total
Asia	5·2	0·8	0·2		2·5	2·4	−0·1	11·0
Europe			1·1			0·5		1·6
Africa			−3·0	1·3	−1·7		0·3	−3·1
North America	2·7		3·8			0·8		7·3
South America	0·1	−0·1	3·3	4·7				8·0
Australia		1·6			−1·1			0·6
Antarctica		0·8		0·2	0·6			1·6
Total	8·0	3·2	5·4	6·1	0·3	3·7	0·2	26·9
$-\nabla_H \cdot \vec{Q}$	7·6	−6·4	−12·3	−8·9	−8·9	2·0		−26·9
Grand total	15·6	−3·2	−6·9	−2·8	−8·6	5·7	0·2	−0·0

consistently. The project will enable us to utilize the atmospheric water balance method for relatively long-term hydrological studies.

Global annual freshwater transportation

The freshwater supply to the ocean has an important effect on the thermohaline circulations of the ocean because it changes the salinity. In this study, the zonal mean discharge $D(\phi)$ is estimated from $-\nabla_H \cdot \vec{Q}$ by the atmospheric water balance [Equation (10)] at the location of river outlet of each basin. Figure 17 shows the annual discharge from the selected 70 river basins estimated by $-\nabla_H \cdot \vec{Q}$. The size of each circle indicates the volume of discharge and the circles are located at the river mouths. The locations of the outlet from the meshes not included in these 70 river basins were also given at the nearest coast.

From this information, the annual freshwater transport from the continents to each ocean can be estimated. Table IV summarizes these results. Some part of water vapour flux convergence remains in the inland basin. There are some negative values in Table V, suggesting that net freshwater transport occurs from the ocean to the continents. It is physically unacceptable, and the detailed discussion of the values shown in Table V may not be meaningful. However, an important point is that such an analysis can be made at least qualitatively using the atmospheric water balance method with geographical information on basin boundaries and the location of river mouths. In so doing it should be noted that the total amount of freshwater transport from the surrounding continents has the same order of magnitude as the freshwater supply that comes directly from the atmosphere expressed by $-\nabla_H \cdot \vec{Q}$.

From Equations (13) and (14), the annual water transport in the latitudinal direction was estimated (Figure 18). Transportation by the atmosphere and by the ocean have almost the same absolute value in each latitude, but with different sign, and the water transport by rivers is about 10% of these other fluxes globally. This number may be underestimated because $-\nabla_H \cdot \vec{Q}$ tends to be smaller than the river discharge observed at the ground surface. The peak at 30°S is mainly carried by the Parana River in South America, and the peak around 10°N is also carried by rivers in South America (the Magdalena, Orinoco, etc.); Russian large rivers (such as the Ob, Yenisey and Lena) carry freshwater towards the north and these transports can be seen around 70°N in Figure 18.

These results indicate that rivers play a significant part in the climate system, not only by the exchange of energy and water within the basin, but also through the freshwater transport by runoff, affecting the hydrological circulation among the atmosphere, oceans and continents.

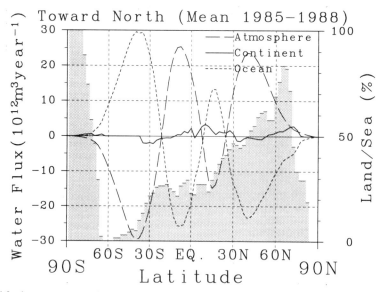

Figure 18. The annual freshwater transport in north–south direction by atmosphere, ocean and rivers (continent). Vapour flux transport of 20×10^{12} m³ year^{-1} corresponds to approximately $1 \cdot 6 \times 10^{15}$ W of latent heat transport

SUMMARY

1. The atmospheric water balance method was introduced, together with an algorithm to compute the vapour flux convergence, $-\nabla_H \cdot \vec{Q}$, using global atmospheric data. Annual runoff can be directly estimated from vertically integrated water vapour flux convergence. Evapotranspiration and the change in total water storage in a drainage basin can be estimated with additional precipitation and discharge data, respectively.

2. The global distribution of vertically integrated vapour flux and its convergence were estimated from 1980 to 1988 using ECMWF four-dimensional assimilation data. For data consistency, the four year mean was calculated using data from 1985 to 1988, and the general characteristics of these estimates were described.

3. The atmospheric water balance method was applied to the Chao Phraya river basin, Thailand, where an intensive regional study is planned under GEWEX/GAME. For a number of reasons, such as the difference in spatial scale and data quality, the values of $-\nabla_H \cdot \vec{Q}$ and river runoff observed at the ground are not the same, but their seasonal change corresponds very well. A correction factor ξ was introduced for $-\nabla_H \cdot \vec{Q}$ to accommodate the observed water balance, and the seasonal change in basin storage and evapotranspiration were estimated by the atmospheric water balance method. The basin storage relates well to the observed discharge and the estimation was found to represent the water balance characteristic of the basin.

4. A comparison between the four year mean $-\nabla_H \cdot \vec{Q}$ and the climatological mean of observed river runoff was made in nearly 70 large river basins. They showed good agreement, especially in mid- and high-latitudes of the northern hemisphere, reflecting the dense and reliable observational network of the atmosphere and runoff in these regions.

5. The estimated $-\nabla_H \cdot \vec{Q}$ was examined globally. Global annual runoff over the earth is estimated to be 165 mm in this study. This value is less than the results from traditional basin water balance method. With the exception of the subtropics, the latitudinal distribution of global runoff corresponds well with previous studies. Individual water balances of oceans and continents were summarized and compared with previous estimates.

6. The mean annual runoff estimated for 70 large river basins in the globe by the atmospheric water balance method was 220 mm. Even though these 70 river basins cover only 55% of the land on earth, approximately 75% of the total $-\nabla_H \cdot \vec{Q}$ for the land is concentrated in these large river basins. From the surface observational data, a similar result can be derived. These results indicate that the areas where large rivers exist have comparatively high runoff. It suggests that the global water balance estimated by the extrapolation of the water balance in large rivers may overestimate the true value.

7. Annual freshwater transport was estimated using the additional geographical information, such as the basin boundaries associated with the location of the river mouth, and using the atmospheric water balance method. The same order of freshwater is supplied to the ocean from the atmosphere and from the surrounding continents through rivers. The rivers carry approximately 10% of the global annual freshwater transport compared with the fluxes by atmosphere and ocean as a zonal mean.

8. To our knowledge, this is the first study to use four-dimensional assimilation global data in estimating the world water balance with validation in near 70 large river basins. Even though atmospheric water balance calculations sometimes give physically unacceptable results, the improvement of data assimilation techniques should make this method a powerful tool for quantifying the global energy and water cycle and for developing and verifying macroscale hydrological models.

ACKNOWLEDGEMENTS

The authors thank the European Centre for Medium-Range Weather Forecasts, the Global Runoff Data Center, the Electricity Generating Authority of Thailand, the Royal Irrigation Department of Thailand, the Thai Meteorological Department and the US National Geophysical Data Center for providing data sets.

The ECMWF data set used here is from the archive in the Center for Climate System Research, University of Tokyo. Special thanks go to Dr M. Sivapalan and anonymous reviewers for their encouragement and helpful advice on the manuscript. A part of this work was supported by the Grant-in-Aid for Scientific Research 'Hydrological characteristics and water management of a river basin in a tropical monsoon region' (representative Professor H. Shiigai, Tsukuba University) of the Ministry of Education, Science and Culture in Japan. Figures were drawn by the Dennou Club Libraries and GMT system.

REFERENCES

Baumgartner, F. and Reichel, E. 1975. *The World Water Balance: Mean Annual Global, Continental and Maritime Precipitation, Evaporation and Runoff*. Ordenbourg, Munich. 179pp.

Brubaker, K. L., Entekhabi, D., and Eagleson, P. S. 1994. 'Atmospheric water vapor transport and continental hydrology over the Americas', *J. Hydrol.*, **155**, 407–428.

Bryan, F. and Oort, A. 1984. 'Seasonal variation of the global water balance based on aerological data', *J. Geophys. Res.*, **89**, 11 717–11 730.

Daley, R. 1991. *Atmospheric Data Analysis*. Cambridge University Press, Cambridge. 457pp.

Edwards, M. H. 1986. 'Digital image processing of local and global bathymetric data', *Master's Thesis*, Washington University, Department of Earth and Planetary Sciences, St. Louis, Missouri, 106pp.

Goff, J. A. and Gratch, S. 1946. 'Low-pressure properties of water — from 160 to 212°F', *Trans. Am. Heat. Vent. Eng.*, **52**, 95–121.

GRDC 1992. *Second Workshop on the Global Runoff Data Centre. Report 1*, Bundesanstalt für Gewässerkunde, Federal Institute of Hydrology, 96pp.

Hoskins, B. J. 1989. 'Diagnostics of the global atmospheric circulation based on ECMWF analyses 1979–1989', *Tech. Rep. WCRP-27, WMO/TD-No.326*, World Meteorological Organization.

Korzun, V. I. (ed.) 1978. *World Water Balance and Water Resources of the Earth. Vol. 25. Studies and Reports in Hydrology*. Unesco, Paris.

Lvovitch, M. I. 1973. 'The global water balance', *Trans. Am. Geophys. Union*, **54**, 28–42.

Masuda, K. 1988a. 'Meridional heat transport by the atmosphere and the ocean; analysis of FGGE data', *Tellus*, **40A**, 285–302.

Masuda, K. 1988b. 'World water balance; analysis of FGGE IIIb data', in Theon, J. S. and N. Fugono (Eds), *Tropical Rainfall Measurements*. A. Deepak, Hampton Virginia, USA. pp. 51–55.

Matsuyama, H. 1992. 'The water budget in the Amazon river basin during the FGGE period', *J. Meteorol. Soc. Jpn*, **70**, 1071–1084.

Matsuyama, H., Oki, T., Shinoda, M., and Masuda, K. 1994. 'The seasonal change of the water budget in the Congo river basin', *J. Meteorol. Soc. Jpn*, **72**, 137–155.

Oki, T. and Musiake, K. 'Seasonal change of the diurnal cycle of precipitation over Japan and Malaysia', *J. Appl. Meteorol.*, Vol. 33 No. 12, 1994 1445–1463.

Oki, R. and Sumi, A. 1993. 'Evaluation of ECMWF latent heat flux in TOGA' in Jones, I. S. F. (Ed.), *Satellite Remote Sensing of the Oceanic Environment*. Seibutsu Kenkyusha, Tokyo. pp. 249–258.

Oki, T., Musiake, K., and Koike, T. 1991a. 'Spatial rainfall distribution at a storm event in mountainous regions estimated by orography and wind direction', *Wat. Resour. Res.*, **27**, 359–369.

Oki, T., Musiake, K., and Shiigai, H. 1991b. 'Water balance using atmospheric data — a case study of Chao Phraya river basin, Thailand' in *Mitteilungsblatt des Hydrographischen Dienstes in Österreich*. Hydographischen Zentralbüro, Vol. 65/66. Vienna. pp. 226–230.

Oki, T., Musiake, K., Masuda, K., and Matsuyama, H. 1993. 'Global runoff estimation by atmospheric water balance using ECMWF data set' in *Macroscale Modelling of the Hydrosphere, IAHS Publ.*, **214**, 163–171.

Peixóto, J. P. 1970. 'Pole to pole divergence of water vapor', *Tellus*, **22**, 17–25.

Peixóto, J. P. and Oort, A. H. 1983. 'The atmospheric branch of the hydrological cycle and climate' in M. Betan and K. Ratcliffe, A. S.-P. (Ed.), *Variations in the Global Water Budget*. D. Reidel, Dordrecht. pp. 5–55.

Phillips, T. J., Gates, W. L., and Arpe, K. 1992. 'The effects of sampling frequency on the climate statistics of the European Centre for Medium-Range Weather Forecasts', *J. Geophys. Res.*, **96**, 20 427–20 436.

Rasmusson, E. M. 1968. 'Atmospheric water vapor transport and the water balance of north America II. Large-scale water balance investigations', *Monthly Weather Rev.*, **96**, 720–734.

Rasmusson, E. M. 1977. 'Hydrological application of atmospheric vapor-flux analyses', *Hydrol. Rep. 11*, WMO, Geneval, 50pp.

Starr, V. P. and Peixóto, J. 1958. 'On the global balance of water vapor and the hydrology of deserts', *Tellus*, **10**, 189–194.

Unesco 1969. *Discharge of Selected Rivers of the World*. Vols I, II and II. Unesco, Paris.

Wijffels, S. E., Schmitt, R. W., Bryden, H. L., and Stigebrandt, A. 1992. 'Transport of freshwater by the oceans', *J. Phys. Oceanogr.*, **22**, 155–162.

25

SCALING OF LAND–ATMOSPHERE INTERACTIONS: AN ATMOSPHERIC MODELLING PERSPECTIVE

RONI AVISSAR

Department of Meterorology and Physical Oceanography, Cook College, Rutgers University, New Brunswick, NJ 08903, USA

ABSTRACT

Investigations of the impact of the ground surface on energy and mass fluxes in the atmosphere as well as on the dynamic response of the atmosphere at different scales are reviewed to emphasize the requirements for hydrological schemes for atmospheric models. Based on these investigations, a discussion on the scaling of the most important land processes and characteristics, from an atmospheric modelling perspective, is provided. It appears that because of the strong, non-linear response of the atmosphere to the spatial variability of land-surface characteristics, the scaling of hydrological processes and parameters is not linear. This implies that appropriate hydrological schemes for atmospheric models need to provide higher statistical moments and characteristic length scales of the spatial distribution of water availability for evapotranspiration.

INTRODUCTION

The microclimate near the ground surface is strongly affected by the ability of the surface to redistribute the radiative energy absorbed from the Sun and the atmosphere into sensible and latent heat fluxes. Indeed, over bare dry land, the absorption of this energy results in a relatively strong heating of the surface, which usually generates a large turbulent sensible heat flux in the atmosphere and a large soil heat flux. In this instance, there is no evaporation (i.e. no latent heat flux). The Bowen ratio, which is the ratio of the sensible heat flux and the latent heat flux, is infinite. However, in wet land, as is common in irrigated agricultural areas and/or after rain events, the incoming radiation is mostly used for evaporation, and the turbulent sensible heat flux and the soil heat flux are usually much smaller than the latent heat flux. As a result, the Bowen ratio is close to zero. When the ground is covered by dense vegetation, water is extracted mostly from the plant root zone by transpiration. In this instance, the latent heat flux is dominant even if the soil surface is dry, but as long as there is enough water available in the root zone and plants are not under stress conditions. As a result, the characteristics of the atmosphere above dry and wet (vegetated) land are significantly different (Avissar, 1992).

Assuming an infinite, homogenous land, Avissar (1992) used an atmospheric model to illustrate this impact. Considering a cloudless mid-summer day at a mid-, northern latitude, he showed that the faster heating rate obtained over a dry land generated a vigorous turbulent mixing and an unstably stratified atmospheric planetary boundary layer (PBL), which expanded up to a height of approximately 3000 m during the afternoon hours. However, the slower heating rate over a wet (vegetated) land limits the development of the PBL to a height of about 1000 m, but evapotranspiration provides a supply of moisture which significantly increases the amount of water in the shallow PBL. He noted that, on average, during the afternoon hours, the temperature of the PBL above the dry land was considerably warmer (about 5 K) than that of the PBL above the wet land, but that during night-time, the strong cooling of the bare and the vegetated surfaces created almost the same atmospheric inversion.

Obviously, these results are specific to the particular conditions that were selected for the simulations (i.e. latitude, day of the year, initial profiles of atmospheric and soil temperatures and humidities, synoptic-scale wind, etc.). Although different results are obtained for different sets of such conditions, the main pattern summarized here is generally reproduced under a broad range of conditions. It is worth mentioning that these results are clearly supported by field observations, e.g. the 1967 Wangara Experiment (Clarke *et al.*, 1971).

Based on these simple simulations, we could assume that if we had a hydrological model that simulates accurately the amount of water available for evapotranspiration at any time, it would be possible to know fairly well how the atmosphere would respond to the land-surface forcing. Unfortunately, the ground is not homogeneous (as was assumed in the above-described simulations) and studies have revealed that the atmospheric response to the ground-surface heterogeneity is strongly non-linear. Therefore, hydrological models that scale linearly from the microscale (at which atmospheric and hydrological processes can be relatively well observed) to the resolvable scale of large-scale atmospheric models (e.g. global climate models, or GCMs) are clearly not appropriate. Nevertheless, this type of model is currently used in most atmospheric models.

The goals of this paper are (i) to review relevant studies that have been conducted to better understand the impact of the land surface on the atmosphere at various scales and (ii) to discuss the relevant information that will need to be produced by hydrological schemes for atmospheric models to improve the simulations of land–atmosphere interactions at various spatial and temporal scales. For this purpose, a definition of the scales commonly used in atmospheric science is presented, and different aspects of subgrid-scale parameterization are also discussed.

ATMOSPHERIC SCALES

The atmosphere is characterized by phenomena, or features, whose typical size and typical lifetime cover a very wide range. The space scales of these phenomena are determined by their typical size (or wavelength) and the time-scales by their typical lifetime (or period). As illustrated in Figure 1, atmospheric features

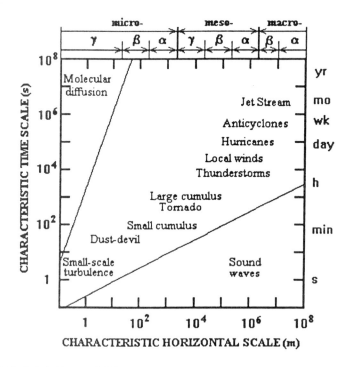

Figure 1. Scale definitions and different processes with characteristic time and horizontal scales

range from turbulence with a space scale of metres and time-scale of seconds up to jet streams with a space scale of thousands of kilometres and time-scale of months.

Even though these phenomena are not discrete but part of a continuum, and attempts to divide them into distinct classes have resulted in disagreement with regard to the scale limits, the classification suggested by Orlanski (1975) (also shown in Figure 1) is accepted as a reasonable consensus. Accordingly, the characteristic horizontal distance scale is divided into a macroscale, which includes phenoma that extend over scales larger than 2000 km, a microscale, which includes features smaller than 2 km, and a mesoscale, which includes the wide range of atmospheric phenomena with scales between the macroscale and the microscale (i.e. between 2 and 2000 km). As indicated in Figure 1, these scales are further subdivided into macro-α (larger than 10 000 km), macro-β (2000–10 000 km), meso-α (200–2000 km), meso-β (20–200 km), meso-γ (2–20 km), micro-α (200 m–2 km), micro-β (20–200 m) and micro-γ (smaller than 20 m).

Classifications according to time-scales are more ambiguous and less frequently used. However, macroscale usually refers to atmospheric features with time-scales of several days or longer and microscale refers to phenomena with time-scales of the order of minutes. Using the classification suggested by Orlanski, meso-α scale corresponds to time scales of one day to one week, meso-β scale to several hours to one day, and meso-γ scale to one-half hour to several hours.

SUBGRID-SCALE PROCESSES

At any location and time, the atmosphere is characterized by a set of conservation principles (i.e. mass, heat, motion, water and other gaseous and aerosol materials) which form a coupled set of relations that must be satisfied simultaneously. This set of relations consists of non-linear partial differential equations, which have no known analytical solution. Thus only a simplified set of relations can be solved analytically, which provides interesting, but limited, analyses of the studied atmospheric process. Alternatively, numerical solutions of this set of equations can be used. These solutions are not perfect either, but do account for non-linear interactions, which have an important impact on many atmospheric processes (e.g. Pielke, 1984).

Because of limited computational resources, the numerical solution of this set of equations is not possible at the spatial and temporal scales needed to apply correctly these conservation principles. For instance, current GCMs typically represent the Earth's surface with a horizontal resolution of 2–8° and the vertical structure of the atmosphere with 2–19 layers. A typical mesoscale model simulates an area of 10^4–10^6 km^2 with a horizontal resolution of 5–50 km and use 5–30 layers to describe the vertical structure of the first 3–15 km of the atmosphere. Large-eddy simulation (LES) models, which are typically used to study microscale processes at a horizontal resolution of 100–250 m currently cannot be used to simulate an area larger than a few square kilometres.

Consequently, all atmospheric numerical models are formulated on the basic concept that atmospheric variables can be separated into a resolved (mean) part and an unresolved (perturbation) part, or subgrid-scale part. The divergences of subgrid-scale fluxes, which result from the averaging procedure of the conservation equations in which atmospheric variables are written as the sum of a mean and a subgrid-scale perturbation, may affect significantly the resolved variables. These subgrid-scale fluxes, which are the covariance of subgrid-scale perturbations, cannot be described by a closed set of equations and, as a result, must be parameterized.

The relative importance of the subgrid-scale fluxes has a major influence on the philosophical approach used to develop the parameterization. For instance, Deardorff (1974) used an LES model with a horizontal resolution of 125 m to demonstrate that in most parts of the PBL, the resolvable fluxes were much more important than the subgrid-scale fluxes. By contrast, in GCMs, clouds and boundary-layer processes (including land–atmosphere interactions at the mesoscale and the microscale) are subgrid-scale processes which considerably affect the simulations produced with these macroscale models. Thus, parameterization of subgrid-scale fluxes might require different degrees of complexity in different numerical models.

LAND-SURFACE IMPACT ON THE ATMOSPHERE

As discussed above, the water availability for evapotranspiration (hereafter referred to as η_{ET}) considerably affects the redistribution of the radiative energy absorbed at the ground surface and, as a result, the entire PBL. However, many hydrological and physiological parameters control η_{ET}. For instance, plant stomata control transpiration through a chain of complex biochemical reactions, which are apparently sensitive to solar radiation, carbon dioxide concentration, temperature, vapour pressure difference between the leaves and their environment and soil water potential in the root zone (e.g. Jarvis and Mansfield, 1981; Martin *et al.*, 1983; Willmer, 1983; Avissar *et al.*, 1985; Zeiger *et al.*, 1987; Collatz *et al.*, 1991; among many others). The soil hydraulic conductivity is one of the parameters that controls the transport of water in the root zone and/or to the evaporating surface and, therefore obviously also affects η_{ET}.

To identify which of the land characteristics have a predominant impact on the redistribution of energy into turbulent* sensible and latent heat fluxes at the ground surface, Collins and Avissar (1994) used a powerful technique, known as the Fourier amplitude sensitivity test (FAST), in conjunction with the land–atmosphere interactions dynamics (LAID), a land-surface scheme developed by Avissar and Mahrer (1982; 1988) and improved by Avissar and Pielke (1991).

The FAST, which was introduced by Cukier *et al.* (1973), is a technique for sensitivity analysis of mathematical models. This is a unique method because it provides information on the model sensitivity to uncertainties in individual parameters. It uses continuous probability distributions of the model input parameters as data, and determines the relative contribution of each parameter to the variances in model outputs. This method has proved to be very accurate and objective. A detailed description of the theory and implementation of the FAST Method and approximations used in computer implementation is given in Collins and Avissar (1994).

The LAID scheme is conceptually similar to the biosphere–atmosphere transfer scheme (BATS) developed by Dickinson *et al.* (1986) or the simple biosphere model (SiB) suggested by Sellers *et al.* (1986) as it is based on the same physical principles and the solution of energy budget equations for soil surface and vegetation. However, some of the mechanisms are parameterized differently in these schemes. For instance, in LAID, heat and moisture conservation equations are solved simultaneously on a multi-layer vertical grid, with a very high resolution near the soil surface (e.g. 1 mm) and progressively less resolution with depth (which varies according to the real soil depth). This more realistic representation of heat and moisture movement in the soil offers some advantages (as compared with BATS and SiB) for the calculation of the hydrological processes.

For their analysis, Collins and Avissar (1994) used the 10 land characteristics provided in Table I, and repeated their experiment for the various environmental conditions summarized in Table II. Furthermore, they considered three different types of distribution of the input parameters (normal, log-normal and rectangular). Their results indicated that land-surface wetness, surface roughness, albedo and, when vegetation covers the ground, leaf area index and plant stomatal conductance, are the most important characteristics which affect the redistribution of energy at the ground surface. Among them, the land-surface wetness, which expresses the relative availability of soil moisture for evaporation in the upper soil layer, was found to have a predominant role in bare land. Correspondingly, in vegetated land, the stomatal conductance, which has a very similar impact on the redistribution of energy into turbulent sensible and latent heat fluxes at the ground surface, was found to be the most important land-surface characteristic.

It is very important to emphasize that in this analysis, the land-surface parameters were allowed to vary independently. In other words, plant stomatal conductance appeared to be one of the most important characteristics because it was not related to, say, the amount of water in the root zone. As far as we understand, however, stomatal conductance is in fact affected by the water potential in the root zone, but this impact was not directly considered in this analysis. This intentional omission was done to prevent the results from being affected by a particular parameterization. Thus, now that it appears that stomatal conductance has a strong impact on the atmosphere, it is important to evaluate which parameter(s) control(s) the behaviour of stomatal conductance at the various scales.

---

* In this paper, turbulence is restricted to microscale perturbations only and, therefore, does not include mesoscale perturbations.

Table I. Land-surface parameters used as input to the FAST. For the analysis, values were chosen within the range between maximum and minimum values (from Collins and Avissar, 1994)

Parameter	Minimum	Maximum
Soil albedo (α_G)	0·05	0·95
Soil emissivity (ϵ_G)	0·80	0·995
Soil thermal conductivity (λ) (W m^{-1} K^{-1})	0·05	3·0
Vegetation albedo (α_v)	0·10	0·30
Vegetation emissivity (ϵ_v)	0·90	0·99
Vegetation extinction coefficient (κ_v)	0·30	2·30
Soil surface wetness (s_w)	0·0	1·0
Relative stomatal conductance (g_{sc})	0·0	1·0
Surface roughness (z_0) (m)	2×10^{-5}	6·0
Leaf area index (l)	0	10

Henderson-Sellers (1993) investigated the same problem using the factorial experiment technique in conjunction with the BATS. In general, her conclusions are similar to those of Collins and Avissar (1994). As part of her study, however, she also tested the most important atmospheric forcings (at the climate time-scale) and found that mean monthly temperature and the interaction between mean monthly temperature and total monthly precipitation have a predominant impact on the behaviour of BATS. She emphasized that fractional cloudiness and other environmental parameters are also important.

It is well known that these five characteristics vary extensively, even at a scale of a few metres. The impact of their spatial variability on the atmosphere at various scales is discussed in the following subsections.

Microscale impact

Li and Avissar (1994) investigated the impact of microscale (or 'patch' scale) variability of these five land-surface characteristics on the atmospheric turbulent heat fluxes near the ground surface. They characterized this variability by 11 different probability density functions (pdfs) for each one of the land-surface characteristics, namely five beta distributions, three log-normal distributions, two normal distributions and one rectangular distribution. All possible combinations of the pdfs of land-surface characteristics and a broad range of atmospheric conditions (totaling 50 625 combinations per pdf) were considered. As an example, Figure 2 presents the histograms of absolute differences obtained between sensible heat flux, latent heat flux and radiative flux emitted by the land surface, calculated with the rectangular distribution and calculated with the corresponding mean of each of the five land-surface characteristics. Each histogram was built with the results from the 50 625 simulated cases. Note that the fluxes were diagnosed for a given set of land-surface characteristics and atmospheric conditions at a given time.

In general, the variability of a particular characteristic affects the three energy fluxes simultaneously, yet not with the same intensity. On average, the latent heat flux is the most sensitive and the radiative flux the

Table II. Input values of environmental variables (T, air temperature; RH, relative humidity; V, wind speed; R_s, solar radiation; and R_L, atmospheric thermal radiation) used with the LAID parameterization (from Collins and Avissar, 1994)

Variable	Minimum	Maximum
T (K)	273	313
RH (%)	20	100
V (m/s)	0·5	6·0
R_s (W/m^2)	0	1000
R_L (W/m^2)	200	400

least sensitive to spatial variability. This is clearly illustrated in Figure 2 by a partition of the frequencies over larger differences for the latent heat fluxes than for the radiative fluxes.

This study showed that the more positively skewed the pdf, the larger the difference between the heat fluxes calculated with the pdf or the mean. Differences as large as 150, 135, 105 and 100 $W\,m^{-2}$ were found between the heat fluxes calculated with the log-normal distributions and with the mean leaf area index, stomatal conductance, soil-surface wetness and surface roughness, respectively. The spatial variability of albedo created differences of not more than 20 $W\,m^{-2}$, indicating a relatively linear relation between this characteristic and land-surface energy fluxes.

A detailed analysis of the land-surface energy fluxes integrated at the patch scale with either the pdfs or the corresponding mean characteristics indicated that, under unstable atmospheric conditions, stomatal conductance and leaf area index variability have the most significant effect on spatially integrated energy

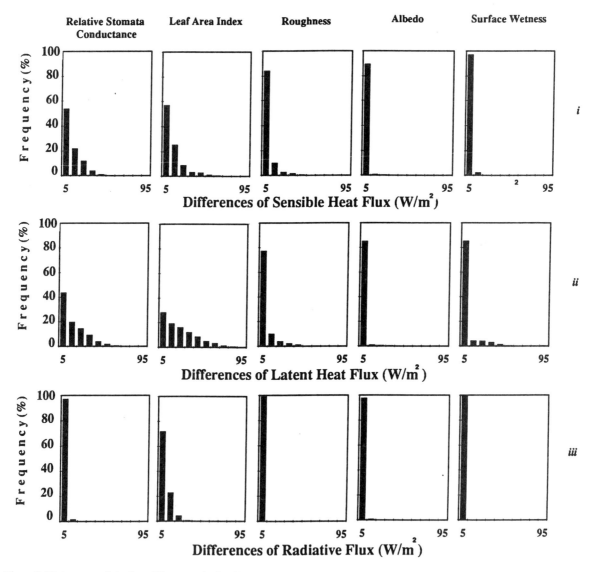

Figure 2. Histograms of absolute differences obtained between (i) sensible heat fluxes, (ii) latent heat flux and (iii) radiative flux emitted by the land surface calculated with a rectangular distribution and calculated with the corresponding mean of relative stomatal conductance, leaf area index, surface roughness, albedo and soil-surface wetness (from Li and Avissar, 1994)

fluxes from vegetated land. In bare land, soil-surface wetness, instead of stomatal conductance, strongly affected the surface fluxes. Under neutral and stable atmospheric conditions, surface roughness appeared to have the predominant effect on the surface fluxes.

In their analysis, however, Li and Avissar (1994) assumed that the impact of the variability of land-surface characteristics did not extend beyond the atmospheric surface layer. This, of course, allowed them to use a relatively simple land-atmosphere model, but limited their study to the patch scale. The impact of the spatial variability of land-surface characteristics at a scale larger than the patch scale, though still considered microscale, can be studied with LES models. This modelling technique, which was pioneered by Deardorff (1972), is accurate and, therefore, very attractive to study the turbulence activity in the PBL. For instance, Deardorff (1974) showed that turbulence statistics obtained with an LES model compared nicely with those observed in the 1967 Wangara Experiment (Clarke *et al.*, 1971). Note that with this modelling technique, only small (less energetic) eddies need to be parameterized and the largest sized eddies are resolved. Numerical experiments with LES by Moeng (1984; 1986; 1987) and, more recently, by Hadfield *et al.* (1991; 1992) and Walko *et al.* (1992) confirmed the high potential of this technique, especially with the powerful computers readily available nowadays.

Unfortunately, because of the extremely large computer resources needed to produce LES, so far only a few investigations have been conducted to better understand the impact of the spatial variability of land-surface characteristics on the PBL. Hadfield *et al.* (1992) simulated the effect of a sinusoidal wave (with two different wavelengths) of surface heat flux with zero or weak wind background conditions. They found that the heat flux variation drives a circulation, which is considerably strengthened with the increase of the wavelength of the perturbation. However, even a light wind weakens this circulation drastically and moves it downwind. Walko *et al.* (1992) studied the impact of small topographical features (assuming horizontal homogeneity of heat fluxes) on the structure of turbulence in the PBL. They found that the horizontal spectra of vertical motion were strongly biased towards the horizontal scales of the terrain. However, both the studies of Hadfield *et al.* (1992) and Walko *et al.* (1992) indicated that vertical profiles of atmospheric variables obtained by horizontal averaging were not affected by the spatial variability of heat flux or the presence of small hills.

More research is needed in this area and with the availability of new, more powerful computers, additional insights are expected to come in the near future. Nevertheless, these preliminary results indicate that as long as the length scale of the surface sensible heat flux perturbation is smaller than about 2 km and the topographical features are not higher than about 200 m, the assumption used by Li and Avissar (1994) seems reasonable. It is interesting to note that, based on theoretical considerations and a simple slab model of the convective boundary layer, Raupach (1993) claimed that this assumption is applicable for a horizontal length scale of 1-5 km.

Mesoscale impact

It is well known that the contrast between land and water generates breezes (i.e. sea, lake and land breezes), which are mesoscale circulations. Several papers and textbooks describe in length the mechanism involved in the generation of these circulations and their impact on the weather and the climate (e.g. Pielke, 1984, among many others). More recent investigations have indicated that other landscape discontinuities, e.g. irrigated land in arid areas, deforestation and afforestation, also provide an environment appropriate for the development of mesoscale circulations. To illustrate the impact of this type of circulation on the atmosphere, Avissar and Chen (in press) produced a 'deforestation' simulation with the Colorado State University (CSU) regional atmospheric modeling system (RAMS). An extensive description of this state of the art model is provided by Pielke *et al.* (1992).

The simulated domain was 250 km wide and was located at latitude 0°. This domain was represented in the model by 50 grid elements with a horizontal grid resolution of 5 km. The deforested area, which was 50 km wide, was located in the middle of the domain. Thus it was flanked by 100 km of forests. The atmosphere was simulated up to a height of 10 km (i.e. roughly the tropopause) and was represented by 18 grid elements with a high grid resolution near the ground surface and progressively less resolution with height in the atmosphere.

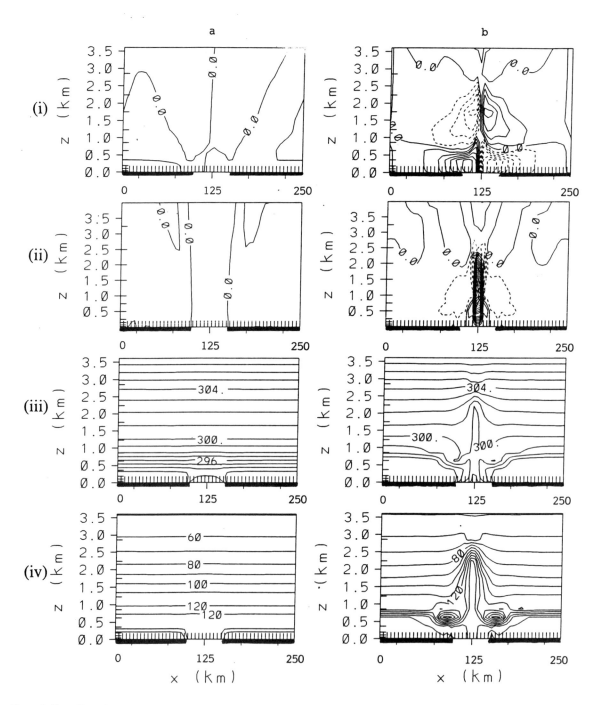

Figure 3. Two-dimensional sections of (i) the horizontal wind component parallel to the domain (u) in m s^{-1}, positive eastward; (ii) the vertical wind component (w) in cm s^{-1}, positive upward; (iii) the potential temperature (θ) in K; and (iv) the mixing ratio ($r \times 10^4$), obtained in the PBL that develops over a deforested area of the Amazonian region during mid-summer time at (a) 9 am, (b) 3 pm, (c) 9 pm and (d) 3 am. Forest is indicated by a dark underbar. Background wind is 0·5 m/s blowing from south. Solid lines indicate positive values and broken lines indicate negative values (from Avissar and Chen, in press)

An initial soil temperature of 300 K, a soil water content at field capacity in the forest area and a soil water content of 4% of the soil water content at field capacity in the deforested area were assumed. A potential temperature lapse of 3·5 K/km, a relative humidity of 95% at all heights in the atmosphere and a southerly background wind of 0·5 m/s were also used for the initialization of the model. Furthermore, the forest was assumed to consist of a very dense canopy, which absorbed all the solar radiation not reflected by the canopy (i.e. zero transmissivity to the soil surface).

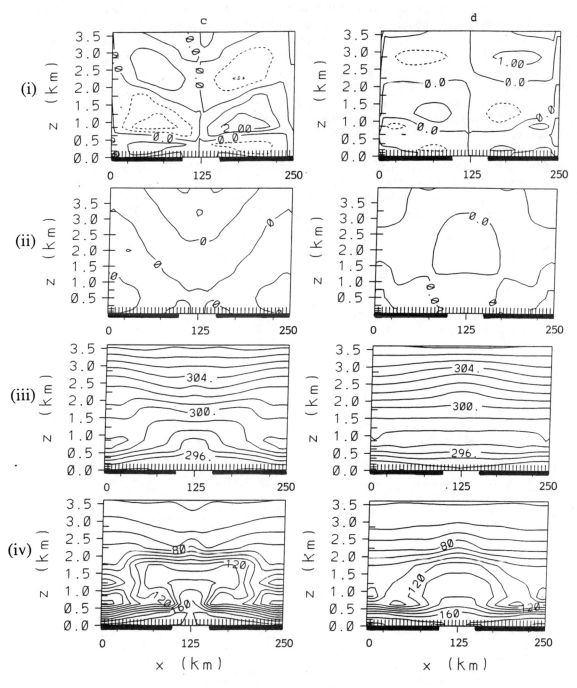

(Figure 3c, d)

The numerical integration for the simulations was started at 6 am (local standard time), which corresponds to the time when the sensible heat flux becomes effective in the development of the convective PBL on sunny summer days. The time step of the numerical integration was 15 s.

Figure 3 presents two-dimensional sections of different meteorological fields obtained in the lower 3·5 km of the atmosphere at different times of the day. The regional circulations depicted by the horizontal and vertical components of the wind result from the differential heating produced by the two very distinct surfaces subjected to the same solar radiation input. In the deforested land (assumed bare and relatively dry), most of the radiative energy received from the Sun and the atmosphere is used to heat the atmosphere and the ground. However, in the forest area (assumed unstressed), a large part of this radiative energy is used for evapotranspiration. The faster heating rate above the deforested land surface generates a vigorous turbulent mixing and an unstable, stratified PBL, as can be seen from the sections of potential temperature in Figure 3. On the contrary, the slower heating rate above the transpiring forest limits the development of the PBL, which remains shallow above the forest. This creates a pressure gradient between the two areas, which generates the circulations from the relatively cold to the relatively warm area.

Because the differential heating is closely related to the redistribution of solar energy absorbed at the ground surface, a diurnal pattern is clearly evident in Figure 3. In the early morning (9 am), the amount of solar radiation reaching the land surface is relatively small and, as a result, only a weak thermal and pressure gradient is created between the forest and the deforested areas. These gradients increase gradually during the morning hours and dissipate during the afternoon and night hours. The time-lag of the maximum intensity of the mesoscale circulation that develops over the domain is mainly due to the time response of the PBL to the heating and cooling, which occur mostly at the ground surface.

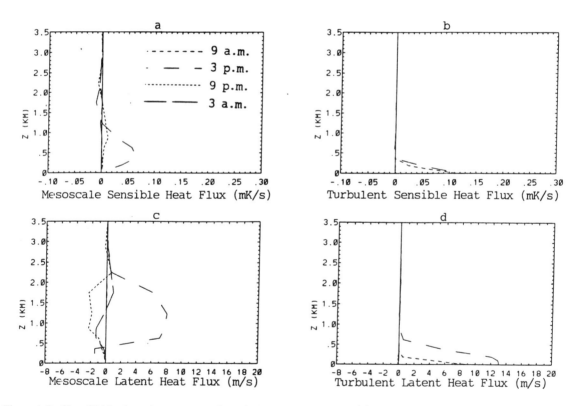

Figure 4. Profiles of (a) horizontal mean mesoscale vertical sensible heat flux ($\langle w'\theta'\rangle$), (b) horizontal mean turbulent vertical sensible heat flux ($\langle \overline{w''\theta''}\rangle$), (c) horizontal mean mesoscale vertical latent heat flux ($\langle w'q'\rangle$) and (d) horizontal mean turbulent vertical latent heat flux ($\langle \overline{w''q''}\rangle$) obtained in the PBL that develops over a deforested area of the Amazonian region during mid-summer time at 9 am, 3 pm, 9 pm and 3 am. Background wind is 0·5 m/s blowing from south (from Avissar and Chen, in press)

The transpiration from the forest provides a supply of moisture which significantly increases the amount of water in the shallow PBL, as can be seen from the two-dimensional sections of the mixing ratio presented in Figure 3. This moisture is advected by the generated mesoscale flow, which strongly converges toward the deforested area, where it is well mixed within the relatively deep convective boundary layer. This process may eventually generate convective clouds and precipitation under appropriate synoptic-scale atmospheric conditions, as discussed in the following.

Figure 4 shows the vertical profiles of mean (i.e. horizontally averaged) mesoscale and turbulent kinematic sensible and latent heat fluxes obtained in the above-described simulation. The mesoscale vertical sensible heat flux is the covariance between the vertical component of the mesoscale circulation (w') and the perturbation of potential temperature (θ'). Thus it is resolved by the model and, as a result, is not subject to the possible flaw of a parameterization. As can be seen in Figure 3, in general, θ' is positive in that region of the PBL where a strong, positive w' has developed (i.e. the lower 2 km of the atmosphere). This results in a positive product $w'\theta'$, indicating a net transport of heat upward. However, in the upper part of the PBL (where the return flow is dominant), the sign of w' and θ' is usually opposite. This results in a negative sensible heat flux, which corresponds to a net transport of heat downward. The general shape of the profile of turbulent sensible heat flux is similar to that observed, simulated and discussed in detail by many others (e.g. Deardorff, 1974; André et al., 1978; Stull, 1988).

The profile of mesoscale vertical latent heat flux (i.e. the covariance between w' and the perturbation of specific humidity, q') is much different in the lower part and the upper part of the PBL. In general, in the lower part of the PBL, positive perturbations of specific humidity are found above the wet land, where w' is typically negative. In this region, however, turbulent fluxes are positive, indicating a net turbulent transport of moisture upward. Thus the mesoscale vertical flow tends to act as a counter-gradient term in this part of the PBL. Above the dry land, where perturbations of specific humidity are negative, w' is mostly positive. Thus, in this part of the PBL, the product $w'q'$ is also negative, indicating a mesoscale transport of relatively dry air upward.

The moisture transported upward in the PBL by the combined effect of turbulence (i.e. microscale) and mesoscale vertical flow, creates a positive mesoscale perturbation above a height of approximately 400 m. As a result, moisture is drained from the lower part of the PBL and dampens the upper part of the PBL.

These mesoscale fluxes are likely to form clouds, which affect the radiation and the precipitation regimes and, as a result, the hydrological cycle. For instance, Bougeault et al. (1991) simulated with the French Weather Service limited-area numerical model (PERIDOT) the formation of clouds next to the boundary between a forest and a crop area. Their results were supported by satellite images and observations of various meteorological parameters in the PBL.

It is interesting to note that turbulence and mesoscale flow complement each other remarkably well to transport heat and moisture upwards in the PBL. As clearly depicted in Figure 4, turbulence is mostly effective close to the ground surface and in the lower part of the PBL, whereas the contribution of the mesoscale flow is mostly effective higher in the PBL. Integrating with height the profiles of mesoscale and turbulent heat fluxes shown in Figure 4, we can see that the mesoscale fluxes are about twice as strong as the turbulent fluxes, emphasizing again the importance of mesoscale fluxes compared with turbulent fluxes.

Chen and Avissar (1994a) investigated in more detail the impact of such landscape discontinuities on clouds and precipitation. They showed that the water availability for evapotranspiration (η_{ET}) significantly affects the timing of onset of clouds, and the intensity and distribution of precipitation. In general, they found that landscape discontinuities enhance shallow convective precipitation. They emphasized that the two mechanisms which are strongly modulated by the spatial distribution of η_{ET}, namely turbulence and mesoscale circulations, also determine the horizontal distribution of maximum precipitation. They found that interactions between shallow cumulus and η_{ET} are highly non-linear and complicated by different factors, such as atmospheric thermodynamic structure, background (large-scale) wind and the horizontal size of the discontinuity.

For instance, Figure 5 shows the accumulated precipitation, at 8 pm, obtained in three different simulations produced with RAMS, in which deforested patches of 25, 100 or 180 km were considered. Initial and boundary conditions for these numerical experiments were similar to those used in the above-described simulation, except that a very high horizontal resolution, namely 250 m, was adopted to resolve explicitly cumulus clouds.

It is interesting to note that the largest precipitation was obtained for the 100 km deforested patch, which corresponds approximately to the local Rossby radius of deformation in these simulations. This relation supports the findings of Dalu *et al.* (1991), who used a set of linearized two-dimensional primitive equations describing thermally forced atmospheric flow to study the impact of land-surface heat flux heterogeneities on the convective boundary layer. In this linear analysis, Dalu *et al.* (1991) obtained a maximum intensity of the mesoscale fluxes for a characteristic length scale of the surface heterogeneity equal to the local Rossby radius of deformation.

To explain this process, it is helpful to examine the two-dimensional sections of the horizontal wind resulting from the 100 km wide patch at 1 pm and 4 pm, which are presented in Figure 6. Two separate mesoscale circulations develop over the two dry-forested land discontinuities in the early afternoon and tend to converge over the central dry area (Figure 6a). Later in the afternoon (Figure 6b), these two meso-scale circulations collide, resulting in a zone of strong convergence, which extends to relatively high elevation and produces strong rainfall. For this particular case, the maximum accumulated precipitation was 41 mm (Figure 5b). When the dry patch is much larger (e.g. in the 180 km wide patch example), the mesoscale circulations generate precipitation before they collide and, as a result, there is no rainfall enhancement. This also explains why two approximately equal peaks of precipitation are obtained in this instance, as

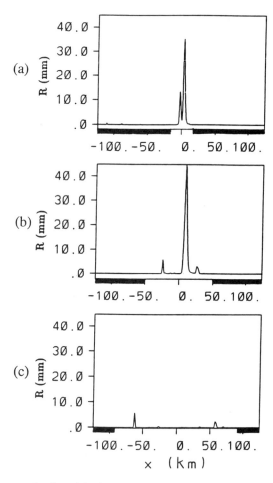

Figure 5. Horizontal distribution of accumulated precipitation at 8 pm for (a) a 25 km wide dry patch, (b) a 100 km wide dry patch and (c) a 180 km wide dry patch. The dry patch is located in the middle of the domain (adapted from Chen and Avissar, 1994a)

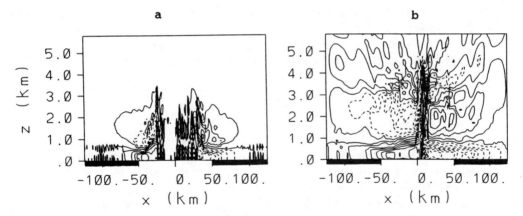

Figure 6. Two-dimensional sections of the horizontal wind component parallel to the domain (u) in m s^{-1} (positive eastward) that develops above a 100 km wide dry patch at (a) 1 pm and (b) 4 pm. The dry patch is located in the middle of the domain (adapted from Chen and Avissar, 1994a)

evident in Figure 5c, whereas in the 25 and 100 km wide patch examples the maximum rainfall concentrates near the centre of the dry patch, where the collision of the two circulations occurs.

Further analyses by Avissar and Chen (1993; in press), Chen and Avissar (1994a, 1994b) and Lynn *et al.* (1995a) indicate that this type of mesoscale circulation (and their associated heat fluxes and precipitation) are affected by other environmental conditions, e.g. the background wind, the thermal and humidity profiles of the atmosphere, the latitude and the day of the year. For instance, an increase in the background wind tends to reduce the importance of mesoscale fluxes, but even with a moderate wind these fluxes remain significant. Although the humidity has only a minor direct impact on the generation of mesoscale circulations, this parameter has dominant role in the formation of cloud and precipitation. As already well known, the latitude and day of the year significantly affect the amount of radiative energy received at the ground surface. Consequently, they affect the turbulent fluxes near the surface, which in turn affect the entire PBL and mesoscale fluxes, as discussed earlier.

Macroscale impact

Although the macroscale variability of the Earth's surface has a significant impact on many atmospheric processes and, as a result, is very relevant to atmospheric modelling, this variability is resolved at the scale of the atmospheric models and, therefore, does not require to be parameterized. Because this paper concentrates on the scaling of land–atmosphere interactions with a perspective of subgrid-scale parameterization, only a brief summary of some of the important studies is given here, just for the purpose of providing interested readers with references that can be used to track other papers on this topic. In addition, the review of Avissar and Verstraete (1990) provides a long list of relevant publications.

One of the first analyses of the impact of the land surface at the macroscale was that by Shukla and Mintz (1982). They compared the global mean precipitation, the surface temperature and the surface pressure obtained from two GCM simulations, which were initialized with identical conditions, except for the constraint placed on the land-surface evaporation. In one instance, the actual rate of evaporation was always set equal to its potential value under the existing climatological conditions. In the other instance, the actual rate of evaporation was permanently set to zero. Significant differences were obtained between the two examples. For instance, over most of North America the mean precipitation for the month of July was about four times less and ground surface temperature was 15–25°C warmer in the zero-evaporation case than in the potential-evaporation case. Similarly, surface pressure was about 5–15 hPa lower over most of the land areas in the zero-evaporation case, which suggests enhanced cyclonic circulations over the continents.

Dickinson and Henderson-Sellers (1988), Henderson-Sellers *et al.* (1988) and Shukla *et al.* (1990) investigated the impact of the deforestation of Amazonia on the global climate with GCMs. Their simulations predicted a net warming, a reduction in precipitation and an extension of the dry season in this region. Xue and Shukla (1993) studied the impact of desertification of Africa on the Sahel drought. They found that rainfall was reduced due to desertification and the rainy season was delayed by about two weeks in this region. On the other hand, rainfall increased south of this region, giving rise to a dipole rainfall anomaly pattern.

DISCUSSION AND CONCLUSIONS

In summary, the review presented in the previous section indicates that five land-surface characteristics, namely stomatal conductance, soil surface wetness, surface roughness, leaf area index and albedo must be described as accurately as possible in atmospheric models. Although using a mean value of albedo does not affect the surface energy fluxes significantly, the spatial variability of the other four characteristics should be considered to calculate the heat fluxes near the ground surface. When the spatial variability of these characteristics results in relatively extended landscape discontinuities, mesoscale circulations may be generated under appropriate environmental conditions. These circulations significantly affect the transport of heat and moisture in the PBL and, as a result, clouds and precipitation. As, in return, precipitation affects the spatial variability of surface wetness, stomatal conductance and albedo, a complex system of interactions with strong non-linear feedbacks develops. These interactions are further complicated by the fact that the dynamic response to precipitation of the atmosphere, of the water in the ground and of the land-surface characteristics, occurs at different time-scales. It is not yet very clear what is the minimum size of a landscape discontinuity needed to generate a significant circulation, but it seems that the maximum effect is obtained for a length scale of the order of 100 km at mid-latitudes, and that significant effect is already produced for a length scale of the order of a few kilometres. Macroscale background atmospheric conditions affect this length scale.

Thus the scaling of hydrological processes from the microscale to the macroscale appears to be non-linear. This implies that it is not appropriate to use a model that was developed and calibrated to operate at a particular scale, for an application at a different scale, just by scaling its parameters. This further implies that it is not possible to use the mean value of a land-surface characteristic as the 'representative' value of even a small area (i.e. at the microscale). Nonetheless, state of the art atmospheric models have adopted parameterizations of land-surface processes based on theories applicable (at best) at the microscale and which are generally calibrated with microscale observations. Most schemes assume spatial homogeneity at the resolvable scale of the hosting atmospheric model. In general, these land-surface schemes are coupled with PBL parameterizations which, in all current atmospheric models, assume horizontal homogeneity at the resolvable scale of the model.

It is worth mentioning that although spatial variability of the land surface induces non-linear feedbacks (e.g. precipitation), this does not mean that this spatial variability must be treated explicitly, as could be done in theory by increasing the spatial resolution of the atmospheric models. In practice, the increase in the spatial resolution of the models is limited by the availability of computer resources. Even though tremendous progress is being achieved constantly in this area, the time when it will be possible to simulate the global climate, or the weather, at a spatial resolution of a few metres is still remote. An alternative approach to treat the variability is by parameterizing its effects, using higher statistical moments.

A few land parameterizations based on this alternative approach have already been suggested. For instance, Avissar and Pielke (1989) proposed to treat the spatial variability of the land as a mosaic of patches (or 'tiles'). With this scheme, the number of patches is not assigned *a priori*, but depends on the heterogeneity of the simulated domain. Assuming that horizontal fluxes between the different patches within a grid element of the hosting model are small compared with the vertical fluxes, patches of the same type located at different places within the grid element can be regrouped into patch classes. Then, for each class, the LAID scheme is applied to evaluate the surface fluxes. A different approach, referred to as the statistical–dynamical approach (e.g. Entekhabi and Eagleson, 1989; Famiglietti and Wood, 1991; Avissar, 1992) has also been proposed. With this approach pdfs are used to describe the variability of the land characteristics found in the real world, at the resolvable scale of interest. However, although these schemes account somehow for the land heterogeneity, the atmosphere above the land is still assumed horizontally homogenous and, therefore, they do not account for the atmospheric feedbacks summarized in the previous section. According to Raupach (1993), this approach is applicable as long as the length scale of the landscape heterogeneity is smaller than 1–5 km.

Lynn et al. (1995b) used the similarity theory approach developed to parameterize turbulence in the PBL (e.g. Stull, 1988) to develop a parameterization of mesoscale fluxes for macroscale models. This method consists of (i) identify the relevant variables to the problem; (ii) organize these variables into dimensionless form; (iii) gather observations or perform experiments to determine their values; and (iv) find an empirical relationship between the dimensionless variables. Ideally, this empirical relationship is 'universal', i.e. it applies for all possible cases.

Lynn et al. (1995b) were able to develop such an empirical relationship between mesoscale fluxes and a combination of a characteristic length scale, the standard deviation of turbulent sensible heat fluxes at the ground surface and other variables representative of the macroscale atmospheric background, e.g. wind intensity, moist static energy, stability and the coriolis force. The characteristic length scale was defined as

$$L = \left(\sum_{i=1}^{n} f_i l_i^m \right)^{\frac{1}{m}} \tag{1}$$

where f_i is the fractional area of a patch i, n is the number of patches found in the domain represented by a single grid element in the model, m is an empirical constant and l is the characteristic length scale of an individual patch, defined as

$$l = \frac{4A}{P} \tag{2}$$

where A and P are the area and the perimeter of the patch, respectively. Note that for $m = 1$, L is the area average of l, and for $m = 2$ (as was adopted by Lynn et al., 1995b), L is the root mean square of l. Also, note that the magnitude of l increases with the 'squareness' of the patch.

This primary parameterization still needs improvements (e.g. the inclusion of moist processes and subgrid-scale topographical features) to be used confidently in macroscale atmospheric models. Yet it is brought into this discussion to emphasize that given a macroscale atmospheric background environment, a characteristic length scale of the land-surface patchiness and the distribution of surface heat fluxes, it is reasonable to assume that a parameterization of subgrid-scale land–atmosphere interactions for macroscale models can be developed. Because a statistical approach is adopted to develop such a parameterization, it would be possible to obtain the higher statistical moments of the product of this parameterization (e.g. mesoscale fluxes, clouds and precipitation). For instance, the distribution of precipitation as characterized by its mean, its standard deviation and maybe its skewness and kurtosis, is likely to be helpful to define landscape patchiness.

Based on this reasonable assumption, we can infer that an important challenge for hydrologists would be to develop a scheme that would be able to provide the characteristic length scale of the land-surface patchiness and the distribution of surface heat fluxes. As to obtain the distribution of surface heat fluxes this patchiness must reflect the time variation of the spatial variability of the five important land characteristics, an appropriate hydrological scheme would probably require high-resolution data sets of topography (which has been shown to affect the redistribution of water in the ground, the surface roughness, the radiation balance and the turbulent heat fluxes in the PBL), soil texture and vegetation type. Clearly, the organization of such data sets at the global scale represents by itself a big challenge.

Although many hydrological processes vary over relatively long time-scales, it is important to mention that the time step of integration of current macroscale atmospheric models is about 20 minutes. Because of numerical stability constraints, an increase in the spatial resolution of the atmospheric model requires a decrease in the time step of integration. For instance, an LES with a horizontal resolution of 100 m requires a time step of integration as small as one second. Thus, in developing an appropriate hydrological scheme for a macroscale atmospheric model (e.g. a GCM), we have to consider that the atmospheric model will require hydrological information at each time step of integration. As discussed in Avissar (1992), however, the statistics of the hydrological processes at the macroscale might be relatively stable over a relatively long period (days to weeks). Clearly, if this appears to be correct, the characteristic length scale required to parameterize land–atmosphere interactions at the macroscale might also be relatively stable, and would be a relatively easier task for hydrologists.

Finally, it is also important to emphasize that hydrologists, atmospheric scientists and ecologists involved in the development of parameterizations at the macroscale are faced with the challenge of collecting appropriate observational data sets, which could be used for the development and the evaluation of these parameterizations. Although field experiments at the microscale, which often involve complex technology and management expertise, are realizable, it is not realistic to assume that the tremendous amount of equipment and human resources needed to carry out a high-resolution field experiment at the macroscale will be available in the foreseeable future. Therefore, the development of new parameterizations will need to rely heavily on numerical and other theoretical analyses, which could provide guidance for the planning of more focused field experiments dedicated to observe only selected important parameterized processes.

In the past, most theories and model parameterizations in hydrology and atmospheric science have been developed based on observations. Because our scientific education has been strongly influenced by this approach, we are reluctant to accept theoretical developments without observational proofs. Nevertheless, we will need to be more receptive to numerical studies (performed with 'robust' models) if progress needs to be achieved. A close collaboration between modellers and experimentalists will be required to better understand what are the key processes that will need to be validated, and how they could be observed.

ACKNOWLEDGEMENTS

This research was partly funded by the National Science Foundation under Grants ATM-9016562 and EAR-9105059, by the National Aeronautics and Space Administration under Grant NAGW-2658, by the US Department of Energy under Grant DE-FG02-92ER61453, and by the National Oceanic and Atmospheric Administration under Grant NA36GP0257. The views expressed herein are those of the author and do not necessarily reflect the views of these agencies.

REFERENCES

André, J.-C., De Moor, G., Lacarrère, P., Therry, G., and du Vachat, R. 1978. 'Modeling the 24-hour evolution of the mean turbulent structures of the planetary boundary layer', *J. Atmos. Sci.*, **35**, 1861–1883.

Avissar, R. 1992. 'Conceptual aspects of a statistical-dynamical approach to represent landscape subgrid-scale heterogeneities in atmospheric models', *J. Geophys. Res.*, **97**, 2729–2742.

Avissar, R. and Chen, F. 1993. 'Development and analysis of prognostic equations for mesoscale kinetic energy and mesoscale (subgrid-scale) fluxes for large scale atmospheric models', *J. Atmos. Sci.*, **50**, 3751–3774.

Avissar, R. and Chen, F. 'An approach to represent mesoscale (subgrid-scale) fluxes in GCM's demonstrated with simulations of local deforestation in Amazonia', *Hydrol. Sci. J.*, in press.

Avissar, R. and Mahrer, Y. 1982. 'Verification study of a numerical greenhouse microclimate model', *Trans. Am. Soc. Agric. Eng.*, **25**, 1711–1720.

Avissar, R. and Mahrer, Y. 1988. 'Mapping frost-sensitive areas with a three-dimensional local scale model. Part I: physical and numerical aspects', *J. Appl. Metereol.*, **27**, 400–413.

Avissar, R. and Pielke, R. A. 1989. 'A parameterization of heterogeneous land-surface for atmospheric numerical models and its impact on regional meteorology', *Monthly Weather Rev.*, **117**, 2113–2136.

Avissar, R. and Pielke, R. A. 1991. 'The impact of plant stomatol control on mesoscale atmospheric circulations', *Agric. Forest Meteorol.*, **54**, 353–372.

Avissar, R. and Verstraete, M. M. 1990. 'The representation of continental surface processes in atmospheric models', *Rev. Geophys.*, **28**, 35–52.

Avissar, R., Avissar, P., Mahrer, Y., and Bravdo, B. A. 1985. 'A model to simulate response of plant stomata to environmental conditions' *Agric. Forest Meteor.*, **34**, 21–29.

Bougeault, P., Bret, B., Lacarrère, P., and Noilhan, J. 1991. 'An experiment with an advanced surface parameterization in a mesobeta-scale model. Part II: the 16 June 1986 simulation', *Monthly Weather Rev.*, **119**, 2374–2392.

Chen, F. and Avissar, R. 1994a. 'Impact of land-surface moisture variability on local shallow convective cumulus and precipitation in large-scale models', *J. Appl. Meteor.*, **33**, 1382–1401.

Chen, F. and Avissar, R. 1994b. 'The impact of land-surface wetness heterogeneity on mesoscale heat fluxes', *J. Appl. Meteor.*, **33**, 1323–1340.

Clarke, R. H., Dyer, A. J., Brook, R. R., Reid, D. G., and Troup, A. J. 'The Wangara experiment: boundary layer data', *Div. Meteorol. Phys. Tech. Pap. No. 19*, CSIRO, Melbourne p. 341.

Collatz, G. J., Grivet, C., Ball, J. T., and Berry, J. A. 1991. 'Physiological and environmental regulation of stomatal conductance, photosynthesis and transpiration: a model that includes a laminar boundary layer', *Agric. Forest Meteorol.*, **54**, 107–136.

Collins, D. and Avissar, R. 1994. 'An evaluation with the Fourier amplitude sensitivity test (FAST) of which land-surface parameters are of greatest importance for atmospheric modelling', *J. Climate*, **7**, 681–703.

Cukier, R. I., Fortuin, C. M., Shuler, K. E., Petschek, A. G., and Schaibly, J. H. 1973. 'Study of the sensitivity of coupled reaction systems to uncertainties in rate coefficients. I. Theory', *J. Chem. Phys.*, **59**, 3873–3878.

Dalu, G. A., Pielke, R. A., Avissar, R., Kallos, G., Baldi, M., and Guerrini, A. 1991. 'Linear impact of thermal inhomogeneities on mesoscale atmospheric flow with zero synoptic wind', *Ann. Geophys.*, **9**, 641–647.

Deardorff, J. W. 1972. 'Numerical investigation of neutral and unstable planetary boundary layers', *J. Atmos. Sci.*, **29**, 91–115.

Deardorff, J. W. 1974. 'Three-dimensional numerical study of the height and mean structure of a heated planetary boundary layer', *Boundary-Layer Meteorol.*, **7**, 81–106.

Dickinson, R. E. and Henderson-Sellers, A. 1988. 'Modelling tropical deforestation, a study of GCM land-surface parameterizations', *Q. J. R. Meteorol. Soc.*, **114**, 439–462.

Dickinson, R. E., Henderson-Sellers, A., Kennedy, P. J., and Wilson, M. F. 1986. 'Biosphere–atmosphere transfer scheme (BATS) for the NCAR community climate model', *NCAR Tech. Note: NCAR/TN-275+STR*, p. 69.

Entekhabi, D. and Eagleson, P. S. 1989. 'Land-surface hydrology parameterization for atmospheric general circulation models including subgrid-scale spatial variability', *J. Climate*, **2**, 816–831.

Famiglietti, J. S. and Wood, E. F. 1991. 'Evapotranspiration and runoff from large land areas: land surface hydrology for atmospheric general circulation models', *Surv. Geophys.*, **12**, 179–204.

Hadfield, M. G., Cotton, W. R. and Pielke, R. A. 1991. 'Large-eddy simulations of thermally-forced circulations in the convective boundary layer. Part I: a small-scale circulation with zero wind', *Boundary-Layer Meteorol.*, **57**, 79–114.

Hadfield, M. G., Cotton, W. R., and Pielke, R. A. 1992. 'Large-eddy simulations of thermally forced circulations in the convective boundary layer. Part II: the effect of changes in wavelength and wind speed', *Boundary-Layer Meteorol.*, **58**, 307–328.

Henderson-Sellers, A. 1993. 'A factorial assessment of the sensitivity of the BATS land-surface parameterization scheme', *J. Climate*, **6**, 227–247.

Henderson-Sellers, A., Dickinson, R. E., and Wilson, M. F. 1988. 'Tropical deforestation, important processes for climate models', *Climatic Change*, **13**, 43–67.

Jarvis, P. G. and Mansfield, T. A. (Eds) 1981. *Stomatal Physiology*. Cambridge University Press, Cambridge. 295 pp.

Li, B. and Avissar, R. 1994. 'The impact of spatial variability of land-surface characteristics on land-surface heat fluxes', *J. Climate*, **7**, 527–537.

Lynn, B. H., Rind, D., and Avissar, R. 1995a. 'The importance of subgrid-scale mesoscale circulations generated by landscape heterogeneity in general circulation models (GCMs)', *J. Climate*, **8**, 191–205.

Lynn, B. H., Abramopolous, F., and Avissar, R. 1995b. 'Using similarity theory to parameterize mesoscale heat fluxes generated by subgrid-scale landscape discontinuities in GCMs', *J. Climate*, **8**, 932–951.

Martin, E. S., Donkin, M. E., and Stevens, R. A. 1983. *Stomata*. Arnold, E. (Ed.). The Camelot Press. p. 60.

Moeng, C.-H. 1984. 'A large-eddy simulation model for the study of planetary boundary-layer turbulence', *J. Atmos. Sci.*, **41**, 2052–2062.

Moeng, C.-H. 1986. 'Large-eddy simulation of a stratus-topped boundary layer, part I: structure and budgets', *J. Atmos. Sci.*, **43**, 2886–2900.

Moeng, C.-H. 1987. 'Large-eddy simulation of a stratus-topped boundary layer, part II: implications for mixed-layer modeling', *J. Atmos. Sci.*, **44**, 1605–1614.

Orlanski, I. 1975. 'A rational subdivision of scales for atmospheric processes', *Bull. Am. Meteorol. Soc.*, **56**, 527–530.

Pielke, R. A. 1984. *Mesoscale Meteorological Modeling*. Academic Press, New York. 612 pp.

Pielke, R. A., Cotton, W. R., Walko, R. L., Tremback, C. J., Lyons, W. A., Grasso, L. D., Nicholls, M. E., Moran, M. D., Wesley, D. A., Lee, T. J., and Copeland, J. H. 1992. 'A comprehensive meteorological modeling system—RAMS', *Meteorol Atmos. Phys.*, **49**, 69–91.

Raupach, M. R. 1993. 'The averaging of surface flux densities in heterogeneous landscapes' in Bolle, H. J., Feddes, R. A., and Kalma, J. D. (Eds), *Exchange Processes at the Land Surface for a Range of Space and Time Scales. IAHS Publ.*, **212**, 343–355.

Sellers, P. J., Mintz, Y., Sud, Y., and Dalcher, A. 1986. 'A simple biosphere model (SiB) for use within general circulation models', *J. Atmos. Sci.*, **6**, 505–531.

Shukla, J. and Mintz, Y. 1982. 'Influence of land-surface evapotranspiration on the Earth's Climate', *Science*, **215**, 1498–1501.

Shukla, J., Nobre, C., and Sellers, P. 1990. 'Amazon deforestation and climate change', *Science*, **247**, 1322–1325.

Stull, R. B. 1988. *An Introduction to Boundary Layer Meteorology*. Kluwer Academic, Dordrecht. 666 pp.

Walko, R. L., Cotton, W. R., and Pielke, R. A. 1992. 'Large-eddy simulations of the effects of hilly terrain on the convective boundary layer', *Boundary-Layer Meteorol.*, **58**, 133–150.

Willmer, C. M. 1983. *Stomata*. Longman, London. 166 pp.

Xue, Y. and Shukla, J. 1993. 'The influence of land surface properties on Sahel climate. Part I: desertification', *J. Climate*, **6**, 2232–2245.

Zeiger, E., Farquhar, G. D., and Cowan, I. R. 1987. *Stomatal Function*. Stanford University Press. 503 pp.

26

EVALUATION OF THE EFFECTS OF GENERAL CIRCULATION MODELS' SUBGRID VARIABILITY AND PATCHINESS OF RAINFALL AND SOIL MOISTURE ON LAND SURFACE WATER BALANCE FLUXES

MURUGESU SIVAPALAN AND ROSS A. WOODS*

Centre for Water Research, Department of Environmental Engineering, The University of Western Australia, Nedlands, WA 6009, Australia

ABSTRACT

Many existing general circulation models (GCMs) use so-called 'bucket algorithms' to represent land-surface hydrology. Wood *et al.* (1992) presented a generalization of the simple bucket representation, based on their variable infiltration capacity (VIC) model. The VIC model, in essence, assumes a statistical distribution of bucket sizes within the grid square. In this way, it provides a simple, computationally efficient and yet physically realistic model of land-surface hydrology. A preliminary attempt is made here, using this simple model, to evaluate the effects of the spatial variability of rainfall intensities and the resulting soil moisture within a hypothetical GCM grid square. Rainfall is assumed to be patchy, with only a fraction of the grid square being wetted by rainfall at any given time. Within the wetted area, however, rainfall is allowed to vary randomly in space. Evaporation during interstorm periods is estimated from the rest of the grid square by a simple non-linear function of the available soil moisture. The soil moisture is also assumed to be patchy and spatially variable due to the antecedent rainfall that caused it. A number of simplifying assumptions have been made in the model about the redistribution of soil moisture at the end of storm and interstorm periods, and the effects of a vegetation canopy have been ignored. The model is applied, under a variety of conditions, to estimate the biases in the modelled water balance fluxes if the assumed spatial heterogeneity is neglected. The model is also used to simulate the long-term water balance dynamics under assumed hypothetical storm and inter-storm climatic inputs, to see how the steady-state hydrological regime is affected by spatial variability. These simulations are relevant to current efforts towards developing simple parameterizations of land-surface hydrology that explicitly incorporate subgrid heterogeneity.

INTRODUCTION

Many existing general circulation models (GCMs) still use variants of the so-called 'bucket model' proposed by Manabe *et al.* (1965) and Manabe (1969) to represent land-surface hydrology. It is now acknowledged that the bucket model is a very simplistic representation of land-surface hydrology—it does not reflect the actual processes of runoff generation and evaporation, it ignores the enormous spatial variability of hydrological processes and it does not explicitly include vegetation.

There has been considerable research in the past decade into developing land-surface hydrological models providing an explicit treatment of the biosphere–atmosphere interactions governing the transfer of energy, mass and momentum between the atmosphere and vegetated land-surfaces. Examples of these include the BATS model (Dickinson *et al.*, 1986) and the SiB model (Sellers *et al.*, 1986). Both the BATS and SiB models have incorporated recent advances in the understanding of plant physiology,

---

* Also at: NIWA Climate, National Institute for Water and Atmospheric Research Ltd, Box 8602, Christchurch, New Zealand

micrometeorology and biosphere–atmosphere interactions. They include a high level of resolution and structure of those surface characteristics that govern the vertical transfer of energy, mass and momentum. Models such as BATS and SiB are often referred to as 'big leaf' models as they assume horizontal homogeneity of the same characteristics within the grid square.

Efforts are currently under way to study the effects of spatial heterogeneity of vegetation and other land-surface parameters. For example, Avissar and Pielke (1989) showed, with a regional mesoscale meteorological model, that spatial heterogeneity in vegetation cover can have significant impacts on land-surface fluxes. A number of approaches have been suggested for developing simple land-surface schemes which incorporate the effects of the subgrid variability of land-surface properties. Avissar (1990) presented a land-surface model based on landscape patchiness heterogeneity. Abramopoulos et al. (1988) used an area-weighted composite scheme for the soil and vegetation parameters. Both the Avissar (1990) and Abramopoulos et al. (1988) approaches effectively involve running, in parallel, big-leaf models for each of a number of 'uniform' land-surface patches and combining the resulting fluxes.

Research carried out by meteorologists, such as by Avissar (1990) and Abramopoulos et al. (1988), concentrated on the effects of heterogeneity on the above-surface processes and did not pay attention to the effects of terrain, soil properties, soil wetness and vegetation on surface and below-surface processes such as runoff generation, soil moisture redistribution and water uptake by plants. In other words, these models incorporated a very high level of complexity in the above-surface parameters and processes, but the subsurface hydrological processes were assumed to be predominantly one dimensional (i.e. vertical) and the relevant state variables and parameters were assumed to be spatially homogeneous. On the other hand, hydrologists such as Entekhabi and Eagleson (1989) and Famiglietti and Wood (1990) concentrated on the heterogeneity of surface and subsurface processes. Entekhabi and Eagleson (1989) prescribed the subgrid variability of soil moisture and precipitation in terms of assumed statistical distributions, and then derived analytical expressions for the resulting hydrological fluxes. In contrast, Famiglietti and Wood (1990) estimated the spatial variability of soil moisture within a catchment, and the hydrological fluxes, using a simple model which relates antecedent soil moisture to the topography–soil index, $\ln(aT_e/T_i\tan\beta)$, presented by Beven and Kirkby (1979) and Sivapalan et al. (1987).

All of the references cited above attempt to capture the effects of spatial variability of land-surface properties in different ways. Approaches developed by Entekhabi and Eagleson (1989), Famiglietti and Wood (1990) and others tend to handle the spatial variability of hydrological variables very well, while using rather simplified treatments of the variability in the vertical direction. In contrast, sophisticated SVAT models developed in recent times (e.g. BATS, SiB, etc.) include considerable vertical resolution and physical realism, whereas the surface and subsurface hydrology are treated in a simple manner and their spatial variability is ignored altogether. There is a continuing need to develop models combining the important aspects of both of these approaches, while at the same time remaining reasonably simple and parsimonious in terms of computational effort and the number of parameters.

Although work in this area is in its early stages, the work of Wood et al. (1992) is an important first step in this direction. Wood et al. present a generalization of the simple bucket representation of Manabe et al. (1965), based on the so-called variable infiltration capacity (VIC) model. With the addition of a single extra parameter, Wood et al. (1992) are able to simulate a distribution of bucket sizes within the grid square. Their model thus provides a simple, but admittedly partial, treatment of subgrid variability, and the resulting hydrological models of runoff generation and evaporation are much more realistic in comparison with the single bucket model. The number of additional parameters is kept to a minimum. Wood et al. (1992) carried out simulations with the VIC model and found that it gave more realistic runoff predictions compared with the single bucket model.

In this paper, the VIC model is used to evaluate the effects of the spatial variability of rainfall intensities and soil moisture within a typical GCM grid square. Rainfall is assumed to be patchy, with only a fraction of the grid square being wetted by rainfall at any given time. Within the wetted area, however, rainfall is assumed to vary spatially, and to be randomly distributed according to the exponential distribution. Evaporation during interstorm periods is assumed to occur at the potential rate from saturated areas and is estimated from the rest of the grid square by means of a simple non-linear function of the available

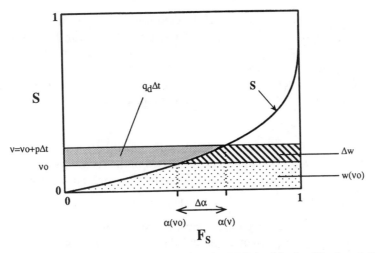

Figure 1. Schematic diagram of variable infiltration capacity (VIC) model, after Wood et al. (1992)

soil moisture. However, soil moisture is also assumed to be patchy and spatially variable due to the antecedent rainfall that caused it.

The model is applied, under a variety of conditions, to estimate the biases in the estimated land-surface hydrological fluxes if the spatial heterogeneity were to be ignored. The model is also used to estimate the long-term water balance dynamics under assumed, hypothetical storm and inter-storm climatic inputs, with a view to simulating the steady-state hydrological regime, and to see how this regime is affected by spatial variability. Finally, we present simple representations of the temporal evolution, during both storm and inter-storm periods, of the spatial variability of soil moisture within the grid square. These simulations are relevant to current efforts towards developing simple parameterizations of land-surface hydrology, explicitly incorporating subgrid variability.

However, the work presented here is only a preliminary attempt to evaluate the impact of subgrid variability of rainfall. The effects of a vegetation canopy, and the effects of subgrid variability of rainfall on interception and throughfall, have been ignored in this study. The effects of spatial locations of the rainfall within the GCM is treated rather simply; so is the problem of redistribution of soil moisture between storm events. These issues have been recognized in the study, and some of their effects elucidated through simulation studies, to draw attention to these problems. However, a full treatment of the problems and possible solutions are beyond the scope of this paper.

VARIABLE INFILTRATION CAPACITY MODEL

This section gives a detailed description of the VIC model used by Wood *et al.* (1992). The VIC model is similar to the earlier Xinanjiang model (Zhao *et al.*, 1980) and the Arno model (Dümenil and Todini, 1992). We continue to use the same terminology as used by Wood *et al.*; however, it should be noted that the infiltration capacity discussed here does not have the same meaning as used by soil physicists. Rather, infiltration capacity is defined as the total volumetric capacity of a vertical soil column (to bedrock) to hold water. Thus the VIC is closely related to the saturation excess mechanism (Dunne-type) of surface runoff generation and not to the infiltration excess (Hortonian-type) mechanism. In fact, it is assumed in this study that runoff by the infiltration excess mechanism is negligible.

Figure 1 presents a schematic of the VIC model, adapted from Wood *et al.* (1992). It describes the cumulative distribution function of the scaled infiltration capacity (scaled by its maximum value)—that is

$$F_S(s) = \mathrm{Prob}\{S \leqslant s\} \tag{1}$$

where the random variable S is the scaled infiltration capacity and $F_S(s)$ is the probability that S takes on values less than or equal to a specified value s. If the depth of the soil profile Z (to bedrock) is known everywhere, with a maximum value of Z_m, and the available porosity $\Delta\theta$ can be assumed to be constant, then $S = Z\Delta\theta)/(Z_m\Delta\theta) = Z/Z_m$. In the context of land-surface parameterizations for GCMs, the storage capacity of the Manabe 'bucket' would be $\bar{Z}\Delta\theta$, where $\bar{Z}$ is the grid-square average of Z. On the other hand, the VIC model can be considered to be equivalent to assuming a distribution of Manabe-type 'buckets', with their sizes, $Z\Delta\theta$, belonging to certain specified statistical distributions.

Xinanjiang distribution

Wood *et al.* (1992) adopted an empirical distribution of infiltration capacities first developed in China for catchment modelling (Zhao *et al.*, 1980) and later also used in the Arno model (Dümenil and Todini, 1992). This empirical distribution, called here the Xinanjiang distribution, can be expressed as

$$F_S(s) = 1 - (1 - s)^\beta \tag{2}$$

where β is an empirical parameter. Figure 1 shows the shape of the Xinanjiang distribution for $\beta = 4$. Wood *et al.* (1992) suggested that when empirical data on soil depths or infiltration capacities do not exist, the parameters β and Z_m could be estimated by calibration, using rainfall–runoff data on representative catchments. In their simulations, Wood *et al.* varied β in the range 0·01–5·0. For GCM land-surface parameterizations, Dümenil and Todini (1992) suggested β values in the range 0·01–0·5, whereas Liang (1994) found a value of $\beta = 0·008$ by calibration on the King's Creek catchment in Kansas. Note that $\beta = 0$ corresponds to a single bucket. To our knowledge, few studies have been conducted where the value of β has been estimated directly from field data.

Based on extensive drilling carried out in the south-west of Western Australia, for the purposes of estimating soil salt storage, considerable information exists on the depth of the A horizon soils that dominate the hydrology of catchments in this region. Surface soils in the region have been classified into a number of distinct landform types. Systematic regionalizations of the distributions of the measured soil depths have been carried out for each of the landform types. For any specific catchment or area of the land-surface, knowing the spatial extent (i.e. the proportion of the surface area) of each of the landforms within it, the composite distribution of the soil depths can then be estimated as the area-weighted combination of the individual distributions.

As an illustration, Figure 2 presents a map of the predominant landforms contained within the Serpentine catchment (661 km² in area), including the estimates of their spatial extent (Brad Harris, pers. comm., 1994). Figure 3 presents the regionalized empirical distributions of the soil depths for six dominant landform types within the south-west region. These are based on data provided by the Water Authority of Western Australia (Leanne Pearce, pers. comm., 1994). The composite, area weighted distribution of the scaled soil depths (all scaled by the maximum depth of 10 m) for the Serpentine catchment is then presented in Figure 4, along with a least-squares fit to the Xinanjiang distribution. It is seen that the Xinanjiang distribution, with $\beta = 4·03$ and $Z_m = 10$ m, provides an excellent fit to the empirical data. Results presented in the remainder of this paper have been obtained using these parameter values for the Xinanjiang distribution. The available porosity used in these calculations is $\Delta\theta = 0·30$, which is assumed spatially and temporally constant.

Note that although the Xinanjiang distribution just presented provides an excellent fit to the empirical distributions of measured soil depths or infiltration capacities in Western Australia, and presumably also in China, its physical basis remains to be established. Note also that the result of $\beta = 4·03$ remains at variance with the values obtained in previous studies from rainfall–runoff model calibration, i.e. 0·01–0·5. Sivapalan and Woods (submitted) has estimated the distributions of soil moisture deficits for a number of catchments assuming that TOPMODEL, the topography based runoff generation model of Beven and Kirkby (1979), applies. They found distributions that were consistent with β values less than unity.

The wide variability of the published β values and the critical importance of the β parameter are important reasons for seeking a physical explanation for the values of the β parameter in different hydrological settings. The available information in Western Australia indicates (see Figure 2) a qualitative connection of

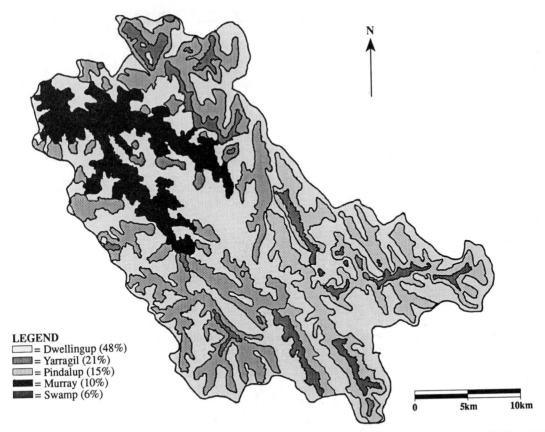

Figure 2. Landform map of Serpentine catchment (661 km² in area) in the south-west of Western Australia (provided by the Water Authority of Western Australia)

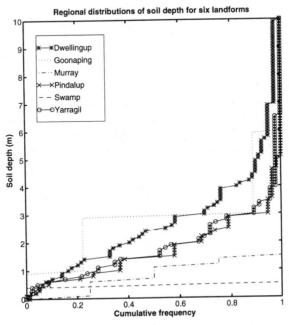

Figure 3. Regionalized cumulative distribution of soil depths for six typical landforms in the south-west of Western Australia (provided by the Water Authority of Western Australia)

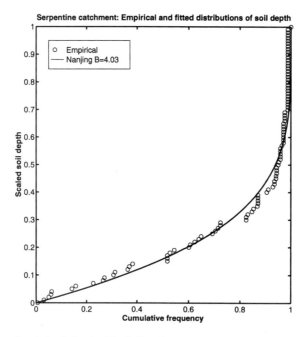

Figure 4. Cumulative distribution of scaled soil depths (also infiltration capacities) for the entire Serpentine catchment (circles) and Xinanjiang distribution with parameter $\beta = 4\cdot03$

the landform types and also of the soil depths (see Figure 3) to the topographic position. However, to have any degree of universality, this connection needs to be expressed quantitatively in terms of a physically based theory. This is a subject of ongoing research (e.g. Dietrich *et al.*, 1995) and useful results can be expected in the future.

Specification of antecedent soil moisture

Following Wood *et al.* (1992), for regions within which the long-term average rainfall can be assumed to be spatially uniform, we express the antecedent soil moisture state of the land surface by an antecedent soil moisture parameter v_0 (see Figure 1), where $v_0 = y_0/Z_m$. This is taken to mean that the soil profile is effectively saturated to a depth of y_0 (above bedrock), except in those parts of the land-surface where the depth of soil is less than y_0, in which instance the land surface is saturated even before the addition of any new rainfall. With the addition of rainfall, the level of soil moisture rises above v_0, and the saturated area expands, consistent with the saturation excess mechanism of runoff generation.

For a given antecedent soil moisture, as specified by v_0, it is possible to estimate the total moisture storage, w, within the soil when the soil depths are distributed according to the Xinanjiang distribution. This storage volume refers to the stippled area in Figure 1 and is denoted by $w(v_0)$. The saturated portion of the total area corresponding to v_0, which is denoted by $\alpha(v_0)$, can be read off Figure 1. Given the parameters β and Z_m, and for any arbitrary soil moisture state denoted by v (assumed constant), the functions $w(v)$ and $\alpha(v)$ can be derived analytically. These functions are also simple enough that they can be inverted, i.e. given w it is possible to analytically estimate both $v(w)$ and $\alpha(w)$. These relationships are

$$w(v) = \frac{1}{(\beta + 1)}[1 - (1 - v)^{\beta+1}] \tag{3}$$

$$\alpha(v) = 1 - (1 - v)^{\beta} \tag{4}$$

$$v(w) = 1 - [1 - (\beta + 1)w]^{1/(\beta+1)} \tag{5}$$

$$\alpha(w) = 1 - [1 - (\beta + 1)w]^{\beta/(\beta+1)} \tag{6}$$

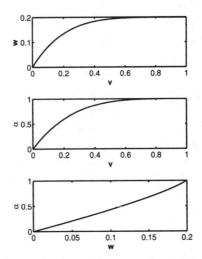

Figure 5. Lumped relationships between scaled soil moisture parameter v, soil moisture storage w and saturated area fraction α, for $\beta = 4$

The maximum value of w (the total area under the S curve) is denoted as w_c and is equal to $w_c = 1/(1 + \beta)$. Figure 5 presents plots of these functions for the Serpentine catchment ($\beta = 4\cdot03$).

The assumption in the VIC model that v_0 is uniform across the catchment may not be justified on physical grounds; however, it leads to simple analytical solutions and may be a sufficient approximation for use in GCMs. Its validity in subsurface runoff dominated parts of the land-surface needs to be investigated.

Estimation of runoff generation

The VIC model incorporates surface runoff generation by the saturation excess mechanism (Dunne-type) and subsurface runoff is included as a simple function of the soil moisture storage w. Infiltration excess runoff (Hortonian-type) is ignored. The main objective of this paper is to investigate the effects of subgrid variability of rainfall on runoff generation and evaporation and on the resulting soil moisture storage. We begin the analysis by presenting expressions for surface and subsurface runoff, for the case where the rainfall can be assumed to be spatially uniform.

The addition of spatially uniform rainfall, assuming it percolates rapidly towards the saturated zone, produces three effects: (i) a uniform rise in the level of the saturated zone, v, except on the already saturated parts of the land-surface; (ii) an increase in groundwater storage, w, and (iii) an expansion of the saturated areas, α. These are illustrated schematically in Figure 1. Let us assume that $p\Delta t$ is the scaled depth of rainfall over a specified time period Δt (i.e. scaled by $Z_m\Delta\theta$), where p is the scaled rainfall intensity. The increase in groundwater storage Δw, the expansion of saturated area fraction $\Delta\alpha$, and the volume of surface runoff generation, $q_d\Delta t$, due to the rainfall depth (and the rise of saturated zone) $p\Delta t$, are presented in Figure 1. All of the above variables are scaled quantities. In this paper, following Wood *et al.* (1992), groundwater runoff, $q_s\Delta t$, is expressed as a linear function of an average value of the soil moisture storage w. However, this may be generalized with more physically based and non-linear functions of the storage. This is left for further study. Note that q_d and q_s, similar to the rainfall intensity p, are scaled runoff intensities.

Using the set of functions given by Equations (3) to (6), all of the above quantities can be expressed analytically as follows:

$$\Delta w = w(v_0 + p\Delta t) - w(v_0) \tag{7}$$

$$\Delta\alpha = \alpha(v_0 + p\Delta t) - \alpha(v_0) \tag{8}$$

Surface runoff: $\qquad q_d\Delta t = p\Delta t - [w(v_0 + p\Delta t) - w(v_0)] \tag{9}$

Subsurface runoff: $\qquad q_s\Delta t = \tfrac{1}{2}k_c[w(v_0 + p\Delta t) + w(v_0)]\Delta t \tag{10}$

where k_c is a constant.

Total runoff intensity, q_T, is then given by

$$\text{Total runoff:} \qquad q_T = q_d + q_s \tag{11}$$

Estimation of evaporation

Even when the level of soil moisture, following rainfall, can be considered to be 'uniform', and equal to $v = v_0 + p\Delta t$ (see Figure 1), the soil moisture deficit in the soil column still exhibits considerable spatial variability. As can be seen in Figure 1, the soil moisture deficit, $d = s - v$, varies from a value of zero over the saturated areas, to a maximum of $1 - v$. This variability will have an impact on the total evapotranspiration from the land-surface and its lumped representation in terms of storage $w = w(v)$. Wood *et al.* (1992), in analogy with Equation (2), proposed a lumped model of evaporation of the following form

$$\frac{e}{e_p} = 1 - \left[1 - \frac{w}{w_c}\right]^{1/\beta_c} \tag{12}$$

where e is the total (lumped) actual evapotranspiration from a 'uniformly' wetted land-surface, e_p is the potential evapotranspiration from the same surface, $w_c = 1/(1 + \beta)$, and β_c is a constant.

However, in this paper, we develop an alternative lumped evapotranspiration model that explicitly incorporates the spatial variability of soil moisture availability. Assuming the land-surface is uniformly wetted to a level of $v = v_0 + p\Delta t$, we develop this lumped model by integrating a point-scale evaporation model over the entire land-surface, using the assumed distribution of infiltration capacities given by the Xinanjiang model [Equation 2].

The point-scale model, adapted from Neghassi (1974), estimates the local actual evapotranspiration e_s (s refers to all locations which have the same value of infiltration capacity), as a function of local soil moisture, as follows:

$$\frac{e_s}{e_p} = \left(\frac{v + \psi_c}{s}\right)^\eta \qquad \text{for } (v + \psi_c) < s \tag{13a}$$

$$\frac{e_s}{e_p} = 1 \qquad \text{for } (v + \psi_c) \geqslant s \tag{13b}$$

where v = level of soil moisture, s = local maximum value of soil moisture (i.e. the infiltration capacity), ψ_c = scaled capillary fringe thickness (scaled by Z_m), and η is a property of the soil and vegetation types, assumed to be constant in space. The total evaporation from the heterogeneous land-surface, wetted by uniform rainfall to a uniform level v, is then given by

$$\frac{e}{e_p}(v, \psi_c) = \int_0^1 \frac{e_s}{e_p}(s; v, \psi_c) f_S(s) \, ds \tag{14}$$

The point-scale model used here [Equation (13)] has been chosen for its simplicity. It can be replaced fairly easily by more physically based functions to reflect the type of soil and vegetation on the land-surface. In the present case, Equation (14) can be simplified as follows:

$$\frac{e}{e_p}(w, \psi_c) = \frac{e}{e_p}(v, \psi_c) = 1 - [1 - (v + \psi_c)]^\beta + \beta G(v + \psi_c | \beta, \eta) \tag{15a}$$

where

$$G(x | \beta, \eta) = x^\eta \int_x^1 s^{-\eta} \int_x^1 s^{-\eta} (1 - s)^{\beta - 1} \, ds \tag{15b}$$

The integral in Equation (15b) is evaluated numerically.

The results of the integration, in terms of lumped e/e_p versus w/w_c relationships, are presented in Figure 6 for four different values of ψ_c. The lumped model proposed by Wood *et al.* (1992) is also presented

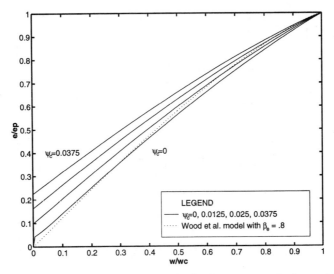

Figure 6. Relationship between e/e_p and w/w_c for $\beta = 4\cdot03$ and $\eta = 3\cdot0$ for the evapotranspiration model developed in this paper (for $\psi_c = 0\cdot0, 0\cdot0125, 0\cdot025, 0\cdot0375$) and for the model proposed by Wood *et al.* (1992) for the case $\beta_e = 0\cdot8$ (dotted line)

for comparison for the case $\beta_e = 0\cdot8$. Although the two models produce similar results, the model developed in this paper has the advantage that it has been obtained from a physically based relation at the point scale. In the rest of this paper the parameters η and ψ_c in Equation (13) have been set at values of $3\cdot0$ and $0\cdot025$, respectively.

It may be feasible to obtain a lumped, closed-form expression relating e/e_p to w, but we have not attempted to do this in this paper. Also, it may be worthwhile generalizing the point-scale models of evapotranspiration to include sophisticated models incorporating plant physiology, soil physics and micrometeorology. Both of these are left for further study. The lumped models of runoff generation and evapotranspiration described here will be later used as the building blocks for estimating the same fluxes when the rainfall wetting is spatially variable.

Estimation of long-term water balance

The lumped models of runoff generation and evapotranspiration presented above, with $\beta = 4\cdot03$, $\eta = 3\cdot0$, $\psi_c = 0\cdot025$ and $Z_m = 10\,\text{m}$, were used to simulate the land-surface responses to the series of hypothetical storm and interstorm climatic inputs ($\bar{p}$ and $e_p = E_p$, a regional value of potential evapotranspiration, assumed constant) shown in Figure 7a. Thus these inputs are characterized by spatial and temporal uniformity of both the rainfall intensities p and the potential evapotranpiration rate e_p. Their magnitudes, and those of the storm duration and inter-storm period, have been chosen such that the land-surface remains in 'hydrological equilibrium'. The resulting 'stationary' temporal variations of total runoff, evaporation rates and the soil moisture storage are shown in Figure 7b and 7c. These variations are used as the baseline conditions against which subsequent results, based on spatially variable rainfall, will be compared.

SIMPLE MODELS OF RAINFALL VARIABILITY

Following Warrilow *et al.* (1986), Shuttleworth (1988) and Thomas and Henderson-Sellers (1991), we assume that rainfall is patchy, falling on a fraction $\mu(\mu \leqslant 1)$ of the GCM grid square. Within the wetted area, however, the scaled rainfall intensity p (rainfall depth per unit time scaled by $Z_m \Delta\theta$) is allowed to vary randomly in space according to the exponential distribution. Given the grid-averaged and similarly scaled rainfall intensity, $\bar{p}$, the probability density function of p (note: this is a conditional distribution

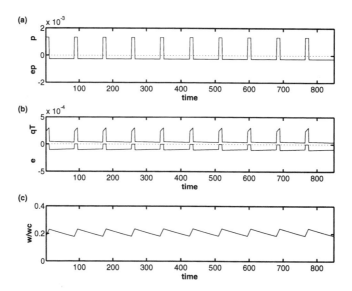

Figure 7. Long-term water balance analysis for spatially uniform rainfall: (a) rainfall (i.e. $\bar{p}$) and potential evapotranspiration (i.e. E_p) sequences; (b) predicted total runoff (i.e. q_T) and actual evapotranspiration rates (i.e., e); and (c) temporal variation of scaled soil moisture storage w/w_c.

and applies within the wetted area only) is then given by

$$f_\mathrm{p}(p|p > 0) + \frac{\mu}{\bar{p}}\exp\left(-\frac{\mu p}{\bar{p}}\right) \tag{16}$$

with conditional mean $= \bar{p}/\mu$ and conditional variance $= (\bar{p}/\mu)^2$. Over the entire grid square (wetted and non-wetted areas), the (unconditional) cumulative distribution function of p is given by

$$F_\mathrm{p}(p) = 0 \qquad\qquad\qquad \text{for } p \leqslant 0 \tag{17a}$$

$$F_\mathrm{p}(p) = (1 - \mu) + \mu\left[1 - \exp\left(-\frac{\mu p}{\bar{p}}\right)\right] \quad \text{for } p > 0 \tag{17b}$$

Distributions other than the exponential have been suggested (Warrilow *et al.*, 1986; Liang, 1994). The attractiveness of the exponential distribution presented here is that it requires only two variables to be prescribed by the atmospheric component of the GCM: $\bar{p}$ and μ. Based on previous climatological studies, it may be possible to estimate the average value of the wetted area fraction μ, given $\bar{p}$ (Thomas and Henderson-Sellers, 1991; Liang, 1994). If more complex distributions are used, however, it may not be so easy to estimate their parameters. This is the reason that we have used the exponential distribution in this study. It should also be remembered that the GCM is not likely to provide the exact location of the rainfall wetting. This will generate an additional degree of bias in the model predictions, which may only be quantified probabilistically. This will be discussed in detail later.

In this paper, we have also assumed that the potential evapotranspiration does not vary with location within the grid square and is unaffected by rainfall wetting. Thus, $e_\mathrm{p} = E_\mathrm{p}$, a constant for the grid square, in time and in space. Also, in the simulations with patchy rainfall, we have assumed that during rainfall periods, no evapotranspiration takes place from anywhere within the grid square, not even from the non-wetted areas.

MODEL OF WATER BALANCE INCLUDING SUBGRID VARIABILITY

The lumped models of evaporation and runoff generation presented in the preceding section assume that the rainfall wetting $p\Delta t$, which is equal to the rise in soil moisture produced by the rainfall, are spatially uniform. In this section we will be concerned with the effects of the spatial variability of rainfall wetting (patchiness as well as random variability) on both large-scale runoff generation and evapotranspiration.

For simplicity, we assume that the length scales of variability of rainfall intensity p, including the random variability, are much larger (being of the order of kilometres) than the length scales of variability of the infiltration capacity s (assumed to be of the order of hundreds of metres). This means that we are effectively ignoring any spatial gradients in s due to systematic geological or geomorphological variations. We are also ignoring any mesoscale circulations or atmospheric feedbacks that might develop as a result of the patchiness of the rainfall wetting.

We begin with a uniform antecedent soil moisture v_0 everywhere on the land-surface. It is then wetted by patchy rainfall described by Equations (16) and (17). To estimate the total runoff generation, saturated area fraction and evapotranspiration, we divide the rainfall depth distribution into classes of uniform precipitation, based on which the land-surface is subdivided into a number of subunits. Within each class, we assume that runoff generation, saturated area fraction, evapotranspiration and groundwater storage can be estimated using the lumped models presented in earlier sections. This is a consequence of the assumption made earlier about the relative length scales of variability of rainfall and infiltration capacity.

Let the number of classes of precipitation values used within the wetted fraction be N ($N = 200$ in this paper). The non-wetted fraction of the grid square forms another class, and thus we have $N + 1$ different classes, or alternatively, land-surface subunits, each of which is assumed to be uniformly wetted. For a typical subunit receiving rainfall intensity p, and subsequently subjected to a potential evapotranspiration demand E_p, we can then apply the lumped water balance model presented earlier. We are thus able to simulate the temporal variations of $q_d(t|p)$, $q_s(t|p)$, $w(t|p)$, $q_T(t|p)$, $\alpha(t|p)$ and $e(t|p)$, for each of the land-surface subunits characterized by p. This procedure can then be continued for any number of such cycles of climatic inputs.

The grid-square average values of these quantities are then estimated by integration over the statistical distribution of rainfall intensities or depths, which is given by Equations (16) and (17). Thus the integrated surface and subsurface runoff generation rates Q_d and Q_s, groundwater storage W, the saturated area fraction A, and the evapotranspiration ratio E/E_p, all of which are grid-square averages, are given by

$$Q_d(t|\bar{p}) = \mu \int_0^\infty q_d(t|p) f_p(p|p > 0) \mathrm{d}p \tag{18}$$

$$Q_s(t|\bar{p}) = \mu \int_0^\infty q_s(t|p) f_p(p|p > 0) \mathrm{d}p + (1 - \mu) q_s(t|0) \tag{19}$$

$$W(t|\bar{p}) = \mu \int_0^\infty w(t|p) f_p(p|p > 0) \mathrm{d}p + (1 - \mu) w(t|0) \tag{20}$$

$$A(t|\bar{p}) = \mu \int_0^\infty \alpha(t|p) f_p(p|p > 0) \mathrm{d}p + (1 - \mu) \alpha(t|0) \tag{21}$$

$$\frac{E}{E_p}(t|\bar{p}) = \mu \int_0^\infty \frac{e}{e_p}(t|p) f_p(p|p > 0) \mathrm{d}p + (1 - \mu) \frac{e}{e_p}(t|0) \tag{22}$$

$$Q_T(t|\bar{p}) = Q_d(t|\bar{p}) + Q_s(t|\bar{p}) \tag{23}$$

The terms $q_s(t|0)$, $w(t|0)$, $\alpha t|0)$, and $e(t|0)$ in Equations (19) to (22) refer to the non-wetted fraction of the grid square, but are estimated in the same way as the other subunits, except that the rainfall depth is zero. The integrals in Equations (18) to (22) are carried out numerically. Note that $e_p = E_p$ in Equation (22). Note also that in Equations (18) to (23) it is assumed that the rainfall distribution does not change with time, i.e. the number of land-surface subunits and their probability weights are assumed constant during all of the storm and inter-storm events.

In an almost parallel development brought to our attention recently (D. Lettenmaier, pers. comm., 1994), Liang (1994) has also studied the effects of subgrid variability of rainfall and patchiness on surface water and energy balances by following an almost identical approach to that adopted in Eqations (18) to (23), and

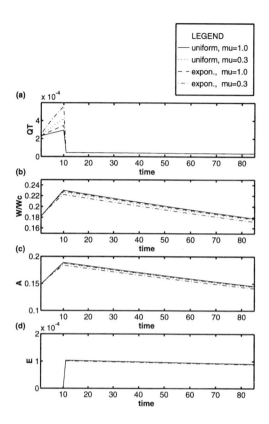

Figure 8. Biases in runoff generation and evapotranspiration due to spatial variability and patchiness of rainfall, during consecutive storm and inter-storm periods, for four rainfall cases: case 1, uniform, widespread ($\mu = 1 \cdot 0$); case 2, uniform, patchy ($\mu = 0 \cdot 3$); case 3, exponential, widespread ($\mu = 1 \cdot 0$); and case 4, exponential, patchy ($\mu = 0 \cdot 3$). (a) Total runoff Q_T; (b) soil moisture storage W/w_c; (c) saturated area fraction A; and (d) evapotranspiration E

then extended this study to include the effects of a vegetation canopy. The results obtained regarding the effects of subgrid variability of rainfall are also similar to those presented next.

Biases in runoff generation and evapotranspiration

The application of Eqations (18) to (23) is a straightforward exercise when we start from a uniform initial moisture v_0, and the land-surface experiences just one climatic cycle of a rainfall period followed by an evapotranspiration period. Running of multi-event sequences of rainfall and evapotranspiration creates problems relating to the locations of subsequent events and to how the soil moisture state is carried through to the subsequent events. Initially we will present the results for a single cycle of rainfall and evapotranspiration periods, thus avoiding these problems.

Here we investigate the biases in runoff generation and evapotranspiration resulting from the 'false' assumption that rainfall is uniform across the entire grid square, when in fact it is not. Figure 8a–d present the results of this investigation for four rainfall cases: case 1, uniform, widespread ($\mu = 1 \cdot 0$); case 2, uniform, patchy ($\mu = 0 \cdot 3$); case 3, exponentially distributed, widespread ($\mu = 1 \cdot 0$); and case 4, exponentially distributed, patchy ($\mu = 0.3$). Figure 8a presents a comparison of the runoff intensities predicted by the model. Runoff generation is highest for case 4 and lowest for case 1, indicating that considerable bias could be caused if rainfall is assumed to be spatially uniform, the bias being greatest if rainfall is patchy. The differences in runoff generated are mirrored in the soil moisture storage at the end of the storm event (Figure 8b), with case 1 having the highest amount of storage. This is also reflected in the saturated area fraction (Figure 8c). This may be counter-intuitive, as we have a situation where the case that has the highest amount of saturated area fraction is also producing the least amount of runoff! However, this result follows directly from the conservation of mass: if we start two grid squares from the same initial storage, and one grid square produces more runoff than the other from the same grid-averaged rainfall, it must have less final moisture storage. The evapotranspiration during the inter-storm periods (Figure 8d) appears to be only marginally affected by spatial variability of rainfall, despite the fact that the beginning storage levels are very different for the three cases. The substantial differences in runoff generation and marginal differences in evapotranspiration mean that the water balance at the end of one cycle of storm and inter-storm periods is substantially different for the four cases. This is discussed in detail in the next section.

BIASES IN LONG-TERM WATER BALANCE

We have seen here that the spatial variability of rainfall causes an increase in runoff, a decrease in groundwater storage and also changes in evapotranspiration during and after individual events. However, for climate studies, it is equally important to know whether ignoring the spatial variability and patchiness of rainfall will lead to substantial biases in the long-term hydrological regime. This is the motivation for this part of the study.

It should be kept in mind that, in this study, we are ignoring the feedbacks between changes in the long-term hydrological regime and any resulting changes in the atmospheric circulation. It may be that land surface–atmospheric feedbacks may enhance or suppress the water balance anomalies that we are attempting to simulate in this paper. These feedbacks need to be investigated carefully as they may have important implications for climate modelling. We are limiting our analyses here to the purely hydrological consequences of rainfall variability. Thus the results obtained may only apply at catchment scales at which the land surface–atmospheric feedbacks may not be critically important.

The modified VIC model presented earlier was run for 10 cycles of storm and interstorm periods consisting of spatially variable (patchy) and temporally constant rainfall intensity p, and spatially and temporally constant potential evapotranspiration rate E_p. These are equivalent to the climatic variables presented in Figure 7a, except that in this instance, rainfall may be patchy and is exponentially distributed within the wetted patch. Recall that the rainfall intensity, potential evapotranspiration rate, storm duration and inter-storm period seen in Figure 7a were chosen such that under spatially constant rainfall wetting, covering the entire grid square, the land-surface will be at hydrological equilibrium (see Figure 7b and 7c). This study can be extended to include more realistic descriptions of rainfall intensity and potential evaporation rate (i.e. temporally variable rainfall, diurnal variations of E_p, random distributions of storm duration, rainfall intensity and inter-storm period, etc.). This is the subject of an extended investigation and will be reported elsewhere

Handling of moisture redistribution

Simulation of long-term water balance, over multi-event sequences of climatic inputs, presents problems which can only be resolved approximately. This has to do with the fact that we do not know precisely whether subsequent rainfall events will have the same wetted area fraction μ, and even if this is so, whether the new rainfall falls on the fraction of the grid square wetted during the previous storm (which is one extreme), or in an area that was not wetted at all during the previous event (which is another extreme). It is likely that there will be some overlap between the wetted area fractions of subsequent storms. A number of factors, such as orography, land-surface changes, regional climatic variations, etc., may significantly affect this question. Even if the wetted fractions are identical, it is not likely that the actual patterns of rainfall variation within them will be identical for all storms.

In this paper, we present simplified treatments of this problem using three alternative, approximate methods, with a view to determining the sensitivity of simulated long-term water balance to the lack of precision of the exact locations of rainfall wetting. All three methods assume that the wetted area fraction μ remains constant for all the storms. An important consideration in the choice of the three methods has been that we do not want to have more than one state variable describing the mean soil moisture state of the grid square. For example, it is impractical in a GCM to have two state variables, describing the states of the wetted and non-wetted fractions.

1. Method 1. Redistribution of the available soil moisture storage at the end of each storm period. In this way, at the end of the storm period the antecedent soil moisture will be uniform, and all memory of the patchiness, and of the location of the patch, are lost.
2. Method 2. Redistribution of the available soil moisture storage at the end of every cycle of storm and inter-storm periods. In this instance, the memory of the patchiness, and of the location of the patch, are explicitly considered during the evapotranspiration period that follows rain, but are lost for any subsequent rainfall events.
3. Method 3. Rainfall wetting takes place repeatedly over the same region, and no redistribution of the soil moisture storage is performed at any time.

We do not believe that any one of these three methods provides for a realistic approximation in all situations. It is easier to argue that, on a statistical basis, storms of random location justify the 'smearing' of the soil moisture between wet and dry areas at the end of a storm or inter-storm period. Some might argue that method 3 is unrealistic, that it is difficult to justify having storms always hit the same part of the grid square. However, in a region affected by orography, such as we find near the Darling Range of Western Australia, it is in fact harder to justify methods 1 and 2 than method 3. We believe the redistribution of soil moisture in a given grid square should be guided by local climatology. In the absence of strong systematic variations of long-term rainfall in a region, method 3 is indeed unrealistic and we next show, using simulations, that it does lead to significant biases in the long-term water balance.

Effects of moisture redistribution

The effects of these three redistribution schemes on the long-term water balance for the sequences of events shown in Figure 7a were investigated first. The results are presented in Figure 9a–d in terms of the temporal variations of groundwater storage and saturated area fraction for two rainfall cases: case 3, exponentially distributed, *widespread* rainfall ($\mu = 1 \cdot 0$); and case 4, exponentially distributed, *patchy* rainfall ($\mu = 0 \cdot 3$). For both cases, methods 1 and 2 give identical results, which is a consequence of the fact that evapotranspiration rates are, relatively, unaffected by the spatial variability of rainfall wetting, as shown previously in Figure 8. On the other hand, the extreme case (method 3), which assumed no redistribution and assumed that rainfall falls repeatedly in the same regions and in the same proportions, produces the highest amount of bias. This is an important result for distributed land-surface modelling.

Effect of patchiness and spatial variability

The results presented in this section, and subsequent results, have all been obtained using method 2 for the redistribution of soil moisture storage for multiple event simulations. Figure 10(a–c) describes the effects of rainfall patchiness and spatial variability on the long-term water balance (for reasons of clarity, only five of the 10 cycles are presented). The same four cases as in Figure 8 are considered: case 1, uniform, $\mu = 1 \cdot 0$; case 2, uniform, $\mu = 0 \cdot 3$; case 3, exponential, $\mu = 1 \cdot 0$; case 4, exponential, $\mu = 0 \cdot 3$. They have been obtained for the same sequence of climatic inputs presented in Figure 7a. The most important result is shown in Figure 10a for soil moisture storage, indicating that ignoring spatial variability and patchiness can lead to a wetter equilibrium than would prevail if the variability is explicitly included. For the numerical values used for patchiness, i.e. $\mu = 0 \cdot 3$, the bias due to patchiness appears to be more important than that due to the random variability within the patch. The results for runoff generation and evapotranspiration are presented in Figure 10(b, c). Together, these results demonstrate the extreme importance of accounting for the spatial variability, especially the patchiness, in long-term water balance models and in GCMs. Without it, the models can lead to consistent underestimation of runoff and overestimation of soil moisture.

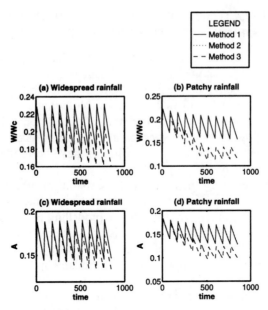

Figure 9. Effects of redistribution of soil moisture during multi-event sequences of *exponentially distributed* rainfall on soil moisture storage W/w_c and saturated area fraction A: method 1 (solid line); method 2 (dotted line); and method 3 (broken line). (a) W/w_c for widespread rainfall ($\mu = 1\cdot0$); (b) W/w_c for patchy rainfall ($\mu = 0\cdot3$); (c) A for widespread rainfall ($\mu = 1\cdot0$); and (d) A for patchy rainfall ($\mu = 0\cdot3$)

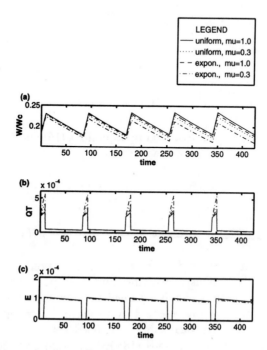

Figure 10. Biases in long-term water balance regime due to spatial variability of rainfall and patchiness, for rainfall cases 1 to 4 discussed in Figure 8, during multi-event sequences of climatic inputs: (a) soil moisture storage, W/w_c; (b) total runoff, Q_T; and (c) evapotranspiration, E

TEMPORAL EVOLUTION OF SOIL MOISTURE HETEROGENEITY: DERIVED DISTRIBUTION APPROACH

If the rainfall wetting, the resulting soil moisture and the potential evapotranspiration are spatially uniform, then the soil moisture storage alone may be sufficient to predict hydrological fluxes at the land-surface. Therefore, all that a land-surface scheme needs to do to accurately predict these fluxes is to continuously monitor the values of the soil moisture storage w. Note that such a scheme, when it is based on the VIC or similar model, is already superior to the simple Manabe bucket as the subgrid variability of infiltration capacities (soil depths or bucket sizes) is already being taken into account. The effects of such variability on runoff generation and evapotranspiration have been effectively 'parameterised' in the lumped models presented in this paper.

However, it is clear from the results of this study that the spatial variability of rainfall, and of the resulting soil moisture storage, can have significant impacts on the land-surface fluxes. How can these effects be handled in a simple and efficient manner for land-surface modelling purposes? The approximate methods presented earlier call for the monitoring of the soil moisture state for a number of land-surface subunits within the grid square for considerable periods of time. Is there a way to monitor the evolution of the sub-grid variability of soil moisture by more parsimonious methods? This question has motivated the results presented in this section.

We present a methodology to predict the continuous evolution of not just the mean soil moisture status, but also its spatial variability, arising from the spatial variability of rainfall, including patchiness. We denote by d the local soil moisture deficit at any point on the land-surface. Referring to Figure 1, we can express d as follows:

$$d = s - v \quad \text{for } s > v \tag{24a}$$

$$d = 0 \quad \text{for } s \leqslant v \tag{24b}$$

Both s and v are random variables with s belonging to the Xinanjiang distribution [Equation (2)]. During rainfall periods only, $v - v_0$ belongs to the same modified exponential distribution as the rainfall depth [Equations (16) and (17)], where v_0 is the starting value of soil moisture.

The distribution of d can be estimated using the convolution relationship given by Mood *et al.* (1974) for deriving the distribution of the difference between two independent random variables. Here we call this method 'the derived distribution approach'. Applying this to the present case, we obtain

$$F_D(d) = \int_0^\infty F_S(v+d)f_V(v)\mathrm{d}v = \int_0^1 F_S(v+d)\mathrm{d}F_V(v) \tag{25}$$

For a rainfall period with patchy rainfall described by Equations (16) and (17), and starting from uniform initial condition $v = v_0$, Equation (25) can be rewritten as

$$F_D = \mu \int_0^\infty F_S(p+v_0+d)f_p(p|p>0)\mathrm{d}p + (1-\mu)F_S(d+v_0)$$

$$= \mu \int_0^1 F_S(p+v_0+d)\mathrm{d}F_P(p|p>0) + (1-\mu)F_S(d+v_0) \tag{26}$$

However, Equation (26) cannot be used to predict the evolution of soil moisture level v during evaporative periods as it is continuously evolving and the rate of decrease of soil moisture is governed by the potential evaporation, which is assumed to be uniform, and the available soil moisture, which varies in space and time. We overcome this problem by monitoring the evolution of soil moisture level v for each of the land-surface subunits defined before [during rainfall periods these are identical to the rainfall classes used in Equations (18) to (23), (25) and (26)]. The model predicts v as a function of time for each land-surface subunit ($v = v_1$ for the subunits inside the wetted fraction, $v = v_2$ for the subunit representing the non-wetted fraction of the grid square). Symbolically these can be

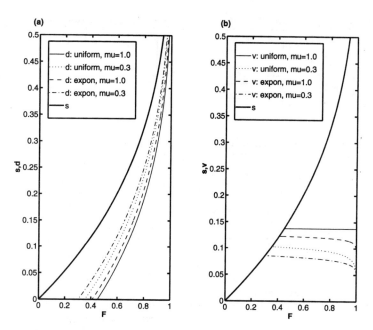

Figure 11. (a) Distributions of the soil moisture deficit d at the end of a rainfall period that produced a grid-square averaged scaled rainfall depth of 0.1 (case 1, solid thin line; case 2, dotted line; case 3 broken line; case 4, dashed–dotted line); and (b) the effective soil moisture profile v that produces the same distribution of soil moisture deficit d as given above

expressed as

$$v_1 = v_1[t|F_P(p > 0)] \qquad \text{(for the wetted fraction)} \tag{27a}$$

$$v_2 = v_2(t) \qquad \text{(for the non-wetted fraction)} \tag{27b}$$

Equation (26) can now be rewritten more generally for any time period during and after randomly distributed (e.g. exponentially distributed) patchy rainfall as follows:

$$F_D(d) = \mu \int_0^1 F_S(v_1 + d)\mathrm{d}F_P(v_1|p > 0) + (1 - \mu)F_S(d + v_2) \tag{28}$$

The integration in Equation (28) is carried out numerically for the four cases presented earlier in Figures (8) and (10): case 1, uniform, $\mu = 1.0$; case 2, uniform, $\mu = 0.3$; case 3, exponential, $\mu = 1.0$; and case 4, exponential, $\mu = 0.3$. The derived distributions of the deficit d can be estimated at any time during both the storm and interstorm periods. For illustration, we present in Figure 11a the cumulative distributions of d at the end of a rainfall period that produced a scaled, grid-square averaged rainfall depth of 0.1.

We next ask the question whether we can specify an effective, antecedent soil moisture (i.e. v) variation within a *uniform* grid square, wetted by *uniform*, grid-square averaged, rainfall of intensity $\bar{p}$, and yet producing the same distribution of soil moisture deficit d, as would be obtained if the rainfall were to be patchy (or widespread) and exponentially distributed. This new soil moisture variation would presumably yield the same runoff, evaporation rate, saturated areas and soil moisture storage as the numerical model presented earlier, but at a fraction of the computational cost. This idea is similar to, and has been motivated by, the concept of 'grid-averaged effective infiltration rate' presented by Shuttleworth (1988). Finding an answer to this question is vital to the development of land-surface schemes for use in GCMs.

Indeed, we can achieve this quite simply. We subtract d from s for each value of F; this is tantamount to subtracting the d curve in Figure 11a from the s curve. The results are presented in Figure 11b. We can clearly see that for a grid square characterized by $F_S(s)$, if the initial moisture profiles v were to be as shown in Figure 11b, then the distributions of soil moisture deficit d within the grid square would be 'preserved'. Note that this does not mean that the v curves presented in Figure 11b are the actual distributions of v. If

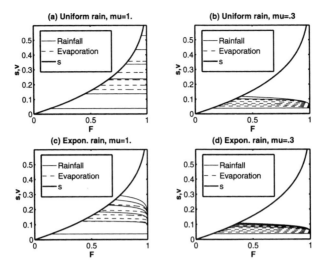

Figure 12. Temporal evolution of effective antecedent soil moisture profile v during rainfall (in steps of $0\cdot1$ in scaled depth units: solid lines) and evapotranspiration periods (equally spaced in time: broken lines) for four rainfall cases: (a) case 1, uniform, widespread ($\mu = 1\cdot0$); (b) case 2, uniform, patchy ($\mu = 0\cdot3$); (c) case 3, exponential, widespread ($\mu = 1\cdot0$); and (d) case 4, exponential, patchy ($\mu = 0\cdot3$)

such a profile of v is available independently or can be assumed on prior knowledge, then there is no longer the need to explicitly consider the spatial variability and patchiness of rainfall. This is an important result for researchers attempting to construct 'smart buckets' that can simulate the subgrid variability of hydrological processes without the use of distributed models.

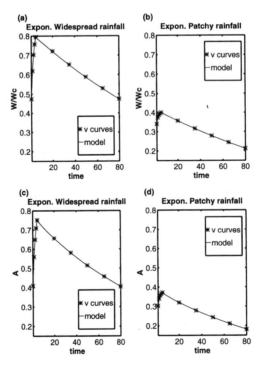

Figure 13. Comparison of soil moisture storages and saturated area fractions, estimated using the profiles of v obtained by the derived distribution approach (stars '*'), and by the numerical integration or brute force approach (solid lines), for consecutive storm and interstorm periods: (a) and (c) case 3, exponential, widespread ($\mu = 1\cdot0$); (b) and (d) case 4, exponential, patchy ($\mu = 0\cdot3$)

The spatial soil moisture variations, i.e. of v, evolve in time in response to changes in the distribution of d. The temporal variations of the 'effective' soil moisture profile v, for the four cases considered in Figures 10 and 11, are presented in Figure 12a–d for one cycle of consecutive rainfall and evapotranspiration periods. During the storm periods, the antecedent soil moisture curves are plotted for increasing rainfall depths, in steps of 0·1 (scaled units). During interstorm periods, the soil moisture curves are plotted for increasing values of time.

During the rainfall period, the v curves are initially widely spaced, reflecting the significant fraction of the rainfall being put into storage, but as the storage fills, more and more rainfall runs off, and storage grows more slowly. During the evapotranspiration period, the curves are widely spaced at first, because much of the catchment is at or near saturation, and is evaporating at near the potential rate. As the storage decreases, evaporation also decreases, so that a considerable period will be required to return to the initial condition. The results are also consistent with our observations in Figure 10. For example, the saturated areas expand more rapidly for case 1 (uniform, widespread) than for case 4 (exponential, patchy). Similarly, the storage is also larger for case 1.

To test the accuracy of the soil moisture variations predicted by the 'derived distribution approach' presented in this section, we compare the variations of the soil moisture storage (i.e. the area under the v curve) and the saturated area fractions during the rainfall and evapotranspiration periods against straightforward estimation of these quantities by the 'brute force' numerical model presented in Equations (18) to (23). These are presented in Figure 13(a–d) for two rainfall cases, case 3 (exponential, $\mu = 1·0$) and case 4 (exponential, $\mu = 0·3$), demonstrating clearly that the derived distribution approach accurately reproduces the variations of these two important state variables. Under what conditions these are sufficient to reproduce important hydrological fluxes, such as surface and subsurface runoff and evaporation, remains to be investigated. For early results, the reader is referred to Sivapalan and Woods (submitted).

It is clear that the v curves incorporate some of the information on the spatial variability of both rainfall and soil moisture necessary to capture a part of the 'bulk' response of the land-surface. For the cases considered, namely groundwater storage and saturated area fraction, if used in conjunction with an appropriate redistribution method, only a brief run with the detailed numerical model is necessary to produce the v curves. From then on, a VIC model with *uniform* rainfall is sufficient for their simulations.

CONCLUSIONS AND SUGGESTIONS FOR FURTHER RESEARCH

The research presented in this paper is of a preliminary nature and has been motivated by the need to develop simple parameterizations for land-surface hydrological processes at large scales. The major simplifying assumptions made in the study are: (i) soil moisture either has no memory from one storm to the next, or storms have perfect memory, i.e. always occur in the same part of the GCM grid square; (ii) the length scale of the variability of precipitation rate is much larger than that of soil moisture capacity, so the distribution of soil moisture capacity remains the same for all precipitation classes; and (iii) there is no vegetation canopy with an interception storage. Furthermore, soil moisture variability in this paper is only an implicit result of soil depth variability. This is a gross simplification, and processes dependent on surface soil moisture content can only be handled by including an additional soil moisture store.

In spite of these simplifying assumptions, and the lack of application to specific catchments with realistic climatic inputs, the following general conclusions can be made, which will have to be kept in mind when developing land-surface boundary parameterizations for climate models:

1. Subgrid variability and patchiness of rainfall wetting can have significant impacts on the estimation of land-surface fluxes and on the long-term hydrological regime.
2. It is possible to develop simple, parsimonious models of the evolution of the space–time variability of antecedent soil moisture. This can considerably reduce the biases in the instantaneous hydrological fluxes and in the long-term hydrological regime.

The derived distribution approach presented in this paper represents an extension of the distributed bucket (i.e. VIC) model, one that allows the effects of subgrid variability of soil moisture capacity (i.e. bucket size) and rainfall wetting to be incorporated into land-surface models without heavy data requirements or computational burden. As such, it holds promise for future research towards parsimonious models of the land-surface. It may also be useful in our attempts to scale up from point-scale soil moisture variations to the corresponding catchment-scale variations.

The work presented in this paper can be considerably improved and extended in a number of directions. These are summarized below:

1. A vegetation canopy layer needs to be included for any realistic simulation of land-surface processes, and the effects of subgrid variability of rainfall on interception and the evaporation of intercepted water need to be explicitly included.
2. As the depth of the soil profile, or the infiltration capacity, is not readily available in most parts of the world, research should be conducted towards finding links between the soil depth or soil moisture capacity and topographic elevation, as topography is the most readily available land-surface property.
3. It is important to improve the treatment of subsurface runoff in the model as large parts of the Earth's land-surface are dominated by subsurface runoff generation. The topographic influences on the antecedent water-table depths should also be included in future studies of subsurface flows. This will lead to a refinement of the treatment of the antecedent soil moisture state used in the present study.
4. The point-scale evapotranspiration model used in this paper can be considerably improved by including aspects of soil physics, plant physiology and micrometeorology. It may be possible to use complex SVAT models of the type of BATS or SiB at the point scale, and then integrate them up to the GCM scale, using assumed spatial distributions (e.g. Xinanjiang) of soil moisture capacities. This has the potential to lead to simple, parsimonious models at the GCM grid-square scale.
5. Simple analytical forms need to be derived for profiles of the antecedent soil moisture, under spatially variable and patchy or widespread rainfall. This will help further reduce the computational burden of the land-surface scheme presented here. Research into these profiles may have the potential to lead to the development of 'smart bucket' models that will allow accurate simulation of subgrid variability without the use of distributed models.
6. The ultimate test of the effects of spatial variability of rainfall and soil moisture, and of the improvements effected to the VIC model in this paper, is to incorporate a land-surface scheme based on this work into a GCM, and to examine whether this produces a substantial improvement in the GCM predictions.

ACKNOWLEDGEMENTS

This research was supported in part by a Special Environmental Fluid Dynamics grant awarded to the first author by the University of Western Australia, and by a grant from the Australian Research Council, Small Grants Scheme, Ref. 04/15/031/254. During the course of this research R. A. Woods was also supported, in part, by the New Zealand Foundation for Research, Science and Technology, under contract CO1319. This support is gratefully acknowledged. The authors are grateful to Leanne Pearce of the Water Authority of Western Australia for providing the landform information and soil depth data for south-west catchments, and to Brad Harris for help with the computations. Centre for Water Research Reference ED 840 MS.

REFERENCES

Abramopoulos, F., Rosenzweig, C., and Choudhury, B. J. 1988. 'Improved ground hydrology calculations for global climate models (GCMs): soil water movement and evapotranspiration', *J. Climate*, **1**, 921–941.
Avissar, R. 1990. 'A statistical-dynamical approach to parameterize subgrid-scale land-surface heterogeneity in climate models', *Surv. Geophys.*, **12**, 155–178.
Avissar, R. and Pielke, R. A. 1989. 'A parameterization of heterogeneous land-surface for atmospheric numerical models and its impact on regional meteorology', *Monthly Weather Rev.*, **117**, 2113–2136.
Beven, K. J. and Kirkby, M. J. 1979. 'A physically-based variable contributing area model of basin hydrology', *Hydrol. Sci. Bull.*, **24**, 43–69.

Dickinson, R. E., Henderson-Sellers, A., Kennedy, P. J., and Wilson, M. F. 1986. 'Biosphere–atmosphere transfer scheme (BATS) for the NCAR community climate model', *NCAR Tech. Note*, National Center for Atmosheric Research, Boulder, Colorado, 69pp.

Dietrich, W. E., Reiss, R., Hsu, M. -L., and Montgomery, D. R. 1995. 'A process-based model for colluvial soil depth and shallow landsliding using digital elevation data', *Hydrol. Process*, **9**, 383–400.

Dümenil, L. and Todini, E. 1992. 'A rainfall–runoff scheme for use in the Hamburg climate model' in O'Kane, J. P. (Ed.), *Advances in Theoretical Hydrology, A Tribute to James Dooge*. Eur. Geophys. Soc. Ser. Hydrol. Sci. Vol. 1. Elsevier, Amsterdam. pp. 129–157,

Entekhabi, D. and Eagleson, P. S. 1989. 'Land surface hydrology parameterization for atmospheric general circulation models including subgrid scale spatial variability', *J. Climate*, **2**, 816–831.

Famiglietti, J. S. and Wood, E. F. 1990. 'Evapotranspiration and runoff from large land areas: land surface hydrology for atmospheric general circulation models', *Surv. Geophys.*, **12**, 179–204.

Liang, X. 1994. 'A two-layer variable infiltration capacity land-surface representation for general circulation models', *Tech. Rep No. 140, Wat Resour. Ser.*, Department of Civil Engineering, University of Washington, Seattle; 208 pp.

Manabe, S. 1969. 'Climate and ocean circulation. I. The atmospheric circulation and the hydrology of the earth's surface', *Monthly Weather Rev.*, **97**, 739–774.

Manabe, S., Smagorinsky, J., and Strickler, R. J. 1965. 'Simulated climatology of a general circulation model with a hydrological cycle', *Monthly Weather Rev.*, **93**, 769–798.

Mood, A. M., Graybill, F. A., and Boes, D. C. 1974. *Introduction to the Theory of Statistics*. McGraw-Hill, New York. 564 pp.

Neghassi, H. M. 1974. 'Crop water use in yield models with limited soil moisture', *PhD dissertation*, Department of Agricultural Engineering., Colorado State University, Fort Collins.

Sellers, P. J., Mintz, Y., Sud, Y. C., and Dalcher, A. 1986.'A simple biosphere model (SiB) for use within general circulation models', *J. Atmos. Sci.*, **43**, 505–531.

Shuttleworth, W. P. 1988. 'Macrohydrology—the new challenge for process hydrology', *J. Hydrol.*, **100**, 31–56.

Sivapalan, M. and Woods, R. A. 1995. 'Variable bucket representation of TOPMODEL incorporating sub-grid variability of rainfall: the meta-catchment concept', *Rep. ED 913 MS, Centre for Water Research*, University of Western Australia, Nedlands, submitted.

Sivapalan, M., Beven, K. J., and Wood, E. F. 1987. 'On hydrologic similarity. 2. A scaled model of storm runoff production', *Wat. Resour. Res.*, **23**, 2266–2278.

Thomas, G. and Henderson-Sellers, A. 1991. 'An evaluation of proposed representation of subgrid hydrologic processes in climate models', *J. Climate*, **4**, 898–910.

Warrilow, D. A., Sangster, A. B. and Slingo, A. 1986. 'Modelling of land-surface processes and their influence on European climate', *Dyn. Climatol. Tech. Note DCTN 38*, UK Meteorological Office, Bracknell, 92 pp.

Wood, E. F., Lettenmaier, D. P., and Zartarian, V. G. 1992. 'A land-surface hydrology parameterization with subgrid variability for general circulation models', *J. Geophys. Res.*, **97**, 2717–2728.

Zhao, R. J., Zhang, Y. L., Fang, L. R., Liu, X. R. and Zhang, Q. S. 1980. 'The Xinanjiang model', in *Hydrological Forecasting, Proceedings of the Oxford Symposium, IAHS Publ.* **129**, 351–356.

27

SIMULATING HETEROGENEOUS VEGETATION IN CLIMATE MODELS. IDENTIFYING WHEN SECONDARY VEGETATION BECOMES IMPORTANT

A. J. PITMAN

School of Earth Sciences, Macquarie University, North Ryde, 2109, Sydney, Australia

ABSTRACT

To parameterize the land surface in global climate models (GCMs) data must be provided at a variety of resolutions. In GCMs, high-resolution data must be aggregated to the coarser resolution of the GCM. The method used for parameter aggregation can lead to the simulation of very different land surface climatologies. It is shown that the most important transition from a simulation of homogeneous tundra to one of homogeneous coniferous forest is the initial increase in forest. The differences between the simulation of tundra and the simulation of tundra with two-ninths coniferous forest are considerable. This suggests that the land surface is sensitive to the aggregated parameters in non-linear ways. It suggests that more care is needed in data aggregation, and that improved algorithms for data aggregation must be developed, because these data sets represent the foundations on which advanced land surface parameterizations are built. Finally, it shows that the influence of relatively small amounts of secondary vegetation should be represented in GCMs.

INTRODUCTION

Global climate models (GCMs) are widely used for predicting the effect of natural and anthropogenically produced perturbations on the Earth's climate. One element of GCMs which has received increasing attention is the land surface. Land surface models range from the simple bucket model (Manabe, 1969) to more complex schemes including the simple biosphere model (SiB; Sellers *et al.*, 1986); the biosphere atmosphere transfer scheme (BATS; Dickinson *et al.*, 1986); the model of Abramopoulos *et al.* (1988) and the bare essentials of surface transfer model (BEST; Cogley *et al.*, 1990). These four models attempt to represent the surface–atmosphere interaction as realistically as possible, within the confines of limited computer time.

Most land surface schemes require data for vegetation parameters (e.g. albedo, roughness length, leaf area index) for each grid square. These data can be derived, for instance, from Matthews (1983) or Wilson and Henderson-Sellers (1985), who provide a global distribution of natural vegetation or land use at 1° × 1° resolution (approximately 100 km × 100 km). There is also the possibility of using satellite products, such as HDVI (Gutman, 1991). These data sets are based on a classification of vegetation type and to use them in a GCM they must be aggregated. The atmosphere does not 'recognize' the surface *type* represented by a GCM. The surface type is only used to specify parameters, which are then used to calculate temperatures and energy fluxes. Therefore, a land surface model does not need 'ecotype', rather it needs parameter values associated with the surface.

Most land surface models assume that the vegetated fraction of the grid square is uniformly covered by a single type of vegetation. This can be defended in terms of computational efficiency, a lack of information about the distribution of vegetation within a grid square and the lack of suitable approaches to accommodate spatial heterogeneity at the surface. Koster and Suarez (1992) proposed a 'mosaic' approach to overcome the limitation of homogeneity in the vegetated fraction of the grid square. In their approach, the

GCM grid square is split into a series of smaller areas (the actual size depends on the GCM resolution and number of mosaic cells prescribed) and fluxes are calculated from each area. These fluxes are areally weighted at the end of each time step.

It is assumed in Koster and Suarez's (1992) approach that the results simulated by a land surface scheme are different if the assumption of homogeneity in the distribution of vegetation (i.e. uniform cover by the primary vegetation type) is changed to one of heterogeneity (i.e. uniform cover of each subarea, but each subarea is covered by different vegetation types). Once a GCM grid square is split into more than one unit, the problem of how small each subunit needs to be must be addressed. There is no equivalent concept to the representative elementary area (Wood *et al.*, 1990) for the atmosphere, and so it is useful to examine the amount of a secondary vegetation type which is needed before it significantly affects the fluxes from an otherwise homogeneous area. In this paper, some simple experiments are reported to determine whether land surface modellers need to be concerned about heterogeneity (at GCM spatial scales). This is investigated by exploring a data aggregation method which retains information about the different vegetation types which make up a GCM grid square. This is a preliminary attempt to determine whether heterogeneity is important at the spatial scales of GCMs (heterogeneity is clearly important at smaller spatial scales; see, for example, Avissar and Pielke, 1989). If heterogeneity is important, then the approaches developed by Koster and Suarez (1992) will need to be implemented in GCMs.

DATA AGGREGATION

Current procedure

Methods for deriving land surface parameters for individual models are ill-defined. Even if the same initial data set was used by every GCM, the aggregated data set at the GCM resolution would vary due to different resolutions and due to the different aggregation criteria used. For instance, the BATS model (Dickinson *et al.*, 1986) uses a combination of the data sets of Matthews (1983), Wilson and Henderson-Sellers (1985) and, where these two data sets differ, an average of these data and a subjective comparison with other data is used to obtain the *dominant* vegetation type for a grid square. This methodology uses subjective analysis, while using the dominant vegetation type leads to a loss of the characteristics of the non-dominant types. In Figure 1, an arbitrary grid square is composed of three vegetation types with indexes of '1', '2' and '3'. Using the method of Dickinson *et al.* (1986), the GCM grid square has a single 'ecotype', the most common, which is index '3'. This index is then used to provide suitable parameter data.

In BEST (Cogley *et al.*, 1990), the data set of Wilson and Henderson-Sellers (1985) is used to derive the canopy index. Although the canopy or vegetation index cannot be aggregated, the quantities in the vegetation look-up tables (Table I) correspond to the vegetation index and are physical quantities which can be aggregated. A simple way to aggregate these data is to average arithmetically. This is probably satisfactory for albedo, and perhaps leaf area index, but, as will be shown below, it is inappropriate for roughness length. Figure 1 shows the results of two different aggregation procedures. In the upper case the most common index is assumed to be representative of the entire grid square and an 'appropriate value' is specified for various parameters (roughness length, albedo, etc.). In the 'proposed method' the same 'appropriate value' is specified for each canopy index before any decision is made about how to aggregate and then the actual values are aggregated to the appropriate resolution for the GCM. The resulting estimates of grid square roughness length, albedo and leaf area index are very different. The next section demonstrates that the differences shown in Figure 1 lead to differences in the simulation of the land surface climatology for selected vegetation types. Note that this approach does not represent different vegetation types simultaneously within a grid square (cf. Koster and Suarez, 1992). Rather the characteristics of different vegetation types are 'lumped' together into a single effective value. This is conceptually less satisfying than the approach of Koster and Suarez (1992), but it is computationally free and is likely to be preferable to assuming that the entire grid square is homogeneously covered by a single vegetation type (e.g. Dickinson *et al.*, 1986).

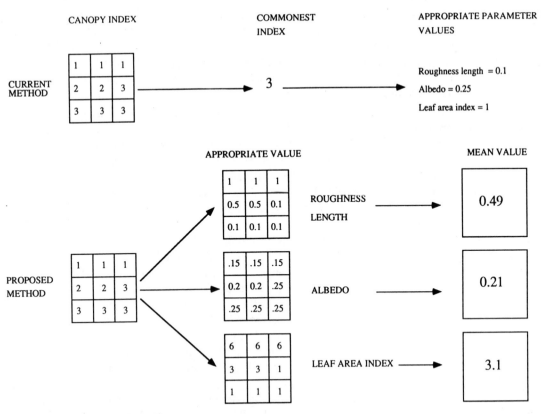

Figure 1. Consequences of the two aggregation approaches discussed in the text. In the 'current method' the most common canopy index is used to assign appropriate parameter values. In the aggregation method, the appropriate value for each parameter for each $1° \times 1°$ area is aggregated, leading to different input parameters in the GCM. Roughness length is expressed in metres, albedo as a fraction and leaf area index in $m^2 m^{-2}$

It is inappropriate to average the roughness length arithmetically. A later section examines three estimates for the effective roughness. These are an arithmetic average, a geometric mean (e.g. Taylor, 1987) and the roughest element (e.g. André and Blondin, 1986; Mason, 1988) to see whether the land surface is sensitive to the type of aggregation methodology used.

SENSITIVITY OF THE SURFACE TO THE SURFACE CHARACTERISTICS

Experiment design

A series of simulations has been performed using BEST in a stand-alone mode driven by prescribed climate forcing to examine the sensitivity of the model to aggregation procedures. The simulations were for 'tundra', 'coniferous forest' and a range of combinations of these two vegetation types. The purpose of these experiments was to identify whether the land surface was sensitive to small amounts of one vegetation type and to determine the scale of heterogeneity at which small vegetation amounts become important.

In the following discussion the experiments where the grid square is composed partly of tundra and partly of coniferous forest are referred to in the following way. Simulations with seven-ninths tundra and two-ninths coniferous forest are referred to as 7T:2C; those with five-ninths tundra and four-ninths coniferous forest as 5T:4C and those with two-ninths tundra and seven-ninths coniferous forest as 2T:7C. The simulations where the whole grid square is composed of either tundra or coniferous forest are referred to as the control simulations. Table I shows the parameter input data used in each of these simulations.

This discussion will be confined to the latent and sensible heat fluxes. The grid average quantities discussed in the following are composites of bare soil, canopy and snow. Heat and moisture fluxes are calculated for each and are averaged according to the fraction of the grid square covered by each. For convenience, the composite of soil, canopy and snow is called the terrestrial surface.

Results of different aggregation procedures

The results of the six simulations (using data from Table I) for evaporation are shown in Figure 2a. The coniferous forest simulation predicts higher latent heat fluxes than the tundra simulation. The higher roughness length and leaf area index, combined with the lower albedo and increased available energy, lead to more evaporation. This result shows that land surface schemes which incorporate vegetation are able to represent the differences in the land surface climatology simulated by different vegetation types. This result also demonstrates that two different evaporation regimes can be simulated from identical climate forcing. This supports the need to represent the natural heterogeneity of the land surface in GCMs (i.e. differentiating between vegetation types).

Although all the curves in Figure 2a (and subsequent figures) are generated using identical climate forcing, the control simulation for tundra and coniferous forest differ by up to $10 \, \text{W m}^{-2}$. In the intermediate experiments, the characteristics of the coniferous forest tend to dominate even if the forest is relatively sparse. For instance, the curve for the 7T:2C simulation shows a result which is more similar to the 4T:5C and coniferous forest control results than to the tundra simulation, even though only two-ninths coniferous forest was included. This implies that small amounts of forest strongly influence the overall simulation. At the other extreme, two-ninths tundra also modifies the control coniferous forest simulation, although adding a small amount of tundra to coniferous forest leads to smaller effects than adding coniferous forest to tundra. Note that adding two-ninths tundra leads to an evaporation flux which is higher than either of the control simulations. This is due to the interaction between high roughness length (in the coniferous forest) and higher moisture holding capacities in the tundra control simulation.

Figure 2b shows an analogous result for the sensible heat flux. Although the results from the control simulation for tundra and coniferous forest form the two extremes the intermediate results do not show the expected pattern during summer (in particular, the results for 4T:5C, 5T:4C and 7T:2C are virtually indistinguishable). The relatively low terrestrial latent heat flux for 7T:2C leads to an increase in the available energy for the sensible heat flux, whereas the relatively high latent heat flux for the 2T:7C simulation leads to a reduced sensible heat flux.

Table I. Input parameters for the six simulations by BEST (parameter values are as prescribed by Dickinson et al., 1986)

Vegetation parameters	Tundra	Composite				Coniferous forest
Variable	9T:0C	7T:2C	5T:4C	4T:5C	2T:7C	0T:9C
$\alpha_{\text{SW,c}}$	0·13	0·12	0·11	0·108	0·099	0·09
$\alpha_{\text{LW,c}}$	0·25	0·237	0·22	0·217	0·203	0·19
S_{AI}	0·5	0·83	1·167	1·33	1·67	2·0
f_{c}	0·01	0·014	0·019	0·02	0·026	0·03
$z_{0\text{c}}$	0·04	0·08	0·17	0·23	0·48	1·0
d_{u}	0·1	0·1	0·1	0·1	0·1	0·1
d_{l}	0·9	1·01	1·12	1·18	1·29	1·4
f_{rootu}	0·9	0·849	0·79	0·77	0·72	0·67
S_{c}	0	0	0	0	0	0
S_{l}	3·5	2·9	2·39	2·11	1·56	1·0
A_{vmax}	0·75	0·81	0·82	0·83	0·867	0·9
LAI_{max}	4	4·44	4·89	5·11	5·56	6

$\alpha_{\text{SW,c}}$ is the visible canopy albedo, $\alpha_{\text{LW,c}}$ is the near-infrared canopy albedo, S_{AI} is the stem area index, f_{c} is a light sensitivity factor, $z_{0\text{c}}$ is the canopy roughness length (m), d is the depth of the upper (u) and lower (l) soil layers (m), f_{rootu} is the fraction of roots contained in d_{u}. S_{c} is the seasonal range of A_{v} (which is the value of A_{vmax} at any time step), S_{l} is the seasonal range of L_{AI}, $A_{v}max$ is the maximum vegetated fraction of the grid square and LAI_{max} is the maximum leaf area index

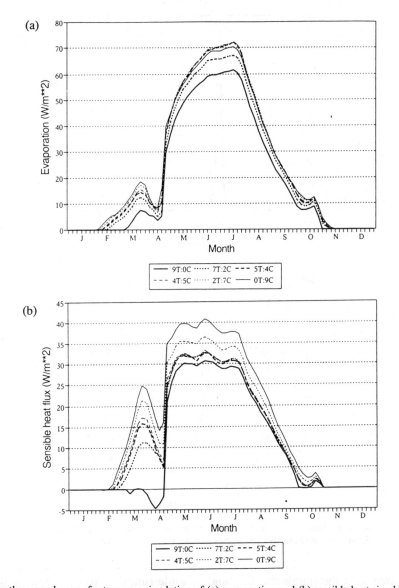

Figure 2. Results from the second year of a two-year simulation of (a) evaporation and (b) sensible heat simulated by BEST for six combinations of tundra and coniferous forest ecotypes. The tundra and coniferous forest control simulations consist of only one ecotype. The remaining simulations consist of combinations of both ecotypes, represented by an aggregation of parameters (see Table I). T7:2C refers to seven-ninths tundra and two-ninths coniferous forest, T5:C4 to five-ninths tundra and four-ninths coniferous forest, T4:C5 to four-ninths tundra and five-ninths coniferous forest and T2:C7 to two-ninths tundra and seven-ninths coniferous forest

In spring, the consequences of spring snow melt on the surface heat balance are clearly shown. All simulations, with the exception of the control simulation for tundra, show similar patterns. The control simulation for tundra shows a very different result, which again indicates that the presence of only two-ninths coniferous forest has a considerable effect of the simulation of the land surface climate (cf. Figure 2a).

Effect of aggregation on runoff simulation

To investigate the response of the land surface in terms of the hydrology, this section briefly considers the effects of data aggregation on runoff. Figure 3a shows that the tundra is characterized by a strong runoff peak in April (11 mm day^{-1}) due to snowmelt, with 0–1·5 mm day^{-1} during most other months (Figure 3b). The coniferous forest is qualitatively similar, but significant differences occur. Runoff in April and June is

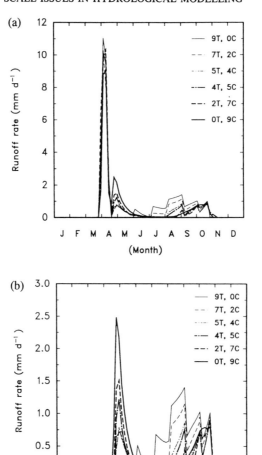

Figure 3. As for Figure 2, but for (a) total runoff and (b) omitting the runoff induced by snowmelt to show the post-snowmelt runoff simulation more clearly. The values plotted are five-day averages in mm day^{-1}

about half compared with the simulation for tundra, while runoff is lower in March. Later in the year, coniferous forest is significantly drier and generates significantly lower runoff.

To examine the lower summer runoff, Figure 3a was redrawn, excluding snowmelt-induced runoff. Figure 3b shows that the late summer reduction in runoff is proportional to the amount of coniferous forest, with the differences increasing linearly until coniferous forest covers seven-ninths, after which there is little further change. The pattern for snowmelt-induced runoff (Figure 3a) is less well defined. Basically, increasing the fraction of coniferous forest to two-ninths leads to an increase in March runoff and a reduction in April runoff. This net change is initially reduced by further increases in coniferous forest until the sign of the change for these two months is reversed for the T2:C7 simulation. Increasing the forest leads to a gradual but consistent increase in May runoff.

In summary, the method used for aggregating the data has a large impact on simulated runoff. These results emphasize the importance of including small amounts of coniferous forest into tundra by using the aggregation method as the T7:C2 simulation is very different to the control simulation for tundra.

Discussion

These results show that small amounts of an aerodynamically rough vegetation type can strongly influence the sensible and latent heat fluxes simulated from an aerodynamically smoother surface. It has also been shown that different combinations of tundra and coniferous forest can lead to non-linear results. The combination of tundra and coniferous forest leads to a variety of different roughness lengths and albedos, which result in different amounts of available energy and exchange rates of turbulent fluxes with the atmosphere. The results in Figure 2a should have led to increasing latent heat fluxes and reduced sensible heat fluxes as the amount of tundra was increased. However, increasing coniferous forest tends to lead to higher latent heat fluxes and lower sensible heat fluxes due to the higher roughness lengths and lower albedos. These factors combine in a variety of ways to lead to a complex interaction resulting in latent heat fluxes exceeding both the control simulations shown in Figure 2a. The increase in roughness length, leaf area index and available energy due to increasing the amount of coniferous forest leads to a tendency for more efficient surface energy exchange. The increase in efficiency is most evident if small amounts of forest are incorporated into tundra.

SENSITIVITY EXPERIMENTS

The roughness length is a parameter which clearly affects the simulation of the atmosphere. However, it is not clear how an effective roughness length should be calculated, hence three simulations have been performed: using a geometric averaging procedure (as suggested by Taylor, 1987); by assuming that the effective roughness length is given by the roughest element contained in any of the nine grid squares (following André and Blondin, 1986; Mason, 1988); and finally by assuming that the effective roughness length is given by the arithmetic average of the nine elements.

In the following section these experiments will be briefly discussed, noting that the results are clearly both model-dependent and affected by the characteristics of the climate forcing used. The results discussed here are intended to show that BEST is sensitive to how the roughness length is aggregated from $1° \times 1°$ to a coarser resolution. Only results for the 5T:4C simulation are discussed.

Results

Figure 4a shows the simulation of evaporation for the three integrations of BEST for the 5T:4C surface. The three curves are for three different methods for aggregating the roughness lengths, by an arithmetic average, a geometric average and by assuming that the effective roughness length is given by the roughest individual element. The three roughness length values used were 0·64, 0·167 and 1·0 m, respectively.

Figure 4a shows for the latent heat flux, and Figure 4b for the sensible heat flux, that changing the roughness length can lead to significant differences, of the order of $15\,\mathrm{W\,m^{-2}}$ for evaporation and $8\,\mathrm{W\,m^{-2}}$ for the sensible heat flux. For most of the winter the snow covering the surface restricts the differences between the simulations as the roughness length is reduced and because energy fluxes are largely suppressed (cf. Figure 2a and 2b). After snowmelt begins (February), the increasing amount of available energy leads to the differences in the roughness length becoming important. However, Figure 4a shows that the differences in the latent heat flux due to the changes in the roughness length are fairly small until May/June. During summer, the three simulations show differences in the latent heat flux of $\sim15\,\mathrm{W\,m^{-2}}$ as the roughness length is increased from 0·167 to 0·64 to 1 m. The increase in the latent heat flux is approximately linear, indicating that errors in prescribing or calculating the roughness length should not grossly affect the simulations.

Figure 4b shows that the simulation of the sensible heat flux is more complex during snowmelt than the latent heat flux. When the 1 m roughness length was used, the sensible heat flux was consistently higher than in the other simulations until snowmelt is completed. Clearly, the higher roughness length aids the onset of snowmelt. For the rest of the year the higher roughness lengths lead to slightly lower sensible heat fluxes ($\sim4\,\mathrm{W\,m^{-2}}$), although the net transfer of turbulent energy is higher as Figure 4a showed that the latent heat flux was higher for longer roughness lengths by $\sim10\,\mathrm{W\,m^{-2}}$.

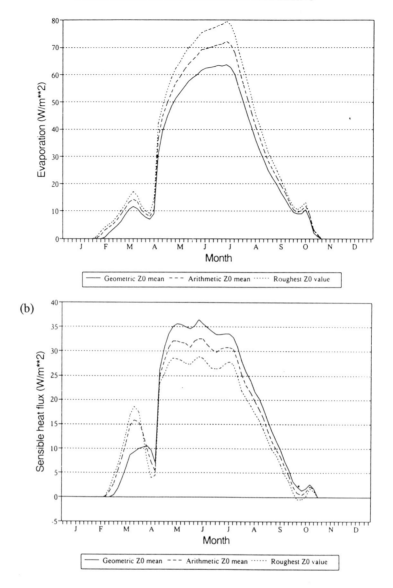

Figure 4. Results from the second year of a two-year simulation of (a) evaporation and (b) sensible heat flux by BEST for three different estimations of the effective roughness length. These results are for the T5:C4 (five-ninths tundra and four-ninths coniferous forest) simulation. The values are five-day averages

DISCUSSION

The results described here have shown that, given constant atmospheric forcing, using parameter data for tundra for the surface, or using parameter data for coniferous forest, produces a different simulation. This should be expected and shows that the type of model structure used by BEST does respond to changes in the input parameters. The results also show that if an aggregation of tundra and coniferous forest vegetation parameters is performed to create an effective ecotype, these data can be used to simulate a different land surface climatology. This new climatology is not an average of the tundra and coniferous forest simulation, but rather is closer to the hydrologically, thermally and aerodynamically dominant type, in this instance coniferous forest.

It is not possible to assume that any one ecotype is appropriate for an GCM grid square as these results have shown that the overall surface–atmosphere system is non-linear. There is no need to determine a single appropriate vegetation type for a grid square as it has been shown here that aggregated parameters for an effective ecotype can be used instead. Calculating these effective parameters from the original $1° \times 1°$ data may require parameter-specific algorithms, but this should be better than just assuming that the entire grid square is covered homogeneously by a single vegetation type. As the simulations described here are sensitive to the aggregation procedure used, the techniques currently used by some GCMs should be reconsidered. The aggregation procedure used for each parameter has to be defined. In these simulations, arithmetic averages have generally been used. However, an arithmetic average is not suitable for roughness length.

Finally, the aggregation method advocated here provides the potential for realizing further flexibility and realism in the provision of data for the land surface. Climate models usually only use the 'primary' vegetation type from the $1° \times 1°$ data sets. One significant advantage of the aggregation procedure discussed here is that as only parameters are aggregated, the methodology need not be restricted to just the primary vegetation type. Wilson and Henderson-Sellers (1985) provide data for 'primary' (defined as the land cover type which occupies $\geqslant 50\%$ of the grid square) and 'secondary' (defined as the land cover type which occupies $< 50\%$ and $\geqslant 25\%$ of the grid square) vegetation type at $1° \times 1°$ resolution. The secondary data set could be aggregated with the primary data set to provide an improved estimation of the grid square effective value. The secondary data would be weighted using the information provided by Wilson and Henderson-Sellers (1985) on the relative fraction of the primary and secondary vegetation types.

CONCLUSIONS

This paper has shown that GCMs need to consider carefully the methodology used for deriving the parameters they require for the land surface from high-resolution data sets. It has been argued that the most realistic aggregation procedure would be to aggregate the parameters themselves rather than an index or by taking the most common ecotype in a grid square. Averaging the individual parameters from the $1° \times 1°$ data set retains information lost by the 'most common' method and implicitly increases the subgrid scale variability as effective rather than absolute quantities are used.

It has also been shown, using stand-alone forcing, that the most important single step in the transition from homogeneous tundra to homogeneous coniferous forest is the initial increase in forest. The addition of two-ninths coniferous forest had a significant effect on the latent and sensible heat fluxes, as well as the runoff. This indicates that even small amounts of heterogeneity can have a significant impact on the simulation of the land surface climate, although the lack of surface–atmospheric feedbacks and advective processes in these simulations means that these conclusions need to be treated with care.

REFERENCES

Abramopoulos, F., Rosenzweig, C., and Choudhury, B. 1988. 'Improved ground hydrology calculations for global climate models (GCMs): soil water movement and evapotranspiration', *J. Climate*, **1**, 921–941.

André, J. C. and Blondin, C. 1986. 'On the effective roughness length for use in numerical three-dimensional models', *Boundary Layer Meteorol.*, **35**, 231–245.

Avissar, R. and Pielke, R. A. 1989. 'A parameterization of heterogeneous land surfaces for atmospheric numerical models and its impact on regional meteorology', *Monthly Weather Rev.*, **117**, 2113–2136.

Cogley, J. G., Pitman, A. J., and Henderson-Sellers, A. 1990. 'A land surface for large-scale climate models', *Trent Climate Note, 90-1*, Trent Univ., Peterborough, Ontario.

Dickinson, R. E., Henderson-Sellers, A., Kennedy, P. J., and Wilson, M. F. 1986. 'Biosphere atmosphere transfer scheme (BATS) for the NCAR Community Climate Model', *NCAR Tech. Note, NCAR, TN275 + STR*, 69pp.

Gutman, G. G. 1991. 'Vegetation indices from AVHRR: an update and future prospects', *Remote Sensing Environ.*, **35**, 121–136.

Koster, R. D. and Suarez, M. J. 1992. 'A comparative analysis of two land surface heterogeneity representations', *J. Climate*, **5**, 1380–1390.

Manabe, S. 1969. 'Climate and the ocean circulation: 1, the atmospheric circulation and the hydrology of the Earths' surface', *Monthly Weather Rev.*, **97**, 739–805.

Mason, P. J. 1988. 'The formation of areally-averaged roughness lengths', *Q. J. Roy. Meteorol. Soc.*, **114**, 399–420.

Matthews, E. 1983. 'Global vegetation and land use: new high-resolution data bases for climate studies', *J. Climate Appl. Meteorol.*, **22**, 474–487.

Sellers, P. J., Mintz, Y., Sud, Y. C., and Dalcher, A. 1986. 'A simple biosphere model (SiB) for use within general circulation models', *J. Atmos. Sci.*, **43**, 505–531.

Taylor, P. A. 1987. 'Comments and further analysis on effective roughness lengths for use in numerical three-dimensional models', *Boundary Layer Meteorol.*, **39,** 403–418.

Wilson, M. F. and Henderson-Sellers, A. 1985. 'A global archive of land cover and soil data sets for use in general circulation climate models', *J. Climatol.*, **5,** 119–143.

Wood, E. F., Sivapalan, M., and Beven, K. 1990. 'Similarity and scale in catchment storm response', *Rev. Geophys.*, **28,** 1–18.

INDEX